AUTISM – A NEURODEVELOPMENTAL JOURNEY FROM GENES TO BEHAVIOUR

Edited by **Valsamma Eapen**

Autism – A Neurodevelopmental Journey from Genes to Behaviour
Edited by Valsamma Eapen

Published by InTech

Janeza Trdine 9, 51000 Rijeka, Croatia

Publishing Process Manager Marina Jozipovic

Technical Editor Teodora Smiljanic

Cover Designer InTech Design Team

Additional hard copies can be obtained from orders@intechopen.com

Autism – A Neurodevelopmental Journey from Genes to Behaviour,
Edited by Valsamma Eapen
p. cm.
ISBN 978-953-307-493-1

For EU product safety concerns:
IN TECH d.o.o., Prolaz Marije Krucifikse Kozulić 3, 51000 Rijeka, Croatia,
info@intechopen.com or visit our website at intechopen.com.

Meet the editor

Valsamma Eapen is Professor & Chair of Infant, Child and Adolescent Psychiatry at the University of New South Wales (UNSW) and Head of the Academic Unit of Child Psychiatry, South West Sydney (AUCS), Australia. Professor Eapen graduated from Kerala, India and continued her postgraduate training in India (NIMHANS) and the UK. She completed her Child Psychiatry training through the Great Ormond Street Hospital Scheme, and PhD work at the National Hospital for Neurology and Neurosurgery, Queen Square, London. She started her academic career at the University College London, UK and then moved to the UAE University, where she was Professor of Child Psychiatry prior to taking up her current position in 2008. As a Child Psychiatrist with special expertise in neurodevelopmental disorders, her research interests include Tourette Syndrome, Autism, and ADHD, and her research has included epidemiological, genetic and clinical aspects of these disorders. Her other areas of research include neurobiology of attachment and separation anxiety disorder as well as developmental trajectory in normal and abnormal development.

Meet the editor

Contents

Preface

The book, "Autism: A Neurodevelopmental Journey from Genes to Behaviour" reflects the wide range of research being conducted into the aetiology, pathogenesis, clinical characteristics, and intervention in autism. While individual chapters cover alterations in genes, neurochemicals, protein complexes, receptors, synaptic connections etc. that contribute to altered neuronal development, together they convey the story of a neurodevelopmental journey that has lost its way and is aberrant in connectivity. In this regard, Dr. Dafna Ben Bashat discusses abnormal developmental trajectories of White Matter (WM) in autism and suggests that disturbed, abnormal and disorganized inter- and intra–cortical connections are one of the core issues in autism, resulting in poorly synchronized and weakly responsive networks, which in turn lead to abnormal cognitive and neurological functioning. While MRI studies have revealed impaired connectivity, and increased brain volume especially early in life, DTI can detect microstructural changes that might help to reveal the neurobiology underpinning autism such as accelerated-myelination at a young age. In the chapter GABAergic dysfunction in autism and epilepsy, Prof. Yuri Bozzi and colleagues explore the role of GABAergic neurons and circuits in autism and review the genetic, cellular, anatomical and neurophysiological studies that support the hypothesis that the imbalance between excitation and inhibition resulting from neurodevelopmental defects in GABAergic circuitry might represent a common cause for autism and epilepsy. Following this theme, Prof. Ricardo Miledi and colleagues in their chapter "GABA and Glutamate Receptors of the Autistic Brain" argue that autism is a developmental synaptic disorder that affects the processing of behavioural relevant information. Given that GABAergic and glutamatergic synapses appear to be convergent nodes of genetic, epigenetic, and probably environmental factors causing the autistic phenotype, they propose that GABA and glutamate receptors may form important targets for pharmacological interventions. In the next chapter Dr. Dayan Goodenowe further elaborates the biochemical basis of autistic behavior focusing on glutamate and mitochondrial toxicity with suggestions for dietary and pharmacological therapies. Here mitochondrial dysfunction is implicated in findings such as abnormal brain growth, seizures, gender bias, and selective autistic pathology involving chronic microglial activation as well as "immunoexcitotoxicity" by glutamate. Serotonergic neurotransmission in autism is the focus of a chapter by Dr. Yoshihiro Takeuchi from Japan while Prof. Susan Masino and colleagues continue this theme by reviewing the actions of Adenosine in the central nervous system, with multiple implications for autism, and the potential for developing new evidence-based therapies. The chapter by Dr. Rene Anand et al using postmortem, genetic, functional, and molecular neurobiological methodologies discuss a rationale that neuronal nicotinic acetylcholine receptor (nAChR) alterations are biomarkers for autism and that specific nAChRs subtypes are likely to be useful therapeutic targets for the treatment of core deficits.

With all the recent attention on the genetic advances, a number of chapters cover gene effects as well as the environmental modulation of these effects. For example, Dr. Abdullah Alqallaf presents an overview for the analysis of genetic variations in the form of DNA copy number changes and suggests that autism is associated with an increased amount of copy number alteration in unstable segments of the genome and that combinations of copy number variations could provide the basis for discriminating autistic and typically developing groups and potentially identifying distinct subgroups within the phenotypic heterogeneity of autism. The chapter "Genome-wide Association Studies of Copy Number Variation in Autism" by Prof. Yeun-Jun Chung reviews the current data on GWAS and argue that the integration of the GWAS data with other resources such as improved bio-imaging, personal whole-genome sequencing, gene-environmental interaction and metagenome analysis data will enable us to get a more comprehensive insight to designing future personalized care of autism. The focus of the chapter by Prof. Haruhiro Higashida et al is the finding of a missense mutation in *CD38* in three autism pedigrees with a proposal that the *CD38* W140 allele could be a risk factor for at least one form of autism by abrogating oxytocin function. On a related topic, Prof. Neumann Inga discusses brain oxytocin as a main regulator of prosocial behaviour and its role in autism.

Although much progress has been made in the recent past in identifying some of the genes that play a crucial role in autism, significant research has also focused on environmental and epigenetic factors involved in the genesis of the disorder. In the chapter "Environmental Factors in the Aetiology of Autism and Lessons from Animals Prenatally Exposed to Valproic Acid", Dr. Tomasz Schneider and colleagues use rodent models to clarify complex relationships between genetic, behavioural and environmental variables to better understand and potentially cure autism. Environmentally induced oxidative stress is the focus of the chapter by Prof. Elizabeth Sajdel-Sulkowska. She discusses disruption of brain thyroid hormone homeostasis with special emphasis on the developmental impact of environmental toxicants, such as herbicides, polychlorinated biphenyls (PCBs), bisphenol A (BPA) and organic mercury compounds that interfere with the thyroid hormone as a possible factor contributing to autistic pathology.

There is increasing evidence indicating the role of immune system in autism and the chapter by Dr. Theoharis Theoharides explores how perinatal immune activation, in the mother and/or the fetus, could adversely affect neurodevelopment. Since mast cell activation during this period by environmental, infectious, neurohormonal and immune triggers appears to be involved in gut-blood-brain barrier disruption and subsequent brain inflammation, the potential use of specific CRH receptor antagonists, as well as drugs that could prevent BBB disruption, or block brain inflammation including the use of mast cell blockers is discussed. On the same theme of immune functions, Prof. Iseri Elvan discusses Neurotrophic Factors with several lines of evidence suggesting that cytokines, growth factors and neurotrophic factors play particular signaling roles within the brain to produce neurochemical, neuroendocrinological and behavioral changes. While autoimmunity in Autism is the focus of the chapter by Dr. Laila AL-Ayadhi, in the next chapter, Dr. Archana Chatterjee provides an analysis of the scientific evidence that autism and vaccines are unlikely to be linked. Immune Dysfunction is again the theme of the chapter contributed by Dr. Anthony Torres and colleagues where they evaluate mounting evidence implicating the involvement of immune molecules, autoimmune associations, maternal/fetal Natural Killer cell population, cytokine levels, and maternal influences in the aetiology of autism. Changes in innate and adaptive immunity and possible impaired oral tolerance in autistic children with food protein induced enterocolitis syndrome (FPIES) is the focus of the chapter by Dr. Harumi Jyonouchi and colleagues. In the chapter on clinical evaluations, Dr. Michaela Dobre reflects on the diagnosis of autism while Dr. Anna Bieniarz relates the phenomenon of loneliness and silence to autism and its implications for psychotherapy.

In the chapter "Mnesic Imbalance and the Neuroanatomy of Autism", Dr. Miguel Ángel Romero-Munguía presents neuropathological, structural and functional imaging data to support mnesic imbalance theory in autism. Behavioral and electrophysiological characterization of induced neural plasticity in the autistic brain is the focus of the chapter by Prof. Jaime Pineda and colleagues. They discuss mirror neurons and EEG neurofeedback in the context of plasticity-inducing rehabilitation training (PIRT), centered on operant conditioning of EEG mu rhythms. Neuroplasticity is again the theme of the next chapter contributed by Dr. Jane Yip who discusses the role of GAD65 and GAD67 in neuronal development and approaching the neuropathology, neuroplasticity, and treatment of autism by studying the transcriptional regulation of GABA. The cerebellar circuitry and conditioned learning are explored alongside implications for behavioral treatment and the use of quantitative electroencephalograph (qEEG) brain map as a tool that holds promise for future research.

Thus the book takes the reader on a unique journey through the neurobiological underpinnings of autism as the different chapters deal with interrelated aetiological theories and causal pathways that explain the diverse manifestations of this complex disorder. The diversity of the findings as elaborated in the 24 chapters of this book is a reflection of the involvement of interconnected neural systems that can be understood through the relationship between functionally relevant anatomic areas and neurochemical pathways, the programming of which are genetically modulated during neurodevelopment and mediated through a range of neuropeptides and interacting neurotransmitter systems. It is hoped that understanding these neurobiological substrates can lead to the design of interventions that accommodate the way the brains of individuals with autism function and may lead to the promotion of more flexible thinking and learning. Furthermore, since genetically mediated deficits and consequent functional impairments involve activity-dependent synapse development that depends on postnatal learning and experience, early intervention can prevent or reduce the risk of these deficits cascading into a trajectory toward full expression of the disorder. Such a model implies the importance of intervening early to prevent downstream effects, and offers an opportunity to interrupt the sequence of events that would otherwise have resulted in an abnormal developmental trajectory, but instead alter the course towards neurotypical development by exploiting the neuronal maturation and brain plasticity especially in the early years of life.

Dr. Valsamma Eapen
Professor & Chair of Infant, Child and Adolescent Psychiatry,
University of New South Wales,
Australia

1

Abnormal Developmental Trajectories of White Matter in Autism - The Contribution of MRI

Dafna Ben Bashat
The Functional Brain Center, The Wohl Institute for Advanced Imaging, Tel Aviv Sourasky Medical Center, Israel

1. Introduction

Autism spectrum disorder (ASD) is a disorder of neuronal connectivity. It has been suggested that disturbed, abnormal and disorganized inter- and intra–cortical connections are one of the core issues in autism[1], resulting in poorly synchronized and weakly responsive networks, which in turn lead to abnormal cognitive and neurological functioning. Evidence accumulated in recent years has led to a shift in the conceptualization of autism, from a localized neurological abnormality to a disorder of distributed networks throughout the brain.

What is connectivity? Two fundamental principles of brain organization have been proposed: functional specialization and functional integration (Friston, 1994, 2002), with the understanding that these two principles are complementary. Functional specialization is usually inferred by the presence of activation foci while functional integration is regarded as a process mediated by connectivity, which reflects the patterns of interaction between neuronal populations, either during the performance of specific tasks or during resting state (Friston, 1994, 2002, 2009a, 2009b; Honey et al., 2009).

Functional integration relies on functional and structural connectivity, while taking into account that these two are not necessarily co-referential (Honey et al., 2009), and that the functional-structural relationship is not straightforward (Damoiseaux & Greicius, 2009). Functional connectivity, as studied by functional magnetic resonance imaging (fMRI), refers to the temporal synchronization of the blood oxygenation level dependent (BOLD) signal of two or more brain areas. Structural connectivity on the other hand, as measured by diffusion tensor imaging (DTI), refers to the physical properties of structural connections - the way in which different brain regions are connected, at the macro level (bundles of axonal tracts) (Mori & van Zijl, 2002).

1.1 The under-connectivity theory in autism

Postmortem and imaging studies support the central role of disordered brain connectivity in autism, and emphasize the importance of studying structural and functional connectivity within the brain. Just et al (Just et al., 2004) formulated the "under-connectivity" theory of

---

[1] In this chapter, the term autism will be equivalent to the acronym ASD, referring collectively to the broader spectrum. When referring solely to the autistic disorder (subgroup), this will be specified.

autism, arguing that "autism is a cognitive and neurobiological disorder marked and caused by underfunctioning integrative circuitry that results in a deficit of integration of information at the neural and cognitive levels". This theory has gained much attention in a large number of studies investigating functional and structural connectivity (for more information see: Belger et al., 2011; Geschwind & Levitt, 2007; Müller, 2007; Wass, 2010).

While there is growing consensus that autism is associated with atypical brain connectivity, there is less agreement regarding the under-connectivity theory, and the location of these abnormal networks. Studies using various methodologies, such as EEG, MEG, fMRI and DTI have reported evidence of under-connectivity in autism (Brock et al., 2002; Castelli, 2002; Just et al., 2004), while others treat the problem as one of over-connectivity (Belmonte & Yurgelun-Todd, 2003; Courchesne & Pierce, 2005; Hutsler & Zhang, 2010).

Several explanations have been proposed to reconcile these two ideas. Belmonte (Belmonte et al., 2004) suggested that "high local connectivity may develop in tandem with low long-range connectivity". This was further supported by other studies reporting a deficiency in the quality of long-range cortico-cortical connections in ASD (Hughes, 2007; Jou et al., 2011) and an increase in short-range connections, as well as connections between subcortical areas and the cortex (Mizuno et al., 2006). In addition, specific methodological characteristics including the choice of tasks were found to affect the results reported in different functional connectivity MRI studies (Müller et al., 2011).

In summary, it is likely that alteration of structural organization underlies functional and behavioral impairment in ASD. As a developmental disorder, we should focus on the trajectories of brain development, in order to better understand the pathology underpinning autism. This will enable better understanding of the nature of this disorder. This chapter will focus on structural connectivity in subjects with autism as detected using MRI with the aim of investigating the integrity and developmental changes of white matter (WM) across the life span. In order to understand abnormal development, a brief review of normal development will first be presented.

2. Normal brain development

The development of the human brain involves extensive structural and neuro-chemical dynamic changes throughout life, with different tissue types, brain structures, and neural circuits exhibiting distinct developmental trajectories. Structural MRI provides information regarding brain development and characterizes age-related changes in brain volume, maturation, cortical thickness and gyrification (Giedd & Judith L. Rapoport, 2010a; Gogtay et al., 2004; Power et al., 2010; Shaw et al., 2008; Vol & Morfologicas, 2000). The focus of this review is WM development therefore the discussion of gray matter (GM) changes will be limited.

2.1 Volumetric changes during brain development

Age-related changes in GM and WM volume have been shown to vary according to sex and brain region. Converging results have been reported by numerous studies, including a large-scale longitudinal study performed at the Child Psychiatry Branch of the National Institute of Mental Health (Giedd et al., 2010; Lenroot & Giedd, 2006). The general pattern for typical brain development in the first 25 years of life is a roughly linear age-dependent increase in WM volume with a steeper increase in males than females. Curves for WM development did not significantly differ between various lobes (Giedd et al., 1999). At the age of 5, 90% of the adult brain volume had already developed (Giedd et al., 1996), and only a small increase in volume was detected later in life (Giedd et al., 1999). The general increase in WM volume throughout childhood and adolescence may reflect greater connectivity and integration of disparate neural circuitry.

In contrast, GM structures show a general pattern of regionally specific inverted U shaped developmental trajectories, with peak volumes occurring in late childhood or early adolescence (Lenroot & Giedd, 2006). Developmental curves for the different cortical regions significantly differed from each other; those for frontal and parietal lobes were the most similar. The absolute size of the cortical GM was approximately 10% larger in boys, and peaked slightly earlier in girls, although the shape of the curves was not significantly different between boys and girls (Giedd et al., 1999).

Brain development including maturation of functional networks and the specific timing and synchronization of the developmental processes across different brain regions should correlate with the well-known temporal sequences of behavior development. Abnormal development of some brain areas will affect the developmental trajectories of networks and cause a failure to acquire normal behavior.

2.2 Diffusion Weighted Imaging (DWI) & Diffusion Tensor Imaging (DTI)

While conventional MRI methods can provide information about changes in the volume and shape of brain structure, diffusion weighted imaging (DWI) can provide additional information characterizing the microstructure of the tissues (Basser & Jones, 2002; Le Bihan, 2003; Moseley et al., 1990). DWI can detect, indirectly, differences between tissue compartments such as size and geometrical shape. Some compartments have isotropic shape (i.e. the water motion is roughly equivalent in all directions, such as in the CSF) while other compartments have anisotropic shape (i.e. the diffusion is more restricted in one axis, which results in anisotropic diffusion, such as in the WM). Diffusion tensor imaging (DTI) can detect information regarding the size and shape of the compartments, by recording the diffusion of water molecules in more than six directions (Basser et al., 1994; Basser & Jones, 2002; Le Bihan & van Zijl, 2002). DTI was previously shown to be a sensitive method for the study of WM connectivity, integrity, development and pathology (Basser et al., 1994; Dubois et al., 2006; Gupta et al., 2005; Huang et al., 2006; Hüppi & Dubois, 2006; Lebihan, 2006; Neil et al., 2002; Wakana et al., 2003).

2.2.1 Diffusion parameters

Several diffusion parameters describe the brain's microstructure, including the three diffusion tensor eigen values (λ_1, λ_2, λ_3), which represent diffusion along the three principal tensor axes, the mean diffusivity (MD) and mathematical measures of anisotropy.

The **mean diffusivity** (MD) is calculated as one third of the trace of the diffusion tensor:

$$MD = \frac{\lambda_1 + \lambda_2 + \lambda_3}{3} \tag{1}$$

MD represents the overall magnitude of water diffusion independent of anisotropy. This parameter provides information on restriction and boundaries (i.e. the extent of packing/ density of cells), and is therefore a sensitive measure of brain maturation and/or injury (Alexander et al., 2007; Dubois et al., 2006).

Water diffusion anisotropy can be described by several parameters (Basser & Pierpaoli, 1996; Uluğ & van Zijl, 1999), of which **fractional anisotropy (FA)** is the most common. The FA parameter is calculated by dividing the magnitude of the anisotropic part of the diffusion tensor, by the magnitude of its isotropic part, resulting in a parameter that describes the degree of water diffusion anisotropy independent of the overall water diffusion coefficient (Basser & Pierpaoli, 1996):

$$FA(\lambda_1,\lambda_2,\lambda_3)=\frac{1}{\sqrt{2}}\sqrt{\frac{(\lambda_1-MD)^2+(\lambda_2-MD)^2+(\lambda_3-MD)^2}{\lambda_1^2+\lambda_2^2+\lambda_3^2}} \quad (2)$$

FA ranges between 0 (for perfectly isotropic diffusion; diffusion that is equal in all directions) and 1 (the hypothetical case of an infinite cylinder, i.e $\lambda_1 \gg\!> \lambda_2=\lambda_3$). A high FA value is detected in a dense and ordered structure such as in WM (Basser & Pierpaoli, 1996; Basser & Jones, 2002). This parameter is considered to reflect fiber density, axonal diameter, and myelination in WM (Alexander et al., 2007; Hüppi & Dubois, 2006).

The FA parameter is frequently used and appears to be quite sensitive to a broad spectrum of pathological conditions. Yet although many studies primarily focus on diffusion anisotropy, this may not be enough to characterize the tissue changes (Alexander et al., 2007). Since the FA value is calculated from the three eigenvalues, different eigenvalue combinations can generate the same FA value. For example, higher eigenvalues may indicate less maturation / myelinization and lower ones may indicate higher maturation, while in both cases the FA value may be reduced. Therefore, looking solely at the FA parameter can occlude trends that may be apparent in a specific eigenvalue. The eigenvalues of the diffusion tensor (axial diffusivity and radial diffusivity), are therefore important for the characterization of changes in the tissue microenvironment.

Axial diffusivity ($Da=\lambda_1$) represents the direction in which water diffusion is highest, which is typically parallel to WM fiber fascicles and is more strongly related to axonal morphology and degradation (Budde et al., 2009). Factors affecting axial diffusivity include buildup of cellular debris, breakdown of axonal structure, disordered microtubule arrangement, aggregation of filaments, and expansion of extracellular space (Schwartz et al., 2005).

Radial diffusivity ($Dr=(\lambda_2+\lambda_3)/2$) is the average value of the two small eigenvalues, and is considered to reflect the diffusivity orthogonal to the axonal bundles and to be mainly affected by the myelin in WM (Song et al., 2002).

2.2.2 Diffusion at high b values

The use of high b values (above > 3000 sec/mm^2) requires a different data analysis approach which results in additional diffusivity parameters (Cohen & Assaf, 2002; Inglis et al., 2001; Ronen et al., 2005). Using high b values, several groups have reported multi-exponential decay of the MRI signal as a function of the b value, detecting signal from different tissue compartments, such as extracellular and intracellular (Inglis et al., 2001; Ronen et al., 2003). High b value DWI seems to be more sensitive to WM integrity than conventional DTI (Ronen et al., 2005) and was used in several studies including investigations of WM development in typically developing children (Ben Bashat et al., 2005; Cihangiroglu et al., 2009) and in young children with autism (Ben Bashat et al., 2007).

2.3 DTI and typical development

DTI has been widely used to describe WM development in children and adolescents (Cascio et al., 2007; Hüppi & Dubois, 2006; Rutherford et al., 1991; Sakuma et al., 1991). Reported changes are consistent across studies with an overall decrease in MD and an increase in FA with age (Barnea-Goraly et al., 2005; Cascio et al., 2007; Mukherjee et al., 2001). A decrease in all three eigen values was reported in newborns and infants from birth to childhood with a much higher rate of decline in the two smaller eigenvalues (Dr) than that of the largest eigenvalue (Da), resulting in an actual increase in FA (Hüppi et al., 1998; Mukherjee et al., 2001; Neil et al., 1998; Song et al., 2002).

The increase in anisotropy with age reflects increased organization of the nerve fibers. Relatively early DTI studies show that a large degree of anisotropy is already present in non-myelinated nerves (Beaulieu & Allen, 1994) or only poorly myelinated fibers such as in the WM of premature newborns (Hüppi et al., 1998; Neil et al., 1998). This increase has been attributed to changes in WM structure which accompany the "premyelinating state" (Wimberger et al., 1995) and are characterized by several histologic changes, including an increase in the number of microtubule-associated proteins in axons, a change in axon caliber, and a significant increase in the number of oligodendrocytes (Hüppi & Dubois, 2006). Following this stage, a continued increase in anisotropy is associated with the histologic appearance of myelin and its maturation (Huppi et al 1998; Huppi and Dubois 2006). These microstructural changes during development are not homogenous throughout the brain, showing considerable regional differences (Dubois et al., 2008; Hüppi & Dubois, 2006).

3. Brain development in autism - MRI findings

MR imaging studies have reported significant changes in GM and WM in subjects with autism compared to age-matched controls. Differences were detected in several brain regions both in the cerebrum and cerebellum (Amaral et al., 2008; Brambilla, 2003). However, it is important to emphasize that to date, all imaging results have been based on group analyses therefore it is not yet possible to make assumptions on an individual level.

3.1 Accelerated brain growth

One of the most consistent findings in autism research is increased brain volume during the first 2-3 years of life. The initial characterization of increased brain size and growth in autism relied heavily on head circumference data, later corroborated by structural MRI studies which established the correlation between head circumference and brain volume (Courchesne et al., 2001). Courchesne and colleagues, using retrospective head circumference records, found that brain volumes appeared normal for all children at birth. However, 2-3 year old children with autism had increased cerebral (18%) and cerebellar (39%) WM, and more cerebral cortical GM (12%) than controls (Courchesne et al., 2001). In contrast, older children and adolescents with autism did not exhibit enlarged gray and WM volumes. Based on these results, the authors hypothesized that overgrowth in autism is restricted to early childhood and followed by a period of abnormally slowed growth. Later studies supported these initial findings (Courchesne et al., 2004; Dementieva et al., 2005), although it was concluded that the increased brain volume was present in only about 70% of children with autism (Lainhart, 2006). This period of accelerated brain growth that occurs in the first years of life is parallel to the emergence of autistic symptoms. Both Courchesne's

early overgrowth theory and current research suggest that overgrowth is not ubiquitous to all regions of the brain. By 2-4 years of age, overgrowth is more evident in some regions and structures than others (Courchesne et al., 2005; Sparks et al., 2002), with the frontal lobes, temporal lobes, and amygdala being the sites of peak overgrowth (Sparks et al., 2002).
A few studies reported differences in older subjects with autism compared to controls. While there is converging evidence that autism is associated with enlarged brain volume early in development, evidence regarding the arrest of this overgrowth abnormality in adulthood are less conclusive demonstrating mixed results. Herbert et al. (Herbert et al., 2004), utilizing a WM parcellation technique, reported an enlargement in the radiate WM in all lobes, particularly in the frontal lobe in high functioning children with autism at a mean age of 9±0.9 years. Aylward (Aylward et al., 2002) reported significantly larger brain volumes in children with autism up to the age of 12 compared to controls, but no differences for individuals older than 12 years. In contrast, Piven and colleagues (Piven et al., 1996) did find significant enlargement in the temporal, parietal, and occipital, but not frontal lobes in 35 subjects with autism, with a mean age of 18 years. Another study reported decreased WM and GM volumes in children with autism at a mean age of 12±1.8, using a voxel-based analysis (McAlonan et al., 2005).

3.2 Volumetric changes in the corpus callosum

In contrast to the increased brain and cerebellar volume, the corpus callosum (CC) seems to be smaller in autism across all age groups (Boger-Megiddo et al., 2006; Stanfield et al., 2008). As the largest fiber connecting the two cerebral hemispheres, the CC has a central role in almost all networks and behaviors, including motor, sensory, visual, cognitive and limbic among other. While most researchers agree on the involvement of the CC in autism, the specific part of the CC which is affected is under debate. Piven and colleagues (Piven et al., 1997) detected smaller size of the body and posterior sub-regions of the CC in individuals with autism, at mean age 47.4 months, and reduced size of the CC only when adjusted for cerebral volume. Vidal et al (Vidal et al., 2006) reported reduction in both the splenium and genu of the CC in subjects with autism, mean age 10±3.3. In a recently published meta-analysis, reduced total CC area was detected in subjects with autism versus healthy controls (Frazier & Hardan, 2009), with the rostral body (Witelson subdivision 3) of the CC demonstrating the largest reduction in volume. Yet, other studies found no significant differences in the CC in adults with high functioning autism (Tepest et al., 2010).
Reduced size of the CC in autism has been demonstrated in contrast to the increased volume of WM which is mainly detected at young ages. Although this finding seems to be consistent in autism, it is non-specific. Reduced volume of the CC was also reported in many other disorders including attention deficit hyperactivity disorder (ADHD) (Giedd et al., 1994) and schizophrenia (Shenton et al., 2001). This leads us to question whether the pathology underlying this imaging abnormality is unique to autism or common to several disorders?

3.3 DTI findings in autism: review of published articles

This review of current published articles on DTI and autism was based on a search in Pubmed.gov performed on the 31st of March, 2011 (Table 1). The search criterion was "DTI and Autism", "DTI and ASD", "DWI and Autism" and "DWI and ASD". Only articles in English and those performed on humans were reviewed, a total of 25 articles. Two additional articles were retrieved from citations in the reviewed articles.

Diffusion Tensor Imaging Studies in ASD

Study	Method	Diagnosis	Group; no.		Group; age, mean (SD) yr		Results				
			Autism	Control	Autism	Control	FA	MD	Da	Dr	Location
Barnea-Goraly et al. 2004	VB	HFA	7	9	14.6±3.4	13.4±2.8	↓	-	-	-	Frontal and temporal WM
Keller et al. 2006	VB	HFA	34	31	18.97±7.3	18.97±6.2	↓	-	-	-	CC and right retrolenticular portion of the IC
Alexander et al. 2007	Semi automated VOI	HFA, PDD-NOS	43	34	16.23±6.70	16.44±5.97	↓	↑	NS	↑	CC
Lee et al. 2007	Semi automated VOI	Autism, PDD-NOS	43	34	16.2±6.7	16.4±6.0	↓	↑	NS	↑	STG and the temporal stem
Ben Bashat et al. 2007	High b value; ROI analysis	Autism	7	41	range: 1.8–3.3y	range: 4m-23y	↑	-	-	-	Genu and splenium - CC, PLIC IC; more in the left hemisphere
Catani et. Al. 2008	Tractography	Asperger	15	16	31±9	35±11	↓	NS	-	-	Short intracerebellar fibres, right superior CP
Thakkar et al. 2008	Surface based analysis	Autism, Asperger, PDD-NOS	12	14	30±11	27±8	↓	-	-	-	Anterior cingulate cortex

Table 1. DTI findings in autism

ASD;autism spectrum disorder,SD;standard deviation, FA;fractional anisotropy, MD;mean diffusivity, Da;axial diffusivity, Dr;radial diffusivity, VB;voxel based, HFA;High functioning autism, WM;white matter, CC;corpus callosum, IC;internal capsule, VOI/ROI;volume/region of interest, PDD-NOS;pervasive developmental disorder-not otherwise specified, NS;not significant, STG/STS;superior temporal gyrus/sulcus, PLIC;Posterior limb of the internal capsule, CP;cerebral peduncle,

Diffusion Tensor Imaging Studies in ASD

Study	Method	Diagnosis	Group; no.		Group; age, mean (SD) yr		Results				
			Autism	Control	Autism	Control	FA	MD	Da	Dr	Location
Sundaram et al. 2008	Tractography	Autism, Asperger, PDD-NOS	50	16	4.79±2.43	6.87±3.45	↓	↑	-	-	Frontal lobe (significant for short range fibers)
Pugliese et al. 2009	Tractography	Asperger	24	42	23±12	25±10	NS	NS	-	-	Limbic pathways
Pardini et al. 2009	Tractography and VB	Autism	10	10	19.7 ± 2.83	19.9 ± 2.64	↓	-	-	-	Orbitofrontal cortex
Lee et al. 2009	VBM	HFA, PDD-NOS	43	34	16.23±6.70	16.44±5.97	↓	↑	-	-	STG, anterior cingulate, thalamus and CC
Ke et al. 2009	VB	HFA	12	10	8.75±2.26	9.40±2.07	↓	-	-	-	Frontal lobe and left temporal lobe
Thomas et al. 2010	Tractography	HFA	12	18	28.5±9.7	22.4±4.1	NS	-	-	-	None
Lange et al. 2010	Semi automated VOI	HFA	30	30	15.78±5.6	15.79±5.5	↓	↑	↑	↑	STG and temporal stem

Table 1. DTI findings in autism

ASD;autism spectrum disorder,SD;standard deviation, FA;fractional anisotropy, MD;mean diffusivity, Da;axial diffusivity, Dr;radial diffusivity, VB;voxel based, HFA;High functioning autism, WM;white matter, CC;corpus callosum, IC;internal capsule, VOI/ROI;volume/region of interest, PDD-NOS;pervasive developmental disorder-not otherwise specified, NS;not significant, STG/STS;superior temporal gyrus/sulcus, PLIC;Posterior limb of the internal capsule, CP;cerebral peduncle

Diffusion Tensor Imaging Studies in ASD

Study	Method	Diagnosis	Group; no.		Group; age, mean (SD) yr		Results				Location
			Autism	Control	Autism	Control	FA	MD	Da	Dr	
Fletcher et al. 2010	Volumetric DTI segmentation	HFA	10	10	14.25±1.92	13.36±1.34	NS	↑	NS	↑	Arcuate fasciculi
Cheng et al. 2010	TBSS	ASD	25	25	13.71±2.54	13.51±2.20	↓or ↑ region dependent	-	↓or ↑ region dependent	↓or ↑ region dependent	Right PLIC, frontal lobe, bilateral insula, right cingulate gyrus, right STG, CP
Noriuchi et al. 2010	VB	HFA, Asperger	7	7	13.96±2.68	13.36±2.74	↓	-	↓	NS	Left DLPFC, posterior STS, right temporal pole, amygdala, SLF, occipitofrontal fasciculus, CC, cingulate
Shukla et al. 2010a	VOIs/ROIs	Autism, Asperger	26	24	12.7±0.6	13.0±0.6	↓	↑	↓	↑	Whole brain, CC, IC, middle CP, IC
Shukla et al. 2010b	TBSS	Autism, Asperger	26	24	12.8±0.6	13.0±0.6	↓	↑	NS	↑	CC, IC, ILF, inferior fronto-occipital fasciculus, SLF, cingulum, anterior thalamic radiation, corticospinal tract
Barnea-Goraly et al. 2010	TBSS	Autism	17	17 siblings 18 controls	10.5±2	8.9±1.9 9.6±2.1	↓	-	↓	NS	Frontal parietal and temporal lobes
Sahyoun et al. 2010	TBSS	HFA	9	12	12.8±1.5	13.3±245	↑	-	-	-	Peripheral WM, including the ventral temporal lobe

Table 1. DTI findings in autism

ASD;autism spectrum disorder,SD;standard deviation,FA;fractional anisotropy, MD;mean diffusivity, Da;axial diffusivity, Dr;radial diffusivity, VB;voxel based, HFA;High functioning autism, WM;white matter, CC;corpus callosum, IC;internal capsule, VOI/ROI;volume/region of interest, PDD-NOS;pervasive developmental disorder-not otherwise specified, NS;not significant, STG/STS;superior temporal gyrus/sulcus, PLIC;Posterior limb of the internal capsule, CP;cerebral peduncle, DLPFC;dorsolateral prefrontal cortex, ILF/SLF;inferior/superior longitudinal fasciculus

Diffusion Tensor Imaging Studies in ASD

Study	Method	Diagnosis	Group; no.		Group; age, mean (SD) yr		Results				
			Autism	Control	Autism	Control	FA	MD	Da	Dr	Location
Bloeman et al. 2010	VB technique	Asperger	13	13	39±9.8	37±9.6	↓	↓	-	↑	WM IC, frontal, temporal, parietal and occipital lobes, cingulum and CC
Shukla et al. 2011	TBSS	Autism, Asperger	26	24	12.8±0.6	13.0±0.6	↓	↑	NS	↑	Frontal,parietal,and temporal lobes
Mengotti 2011	DWI VB and ROI	Autism	20	22	7±2.75	7.68±2.03	-	↓or ↑ Age dependent	-	-	Bilateral frontal cortex and in the left side of the genu of the CC
Weinstein et al. 2011	Tractography and TBSS	Autism	22	32	3.2±1.1	3.4±1.3	↑	NS	NS	↓	Genu and body of the CC, left SLF and right and left cingulum
Groen et al. 2011	Kurtosis VB	HFA	17	25	14.4±1.6	15.5±1.8	NS	↑	-	-	Whole WM and GM
Jou et al. 2011	Tractography and VB	ASD	10	10	13.06 ±9.85	13.94±4.23	↓	-	-	-	ILF/inferior fronto-occipital fasciculus, SLF,and CC/cingulum

Table 1. DTI findings in autism

ASD;autism spectrum disorder,SD;standard deviation,FA;fractional anisotropy, MD;mean diffusivity, Da;axial diffusivity, Dr;radial diffusivity, VB;voxel based, HFA;High functioning autism, WM;white matter, CC;corpus callosum, IC;internal capsule, VOI/ROI;volume/region of interest, PDD-NOS;pervasive developmental disorder-not otherwise specified, NS;not significant, STG/STS;superior temporal gyrus/sulcus, PLIC;Posterior limb of the internal capsule, CP;cerebral peduncle, DLPFC;dorsolateral prefrontal cortex, ILF/SLF;inferior/superior longitudinal fasciculus

Barnea Goraly and colleagues (Barnea-Goraly, 2004) were the first to apply DTI to a small number of children with autism and a control group using a voxel-based approach. They reported reduced FA in the CC and in the WM of the ventromedial prefrontal cortices, anterior cingulated gyri and temporoparietal junctions, indicating a reduction in WM integrity in the autism group. Following this auspicious start, most autism research to date has avoided early childhood studies, focusing instead on high functioning adolescents or adults with ASD, and focused mainly on FA and / or MD, to the exclusion of other diffusivity parameters. Most studies in these age groups reported reduced integrity of the WM in several brain regions including the limbic system, temporal and frontal lobes and CC.

Many studies were conducted in subjects from the entire ASD spectrum, including autism, Asperger's and PDD-NOS, i.e. a very heterogeneous group. Consistent findings in these studies were reduced FA in the CC (Alexander et al., 2007; Jou et al., 2011; Lee et al., 2009; Shukla et al., 2010, 2011b) in the superior temporal gyrus (Cheung et al., 2009; Lee et al., 2007, 2009; Shukla et al., 2011a); the anterior cingulate cortex (Cheng et al., 2010; Thakkar et al., 2008); the frontal lobe (Cheng et al., 2010; Shukla etl al., 2011a; Sundaram et al., 2008); the thalamus (Lee et al., 2009) and the interior and posterior limbs of the internal capsule (Cheng et al., 2010; Shukla et al., 2010).

Most studies conducted on subjects with high functioning autism or with Asperger's syndrome reported reductions in FA in the frontal and temporal lobes (Barnea-Goraly, 2004; Bloemen et al., 2010; Ke et al., 2009; Lange et al., 2010; Noriuchi et al., 2010); the limbic system (Bloemen et al., 2010; Noriuchi et al., 2010; Pugliese et al., 2009); the superior longitudinal fasciculus (Noriuchi et al., 2010); the CC (Bloemen et al., 2010; Keller et al., 2006; Noriuchi et al., 2010); and cerebellum (Catani et al., 2008). One study reported areas of reduced FA in children with high functioning autism (mean age 12.8 years) compared to controls, within the frontal WM and the superior longitudinal fasciculus, and increased FA within peripheral WM (Sahyoun et al., 2010). Three studies did not find any significant differences in FA values (Groen et al., 2011; Fletcher et al., 2010; Thomas et al., 2010).

And finally, studies conducted only in subjects with autism (subgroup, not the entire spectrum) reported reductions in FA in the frontal, parietal and temporal lobes (Barnea-Goraly et al., 2010); the frontal lobe (Mengotti et al., 2011; Pardini et al., 2009); and CC (Mengotti et al., 2011).

Nine of eleven studies that investigated axial and radial diffusivity, reported increased Dr along with increased MD and reduced FA in subjects with autism compared to controls (one study reported mixed results, region dependent). Five of these articles did not find significant results in Da, while another five reported mixed results (3 reported reduced Da, one region dependent, and one increased Da). Two studies reported reduced Da without significant results in Dr (see Table 1).

A few diffusion studies reported results in young children with autism (<6 years) (Ben Bashat et al., 2007; Mengotti et al., 2011; Sundaram et al., 2008; Weinstein et al., 2011). Two of these studies reported an opposite trend of increased FA values, in several brain regions. The first study, using high b value DWI, demonstrated an increase in FA values in the frontal lobe of 1.8-3.3 year old children with autism (Ben Bashat et al., 2007). Higher restriction was more dominant in the left hemisphere and was mainly detected in the frontal lobe, indicating abnormal density with regards to age. Higher restriction and increased FA values were also detected in the genu and splenium of the CC (the body of the CC was not studied). In this study, it was suggested that early and accelerated abnormal maturation occurring in young subjects with autism supports brain overgrowth at these ages. In the second study, increased FA was detected in the genu and body of the CC (Witelson subdivision 3), left superior longitudinal fasiculus and right cingulum compared with age matched controls (Weinstein et al., 2011). Changes in FA reported in this study were mainly driven by a decrease in Dr. A third study (Mengotti et al., 2011) used DWI, and reported reduced MD in a restricted group of 7 children with autism (mean age 7.28 years old) in the frontal cortex, the genu and splenium of the CC. The fourth study performed at young ages included children from the entire spectrum, and reported inverse results of reduced FA in short range association fibers (Sundaram et al., 2008).

To sum up, several brain regions show structural differences between subjects with autism and controls. The reported regions form part of major networks relating to several behaviors that are recognized as core deficits in autism. These findings support the conception of autism as a connectivity disorder. The diversity of findings might be due to the numerous issues inherent in autism research in general, and imaging studies in particular, as well as age-related differences, which will be further discussed.

4. Abnormal developmental trajectories

Studies of young subjects with autism indicate increased FA which contrasts with findings of reduced FA in adolescents and young adults. This inconsistency seems to be age-related. The majority of autism imaging studies to date have been conducted on adolescents and young adults. Some studies present developmental curves of FA according to age, and extrapolation of this data points to increased FA at young ages, although most authors did not discuss these findings. A consideration of the data presented in the study by Lee et al, (Lee et al., 2007) (see Figure 2 in that manuscript), suggests that at younger ages (<12 years) there could be increased FA in the temporal stem relative to normal controls. A similar trend can be seen in another study (Shukla et al., 2011a) with extrapolated increased FA and reduced MD in children with autism younger than 8 years, in whole brain WM skeleton compared to controls (see Figure 3 in that manuscript). In a study performed by Cheng et al, (Cheng et al., 2010), FA was higher in children with autism below the age of ~13 years in several brain regions: the right paracentral lobule, right superior frontal gyrus and left superior longitudinal fasciculus (see Figure 3 in that manuscript). Mengotti et al (Mengotti et al., 2011) detected reduced MD in a restricted group of 7 children within the autism subgroup, mean age of 7 years, in the bilateral frontal cortex and in the left side of the genu of the CC.

A similar concern relating to the possibility of age dependency of other imaging measurements in autism can be seen in volumetric measurement, both of WM volume, such as in the frontal lobe and several GM structures. In both cases enlargement was found in young children with autism, while these differences were either not present or reversed in older subjects. GM volume in younger individuals with autism was larger for the amygdala (Sparks et al., 2002) and smaller in the cerebellar vermal lobules compared to controls. These findings were either less pronounced or were not present in older groups (Stanfield et al., 2008). Similar differences were detected in studies using event-related potentials (ERPs), detecting higher amplitude of response to unexpected novel event in subjects with autism compared to normal controls during childhood, and the opposite during young adulthood (Ferri, 2003).

In summary, higher FA and reduced MD were detected in young subjects with autism compared to controls. This pattern seems to be reversed above the age of 7-13 years when reduced FA and increased MD are detected in subjects with autism compared to controls. In the age range of 7-13 there seems to be a period of "pseudo-normalization" of the FA and MD. Longitudinal studies are still needed to confirm this assumption, yet, studies that aim to compare subjects with autism to controls, at a specific age, are recommended not to focus on this age range, since significant results are less likely to be detected.

4.1 What is the pathology underlying abnormal white matter development?

Accelerated brain growth in young children with autism, as measured by volumetric studies, seems to coincide with increased FA and reduced MD, as detected by DTI. What leads to this deviant developmental trajectory? Could excessive prenatal neurogenesis, abnormal pruning or excessive dendrite growth be involved, resulting in aberrant connectivity and overall brain enlargement after birth? Or is there perhaps an increase of myelination or inflammatory response leading to excessive microglial activation? (Schumann & Nordahl, 2010). Post mortem and genetic studies support the combination of several cellular factors that account for autism pathology during early development (Morgan et al., 2010; Rubenstein, 2010; Schumann & Nordahl, 2010). Genetic studies reveal that a number of mutations converge on a common neurodevelopmental pathway involved in neurogenesis, axon guidance and synapse formation, all of which are critical for proper neural connectivity (Benvenuto et al., 2009; Geschwind, 2009). A recent post mortem study, detected microglial activation and increased microglial density in two thirds of their sample of young children (n=5, age < 6 years) (Morgan et al., 2010). An over expression of some or of a combination of these processes can explain the reduced MD and increased FA in young children with autism. Although FA seems to be highly sensitive to microstructural changes, it is less sensitive to the type of changes (Alexander et al., 2007). Changes in FA should therefore be interpreted with caution, and examining other diffusivity values may improve our understanding. Reduced Dr without significant change of Da was detected at a young age, accounting for the reduced MD and increased FA. Normal developmental studies, related reduction in Dr with the myelination process (Song et al., 2002). It is postulated therefore, that the reduced Dr in autism may express over-myelination at a young age. Another imaging study supports this finding (using T2 weighted), showing overdevelopment of WM in several brain regions in children with autism, which were considered to reflect myelination changes (Carmody & Lewis, 2010).

Therefore, it is hypothesized that accelerated-myelination contributes to brain overgrowth in young children with autism, probably along with other developmental processes. Is this finding unique to autism? Once again, findings are mixed. Previous studies in children with developmental delay showed reduced myelination, (Pujol et al., 2004). Reduced WM volume was detected in subjects with ADHD (Castellanos et al., 2002). In contrast, a study of children with developmental language disorder showed increased volume and later or longer-myelinating regions compared to controls, similar to autism findings (but not in all brain regions) (Herbert et al., 2004).

5. Why have we failed so far to find imaging biomarkers?

5.1 Imaging research in autism – only the beginning

Despite the developmental nature of the disorder, it seems that autism research focused on early childhood, while critical to our understanding of the nature of abnormal development, is still in the early stages. While evidence of accelerated brain growth during the first years of life has been accumulating for over a decade, there has not been much progress in the interpretation of this finding. Most DTI studies in autism were performed on adults and adolescents, with just a few studies performed at young ages (<6 years). In addition, there are currently no post mortem studies on young children focusing on developmental trajectories which could shed light on the underlying pathology. Hence, future neuroimaging studies can contribute substantially to the understanding of the neurobiology of autism and, in particular, to the understanding of the important distinction between congenital pathology and acquired impairments.

A reduction in the age of diagnosis of ASDs, access to services and early intensive intervention are crucial for improving developmental outcomes (Dawson et al., 2010). Identification of imaging markers that may assist in early diagnosis is therefore of the utmost importance. In addition, studies of autism at young ages may be more sensitive to the origin of the disorder and may be less confounded by developmental changes, medication, seizures and more. Future studies should aim to identify imaging biomarkers with an emphasis on young ages.

5.2 Inherent problems in autism research

In most autism research, study populations vary widely, due to the heterogeneity of ASDs, as well as differences in diagnostic criteria, subject characteristics (including age, IQ, etc.) and research methodologies. These core issues no doubt account for some of the diversity in results reported in the literature and the failure to find any biomarker, as yet.

The definition of autism is the first and probably the major problem. ASD is a category that includes autistic disorder, pervasive developmental disorder not otherwise specified (PDD-NOS), Asperger's syndrome, disintegrative disorder and Rett syndrome. Considerable debate exists as to whether conditions at the higher functioning end of the autistic spectrum (i.e., high functioning autism, Asperger's syndrome, and PDD-NOS) are separate disorders or simply different expressions of the same underlying condition (Macintosh & Dissanayake, 2004; Matson, 2007).

In addition, there is a wide developmental and behavioral heterogeneity within each group in the spectrum with wide ranging symptoms (S.E. Levy et al, 2009). Some children display signs of developmental delay within the first 18 months of life, however, 25-40% of children with autism initially demonstrate near-normal development until 18-24 months, when they regress into an autism that is generally indistinguishable from early onset autism (Ozonoff et al., 2008). Some children develop language while others remain non verbal; some display interest in social interaction while others remain secluded; some manifest repetitive or obsessive compulsive behavior patterns while others do not; some respond well to therapeutic intervention while others show limited progress (Ben-Itzchak & Zachor, 2007; Charman et al., 2011; Luyster et al., 2008; Pelphrey et al., 2011).

Furthermore, co-morbidity is highly prevalent in ASD. For example, 40-75% of children with ASD are mentally retarded (Fombonne, 2003; Newschaffer et al., 2007) and epilepsy is reported in up to one third of children (Jeste, 2011). Several genetic diseases have also been associated with autism including Fragile-X and Tuberous sclerosis complex (Benvenuto at al., 2009).

Protocol parameters and the various image processing approaches used in imaging studies, may also affect results. Some studies used volume of interest definition, either by manual selection or using reconstruction of a specific fiber bundle (i.e. fiber tractography) while others used whole brain comparison (including voxel based DTI analysis and tract-based special statistics (TBSS)). Several inherent differences between these two approaches may account for some of the contrasting results reported in the literature (Alexander et al., 2007).

5.3 Is studying autism as a syndrome the right direction?

Autism has diverse clinical manifestations, behavioral phenotypes, developmental dimensions and genetic origins, all of which complicate research and clinical practice with regard to etiology, the selection of appropriate interventions and the search for biomarkers. Most DTI studies in autism were performed on high functioning subjects, since scanning subjects with low functioning disorders is more difficult. It is therefore debatable whether similar findings can be expected among these heterogeneous groups and whether any conclusions can be drawn about the whole spectrum or generalized to other groups, based on findings in one particular group.

Recently Happe et al (Happé et al., 2006) argued that attempts to propose a unified account of autistic symptoms failed at all levels of analysis - genetic, imaging and behavioral. Bearden & Freimer (Bearden & Freimer, 2006) claim that the inherent imprecision of behavioral phenotyping is probably the most important factor contributing to the failure to discover the biological factors involved in psychiatric and neurodevelopmental disorders. In a recent review article, Levy and Ebstein (Y. Levy & Ebstein, 2009) argue that syndrome heterogeneity, cross-syndrome similarities and syndrome comorbidities challenge the relevance of syndromes to biological research, and that cohort selection based on cross-syndrome trait classification would be more accurate than based on syndromic groups.

6. Conclusions

In summary, current theories of neural deficiencies in autism emphasize the first few years of life as a key period when abnormalities in the development of neural circuitry occur, along with the first behavioral signs of autism. These abnormalities, which can be detected on the basis of group differences, persist into adulthood. Despite recent advances in autism research, early childhood neuroimaging studies are few and far between, hindering investigation of the developmental nature of autism. Thus the specific relationship between etiology, mechanisms, genetic and imaging markers and the ensuing behavioral abnormalities remains unclear.

Abnormal trajectories in WM development in autism, resulting in impaired connectivity, have been demonstrated by imaging studies. While several mechanisms may account for the increased brain volume in young children, DTI can detect microstructural changes and might help to reveal the neurobiology underpinning autism. Based on recent findings, it is suggested that accelerated myelination might be one of the processes occurring at a young age in subjects with autism.

Subjects with autism exhibit changes in diffusivity values with age. Higher FA values along with reduced MD were detected in young children with autism, while reduced FA and increased MD were reported at older ages. The shift of FA from higher to lower values results in a period of suggested "pseudo-normalization" which seems to occur between the ages of 7-13 years. This hypothesis accounts for the seemingly controversial results detected in young children versus adolescents and young adults.

There are many contrasting reports regarding the location of abnormalities within the brain of subjects with autism, both during adolescence and at younger ages. This might support asynchrony in maturational processes in different brain regions which may be the basis for abnormal connectivity and behavior. Studies performed at young ages may be able to detect congenital neurobiological pathologies and distinguish these from acquired impairments that are more likely to be detected at older ages.

Longitudinal studies in a large cohort may be the best way to solve the autism puzzle. Integrating several approaches, including genetic, postmortem and imaging, may be the only way to provide answers concerning the neuropathology of autism (Schumann & Nordahl, 2010). In addition, a multimodal approach in imaging studies, which has demonstrated major advantages in several brain pathologies and in a recently published study of autism (Ecker et al., 2010), should be incorporated in future studies.

Future research should focus on subgroups with specific traits of the autistic disorder, or endophenotypes such as language impairment, in order to provide promising avenues for understanding the neurobiological processes underlying autism. Future studies will reveal whether differences are detectable on an individual basis; whether imaging results can be powerful enough to be included in the diagnostic criteria of autism; and whether reported imaging findings are specific enough to differentiate autism from other developmental disorders.

7. Acknowledgment

I am indebted to Maya Weinstein, Moran Artzi, and Reut Granot Halevi, for their helpful assistance in preparing this chapter; Vicki Myers for editorial assistance; Yonata Levy, Liat Ben Sira, Talma Hendler, Ditza Zachor and Esti Ben Itzhak for being part of our group searching for imaging clues in the puzzle of autism; March of Dimes for partially supporting our research; and last but not least, to the children and their families who so willingly collaborated.

8. References

Alexander, A. L., Lee, J. E., Lazar, M., Boudos, R., Dubray, M. B., Oakes, T. R., Miller, J. N., Lu, J., Jeong, E-K., McMahon, W. M., Bigler, E. D., & Lainhart, J. E. (2007). Diffusion tensor imaging of the corpus callosum in Autism. *NeuroImage*, 34, 61 - 73.

Alexander, A. L., Lee, J. E., Lazar, M., & Field, A. S. (2007). Diffusion Tensor Imaging of the Brain. *Neurotherapeutics.*, 4(3), 316-329.

Amaral, D. G., Schumann, C. M., & Nordahl, C. W. (2008). Neuroanatomy of autism. *Trends in neurosciences*, 31(3), 137-45.

Aylward, E. H., Minshew, N. J, Field, K., Sparks, B. F., & Singh, N. (2002). Effects of age on brain volume and head circumference in autism. *Neurology*, 59(2), 175-83.

Barnea-Goraly, N. (2004). White matter structure in autism: preliminary evidence from diffusion tensor imaging. *Biological Psychiatry*, 55(3), 323-326.

Barnea-Goraly, N., Lotspeich, L. J., & Reiss, A. L. (2010). Similar white matter aberrations in children with autism and their unaffected siblings: a diffusion tensor imaging study using tract-based spatial statistics. *Archives of general psychiatry*, 67(10), 1052-60.

Barnea-Goraly, N., Menon, V., Eckert, M., Tamm, L., Bammer, R., Karchemskiy, A., Dant, C., & Reiss, A. L. (2005). White matter development during childhood and adolescence: a cross-sectional diffusion tensor imaging study. *Cerebral cortex*, 15(12), 1848-54.

Basser, P. J, Mattiello, J., & LeBihan, D. (1994). Estimation of the effective self-diffusion tensor from the NMR spin echo. *Journal of magnetic resonance. Series B*, 103(3), 247-54.

Basser, P J, & Pierpaoli, C. (1996). Microstructural and physiological features of tissues elucidated by quantitative-diffusion-tensor MRI. *Journal of magnetic resonance. Series B*, 111(3), 209-19.

Basser, Peter J, & Jones, D. K. (2002). Diffusion-tensor MRI: theory, experimental design and data analysis - a technical review. *NMR in biomedicine*, 15(7-8), 456-67.

Basser, Peter J, & Jones, D. K. (2002). Diffusion-tensor MRI: theory, experimental design and data analysis - a technical review. *NMR in biomedicine*, 15(7-8), 456-67.

Bearden, C. E., & Freimer, N. B. (2006). Endophenotypes for psychiatric disorders: ready for primetime? *Trends in genetics* : TIG, 22(6), 306-13.

Beaulieu, C., & Allen, P. S. (1994). Determinants of anisotropic water diffusion in nerves. *Magnetic Resonance in Medicine*, 31(4), 394-400.

Belger, A., Carpenter, K. L. H., Yucel, G. H., Cleary, K. M., & Donkers, F. C. L. (2011). The Neural Circuitry of Autism. *Neurotoxicity research*.

Belmonte, M. K., Allen, G., Beckel-Mitchener, A., Boulanger, L. M., Carper, R. a, & Webb, S. J. (2004). Autism and abnormal development of brain connectivity. *The Journal of neuroscience : the official journal of the Society for Neuroscience*, 24(42), 9228-31.

Belmonte, M. K., & Yurgelun-Todd, D. A. (2003). Functional anatomy of impaired selective attention and compensatory processing in autism. *Brain research. Cognitive brain research*, 17(3), 651-64.

Ben Bashat, D., Ben Sira, L., Graif, M., Pianka, P., Hendler, T., Cohen, Y & Assaf, Y. (2005). Normal white matter development from infancy to adulthood: comparing diffusion tensor and high b value diffusion weighted MR images. *Journal of magnetic resonance imaging*, 21(5), 503-11.

Ben Bashat, D., Kronfeld-Duenias, V., Zachor, D. a, Ekstein, P. M., Hendler, T., Tarrasch, R., Even, A., Levy, Y., & Ben Sira, L. (2007). Accelerated maturation of white matter in young children with autism: a high b value DWI study. *NeuroImage*, 37(1), 40-7.

Ben-Itzchak, E., & Zachor, D. A. (2007). The effects of intellectual functioning and autism severity on outcome of early behavioral intervention for children with autism. *Research in developmental disabilities*, *28*(3), 287-303.

Benvenuto, A., Moavero, R., Alessandrelli, R., Manzi, B., & Curatolo, P. (2009). Syndromic autism: causes and pathogenetic pathways. *World journal of pediatrics*, 5(3), 169-76.

Bloemen, O. J. N., Deeley, Q., Sundram, F., Daly, E. M., Barker, G. J., Jones, D. K., van Amelsvoort, T. A. M. J., Schmitz, N., Robertson, D., Murphy, K. C., & Murphy, D. G. M. (2010). White matter integrity in Asperger syndrome: a preliminary diffusion tensor magnetic resonance imaging study in adults. *Autism research*, (July), 203-213.

Boger-Megiddo, I., Shaw, D. W. W., Friedman, S. D., Sparks, B. F., Artru, A. A., Giedd, J. N., Dawson, G., & Dager, S. R. (2006). Corpus callosum morphometrics in young children with autism spectrum disorder. *Journal of autism and developmental disorders*, 36(6), 733-9.

Brambilla, P. (2003). Brain anatomy and development in autism: review of structural MRI studies. *Brain Research Bulletin*, 61(6), 557-569.

Brock, J., Brown, C. C., Boucher, J., & Rippon, G. (2002). The temporal binding deficit hypothesis of autism. *Development and psychopathology*, 14(2), 209-24.

Budde, M. D., Xie, M., Cross, A. H., & Song, S-K. (2009). Axial diffusivity is the primary correlate of axonal injury in the experimental autoimmune encephalomyelitis spinal cord: a quantitative pixelwise analysis. *The Journal of neuroscience*, 29(9), 2805-13.

Carmody, D. P., & Lewis, M. (2010). Regional white matter development in children with autism spectrum disorders. *Developmental psychobiology*, 52(8), 755-63.

Cascio, C. J., Gerig, G., & Piven, J. (2007). Diffusion tensor imaging: Application to the study of the developing brain. *Journal of the American Academy of Child and Adolescent Psychiatry*, 46(2), 213-23.

Castellanos, X. F., Lee, P. P., Sharp, W., Jeffries, N. O., Greenstein, D. K., Clasen, L. S., Blumenthal, J. D., James, R. S., Ebens, C. L., Walter, J. M., Zijdenbos, A., Evans, A. C., Giedd, J. N., & Rapoport, J. L. (2002). Developmental Trajectories of Brain Volume Abnormalities in Children and Adolescents With Attention-Deficit / Hyperactivity Disorder. *JAMA*, 288(14), 1740-1748.

Castelli, F. (2002). Autism, Asperger syndrome and brain mechanisms for the attribution of mental states to animated shapes. *Brain*, 125(8), 1839-1849.

Catani, M., Jones, D. K., Daly, E., Embiricos, N., Deeley, Q., Pugliese, L., Curran, S., Robertson, D., & Murphy, D. G. M. (2008). Altered cerebellar feedback projections in Asperger syndrome. *NeuroImage*, 41(4), 1184-91.

Charman, T., Pickles, a, Simonoff, E., Chandler, S., Loucas, T., & Baird, G. (2011). IQ in children with autism spectrum disorders: data from the Special Needs and Autism Project (SNAP). *Psychological medicine*, 41(3), 619-27.

Cheng, Y., Chou, K.-H., Chen, I.-Y., Fan, Y.-T., Decety, J., & Lin, C.-P. (2010). Atypical development of white matter microstructure in adolescents with autism spectrum disorders. *NeuroImage*, 50(3), 873-82.

Cheung, C., Chua, S. E., Cheung, V., Khong, P. L., Tai, K. S., Wong, T. K. W., Ho, T. P., & McAlonan, G. M. (2009). White matter fractional anisotrophy differences and correlates of diagnostic symptoms in autism. *Journal of child psychology and psychiatry, and allied disciplines*, 50(9), 1102-12.

Cihangiroglu, M., Uluğ, A. M., Firat, Z., Bayram, A., Kovanlikaya, A., & Kovanlikaya, I. (2009). High b-value diffusion-weighted MR imaging of normal brain at 3T. *European journal of radiology*, 69(3), 454-8.

Cohen, Yoram, & Assaf, Y. (2002). High b-value q-space analyzed diffusion-weighted MRS and MRI in neuronal tissues - a technical review. *NMR in biomedicine*, 15(7-8), 516-42.

Courchesne, E., & Pierce, Karen. (2005). Why the frontal cortex in autism might be talking only to itself: local over-connectivity but long-distance disconnection. *Current opinion in neurobiology*, 15(2), 225-30.

Courchesne, E., Redcay, E., & Kennedy, D. P. (2004). The autistic brain: birth through adulthood. *Current Opinion in Neurology*, 17(4), 489-496.

Courchesne, E., Redcay, E., Morgan, J. T., & Kennedy, D. P. (2005). Autism at the beginning: microstructural and growth abnormalities underlying the cognitive and behavioral phenotype of autism. *Development and psychopathology*, 17(3), 577-97.

Courchesne, E., Karns, C. M., Davis, H. R., Ziccardi, R., Carper, R. A., Tigue, Z. D., Chisum, H. J., Moses, P., Pierce, K., Lord, C., Lincoln, A. J., Pizzo, S., Schreinbman, L., Hass, R. H., Akshoomoff, N. A., & Courchesne, R. Y. (2001). Unusual brain growth patterns in early An MRI study. *Neurology*, 57, 245-254.

Damoiseaux, J. S., & Greicius, M. D. (2009). Greater than the sum of its parts: a review of studies combining structural connectivity and resting-state functional connectivity. *Brain structure & function*, 213(6), 525-33.

Dawson, G., Rogers, S., Munson, J., Smith, M.., Winter, J., Greenson, J., Donaldson, A & Varkey J. (2010). Randomized, controlled trial of an intervention for toddlers with autism: the Early Start Denver Model. *Pediatrics*, 125(1), e17-23.

Dementieva, Y. a, Vance, D. D., Donnelly, S. L., Elston, L. a, Wolpert, C. M., Ravan, S. A., DeLong, G. R., Abramson, R. K., Wright, H. H., & Cuccaro, M. L. (2005). Accelerated head growth in early development of individuals with autism. *Pediatric neurology*, 32(2), 102-8.

Dubois, J, Hertz-Pannier, L, Dehaene-Lambertz, G, Cointepas, Y, & Le Bihan, D. (2006). Assessment of the early organization and maturation of infants' cerebral white matter fiber bundles: a feasibility study using quantitative diffusion tensor imaging and tractography. *NeuroImage*, 30(4), 1121-32.

Dubois, J., Dehaene-Lambertz, G., Perrin, M., Mangin, J-F., Cointepas, Y., Duchesnay, E., Le Bihan, D., & Hertz-Pannier, L. (2008). Asynchrony of the early maturation of white matter bundles in healthy infants: quantitative landmarks revealed noninvasively by diffusion tensor imaging. *Human brain mapping*, 29(1), 14-27.

Ecker, C., Marquand, a, Mourao-Miranda, J., Johnston, P., Daly, E. M., Brammer, M. J., Maltezos, S., Murphy, C. M., Robertson, D., Williams, S. C., & Murphy, D. G. M. (2010). Describing the Brain in Autism in Five Dimensions--Magnetic Resonance Imaging-Assisted Diagnosis of Autism Spectrum Disorder Using a Multiparameter Classification Approach. *The Journal of Neuroscience*, 30(32), 10612-10623.

Ferri, R. (2003). The mismatch negativity and the P3a components of the auditory event-related potentials in autistic low-functioning subjects. *Clinical Neurophysiology*, 114(9), 1671-1680.

Fletcher, P. T., Whitaker, R. T., Tao, R., DuBray, M. B., Froehlich, A., Ravichandran, C., Alexander, A. L., Bigler, E. D., Lange, N., Lainhart, J. E. (2010). Microstructural connectivity of the arcuate fasciculus in adolescents with high-functioning autism. *NeuroImage*, 51(3), 1117-25.

Fombonne, E. (2003). Epidemiological surveys of autism and other pervasive developmental disorders: an update. *Journal of autism and developmental disorders*, 33(4), 365-82.

Frazier, T. W., & Hardan, A. Y. (2009). A meta-analysis of the corpus callosum in autism. *Biological psychiatry*, 66(10), 935-41.

Friston, K. J. (1994). Functional and effective connectivity in neuroimaging: A synthesis. *Human Brain Mapping*, 2(1-2), 56-78.

Friston, K. J. (2002). Functional integration and inference in the brain. *Progress in neurobiology*, *68*(2), 113-43.

Friston, K. J. (2009a). Causal modelling and brain connectivity in functional magnetic resonance imaging. *PLoS biology*, 7(2), e33.

Friston, K. J. (2009b). Modalities, modes, and models in functional neuroimaging. *Science*, 326(5951), 399-403.

Geschwind, D. H., & Levitt, P. (2007). Autism spectrum disorders: developmental disconnection syndromes. *Current opinion in neurobiology*, 17(1), 103-11.

Geschwind, D. H. (2009). Advances in autism. *Annual review of medicine*, 60, 367-80.

Giedd, J N, Blumenthal, J., Jeffries, N. O., Castellanos, F. X., Liu, H., Zijdenbos, A., Paus, T., Evans, A. C., & Rapoport, J. L. (1999). Brain development during childhood and adolescence: a longitudinal MRI study. *Nature neuroscience*, 2(10), 861-3.

Giedd, J N, Snell, J. W., Lange, N, Rajapakse, J. C., Casey, B. J., Kozuch, P. L., Vaituzis, A. C., Vauss, Y. C., Hamburger, S. D., Kaysen, D., & Rapoport, J. L. (1996). Quantitative magnetic resonance imaging of human brain development: ages 4-18. *Cerebral cortex*, 6(4), 551-60.

Giedd, Jay N., & Rapoport, Judith L. (2010). Structural MRI of Pediatric Brain Development: What Have We Learned and Where Are We Going? *Neuron*, 67(5), 728-734.

Giedd, J., Castellanos, F., Casey, B., Kozuch, P., King, A., Hamburger, S., & Rapoport, J. L. (1994). Quantitative morphology of the corpus callosum in attention deficit hyperactivity disorder. *Am J Psychiatry*, 151(5), 665-669.

Gogtay, N., Giedd, Jay N, Lusk, L., Hayashi, K. M., Greenstein, D., Vaituzis, A. C., Nugen, T. F., Herman, D. H., Clasen, L. S., Toga, A. W., Rapoport, J. L., & Thompson, P. M. (2004). Dynamic mapping of human cortical development during childhood through early adulthood. *Proceedings of the National Academy of Sciences of the United States of America*, 101(21), 8174-9.

Groen, W. B., Buitelaar, J. K., Gaag, R. J. van der, & Zwiers, M. P. (2011). Pervasive microstructural abnormalities in autism: a DTI study. *Journal of psychiatry & neuroscience*, 36(1), 32-40.

Gupta, R. K., Hasan, K. M., Trivedi, R., Pradhan, M., Das, V., Parikh, N., & Narayana, A., (2005). Diffusion tensor imaging of the developing human cerebrum. *Journal of neuroscience research*, 81(2), 172-8.

Happé, F., Ronald, A., & Plomin, R. (2006). Time to give up on a single explanation for autism. *Nature neuroscience*, 9(10), 1218-20.

Herbert, M. R., Ziegler, D. A., Makris, N., Filipek, P. A., Kemper, T. L., Normandin, J. J., Sanders, H. A., Kennedy, D. N., & Caviness, C. S. (2004). Localization of white matter volume increase in autism and developmental language disorder. *Annals of neurology*, 55(4), 530-40.

Honey, C. J., Sporns, O., Cammoun, L., Gigandet, X., Thiran, J. P., Meuli, R., & Hagmann, P. (2009). Predicting human resting-state functional connectivity from structural connectivity. *Proceedings of the National Academy of Sciences of the United States of America*, 106(6), 2035-40.

Huang, H., Zhang, J., Wakana, S., Zhang, W., Ren, T., Richards, L. J., Yarowsky, P., Donohue, P., Graham, E., van Zijl, P. C. M., & Mori, S. (2006). White and gray matter development in human fetal, newborn and pediatric brains. *NeuroImage*, 33(1), 27-38.

Hughes, J. R. (2007). Autism: the first firm finding = underconnectivity? *Epilepsy & behavior : E&B*, 11(1), 20-4.

Hutsler, J. J., & Zhang, H. (2010). Increased dendritic spine densities on cortical projection neurons in autism spectrum disorders. *Brain research*, 1309, 83-94.

Hüppi, P. S., Maier, S. E., Peled, S., Zientara, G. P., Barnes, P. D., Jolesz, F. A., & Volpe, J. J. (1998). Microstructural development of human newborn cerebral white matter assessed in vivo by diffusion tensor magnetic resonance imaging. *Pediatric research*, 44(4), 584-90.

Hüppi, P. S., & Dubois, J. (2006). Diffusion tensor imaging of brain development. *Seminars in fetal & neonatal medicine*, 11(6), 489-97.

Inglis, B. A,, Bossart, E. L., Buckley, D. L., Wirth, E. D., & Mareci, T. H. (2001). Visualization of neural tissue water compartments using biexponential diffusion tensor MRI. *Magnetic resonance in medicine*, 45(4), 580-7.

Jeste, S. S. (2011). The neurology of autism spectrum disorders. *Current opinion in neurology*, 24(2), 132-9.

Jou, R. J., Jackowski, A. P., Papademetris, X., Rajeevan, N., Staib, L. H., & Volkmar, F. R. (2011). Diffusion tensor imaging in autism spectrum disorders: preliminary evidence of abnormal neural connectivity. *The Australian and New Zealand journal of psychiatry*, 45(2), 153-62.

Jou, R. J., Mateljevic, N., Minshew, Nancy J., Keshavan, M. S., & Hardan, A. Y. (2011). Reduced central white matter volume in autism: Implications for long-range connectivity. *Psychiatry and Clinical Neurosciences*, 651, 98-101.

Just, M. A., Cherkassky, V. L., Keller, T. a, & Minshew, Nancy J. (2004). Cortical activation and synchronization during sentence comprehension in high-functioning autism: evidence of underconnectivity. *Brain*, 127(8), 1811-21.

Ke, X., Tang, T., Hong, S., Hang, Y., Zou, B., Li, H., et al. (2009). White matter impairments in autism, evidence from voxel-based morphometry and diffusion tensor imaging. *Brain research*, 1265, 171-7.

Keller, T. A., Kana, R. K., & Just, M. A. (2006). A developmental study of the structural integrity of white matter in autism. *Neuroreport*, 18(1), 23-7.

Lainhart, J. E. (2006). Advances in Autism Neuroimaging Research for the Clinician and Geneticist. *American Journal of Medical Genetics*, 142(C), 33 - 39.

Lange, Nicholas, Dubray, M. B., Lee, J. E., Froimowitz, M. P., Froehlich, A., Adluru, N., Wright, B., Ravichandran, C., Fletcher, P. T., Bigler, E. D., Alexander, A. L. & Lainhart, J. E. (2010). Atypical diffusion tensor hemispheric asymmetry in autism. *Autism research*, (December), 350-358.

Le Bihan, Denis. (2003). Looking into the functional architecture of the brain with diffusion MRI. *Nature reviews*. 4(6), 469-80.

Le Bihan, Denis, & Zijl, P. van. (2002). From the diffusion coefficient to the diffusion tensor. *NMR in biomedicine*, 15(7-8), 431-4.

Lebihan, D. (2006). Looking into the functional architecture of the brain with diffusion MRI☆. *International Congress Series*, 1290, 1-24.

Lee, J. E., Bigler, E. D., Alexander, A. L., Lazar, M., DuBray, M. B., Chung, M. K., Johnson, M. Morgan, J., Miller, J. N., McMahon, W. M., Lu. J., Jeong. E-K., & Lainhart, J. E. (2007). Diffusion tensor imaging of white matter in the superior temporal gyrus and temporal stem in autism. *Neuroscience letters*, 424(2), 127-32.

Lee, J. E., Chung, M. K., Lazar, M., DuBray, M. B., Kim, J., Bigler, E. D., Lainhart, J. E., & Alexander, A. L. (2009). A study of diffusion tensor imaging by tissue-specific, smoothing-compensated voxel-based analysis. *NeuroImage*, 44(3), 870-83.

Lenroot, R. K., & Giedd, J. N. (2006). Brain development in children and adolescents: insights from anatomical magnetic resonance imaging. *Neuroscience and biobehavioral reviews*, 30(6), 718-29.

Levy, S. E., Mandell, D. S., & Schultz, R. t. (2009). Autism. *lancet*, 374(9701), 1627-1638.

Levy, Y., & Ebstein, R. P. (2009). Research review: crossing syndrome boundaries in the search for brain endophenotypes. *Journal of child psychology and psychiatry, and allied disciplines*, 50(6), 657-68.

Luyster, R. J., Kadlec, M. B., Carter, A., & Tager-Flusberg, H. (2008). Language assessment and development in toddlers with autism spectrum disorders. *Journal of autism and developmental disorders*, 38(8), 1426-38.

Macintosh, K. E., & Dissanayake, C. (2004). Annotation: The similarities and differences between autistic disorder and Asperger's disorder: a review of the empirical evidence. *Journal of Child Psychology and Psychiatry*, 45(3), 421-434.

Matson, J. L. (2007). Current status of differential diagnosis for children with autism spectrum disorders. *Research in developmental disabilities*, 28(2), 109-18.

Mengotti, P., D'Agostini, S., Terlevic, R., De Colle, C., Biasizzo, E., Londero, D., Ferro, A ., Rambaldelli, G., Balestrieri, M., Zanini, S., Fabbro, F., Molteni, M., & Brambilla, P. (2011). Altered white matter integrity and development in children with autism: A combined voxel-based morphometry and diffusion imaging study. *Brain research bulletin*, 84(2), 189-195.

Mizuno, A., Villalobos, M. E., Davies, M. M., Dahl, B. C., & Müller, R-A. (2006). Partially enhanced thalamocortical functional connectivity in autism. *Brain research*, 1104(1), 160-74.

Morgan, J. T., Chana, G., Pardo, C. a, Achim, C., Semendeferi, K., Buckwalter, J., Courchesne, E., & Everall, I. P. (2010). Microglial activation and increased microglial density observed in the dorsolateral prefrontal cortex in autism. *Biological psychiatry*, 68(4), 368-76.

Mori, S., & Zijl, P. C. M. van. (2002). Fiber tracking: principles and strategies - a technical review. *NMR in biomedicine*, 15(7-8), 468-80.

Moseley, M. E., Cohen, Y, Kucharczyk, J., Mintorovitch, J., Asgari, H. S., Wendland, M. F., Tsuruda, J., & Norman, D. (1990). Diffusion-weighted MR imaging of anisotropic water diffusion in cat central nervous system. *Radiology*, 176(2), 439-45.

Mukherjee, P., Miller, J H, Shimony, J S, Conturo, T E, Lee, B. C., Almli, C R, & McKinstry, R. C. (2001). Normal brain maturation during childhood: developmental trends characterized with diffusion-tensor MR imaging. *Radiology, 221*(2), 349-58.

Müller, R-A. (2007). The study of autism as a ditributed disorder. *Mental Retardation and developmental disabilities*, 13(225), 85 - 95.

Müller, R.-A., Shih, P., Keehn, B., Deyoe, J. R., Leyden, K. M., & Shukla, D. K. (2011). Underconnected, but How? A Survey of Functional Connectivity MRI Studies in Autism Spectrum Disorders. *Cerebral cortex*.

Neil, J. J., Shiran, S. I., McKinstry, R C, Schefft, G. L., Snyder, A. Z., Almli, C R, Akbudak, E., Aronovitz, J. A., Miller, J. P., & Lee, B. C., & Conturo, T. E. (1998). Normal brain in human newborns: apparent diffusion coefficient and diffusion anisotropy measured by using diffusion tensor MR imaging. *Radiology*, 209(1), 57-66.

Neil, J., Miller, J., Mukherjee, P., & Hüppi, P S. (2002). Diffusion tensor imaging of normal and injured developing human brain - a technical review. *NMR in biomedicine*, 15(7-8), 543-52.

Newschaffer, C. J., Croen, L. a, Daniels, J., Giarelli, E., Grether, J. K., Levy, S. E., et al. (2007). The epidemiology of autism spectrum disorders. *Annual review of public health*, 28, 235-58.

Noriuchi, M., Kikuchi, Y., Yoshiura, T., Kira, R., Shigeto, H., Hara, T., Tobimatsu, S., & Kamio, Y. (2010). Altered white matter fractional anisotropy and social impairment in children with autism spectrum disorder. *Brain research, 1362*, 141-149.

Ozonoff, S., Heung, K., Byrd, R., Hansen, R., & Hertz-Picciotto, I. (2008). The onset of autism: patterns of symptom emergence in the first years of life. *Autism research*, 1(6), 320-8.

Pardini, M., Garaci, F. G., Bonzano, L., Roccatagliata, L., Palmieri, M. G., Pompili, E., Coniglione, F., Krueger, F., Ludovici, A., Floris, R., Benassi, F., & Emberti Gialloreti, L. (2009). White matter reduced streamline coherence in young men with autism and mental retardation. *European journal of neurology*, 16(11), 1185-90.

Pelphrey, K. a, Shultz, S., Hudac, C. M., & Vander Wyk, B. C. (2011). Research Review: Constraining heterogeneity: the social brain and its development in autism spectrum disorder. *Journal of child psychology and psychiatry, and allied disciplines.*

Piven, J, Arndt, S., Bailey, J., & Andreasen, N. (1996). Regional brain enlargement in autism: a magnetic resonance imaging study. *Journal of the American Academy of Child and Adolescent Psychiatry*, 35(4), 530-6.

Piven, J, Bailey, J., Ranson, B. J., & Arndt, S. (1997). An MRI study of the corpus callosum in autism. *The American journal of psychiatry*, 154(8), 1051-6.

Power, J. D., Fair, D. a, Schlaggar, B. L., & Petersen, S. E. (2010). The Development of Human Functional Brain Networks. *Neuron*, 67(5), 735-748. Elsevier Inc.

Pugliese, L., Catani, M., Ameis, S., Dell'Acqua, F., Thiebaut de Schotten, M., Murphy, C., Robertson, D., Deeley, Q., Daly, E., & Murphy, D. G. M. (2009). The anatomy of extended limbic pathways in Asperger syndrome: a preliminary diffusion tensor imaging tractography study. *NeuroImage*, 47(2), 427-34.

Pujol, J., López-Sala, A., Sebastián-Gallés, N., Deus, J., Cardoner, N., Soriano-Mas, C., Moreno, A., & Sans, A. (2004). Delayed myelination in children with developmental delay detected by volumetric MRI. NeuroImage, 22(2), 897-903.

Ronen, I., Kim, K.-H., Garwood, M., Ugurbil, K., & Kim, D.-S. (2003). Conventional DTI vs. slow and fast diffusion tensors in cat visual cortex. *Magnetic resonance in medicine, 49*(5), 785-90.

Ronen, I., Ugurbil, K., & Kim, D.-S. (2005). How does DWI correlate with white matter structures? *Magnetic resonance in medicine, 54*(2), 317-23.

Rubenstein, J. L. R. (2010). Three hypotheses for developmental defects that may underlie some forms of autism spectrum disorder. Current opinion in neurology, 23(2), 118-23.

Rutherford, M. A., Cowan, F. M., Manzur, A. Y., Dubowitz, L. M., Pennock, J. M., Hajnal, J. V., Young, I. R., & Bydder, G. M. (1991). MR imaging of anisotropically restricted diffusion in the brain of neonates and infants. *Journal of computer assisted tomography*, 15(2), 188-98.

Sahyoun, C. P., Belliveau, J. W., & Mody, M. (2010). White matter integrity and pictorial reasoning in high-functioning children with autism. *Brain and cognition*, 73(3), 180-8.

Sakuma, H., Nomura, Y., Takeda, K., Tagami, T., Nakagawa, T., Tamagawa, Y., Ishii, Y., & Tsukamoto, T. (1991). Adult and neonatal human brain: diffusional anisotropy and myelination with diffusion-weighted MR imaging. *Radiology*, 180(1), 229-33.

Schumann, C. M., & Nordahl, C. W. (2010). Bridging the Gap between MRI and Postmortem Research in Autism. *Brain research*, 1380, 175-186.

Schwartz, E. D., Cooper, E. T., Fan, Y., Jawad, A. F., Chin, C-liang, Nissanov, J., & Hackney, D. B. (2005). MRI diffusion coefficients in spinal cord correlate with axon morphometry. *NeuroReport*, 16(1), 73-76.

Shaw, P., Kabani, N. J., Lerch, J. P., Eckstrand, K., Lenroot, R., Gogtay, N., Greenstein, D., & Clasen, L., Evans, A. (2008). Neurodevelopmental trajectories of the human cerebral cortex. *The Journal of neuroscience*, 28(14), 3586-94.

Shenton, M. E., Dickey, C. C., Frumin, M., & McCarley, R. W. (2001). A review of MRI findings in schizophrenia. *Schizophrenia research*, 49(1-2), 1-52.

Shukla, D. K., Keehn, B., Lincoln, Alan J, & Müller, R.-A. (2010). White matter compromise of callosal and subcortical fiber tracts in children with autism spectrum disorder: a diffusion tensor imaging study. *Journal of the American Academy of Child and Adolescent Psychiatry*, 49(12), 1269-78.

Shukla, D. K., Keehn, B., Smylie, D. M., & Müller, R. A. (2011a). Microstructural abnormalities of short-distance white matter fiber tracts in autism spectrum disorder. *Neuropsychologia*, 1-5.

Shukla, D. K., Keehn, B., & Müller, R. A. (2011b). Tract-specific analyses of diffusion tensor imaging show widespread white matter compromise in autism spectrum disorder. *Journal of Child Psychology and Psychiatry*, 3, 286-295.

Song, S., Sun, S., Ramsbottom, M., Chang, C., Russell, J., & Cross, A. (2002). Dysmyelination Revealed through MRI as Increased Radial (but Unchanged Axial) Diffusion of Water. *NeuroImage*, 17(3), 1429-1436.

Sparks, B F, Friedman, S D, Shaw, D. W., Aylward, E. H., Echelard, D., Artru, A. A., Maravilla, K. R., Giedd, J. N., Munson, J., Dawson, G., & Dager, S. R. (2002). Brain structural abnormalities in young children with autism spectrum disorder. *Neurology*, 59(2), 184-92.

Stanfield, A. C., McIntosh, A. M., Spencer, M. D., Philip, R., Gaur, S., & Lawrie, S. M. (2008). Towards a neuroanatomy of autism: a systematic review and meta-analysis of structural magnetic resonance imaging studies. *European psychiatry*, 23(4), 289-99.

Sundaram, S. K., Kumar, A., Makki, M. I., Behen, M. E., Chugani, H. T., & Chugani, D. C. (2008). Diffusion tensor imaging of frontal lobe in autism spectrum disorder. *Cerebral cortex*, 18(11), 2659-65.

Tepest, R., Jacobi, E., Gawronski, A., Krug, B., Möller-Hartmann, W., Lehnhardt, F. G., & Vogeley, K. (2010). Corpus callosum size in adults with high-functioning autism and the relevance of gender. *Psychiatry research*, 183(1), 38-43.

Thakkar, K. N., Polli, F. E., Joseph, R. M., Tuch, D. S., Hadjikhani, N., Barton, J. J. S., & Manoach, D. S. (2008). Response monitoring, repetitive behaviour and anterior cingulate abnormalities in autism spectrum disorders (ASD). *Brain*, 131(Pt 9), 2464-78.

Thomas, C., Humphreys, K., Jung, K.-J., Minshew, N., & Behrmann, M. (2010). The anatomy of the callosal and visual-association pathways in high-functioning autism: A DTI tractography study. *Cortex*, 1-11.

Uluğ, a M., & van Zijl, P. C. M. (1999). Orientation-independent diffusion imaging without tensor diagonalization: anisotropy definitions based on physical attributes of the diffusion ellipsoid. *Journal of magnetic resonance imaging*, 9(6), 804-13.

Vidal, C. N., Nicolson, R., DeVito, T. J., Hayashi, K. M., Geaga, J. a, Drost, D. J., Williamson, P. C. Rajakumar, N., Sui, Y., Dutton, R. A., Toga, A. W. & Thompson, P. M. (2006). Mapping corpus callosum deficits in autism: an index of aberrant cortical connectivity. *Biological psychiatry*, 60(3), 218-25.

Vol, H. E., & Morfologicas, C. (2000). A. Pefia-Melian. *Human Evolution, 15.*

Wakana, S., Jiang, H., & van Zijl, P. C. M. (2003). Radiology Fiber Tract – based Atlas of. *Radiology*, 21-29.

Wass, S. (2010). Distortions and disconnections: Disrupted brain connectivity in autism. *Brain and cognition*, 75(1), 18-28.

Weinstein, M., Ben-Sira, L., Levy, Y., Zachor, D. A., Itzhak, E. B., Artzi, M., et al. (2011). Abnormal white matter integrity in young children with autism. *Human Brain Mapping*, 32(4), 534-543.

Wimberger, D. M., Roberts, T. P., Barkovich, A. J., Prayer, L. M., Moseley, M. E., & Kucharczyk, J. (1995). Identification of "premyelination" by diffusion-weighted MRI. *Journal of computer assisted tomography*, 19(1), 28-33.

2

GABAergic Dysfunction in Autism and Epilepsy

Sacha Genovesi[1], Giovanni Provenzano[1],
Mark Dunleavy[1,2], Paola Sgadò[1] and Yuri Bozzi[1,3]
[1]Laboratory of Molecular Neuropathology, Centre for Integrative Biology (CIBIO), University of Trento,
[2]Department of Physiology and Medical Physics, Royal College of Surgeons in Ireland, Dublin,
[3]CNR Neuroscience Institute, Pisa,
[1,3]Italy
[2]Ireland

1. Introduction

Autism spectrum disorders (ASD) and epilepsy are among the most devastating and common neurological disorders of childhood, with an estimated incidence of about 0.5 - 1% in worldwide population. Autism and epilepsy are often associated: about 30% of autistic patients develop epilepsy, and a relevant percentage of epileptic patients in paediatric age shows ASD symptoms. This suggests that - at least in certain cases - common neurodevelopmental bases may exist for these two diseases (Brooks-Kayal, 2010).

The neurodevelopmental bases of both autism and epilepsy have been clearly showed by a number of clinical, neuroimaging and neuropathological studies. A large series of evidence also indicates that both autism and epilepsy have a primarily genetic origin. A wide variety of genes have been associated to these diseases, including genes regulating brain development, gene transcription, synaptic scaffolding, neurotransmission and signal transduction. Indeed, genetic heterogeneity is recognised as a typical feature of both autism and epilepsy, meaning that different mutations may result in similar disease phenotypes. Since autism and epilepsy are neurological disorders involving multiple genes and resulting in complex pathological traits, understanding the underlying mechanisms is a very difficult task. Moreover, even though the two diseases may have a common neurodevelopmental origin, a precise link between these two pathologies still remains to be determined.

In recent years, inhibitory circuit dysfunction gained increasing attention in ASD research. γ-aminobutyric acid (GABA) is the major inhibitory neurotransmitter in the brain, and human genetics studies clearly indicate an association between ASD and genes for GABA receptor subunits as well as genes controlling GABAergic neuron development or GABAergic synapse structure. Moreover, recent studies, performed on both animal models and postmortem human samples, suggest that GABAergic neurons and circuits may be altered in ASD. It is likely that the imbalance between excitation and inhibition resulting from neurodevelopmental defects in GABAergic circuitry might represent a common cause for ASD and epilepsy. Here, we will review the genetic, cellular, anatomical and neurophysiological studies that support this hypothesis.

2. Genetic determinants of ASD

ASD represent a group of very heterogeneous group of neurodevelopmental disabilities of proven genetic origin, with an incidence of about 60-70/10,000 (Fombonne, 2009). A gender distortion is observed in ASD (4:1 males to females ratio; Abrahams & Geschwind, 2008), reflecting a possible involvement of the X chromosome or imprinting mechanisms. The genetic factors play an important role in the pathogenesis of these diseases (Persico et al., 2006), as documented by the recurrence risk in families and twin studies. These studies show a concordance rate of 82–92% in monozygotic versus 1–10% in dizygotic twins. Heritability is estimated above 90% and sibling recurrence risk is above 6–20% (Abrahams & Geschwind, 2008; Toro et al., 2010). The genetic architecture of ASDs is complex; only 10–20% of ASD patients have an identified genetic etiology, whereas in the majority of patients the origin of the disorder remains unknown. Genetic forms of ASD include monogenic and complex disorders, as well as chromosomal abnormalities. Monogenic disorders include neurofibromatosis (NF1), fragile X syndrome (FMR1), tuberous sclerosis (TSC1, TSC2), Angelman syndrome (UBEA3A) and Rett syndrome (MECP2), covering only the 2–5% of the ASD cases (Hatton et al., 2006). Cytogenetic investigations and genome-wide scans have been performed to identify chromosomal regions containing ASD susceptibility genes. Results from these genome-wide linkage scans indicate potential susceptibility regions that spread across the entire genome, but only a few loci (Freitag, 2007; Yang & Gill, 2007). The most common chromosomal rearrangement is the maternal duplication of 15q11-q13, which accounts for approximately 1-2% of ASD cases (Vorstman et al. 2006). Recent genome-wide association (GWA) studies have identified novel candidate loci between the cadherin genes CDH9 and CDH10 (5p14.1; Wang et al., 2009) and between the SEMA5A and TAS2R1 genes (5p15.2 ; Weiss et al., 2009). In addition, the Autism Genome Project (AGP) Consortium has genotyped 1,558 ASD families for one million single nucleotide polymorphisms (SNPs), identifying a novel locus near the gene MACROD2 (20p12.1; Anney et al., 2010). Syndromic forms of ASD have been associated with both copy number variations (CNVs) and rare mutations in several genes, including SHANK3, NLGN3, NLGN4, NRXN1 and HOXA1 (Lintas & Persico, 2009). A recent extensive metanalysis of the literature (Betancur, 2011) allowed identification of 103 disease genes and 44 genomic loci reported in subjects with ASD or autistic behaviour. It is interesting to note that the vast majority of these genes and loci have been also causally implicated in epilepsy (including genes regulating brain development, gene transcription, synaptic scaffolding, neurotransmission and signal transduction), suggesting that these two neurodevelopmental disorders share common genetic bases (Betancur, 2011). Specifically, several evidences suggest that an impairment of inhibitory neurotransmission may constitute a fundamental event in the development of both ASD and epilepsy (Rubenstein & Merzenich, 2003). Since GABA is the major inhibitory neurotransmitter in the brain, here we briefly summarize the genes involved in GABAergic dysfunction and their relation with susceptibility to ASD and epilepsy.

2.1 Genes regulating GABAergic neuron development

Several genes control the process of development of GABAergic neurons, including DLX1, DLX2, MASH1 and RELN (Wonders & Anderson, 2006). The DLX1 and DLX2 genes encode homeodomain-containing transcription factors and are located head-to-head on chromosome 2q31, a region previously associated to autism susceptibility in several genome-wide linkage studies. Two studies examining SNPs in the DLX1 and DLX2 genes have found an association with ASD, suggesting that common genetic variations in these genes play a critical role in the disease (Liu et al., 2009; Chang et al., 2010). GABAergic neuron development dysfunction may also occur in conjunction with abnormalities in the RELN gene, coding for the extracellular matrix glycoprotein Reelin which is involved in neuronal migration and lamination of the cerebral cortex during embryogenesis (Forster et al., 2002). RELN maps to 7q22 human chromosome (De Silva et al., 1997). Linkage in this region is among the most robust genetic findings in ASD. In family-based and case–control studies the 5′-untranslated region (5′-UTR) GGC repeat alleles was associated with ASD (Persico et al., 2001). Importantly, Reelin (the product of RELN gene) is expressed in GABAergic neurons the adult brain (van Kooten et al., 2005).

Among the numerous ASD associated genes, EN2 (coding for the homeobox-containing transcription factor Engrailed-2) was originally shown to be involved in posterior brain (mesencephalon/hidbrain) embryonic development (Joyner et al., 1991). Recent studies on En2 null mice suggest that En2 deletion may alter GABAergic circuitry in the adult brain (Tripathi et al., 2009; see section 3.3.2). EN2 maps to a region of chromosome 7 implicated in ASD susceptibility, and GWA studies indicated EN2 as a candidate gene for ASD (Benayed et al., 2009). Namely, two SNPs in the human EN2 gene have been associated to ASD, one of which (rs1861973, A-C haplotype) is functional: when tested in a luciferase reporter assay in rat, mouse and human cell lines, this SNP markedly affected EN2 promoter activity (Benayed et al., 2009).

2.2 Genes coding for $GABA_A$ receptor subunits

Genes coding for GABA receptors has been extensively studied to evaluate their role in the pathogenesis of ASD. Three classes of GABA receptors exist in the mature mammalian brain, named as $GABA_A$, $GABA_B$ and $GABA_A$-rho ($GABA_C$). $GABA_A$ receptor is a ionotropic receptor that allows chloride ion fluxes through the neuronal membrane. Nearly 20 $GABA_A$ subunits have been reported in humans (Olsen & Sieghart, 2009). Mutations in $GABA_A$ receptor subunits genes have been associated to ASD. The chromosome 15q11, containing the genes coding the three $GABA_A$ receptor subunits $\alpha5$, $\beta3$ and $\gamma3$ (GABRA5, GABRB3 and GABRG3, respectively) has been linked to ASD, and SNPs in the above mentioned genes have been associated to ASD (Buxbaum et al., 2002; Hogart et al., 2007, 2009). It is interesting to note that, among these genes, GABRB3 has also been indicated as susceptibility gene for childhood absence epilepsy (Urak et al., 2006).

2.3 Genes involved in GABAergic synapse structure and function

Different genes involved in the development and function of inhibitory GABAergic synapses have been associated to ASD. Neurexins (NRXNs) are presynaptic proteins, binding postsynaptic neuroligins. This interaction is thought to trigger postsynaptic differentiation and control the balance of inhibitory GABAergic and excitatory glutamatergic inputs (Graf et al., 2004; Scheiffele et al., 2000). There are three NRXN genes (NRXN 1-3) in mammals; among these, mutations and chromosomal rearrangements in NRXN1 has been associated with ASD (Feng et al., 2006; Kim et al., 2008; Wisniowiecka-Kowalnik et al., 2010). Recently it has been shown that NRXNs can bind not only Neuroligins (NLGNs) but also $GABA_A$ receptors (Zhang et al., 2010). The effect of this ligand-receptor interaction decreases GABAergic transmission. NLGNs are neural cell adhesion molecules, which act as ligands for neurexins (Graf et al., 2004; Scheiffele et al., 2000). NLGNs play a key role in the formation, organization, and remodeling of synapses and different isoforms are associated with different synaptic types. NLGN1, NLGN4X and NLGN4Y are localized at glutamatergic synapses (Persico et al., 2006; Craig & Kang, 2007). NLGN3 is present in both excitatory and inhibitory synapses (Chih et al., 2005; Budreck & Scheiffele, 2007), whereas NLGN2 is located in GABAergic synapses (Craig & Kang, 2007; Persico et al., 2006; Varoqueaux et al., 2006). Mutations in NLGN1, 3 and 4X genes have been identified in patients with familial ASD (Jamain et al., 2003; Laumonnier et al., 2004; Lawson-Yuen et al., 2008).

The MECP2 gene, coding for the epigenetic regulator methyl-CpG-binding protein 2, is the causative gene for Rett syndrome, which belong to the family of ASD. Rett syndrome is characterized by loss of language capability, motor stereotyped behaviors, severe mental retardation and seizures (Chahrour & Zoghby, 2007). Recent studies indicate that Mecp2 dysfunction in GABAergic interneurons severely impacts GABA signaling and results in ASD-like phenotypes in the mouse (Chao et al., 2010; see section 3.3.6).

Fragile X syndrome (FXS) is one of the disorders included in ASD. FXS is caused by an expansion of the trinucleotide repeat in the promoter region of the fragile X mental retardation 1 (FMR1) gene (Verkerk et al., 1991). FMR1 gene encodes the fragile X mental retardation protein (FMRP), a mRNA binding protein with a key role in the intracellular transport and translation of 4–8% of synaptic proteins (Bassell & Warren, 2008). Recent evidence also indicates the involvement of the GABAergic system in the pathogenesis of FXS (D'Hulst & Kooy, 2007, 2009; Olmos-Serrano et al., 2010; see also 3.3.7).

3. Deficits of GABAergic neurons and circuits in ASD

From the literature data reported in the previous chapter, it is evident that genetic defects in genes regulating GABAergic neuron development as well as GABAergic synapse structure and function are crucially involved in ASD pathogenesis. Alterations of GABAergic neurons and circuits have been reported in postmortem brain tissue samples from ASD patients, as well as in mouse models of the disease. Here we will review the more significant findings from these studies.

3.1 GABAergic neuron defects in the brain of ASD patients

The analysis of postmortem tissues revealed that many brain regions are affected in ASD patients, including the cerebral cortex, limbic system and cerebellum. Minicolumns represent the cellular and functional organization of glutamatergic and GABAergic neurons in the cerebral cortex (De Felipe et al., 1990; Mountcastle, 1997; Polleux & Lauder, 2004). Minicolumns are anatomically characterized by vertical arrays of pyramidal neurons with their dentrides and axon projections. Pyramidal cells arrays are accompanied by their GABAergic interneurons that establish synapses with pyramidal cells bodies, their axons emergences and dentrites. A narrowing of cortical minicolumns (namely, a reduced distance between columns) has been shown in ASD patients (Casanova et al., 2002). This reduced intercolumnar distance was proposed to depend on structural/anatomical defects in GABAergic interneurons surrounding principal pyramidal cortical neurons (Casanova 2003; Casanova & Trippe 2009, Raghanti et al., 2010). An increased cell density and a reduced cell size have long been reported in the limbic system of ASD patients (Kemper & Bauman, 1993). More recently, a significant increase in the number of parvalbumin-, calbindin- and calretinin-positive interneurons has been reported in the hippocampus of ASD patients (Lawrence et al., 2010). In the cerebellum, a clear reduction in the number of inhibitory Purkinje neurons has long been reported in ASD postmortem tissues (Kemper & Bauman, 1993). More recent studies confirmed the selective loss of calbindin-positive Purkinje cells but not parvalbumin-positive stellate and basket interneurons in the ASD cerebellum (Whitney et al., 2008, 2009).

3.2 GABAergic signaling deficits in the brain of ASD patients

Several studies suggest a GABAergic signaling dysfunction in ASD, mainly due to altered levels of the GABA synthetic enzyme (glutamic acid decarboxylase, GAD) and GABA receptors. Two GAD isoforms exist, named GAD65 (GAD2) and GAD67 (GAD1), respectively localized on chromosome 10 and 2 in humans (Karlsen et al., 1991; Kaufman et al., 1991; Martin & Rimvall, 1993). Several studies were conducted in post-mortem brains of autistic patients focusing on parietal cortices and cerebellar alterations in terms of GAD proteins amount and Purkinje cells quantification. A 50% reduction in GAD65/67 proteins levels was reported in the cerebellum and parietal cortex from ASD patients (Fatemi et al., 2002). Accordingly, reduced levels of GAD67 and GAD65 mRNAs were also detected in Purkinje cells and dentate nuclei neurons in the cerebellum from ASD cases (Yip et al., 2007, 2008). Interestingly, the same author also reported an increase of GAD67 mRNA levels in GABAergic basket cells in the cerebellar molecular layer that was interpreted as a compensatory up-regulation to supply the loss of Purkinje cells in ASD brains (Yip et al., 2008). Several studies showed a significant decrease in $GABA_A$ receptor $\alpha 4$, $\alpha 5$, $\beta 1$ and $\beta 3$ subunits (Blatt et al. 2001; Fatemi et al., 2010, Samaco et al. 2005), as well as a significat reduction of benzodiazepine (the $GABA_A$ receptor ligand) binding sites (Oblak et al., 2011) in ASD brains. $GABA_B$ receptors were also reduced in restricted regions of the cerebral cortex from ASD patients (Oblak et al., 2010). Importantly, a recent preliminary study performed by transcranial magnetic stimulation allowed to detect a reduced cortical inhibition (interpreted as a possible disruption of $GABA_A$ receptor activity) in the brain of a subset of ASD patients (Enticott et al., 2010). Taken together, these data support the hypothesis of a GABAergic signaling deficit in ASD.

3.3 Evidence from animal studies

Impaired function of inhibitory circuits has been proposed as a major pathogenic cause of ASD (Rubenstein & Merzenich, 2003). Evidence is essentially limited to human genetic association studies as well as expression analyses performed on postmortem samples from ASD and control patients (see above). Conversely, this hypothesis is strongly supported by a vast number of studies performed on animal models recapitulating different aspects of the ASD pathology. So far, several transgenic and pharmacological mouse models have been developed to mimic different features of ASD-like syndromes. These include mutant mice for DLX1/2, EN2, GABRB3, RELN, NLGN3, MECP2 and FMR1.

Different anatomical and functional deficits of the GABAergic system were discovered in all these mouse models, and loss of parvalbumin (PV) expressing interneurons seems to be a hallmark of ASD-like dysfunctions in all models analyzed. The physiological formation of synaptic connections between PV-positive interneurons and principal pyramidal neurons has been implicated in functional maturation of the postnatal cerebral cortex, and deficits in this process have been proposed as a pathogenic mechanism of ASD (Di Cristo, 2007). PV-positive interneurons approximately represent the 40% of GABAergic interneurons of the cerebral cortex, and comprise basket and chandelier fast spiking cells (Rudy et al., 2010). The other two principal groups of cortical interneurons are somatostatin (SST) positive neurons (about 30%) and neurons expressing the 5HT3a serotonin receptor (about 30%); other less-represented cortical interneuron subtypes are those expressing the calcium binding proteins calretinin (CR) and calbindin (CB) and neuropeptide Y (NPY) (Rudy et al., 2010). Defects in different interneuron subtypes, and more generally in the anatomical organization and physiological function of the GABAergic system, have been reported in several mouse models of ASD. The principal findings are described in the following paragraphs.

3.3.1 Dlx1/2 knockout mice

The family of Dlx homeobox transcription factors regulates the development of inhibitory interneurons; members of this family, namely Dlx1, Dlx2, Dlx5, and Dlx6, control the differentiation of GABAergic neurons in basal ganglia and cerebral cortex (Pleasure, 2000). The principal finding reported in Dlx1/2 mutant mice is a migration defect of GABAergic interneurons into the cerebral cortex (Anderson et al., 1997a, 1997b). Dlx1 null mice displayed a selective loss of SST-, NPY-, CR- and reelin-expressing interneurons accompanied by reduced GABAergic inhibitory transmission and late-onset epilepsy (Cobos et al., 2005). More recently, additional behavioural abnormalities (such as conditioned fear response) linked to impairment of GABAergic systems were described in Dlx1-null mice (Mao et al., 2009).

3.3.2 Engrailed2 knockout mice

En2 null mice have been proposed as a model for ASD, due to their complex anatomical and behavioural phenotype. En2 null mice display cerebellar hypoplasia and a reduced number of Purkinje cells (Joyner et al., 1991; Kuemerle et al., 1997). These abnormalities resemble some of those reported in ASD (see section 3.1). Importantly, ASD-like behaviours such as decreased play, reduced sociality and impaired spatial learning and memory were described in these mutants (Cheh et al., 2006). Recently we showed an increased susceptibility to seizures in En2 null mice, that was accompanied by reduced PV immunostaining on cell bodies of CA3 pyramidal neurons, and reduced SST immunostaining in the the stratum lacunosum moleculare of the hippocampal formation (Tripathi et al., 2009). These findings suggest that the En2 gene may be involved in GABAergic system development and maintenance, and altered En2 function may be a common cause of ASD and seizures.

3.3.3 GABRB3 knockout mice

Mice lacking the $GABA_A$ receptor subunit β3 (GABRB3) display high mortality rate and symptoms consistent to Angelman's syndrome, including learning and memory deficits, poor motor skills, stereotyped behaviours and seizures suscetibility (Homanics et al., 1997; DeLorey et al., 1998). More recently, GABRB3 gene deficient mice have been shown to exhibit impaired social and exploratory behaviors, deficits in non-selective attention and hypoplasia of cerebellar vermal lobules, thus resempling a wide renge of ASD phenotypes (DeLorey et al., 2008).

3.3.4 Reeler mice

Reelin is an extracellular glycoprotein belonging to the family of serine proteases (Fatemi et al., 2005). Reelin binding to membrane receptors enhances signal transduction pathways leading to synaptic plasticity and axonal growth (Beffert, 2005). Reeler mice are lacking the Reelin gene. In these mice, neuronal migration in the cerebral cortex is dramatically impaired. This results a disorganization of laminated brain regions as cerebral cortices and cerebellum (Curran & D'Arcangelo, 1998). Reeler also mice show a decrease of dentritic spine density and a decreased GABA metabolism turnover (Carboni, 2004). ASD-like behaviours and loss of PV interneurons was recently reported in Reeler mice (Macrì et al., 2010).

3.3.5 NLGN3 knock-in mice

Mice carrying the R451C mutation in the NLGN3 gene show behavioural phenotypes related to ASD (lack of social behaviours, reduced ultrasound vocalization; Radyushkin et al., 2009; Tabuchi et al., 2007). In addition, Nlgn3 R451C knock-in mice present an increase in the number of GABAergic synapses (as evaluated by vesicular GABA transporter and gephyrin immunostaining) and in the amplitude of inhibitory currents, suggesting that the R451C mutation switches Nlgn3 synaptic specificity from glutamatergic to GABAergic (Tabuchi et al., 2007). Further characterization of these mutants demonstrated that loss of parvalbumin-positive basket cells is detectable across the two hemispheres in these mice (Gogolla et al., 2009).

3.3.6 Mecp2 knockout mice

Several conditional MeCp2 mutants mice were generated in order to remove Mecp2 from distinct neuronal populations. With respect to ASD, the most interesting data were recently obtained in conditional mutants lacking Mecp2 in inhibitory neurons expressing Viiat (Vesicular inhibitory aminoacid transporter, required to load GABA and glycine into synaptic vesicles) (Chao et al., 2010). Viaat-Mecp2 conditional mutants started to exhibit ASD-like repetitive and stereotyped behaviours, developing also self-injury behaviours. Interneurons immunolabelling in Viaat-Mecp2 mutants also showed a reduction of GAD65 and GAD67 mRNA in the cerebral cortex. Mecp2 loss in inhibitory neurons also resulted in EEG abnormalities and seizures. Moreover, electrophysiological recordings showed decreased miniature inhibitory post-synaptic currents (mIPSC) in cortical slices of Viaat-Mecp2 mutants, demonstrating that Mecp2 deficiency in GABAergic neurons determines a reduction of GABA neurotransmitter release due to a reduction of GAD amount in presynaptic terminals. Thus, loss of Mecp2 in inhibitory neurons might be a crucial determinant of severe forms of ASD.

3.3.7 Fmr1 knockout mice

Several studies show a strong reduction in the expression of $GABA_A$ receptor subunit mRNAs and proteins in adult Fmr1 knockout mice (Adusei et al., 2010; D'Hulst et al., 2009), that is accompanied by abnormal GABAergic transmission (Centonze et al., 2008; Curia et al., 2009), deficits of PV (but not CB- or CR-) cortical interneurons (Selby et al., 2007) and increased audiogenic seizure susceptibility (Musumeci et al., 2007). Table 1 summarizes the major GABAergic deficits described in ASD mouse models.

Mouse	Genetics	Signaling	Anatomy	Behaviour	Seizures	Sections in text
Dlx1/2 null	SNPs in Dlx1/2 genes found in ASD patients	-	Interneuron migration defects Loss of SST, NPY, CR and reeling interneurons	Impaired conditioned fear response	Late-onset epilepsy	2.1 3.3.1
En2 null	SNP found in ASD patients	-	Cerebellar hypoplasia Reduced PV and SST staining in hippocampus	Learning and memory deficit Reduced social behaviours	Increased susceptibility to kainic acid	2.1 3.3.2
Gabrb3 null	Mutation in Gabrb3 linked to ASD	-	Cerebellar hypoplasia	Learning and memory deficit Stereotypies Attention deficits	Yes	2.2 3.3.3 4.2.1
Reeler	GGC repeat in 5'UTR associated with ASD	Decreased GABA metabolism turnover	Neuronal migration defects Abnormal lamination of cerebral cortex and cerebellum Loss of PV interneurons	ASD-like behaviours	Yes	2.1 3.3.4
Nlgn3 R451C knock-in	R451C mutation is causative of ASD	Increased IPSCs	Increased mumber of GABAergic synapses Loss of PV interneurons	Lack of social behaviours Reduced ultrasound vocalization	-	2.3 3.3.5
Mecp2 knockout in GABA neurons	Mecp2 mutation is causative of Rett syndrome	Decreased GAD mRNA Decreased IPSCs	Loss of GABA interneurons (cerebral cortex)	ASD-like behaviours	Yes	2.3 3.3.6
Fmr1 null	Fmr1 mutation is causative of Fragile X syndrome	Reduced $GABA_A$ receptor subunit mRNAs and GABA transmission	Loss of PV interneurons (cerebral cortex)	-	Audiogenic seizures	2.3 3.3.7

Table 1. GABAergic defects in genetic mouse models of ASD.

4. GABAergic dysfunction in autism and epilepsy

Epilepsy is a neurological disorder characterised by spontaneous recurrent seizures. Epilepsy, similar to ASD, is increasingly being considered as a spectrum disorder due to the range of pathologies, seizures and behavioural and cognitive deficits associated with it (Jensen, 2011). As with ASD, epileptic seizures are considered to be as a result of imbalance between excitation and inhibition in the brain (Bradford, 1995; Olsen & Avoli, 1997). It is of particular relevance to developmental disorders, since seizures are the most common neurological emergency in children and occur most frequently in the first two years of life (Hauser, 1990; Chin et al., 2006), a critical period in brain development.
A strong association has been shown between epilepsy and ASD. The incidence of epilepsy in ASD has been reported to be between 5 - 40% (Canitano, 2007). Factors such as referral criteria, age and severity of cognitive impairments all contribute to the variability in report rate (Canitano, 2007; Tuchman et al., 2009). Children that co-express autism and epilepsy show a poorer outcome in cognitive and adaptive behaviour than those without epilepsy (Danielsson et al., 2005; Hara, 2007). The severity of the epilepsy phenotype seems to be closely related to the severity of ASD and is not associated with one particular type of seizure; simple and complex partial seizures, atypical absence, tonic-clonic and myoclonic seizures have all been reported. In particular, coexpression of mental retardation with autism is a risk factor for epilepsy (Volkmar & Nelson, 1990). Early onset of seizure is also an indicator of poor outcome in children with ASDs with more developmental disorders and greater seizure intractability reported (Wong, 1993; Bombardieri et al., 2010). "Seizures beget seizures" was a phrase coined by Sir William Gowers in 1881 and it may go some way to explaining this poorer outcome associated with early onset of ASD and seizures. Where seizures are initially a manifestation of the underlying imbalance in excitation/inhibition they may ultimately contribute to progressive increase in seizure severity and secondarily, behavioural and cognitive deficits. This is particularly evident in the developing brain where susceptibility to seizure-induced neuropathology leads to epilepsy and further cognitive deficits in later life (Ben-Ari, 2006; Ben-Ari & Holmes, 2006). Here we examine the link between ASD and epilepsy with particular focus on the role of GABAergic dysfunction of the pathogenesis of these diseases. In particular, we will describe some of the key human and animal studies further outlining the link between epilepsy and ASD (e.g., the presence of mental retardation) and the potential mechanistic role of GABA dysfunction.

4.1 Clinical evidence of a role for GABAergic involvement in epilepsy-autism disorders

An increasing number of studies have implicated the GABAergic system dysfunction in epilepsy and ASD. Chromosome 15q, which contains genes coding for GABA receptor subunits, has been reported to be a common site for mutations in ASDs (Schroer et al., 1998). This data is further supported by association studies linking GABA receptor subunit genes and SNPs associated with autism and seizures (Collins et al., 2006). An increasing body of evidence suggests a downregulation of GABAergic function is critical in ASD-associated epilepsy. Quantitative receptor autoradiographic studies examining the density and distribution of GABAergic subunits indicated a downregulation of GABAergic function in the hippocampus of ASD patients with seizures (Blatt et al., 2001). Furthermore, altered packing of GABAergic interneurons in the CA1 and CA3 hippocampal subfields where malformations are associated with the generation of seizures (Bauman & Kemper, 2005).

Supporting this human studies have demonstrated that there is a loss of inhibitory interneurons in the epileptic brain (Zhu et al., 1997; Wittner et al., 2001).
As outlined above, FXS is one such autism-related disorder that is a leading cause of mental retardation (Bardoni et al., 2006). It is also strongly associated with abnormal EEG activity with a mean epilepsy prevalence of between 22 - 25% (Wisniewski et al., 1991; El Idrissi et al., 2005). Fragile X appears to have a wide profile with some reporting a benign condition with generalised seizures responding well to antiepileptic drugs (AEDs) treatment (Wisniewski et al., 1991) that disappear after childhood. Other groups report long lasting generalised and partial epilepsy and EEG abnormalities in adults with fragile X (Sabaratnam, 2000). Furthermore, Gauthey et al describe a more severe phenotype with children with fragile X presenting with status epilepticus (seizures lasting > 30 min) at their initial seizure with recurrent prolonged seizures on follow-up (Gauthey et al., 2010). Status epilepticus in the developing brain is known to significantly increase the risk of epilepsy, hippocampal sclerosis and further behavioural and cognitive deficits in later life in human and animals studies (Raspall-Chaure et al., 2006; Dunleavy et al., 2010). There is a large amount of data from animal studies (described above) to support a role for disruption of normal GABAergic function in seizure generation in FXS.
Rett syndrome is a postnatal neurodevelopmental disorder typically emerging between 6 - 18 months of age consisting of progressive loss of cognitive and motor function and the emergence of epilepsy (Chahrour and Zoghbi, 2007). Seizures have been reported to be present in between 50 - 90% of patients (Witt Engerstrom, 1992; Steffenburg et al., 2001). As with FXS, epilepsy is most severe through childhood and young adulthood while the phenotype ranges from mild seizures that are well controlled with AEDs to refractory epilepsy, most common types being partial complex and tonic-clonic seizures (Steffenburg et al., 2001; Jian et al., 2006).

4.2 Experimental evidence of a role for GABAergic involvement in epilepsy-autism disorders

There are two main aspects to the role of GABA dysfunction in the pathogenesis of epilepsy in autism. Firstly, absence of GABA signaling results in loss of inhibitory neuronal firing that normally prevents the spread of paroxysmal discharge. Furthermore, normal GABAergic function is integral in the brain development alteration in this function can have significant effects on neuronal migration, differentiation, synaptogenesis and circuit formation. Presently, we will outline the current data from animal studies and examine the mechanisms involved in these processes.

4.2.1 Reduced GABA transmission

GABAergic inhibition can be affected in two ways, presynaptically by a reduction in GABA release into the synapse or postsynaptically by an alteration in GABA receptor function. There is some evidence that both of these situations may contribute in epilepsy-autism disorders. GAD65 is one of two glutamate decarboxylase isoforms that synthesis GABA in the brain. Previously linked to animal models of ASD, GAD65 knockout mice have also been shown to display an epileptic phenotype, with animals undergoing spontaneous seizures involving the limbic system (Kash et al., 1997). The presence of the seizures and altered behaviour was attributed to the loss of tonic inhibition to prevent hyperexcitability in the developing nervous system (Stork et al., 2000). $GABA_A$ receptor dysfunction has been well documented the hippocampus and neocortex in human epilepsy (Loup et al., 2000, 2006). Animal models of temporal lobe epilepsy (Pirker et al., 2003) and absence seizures (Li et al., 2006) suggest alterations in receptor subunit expression and receptor localization as potential mechanisms. Data is limited for epilepsy-autism disorders, however mice lacking the $GABA_A$ receptor subunit β3 (see above) displays altered EEG along with a reduced threshold to chemoconvulsant seizures (DeLorey et al., 1998; Liljelund et al., 2005).

4.2.2 Altered brain development as a result of GABAergic dysfunction

In addition to the direct effect of altered GABA system on the ability of interneurons to inhibit the generation of synchronized discharges, there are a vast array of ASD candidate genes involved in secondary regulation of the GABAergic system during development that may play a role in the pathogenesis of epilepsy-autism disorders.

The effects of Mecp2 and En2 mutations on interneuron and seizure susceptibility has been described in the previous sections. Similarly, deficits in inhibitory interneurons and reduced seizure threshold were observed in neuropilin 2 (NPN2) deficient mice (Gant et al., 2009). The gene for NPN2 (also known as NRP2) is coded for at 2q34, a region known to be strongly associated with autism (Wu et al., 2007). NPN2 functions as a chemorepulsive receptor for the axon guidance molecule Semaphorin 3F, and together regulate neuronal migration and differentiation, contributing to brain development and network formation. NPN2 deficient mice had shorter seizure latencies, increased vulnerability to seizure-induced neuronal death and developed chemically-induced spontaneous recurrent seizures (Gant et al., 2009). Importantly, NPN2 null mice had a reduced number of GABA, PV and NPY interneurons (Gant et al., 2009).

As described in the table above, Fmr1 mutant mice also showed increased susceptibility to audiogenic seizures (Musumeci et al., 2000) but not chemoconvulsants (Chen & Toth, 2001). An imbalance in the inhibition-excitation system (mainly due to reduced $GABA_A$ receptor expression; El Idrissi et al., 2005) is thought to be the major cause of increased susceptibility to seizures in these mutants. Importantly, hyperexcitibility of Fmr1 mutant mice was shown to be reduced by pharmacological intervention with a GABAergic agonist, augmenting tonic inhibitory tone (Olmos-Serrano et al., 2010).

Taken together, all these data provide a sound rationale for proposing GABA dysfunction, primarily through loss of GABA transmission, and secondarily, through altered circuit formation in development, as a potential link between epilepsy and autism, possibly even a common pathology. The range of genes involved may reflect the spectrum of pathologies associated with autism and epilepsy and warrant more detailed investigation. Overall, these data indicate that the ASD-epilepsy condition is a spectrum disorder within itself. It appears that the severity of the autism condition, presence or absence of mental retardation, is closely associated with the epilepsy phenotype, in seizure frequency, severity and intractability. Early diagnosis and suitable treatment protocols are vital for successful outcomes (Tuchman, 2000; Bombardieri et al., 2010). The Australian Rett Syndrome Database (Laurvick et al., 2006) is an ongoing longitudinal study profiling the progression of the disease in a growing cohort of cases. Similar properly constructed prospective clinical studies throughout ASD could provide vital insights required to develop successful therapeutic approaches for epilepsy in ASD. Key to advancing the experimental models will be temporal and spatial conditional deletions of various levels of the signaling system will allow greater insight into the contributions of the various genes in migration, differentiation and circuit formation.

4.3 Antiepileptic drugs in epilepsy-autism disorders

Due to the clinical heterogeneity of epilepsy and ASD significant issues exist regarding the treatment of epilepsy in children with ASD. Current treatment of epilepsy in ASD patients is based on the existing strategies for treating childhood epilepsy. Treatment with traditional AEDs such as phenobarbital, valproic acid and lamotrigine have all been shown to reduce abnormal EEG discharge (Depositario-Cabacar & Zelleke, 2010). Although, despite some studies reporting improved cognitive performance following treatment, the results from other studies are equivocal. It is clear, however, that there is an increased risk of the patient being refractory to treatment where developmental disabilities coexist (Alvarez et al., 1998; Airaksinen et al., 2000).

Studies of treatment on non autism-related epilepsies in children also raise a number of concerns about effects of AED treatment which are relevant to the current discussion. GABA is the principal inhibitory neurotransmitter in the adult brain. However, during embryonic development and up until early postnatal development in both humans and rodents, high expression of the $Na^+/K^+/2Cl^-$ (NKCC1) cotransporter and weak expression of the K^+/Cl^- cotransporter (KCC2) are present. As a result, opening of the GABAergic channels leads to depolarization and excitation of the neuron (Ben-Ari, 2002). There is an increasing body of evidence that this property of GABAergic neurons in the developing brain plays a crucial role in the development of normal neuronal circuitry, facilitating the development of both inhibitory and excitatory synapses (Akerman & Cline, 2007). In this respect, the use of GABAergic agonists to treat seizures in ASD children during developmental age might increase excitation and could result detrimental. In addition, approximately 30% of children are refractory to AED treatment (Treiman et al., 1998; Lowenstein, 2006). Furthermore, when epileptic seizures are successfully suppressed, AED treatment is associated with decreased cognitive and behavioural developmental outcomes (Loring et al., 2007), which may play a role in exacerbating the autistic condition, or at least moderate the positive outcome from absence of seizures (Tuchman, 2000). Careful monitoring for behavioural and cognitive side-effects is required since they still have the potential to exacerbate the existing ASD-related deficits. Elucidating the role of GABA dysfunction in autism-epilepsy disorders will provide greater insight into the pathogenesis of these diseases and hopefully facilitate more targeted approach producing improved outcomes in both disorders.

5. Conclusion

ASD and epilepsy both have a clear neurodevelopmental origin, and are characterized by a high degree of genetic heterogeneity. Genes regulating brain development, gene transcription, synaptic scaffolding, neurotransmission and signal transduction have been implicated in their pathogenesis, indicating that these two neurodevelopmental disorders may share common genetic bases. Indeed, epilepsy and ASD are often associated. Typically, severe forms of ASD and ASD-related pathologies always present seizures. Defects in the development, maintenance and function of GABAergic interneurons in the cerebral cortex and other brain areas have been postulated as a pathogenic mechanism of ASD-epilepsy syndromes. However, a direct, causal demonstration of a defect of GABAergic neurotransmission in restricted brain areas of ASD patients is still lacking. Conversely, evidence from several genetic mouse models of ASD strongly supports the hypothesis of GABAergic dysfunction in ASD-epilepsy. In the near future, it will be crucial to use these models to test the efficacy of GABAergic drugs to rescue ASD-like anatomical, physiological and behavioural deficits in preclinical studies. If successful, these studies might contribute to developing novel therapies against human ASD.

6. Acknowledgment

S.G., G.P. and M.D. equally contributed to this study. M.D. and P.S. are postodctoral fellows, respectively supported by IRCSET (Ireland) and Provincia Autonoma di Trento (Italy) under the Marie Curie-People cofunding action of the European Community. This work was funded by the Italian Ministry of University and Research (PRIN 2008 grant # 200894SYW2_002) and the University of Trento (CIBIO start-up grant). Correspondence should be addressed to Yuri Bozzi, Laboratory of Molecular Neuropathology, Centre for Integrative Biology (CIBIO), University of Trento, Italy (bozzi@science.unitn.it).

7. References

Abrahams, B. S. & Geschwind, D. H. (2008). Advances in Autism Genetics: On the Threshold of a New Neurobiology. *Nat Rev Genet*, Vol.9, No.5, May, pp.341-355,1471-0064

Adusei, D. C., Pacey, L. K., Chen, D. & Hampson, D. R. (2010). Early Developmental Alterations in Gabaergic Protein Expression in Fragile X Knockout Mice. *Neuropharmacology*, Vol.59, No.3, Sep, pp.167-171,1873-7064

Airaksinen, E. M., Matilainen, R., Mononen, T., Mustonen, K., Partanen, J., Jokela, V. & Halonen, P. (2000). A Population-Based Study on Epilepsy in Mentally Retarded Children. *Epilepsia*, Vol.41, No.9, Sep, pp.1214-1220,0013-9580

Akerman, C. J. & Cline, H. T. (2007). Refining the Roles of Gabaergic Signaling During Neural Circuit Formation. *Trends Neurosci*, Vol.30, No.8, Aug, pp.382-389,0166-2236

Alvarez, N., Besag, F. & Iivanainen, M. (1998). Use of Antiepileptic Drugs in the Treatment of Epilepsy in People with Intellectual Disability. *J Intellect Disabil Res*, Vol.42 Suppl 1, Dec, pp.1-15,0964-2633

Anderson, S. A., Eisenstat, D. D., Shi, L. & Rubenstein, J. L. (1997a). Interneuron Migration from Basal Forebrain to Neocortex: Dependence on Dlx Genes. *Science*, Vol.278, No.5337, Oct 17, pp.474-476,0036-8075

Anderson, S. A., Qiu, M., Bulfone, A., Eisenstat, D. D., Meneses, J., Pedersen, R. & Rubenstein, J. L. (1997b). Mutations of the Homeobox Genes Dlx-1 and Dlx-2 Disrupt the Striatal Subventricular Zone and Differentiation of Late Born Striatal Neurons. *Neuron*, Vol.19, No.1, Jul, pp.27-37,0896-6273

Anney, R.,Klei, L.,Pinto, D.,Regan, R.,Conroy, J.,Magalhaes, T. R.,Correia, C.,Abrahams, B. S.,Sykes, N.,Pagnamenta, A. T.,Almeida, J.,Bacchelli, E.,Bailey, A. J.,Baird, G.,Battaglia, A.,Berney, T.,Bolshakova, N.,Bolte, S.,Bolton, P. F.,Bourgeron, T.,Brennan, S.,Brian, J.,Carson, A. R.,Casallo, G.,Casey, J.,Chu, S. H.,Cochrane, L.,Corsello, C.,Crawford, E. L.,Crossett, A.,Dawson, G.,de Jonge, M.,Delorme, R.,Drmic, I.,Duketis, E.,Duque, F.,Estes, A.,Farrar, P.,Fernandez, B. A.,Folstein, S. E.,Fombonne, E.,Freitag, C. M.,Gilbert, J.,Gillberg, C.,Glessner, J. T.,Goldberg, J.,Green, J.,Guter, S. J.,Hakonarson, H.,Heron, E. A.,Hill, M.,Holt, R.,Howe, J. L.,Hughes, G.,Hus, V.,Igliozzi, R.,Kim, C.,Klauck, S. M.,Kolevzon, A.,Korvatska, O.,Kustanovich, V.,Lajonchere, C. M.,Lamb, J. A.,Laskawiec, M.,Leboyer, M.,Le Couteur, A.,Leventhal, B. L.,Lionel, A. C.,Liu, X. Q.,Lord, C.,Lotspeich, L.,Lund, S. C.,Maestrini, E.,Mahoney, W.,Mantoulan, C.,Marshall, C. R.,McConachie,

H.,McDougle, C. J.,McGrath, J.,McMahon, W. M.,Melhem, N. M.,Merikangas, A.,Migita, O.,Minshew, N. J.,Mirza, G. K.,Munson, J.,Nelson, S. F.,Noakes, C.,Noor, A.,Nygren, G.,Oliveira, G.,Papanikolaou, K.,Parr, J. R.,Parrini, B.,Paton, T.,Pickles, A.,Piven, J.,Posey, D. J.,Poustka, A.,Poustka, F.,Prasad, A.,Ragoussis, J.,Renshaw, K.,Rickaby, J.,Roberts, W.,Roeder, K.,Roge, B.,Rutter, M. L.,Bierut, L. J.,Rice, J. P.,Salt, J.,Sansom, K.,Sato, D.,Segurado, R.,Senman, L.,Shah, N.,Sheffield, V. C.,Soorya, L.,Sousa, I.,Stoppioni, V.,Strawbridge, C.,Tancredi, R.,Tansey, K.,Thiruvahindrapduram, B.,Thompson, A. P.,Thomson, S.,Tryfon, A.,Tsiantis, J.,Van Engeland, H.,Vincent, J. B.,Volkmar, F.,Wallace, S.,Wang, K.,Wang, Z.,Wassink, T. H.,Wing, K.,Wittemeyer, K.,Wood, S.,Yaspan, B. L.,Zurawiecki, D.,Zwaigenbaum, L.,Betancur, C.,Buxbaum, J. D.,Cantor, R. M.,Cook, E. H.,Coon, H.,Cuccaro, M. L.,Gallagher, L.,Geschwind, D. H.,Gill, M.,Haines, J. L.,Miller, J.,Monaco, A. P.,Nurnberger, J. I., Jr.,Paterson, A. D.,Pericak-Vance, M. A.,Schellenberg, G. D.,Scherer, S. W.,Sutcliffe, J. S.,Szatmari, P.,Vicente, A. M.,Vieland, V. J.,Wijsman, E. M.,Devlin, B.,Ennis, S. & Hallmayer, J. (2010). A Genome-Wide Scan for Common Alleles Affecting Risk for Autism. *Hum Mol Genet*, Vol.19, No.20, Oct 15, pp.4072-4082,1460-2083

Bardoni, B., Davidovic, L., Bensaid, M. & Khandjian, E. W. (2006). The Fragile X Syndrome: Exploring Its Molecular Basis and Seeking a Treatment. *Expert Rev Mol Med*, Vol.8, No.8, pp.1-16,1462-3994

Bassell, G. J. & Warren, S. T. (2008). Fragile X Syndrome: Loss of Local Mrna Regulation Alters Synaptic Development and Function. *Neuron*, Vol.60, No.2, Oct 23, pp.201-214,1097-4199

Bauman, M. L. & Kemper, T. L. (2005). Neuroanatomic Observations of the Brain in Autism: A Review and Future Directions. *Int J Dev Neurosci*, Vol.23, No.2-3, Apr-May, pp.183-187,0736-5748

Beffert, U., Weeber, E. J., Durudas, A., Qiu, S., Masiulis, I., Sweatt, J. D., Li, W. P., Adelmann, G., Frotscher, M., Hammer, R. E. & Herz, J. (2005). Modulation of Synaptic Plasticity and Memory by Reelin Involves Differential Splicing of the Lipoprotein Receptor Apoer2. *Neuron*, Vol.47, No.4, Aug 18, pp.567-579,0896-6273

Ben-Ari, Y. (2002). Excitatory Actions of Gaba During Development: The Nature of the Nurture. *Nat Rev Neurosci*, Vol.3, No.9, Sep, pp.728-739,1471-003X

Ben-Ari, Y. (2006). Seizures Beget Seizures: The Quest for Gaba as a Key Player. *Crit Rev Neurobiol*, Vol.18, No.1-2, pp.135-144,0892-0915

Ben-Ari, Y. & Holmes, G. L. (2006). Effects of Seizures on Developmental Processes in the Immature Brain. *Lancet Neurol*, Vol.5, No.12, Dec, pp.1055-1063,1474-4422

Benayed, R., Choi, J., Matteson, P. G., Gharani, N., Kamdar, S., Brzustowicz, L. M. & Millonig, J. H. (2009). Autism-Associated Haplotype Affects the Regulation of the Homeobox Gene, Engrailed 2. *Biol Psychiatry*, Vol.66, No.10, Nov 15, pp.911-917,1873-2402

Betancur, C. (2011). Etiological Heterogeneity in Autism Spectrum Disorders: More Than 100 Genetic and Genomic Disorders and Still Counting. *Brain Res*, Vol.1380, Mar 22, pp.42-77,1872-6240

Blatt, G. J., Fitzgerald, C. M., Guptill, J. T., Booker, A. B., Kemper, T. L. & Bauman, M. L. (2001). Density and Distribution of Hippocampal Neurotransmitter Receptors in

Autism: An Autoradiographic Study. *J Autism Dev Disord*, Vol.31, No.6, Dec, pp.537-543,0162-3257

Bombardieri, R., Pinci, M., Moavero, R., Cerminara, C. & Curatolo, P. (2010). Early Control of Seizures Improves Long-Term Outcome in Children with Tuberous Sclerosis Complex. *Eur J Paediatr Neurol*, Vol.14, No.2, Mar, pp.146-149,1532-2130

Bradford, H. F. (1995). Glutamate, Gaba and Epilepsy. *Prog Neurobiol*, Vol.47, No.6, Dec, pp.477-511,0301-0082

Brooks-Kayal, A. (2010). Epilepsy and Autism Spectrum Disorders: Are There Common Developmental Mechanisms? *Brain Dev*, Vol.32, No.9, Oct, pp.731-738,1872-7131

Budreck, E. C. & Scheiffele, P. (2007). Neuroligin-3 Is a Neuronal Adhesion Protein at Gabaergic and Glutamatergic Synapses. *Eur J Neurosci*, Vol.26, No.7, Oct, pp.1738-1748,0953-816X

Buxbaum, J. D., Silverman, J. M., Smith, C. J., Greenberg, D. A., Kilifarski, M., Reichert, J., Cook, E. H., Jr., Fang, Y., Song, C. Y. & Vitale, R. (2002). Association between a Gabrb3 Polymorphism and Autism. *Mol Psychiatry*, Vol.7, No.3, pp.311-316,1359-4184

Canitano, R. (2007). Epilepsy in Autism Spectrum Disorders. *Eur Child Adolesc Psychiatry*, Vol.16, No.1, Feb, pp.61-66,1018-8827

Carboni, G., Tueting, P., Tremolizzo, L., Sugaya, I., Davis, J., Costa, E. & Guidotti, A. (2004). Enhanced Dizocilpine Efficacy in Heterozygous Reeler Mice Relates to Gaba Turnover Downregulation. *Neuropharmacology*, Vol.46, No.8, Jun, pp.1070-1081,0028-3908

Casanova, M. & Trippe, J. (2009). Radial Cytoarchitecture and Patterns of Cortical Connectivity in Autism. *Philos Trans R Soc Lond B Biol Sci*, Vol.364, No.1522, May 27, pp.1433-1436,1471-2970

Casanova, M. F., Buxhoeveden, D. & Gomez, J. (2003). Disruption in the Inhibitory Architecture of the Cell Minicolumn: Implications for Autisim. *Neuroscientist*, Vol.9, No.6, Dec, pp.496-507,1073-8584

Casanova, M. F., Buxhoeveden, D. P., Switala, A. E. & Roy, E. (2002). Minicolumnar Pathology in Autism. *Neurology*, Vol.58, No.3, Feb 12, pp.428-432,0028-3878

Centonze, D., Rossi, S., Mercaldo, V., Napoli, I., Ciotti, M. T., De Chiara, V., Musella, A., Prosperetti, C., Calabresi, P., Bernardi, G. & Bagni, C. (2008). Abnormal Striatal Gaba Transmission in the Mouse Model for the Fragile X Syndrome. *Biol Psychiatry*, Vol.63, No.10, May 15, pp.963-973,1873-2402

Chahrour, M. & Zoghbi, H. Y. (2007). The Story of Rett Syndrome: From Clinic to Neurobiology. *Neuron*, Vol.56, No.3, Nov 8, pp.422-437,0896-6273

Chang, S. C., Pauls, D. L., Lange, C., Sasanfar, R. & Santangelo, S. L. (2010). Common Genetic Variation in the Gad1 Gene and the Entire Family of Dlx Homeobox Genes and Autism Spectrum Disorders. *Am J Med Genet B Neuropsychiatr Genet*, Dec 16,1552-485X

Chao, H. T., Chen, H., Samaco, R. C., Xue, M., Chahrour, M., Yoo, J., Neul, J. L., Gong, S., Lu, H. C., Heintz, N., Ekker, M., Rubenstein, J. L., Noebels, J. L., Rosenmund, C. & Zoghbi, H. Y. (2010). Dysfunction in Gaba Signalling Mediates Autism-Like Stereotypies and Rett Syndrome Phenotypes. *Nature*, Vol.468, No.7321, Nov 11, pp.263-269,1476-4687

Cheh, M. A., Millonig, J. H., Roselli, L. M., Ming, X., Jacobsen, E., Kamdar, S. & Wagner, G. C. (2006). En2 Knockout Mice Display Neurobehavioral and Neurochemical Alterations Relevant to Autism Spectrum Disorder. *Brain Res*, Vol.1116, No.1, Oct 20, pp.166-176,0006-8993

Chen, L. & Toth, M. (2001). Fragile X Mice Develop Sensory Hyperreactivity to Auditory Stimuli. *Neuroscience*, Vol.103, No.4, pp.1043-1050,0306-4522

Chih, B., Engelman, H. & Scheiffele, P. (2005). Control of Excitatory and Inhibitory Synapse Formation by Neuroligins. *Science*, Vol.307, No.5713, Feb 25, pp.1324-1328,1095-9203

Chin, R. F., Neville, B. G., Peckham, C., Bedford, H., Wade, A. & Scott, R. C. (2006). Incidence, Cause, and Short-Term Outcome of Convulsive Status Epilepticus in Childhood: Prospective Population-Based Study. *Lancet*, Vol.368, No.9531, Jul 15, pp.222-229,1474-547X

Cobos, I., Calcagnotto, M. E., Vilaythong, A. J., Thwin, M. T., Noebels, J. L., Baraban, S. C. & Rubenstein, J. L. (2005). Mice Lacking Dlx1 Show Subtype-Specific Loss of Interneurons, Reduced Inhibition and Epilepsy. *Nat Neurosci*, Vol.8, No.8, Aug, pp.1059-1068,1097-6256

Collins, A. L., Ma, D., Whitehead, P. L., Martin, E. R., Wright, H. H., Abramson, R. K., Hussman, J. P., Haines, J. L., Cuccaro, M. L., Gilbert, J. R. & Pericak-Vance, M. A. (2006). Investigation of Autism and Gaba Receptor Subunit Genes in Multiple Ethnic Groups. *Neurogenetics*, Vol.7, No.3, Jul, pp.167-174,1364-6745

Craig, A. M. & Kang, Y. (2007). Neurexin-Neuroligin Signaling in Synapse Development. *Curr Opin Neurobiol*, Vol.17, No.1, Feb, pp.43-52,0959-4388

Curia, G., Papouin, T., Seguela, P. & Avoli, M. (2009). Downregulation of Tonic Gabaergic Inhibition in a Mouse Model of Fragile X Syndrome. *Cereb Cortex*, Vol.19, No.7, Jul, pp.1515-1520,1460-2199

Curran, T. & D'Arcangelo, G. (1998). Role of Reelin in the Control of Brain Development. *Brain Res Brain Res Rev*, Vol.26, No.2-3, May, pp.285-294

D'Hulst, C., Heulens, I., Brouwer, J. R., Willemsen, R., De Geest, N., Reeve, S. P., De Deyn, P. P., Hassan, B. A. & Kooy, R. F. (2009). Expression of the Gabaergic System in Animal Models for Fragile X Syndrome and Fragile X Associated Tremor/Ataxia Syndrome (Fxtas). *Brain Res*, Vol.1253, Feb 9, pp.176-183,1872-6240

D'Hulst, C. & Kooy, R. F. (2007). The Gabaa Receptor: A Novel Target for Treatment of Fragile X? *Trends Neurosci*, Vol.30, No.8, Aug, pp.425-431,0166-2236

D'Hulst, C. & Kooy, R. F. (2009). Fragile X Syndrome: From Molecular Genetics to Therapy. *J Med Genet*, Vol.46, No.9, Sep, pp.577-584,1468-6244

Danielsson, S., Gillberg, I. C., Billstedt, E., Gillberg, C. & Olsson, I. (2005). Epilepsy in Young Adults with Autism: A Prospective Population-Based Follow-up Study of 120 Individuals Diagnosed in Childhood. *Epilepsia*, Vol.46, No.6, Jun, pp.918-923,0013-9580

DeFelipe, J., Hendry, S. H., Hashikawa, T., Molinari, M. & Jones, E. G. (1990). A Microcolumnar Structure of Monkey Cerebral Cortex Revealed by Immunocytochemical Studies of Double Bouquet Cell Axons. *Neuroscience*, Vol.37, No.3, pp.655-673,0306-4522

DeLorey, T. M., Handforth, A., Anagnostaras, S. G., Homanics, G. E., Minassian, B. A., Asatourian, A., Fanselow, M. S., Delgado-Escueta, A., Ellison, G. D. & Olsen, R. W.

(1998). Mice Lacking the Beta3 Subunit of the Gabaa Receptor Have the Epilepsy Phenotype and Many of the Behavioral Characteristics of Angelman Syndrome. *J Neurosci*, Vol.18, No.20, Oct 15, pp.8505-8514,0270-6474

DeLorey, T. M., Sahbaie, P., Hashemi, E., Homanics, G. E. & Clark, J. D. (2008). Gabrb3 Gene Deficient Mice Exhibit Impaired Social and Exploratory Behaviors, Deficits in Non-Selective Attention and Hypoplasia of Cerebellar Vermal Lobules: A Potential Model of Autism Spectrum Disorder. *Behav Brain Res*, Vol.187, No.2, Mar 5, pp.207-220,0166-4328

Depositario-Cabacar, D. F. & Zelleke, T. G. (2010). Treatment of Epilepsy in Children with Developmental Disabilities. *Dev Disabil Res Rev*, Vol.16, No.3, pp.239-247,1940-5529

DeSilva, U., D'Arcangelo, G., Braden, V. V., Chen, J., Miao, G. G., Curran, T. & Green, E. D. (1997). The Human Reelin Gene: Isolation, Sequencing, and Mapping on Chromosome 7. *Genome Res*, Vol.7, No.2, Feb, pp.157-164,1088-9051

Di Cristo, G. (2007). Development of Cortical Gabaergic Circuits and Its Implications for Neurodevelopmental Disorders. *Clin Genet*, Vol.72, No.1, Jul, pp.1-8,0009-9163

Dunleavy, M., Shinoda, S., Schindler, C., Ewart, C., Dolan, R., Gobbo, O. L., Kerskens, C. M. & Henshall, D. C. (2010). Experimental Neonatal Status Epilepticus and the Development of Temporal Lobe Epilepsy with Unilateral Hippocampal Sclerosis. *Am J Pathol*, Vol.176, No.1, Jan, pp.330-342,1525-2191

El Idrissi, A., Ding, X. H., Scalia, J., Trenkner, E., Brown, W. T. & Dobkin, C. (2005). Decreased Gaba(a) Receptor Expression in the Seizure-Prone Fragile X Mouse. *Neurosci Lett*, Vol.377, No.3, Apr 4, pp.141-146,0304-3940

Enticott, P. G., Rinehart, N. J., Tonge, B. J., Bradshaw, J. L. & Fitzgerald, P. B. (2010). A Preliminary Transcranial Magnetic Stimulation Study of Cortical Inhibition and Excitability in High-Functioning Autism and Asperger Disorder. *Dev Med Child Neurol*, Vol.52, No.8, Aug, pp.e179-e183,0012-1622

Fatemi, S. H., Halt, A. R., Stary, J. M., Kanodia, R., Schulz, S. C. & Realmuto, G. R. (2002). Glutamic Acid Decarboxylase 65 and 67 Kda Proteins Are Reduced in Autistic Parietal and Cerebellar Cortices. *Biol Psychiatry*, Vol.52, No.8, Oct 15, pp.805-810,0006-3223

Fatemi, S. H., Reutiman, T. J., Folsom, T. D., Rooney, R. J., Patel, D. H. & Thuras, P. D. (2010). Mrna and Protein Levels for Gabaaalpha4, Alpha5, Beta1 and Gababr1 Receptors Are Altered in Brains from Subjects with Autism. *J Autism Dev Disord*, Vol.40, No.6, Jun, pp.743-750,1573-3432

Fatemi, S. H., Snow, A. V., Stary, J. M., Araghi-Niknam, M., Reutiman, T. J., Lee, S., Brooks, A. I. & Pearce, D. A. (2005). Reelin Signaling Is Impaired in Autism. *Biol Psychiatry*, Vol.57, No.7, Apr 1, pp.777-787,0006-3223

Feng, J., Schroer, R., Yan, J., Song, W., Yang, C., Bockholt, A., Cook, E. H., Jr., Skinner, C., Schwartz, C. E. & Sommer, S. S. (2006). High Frequency of Neurexin 1beta Signal Peptide Structural Variants in Patients with Autism. *Neurosci Lett*, Vol.409, No.1, Nov 27, pp.10-13,0304-3940

Fombonne, E. (2009). Epidemiology of Pervasive Developmental Disorders. *Pediatr Res*, Vol.65, No.6, Jun, pp.591-598,1530-0447

Forster, E., Tielsch, A., Saum, B., Weiss, K. H., Johanssen, C., Graus-Porta, D., Muller, U. & Frotscher, M. (2002). Reelin, Disabled 1, and Beta 1 Integrins Are Required for the

Formation of the Radial Glial Scaffold in the Hippocampus. *Proc Natl Acad Sci U S A*, Vol.99, No.20, Oct 1, pp.13178-13183,0027-8424

Freitag, C. M. (2007). The Genetics of Autistic Disorders and Its Clinical Relevance: A Review of the Literature. *Mol Psychiatry*, Vol.12, No.1, Jan, pp.2-22,1359-4184

Gant, J. C., Thibault, O., Blalock, E. M., Yang, J., Bachstetter, A., Kotick, J., Schauwecker, P. E., Hauser, K. F., Smith, G. M., Mervis, R., Li, Y. & Barnes, G. N. (2009). Decreased Number of Interneurons and Increased Seizures in Neuropilin 2 Deficient Mice: Implications for Autism and Epilepsy. *Epilepsia*, Vol.50, No.4, Apr, pp.629-645,1528-1167

Gauthey, M., Poloni, C. B., Ramelli, G. P., Roulet-Perez, E. & Korff, C. M. (2010). Status Epilepticus in Fragile X Syndrome. *Epilepsia*, Vol.51, No.12, Dec, pp.2470-2473,1528-1167

Gogolla, N., Leblanc, J. J., Quast, K. B., Sudhof, T., Fagiolini, M. & Hensch, T. K. (2009). Common Circuit Defect of Excitatory-Inhibitory Balance in Mouse Models of Autism. *J Neurodev Disord*, Vol.1, No.2, Jun 1, pp.172-181,1866-1947

Graf, E. R., Zhang, X., Jin, S. X., Linhoff, M. W. & Craig, A. M. (2004). Neurexins Induce Differentiation of Gaba and Glutamate Postsynaptic Specializations Via Neuroligins. *Cell*, Vol.119, No.7, Dec 29, pp.1013-1026,0092-8674

Hara, H. (2007). Autism and Epilepsy: A Retrospective Follow-up Study. *Brain Dev*, Vol.29, No.8, Sep, pp.486-490,0387-7604

Hatton, D. D., Sideris, J., Skinner, M., Mankowski, J., Bailey, D. B., Jr., Roberts, J. & Mirrett, P. (2006). Autistic Behavior in Children with Fragile X Syndrome: Prevalence, Stability, and the Impact of Fmrp. *Am J Med Genet A*, Vol.140A, No.17, Sep 1, pp.1804-1813,1552-4825

Hauser, W. A. (1990). Status Epilepticus: Epidemiologic Considerations. *Neurology*, Vol.40, No.5 Suppl 2, May, pp.9-13,0028-3878

Hogart, A., Leung, K. N., Wang, N. J., Wu, D. J., Driscoll, J., Vallero, R. O., Schanen, N. C. & LaSalle, J. M. (2009). Chromosome 15q11-13 Duplication Syndrome Brain Reveals Epigenetic Alterations in Gene Expression Not Predicted from Copy Number. *J Med Genet*, Vol.46, No.2, Feb, pp.86-93,1468-6244

Hogart, A., Nagarajan, R. P., Patzel, K. A., Yasui, D. H. & Lasalle, J. M. (2007). 15q11-13 Gabaa Receptor Genes Are Normally Biallelically Expressed in Brain yet Are Subject to Epigenetic Dysregulation in Autism-Spectrum Disorders. *Hum Mol Genet*, Vol.16, No.6, Mar 15, pp.691-703,0964-6906

Homanics, G. E., DeLorey, T. M., Firestone, L. L., Quinlan, J. J., Handforth, A., Harrison, N. L., Krasowski, M. D., Rick, C. E., Korpi, E. R., Makela, R., Brilliant, M. H., Hagiwara, N., Ferguson, C., Snyder, K. & Olsen, R. W. (1997). Mice Devoid of Gamma-Aminobutyrate Type a Receptor Beta3 Subunit Have Epilepsy, Cleft Palate, and Hypersensitive Behavior. *Proc Natl Acad Sci U S A*, Vol.94, No.8, Apr 15, pp.4143-4148,0027-8424

Jamain, S., Quach, H., Betancur, C., Rastam, M., Colineaux, C., Gillberg, I. C., Soderstrom, H., Giros, B., Leboyer, M., Gillberg, C. & Bourgeron, T. (2003). Mutations of the X-Linked Genes Encoding Neuroligins Nlgn3 and Nlgn4 Are Associated with Autism. *Nat Genet*, Vol.34, No.1, May, pp.27-29,1061-4036

Jensen, F. E. (2011). Epilepsy as a Spectrum Disorder: Implications from Novel Clinical and Basic Neuroscience. *Epilepsia*, Vol.52 Suppl 1, Jan, pp.1-6,1528-1167

Jian, L., Nagarajan, L., de Klerk, N., Ravine, D., Bower, C., Anderson, A., Williamson, S., Christodoulou, J. & Leonard, H. (2006). Predictors of Seizure Onset in Rett Syndrome. *J Pediatr*, Vol.149, No.4, Oct, pp.542-547,0022-3476

Joyner, A. L., Herrup, K., Auerbach, B. A., Davis, C. A. & Rossant, J. (1991). Subtle Cerebellar Phenotype in Mice Homozygous for a Targeted Deletion of the En-2 Homeobox. *Science*, Vol.251, No.4998, Mar 8, pp.1239-1243,0036-8075

Karlsen, A. E., Hagopian, W. A., Grubin, C. E., Dube, S., Disteche, C. M., Adler, D. A., Barmeier, H., Mathewes, S., Grant, F. J., Foster, D. & et al. (1991). Cloning and Primary Structure of a Human Islet Isoform of Glutamic Acid Decarboxylase from Chromosome 10. *Proc Natl Acad Sci U S A*, Vol.88, No.19, Oct 1, pp.8337-8341,0027-8424

Kash, S. F., Johnson, R. S., Tecott, L. H., Noebels, J. L., Mayfield, R. D., Hanahan, D. & Baekkeskov, S. (1997). Epilepsy in Mice Deficient in the 65-Kda Isoform of Glutamic Acid Decarboxylase. *Proc Natl Acad Sci U S A*, Vol.94, No.25, Dec 9, pp.14060-14065,0027-8424

Kaufman, D. L., Houser, C. R. & Tobin, A. J. (1991). Two Forms of the Gamma-Aminobutyric Acid Synthetic Enzyme Glutamate Decarboxylase Have Distinct Intraneuronal Distributions and Cofactor Interactions. *J Neurochem*, Vol.56, No.2, Feb, pp.720-723,0022-3042

Kemper, T. L. & Bauman, M. L. (1993). The Contribution of Neuropathologic Studies to the Understanding of Autism. *Neurol Clin*, Vol.11, No.1, Feb, pp.175-187,0733-8619

Kim, H. G., Kishikawa, S., Higgins, A. W., Seong, I. S., Donovan, D. J., Shen, Y., Lally, E., Weiss, L. A., Najm, J., Kutsche, K., Descartes, M., Holt, L., Braddock, S., Troxell, R., Kaplan, L., Volkmar, F., Klin, A., Tsatsanis, K., Harris, D. J., Noens, I., Pauls, D. L., Daly, M. J., MacDonald, M. E., Morton, C. C., Quade, B. J. & Gusella, J. F. (2008). Disruption of Neurexin 1 Associated with Autism Spectrum Disorder. *Am J Hum Genet*, Vol.82, No.1, Jan, pp.199-207,1537-6605

Kuemerle, B., Zanjani, H., Joyner, A. & Herrup, K. (1997). Pattern Deformities and Cell Loss in Engrailed-2 Mutant Mice Suggest Two Separate Patterning Events During Cerebellar Development. *J Neurosci*, Vol.17, No.20, Oct 15, pp.7881-7889,0270-6474

Laumonnier, F., Bonnet-Brilhault, F., Gomot, M., Blanc, R., David, A., Moizard, M. P., Raynaud, M., Ronce, N., Lemonnier, E., Calvas, P., Laudier, B., Chelly, J., Fryns, J. P., Ropers, H. H., Hamel, B. C., Andres, C., Barthelemy, C., Moraine, C. & Briault, S. (2004). X-Linked Mental Retardation and Autism Are Associated with a Mutation in the Nlgn4 Gene, a Member of the Neuroligin Family. *Am J Hum Genet*, Vol.74, No.3, Mar, pp.552-557,0002-9297

Laurvick, C. L., De Klerk, N., Bower, C., Christodoulou, J., Ravine, D., Ellaway, C., Williamson, S. & Leonard, H. (2006). Rett Syndrome in Australia: A Review of the Epidemiology. *Journal of Pediatrics*, Vol.148, No.3, Mar, pp.347-352,0022-3476

Lawrence, Y. A., Kemper, T. L., Bauman, M. L. & Blatt, G. J. (2010). Parvalbumin-, Calbindin-, and Calretinin-Immunoreactive Hippocampal Interneuron Density in Autism. *Acta Neurol Scand*, Vol.121, No.2, Feb, pp.99-108,1600-0404

Lawson-Yuen, A., Saldivar, J. S., Sommer, S. & Picker, J. (2008). Familial Deletion within Nlgn4 Associated with Autism and Tourette Syndrome. *Eur J Hum Genet*, Vol.16, No.5, May, pp.614-618,1018-4813

Li, H., Kraus, A., Wu, J., Huguenard, J. R. & Fisher, R. S. (2006). Selective Changes in Thalamic and Cortical Gabaa Receptor Subunits in a Model of Acquired Absence Epilepsy in the Rat. *Neuropharmacology*, Vol.51, No.1, Jul, pp.121-128,0028-3908

Liljelund, P., Handforth, A., Homanics, G. E. & Olsen, R. W. (2005). Gabaa Receptor Beta3 Subunit Gene-Deficient Heterozygous Mice Show Parent-of-Origin and Gender-Related Differences in Beta3 Subunit Levels, Eeg, and Behavior. *Brain Res Dev Brain Res*, Vol.157, No.2, Jun 30, pp.150-161,0165-3806

Lintas, C. & Persico, A. M. (2009). Autistic Phenotypes and Genetic Testing: State-of-the-Art for the Clinical Geneticist. *J Med Genet*, Vol.46, No.1, Jan, pp.1-8,1468-6244

Liu, X., Novosedlik, N., Wang, A., Hudson, M. L., Cohen, I. L., Chudley, A. E., Forster-Gibson, C. J., Lewis, S. M. & Holden, J. J. (2009). The Dlx1and Dlx2 Genes and Susceptibility to Autism Spectrum Disorders. *Eur J Hum Genet*, Vol.17, No.2, Feb, pp.228-235,1476-5438

Loring, D. W., Marino, S. & Meador, K. J. (2007). Neuropsychological and Behavioral Effects of Antiepilepsy Drugs. *Neuropsychol Rev*, Vol.17, No.4, Dec, pp.413-425,1040-7308

Loup, F., Picard, F., Andre, V. M., Kehrli, P., Yonekawa, Y., Wieser, H. G. & Fritschy, J. M. (2006). Altered Expression of Alpha3-Containing Gabaa Receptors in the Neocortex of Patients with Focal Epilepsy. *Brain*, Vol.129, No.Pt 12, Dec, pp.3277-3289,1460-2156

Loup, F., Wieser, H. G., Yonekawa, Y., Aguzzi, A. & Fritschy, J. M. (2000). Selective Alterations in Gabaa Receptor Subtypes in Human Temporal Lobe Epilepsy. *J Neurosci*, Vol.20, No.14, Jul 15, pp.5401-5419,0270-6474

Lowenstein, D. H. (2006). The Management of Refractory Status Epilepticus: An Update. *Epilepsia*, Vol.47 Suppl 1, pp.35-40,0013-9580

Macri, S., Biamonte, F., Romano, E., Marino, R., Keller, F. & Laviola, G. (2010). Perseverative Responding and Neuroanatomical Alterations in Adult Heterozygous Reeler Mice Are Mitigated by Neonatal Estrogen Administration. *Psychoneuroendocrinology*, Vol.35, No.9, Oct, pp.1374-1387,1873-3360

Mao, R., Page, D. T., Merzlyak, I., Kim, C., Tecott, L. H., Janak, P. H., Rubenstein, J. L. & Sur, M. (2009). Reduced Conditioned Fear Response in Mice That Lack Dlx1 and Show Subtype-Specific Loss of Interneurons. *J Neurodev Disord*, Vol.1, No.3, Sep, pp.224-236,1866-1947

Martin, D. L. & Rimvall, K. (1993). Regulation of Gamma-Aminobutyric Acid Synthesis in the Brain. *J Neurochem*, Vol.60, No.2, Feb, pp.395-407,0022-3042

Mountcastle, V. B. (1997). The Columnar Organization of the Neocortex. *Brain*, Vol.120 (Pt 4), Apr, pp.701-722,0006-8950

Musumeci, S. A., Bosco, P., Calabrese, G., Bakker, C., De Sarro, G. B., Elia, M., Ferri, R. & Oostra, B. A. (2000). Audiogenic Seizures Susceptibility in Transgenic Mice with Fragile X Syndrome. *Epilepsia*, Vol.41, No.1, Jan, pp.19-23,0013-9580

Musumeci, S. A., Calabrese, G., Bonaccorso, C. M., D'Antoni, S., Brouwer, J. R., Bakker, C. E., Elia, M., Ferri, R., Nelson, D. L., Oostra, B. A. & Catania, M. V. (2007). Audiogenic Seizure Susceptibility Is Reduced in Fragile X Knockout Mice after Introduction of Fmr1 Transgenes. *Exp Neurol*, Vol.203, No.1, Jan, pp.233-240,0014-4886

Oblak, A. L., Gibbs, T. T. & Blatt, G. J. (2010). Decreased Gaba(B) Receptors in the Cingulate Cortex and Fusiform Gyrus in Autism. *J Neurochem*, Vol.114, No.5, Sep 1, pp.1414-1423,1471-4159

Oblak, A. L., Gibbs, T. T. & Blatt, G. J. (2011). Reduced Gaba(a) Receptors and Benzodiazepine Binding Sites in the Posterior Cingulate Cortex and Fusiform Gyrus in Autism. *Brain Res*, Vol.1380, Mar 22, pp.218-228,1872-6240

Olmos-Serrano, J. L., Paluszkiewicz, S. M., Martin, B. S., Kaufmann, W. E., Corbin, J. G. & Huntsman, M. M. (2010). Defective Gabaergic Neurotransmission and Pharmacological Rescue of Neuronal Hyperexcitability in the Amygdala in a Mouse Model of Fragile X Syndrome. *J Neurosci*, Vol.30, No.29, Jul 21, pp.9929-9938,1529-2401

Olsen, R. W. & Avoli, M. (1997). Gaba and Epileptogenesis. *Epilepsia*, Vol.38, No.4, Apr, pp.399-407,0013-9580

Olsen, R. W. & Sieghart, W. (2009). Gaba a Receptors: Subtypes Provide Diversity of Function and Pharmacology. *Neuropharmacology*, Vol.56, No.1, Jan, pp.141-148,0028-3908

Persico, A. M. & Bourgeron, T. (2006). Searching for Ways out of the Autism Maze: Genetic, Epigenetic and Environmental Clues. *Trends Neurosci*, Vol.29, No.7, Jul, pp.349-358,0166-2236

Persico, A. M., D'Agruma, L., Maiorano, N., Totaro, A., Militerni, R., Bravaccio, C., Wassink, T. H., Schneider, C., Melmed, R., Trillo, S., Montecchi, F., Palermo, M., Pascucci, T., Puglisi-Allegra, S., Reichelt, K. L., Conciatori, M., Marino, R., Quattrocchi, C. C., Baldi, A., Zelante, L., Gasparini, P. & Keller, F. (2001). Reelin Gene Alleles and Haplotypes as a Factor Predisposing to Autistic Disorder. *Mol Psychiatry*, Vol.6, No.2, Mar, pp.150-159,1359-4184

Pirker, S., Schwarzer, C., Czech, T., Baumgartner, C., Pockberger, H., Maier, H., Hauer, B., Sieghart, W., Furtinger, S. & Sperk, G. (2003). Increased Expression of Gaba(a) Receptor Beta-Subunits in the Hippocampus of Patients with Temporal Lobe Epilepsy. *J Neuropathol Exp Neurol*, Vol.62, No.8, Aug, pp.820-834,0022-3069

Pleasure, S. J., Anderson, S., Hevner, R., Bagri, A., Marin, O., Lowenstein, D. H. & Rubenstein, J. L. (2000). Cell Migration from the Ganglionic Eminences Is Required for the Development of Hippocampal Gabaergic Interneurons. *Neuron*, Vol.28, No.3, Dec, pp.727-740,0896-6273

Polleux, F. & Lauder, J. M. (2004). Toward a Developmental Neurobiology of Autism. *Ment Retard Dev Disabil Res Rev*, Vol.10, No.4, pp.303-317,1080-4013

Radyushkin, K., Hammerschmidt, K., Boretius, S., Varoqueaux, F., El-Kordi, A., Ronnenberg, A., Winter, D., Frahm, J., Fischer, J., Brose, N. & Ehrenreich, H. (2009). Neuroligin-3-Deficient Mice: Model of a Monogenic Heritable Form of Autism with an Olfactory Deficit. *Genes Brain Behav*, Vol.8, No.4, Jun, pp.416-425,1601-183X

Raghanti, M. A., Spocter, M. A., Butti, C., Hof, P. R. & Sherwood, C. C. (2010). A Comparative Perspective on Minicolumns and Inhibitory Gabaergic Interneurons in the Neocortex. *Front Neuroanat*, Vol.4, pp.3,1662-5129

Raspall-Chaure, M., Chin, R. F., Neville, B. G. & Scott, R. C. (2006). Outcome of Paediatric Convulsive Status Epilepticus: A Systematic Review. *Lancet Neurol*, Vol.5, No.9, Sep, pp.769-779,1474-4422

Rubenstein, J. L. & Merzenich, M. M. (2003). Model of Autism: Increased Ratio of Excitation/Inhibition in Key Neural Systems. *Genes Brain Behav*, Vol.2, No.5, Oct, pp.255-267,1601-1848

Rudy, B., Fishell, G., Lee, S. & Hjerling-Leffler, J. (2010). Three Groups of Interneurons Account for Nearly 100% of Neocortical Gabaergic Neurons. *Dev Neurobiol*, Nov 30,1932-846X

Sabaratnam, M. (2000). Pathological and Neuropathological Findings in Two Males with Fragile-X Syndrome. *J Intellect Disabil Res*, Vol.44 (Pt 1), Feb, pp.81-85,0964-2633

Samaco, R. C., Hogart, A. & LaSalle, J. M. (2005). Epigenetic Overlap in Autism-Spectrum Neurodevelopmental Disorders: Mecp2 Deficiency Causes Reduced Expression of Ube3a and Gabrb3. *Hum Mol Genet*, Vol.14, No.4, Feb 15, pp.483-492,0964-6906

Scheiffele, P., Fan, J., Choih, J., Fetter, R. & Serafini, T. (2000). Neuroligin Expressed in Nonneuronal Cells Triggers Presynaptic Development in Contacting Axons. *Cell*, Vol.101, No.6, Jun 9, pp.657-669,0092-8674

Schroer, R. J., Phelan, M. C., Michaelis, R. C., Crawford, E. C., Skinner, S. A., Cuccaro, M., Simensen, R. J., Bishop, J., Skinner, C., Fender, D. & Stevenson, R. E. (1998). Autism and Maternally Derived Aberrations of Chromosome 15q. *Am J Med Genet*, Vol.76, No.4, Apr 1, pp.327-336,0148-7299

Selby, L., Zhang, C. & Sun, Q. Q. (2007). Major Defects in Neocortical Gabaergic Inhibitory Circuits in Mice Lacking the Fragile X Mental Retardation Protein. *Neurosci Lett*, Vol.412, No.3, Feb 2, pp.227-232,0304-3940

Steffenburg, U., Hagberg, G. & Hagberg, B. (2001). Epilepsy in a Representative Series of Rett Syndrome. *Acta Paediatr*, Vol.90, No.1, Jan, pp.34-39,0803-5253

Stork, O., Ji, F. Y., Kaneko, K., Stork, S., Yoshinobu, Y., Moriya, T., Shibata, S. & Obata, K. (2000). Postnatal Development of a Gaba Deficit and Disturbance of Neural Functions in Mice Lacking Gad65. *Brain Res*, Vol.865, No.1, May 19, pp.45-58,0006-8993

Tabuchi, K., Blundell, J., Etherton, M. R., Hammer, R. E., Liu, X., Powell, C. M. & Sudhof, T. C. (2007). A Neuroligin-3 Mutation Implicated in Autism Increases Inhibitory Synaptic Transmission in Mice. *Science*, Vol.318, No.5847, Oct 5, pp.71-76,1095-9203

Toro, R., Konyukh, M., Delorme, R., Leblond, C., Chaste, P., Fauchereau, F., Coleman, M., Leboyer, M., Gillberg, C. & Bourgeron, T. (2010). Key Role for Gene Dosage and Synaptic Homeostasis in Autism Spectrum Disorders. *Trends Genet*, Vol.26, No.8, Aug, pp.363-372,0168-9525

Treiman, D. M., Meyers, P. D., Walton, N. Y., Collins, J. F., Colling, C., Rowan, A. J., Handforth, A., Faught, E., Calabrese, V. P., Uthman, B. M., Ramsay, R. E. & Mamdani, M. B. (1998). A Comparison of Four Treatments for Generalized Convulsive Status Epilepticus. Veterans Affairs Status Epilepticus Cooperative Study Group. *N Engl J Med*, Vol.339, No.12, Sep 17, pp.792-798,0028-4793

Tripathi, P. P., Sgado, P., Scali, M., Viaggi, C., Casarosa, S., Simon, H. H., Vaglini, F., Corsini, G. U. & Bozzi, Y. (2009). Increased Susceptibility to Kainic Acid-Induced Seizures in Engrailed-2 Knockout Mice. *Neuroscience*, Vol.159, No.2, Mar 17, pp.842-849,0306-4522

Tuchman, R. (2000). Treatment of Seizure Disorders and Eeg Abnormalities in Children with Autism Spectrum Disorders. *J Autism Dev Disord*, Vol.30, No.5, Oct, pp.485-489,0162-3257

Tuchman, R., Moshe, S. L. & Rapin, I. (2009). Convulsing toward the Pathophysiology of Autism. *Brain Dev*, Vol.31, No.2, Feb, pp.95-103,1872-7131

Urak, L., Feucht, M., Fathi, N., Hornik, K. & Fuchs, K. (2006). A Gabrb3 Promoter Haplotype Associated with Childhood Absence Epilepsy Impairs Transcriptional Activity. *Hum Mol Genet,* Vol.15, No.16, Aug 15, pp.2533-2541,0964-6906

van Kooten, I.A.J., Hof, P.R., van Engeland, H., Steinbusch, H.W.M., Patterson, P.H., & Schmitz, C. (2005). Autism: Neuropathology, Alterations of the GABAergic System, and Animal Models, In: *International Review of Neurobiology.* Vol. 71, pp. 2-17, Elsevier Inc., 0074-7742

Varoqueaux, F., Aramuni, G., Rawson, R. L., Mohrmann, R., Missler, M., Gottmann, K., Zhang, W., Sudhof, T. C. & Brose, N. (2006). Neuroligins Determine Synapse Maturation and Function. *Neuron,* Vol.51, No.6, Sep 21, pp.741-754,0896-6273

Verkerk, A. J., Pieretti, M., Sutcliffe, J. S., Fu, Y. H., Kuhl, D. P., Pizzuti, A., Reiner, O., Richards, S., Victoria, M. F., Zhang, F. P. & et al. (1991). Identification of a Gene (Fmr-1) Containing a Cgg Repeat Coincident with a Breakpoint Cluster Region Exhibiting Length Variation in Fragile X Syndrome. *Cell,* Vol.65, No.5, May 31, pp.905-914,0092-8674

Volkmar, F. R. & Nelson, D. S. (1990). Seizure Disorders in Autism. *J Am Acad Child Adolesc Psychiatry,* Vol.29, No.1, Jan, pp.127-129,0890-8567

Vorstman, J. A., Staal, W. G., van Daalen, E., van Engeland, H., Hochstenbach, P. F. & Franke, L. (2006). Identification of Novel Autism Candidate Regions through Analysis of Reported Cytogenetic Abnormalities Associated with Autism. *Mol Psychiatry,* Vol.11, No.1, Jan, pp.1, 18-28,1359-4184

Wang, K., Zhang, H., Ma, D., Bucan, M., Glessner, J. T., Abrahams, B. S., Salyakina, D., Imielinski, M., Bradfield, J. P., Sleiman, P. M., Kim, C. E., Hou, C., Frackelton, E., Chiavacci, R., Takahashi, N., Sakurai, T., Rappaport, E., Lajonchere, C. M., Munson, J., Estes, A., Korvatska, O., Piven, J., Sonnenblick, L. I., Alvarez Retuerto, A. I., Herman, E. I., Dong, H., Hutman, T., Sigman, M., Ozonoff, S., Klin, A., Owley, T., Sweeney, J. A., Brune, C. W., Cantor, R. M., Bernier, R., Gilbert, J. R., Cuccaro, M. L., McMahon, W. M., Miller, J., State, M. W., Wassink, T. H., Coon, H., Levy, S. E., Schultz, R. T., Nurnberger, J. I., Haines, J. L., Sutcliffe, J. S., Cook, E. H., Minshew, N. J., Buxbaum, J. D., Dawson, G., Grant, S. F., Geschwind, D. H., Pericak-Vance, M. A., Schellenberg, G. D. & Hakonarson, H. (2009). Common Genetic Variants on 5p14.1 Associate with Autism Spectrum Disorders. *Nature,* Vol.459, No.7246, May 28, pp.528-533,1476-4687

Weiss, L. A., Arking, D. E., Daly, M. J. & Chakravarti, A. (2009). A Genome-Wide Linkage and Association Scan Reveals Novel Loci for Autism. *Nature,* Vol.461, No.7265, Oct 8, pp.802-808,1476-4687

Whitney, E. R., Kemper, T. L., Rosene, D. L., Bauman, M. L. & Blatt, G. J. (2008). Calbindin-D28k Is a More Reliable Marker of Human Purkinje Cells Than Standard Nissl Stains: A Stereological Experiment. *J Neurosci Methods,* Vol.168, No.1, Feb 15, pp.42-47,0165-0270

Whitney, E. R., Kemper, T. L., Rosene, D. L., Bauman, M. L. & Blatt, G. J. (2009). Density of Cerebellar Basket and Stellate Cells in Autism: Evidence for a Late Developmental Loss of Purkinje Cells. *J Neurosci Res,* Vol.87, No.10, Aug 1, pp.2245-2254,1097-4547

Wisniowiecka-Kowalnik, B., Nesteruk, M., Peters, S. U., Xia, Z., Cooper, M. L., Savage, S., Amato, R. S., Bader, P., Browning, M. F., Haun, C. L., Duda, A. W., 3rd, Cheung, S. W. & Stankiewicz, P. (2010). Intragenic Rearrangements in Nrxn1 in Three Families with Autism Spectrum Disorder, Developmental Delay, and Speech Delay. *Am J Med Genet B Neuropsychiatr Genet*, Vol.153B, No.5, Jul, pp.983-993,1552-485X

Wisniewski, K. E., Segan, S. M., Miezejeski, C. M., Sersen, E. A. & Rudelli, R. D. (1991). The Fra(X) Syndrome: Neurological, Electrophysiological, and Neuropathological Abnormalities. *Am J Med Genet*, Vol.38, No.2-3, Feb-Mar, pp.476-480,0148-7299

Witt Engerstrom, I. (1992). Age-Related Occurrence of Signs and Symptoms in the Rett Syndrome. *Brain Dev*, Vol.14 Suppl, May, pp.S11-20,0387-7604

Wittner, L., Magloczky, Z., Borhegyi, Z., Halasz, P., Toth, S., Eross, L., Szabo, Z. & Freund, T. F. (2001). Preservation of Perisomatic Inhibitory Input of Granule Cells in the Epileptic Human Dentate Gyrus. *Neuroscience*, Vol.108, No.4, pp.587-600,0306-4522

Wonders, C. P. & Anderson, S. A. (2006). The Origin and Specification of Cortical Interneurons. *Nat Rev Neurosci*, Vol.7, No.9, Sep, pp.687-696,1471-003X

Wong, V. (1993). Epilepsy in Children with Autistic Spectrum Disorder. *J Child Neurol*, Vol.8, No.4, Oct, pp.316-322,0883-0738

Wu, S., Yue, W., Jia, M., Ruan, Y., Lu, T., Gong, X., Shuang, M., Liu, J., Yang, X. & Zhang, D. (2007). Association of the Neuropilin-2 (Nrp2) Gene Polymorphisms with Autism in Chinese Han Population. *Am J Med Genet B Neuropsychiatr Genet*, Vol.144B, No.4, Jun 5, pp.492-495,1552-4841

Yang, M. S. & Gill, M. (2007). A Review of Gene Linkage, Association and Expression Studies in Autism and an Assessment of Convergent Evidence. *Int J Dev Neurosci*, Vol.25, No.2, Apr, pp.69-85,0736-5748

Yip, J., Soghomonian, J. J. & Blatt, G. J. (2007). Decreased Gad67 Mrna Levels in Cerebellar Purkinje Cells in Autism: Pathophysiological Implications. *Acta Neuropathol*, Vol.113, No.5, May, pp.559-568,0001-6322

Yip, J., Soghomonian, J. J. & Blatt, G. J. (2008). Increased Gad67 Mrna Expression in Cerebellar Interneurons in Autism: Implications for Purkinje Cell Dysfunction. *J Neurosci Res*, Vol.86, No.3, Feb 15, pp.525-530,1097-4547

Zhang, C., Atasoy, D., Arac, D., Yang, X., Fucillo, M. V., Robison, A. J., Ko, J., Brunger, A. T. & Sudhof, T. C. (2010). Neurexins Physically and Functionally Interact with Gaba(a) Receptors. *Neuron*, Vol.66, No.3, May 13, pp.403-416,1097-4199

Zhu, Z. Q., Armstrong, D. L., Hamilton, W. J. & Grossman, R. G. (1997). Disproportionate Loss of Ca4 Parvalbumin-Immunoreactive Interneurons in Patients with Ammon's Horn Sclerosis. *J Neuropathol Exp Neurol*, Vol.56, No.9, Sep, pp.988-998,0022-3069

3

GABA and Glutamate Receptors of the Autistic Brain

Agenor Limon, Jorge M. Reyes-Ruiz and Ricardo Miledi
University of California at Irvine
United States of America

1. Introduction

Autism is a severe neuropsychiatric disorder characterized by impaired communication, significant reduction in social interaction, and repetitive and stereotyped behaviour. It is highly hereditable (Hoekstra et al., 2007); however, genomic alterations associated to autism have been found only in less than a fifth of the total number of cases. How those alterations ultimately cause the autistic phenotype is still very poorly understood. Besides genomic abnormalities, environmental and epigenetic factors may also increase the risk of developing autism or autistic traits. Prenatal exposure to rubella virus, cytomegalovirus, or to the chemical substances thalidomide and valproate are among the non-genetic causes linked to autism (Persico & Bourgeron, 2006), however causal relationships are not established. Regardless of the origins of autism, neuropathological observations are consistently found in several areas of autistic brains (Bauman & Kemper, 1985; Ritvo et al., 1986), and abnormal patterns of synaptic connectivity are thought to be at the core of the autistic disorder (Belmonte et al., 2004). Indeed, many of the genes associated with high risk for autism and those increasing susceptibility are directly, or indirectly, involved in axon guidance, neuronal signalling, metabolism, cell differentiation and synaptic homeostasis (Weiss et al., 2009; Autism genome project consortium, 2007; Tabuchi et al., 2007; Toro et al., 2010). Therefore, along with an early diagnosis (Limon 2007), the prevention and correction of the abnormal connectivity, and the modulation of the synaptic function are the main goals of current and future treatments of the pathological characteristics of the autistic disorders.

Glutamate and GABA are the main excitatory and inhibitory neurotransmitters in the human brain and both have important roles during early development of the nervous system, an ontological stage when the evidence indicates that autism begins. Therefore, it is important to analyse the functional status of glutamatergic and GABAergic neurotransmission in the autistic brain. Cumulative evidence indicates that dysfunctional excitatory and inhibitory synaptic activities underlie several of the characteristics of autism and are, consequently, important targets of pharmacological intervention. In this chapter we describe how glutamate and GABA receptors may participate in the aetiology of autistic disorders and will discuss some of the methods that we have developed to study functional and pharmacological properties of human membrane receptors. We include information demonstrating that functional studies of GABA and glutamate receptors from autistic tissue are feasible and important for the development of new drugs aimed at the manipulation of the GABAergic or glutamatergic systems in the autistic brain.

2. Evidence of glutamatergic dysfunction in autism

Glutamate is the main excitatory neurotransmitter in the vertebrate brain and its effects are exerted mainly through metabotropic (mGlu) and ionotropic (iGlu) glutamate receptors localized in the cellular membranes of neurons and glia. The iGlu receptors are tetrameric proteins that mediate fast synaptic transmission and are grouped into three classes, according to their differential affinity for the agonists N-methyl-D-aspartate (NMDA), kainate (KA), and α-amino-3-hydroxy-5-methyl-4-isoxazole propionic acid (AMPA) (Dingledine et al., 1999). The metabotropic receptors belong to the G-protein-coupled receptor superfamily, and can be divided in the group I (mGluR1 and mGluR5), group II (mGluR2 and mGluR3) and group III (mGluR4 and mGluR6-8) according to their agonist pharmacology, primary sequence and G-protein effector coupling (Fagni et al., 2004). The increased probability of epilepsy in the autistic population (Tuchman & Rapin, 2002) suggests that abnormally enhanced glutamatergic signalling may contribute to some of the autistic characteristics. Indeed, there is a slight positive correlation between plasma levels of glutamate and the severity of autism that supports this hypothesis (Shinohe et al., 2006). Moreover, the cerebellum of autistic patients had increased expression of mRNAs encoding the excitatory amino acid transporter 1 (EAAT 1) and the AMPA 1 receptor. Increments in those proteins would elevate the extracellular concentration of glutamate and enhance the postsynaptic activity of glutamatergic synapses (Puercell et al., 2001). Genome studies have found associations between autism and *GRM8,* which encodes the metabotropic glutamate receptor 8 (mGluR8) (Serajee et al., 2003). This receptor is localized at or near presynaptic sites and when activated, it negatively modulates the release of glutamate from the presynaptic terminal (Cartmell & Schoepp, 2000); therefore, mGluR8 dysfunction may lead to increased neuronal activity. Interestingly, regional subsets of presynaptic GABAergic and glutamatergic terminals innervating GABAergic interneurons are highly enriched with mGluR8 receptors in rat hippocampus (Ferraguti et al., 2005), suggesting that, if the same pattern is present in the human hippocampus, abnormally expressed mGluR8 would produce aberrant GABAergic activity and disrupt the temporal patterns of synchronic activity generated by inhibitory interneurons. Another important evidence suggesting a central role of glutamate receptors in autism comes from animal models of Fragile X syndrome, a neurological disorder with high prevalence of autism that is caused by silencing of the *FMR1* gene and its translated protein, the fragile X mental retardation protein FMRP (Dölen et al., 2010). This protein inhibits the synthesis of a wide range of proteins by binding to actively translating mRNAs. FMRP is a counterbalance to the synthesis of proteins during mGluR5-mediated neuronal plasticity, therefore, when FMRP is dysfunctional, excessive protein synthesis mediated by activation of mGluR5 leads to synaptic dysfunction in *Frm1* KO mice. Importantly, the phenotype of *Frm1* KO mice can be rescued by 50% reduction of the expression of mGluR5, suggesting that specific antagonists of mGluR5 will have beneficial effects on humans with Fragile X syndrome (Dölen et al., 2010) and encourages the search of pharmacological approaches to treat idiopathic autism. Genome studies have also found several single nucleotide polymorphisms (SNPs) of *GRIK2* associated with risk of autism (Jamain et al., 2002; Shuang et al., 2004). *GRIK2* produces the kainate receptor GluK2 (previously known as GluR6) that participates in processes of learning and memory at postsynaptic sites and at presynaptic ones, modulates the release of glutamate and GABA from synaptic terminals (Pinheiro & Mulle, 2006). Moreover, one of the reported SNPs changes a methionine to an isoleucine in position 867 (M867I) of the intracellular C-terminus, producing a structural change of the receptor that may be related to the aetiology of autism. Electrophysiological experiments have shown that *Xenopus* oocytes expressing the rat version of M867I-GluK2 elicit larger ion currents than oocytes expressing the wild type receptor, and the increment of current correlates with an enhanced density of the mutated receptors on the plasma membrane of the oocytes (Strutz-Seebohm et al., 2006). Except for a slower rate of desensitization, no major changes in the kinetic properties of the human or rat version of the M867I-GluK2 could account for the observed increment in the ion currents (Han et al., 2010); therefore, the potential link between the M867I and the autistic phenotype could be related mostly to alterations in the normal trafficking of GluK2 in and out of presynaptic and postsynaptic terminals. Interestingly, despite a 99% homology between rat and human GluK2, both receptors had important kinetic and potency differences, highlighting the importance of the validation of animal models with data obtained from the human brain (Halladay et al., 2009; Limon et al., 2011).

3. Evidence of GABAergic dysfunction in autism

GABA is the most abundant and versatile neurotransmitter in the Central Nervous System (CNS). GABA is excitatory in the immature brain and inhibitory in the mature one (Ben-Ari, 2008). At early stages of neural development GABA has a paracrine action on immature neurons. It modulates neuronal migration, stimulating developing networks and exerting a wide range of trophic actions that lead to the correct establishment of neural circuits (Ben-Ari, 2008). In the adult brain, GABA participates in the generation of synchronous rhythms of cortical assemblies. This allows local time-precise communication among neurons and coherent communication with other cerebral centres, thus creating behavioural relevant processes (Somogyi et al., 2005). GABA actions on the membrane potential are mediated mostly by ionotropic receptors. Ionotropic GABA receptors in the CNS are pentameric channels made up by the combination of α (1-6), β (2-3), γ (1-3) and δ subunits in a γ–β–α–β–α arrangement, with δ substituting γ in some extrasynaptic receptors (Olsen & Sieghart, 2008). They are permeable to chloride ions and whether they depolarize or hyperpolarize the cell membrane potential depends on the developmentally regulated expression of the $Na^+/K^+/Cl^-$-cotransporter, NKCC1, and the K^+/Cl^--cotransporter, KCC2. NKCC1 is highly expressed in immature neurons and transports chloride into the neuron producing a high intracellular concentration of chloride, thus making GABA depolarizing. In mature neurons the increased expression of KCC2 and reduction of NKCC1 lowers the concentration of intracellular chloride and makes GABA hyperpolarizing (Mathews, 2007).

Due to the concerted and intertwined activity of GABAergic and glutamatergic neurotransmission, even small net deviances of GABAergic activity could affect the excitation-inhibition balance in the autistic brain. Such an imbalance would reduce the ratio signal to noise of the sensory and procedural information in mild cases (Casanova et al., 2006) and, in the extreme ones, it could lead to epilepsy. Actually, the incidence of epilepsy in one third of people with autism (Tuchman & Rapin, 2002) and the presence of paroxystic activity in the electroencephalogram (EEG) of approximately 68% of autistic people (Kim et al., 2006) is consistent with this hypothesis. These gradual alterations of the EEG traces suggest that aberrant electrical activity in the autistic brain is expressed as a continuum and it may be present even in cases of autism with normal EEG recordings. Although abnormal electrical activity could have scores of causes, genetic association studies have implicated genes coding for the subunits γ1, α2, α4, β1 and β3 of $GABA_A$ receptors as likely contributors for autism (Blatt et al., 2001; Hussman, 2001; Cook et al., 1998; Ma et al., 2005; Vincent et al., 2006; Kakinuma & Sato, 2008). Indeed, evidence from disorders that share overlapping autistic characteristics like Prader-Willi/Angelman syndrome (AS) and Rett syndrome, supports the idea that GABA receptors are convergent nodes in autistic phenotypes of different genetic origins. Angelman syndrome is an imprinted disorder caused by a maternal deficiency of chromosome 15q11-q13 (Magenis et al., 1987; Lalande, 1996) that includes autistic characteristics and developmental delay, seizures and stereotyped behaviours (Samaco et al., 2005). Because the 15q11-q13 region contains genes for the β3, α5 and γ3 subunits of $GABA_A$ receptors, the chromosomal deficiencies in Angelman syndrome may produce alterations in the expression of these GABA subunits. Knockout mice with deletions in the $GABA_A$ α5 and γ3 subunits did not show a drastic phenotype; but a deletion in the gene for the $GABA_A$ β3 subunit produced a neonatal mortality of 90-95% and the survivors displayed a phenotype resembling severe forms of AS (Sinkkonen et al., 2003). These alterations were associated to a decreased number of GABA receptors in areas like the hippocampus (Sinkkonen et al., 2003) which is commonly affected in autism. Rett syndrome is a pervasive disorder classified within the autism spectrum category. Patients diagnosed with Rett syndrome, in addition to the triad of autistic characteristics, also show cognitive deficits, apraxia, ataxia, seizures and respiratory abnormalities (Chahrour & Zoghbi, 2007). Rett syndrome is the result of mutations of *MECP2*, a gene encoding the transcriptional regulator methyl-CpG-binding protein 2 (MeCp2) (Amir et al., 1999). Mice genetically modified to reduce the function of MeCp2 reproduce much of the phenotype of Rett syndrome. Interestingly these mice also show defects in the expression of the $GABA_A$ β3 subunit (Samaco et al., 2005). Recent studies have shown that selective deletion of *MECP2* in GABAergic neurons reproduces some of the characteristics of regressive autism. Initially, mice with *MECP2* deficiency were indistinguishable from the normal mice; however, after few weeks they started to develop stereotyped movements, compulsive grooming, impaired motor coordination, EEG hyperexcitability and abnormal behavioural patterns (Chao et al., 2010). The EEG abnormalities were tracked down to an impaired production of intracellular GABA[1] and a consequent reduction in the quantal GABA released (Chao et al., 2010). Interestingly, MeCp2 reductions in frontal cortex are frequent in autism and other mental disorders (Nagarajan et al., 2006). Monoallelic or skewed expression of *GABRB3* and a subsequent decrease of $GABA_A$ β3 subunit has been found in 4 out of 8 autistic patients (Hogart et al., 2007), indicating that origins of autism are not exclusively rooted in the genome but any epigenetic, or environmental factor that is able to modify the function of GABA receptors during the development of the nervous system is prone to cause neurodevelopmental disorders and, particularly in the case of GABA β3 subunits, it may lead to autistic phenotypes. Actually, there is strong evidence implicating GABAergic dysfunction in

[1]GABA is produced by the enzymes Gad65 and Gad67 which are respectively coded by the genes *Gad1* and *Gad2*. In *MECP2* deficient mice the expression of *Gad1* and *Gad2* and the immunoreactivity to GABA in interneurons were importantly reduced. These reductions were associated with smaller miniature postsynaptic events (mIPSC) without a change in their frequency (Chao et al., 2010).

anxiety, a common feature of autistic disorders (Amaral & Corbett., 2003). The hippocampus, lateral septum, periaqueductal gray matter and amygdala, which are all cortico-limbic structures involved in modulating anxiety states and have neuroanatomical changes in autism, contain major networks of GABAergic interneurons (Millan, 2003). Many studies focus on $GABA_A$ receptors and anxiety (Yilmazer-Hanke, 2003), but even $GABA_B$ (Mombereau et al., 2004) and $GABA_C$ receptors (Flores-Gracia et al., 2010) have been reported to be involved in anxiety-like behaviours.

The wide number of factors with effects on GABA receptors and the different degrees of changes in GABAergic signalling may explain, at least in part, the high heterogeneity found in the clinical phenotypes within the autism spectrum. However whether changes in animal models apply to humans and what other qualitative changes are present in the human brain is a matter of current research. Blatt et al. (2001) reported a reduced binding of benzodiazepines and muscimol in the hippocampus of autistic brains, suggesting a decrease in the number of $GABA_A$ receptors. Posterior studies using different concentrations of $[H^3]$flunitrazepan showed that the decrement of benzodiazepine binding was due to reductions of binding sites with no changes in the binding affinity of the receptors (Guptill et al., 2007). Western blot analyses of four $GABA_A$ subunits (α1-3 and β3) showed reductions of all subunits in parietal cortex, of α1 in frontal cortex and of α1 and β3 in cerebellum of post-mortem autistic brains (Fatemi et al., 2009). And recent studies showed reductions in the binding sites to muscimol and benzodiazepines in cingulate cortex and fusiform gyrus of autistic brains (Oblak et al., 2009; Oblak et al., 2011). These authors also found a reduction in the affinity of the binding sites to muscimol, suggesting pharmacological changes in the receptors due to changes in the properties of the same receptors or switching of $GABA_A$ subunits. Indeed, an increment in the expression of α5 has been reported in the autistic brain (Purcell et al., 2001), and a potential remodelling of GABA subunits may explain the several reports of paradoxical benzodiazepine-based sedatives on severe autistic individuals with mental retardation (Marrosu et al., 1987; Sandman & Barron; 1992; Aman & Langworthy, 2000).

Undoubtedly binding experiments are, and will continue, providing important information about the status of GABA receptors in the autistic brain, particularly about density and tissue localization. However, their resolution is limited and the absence of functional information on the receptors is a great drawback compared with electrophysiological studies, where even the ion current through an activated single channel can be detected. Therefore, important differences between $GABA_A$ receptors might be overlooked. Another concern is that the benzodiazepine binding site is not present in all $GABA_A$ receptor isoforms and alterations in GABA receptors containing δ, α4 and α6 subunits, which do not bind classical benzodiazepines but are highly expressed in cerebellum (Sieghart, 2006), have not yet been appropriately addressed. It is worth noting that because of the duplication of the 4p12 chromosome (Ma et al., 2005), a gain of function of the α4 subunits is expected, but not yet explored. Also important is the fact that normal binding does not necessarily mean normal activation; therefore, even though the agonists and antagonists can bind to the receptor, determining whether the receptors are functional or not requires a multidisciplinary approach, using binding, biochemical and electrophysiological methodologies. Because GABA and glutamate are main targets of pharmacological intervention, a detailed analyses of their kinetic and biophysical properties will help to evaluate new drugs and therapeutic treatments.

4. Pharmacology of human receptors beyond binding studies

We have developed two methods that allow in-depth studies of neurotransmitter receptors and ion channels of the **human** brain. The first, now widely used, involves the heterologous expression of human receptors in *Xenopus* oocytes, which, because of their size, sturdiness, and availability, permit experiments that would be very difficult, or impossible, to carry out in the human brain. The second method is the microtransplantation of native receptors that allows the study of native receptors **still embedded in their own lipids and with their associated proteins.**

4.1 Heterologous expression of human receptors

The *Xenopus* oocyte system is a very convenient model for studying the electrophysiological properties of membrane receptors (Kusano et al. 1977). G protein-coupled receptors (GPCRs) that are activated through the phosphatidylinositol signal pathway, and ligand gated ion channels, are particularly easy to study in this system. GPCRs activated through the cAMP signal pathway are also possible, although they are more difficult to study and require co-expression of some of the pathway elements (Gether et al., 2002). Heterologous expression in *Xenopus* oocytes is now a classic method to study the functional impact of mutations in membrane receptors, and has a long tradition on the study of neurological disorders, such as Alzheimer's disease and epilepsy (Dauch et al., 1997; Green et al., 2008; Palma et al. 2002). In autism research, the use of *Xenopus* oocytes is just at the beginning. However, it has already helped to demonstrate the gain of function of the M867I-GluK2 mutation associated with autism (Strutz-Seebohm et al., 2006) and the functional impact of mutations in voltage-gated calcium channels that lead to autistic traits in Timothy syndrome (Splawski et al., 2004). Timothy syndrome is a multisystem disorder characterized by syndactyly, arrhythmias and low survival. A high percentage of the children that survive the first years develop autism (Splawski et al., 2004). Timothy syndrome is caused by a substitution of glycine by arginine at residue 406 (G640R) of the cardiac L-type $Ca_v1.2$ channel and it is expressed in several organs including the brain. The expression of wild type and the mutant versions of $Ca_v1.2$ channels in *Xenopus* oocytes showed that the mutation dramatically impairs the voltage-dependent inactivation of the channel and leads to larger and longer calcium currents (Splawski et al., 2004). The gain of function of the $Ca_v1.2$ channel suggests that impairments of calcium signalling participate in the aetiology of autism. Such central role of calcium in autism is supported by the recent finding that increased levels of calcium enhance the activity of the mitochondrial aspartate/glutamate carrier AGC1, a protein coded by the autism susceptibility gene *SLC25A12* (Palmieri et al., 2010).

The evidence that single mutations of glutamate- and calcium channels can lead to autism indicates that abnormal network excitability and intracellular signalling are critical factors in the generation of autistic phenotypes; and highlights the importance of understanding the electrophysiological properties of the receptors and channels, even in cases where no mutations have been reported. The expression of mRNA or cloned human receptors in *Xenopus* oocytes is a potent tool to evaluate the functional properties of GABA-, glutamate - and calcium-channels in the autistic brain (Limon et al., 2008). Next we describe the method we use to express human receptors in *Xenopus* oocytes.

4.1.1 Methodological insight into heterologous expression in *Xenopus* oocytes

In order to study ion channels and other membrane proteins via the expression approach, the gene must be available in a plasmid. Nuclear injections of plasmidic DNAs can be done, provided a relative high purity sample is used. We have found that plasmidic DNA isolated with spin columns is good enough to express receptors. The human cytomegalovirus (CMV) promoter is our promoter of choice. We routinely inject 14 nL at a concentration of about 200 ng/mL, aiming at the animal pole and injecting deep to increase the likelihood of reaching the nucleus. Notwithstanding, the rate of successful nuclear injection is not very high, ending always with some oocytes that do not express the protein and with increased oocyte "mortality". We also use the T7 promoter, but the level of expression is considerably lower (see also Geib et al., 2001). Another factor to consider is our observation that the receptor induced membrane currents achieved in the oocytes successfully injected with DNA are lower than those achieved after injecting cRNA. Therefore, our preferred choice for expression of channels is the cytoplasmic injection of synthetic cRNA. If the gene of interest is placed after any RNA Polymerase promoter such as T7, T3, SP6, synthetic cRNA can be produced in large quantities and with a desirable level of purity. One can use the template DNA and add the Polymerase, rNTPs and a reaction of a couple of hours will yield microgram quantities of RNA (Krieg et al., 1984). The ready to use kits are very convenient and even incorporate a 5′ cap analog and have the option of adding an enzyme that will incorporate a Poly(A)+ tail at the 3′ end to resemble more closely a natural RNA.

Of all the kits we have tested, the mMessage mMachine kit from Ambion is the one that has given us the best results. That kit uses ARCA (Anti Reverse Cap Analog) 5′ cap analog (Stepinski et al., 2001), which prevents incorporation in the reverse orientation and maximizes translation efficiency. In such a way, we obtain RNAs that have strong **expressional potency** after injecting 50 nL (at ca. 1 mg/mL) into the equator of the oocyte (Limon et al., 2007; Reyes-Ruiz et al., 2010; Limon et al., 2010). Besides its use for the expression of recombinant proteins, *Xenopus* oocytes can also express membrane proteins after injection of mRNAs directly isolated from human biopsy tissue or from post-mortem brains (Gundersen et al., 1984; Palma et al., 2002; Limon et al., 2008). For the expression of mRNA, the quality of the starting material is important, specially making sure the tissue is not fixed, but snap frozen in liquid nitrogen or CO_2. We have isolated mRNAs that were able to express functional receptors and ion channels from tissue that was obtained several hours post-mortem and was stored frozen for several years (Limon et al., 2008). For the mRNA isolation procedure, it is preferable to start with at least one gram of tissue. We use the TRIzol method for total RNA isolation. The Poly(A)$^+$ RNA is then isolated using an Oligo (dT) resin. A chromatography column is our method of choice because it is more convenient for larger volumes. After overnight precipitation with NaCl, the mRNA is thoroughly washed with 70% ethanol and resuspended in RNase free water at a concentration of 1 mg/mL or higher if possible. By starting with 1 g of tissue we generally recover about 20 or 30 μg of mRNA, suitable for at least 10 injections of 20 oocyte batches. Even though this is the method that has given us the best results, very often the mRNA obtained fails to express large currents after injection, independently of the yield or A260/A280 ratio, although that mRNA is usually good enough for qPCR assays. This methodology extends the amount of information yielded by the mRNA, because functional data can be cross-correlated with data from qPCR analyses of GABA subunits and synaptic markers.

4.2 Microtransplantation of human receptors

Proteins expressed *de novo* in *Xenopus* oocytes have post-translational modifications and membrane lipids that are specific for the oocyte, not for the human brain. Moreover, microenvironmental effects on the receptor or epigenetic modifications that may have arisen through complex cellular signalling are not necessarily reproduced by the expression of mRNA. To overcome those problems, we have been studying native human receptors using the microtransplantation method which is conceptually depicted in Figure 1. To microtransplant native membranes from human brains, part of the brain is homogenized and the cell membranes are isolated by centrifugation (Miledi et al., 2004).

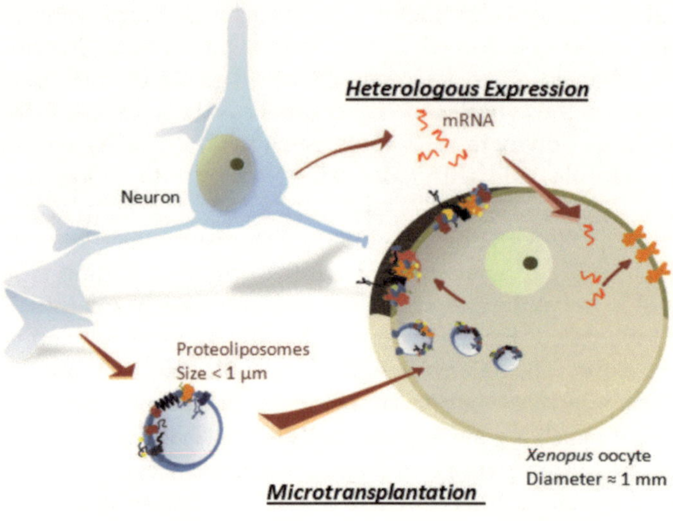

Fig. 1. Methods for evaluating the electrophysiological properties of human neurotransmitter receptors. Diagram showing the heterologous expression of proteins in *Xenopus* oocytes (top) and the microtransplantation of native cell membranes to oocytes (bottom). For the microtransplantation, proteoliposomes isolated from brain tissue specimens, containing the original native receptors, are injected into the oocytes. Within a few hours after injection the proteoliposomes fuse with the oocyte's plasma membrane and expose the native receptors to pharmacological and biophysical experimentation. For the expression of *de novo* the human receptors, mRNA is isolated and injected into the oocytes; and the newly expressed receptors can then be studied in great detail.

Cell membranes, mostly in the form of small vesicles, are adjusted to a protein concentration of 1-2 mg/mL and injected into an oocyte. Within a few hours the membranes, carrying their original neurotransmitter receptors and ion channels, begin to fuse with the oocyte plasma membrane. Voltage-clamp recording is then used to study the functional characteristics of the transplanted receptors. Oocytes with transplanted receptors can be studied up to several days post-injection (Miledi et al., 2004; Limon et al., 2008).

We have assessed whether autistic brains with long post-mortem intervals still contain functional neurotransmitter receptors and voltage-operated ion channels that could be microtransplanted into *Xenopus* oocytes. For that purpose, we chose the cerebellum and temporal cortex, because neuroanatomical and biochemical studies have shown abnormalities in the neuronal organization of those areas (Bauman & Kemper, 1985; Bailey et al., 1998; Purcell et al., 2001). Our initial studies of six autistic cerebella showed that the amounts of microtransplanted receptors differed from their respective controls in all cases.

Four of them yielded responses to GABA, kainate, and glutamate that were smaller than their respective paired controls. The other two cases generated larger responses. In contrast, when the receptors of the temporal cortex were microtransplanted, the amplitudes of the currents induced by kainate, glutamate, and GABA by the receptors from the autistic membranes were larger than their controls in two of the three autistic cases (Limon et al., 2008). Interestingly, even though the number of patients tested is low, deviations from the controls were already found, supporting the hypothesis of an altered GABAergic signalling in the autistic brain, thus encouraging further studies.

It will be interesting to microtransplant receptors from other areas of the brain. The hippocampus is important in memory and learning processes; and together with the amygdala and prefrontal cortex participates in the negative feedback of the hypothalamic-pituitary-adrenal (HPA) axis which in turn controls the body's response to stress (Morris, 2007). The hippocampus and amygdala have clear neuroanatomical alterations in autistic brains (Bauman & Kemper, 2005) and there is a hypothesis that prenatal stress may affect the HPA axis during development and increase the risk of developing autism (O'Donnell et al., 2009). Accordingly we evaluated if neurotransmitter receptors from hippocampi with long post-mortem intervals can be transplanted to *Xenopus* oocytes.

4.2.1 Microtransplantation of native receptors from the hippocampus of autistic brains

For these experiments we used the anterior hippocampus from two autistic brains and two matching control brains (Table 1). Cell membranes were prepared as previously reported (Limon et al., 2008) and then injected into the equator of *Xenopus* oocytes. Ligand-activated ion currents were observed after 24 h post-injection in oocytes voltage-clamped at -80 mV, clearly indicating the successful transplantation of functional GABA and glutamate receptors (Fig. 2). Oocytes injected with cell membranes derived from both autistic brains showed smaller GABA-currents than their respective controls (Fig. 3); suggesting that the number of GABA receptors is decreased in the hippocampus of autistic brains. However, for the moment we have not discounted the possibility that changes in the properties of the receptors, or in the efficiency of membrane fusion, produce a decrease of the currents without a decrease of receptor density. Kainate induced currents were reduced in the autistic sample from the older patient and unaffected in the other autistic sample. Evidently further studies are needed including a statistically meaningful number of samples. However, it is already evident that the microtransplantation of receptors from hippocampus, even with long post-mortem intervals is possible.

Case	Dx	Age (years)	Gender	PMI (h)
B6076	Control	38	Male	25.47
B6401	Autism	39	Male	13.95
B6207	Control	16	Male	26.16
B5666	Autism	8	Male	22.16

Table 1. Characteristics of the tissue used. Dx, diagnosis; PMI, post-mortem interval.

An important advantage of the microtransplantation method is that the biophysical studies are done directly on native receptors that were once in the human brain and are still embedded in their original lipids and with their own cohort of associated proteins. Another advantage is the rapid and high yield of functional and pharmacological information, obtained from minimal amounts of protein. For comparative purposes, consider for example that to prepare a 2-dimensional gel ranges of 2-2000 μg of protein are needed (Adams & Gallagher, 2005); for a Western Blot about 40 μg is recommended. Even for mass spectrometry-based proteomics few μg of total protein is used. In contrast, with the microtransplantation method 50 ng of protein are injected into a single oocyte and a full dose response plus several pharmacological experiments can be done with such a small amount of protein.

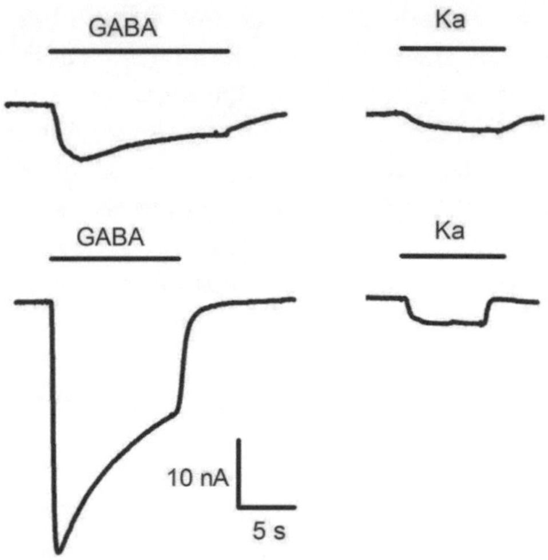

Fig. 2. Sample responses to 1 mM GABA and 100 μM kainate of an oocyte injected with membranes from an autistic hippocampus (above; case B6076, PMI = 25.47 h) and of another oocyte injected with its matching control (below, case B6401, PMI = 13.95 h).

5. Conclusion

Strong evidence indicates that autism is a developmental synaptic disorder that affects the processing of behavioural relevant information. Although the causes of autism are still not known, GABAergic and glutamatergic synapses appear to be convergent nodes of genetic, epigenetic, and probably environmental factors causing the autistic phenotype. Even small alterations in GABAergic or glutamatergic signalling produce autistic characteristics in animal models, and it is highly probable that a similar phenomenon is present in the human brain. GABA and glutamate receptors are also important targets of pharmacological interventions and a detailed knowledge of their function in the human brain will improve the use and design of molecules with therapeutic activity. The microtransplantation of the original receptors from postmortem brains coupled to the expression of receptors by post-mortem mRNAs will help to determine in great detail the type, number, and functional properties of autistic neurotransmitter receptors and channels. These procedures will help to decipher the functional impact of genetic and epigenetic factors in autism. Furthermore, the microtransplantation method will help to determine the mode of action of the medicines presently used to treat autism, and help to develop new medicines and evaluate their pharmacological activity.

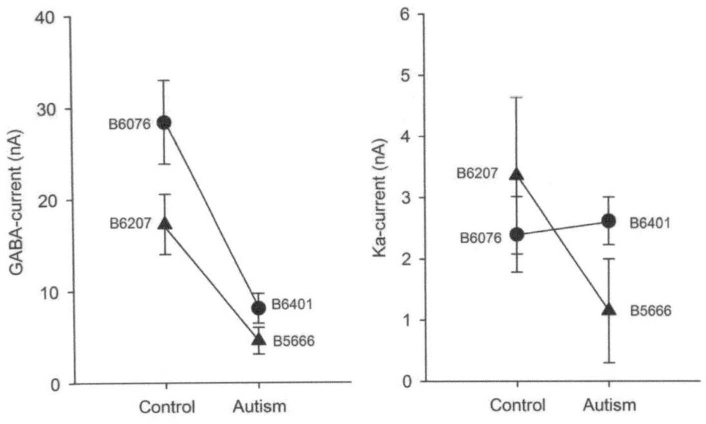

Fig. 3. Responses to 1 mM GABA and 100 μM kainate in oocytes injected with membrane preparations from anterior hippocampus of autistic and control brains (n=5-6 oocytes/case). Data is mean ± SEM.

6. Acknowledgment

We are extremely grateful to all the persons that donated their brains for scientific research. Human brain tissue samples were generously provided by the Autism Tissue Program via the Harvard Brain Tissue Resource Center (Belmont, MA). This work was supported by the King Abdul Aziz City for Science and Technology, (Riyadh, SA; Grant KACST-46749).

7. References

Adams, L. D., and S. R. Gallagher. "Two-Dimensional Gel Electrophoresis." Curr Protoc Immunol Chapter 8 (2005): Unit 8 5.

Aman, M. G., and K. S. Langworthy. "Pharmacotherapy for Hyperactivity in Children with Autism and Other Pervasive Developmental Disorders." J Autism Dev Disord 30, no. 5 (2000): 451-9.

Amaral, D. G., and B. A. Corbett. "The Amygdala, Autism and Anxiety." Novartis Found Symp 251 (2003): 177-87; discussion 87-97, 281-97.

Amir, R. E., I. B. Van den Veyver, M. Wan, C. Q. Tran, U. Francke, and H. Y. Zoghbi. "Rett Syndrome Is Caused by Mutations in X-Linked MeCp2, Encoding Methyl-Cpg-Binding Protein 2." Nat Genet 23, no. 2 (1999): 185-8.

Autism genome project consortium. "Mapping Autism Risk Loci Using Genetic Linkage and Chromosomal Rearrangements." Nat Genet 39, no. 3 (2007): 319-28.

Bailey, A., P. Luthert, A. Dean, B. Harding, I. Janota, M. Montgomery, M. Rutter, and P. Lantos. "A Clinicopathological Study of Autism." Brain 121 (Pt 5) (1998): 889-905.

Bauman, M., and T. L. Kemper. "Histoanatomic Observations of the Brain in Early Infantile Autism." Neurology 35, no. 6 (1985): 866-74.

Belmonte, M. K., E. H. Cook, Jr., G. M. Anderson, J. L. Rubenstein, W. T. Greenough, A. Beckel-Mitchener, E. Courchesne, L. M. Boulanger, S. B. Powell, P. R. Levitt, E. K. Perry, Y. H. Jiang, T. M. DeLorey, and E. Tierney. "Autism as a Disorder of Neural

Information Processing: Directions for Research and Targets for Therapy." Mol Psychiatry 9, no. 7 (2004): 646-63.

Ben-Ari, Y. "Neuro-Archaeology: Pre-Symptomatic Architecture and Signature of Neurological Disorders." Trends Neurosci 31, no. 12 (2008): 626-36.

Blatt, G. J., C. M. Fitzgerald, J. T. Guptill, A. B. Booker, T. L. Kemper, and M. L. Bauman. "Density and Distribution of Hippocampal Neurotransmitter Receptors in Autism: An Autoradiographic Study." J Autism Dev Disord 31, no. 6 (2001): 537-43.

Bauman, M. L., and T. L. Kemper. "Neuroanatomic Observations of the Brain in Autism: A Review and Future Directions." Int J Dev Neurosci 23, no. 2-3 (2005): 183-7.

Cartmell, J., and D. D. Schoepp. "Regulation of Neurotransmitter Release by Metabotropic Glutamate Receptors." J Neurochem 75, no. 3 (2000): 889-907.

Casanova, M. F., I. A. van Kooten, A. E. Switala, H. van Engeland, H. Heinsen, H. W. Steinbusch, P. R. Hof, J. Trippe, J. Stone, and C. Schmitz. "Minicolumnar Abnormalities in Autism." Acta Neuropathol 112, no. 3 (2006): 287-303.

Chahrour, M., and H. Y. Zoghbi. "The Story of Rett Syndrome: From Clinic to Neurobiology." Neuron 56, no. 3 (2007): 422-37.

Chao, H. T., H. Chen, R. C. Samaco, M. Xue, M. Chahrour, J. Yoo, J. L. Neul, S. Gong, H. C. Lu, N. Heintz, M. Ekker, J. L. Rubenstein, J. L. Noebels, C. Rosenmund, and H. Y. Zoghbi. "Dysfunction in Gaba Signalling Mediates Autism-Like Stereotypies and Rett Syndrome Phenotypes." Nature 468, no. 7321 (2010): 263-9.

Cook, E. H., Jr., R. Y. Courchesne, N. J. Cox, C. Lord, D. Gonen, S. J. Guter, A. Lincoln, K. Nix, R. Haas, B. L. Leventhal, and E. Courchesne. "Linkage-Disequilibrium Mapping of Autistic Disorder, with 15q11-13 Markers." Am J Hum Genet 62, no. 5 (1998): 1077-83.

Dauch, P., G. Champigny, J. E. Ricci, and F. Checler. "Lack of Effect of Presenilin 1, Betaapp and Their Alzheimer's Disease-Related Mutated Forms on Xenopus Oocytes Membrane Currents." Neurosci Lett 221, no. 2-3 (1997): 85-8.

Dingledine, R., K. Borges, D. Bowie, and S. F. Traynelis. "The Glutamate Receptor Ion Channels." Pharmacol Rev 51, no. 1 (1999): 7-61.

Dölen, G., R. L. Carpenter, T. D. Ocain, and M. F. Bear. "Mechanism-Based Approaches to Treating Fragile X." Pharmacol Ther 127, no. 1: 78-93.

Fagni, L., F. Ango, J. Perroy, and J. Bockaert. "Identification and Functional Roles of Metabotropic Glutamate Receptor-Interacting Proteins." Semin Cell Dev Biol 15, no. 3 (2004): 289-98.

Fatemi, S. H., T. J. Reutiman, T. D. Folsom, and P. D. Thuras. "GABA(A) Receptor Downregulation in Brains of Subjects with Autism." J Autism Dev Disord 39, no. 2 (2009): 223-30.

Ferraguti, F., T. Klausberger, P. Cobden, A. Baude, J. D. Roberts, P. Szucs, A. Kinoshita, R. Shigemoto, P. Somogyi, and Y. Dalezios. "Metabotropic Glutamate Receptor 8-Expressing Nerve Terminals Target Subsets of Gabaergic Neurons in the Hippocampus." J Neurosci 25, no. 45 (2005): 10520-36.

Flores-Gracia, C., A. Nuche-Bricaire, M. Crespo-Ramirez, R. Miledi, K. Fuxe, and M. Perez de la Mora. "GABA(A) Rho Receptor Mechanisms in the Rat Amygdala and Its Role in the Modulation of Fear and Anxiety." Psychopharmacology (Berl) 212, no. 4 (2010): 475-84.

Geib, S., G. Sandoz, E. Carlier, V. Cornet, V. Cheynet-Sauvion, and M. De Waard. "A Novel Xenopus Oocyte Expression System Based on Cytoplasmic Coinjection of T7-Driven Plasmids and Purified T7-Rna Polymerase." Receptors Channels 7, no. 5 (2001): 331-43.

Gether, U., F. Asmar, A. K. Meinild, and S. G. Rasmussen. "Structural Basis for Activation of G-Protein-Coupled Receptors." Pharmacol Toxicol 91, no. 6 (2002): 304-12.

Green, K. N., A. Demuro, Y. Akbari, B. D. Hitt, I. F. Smith, I. Parker, and F. M. LaFerla. "Serca Pump Activity Is Physiologically Regulated by Presenilin and Regulates Amyloid Beta Production." J Cell Biol 181, no. 7 (2008): 1107-16.

Guptill, J. T., A. B. Booker, T. T. Gibbs, T. L. Kemper, M. L. Bauman, and G. J. Blatt. "[3h]-Flunitrazepam-Labeled Benzodiazepine Binding Sites in the Hippocampal Formation in Autism: A Multiple Concentration Autoradiographic Study." J Autism Dev Disord 37, no. 5 (2007): 911-20.

Gundersen, C. B., R. Miledi, and I. Parker. "Messenger Rna from Human Brain Induces Drug- and Voltage-Operated Channels in Xenopus Oocytes." Nature 308, no. 5958 (1984): 421-4.

Halladay, A. K., D. Amaral, M. Aschner, V. J. Bolivar, A. Bowman, E. DiCicco-Bloom, S. L. Hyman, F. Keller, P. Lein, I. Pessah, L. Restifo, and D. W. Threadgill. "Animal Models of Autism Spectrum Disorders: Information for Neurotoxicologists." Neurotoxicology 30, no. 5 (2009): 811-21.

Han, Y., C. Wang, J. S. Park, and L. Niu. "Channel-Opening Kinetic Mechanism for Human Wild-Type Gluk2 and the M867i Mutant Kainate Receptor." Biochemistry 49, no. 43 (2010): 9207-16.

Hoekstra, R. A., M. Bartels, C. J. Verweij, and D. I. Boomsma. "Heritability of Autistic Traits in the General Population." Arch Pediatr Adolesc Med 161, no. 4 (2007): 372-7.

Hogart, A., R. P. Nagarajan, K. A. Patzel, D. H. Yasui, and J. M. Lasalle. "15q11-13 $GABA_A$ Receptor Genes Are Normally Biallelically Expressed in Brain yet Are Subject to Epigenetic Dysregulation in Autism-Spectrum Disorders." Hum Mol Genet 16, no. 6 (2007): 691-703.

Hussman, J. P. "Suppressed Gabaergic Inhibition as a Common Factor in Suspected Etiologies of Autism." J Autism Dev Disord 31, no. 2 (2001): 247-8.

Jamain, S., C. Betancur, H. Quach, A. Philippe, M. Fellous, B. Giros, C. Gillberg, M. Leboyer, and T. Bourgeron. "Linkage and Association of the Glutamate Receptor 6 Gene with Autism." Mol Psychiatry 7, no. 3 (2002): 302-10.

Kakinuma, H., and H. Sato. "Copy-Number Variations Associated with Autism Spectrum Disorder." Pharmacogenomics 9, no. 8 (2008): 1143-54.

Kim, H. L., J. H. Donnelly, A. E. Tournay, T. M. Book, and P. Filipek. "Absence of Seizures Despite High Prevalence of Epileptiform Eeg Abnormalities in Children with Autism Monitored in a Tertiary Care Center." Epilepsia 47, no. 2 (2006): 394-8.

Krieg, P. A., and D. A. Melton. "Functional Messenger Rnas Are Produced by Sp6 in Vitro Transcription of Cloned Cdnas." Nucleic Acids Res 12, no. 18 (1984): 7057-70.

Kusano, K., R. Miledi, and J. Stinnakre. "Acetylcholine Receptors in the Oocyte Membrane." Nature 270, no. 5639 (1977): 739-41.

Lalande, M. "Parental Imprinting and Human Disease." Annu Rev Genet 30 (1996): 173-95.

Limon, A. "Importance of Early Detection in Autism Spectrum Disorder." Gac Med Mex 143, no. 1 (2007): 73-8.

Limon, A., J. M. Reyes-Ruiz, F. Eusebi, and R. Miledi. "Properties of Glur3 Receptors Tagged with GFP at the Amino or Carboxyl Terminus." Proc Natl Acad Sci U S A 104, no. 39 (2007): 15526-30.

Limon, A., J. M. Reyes-Ruiz, and R. Miledi. "Microtransplantation of Neurotransmitter Receptors from Postmortem Autistic Brains to Xenopus Oocytes." Proc Natl Acad Sci U S A 105, no. 31 (2008): 10973-7.

Limon, A., J. M. Reyes-Ruiz, R. G. Vaswani, A. R. Chamberlin, and R. Miledi. "Kaitocephalin Antagonism of Glutamate Receptors Expressed in Xenopus Oocytes." ACS Chem Neurosci 1, no. 3 (2010): 175-81.

Limon, A., J. M. Reyes-Ruiz, and R. Miledi. "GABAergic drugs and Alzheimer's disease." Fut Med Chem 3, no. 2 (2011): 149-53.

Ma, D. Q., P. L. Whitehead, M. M. Menold, E. R. Martin, A. E. Ashley-Koch, H. Mei, M. D. Ritchie, G. R. Delong, R. K. Abramson, H. H. Wright, M. L. Cuccaro, J. P. Hussman, J. R. Gilbert, and M. A. Pericak-Vance. "Identification of Significant Association and Gene-Gene Interaction of GABA Receptor Subunit Genes in Autism." Am J Hum Genet 77, no. 3 (2005): 377-88.

Magenis, R. E., M. G. Brown, D. A. Lacy, S. Budden, and S. LaFranchi. "Is Angelman Syndrome an Alternate Result of Del(15)(Q11q13)?" Am J Med Genet 28, no. 4 (1987): 829-38.

Marrosu, F., G. Marrosu, M. G. Rachel, and G. Biggio. "Paradoxical Reactions Elicited by Diazepam in Children with Classic Autism." Funct Neurol 2, no. 3 (1987): 355-61.

Mathews, G. C. "The Dual Roles of GABA in Seizures and Epilepsy Generate More Excitement." Epilepsy Curr 7, no. 1 (2007): 28-30.

Miledi, R., Z. Duenas, A. Martinez-Torres, C. H. Kawas, and F. Eusebi. "Microtransplantation of Functional Receptors and Channels from the Alzheimer's Brain to Frog Oocytes." Proc Natl Acad Sci U S A 101, no. 6 (2004): 1760-3.

Millan, M. J. "The Neurobiology and Control of Anxious States." Prog Neurobiol 70, no. 2 (2003): 83-244.

Mombereau, C., K. Kaupmann, W. Froestl, G. Sansig, H. van der Putten, and J. F. Cryan. "Genetic and Pharmacological Evidence of a Role for Gaba(B) Receptors in the Modulation of Anxiety- and Antidepressant-Like Behavior." Neuropsychopharmacology 29, no. 6 (2004): 1050-62.

Morris, R. "Stress and the hippocampus.", In: *The hippocampus book*, P. Andersen, R. Morris, D. Amaral, T. Bliss, J O'Keefe. (Ed), (2007): pp. 751-68.

Nagarajan, R. P., A. R. Hogart, Y. Gwye, M. R. Martin, and J. M. LaSalle. "Reduced MeCp2 Expression Is Frequent in Autism Frontal Cortex and Correlates with Aberrant Mecp2 Promoter Methylation." Epigenetics 1, no. 4 (2006): e1-11.

Oblak, A., T. T. Gibbs, and G. J. Blatt. "Decreased $GABA_A$ Receptors and Benzodiazepine Binding Sites in the Anterior Cingulate Cortex in Autism." Autism Res 2, no. 4 (2009): 205-19.

Oblak, A. L., T. T. Gibbs, and G. J. Blatt. "Reduced GABA(A) Receptors and Benzodiazepine Binding Sites in the Posterior Cingulate Cortex and Fusiform Gyrus in Autism." Brain Res 1380 (2011): 218-28.

O'Donnell, K., T. G. O'Connor, and V. Glover. "Prenatal Stress and Neurodevelopment of the Child: Focus on the HPA Axis and Role of the Placenta." Dev Neurosci 31, no. 4 (2009): 285-92.

Olsen, R. W., and W. Sieghart. "International Union of Pharmacology. Lxx. Subtypes of Gamma-Aminobutyric Acid(a) Receptors: Classification on the Basis of Subunit Composition, Pharmacology, and Function. Update." Pharmacol Rev 60, no. 3 (2008): 243-60.

Palma, E., V. Esposito, A. M. Mileo, G. Di Gennaro, P. Quarato, F. Giangaspero, C. Scoppetta, P. Onorati, F. Trettel, R. Miledi, and F. Eusebi. "Expression of Human Epileptic Temporal Lobe Neurotransmitter Receptors in Xenopus Oocytes: An Innovative Approach to Study Epilepsy." Proc Natl Acad Sci U S A 99, no. 23 (2002): 15078-83.

Palmieri, L., V. Papaleo, V. Porcelli, P. Scarcia, L. Gaita, R. Sacco, J. Hager, F. Rousseau, P. Curatolo, B. Manzi, R. Militerni, C. Bravaccio, S. Trillo, C. Schneider, R. Melmed, M. Elia, C. Lenti, M. Saccani, T. Pascucci, S. Puglisi-Allegra, K. L. Reichelt, and A. M. Persico. "Altered Calcium Homeostasis in Autism-Spectrum Disorders: Evidence from Biochemical and Genetic Studies of the Mitochondrial Aspartate/Glutamate Carrier Agc1." Mol Psychiatry 15, no. 1: 38-52.

Persico, A. M., and T. Bourgeron. "Searching for Ways out of the Autism Maze: Genetic, Epigenetic and Environmental Clues." Trends Neurosci 29, no. 7 (2006): 349-58.

Pinheiro, P., and C. Mulle. "Kainate Receptors." Cell Tissue Res 326, no. 2 (2006): 457-82.

Purcell, A. E., O. H. Jeon, A. W. Zimmerman, M. E. Blue, and J. Pevsner. "Postmortem Brain Abnormalities of the Glutamate Neurotransmitter System in Autism." Neurology 57, no. 9 (2001): 1618-28.

Reyes-Ruiz, J. M., L. D. Ochoa-de la Paz, A. Martinez-Torres, and R. Miledi. "Functional Impact of Serial Deletions at the C-Terminus of the Human GABArho1 Receptor." Biochim Biophys Acta 1798, no. 5: 1002-7.

Ritvo, E. R., B. J. Freeman, A. B. Scheibel, T. Duong, H. Robinson, D. Guthrie, and A. Ritvo. "Lower Purkinje Cell Counts in the Cerebella of Four Autistic Subjects: Initial Findings of the Ucla-Nsac Autopsy Research Report." Am J Psychiatry 143, no. 7 (1986): 862-6.

Samaco, R. C., A. Hogart, and J. M. LaSalle. "Epigenetic Overlap in Autism-Spectrum Neurodevelopmental Disorders: MeCp2 Deficiency Causes Reduced Expression of UBE3A and GABRB3." Hum Mol Genet 14, no. 4 (2005): 483-92.

Sandman, C.A., and J. Barron. "Paradoxical response to sedative/hypnotics in subjects with self-injurious behavior and stereotypy". J Dev Phys Disabil 4, no. 4 (1992):307-16.

Serajee, F. J., H. Zhong, R. Nabi, and A. H. Huq. "The Metabotropic Glutamate Receptor 8 Gene at 7q31: Partial Duplication and Possible Association with Autism." J Med Genet 40, no. 4 (2003): e42.

Sinkkonen, S. T., G. E. Homanics, and E. R. Korpi. "Mouse Models of Angelman Syndrome, a Neurodevelopmental Disorder, Display Different Brain Regional GABA(A) Receptor Alterations." Neurosci Lett 340, no. 3 (2003): 205-8.

Shinohe, A., K. Hashimoto, K. Nakamura, M. Tsujii, Y. Iwata, K. J. Tsuchiya, Y. Sekine, S. Suda, K. Suzuki, G. Sugihara, H. Matsuzaki, Y. Minabe, T. Sugiyama, M. Kawai, M. Iyo, N. Takei, and N. Mori. "Increased Serum Levels of Glutamate in Adult Patients with Autism." Prog Neuropsychopharmacol Biol Psychiatry 30, no. 8 (2006): 1472-7.

Shuang, M., J. Liu, M. X. Jia, J. Z. Yang, S. P. Wu, X. H. Gong, Y. S. Ling, Y. Ruan, X. L. Yang, and D. Zhang. "Family-Based Association Study between Autism and Glutamate

Receptor 6 Gene in Chinese Han Trios." Am J Med Genet B Neuropsychiatr Genet 131B, no. 1 (2004): 48-50.

Sieghart, W. "Structure, Pharmacology, and Function of $GABA_A$ Receptor Subtypes." Adv Pharmacol 54 (2006): 231-63.

Somogyi, P., and T. Klausberger. "Defined Types of Cortical Interneurone Structure Space and Spike Timing in the Hippocampus." J Physiol 562, no. Pt 1 (2005): 9-26.

Splawski, I., K. W. Timothy, L. M. Sharpe, N. Decher, P. Kumar, R. Bloise, C. Napolitano, P. J. Schwartz, R. M. Joseph, K. Condouris, H. Tager-Flusberg, S. G. Priori, M. C. Sanguinetti, and M. T. Keating. "Ca(V)1.2 Calcium Channel Dysfunction Causes a Multisystem Disorder Including Arrhythmia and Autism." Cell 119, no. 1 (2004): 19-31.

Stepinski, J., C. Waddell, R. Stolarski, E. Darzynkiewicz, and R. E. Rhoads. "Synthesis and Properties of Mrnas Containing the Novel "Anti-Reverse" Cap Analogs 7-Methyl(3'-O-Methyl)Gpppg and 7-Methyl (3'-Deoxy)Gpppg." RNA 7, no. 10 (2001): 1486-95.

Strutz-Seebohm, N., G. Korniychuk, R. Schwarz, R. Baltaev, O. N. Ureche, A. F. Mack, Z. L. Ma, M. Hollmann, F. Lang, and G. Seebohm. "Functional Significance of the Kainate Receptor Glur6(M836i) Mutation That Is Linked to Autism." Cell Physiol Biochem 18, no. 4-5 (2006): 287-94.

Tabuchi, K., J. Blundell, M. R. Etherton, R. E. Hammer, X. Liu, C. M. Powell, and T. C. Sudhof. "A Neuroligin-3 Mutation Implicated in Autism Increases Inhibitory Synaptic Transmission in Mice." Science 318, no. 5847 (2007): 71-6.

Tuchman, R., and I. Rapin. "Epilepsy in Autism." Lancet Neurol 1, no. 6 (2002): 352-8.

Toro, R., M. Konyukh, R. Delorme, C. Leblond, P. Chaste, F. Fauchereau, M. Coleman, M. Leboyer, C. Gillberg, and T. Bourgeron. "Key Role for Gene Dosage and Synaptic Homeostasis in Autism Spectrum Disorders." Trends Genet 26, no. 8 (2010): 363-72.

Vincent, J. B., S. I. Horike, S. Choufani, A. D. Paterson, W. Roberts, P. Szatmari, R. Weksberg, B. Fernandez, and S. W. Scherer. "An Inversion Inv(4)(P12-P15.3) in Autistic Siblings Implicates the 4p Gaba Receptor Gene Cluster." J Med Genet 43, no. 5 (2006): 429-34.

Weiss, L. A., D. E. Arking, M. J. Daly, and A. Chakravarti. "A Genome-Wide Linkage and Association Scan Reveals Novel Loci for Autism." Nature 461, no. 7265 (2009): 802-8.

Yilmazer-Hanke, D. M., T. Roskoden, K. Zilles, and H. Schwegler. "Anxiety-Related Behavior and Densities of Glutamate, Gabaa, Acetylcholine and Serotonin Receptors in the Amygdala of Seven Inbred Mouse Strains." Behav Brain Res 145, no. 1-2 (2003): 145-59.

4

The Biochemical Basis of Autistic Behavior and Pathology

Dayan Goodenowe and Elodie Pastural
Phenomenome Discoveries Inc.
Canada

1. Introduction

Autism and autism spectrum disorders (ASD) are diseases which are characterized by physical (neurological function and pathology) and behavioral (social interaction) abnormalities that are most commonly diagnosed in children (predominately males) between the ages of 2 and 10 years. Autism is a neurodevelopmental disorder. In autism the Central Nervous System (CNS) cells preferentially affected are the GABAergic Purkinje neurons of the cerebellum. However unlike classical neurodegenerative diseases, in autism, progressive decline is uncommon and there is a concomitant superimposition of neuronal hypersensitivity. This apparent contradiction can be best explained by metabolic abnormalities which have differing effects on specific cell types. Identifying the root of these pleiotropic events has been a topic of intense investigation. From a biochemical perspective, the preponderance of evidence implicates the breakdown of mitochondrial function. From a genetics perspective, genetic mutations targeting mitochondrial function collectively account for more than 10% of all cases, by far the largest single site genetic contribution. However, despite the evidence implicating impaired mitochondrial function, the extra-mitochondrial biochemical implications and consequences of an impaired mitochondrial system have not been thoroughly investigated. The following chapter outlines the causes and implications of mitochondrial impairment in autism.

2. Overview of mitochondrial dysfunction in ASD

Mitochondrial dysfunction in autism has been implicated by several research groups [1-3]. Elevated plasma lactate, a commonly used indicator of mitochondrial dysfunction, has been observed to present in 20 [4] to 40 percent of ASD subjects [5]. Levels of carnitine, the required fatty acid carrier from the cytosol to the mitochondria, have been reported to be low in ASD subjects' serum [6]. The activity of the mitochondrial electron transport chain (ETC), complexes I and III, has been reported to be decreased [7-9]. Glutathione, the key intramitochondrial reactive oxygen species (ROS) neutralizer, is decreased in ASD [2,10-13]. In addition, lipid peroxidation, a down-stream effect of reduced ROS deactivation, is increased in autistic children [14,15]. Extra-mitochondrial processing of palmitate, the key energy source for mitochondria, was observed to be universally increased in ASD [2].

Subjects with definite mitochondrial disease (according to the criteria defined by [16]) have a higher occurrence of autism than expected by chance in the average population [1], especially of regressive autism [17]. Stressors such as fever can lead to the appearance of autistic phenotypic traits in individuals with mitochondrial disease [17]. Since neurodegeneration is a feature of mitochondrial disease, Richard Haas hypothesized that ASD subjects children who undergo regression and/or with symptoms of multisystem disorders are the populations with the highest mitochondrial disease occurrence [1].

Despite tremendous efforts dedicated to the identification of loci associated to autism susceptibility, the numbers of genes and genomic regions involved in ASD families can still not account for the majority of autism cases, with an estimated 10% to 20% of all ASD explained by genetic defects [18]. However the occurrence of genetic impairments in mitochondrial genome and, as importantly, in the nuclear DNA coding for the estimated 1,500 mitochondrial proteins, represents a high fraction of this estimate. A study on a Portuguese autism population estimated that as high as 7% of autistic cases could be attributed to mitochondrial respiratory chain disorders, suggesting that this might be one of the most common disorders associated with autism, especially since not all of the children had been tested for these disorders [19]. Anecdotally, a study on Copy Number Variations reported a copy number gain in the *SUCLG2* gene encoding the beta subunit of succinyl-CoA synthetase ligase, involved in the tricarboxylic acid (TCA) cycle, and in *NDUFA11* and *ATP5J*, both involved in oxidative phosphorylation, in three autistic patients [20]. In addition, many linkage analyses pointed to chromosomal regions that contain mitochondria- related genes. At least two studies reported an association between autism and *SLC25A12* gene, which encodes the mitochondrial aspartate-glutamate carrier AGC1 [21,22] and whose expression has been shown to be up-regulated in autistic prefrontal cortex [23]; recombinant expression of *SLC25A12* had been reported to increase mitochondrial metabolism [24]. One of the most commonly identified abnormalities in ASD, the inverted duplication of chromosome 15q11-q13, displayed mitochondrial dysfunction, with mitochondrial proliferation, partial deficiency of respiratory complex III, and moderate acid lactosis in two initially studied autistic patients [25].

Conversely, mitochondrial DNA mutations have been associated to autistic features [8]. Mitochondrial encephalopathy with lactic acidosis and stroke-like episodes (MELAS) is frequently caused by the A3243G mutation in the mitochondrial $tRNA^{Leu}$ gene and has been associated to autistic clinical traits [8]; another mutation of the mitochondrial genome, A3260G, usually associated with cardiomyopathy and myopathy, has been recently reported to be also associated to MELAS and autism [26]. Mitochondrial DNA depletion syndrome (MDS) is characterized by a reduction of the mtDNA copy number in affected tissues. The same mitochondrial mutation A3243G has been identified in autistic children with MDS [8].

Mitochondrial DNA mutations may be far more frequent in ASD than anticipated, with a very interesting study on neonatal cord blood samples showing a frequency of 1/200 of common point mutations in mitochondrial DNA [27]; the authors state that "at least one in 200 healthy humans harbors a pathogenic mtDNA mutation that potentially causes disease in the offspring of female carriers" [27]. The offspring of the transmitting females would indeed inherit some of the mutant mitochondria by heteroplasmy.

The contribution of environmental factors to the elevation of autism prevalence has now been clearly established [28]. Some mechanisms underlying these environmental toxic insults have been identified and largely point to mitochondrial dysfunction. One of the most detailed examples is propionic acid, by-product of *Clostridium Difficile* and now a common food preservative [29]. Propionate infused rats are an appropriate animal model of autism not only because of their very similar behaviour to autistic children's but also because of their similar biochemical profiles [30]; for instance, carnitine levels and acyl-carnitine levels are respectively lower and higher in the brain of rats infused intraventricularly with propionate than with PBS [30]. Oxidative stress is increased in the infused rats' brain, which according to the authors may lead to mitochondrial failure, or alternatively, is a result of mitochondrial failure induced by propionate [29]. Among the anticonvulsant drugs whose prenatal exposure has been associated to ASD development, valproate sodium seems to present the highest risk with 9% of ASD or Asperger syndrome diagnoses among the exposed fetuses as reported in a Scottish study [31]. Valproic acid (VPA) is a well-known carnitine inhibitor, with a mechanism of action involving the TCA cycle component α-ketoglutarate [32].

It is also tempting to associate the increase in autism prevalence and the expansion of microwave radiation sources in our environment. Early in the eighties it was already shown that rat brain exposure to microwave radiation inhibited mitochondrial electron transport chain function, which resulted in decreased ATP and creatine phosphate levels [33]. Particulate matters from air or water pollution, which are now commonly found in our environment, have been shown to induce microglial activation as reviewed by Block and Calderón-Garcidueñas [34]. Microgliosis is a critical process in neurodegenerative disorders but also and as importantly in ASD [35]. Examples of particulate matters include manufactured aluminum oxide particles, the treatment of which alters mitochondrial potential in human brain microvascular endothelial cells [36], or the pesticide rotenone, which inhibits the ETC complex I and deprives cells of ATP, although not directly inducing microglial activation [37].

Biochemical abnormalities in autism are the norm, not the exception. Genetic abnormalities in mitochondrial processes are the most prevalent in ASD and are likely underestimated, mainly because of poor testing. The list of environmental toxicants clearly associated to ASD is expected to be growing as mitochondrial toxicology is a rapidly emerging field [38]. Biochemical, genetic and environmental data in ASD all point to a very likely role of mitochondria dysfunction in the aetiology of autism, or at least as an autism phenotype [39]. The causes and the effects of these abnormalities are topics of heated discussion within the research community. The focus of this chapter is to describe in greater detail the intra and intercellular role of mitochondria and the consequences of impaired mitochondrial function.

3. Glutamate, mitochondrial toxicity and selective autistic neuropathology

Reduced cerebellar Purkinje neuron density is the key neuropathological observation in autism [40,41]. Purkinje neurons are glutamate receiving (from excitatory climbing fibers) and GABA transmitting (to the deep cerebellar nuclei) neurons. Purkinje neurons coexist with specialized astrocytes (Bergman glia), which protect the neurons. Subjects with ASD have activated microglia [35], which export copious amounts of glutamate [42]. Glutamate is a mitochondrial toxin [43-45], and is selectively toxic to neurons [46-48]. Could glutamate be the cause of the mitochondrial dysfunction and selective Purkinje neuron degeneration observed in autism?

4. Extracellular glutamate transport and receptor activation

There are two sources of extracellular glutamate near Purkinje neurons. The first is glutamate arising from pre-synaptic depolarization of climbing fiber neurons. This mechanism is estimated to result in synaptic glutamate concentrations in the low mM range, sufficient to activate all known glutamate receptors [49]. The second source of glutamate can arise from synaptic spillover, which is estimated to be in the 100-200μM range [49], or microglial activation which can result in mM levels of glutamate [50,51]. *In situ*, glutamate released during neurotransmission is primarily transported into astrocytes [52]. Accordingly, astrocytes are neuroprotective [53,54] against neuronal glutamate toxicity. Although the average intracellular concentration of glutamate in the central nervous system (CNS) is greater than 10mM [55], extracellular concentrations in the low μM range are toxic [56]. Therefore all major CNS cell types contain high affinity glutamate transport mechanisms [57] which maintain extracellular concentrations at less than 1 μM [55].

There is considerable regional and cell type variability of glutamate transporters. Cerebellar astrocytes have similar K_m values for glutamate uptake relative to other brain regions (~50 μM), however the Vmax of these astrocytes is the lowest of all brain regions studied (2.2 nmol/min/mg protein vs. >10nmol/min/mg in cortical regions) [58]. The key glutamate transporters in this region and their relative contributions to glutamate uptake have been recently studied extensively [59-63]. The general consensus is that although Bergmann glia express both EAAT2 (GLT-1) and EAAT1 (GLAST), EAAT1 is responsible for the majority of glutamate uptake [63]. Likewise, although Purkinje neurons express both EAAT4 and EAAT3 (EAAC), EAAT4 is the principal glutamate uptake transporter [64]. It has been estimated that <10% of glutamate arising from climbing fiber depolarization is taken up by Purkinje neurons [64]. However, it has also been proposed that the neuronal EAAT4 transporter is responsible for maintaining low extracellular glutamate levels in between neuronal firing events [59] and that this transporter has a 20-fold greater affinity for glutamate (2.5 μM) [65] vs. EAAT1 (48 μM), EAAT2 (97 μM), or EAAT3 (62 μM) [66]. Although EAAT4 is also present in astrocytes [67], cerebellar Purkinje neurons express the highest levels of EAAT4 in the human brain [68]. These results suggest that there is an inherent weakness in astrocytic glutamate uptake in cerebellar regions These data collectively suggest that chronic low-level exposure to extracellular glutamate would be expected to have a disproportionate effect on these neurons.

Time course studies of glutamate toxicity on neurons [69] and astrocytes [70] suggest that toxicity resulting from chronic exposure to glutamate is mediated by intracellular metabolic disturbances, most notably an increase in oxidative stress and depleted glutathione levels. However, both receptor-dependent and receptor-independent mechanisms of glutamate toxicity have been reported. The N-methyl-D-Aspartate (NMDA) and the α-amino-3-hydroxy-5-methyl-4-isoxazole-proprionic acid (AMPA) receptors are the most commonly associated with glutamate toxicity. Glutamate activates NMDA receptors with a EC50 of 2.3 μM and AMPA receptors with a EC50 of 480 μM [71]. In neuron-astrocyte co-cultures, the blockade of neuronal NMDA receptors reduces glutamate toxicity whereas the blockade of glutamate transport into astrocytes increases neuronal glutamate toxicity [56]. These results suggest that the acute toxic effect of extracellular glutamate on neurons is primarily mediated via glutamate receptors and that glutamate uptake, primarily into astrocytes, is the principal deactivation / neuroprotection mechanism [72]. Studies involving co-cultures of astrocytes and neurons reveal that astrocytes are neuroprotective and that the uptake and metabolic deactivation of glutamate is a key factor in their neuroprotection [48,53,54]. Receptor independent mechanisms (i.e. intracellular transport) of extracellular glutamate toxicity are also well documented [69,70,73]. The chronic toxic effects of glutamate are most likely mediated via these mechanisms. In particular, Purkinje neuron viability is dependent upon functional glutamate uptake and metabolism in Bergmann glia [60,61]. The regulation of extracellular glutamate levels and their toxicity to Purkinje neurons are determined by the collective ability of these cells reduce extracellular glutamate levels via transport and then to detoxify glutamate via intracellular metabolic deactivation.

5. Glutamate transport and intracellular metabolism in Purkinje neurons and Bergmann glia

Each molecule of glutamate transported into the cell is co-transported with 3 sodium (Na^+) ions, one hydroxyl (OH^-) or chloride (Cl^-) ion, and one proton (H^+) with one potassium (K^+) ion being transported out, resulting in a net import of one positive charge (glutamate having a negative charge) and thus depolarization of the cell. Both Purkinje neurons and Bergmann glia express Na^+/K^+-ATPase [74], which restores the sodium gradient by exporting three Na^+ ions and importing two K^+ ions. Therefore, glutamate transport consumes one ATP per glutamate transported. In regards to restoring cytosolic ATP levels both astrocytes and neurons rely upon glucose as the first response [the rate of glycolysis is regulated by cytosolic ATP, see [75] for an excellent review on this topic]. Glucose utilization appears to be roughly equal in both Purkinje neurons and cerebellar astrocytes [76,77]. Predictably, glutamate uptake is therefore a secondary activator of glycolysis [78] and results in the stoichiometric utilization of glucose [79].

Pyruvate is the metabolic product of glycolysis. Studies using (2-^{14}C)-pyruvate (TCA and non-TCA metabolism) indicate that both neurons and astrocytes primarily process pyruvate via the TCA cycle and that glutamate (50%) and aspartate (20%) are the two key metabolites formed. However, astrocytes also generate significant amounts of alanine with less than 10% of the label unaccounted for whereas in neurons a small amount of GABA is formed and 20% of the label is unaccounted for [80]. Studies using (1-^{14}C)-pyruvate (non-TCA metabolism only) indicate that in Purkinje neurons, most of the label ends up in Asp, whereas in cerebellar astrocytes most of the label ends up in alanine [80]. Since the synthesis of aspartate from pyruvate proceeds via oxaloacetate (OAA), mitochondrial pyruvate carboxylation must be active in neurons, as previously reported [81,82]. In contrast, non-TCA cycle processing of pyruvate in astrocytes occurs primarily via alanine aminotransferase. These data clearly indicate that under normal conditions, extra pyruvate is preferentially processed via the first half of the TCA cycle and that glutamate is the predominant non-CO_2 product in both neurons and astrocytes. The above data indicate that lactate is definitely not a significant product of pyruvate metabolism in astrocytes, but the possibility exists that up to 20% of pyruvate could be converted to lactate via lactate dehydrogenase (LDH) in Purkinje neurons. It is interesting to note that Purkinje neurons have a disproportionately high LDH activity relative to other CNS neurons [83].

Studies involving purified mitochondria and cell cultures have repeatedly shown that, in the brain, glutamate is stoichiometrically converted to aspartate [84,85]. However more recent cell culture studies utilizing ^{15}N-glutamate indicate that in astrocytes, glutamate nitrogen is almost exclusively converted to glutamine along with a small but significant formation of alanine [86,87] whereas in neurons, the glutamate nitrogen is almost exclusively converted into aspartate [88,89]. In addition, *in vitro* tracer studies of (U-^{14}C)-L-glutamate reveal that both aspartate and glutamine are labeled with the synaptosomal ratio being 2:1 in favor of aspartate but the astrocytic fraction being just over 1:1 in favor of glutamine [80]. Enzyme activity studies indicate that Purkinje neurons predominantly express aspartate aminotransferase (AAT), a little glutamine synthetase (GS) and almost no glutamate dehydrogenase (GDH), whereas astrocytes and Bergmann glia strongly express GDH and GS with only a minor amount of AAT [90-94]. Clearly, astrocytes and neurons metabolize glutamate differently. However the underlying reason has been difficult to understand until recently. AAT, the favored glutamate-metabolizing enzyme, is ubiquitously distributed in the brain and exists in both the cytosol and mitochondria [89]. The metabolic flux through AAT, especially in the mitochondria, has long been shown to be heavily controlled by the electrogenic mitochondrial aspartate/glutamate carrier (AGC) [95,96], as well as by metabolite substrate availability where the addition of pyruvate (which drives OAA through citrate synthase) or α-ketoglutarate (α-KG) reduces glutamate flux, unlike malate (which increases OAA availability) [85,97]. These effects are dramatically reduced in liver mitochondria which have high GDH activity, exemplifying the brain's reliance on AAT [85,97]. Recently, three independent groups have confirmed that, *in situ,* only neurons express AGC [98-100]. The lack of this carrier in astrocytes explains the increased mitochondrial flux of glutamate into glutamine [101] and aspartate into OAA [102], and the lack of deamination of aspartate in neurons [102].

The subsequent metabolism of aspartate in neurons and astrocytes is even more specialized. First of all, the two key glutamate transporters expressed in astrocytes (EAAT1, EAAT2) have a two times higher affinity for L-aspartate vs. L-glutamate [66], yet the intracellular/extracellular ratio for glutamate in astrocytes is significantly higher [86], which is suggestive of a very rapid intracellular aspartate metabolic rate. Comparison of ^{15}N-aspartate with ^{15}N-glutamate metabolism in cultured astrocytes in the presence of adequate glucose reveals that ^{15}N-aspartate flux is almost two times that of ^{15}N-glutamate [86,101]. In these two studies, it is clear that the majority of aspartate and glutamate metabolism in astrocytes is occurring in the cytosol and that the principal products are arginine and glutamine, respectively. Perhaps more importantly is that it appears that glutamate transfers its nitrogen to glutamine via deamination in the mitochondria, not transamination, and that the resulting α-KG is metabolized via the TCA cycle [103]. These findings are remarkably consistent with [102] where aspartate was found to be deaminated in astrocytes. The most likely explanation for these findings stems from the work of Fahien *et al.* [104,105] where it was found that GDH-AAT complexes resulted in the oxidative deamination of aspartate. When the membrane potential gets above -20nA in neurons, EAAT4 actually exports aspartate [65], a property not shared by either EAAT1 [106] or EAAT2 [107], the primary astrocyte transporters. These data strongly suggest that neurons primarily convert glutamate to aspartate and export aspartate to the extracellular space where it can be taken up by astrocytes (Figure 1). Astrocytes, on the other hand primarily neutralize glutamate and aspartate by converting glutamate to glutamine and aspartate to arginine with the excess being metabolized via the TCA cycle.

As discussed above, the preferred fate of glutamate in neurons is the transport of glutamate via the AGC into the mitochondria, conversion to aspartate via AAT, and the export of aspartate via AGC to the cytosol and finally out of the cell via EAAT4. Therefore the maintenance of high flux (and thus detoxification) of glutamate in neurons is principally dependent upon mitochondrial aspartate efflux via AGC. This efflux is directly regulated by the mitochondrial ETC proton gradient and indirectly regulated by precursor availability for AAT (i.e. OAA). Aspartate efflux requires that the cytosol is acidic relative to mitochondrial matrix. Glutamate is transported into the mitochondria in its protonated state and aspartate is transported out of the mitochondria in its de-protonated state. Therefore the exchange process results in the net import of one proton per glutamate. This leads to acidification of the mitochondria, reduction in the membrane proton gradient, and thus reduction in aspartate efflux. To maintain the proton gradient, it is essential that the ETC is operating efficiently as it is this process that ejects protons from the mitochondria and maintains the proton gradient. Therefore to restore the electrogenic balance during glutamate exposure, neuronal mitochondria need to convert one NADH to NAD^+ per molecule of glutamate imported. The standard hypothesis is that the α-KG formed from AAT is exported out in exchange for cytosolic malate. This malate is then converted to OAA via malate dehydrogenase which consumes one NAD^+, producing one NADH which is then processed by complex I of the ETC. The importance of cytosolic malate in maintaining efficient mitochondrial processing of glutamate cannot be overstated (see [85]). The exported α-KG and aspartate are then converted to glutamate and OAA by cytosolic AAT and the resultant OAA converted to malate in the cytosol. However, under a toxic glutamate load, the cytosolic environment would not favor this reaction. Instead, cytosolic AAT will be driven to aspartate, not glutamate, and since neurons have no other major metabolic pathway for aspartate, the export of aspartate via EAAT4 would be its logical fate. Just as the export of aspartate from mitochondria is the rate limiting step of AAT, total cellular export of aspartate would be rate limiting for the metabolic detoxification of glutamate to aspartate in neurons (Figure 1). Increased levels of pyruvate from stimulated glycolysis combined with reduced mitochondrial capacity for acetyl-CoA utilization (due to mitochondrial OAA being

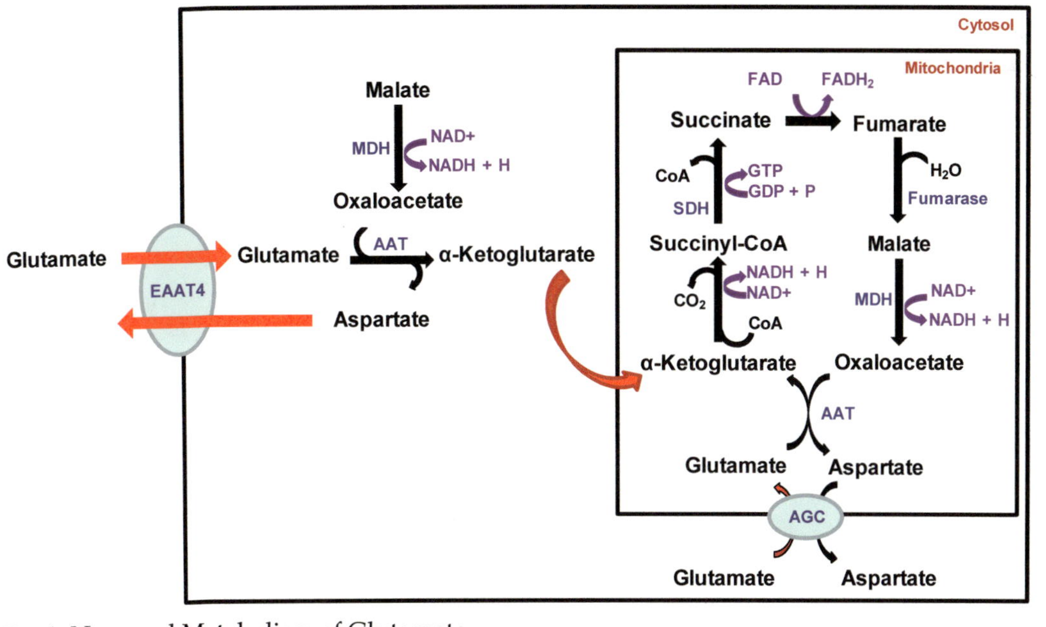

Fig. 1. Neuronal Metabolism of Glutamate

used to create aspartate instead of citrate) would result in pyruvate being driven to lactate via lactate dehydrogenase (LDH). The combination of high acidity (from glutamate) and lactate (from glycolysis) would favor the export of lactate via the monocarboxylate transporter 2 (MCT2). Therefore the export of lactate via MCT2 becomes rate limiting for the conversion of pyruvate to lactate via LDH and the cytosolic regeneration of NAD^+ (Figure 2). As lactate builds up, the regeneration of NAD^+ from LDH will go down. In the mitochondria, as the AGC gets overwhelmed, intra-mitochondrial α-KG goes up and NADH goes up. This results in α-KG dehydrogenase switching from succinate formation to peroxide formation [108]. The principal mechanism of detoxifying peroxide is via glutathione. The oxidation of GSH is one of the first toxic metabolic consequences of glutamate toxicity in neurons [69], and decreased GSH is a common observation in autism [2,11-13,109].

Unlike neurons, which have metabolic mechanisms that enable it to rapidly cycle glutamate, astrocytes must process the aspartate and glutamate they import. Astrocytes highly express EAAT1 which transports both aspartate and glutamate with high affinity. Astrocytes are far

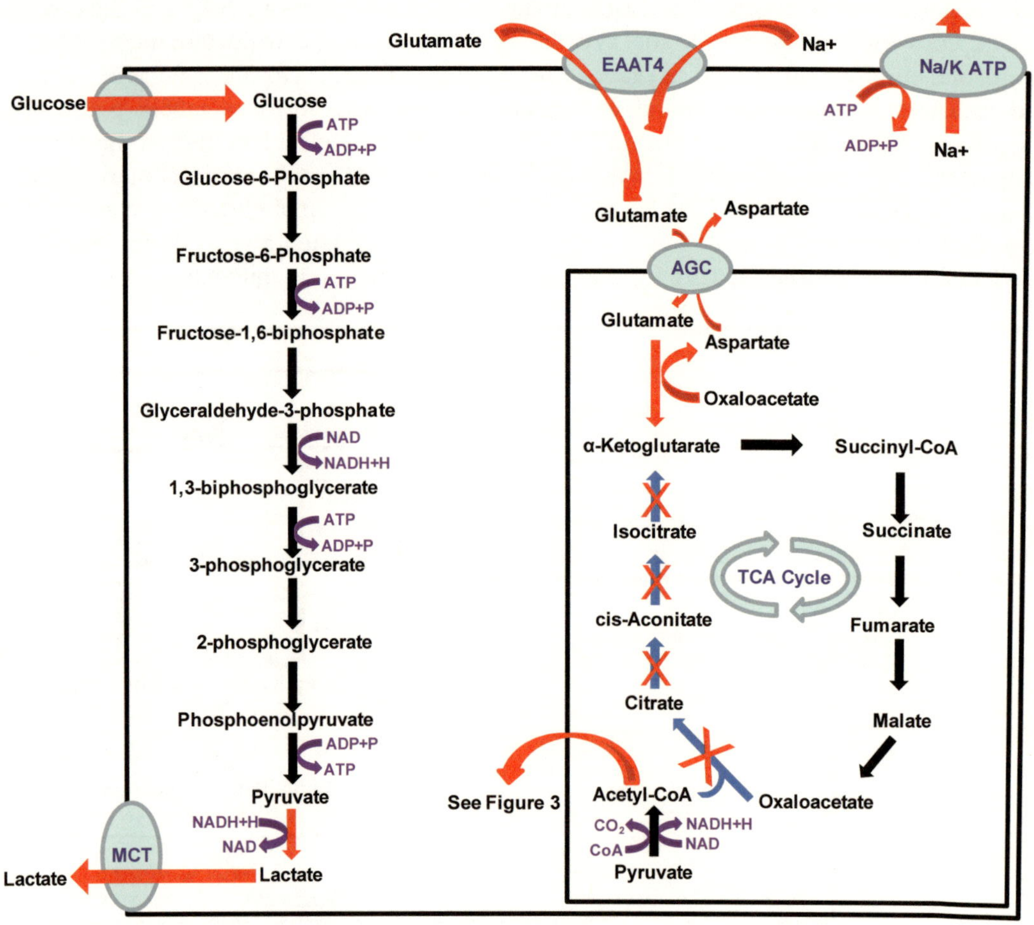

Fig. 2. Effect of Glutamate Transport on Glycolysis and Pyruvate

more reliant on the mitochondrial TCA cycle than neurons due to the lack of the AGC. Glutamate can enter the mitochondria via the dicarboxylic acid carrier [98]. Once in the mitochondrial matrix, two glutamate molecules can be converted to one α-KG and one glutamine via GDH and GS respectively. α-KG can then directly enter the TCA cycle. Therefore the production of α-KG from glutamate via GDH can completely bypass glucose metabolism. Normally, the primary TCA energy source is acetyl-CoA which can come from either glycolysis (pyruvate) or fatty acid oxidation (palmitate, 16:0) and this acetyl-CoA enters the TCA cycle via citrate synthase, which condenses one OAA with one acetyl-CoA to form citrate, which then goes through the cycle releasing two CO_2 molecules ultimately leading back to OAA. Since α-KG can be derived from either citrate or glutamate, the oxidative capacity of the TCA cycle can be broken into two parts: OAA to α-KG via citrate and then α-KG to OAA, independent of citrate (or acetyl-CoA for that matter). It turns out that this latter part of the TCA cycle has three times the capacity vs. the former [110]. So, for every one acetyl-CoA entering via glycolysis or fatty acid oxidation, the TCA cycle can accept two additional α-KG molecules. Since the primary metabolic route of glutamate directly creates α-KG, this provides an effective means of detoxifying glutamate in astrocytes.

The TCA cycle has multiple regulatory systems. One of the more important ones is the succinyl-CoA/acetyl-CoA ratio. As this ratio goes up both citrate synthase and α-KG dehydrogenase are inhibited. However, the inhibition of α-KG dehydrogenase can be over-ridden by high levels of α-KG, as is the case when glutamate is present in abundance. So the net effect of glutamate loading is to inhibit citrate synthase, which shuts down both aerobic glycolysis and fatty acid oxidation. However, this intra-mitochondrial glutamate detoxification pathway consumes two NAD^+, which need to be restored by the ETC. When the succinyl-CoA:acetyl-CoA ratio gets too high, respiration is shut down. When cytosolic glutamate accumulates, cytosolic conversion to aspartate occurs resulting in α-KG. Unlike the aspartate shuttle which energetically operates in only one direction, the α-KG transporter is completely reversible, which enables cytosolic α-KG to equilibrate with mitochondrial levels; this leads to increased mitochondrial α-KG, which will over-ride the α-KG dehydrogenase inhibition caused by a high succinyl-CoA:acetyl-CoA ratio. Since these conditions also create conditions of low NAD^+/NADH ratio, the α-KG dehydrogenase reaction switches from creating succinate to creating hydrogen peroxide [108]. This hydrogen peroxide must be detoxified by GSH, which is why decreased GSH occurs rapidly upon glutamate loading. Furthermore, GSH is exclusively synthesized in the cytosol and then transported to the mitochondria, a process that is inhibited by glutamate [111]. The shut-down of mitochondrial oxidative phosphorylation results in a decrease in the ATP:ADP ratio which further turns on hexokinase, and since pyruvate is blocked from entering the TCA cycle, the anaerobic pathway becomes a critical short-term source of ATP. However, for anaerobic glycolysis not to become self-limiting, the cell needs to export lactate. Astrocytes export lactate via MCT1 [112].

6. Effect of glutamate on neuronal and astrocytic glutathione metabolism

Oxidative metabolism generates reactive oxygen species (ROS). The primary intracellular neutralizer of ROS is GSH. The cystine-glutamate antiporter, which is highly active in microglia [42], is also highly expressed on both astrocytes in the granular layer and on Bergmann glia in the molecular layer, but not on oligodendrocytes or Purkinje neurons [113]. This transporter is an energy-neutral ion exchange protein that operates according to the relative intracellular and extracellular concentrations of glutamate and cystine [114]. The fact that intracellular glutamate levels are orders of magnitude greater than extracellular glutamate means that this transporter's primary purpose in astrocytes is for the import of cystine and does not contribute significantly under resting conditions to the uptake of glutamate. Astrocytes contain high concentrations of glutathione relative to neurons (more than 20 times higher [115]). In addition, resistance to glutamate toxicity in astrocytes is primarily mediated by glutathione [70]. Glutamate-derived oxidative phosphorylation in astrocytes is about twenty times that in neurons [116] and >80% of extracellular glutamate is transported into astrocytes. Since intracellular glutamate stimulates both glutathione synthesis [2,117] and the inward flow of cystine [114], this anti-porter provides astrocytes with a glutamate-dependent means of maintaining high GSH levels. Neurons, on the other hand, utilize EAAT2 and EAAT3 to import cysteine for the synthesis of GSH and this import process is competitively inhibited by glutamate [118]. Therefore the net effect of high levels of extracellular glutamate arising from activated microglia would be to preferentially starve neurons of cysteine in favor of ensuring adequate astrocytic levels of GSH. This observation is consistent with both *in vitro* studies that show decreased GSH levels in neurons as a result of glutamate treatment [69,119] and clinical studies that show that GSH levels are reduced in autism [2,11-14]. Furthermore, the metabolic precursors of GSH, methionine and cysteine, are also reduced in autism [2,12,13]. Collectively, these data suggest that glutamate toxicity resulting from activated microglia would simultaneously increase astrocytic GSH synthesis and oxidation, which would be expected to result in decreased levels of both GSH precursors and GSH, conditions shown to be present in autistic children.

7. Impaired mitochondrial fatty acid oxidation in autism

Glutamate-induced mitochondrial dysfunction indirectly and selectively suppresses mitochondrial fatty acid β-oxidation. The formation of aspartate from glutamate via the transaminase reaction outcompetes citrate synthase for OAA resulting in a dramatic decrease in citrate and effectively shutting down mitochondrial processing of acetyl-CoA [2]. High levels of acetyl-CoA then feedback inhibit mitochondrial β-oxidation. Indirectly, the energetic outward transport of aspartate leads to an increased flux through malate dehydrogenase, which causes an increase in the mitochondrial $NADH/NAD^+$ ratio, which inhibits β-oxidation at the NAD^+-linked β-hydroxyacyl-CoA dehydrogenase reaction [120]. The extra-mitochondrial effects of disrupted mitochondrial fatty acid β-oxidation are related to the carnitine-dependency of this system. Carnitine performs two essential metabolic functions. Its primary and most widely recognized function is to shuttle fatty acids (palmitate (16:0) and stearate (18:0)) from the cytosol into the mitochondrial matrix where it can be β-oxidized to acetyl-CoA. Its secondary, less recognized function is to shuttle excess acetyl-CoA out of the mitochondrial matrix to the cytosol (for reviews see [121-124]).

If mitochondrial acetyl-CoA metabolism is impaired, carnitine-fatty acid cycling is impaired and carnitine usage is shifted to acetyl-carnitine from palmitoyl-carnitine (Figure 3). This results in a build-up of palmitate in the cytosol. Normally, peroxisomes only oxidize 20-30% of cellular palmitate [125,126]. Fatty acid transport into peroxisomes occurs via their CoA esters, not carnitine. Peroxisomes are designed to consume excess cytosolic fatty acids. However, unlike mitochondria fatty acid β-oxidation, which is a catabolic, energy generating process [127], peroxisomal β-oxidation plays primarily an anabolic role where imported fatty acid-CoA is partially β-oxidized to acetyl-CoA and medium-chain fatty acids [128]. Within the peroxisome, this acetyl-CoA is used for the synthesis of the fatty alcohol that ultimately becomes the sn-1 ether in plasmalogens [129-132]. In addition to the synthesis of the 1-O-alkyl bond of plasmalogens, the synthesis of docosahexaenoic acid (DHA) also involves a peroxisomal component. Following the synthesis of 24:6 (tetracosahexaenoic acid) via fatty acid elongation and desaturation of 18:3 (α-linolenic acid) in the endoplasmic reticulum, 24:6 is transported to the peroxisome where it is β-oxidized to 22:6 (DHA) [133]. Peroxisomal acetyl-CoA is normally the major source of cytosolic acetyl-CoA. Within the cytosol, this acetyl-CoA is used for cholesterol synthesis [126,132] and other lipogenic processes such as VLCFA synthesis [134]. Findings of elevated DHA [135], PlsEtn,

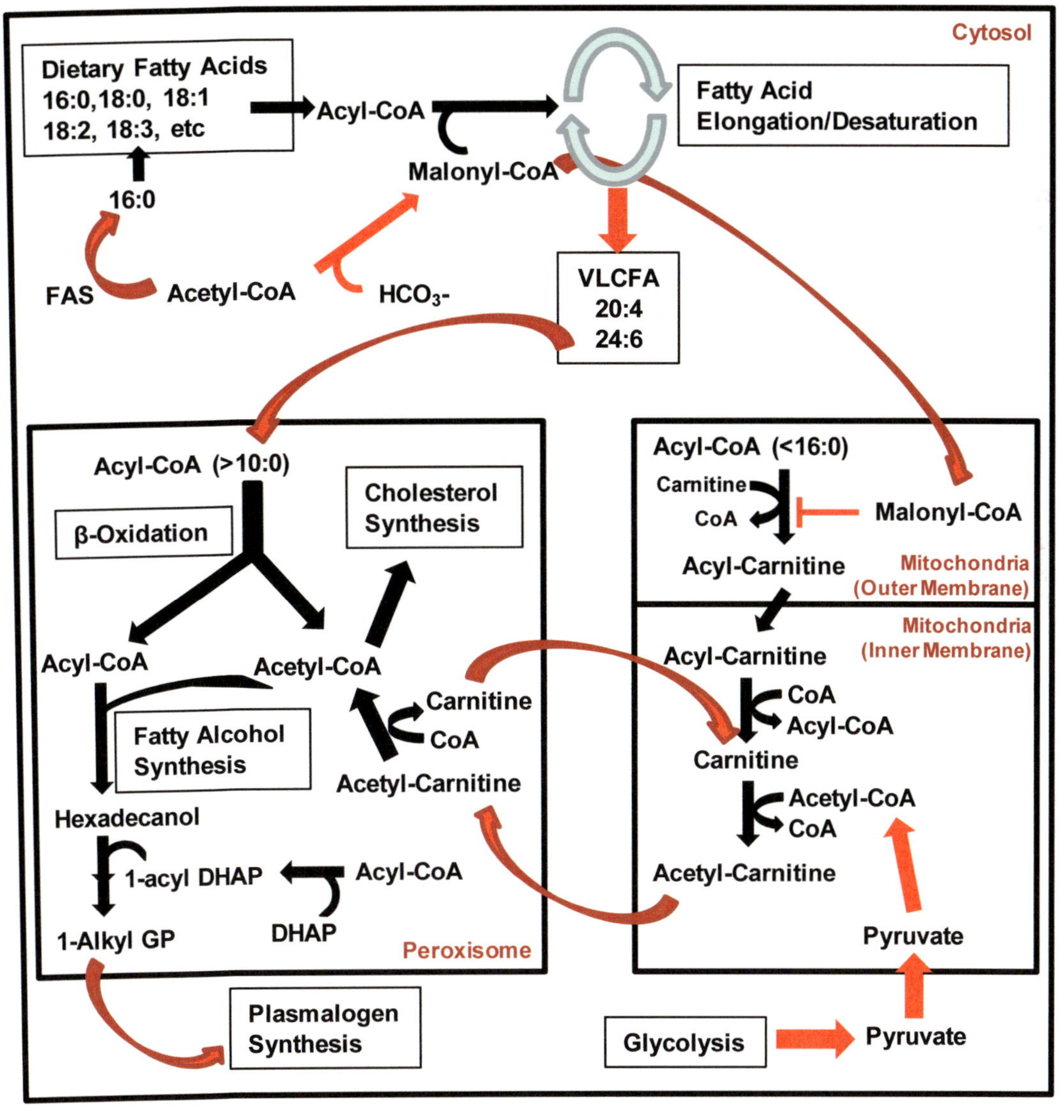

Fig. 3. Metabolic Consequences of Impaired Mitochondrial Tricarboxylic Acid Cycle

and VLCFA levels [2] in autism plasma are consistent with a combined decrease in mitochondrial β-oxidation and an increase in peroxisomal β-oxidation. *In vitro* cell culture assays performed on hepatocytes, neurons and astrocytes revealed that glutamate exposure decreased 16:0 mitochondrial β-oxidation and increased peroxisomal processing of 16:0 and cytosolic fatty acid elongation/desaturation [2].

It is highly speculative but nonetheless tempting to associate some of the epigenetic findings in autism to the recently identified role of mitochondrial acetylcarnitine in nuclear histone acetylation [136]. The acetylcarnitine formed in mitochondria from acetyl-CoA is indeed translocated to cytosol as commonly known, but also to the nucleus where it is converted to acetyl-CoA, which is then used as a main source of acetyl groups for histone acetylation [136]. The situation of high cytosolic acetylcarnitine described above would therefore be expected to enhance histone acetylation, since the other putative source of nuclear acetyl groups, the ATP-citrate lyase pathway, would similarly face high citrate. On the other hand, acetylation and methylation are altered in Rett syndrome, an autism spectrum disorder caused by mutations in MeCP2, a global transcriptional repressor of methylated promoters during postnatal brain development [137]. Brains from MeCP2$^{308/y}$ mutant mice exhibit elevated histone H3 acetylation [138], similarly to brains from Rett syndrome patients with mutant MeCP2 or autistic brains with MeCP2 deficiency [139], which can be explained by the role of MeCP2 in HDAC's recruitment [139]. Higher histone acetylation in Rett syndrome and autism with MeCP2 deficiency may therefore be correlated to the putative higher histone acetylation resulting from the mitochondrial dysfunction model described above.

8. Mitochondrial dysfunction and microglial activation

As exhaustively reviewed by Chauhan and Chauhan [10], the brain is highly vulnerable to oxidative stress, particularly during the early part of development. The reactive oxygen species (ROS) generated by oxidative stress is a cause of lipid peroxidation, which has been reported to be increased in serum and urine from autistic children [14,15]. Lipid peroxidation is a well established cause of reactive aldehyde generation, which plays a key role in apoptotic mechanisms leading to both neuronal and glial cell death [140]. Damaged cells further stimulate microglial activation, which also contributes to free radical production [141].

VLCFA accumulation is also a cause of microgliosis, as evidenced by microglial inflammation in subcortical region in X-adrenoleukodystrophy [142]. The increase in plasmalogens is expected to aggravate the damage caused by VLCFA accumulation since, as recently reported, VLCFA-induced microgliosis seems to be dependent on plasmalogens [143]. The decrease in GSH and the increase in VLCFA and/or plasmalogens observed in autistic subjects are therefore expected to contribute to continuous microglial activation in a positive feedback [2].

The decrease in GSH and the increase in VLCFA and/or plasmalogens observed in autistic subjects [2] are therefore expected to both contribute to continuous microglial activation in a positive feedback or "vicious cycle" [144]. Microglial activation and subsequent "immunoexcitotoxicity" by glutamate are growingly proposed as a central causative model in autism [145].

9. Mitochondrial dysfunction and the observed gender bias in autism

There is no escaping the irrefutable epidemiological fact that autism exhibits a marked gender bias with approximately four times more males diagnosed as females [146-148].

Prepubertal boys and girls are different, not just genetically, but biochemically as well. Prepubertal boys and girls have similar testosterone levels (3.9 vs. 4.7 ng/dL in [149] and 0.41 vs. 0.45 nM in [150] for 17 ß-testosterone), but prepubertal girls have 4-fold higher levels of estrogen than prepubertal boys on average (5.9 +/- 9.7 pmol /L vs. 1.5 +/- 4.1) [151] or at least 3 times higher (9.6 pmol/L vs. <3.7) [150] for ß-estradiol; this difference masks a very high heterogeneity in the level distribution, with undetectable levels for some girls and elevated levels for some boys [150,151,161]. Estrogen is a well known neuroprotectant, especially in glutamate induced neurotoxicity [152-160]. Interestingly, the serum distribution levels in the Courant *et al.* study [150] reveal that if 5 pmol/L were arbitrarily selected as the minimal protective cut-off value, less than 1/3 of girls would be at risk versus 3/4 of boys, a gender bias identical to that observed in autism.
Multiple mechanisms for ß-estradiol neuroprotection have been demonstrated and most involve the mitochondria [162,163]. The chemical structure of estrogens, with the presence of a phenolic A-ring, directly participates in neuroprotection as the "chemical shield" scavenges reactive oxygen species [164]. Another interesting mechanism seems to be structural as estradiol intercalates within cell membranes, preserving mitochondrial integrity [162].
But the most powerful support for estrogen protection as the cause of the gender bias in autism comes from the work of Djouadi and colleagues [165]. The authors studied the simultaneous inhibition of mitochondrial 16:0 processing via an irreversible pharmacological CPT-I inhibitor (Etomoxir) and peroxisome proliferation via PPARα double knockout (-/-). They observed an unexpected gender effect. 100% of the male mice died but only 25% of the female mice died. 100% protection of the male mice was afforded by pretreatment of the mice with β-estradiol. Clearly, β-estradiol is protective against complications arising from impaired extra mitochondrial processing of 16:0. In addition, it was observed that blood glucose levels of female PPARα (-/-) mice recovered relatively quickly to Etomoxir-induced hypoglycemia but that male PPARα (-/-) did not. β-Estradiol pretreated male PPARα (-/-) exhibited a similar result as female PPARα (-/-) mice. These data are consistent with the data of [166] in that recovery from insulin-induced hypoglycemia was significantly slower in autistic children versus non-autistic children. This hypoglycemic response is relevant in that glutamate toxicity creates localized hypoglycemia presumably due to increased glucose uptake [156]. In addition, β-estradiol increases lactate dehydrogenase activity and synthesis [167] as well as lactate export [156], which increases glycolysis flux capacity. β-estradiol also increases cytosolic acetyl-CoA utilization by increasing fatty acid synthase and acetyl-CoA carboxylase activity [168], which would irreversibly remove acetyl-CoA from the cytosolic pool and free up cytosolic carnitine, which would enhance mitochondrial processing of 16:0.

10. Mitochondrial dysfunction and abnormal brain growth in autism

Abnormal brain growth, particularly in cerebellar white matter [169-171], has been observed in autism. No direct link between mitochondrial dysfunction and abnormal brain growth has been proposed. However, the pro-osmotic effect of glutamate import in astrocytes [2], has been proposed as a putative mechanism. As mentioned earlier, the import of a glutamate molecule results in a net import of one positive charge, which in astrocytes results in significant swelling (up to 9% increase in volume) [172]; the swelling ceases as extracellular glutamate levels decrease [172]. It is therefore possible that the increased circumference observed in autism is the result of continuously high extracellular concentrations of glutamate.

11. Mitochondrial dysfunction and seizures in autism

Epilepsy is a common clinical feature associated to autism; it is conservatively estimated that 20-25% children with ASD present with seizures, the most frequent type being complex partial seizures [173]. Oxidative stress [174] and subsequent mitochondrial dysfunction [175] are growingly recognized as being linked to seizure susceptibility, and very interestingly, may actually be contributing factors to epileptic susceptibility, at least in the case of acquired epilepsy, as after brain injury [175]. In their excellent review of the association between mitochondrial dysfunction and temporal lobe epilepsy, Waldbaum and Patel remind that the first suggestion of mitochondrial dysfunction in epilepsy arose from the observation that epilepsy is frequent in inherited mitochondrial disorders such as those associated with childhood encephalopathies [175]. Suggested causative mechanisms underlying a mitochondrial role in epilepsy are imbalances in glutamate and/or calcium signalling [176,177], or respiratory chain complex I dysfunction [178].

12. Mitochondrial dysfunction and visual acuity

There is still a debate as to whether children with ASD display a higher visual acuity than normally developing children, but numerous reports, even anecdotic, seem to support this common observation from parents [179,180]. The importance of dietary long chain polyunsaturated fatty acids, and of DHA particularly, for visual development has been demonstrated by several clinical trials in infants, showing benefits for DHA-rich formulas fed on longer periods [181]. It is therefore tempting to associate the "eagle-eyed" visual acuity detected in children with ASD [179] to the higher content of docosahexaenoic acid detected in plasma [2,135] and presumably present in retina. This assumption must however be nuanced by the fact that benefits on visual acuity were reported for DHA fed and monitored through the diet, with few studies monitoring the omega-3 body burden [181,182]; the umbilical cord DHA content has however been found to be positively correlated with visual system function [182]. Another nuance is the limit of the benefits to visual acuity contributed by DHA: in the DIAMOND study, which monitored the dose effect of DHA supplementation in infant formulas, DHA supplementation improved visual acuity but this improvement did not show a dose response as visual acuity did not improve further with higher DHA content, even if this higher dietary DHA concentration was reflected in higher DHA concentration in red blood cells [181]. Overall, it seems that DHA concentration is positively correlated with future visual function during gestation and infancy, with the effect of dose and endogenous/dietary balance requiring further investigation.

13. Conclusive remarks

In summary, there is a significant amount of direct and indirect evidence of mitochondrial dysfunction in autism. The metabolic cascade observed in autism due to mitochondrial dysfunction can be reproduced with glutamate. Glutamate is the most obvious perpetrator due to its role in microgliosis and the selective glutamergic architecture of Purkinje neurons and Bergmann glia. The data and research performed to date regarding autism and mitochondrial dysfunction however, are unable to ascertain whether subjects who suffer from autism have a yet undetermined mitochondrial weakness that leads to mitochondrial dysfunction under circumstances that would be tolerated by non-autistic subjects, or whether subjects with autism have some other abnormality that results in chronic microglial activation and subsequent "immunoexcitotoxicity" by glutamate [145], but with otherwise normal mitochondrial function. Regardless of the cause of the mitochondrial dysfunction and related consequences, the female gender protection from autism appears to be due to circulating β-estradiol levels and its buffering effects on glucose metabolism. Accordingly, dietary and pharmacological therapies directed at the treatment of autism should thus focus on reducing intracellular mitochondrial demand by either reducing cellular uptake (or extracellular production) of mitochondrial demanding substrates (i.e. glutamate or glucose) or increasing cellular export of mitochondrial or glycolysis products (i.e. aspartate, pyruvate or lactate).

14. References

[1] Haas RH. Autism and mitochondrial disease. Dev Disabil Res Rev 2010;16 (2):144-53.

[2] Pastural E, Ritchie S, Lu Y, Jin W, Kavianpour A, Su-Myat KK, Heath D, Wood PL, Fisk M, Goodenowe DB. Novel Plasma Phospholipid Biomarkers of Autism: Mitochondrial Dysfunction as a Putative Causative Mechanism. Prostaglandins Leukot Essent Fatty Acids 2009;81 (4):253-64.

[3] Rossignol DA, Frye RE. Mitochondrial dysfunction in autism spectrum disorders: a systematic review and meta-analysis. Mol Psychiatry 2011.

[4] Oliveira G, Diogo L, Grazina M, Garcia P, Ataide A, Marques C, Miguel T, Borges L, Vicente AM, Oliveira CR. Mitochondrial dysfunction in autism spectrum disorders: a population-based study. Dev Med Child Neurol 2005;47 (3):185-9.

[5] Laszlo A, Horvath E, Eck E, Fekete M. Serum serotonin, lactate and pyruvate levels in infantile autistic children. Clin Chim Acta 1994;229 (1-2):205-7.

[6] Filipek PA, Juranek J, Nguyen MT, Cummings C, Gargus JJ. Relative carnitine deficiency in autism. J Autism Dev Disord 2004;34 (6):615-23.

[7] Giulivi C, Zhang YF, Omanska-Klusek A, Ross-Inta C, Wong S, Hertz-Picciotto I, Tassone F, Pessah IN. Mitochondrial dysfunction in autism. JAMA 2010;304 (21):2389-96.

[8] Pons R, Andreu AL, Checcarelli N, Vila MR, Engelstad K, Sue CM, Shungu D, Haggerty R, de Vivo DC, DiMauro S. Mitochondrial DNA abnormalities and autistic spectrum disorders. J Pediatr 2004;144 (1):81-5.

[9] Weissman JR, Kelley RI, Bauman ML, Cohen BH, Murray KF, Mitchell RL, Kern RL, Natowicz MR. Mitochondrial disease in autism spectrum disorder patients: a cohort analysis. PLoS ONE 2008;3 (11):e3815.

[10] Chauhan A, Chauhan V. Oxidative stress in autism. Pathophysiology 2006;13 (3):171-81.

[11] Geier DA, Kern JK, Garver CR, Adams JB, Audhya T, Geier MR. A Prospective Study of Transsulfuration Biomarkers in Autistic Disorders. Neurochem Res 2008.

[12] James SJ, Cutler P, Melnyk S, Jernigan S, Janak L, Gaylor DW, Neubrander JA. Metabolic biomarkers of increased oxidative stress and impaired methylation capacity in children with autism. Am J Clin Nutr 2004;80 (6):1611-7.

[13] Kern JK, Jones AM. Evidence of toxicity, oxidative stress, and neuronal insult in autism. J Toxicol Environ Health B Crit Rev 2006;9 (6):485-99.

[14] Chauhan A, Chauhan V, Brown WT, Cohen I. Oxidative stress in autism: increased lipid peroxidation and reduced serum levels of ceruloplasmin and transferrin--the antioxidant proteins. Life Sci 2004;75 (21):2539-49.
[15] Ming X, Stein TP, Brimacombe M, Johnson WG, Lambert GH, Wagner GC. Increased excretion of a lipid peroxidation biomarker in autism. Prostaglandins Leukot Essent Fatty Acids 2005;379 (5):379-84.
[16] Bernier FP, Boneh A, Dennett X, Chow CW, Cleary MA, Thorburn DR. Diagnostic criteria for respiratory chain disorders in adults and children. Neurology 2002;59 (9):1406-11.
[17] Shoffner J, Hyams L, Langley GN, Cossette S, Mylacraine L, Dale J, Ollis L, Kuoch S, Bennett K, Aliberti A, Hyland K. Fever plus mitochondrial disease could be risk factors for autistic regression. J Child Neurol;25 (4):429-34.
[18] Abrahams BS, Geschwind DH. Advances in autism genetics: on the threshold of a new neurobiology. Nat Rev Genet 2008;9 (5):341-55.
[19] Oliveira GD, L; Grazina, M; Garcia, P; Ataíde, A; Marques, C; Miguel, T; Borges, L; Vicente, AM; Oliveira, CR. Mitochondrial dysfunction in autism spectrum disorders: a population-based study. Dev Med Child Neurol 2005;47 (3):185-9.
[20] Autism Genome Project Consortium SP, Paterson AD, Zwaigenbaum L, Roberts W, Brian J, Liu XQ, Vincent JB, Skaug JL, Thompson AP, Senman L, Feuk L, Qian C, Bryson SE, Jones MB, Marshall CR, Scherer SW, Vieland VJ, Bartlett C, Mangin LV, Goedken R, Segre A, Pericak-Vance MA, Cuccaro ML, Gilbert JR, Wright HH, Abramson RK, Betancur C, Bourgeron T, Gillberg C, Leboyer M, Buxbaum JD, Davis KL, Hollander E, Silverman JM, Hallmayer J, Lotspeich L, Sutcliffe JS, Haines JL, Folstein SE, Piven J, Wassink TH, Sheffield V, Geschwind DH, Bucan M, Brown WT, Cantor RM, Constantino JN, Gilliam TC, Herbert M, Lajonchere C, Ledbetter DH, Lese-Martin C, Miller J, Nelson S, Samango-Sprouse CA, Spence S, State M, Tanzi RE, Coon H, Dawson G, Devlin B, Estes A, Flodman P, Klei L, McMahon WM, Minshew N, Munson J, Korvatska E, Rodier PM, Schellenberg GD, Smith M, Spence MA, Stodgell C, Tepper PG, Wijsman EM, Yu CE, Rogé B, Mantoulan C, Wittemeyer K, Poustka A, Felder B, Klauck SM, Schuster C, Poustka F, Bölte S, Feineis-Matthews S, Herbrecht E, Schmötzer G, Tsiantis J, Papanikolaou K, Maestrini E, Bacchelli E, Blasi F, Carone S, Toma C, Van Engeland H, de Jonge M, Kemner C, Koop F, Langemeijer M, Hijmans C, Staal WG, Baird G, Bolton PF, Rutter ML, Weisblatt E, Green J, Aldred C, Wilkinson JA, Pickles A, Le Couteur A, Berney T, McConachie H, Bailey AJ, Francis K, Honeyman G, Hutchinson A, Parr JR, Wallace S, Monaco AP, Barnby G, Kobayashi K, Lamb JA, Sousa I, Sykes N, Cook EH, Guter SJ, Leventhal BL, Salt J, Lord C, Corsello C, Hus V, Weeks DE, Volkmar F, Tauber M, Fombonne E, Shih A, Meyer KJ. Mapping autism risk loci using genetic linkage and chromosomal rearrangements. Nature Genetics 2007;39 (3):319-28.
[21] Ramoz N, Reichert J, Smith CJ, Silverman JM, Bespalova IN, Davis KL, Buxbaum JD. Linkage and association of the mitochondrial aspartate/glutamate carrier SLC25A12 gene with autism. AM J Psychiatry 2004;161 (4):662-9.
[22] Segurado R, Conroy J, Meally E, Fitzgerald M, Gill M, Gallagher L. Confirmation of association between autism and the mitochondrial aspartae/glutamate carrier

SLC25A2 gene on chromosome 2q31. American Journal of Psychiatry 2005;162:2182-4.

[23] Lepagnol-Bestel AM, Maussion G, Boda B, Cardona A, Iwayama Y, Delezoide A, Moalic J, Muller D, Dean B, Yoshikawa T, Gorwood P, Buxbaum J, Ramoz N, Simmoneau M. *SLC25A12* expression is associated with neurite outgrowth and is upregulated in the prefontal cortex of autistic subjects. Molecular Psychiatry 2008.

[24] Lasorsa FM, Pinton P, Palmieri L, Fiermonte G, Rizzuto R, F. P. Recombinant expression of the Ca(2+)-sensitive aspartate/glutamate carrier increases mitochondrial ATP production in agonist-stimulated Chinese hamster ovary cells. J Biol Chem 2003;278 (40):38686-92.

[25] Filipek PA, Juranek J, Smith M, Mays LZ, Ramos ER, Bocian M, Masser-Frye D, Laulhere TM, Modahl C, Spence MA, Gargus JJ. Mitochondrial dysfunction in autistic patients with 15q inverted duplication. Ann Neurol 2003;53 (6):801-4.

[26] Connolly BS, Feigenbaum AS, Robinson BH, Dipchand AI, Simon DK, Tarnopolsky MA. MELAS syndrome, cardiomyopathy, rhabdomyolysis, and autism associated with the A3260G mitochondrial DNA mutation. Biochem Biophys Res Commun 2010.

[27] Elliott HR, Samuels DC, Eden JA, Relton CL, Chinnery PF. Pathogenic mitochondrial DNA mutations are common in the general population. Am J Hum Genet 2008;83 (2):254-60.

[28] Herbert MR. Contributions of the environment and environmentally vulnerable physiology to autism spectrum disorders. Curr Opin Neurol;23 (2):103-10.

[29] MacFabe DF, Cain DP, Rodriguez-Capote K, Franklin AE, Hoffman JE, Boon F, Taylor AR, Kavaliers M, Ossenkopp KP. Neurobiological effects of intraventricular propionic acid in rats: possible role of short chain fatty acids on the pathogenesis and characteristics of autism spectrum disorders. Behav Brain Res 2007;176 (1):149-69.

[30] Thomas RH, Foley KA, Mepham JR, Tichenoff LJ, Possmayer F, MacFabe DF. Altered brain phospholipid and acylcarnitine profiles in propionic acid infused rodents: further development of a potential model of autism spectrum disorders. J Neurochem 2010;113 (2):515-29.

[31] Rasalam AD, Hailey H, Williams JH, Moore SJ, Turnpenny PD, Lloyd DJ, Dean JC. Characteristics of fetal anticonvulsant syndrome associated autistic disorder. Dev Med Child Neurol 2005;47 (8):551-5.

[32] Farkas V, Bock I, Cseko J, Sandor A. Inhibition of carnitine biosynthesis by valproic acid in rats--the biochemical mechanism of inhibition. Biochem Pharmacol 1996;52 (9):1429-33.

[33] Sanders AP, Schaefer DJ, Joines WT. Microwave effects on energy metabolism of rat brain. Bioelectromagnetics 1980;1 (2):171-81.

[34] Block ML, Calderon-Garciduenas L. Air pollution: mechanisms of neuroinflammation and CNS disease. Trends Neurosci 2009;32 (9):506-16.

[35] Vargas DL, Nascimbene C, Krishnan C, Zimmerman AW, Pardo CA. Neuroglial activation and neuroinflammation in the brain of patients with autism. Ann Neurol 2005;57 (1):67-81.

[36] Chen L, Yokel RA, Hennig B, Toborek M. Manufactured aluminum oxide nanoparticles decrease expression of tight junction proteins in brain vasculature. J Neuroimmune Pharmacol 2008;3 (4):286-95.

[37] Klintworth H, Garden G, Xia Z. Rotenone and paraquat do not directly activate microglia or induce inflammatory cytokine release. Neurosci Lett 2009;462 (1):1-5.

[38] Schmidt CW. Mito-Conundrum: Unraveling Environmental Effects on Mitochondria. Environ Health Perspect 2010;118:a292-a7.

[39] Palmieri L, Persico AM. Mitochondrial dysfunction in autism spectrum disorders: cause or effect? Biochim Biophys Acta;1797 (6-7):1130-7.

[40] Palmen SJ, van Engeland H, Hof PR, Schmitz C. Neuropathological findings in autism. Brain 2004;127 (Pt 12):2572-83.

[41] Courchesne E. Brainstem, cerebellar and limbic neuroanatomical abnormalities in autism. Curr Opin Neurobiol 1997;7 (2):269-78.

[42] Domercq M, Sanchez-Gomez MV, Sherwin C, Etxebarria E, Fern R, Matute C. System xc- and glutamate transporter inhibition mediates microglial toxicity to oligodendrocytes. J Immunol 2007;178 (10):6549-56.

[43] Khodorov BI. Mechanisms of destabilization of Ca2+-homeostasis of brain neurons caused by toxic glutamate challenge. Membr Cell Biol 2000;14 (2):149-62.

[44] Pereira CF, Oliveira CR. Oxidative glutamate toxicity involves mitochondrial dysfunction and perturbation of intracellular Ca2+ homeostasis. Neurosci Res 2000;37 (3):227-36.

[45] Stout AK, Raphael HM, Kanterewicz BI, Klann E, Reynolds IJ. Glutamate-induced neuron death requires mitochondrial calcium uptake. Nat Neurosci 1998;1 (5):366-73.

[46] Liang J, Takeuchi H, Doi Y, Kawanokuchi J, Sonobe Y, Jin S, Yawata I, Li H, Yasuoka S, Mizuno T, Suzumura A. Excitatory amino acid transporter expression by astrocytes is neuroprotective against microglial excitotoxicity. Brain Res 2008;1210:11-9.

[47] Piani D, Fontana A. Involvement of the cystine transport system xc- in the macrophage-induced glutamate-dependent cytotoxicity to neurons. J Immunol 1994;152 (7):3578-85.

[48] Zhao W, Xie W, Le W, Beers DR, He Y, Henkel JS, Simpson EP, Yen AA, Xiao Q, Appel SH. Activated microglia initiate motor neuron injury by a nitric oxide and glutamate-mediated mechanism. J Neuropathol Exp Neurol 2004;63 (9):964-77.

[49] Dzubay JA, Jahr CE. The concentration of synaptically released glutamate outside of the climbing fiber-Purkinje cell synaptic cleft. J Neurosci 1999;19 (13):5265-74.

[50] McNaught KS, Jenner P. Extracellular accumulation of nitric oxide, hydrogen peroxide, and glutamate in astrocytic cultures following glutathione depletion, complex I inhibition, and/or lipopolysaccharide-induced activation. Biochem Pharmacol 2000;60 (7):979-88.

[51] Yawata I, Takeuchi H, Doi Y, Liang J, Mizuno T, Suzumura A. Macrophage-induced neurotoxicity is mediated by glutamate and attenuated by glutaminase inhibitors and gap junction inhibitors. Life Sci 2008;82 (21-22):1111-6.

[52] McLennan H. The autoradiographic localization of L-[3h]glutamate in rat brain tissue. Brain Res 1976;115 (1):139-44.

[53] Brown DR. Neurons depend on astrocytes in a coculture system for protection from glutamate toxicity. Mol Cell Neurosci 1999;13 (5):379-89.

[54] Rosenberg PA, Aizenman E. Hundred-fold increase in neuronal vulnerability to glutamate toxicity in astrocyte-poor cultures of rat cerebral cortex. Neurosci Lett 1989;103 (2):162-8.

[55] Bittigau P, Ikonomidou C. Glutamate in neurologic diseases. J Child Neurol 1997;12 (8):471-85.

[56] Amin N, Pearce B. Glutamate toxicity in neuron-enriched and neuron-astrocyte co-cultures: effect of the glutamate uptake inhibitor L-trans-pyrrolidine-2,4-dicarboxylate. Neurochem Int 1997;30 (3):271-6.

[57] Sheldon AL, Robinson MB. The role of glutamate transporters in neurodegenerative diseases and potential opportunities for intervention. Neurochem Int 2007;51 (6-7):333-55.

[58] Drejer J, Larsson OM, Schousboe A. Characterization of L-glutamate uptake into and release from astrocytes and neurons cultured from different brain regions. Exp Brain Res 1982;47 (2):259-69.

[59] Takayasu Y, Iino M, Kakegawa W, Maeno H, Watase K, Wada K, Yanagihara D, Miyazaki T, Komine O, Watanabe M, Tanaka K, Ozawa S. Differential roles of glial and neuronal glutamate transporters in Purkinje cell synapses. J Neurosci 2005;25 (38):8788-93.

[60] Slemmer JE, De Zeeuw CI, Weber JT. Don't get too excited: mechanisms of glutamate-mediated Purkinje cell death. Prog Brain Res 2005;148:367-90.

[61] Yamashita A, Makita K, Kuroiwa T, Tanaka K. Glutamate transporters GLAST and EAAT4 regulate postischemic Purkinje cell death: an in vivo study using a cardiac arrest model in mice lacking GLAST or EAAT4. Neurosci Res 2006;55 (3):264-70.

[62] Takatsuru Y, Takayasu Y, Iino M, Nikkuni O, Ueda Y, Tanaka K, Ozawa S. Roles of glial glutamate transporters in shaping EPSCs at the climbing fiber-Purkinje cell synapses. Neurosci Res 2006;54 (2):140-8.

[63] Takatsuru Y, Iino M, Tanaka K, Ozawa S. Contribution of glutamate transporter GLT-1 to removal of synaptically released glutamate at climbing fiber-Purkinje cell synapses. Neurosci Lett 2007;420 (1):85-9.

[64] Huang YH, Dykes-Hoberg M, Tanaka K, Rothstein JD, Bergles DE. Climbing fiber activation of EAAT4 transporters and kainate receptors in cerebellar Purkinje cells. J Neurosci 2004;24 (1):103-11.

[65] Fairman WA, Vandenberg RJ, Arriza JL, Kavanaugh MP, Amara SG. An excitatory amino-acid transporter with properties of a ligand-gated chloride channel. Nature 1995;375 (6532):599-603.

[66] Arriza JL, Fairman WA, Wadiche JI, Murdoch GH, Kavanaugh MP, Amara SG. Functional comparisons of three glutamate transporter subtypes cloned from human motor cortex. J Neurosci 1994;14 (9):5559-69.

[67] Hu WH, Walters WM, Xia XM, Karmally SA, Bethea JR. Neuronal glutamate transporter EAAT4 is expressed in astrocytes. Glia 2003;44 (1):13-25.

[68] Furuta A, Martin LJ, Lin CL, Dykes-Hoberg M, Rothstein JD. Cellular and synaptic localization of the neuronal glutamate transporters excitatory amino acid transporter 3 and 4. Neuroscience 1997;81 (4):1031-42.

[69] Tirosh O, Sen CK, Roy S, Packer L. Cellular and mitochondrial changes in glutamate-induced HT4 neuronal cell death. Neuroscience 2000;97 (3):531-41.

[70] Chen CJ, Liao SL, Kuo JS. Gliotoxic action of glutamate on cultured astrocytes. J Neurochem 2000;75 (4):1557-65.

[71] Patneau DK, Mayer ML. Structure-activity relationships for amino acid transmitter candidates acting at N-methyl-D-aspartate and quisqualate receptors. J Neurosci 1990;10 (7):2385-99.

[72] Camacho A, Massieu L. Role of glutamate transporters in the clearance and release of glutamate during ischemia and its relation to neuronal death. Arch Med Res 2006;37 (1):11-8.

[73] Greenwood SM, Mizielinska SM, Frenguelli BG, Harvey J, Connolly CN. Mitochondrial dysfunction and dendritic beading during neuronal toxicity. J Biol Chem 2007;282 (36):26235-44.

[74] Atterwill CK, Cunningham VJ, Balazs R. Characterization of Na+,K+-ATPase in cultured and separated neuronal and glial cells from rat cerebellum. J Neurochem 1984;43 (1):8-18.

[75] Chih CP, Roberts Jr EL. Energy substrates for neurons during neural activity: a critical review of the astrocyte-neuron lactate shuttle hypothesis. J Cereb Blood Flow Metab 2003;23 (11):1263-81.

[76] Akabayashi A, Kato T. Glucose utilization rates in single neurons and neuropil determined by injecting nontracer amounts of 2-deoxyglucose. J Neurochem 1993;60 (3):931-5.

[77] Itoh Y, Abe T, Takaoka R, Tanahashi N. Fluorometric determination of glucose utilization in neurons in vitro and in vivo. J Cereb Blood Flow Metab 2004;24 (9):993-1003.

[78] Pellerin L, Magistretti PJ. Glutamate uptake into astrocytes stimulates aerobic glycolysis: a mechanism coupling neuronal activity to glucose utilization. Proc Natl Acad Sci U S A 1994;91 (22):10625-9.

[79] Sibson NR, Dhankhar A, Mason GF, Rothman DL, Behar KL, Shulman RG. Stoichiometric coupling of brain glucose metabolism and glutamatergic neuronal activity. Proc Natl Acad Sci U S A 1998;95 (1):316-21.

[80] Shank RP, Campbell GL. Amino acid uptake, content, and metabolism by neuronal and glial enriched cellular fractions from mouse cerebellum. J Neurosci 1984;4 (1):58-69.

[81] Hassel B. Carboxylation and anaplerosis in neurons and glia. Mol Neurobiol 2000;22 (1-3):21-40.

[82] Hassel B. Pyruvate carboxylation in neurons. J Neurosci Res 2001;66 (5):755-62.

[83] Kato T, Lowry OH. Enzymes of energy-converting systems in individual mammalian nerve cell bodies. J Neurochem 1973;20 (1):151-63.

[84] Dennis SC, Clark JB. The pathway of glutamate metabolism in rat brain mitochondria. Biochem J 1977;168 (3):521-7.

[85] Balazs R. Control of Glutamate Metabolism. The Effect of Pyruvate. J Neurochem 1965;12:63-76.

[86] Yudkoff M, Nissim I, Hummeler K, Medow M, Pleasure D. Utilization of [15N]glutamate by cultured astrocytes. Biochem J 1986;234 (1):185-92.

[87] Sonnewald U, Westergaard N, Schousboe A. Glutamate transport and metabolism in astrocytes. Glia 1997;21 (1):56-63.

[88] Erecinska M, Zaleska MM, Nissim I, Nelson D, Dagani F, Yudkoff M. Glucose and synaptosomal glutamate metabolism: studies with [15N]glutamate. J Neurochem 1988;51 (3):892-902.
[89] Erecinska M, Silver IA. Metabolism and role of glutamate in mammalian brain. Prog Neurobiol 1990;35 (4):245-96.
[90] McKenna MC, Hopkins IB, Lindauer SL, Bamford P. Aspartate aminotransferase in synaptic and nonsynaptic mitochondria: differential effect of compounds that influence transient hetero-enzyme complex (metabolon) formation. Neurochem Int 2006;48 (6-7):629-36.
[91] Daikhin Y, Yudkoff M. Compartmentation of brain glutamate metabolism in neurons and glia. J Nutr 2000;130 (4S Suppl):1026S-31S.
[92] McKenna MC, Stevenson JH, Huang X, Hopkins IB. Differential distribution of the enzymes glutamate dehydrogenase and aspartate aminotransferase in cortical synaptic mitochondria contributes to metabolic compartmentation in cortical synaptic terminals. Neurochem Int 2000;37 (2-3):229-41.
[93] Wurdig S, Kugler P. Histochemistry of glutamate metabolizing enzymes in the rat cerebellar cortex. Neurosci Lett 1991;130 (2):165-8.
[94] Subbalakshmi GY, Murthy CR. Isolation of astrocytes, neurons, and synaptosomes of rat brain cortex: distribution of enzymes of glutamate metabolism. Neurochem Res 1985;10 (2):239-50.
[95] LaNoue KF, Meijer AJ, Brouwer A. Evidence for electrogenic aspartate transport in rat liver mitochondria. Arch Biochem Biophys 1974;161 (2):544-50.
[96] LaNoue KF, Bryla J, Bassett DJ. Energy-driven aspartate efflux from heart and liver mitochondria. J Biol Chem 1974;249 (23):7514-21.
[97] Balazs R. Control of Glutamate Oxidation in Brain and Liver Mitochondrial Systems. Biochem J 1965;95:497-508.
[98] Berkich DA, Ola MS, Cole J, Sweatt AJ, Hutson SM, LaNoue KF. Mitochondrial transport proteins of the brain. J Neurosci Res 2007;85 (15):3367-77.
[99] Ramos M, del Arco A, Pardo B, Martinez-Serrano A, Martinez-Morales JR, Kobayashi K, Yasuda T, Bogonez E, Bovolenta P, Saheki T, Satrustegui J. Developmental changes in the Ca2+-regulated mitochondrial aspartate-glutamate carrier aralar1 in brain and prominent expression in the spinal cord. Brain Res Dev Brain Res 2003;143 (1):33-46.
[100] Xu Y, Ola MS, Berkich DA, Gardner TW, Barber AJ, Palmieri F, Hutson SM, LaNoue KF. Energy sources for glutamate neurotransmission in the retina: absence of the aspartate/glutamate carrier produces reliance on glycolysis in glia. J Neurochem 2007;101 (1):120-31.
[101] Yudkoff M, Nissim I, Pleasure D. [15N]aspartate metabolism in cultured astrocytes. Studies with gas chromatography-mass spectrometry. Biochem J 1987;241 (1):193-201.
[102] Bakken IJ, White LR, Aasly J, Unsgard G, Sonnewald U. [U-13C] aspartate metabolism in cultured cortical astrocytes and cerebellar granule neurons studied by NMR spectroscopy. Glia 1998;23 (3):271-7.
[103] Westergaard N, Drejer J, Schousboe A, Sonnewald U. Evaluation of the importance of transamination versus deamination in astrocytic metabolism of [U-13C]glutamate. Glia 1996;17 (2):160-8.

[104] Fahien LA, Hsu SL, Kmiotek E. Effect of aspartate on complexes between glutamate dehydrogenase and various aminotransferases. J Biol Chem 1977;252 (4):1250-6.
[105] Fahien LA, Smith SE. The enzyme-enzyme complex of transaminase and glutamate dehydrogenase. J Biol Chem 1974;249 (9):2696-703.
[106] Klockner U, Storck T, Conradt M, Stoffel W. Functional properties and substrate specificity of the cloned L-glutamate/L-aspartate transporter GLAST-1 from rat brain expressed in Xenopus oocytes. J Neurosci 1994;14 (10):5759-65.
[107] Otis TS, Kavanaugh MP. Isolation of current components and partial reaction cycles in the glial glutamate transporter EAAT2. J Neurosci 2000;20 (8):2749-57.
[108] Tretter L, Adam-Vizi V. Generation of reactive oxygen species in the reaction catalyzed by alpha-ketoglutarate dehydrogenase. J Neurosci 2004;24 (36):7771-8.
[109] Geier DA, Geier MR. A clinical and laboratory evaluation of methionine cycle-transsulfuration and androgen pathway markers in children with autistic disorders. Horm Res 2006;66 (4):182-8.
[110] Yudkoff M, Nelson D, Daikhin Y, Erecinska M. Tricarboxylic acid cycle in rat brain synaptosomes. Fluxes and interactions with aspartate aminotransferase and malate/aspartate shuttle. J Biol Chem 1994;269 (44):27414-20.
[111] Dringen R. Metabolism and functions of glutathione in brain. Prog Neurobiol 2000;62 (6):649-71.
[112] Pierre K, Pellerin L. Monocarboxylate transporters in the central nervous system: distribution, regulation and function. J Neurochem 2005;94 (1):1-14.
[113] Pow DV. Visualising the activity of the cystine-glutamate antiporter in glial cells using antibodies to aminoadipic acid, a selectively transported substrate. Glia 2001;34 (1):27-38.
[114] Bannai S, Tateishi N. Role of membrane transport in metabolism and function of glutathione in mammals. J Membr Biol 1986;89 (1):1-8.
[115] Raps SP, Lai JC, Hertz L, Cooper AJ. Glutathione is present in high concentrations in cultured astrocytes but not in cultured neurons. Brain Res 1989;493 (2):398-401.
[116] Hertz L, Peng L. Energy metabolism at the cellular level of the CNS. Can J Physiol Pharmacol 1992;70 Suppl:S145-57.
[117] Rimaniol AC, Mialocq P, Clayette P, Dormont D, Gras G. Role of glutamate transporters in the regulation of glutathione levels in human macrophages. Am J Physiol Cell Physiol 2001;281 (6):C1964-70.
[118] Chen Y, Swanson RA. The glutamate transporters EAAT2 and EAAT3 mediate cysteine uptake in cortical neuron cultures. J Neurochem 2003;84 (6):1332-9.
[119] Tan S, Schubert D, Maher P. Oxytosis: A novel form of programmed cell death. Curr Top Med Chem 2001;1 (6):497-506.
[120] Bremer J, Wojtczak AB. Factors controlling the rate of fatty acid -oxidation in rat liver mitochondria. Biochim Biophys Acta 1972;280 (4):515-30.
[121] Pettegrew JW, Levine J, McClure RJ. Acetyl-L-carnitine physical-chemical, metabolic, and therapeutic properties: relevance for its mode of action in Alzheimer's disease and geriatric depression. Mol Psychiatry 2000;5 (6):616-32.
[122] Nalecz KA, Nalecz MJ. Carnitine--a known compound, a novel function in neural cells. Acta Neurobiol Exp (Wars) 1996;56 (2):597-609.
[123] Bremer J. The role of carnitine in intracellular metabolism. J Clin Chem Clin Biochem 1990;28 (5):297-301.

[124] Bieber LL. Carnitine. Annu Rev Biochem 1988;57:261-83.

[125] Aoyama T, Peters JM, Iritani N, Nakajima T, Furihata K, Hashimoto T, Gonzalez FJ. Altered constitutive expression of fatty acid-metabolizing enzymes in mice lacking the peroxisome proliferator-activated receptor alpha (PPARalpha). J Biol Chem 1998;273 (10):5678-84.

[126] Kondrup J, Lazarow PB. Flux of palmitate through the peroxisomal and mitochondrial beta-oxidation systems in isolated rat hepatocytes. Biochim Biophys Acta 1985;835 (1):147-53.

[127] Bartlett K, Eaton S. Mitochondrial beta-oxidation. Eur J Biochem 2004;271 (3):462-9.

[128] Skorin C, Necochea C, Johow V, Soto U, Grau AM, Bremer J, Leighton F. Peroxisomal fatty acid oxidation and inhibitors of the mitochondrial carnitine palmitoyltransferase I in isolated rat hepatocytes. Biochem J 1992;281 (Pt 2):561-7.

[129] Hayashi H, Hara M. 1-Alkenyl group of ethanolamine plasmalogen derives mainly from de novo-synthesized fatty alcohol within peroxisomes, but not extraperoxisomal fatty alcohol or fatty acid. J Biochem (Tokyo) 1997;121 (5):978-83.

[130] Hayashi H, Oohashi M. Incorporation of acetyl-CoA generated from peroxisomal beta-oxidation into ethanolamine plasmalogen of rat liver. Biochim Biophys Acta 1995;1254 (3):319-25.

[131] Hayashi H, Sato A. Fatty alcohol synthesis accompanied with chain elongation in liver peroxisomes. Biochim Biophys Acta 1997;1346 (1):38-44.

[132] Hayashi H, Takahata S. Role of peroxisomal fatty acyl-CoA beta-oxidation in phospholipid biosynthesis. Arch Biochem Biophys 1991;284 (2):326-31.

[133] Sprecher H, Luthria DL, Mohammed BS, Baykousheva SP. Reevaluation of the pathways for the biosynthesis of polyunsaturated fatty acids. J Lipid Res 1995;36 (12):2471-7.

[134] Reszko AE, Kasumov T, David F, Jobbins KA, Thomas KR, Hoppel CL, Brunengraber H, Des Rosiers C. Peroxisomal fatty acid oxidation is a substantial source of the acetyl moiety of malonyl-CoA in rat heart. J Biol Chem 2004;279 (19):19574-9.

[135] Sliwinski S, Croonenberghs J, Christophe A, Deboutte D, Maes M. Polyunsaturated fatty acids: do they have a role in the pathophysiology of autism? Neuro Endocrinol Lett 2006;27 (4):465-71.

[136] Madiraju P, Pande SV, Prentki M, Madiraju SR. Mitochondrial acetylcarnitine provides acetyl groups for nuclear histone acetylation. Epigenetics 2009;4 (6):399-403.

[137] Lasalle JM, Yasui DH. Evolving role of MeCP2 in Rett syndrome and autism. Epigenomics 2009;1 (1):119-30.

[138] Shahbazian M, Young J, Yuva-Paylor L, Spencer C, Antalffy B, Noebels J, Armstrong D, Paylor R, Zoghbi H. Mice with truncated MeCP2 recapitulate many Rett syndrome features and display hyperacetylation of histone H3. Neuron 2002;35 (2):243-54.

[139] Thatcher KN, LaSalle JM. Dynamic changes in Histone H3 lysine 9 acetylation localization patterns during neuronal maturation require MeCP2. Epigenetics 2006;1 (1):24-31.

[140] Wood PL, Khan MA, Kulow SR, Mahmood SA, Moskal JR. Neurotoxicity of reactive aldehydes: the concept of "aldehyde load" as demonstrated by neuroprotection with hydroxylamines. Brain Res 2006;1095 (1):190-9.

[141] Dringen R. Oxidative and Antioxidative Potential of Brain Microglial Cells. Antioxidants & Redox Signaling 2005;7 (9-10):1223-33.
[142] Eichler FS, Ren JQ, Cossoy M, Rietsch AM, Nagpal S, Moser AB, Frosch MP, Ransohoff RM. Is microglial apoptosis an early pathogenic change in cerebral X-linked adrenoleukodystrophy? Ann Neurol 2008;63 (6):729-42.
[143] Brites P, Mooyer PAW, el Mrabet L, Waterham HR, Wanders RJA. Plasmalogens participate in very-long-chain fatty acid-induced pathology. Brain 2009;132:482-92.
[144] Block ML, Hong JS. Microglia and inflammation-mediated neurodegeneration: multiple triggers with a common mechanism. Prog Neurobiol 2005;76 (2):77-98.
[145] Blaylock RL. A possible central mechanism in autism spectrum disorders, part I. Altern Ther Health Med 2008;14 (6):46-53.
[146] Fombonne E. Epidemiology of autistic disorder and other pervasive developmental disorders. J Clin Psychiatry 2005;66 Suppl 10:3-8.
[147] Fombonne E. The epidemiology of autism: a review. Psychol Med 1999;29 (4):769-86.
[148] Ritvo ER, Freeman BJ, Pingree C, Mason-Brothers A, Jorde L, Jenson WR, McMahon WM, Petersen PB, Mo A, Ritvo A. The UCLA-University of Utah epidemiologic survey of autism: prevalence. Am J Psychiatry 1989;146 (2):194-9.
[149] Horlick MB, Rosenbaum M, Nicolson M, Levine LS, Fedun B, Wang J, Pierson RN, Jr., Leibel RL. Effect of puberty on the relationship between circulating leptin and body composition. J Clin Endocrinol Metab 2000;85 (7):2509-18.
[150] Courant F, Aksglaede L, Antignac JP, Monteau F, Sorensen K, Andersson AM, Skakkebaek NE, Juul A, Bizec BL. Assessment of circulating sex steroid levels in prepubertal and pubertal boys and girls by a novel ultrasensitive gas chromatography-tandem mass spectrometry method. J Clin Endocrinol Metab;95 (1):82-92.
[151] Janfaza M, Sherman TI, Larmore KA, Brown-Dawson J, Klein KO. Estradiol levels and secretory dynamics in normal girls and boys as determined by an ultrasensitive bioassay: a 10 year experience. J Pediatr Endocrinol Metab 2006;19 (7):901-9.
[152] Goodman Y, Bruce AJ, Cheng B, Mattson MP. Estrogens attenuate and corticosterone exacerbates excitotoxicity, oxidative injury, and amyloid beta-peptide toxicity in hippocampal neurons. J Neurochem 1996;66 (5):1836-44.
[153] Harms C, Lautenschlager M, Bergk A, Katchanov J, Freyer D, Kapinya K, Herwig U, Megow D, Dirnagl U, Weber JR, Hortnagl H. Differential mechanisms of neuroprotection by 17 beta-estradiol in apoptotic versus necrotic neurodegeneration. J Neurosci 2001;21 (8):2600-9.
[154] Hilton GD, Ndubuizu AN, McCarthy MM. Neuroprotective effects of estradiol in newborn female rat hippocampus. Brain Res Dev Brain Res 2004;150 (2):191-8.
[155] McClean J, Nuñez JL. 17alpha-Estradiol is neuroprotective in male and female rats in a model of early brain injury. Exp Neurol 2008;210 (1):41-50.
[156] Mendelowitsch A, Ritz MF, Ros J, Langemann H, Gratzl O. 17beta-Estradiol reduces cortical lesion size in the glutamate excitotoxicity model by enhancing extracellular lactate: a new neuroprotective pathway. Brain Res 2001;901 (1-2):230-6.
[157] Nilsen J, Diaz Brinton R. Mechanism of estrogen-mediated neuroprotection: regulation of mitochondrial calcium and Bcl-2 expression. Proc Natl Acad Sci U S A 2003;100 (5):2842-7.

[158] Singer CA, Rogers KL, Strickland TM, Dorsa DM. Estrogen protects primary cortical neurons from glutamate toxicity. Neurosci Lett 1996;212 (1):13-6.

[159] Vavaiya KV, Briski KP. Effects of caudal hindbrain lactate infusion on insulin-induced hypoglycemia and neuronal substrate transporter glucokinase and sulfonylurea receptor-1 gene expression in the ovariectomized female rat dorsal vagal complex: Impact of estradiol. J Neurosci Res 2008;86 (3):694-701.

[160] Zhang Y, Lu X, Bhavnani BR. Equine estrogens differentially inhibit DNA fragmentation induced by glutamate in neuronal cells by modulation of regulatory proteins involved in programmed cell death. BMC Neurosci 2003;4:32.

[161] Klein KO, Baron J, Colli MJ, McDonnell DP, Cutler GB, Jr. Estrogen levels in childhood determined by an ultrasensitive recombinant cell bioassay. J Clin Invest 1994;94 (6):2475-80.

[162] Simpkins JW, Wang J, Wang X, Perez E, Prokai L, Dykens JA. Mitochondria play a central role in estrogen-induced neuroprotection. Curr Drug Targets CNS Neurol Disord 2005;4 (1):69-83.

[163] Simpkins JW, Yi KD, Yang SH, Dykens JA. Mitochondrial mechanisms of estrogen neuroprotection. Biochim Biophys Acta 2010;1800 (10):1113-20.

[164] Prokai L, Prokai-Tatrai K, Perjesi P, Zharikova AD, Perez EJ, Liu R, Simpkins JW. Quinol-based cyclic antioxidant mechanism in estrogen neuroprotection. Proc Natl Acad Sci U S A 2003;100 (20):11741-6.

[165] Djouadi F, Weinheimer CJ, Saffitz JE, Pitchford C, Bastin J, Gonzalez FJ, Kelly DP. A gender-related defect in lipid metabolism and glucose homeostasis in peroxisome proliferator- activated receptor alpha- deficient mice. J Clin Invest 1998;102 (6):1083-91.

[166] Maher KR, Harper JF, Macleay A, King MG. Peculiarities in the endocrine response to insulin stress in early infantile autism. J Nerv Ment Dis 1975;161 (3):180-4.

[167] Nagy I, Hirka G, Kurcz M, Anda E, Baranyai P. The role of estrogens in the regulation of lactate dehydrogenase activity and its submolecular organization in rat anterior pituitary. Endokrinologie 1978;71 (1):1-12.

[168] Mandour T, Kissebah AH, Wynn V. Mechanism of oestrogen and progesterone effects on lipid and carbohydrate metabolism: alteration in the insulin: glucagon molar ratio and hepatic enzyme activity. Eur J Clin Invest 1977;7 (3):181-7.

[169] Courchesne E, Redcay E, Kennedy DP. The autistic brain: birth through adulthood. Curr Opin Neurol 2004;17 (4):489-96.

[170] Courchesne E, Saitoh O, Yeung-Courchesne R, Press GA, Lincoln AJ, Haas RH, Schreibman L. Abnormality of cerebellar vermian lobules VI and VII in patients with infantile autism: identification of hypoplastic and hyperplastic subgroups with MR imaging. AJR Am J Roentgenol 1994;162 (1):123-30.

[171] Herbert MR, Ziegler DA, Makris N, Filipek PA, Kemper TL, Normandin JJ, Sanders HA, Kennedy DN, Caviness VS, Jr. Localization of white matter volume increase in autism and developmental language disorder. Ann Neurol 2004;55 (4):530-40.

[172] Schneider GH, Baethmann A, Kempski O. Mechanisms of glial swelling induced by glutamate. Can J Physiol Pharmacol 1992;70 Suppl:S334-43.

[173] Canitano R. Epilepsy in autism spectrum disorders. Eur Child Adolesc Psychiatry 2007;16 (1):61-6.

[174] Devi PU, Manocha A, Vohora D. Seizures, antiepileptics, antioxidants and oxidative stress: an insight for researchers. Expert Opin Pharmacother 2008;9 (18):3169-77.
[175] Waldbaum S, Patel M. Mitochondrial dysfunction and oxidative stress: a contributing link to acquired epilepsy? J Bioenerg Biomembr 2010;42 (6):449-55.
[176] Gargus JJ. Genetic calcium signaling abnormalities in the central nervous system: seizures, migraine, and autism. Ann N Y Acad Sci. 2009;1151:133-56.
[177] Palmieri F. Diseases caused by defects of mitochondrial carriers: a review. Biochim Biophys Acta 2008;1777 (7-8):564-78.
[178] Chen SD, Chang AY, Chuang YC. The potential role of mitochondrial dysfunction in seizure-associated cell death in the hippocampus and epileptogenesis. J Bioenerg Biomembr 2010;42 (6):461-5.
[179] Ashwin E, Ashwin C, Rhydderch D, Howells J, Baron-Cohen S. Eagle-eyed visual acuity: an experimental investigation of enhanced perception in autism. Biol Psychiatry 2009;65 (1):17-21.
[180] Simmons DR, Robertson AE, McKay LS, Toal E, McAleer P, Pollick FE. Vision in autism spectrum disorders. Vision Res 2009;49 (22):2705-39.
[181] Birch EE, Carlson SE, Hoffman DR, Fitzgerald-Gustafson K, Fu VL, Drover JR, Castañeda YS, Minns L, Wheaton DK, Mundy D, Marunycz J, Diersen-Schade DA. The DIAMOND (DHA Intake And Measurement Of Neural Development) Study: a double-masked, randomized controlled clinical trial of the maturation of infant visual acuity as a function of the dietary level of docosahexaenoic acid. Am J Clin Nutr 2010;91 (4):848-59.
[182] Jacques C, Levy E, Muckle G, Jacobson SW, Bastien C, Dewailly E, Ayotte P, Jacobson JL, Saint-Amour D. Long-term effects of prenatal omega-3 fatty acid intake on visual function in school-age children. J Pediatr 2011;158 (1):83-90.

5

Serotonergic Neurotransmission in Autism Spectrum Disorders

Yoshihiro Takeuchi
Shiga University of Medical Science
Japan

1. Introduction

Serotonin (5-hydroxytryptamine, 5-HT) was isolated and characterized during the late 1930s through the 1950s. Since then, serotonin has been shown to play a key role in a range of behaviors and processes, including sensory gating and processing, behavioral inhibition, appetite, aggression, sleep, mood, and neuroendocrine secretion (Anderson, 2005). The serotonin neuron system stretches abundant branches from a limited number of cells in the brain stem and widely and densely projects the brain (Takeuchi, 1988). Therefore, the serotonin neuron system is considered to be "the total control system". During development of the brain, the serotonin neuron system is not only essential for formation and maintenance of synapses (Lauder, 1990), but also is affected by a variety of environmental factors. These findings are crucial for understanding the pathogenesis of many developmental disorders. This paper reviews the involvement of the serotonin neuron system in neurotransmission in autism spectrum disorders.

2. Anatomical characteristics of the serotonin neuron system

The distribution of the serotonin neuron system overwhelms that of other neuron systems (Nieuwenhuys, 1985; Takeuchi, 1988). The serotonin neuron system is a slow synaptic potential system with a long time lapse, similar to the physiology of noradrenalin and dopamine neuron systems. In this review, we examine the central nervous system (CNS) in monkeys (macaca fuscata) describing the anatomical characteristics of the serotonin neuron system, in order to understand the diversity of functions of serotonin in the brain and the pathogenesis of developmental disorders.

2.1 Cell bodies of serotonin neurons are localized in the brain stem

The cell bodies of serotonin neurons are localized from the caudal part of the midbrain red nucleus to the decussation of the pyramid of the medulla. This pattern is similar among species. Approximately 65% of the cell bodies are localized near the raphe nuclei group at the center of the brain stem, including the dorsal raphe nucleus (B7), superior central nucleus (B8), pontine raphe nucleus (B5), raphe magnus nucleus (B3), obscurus raphe nucleus (B2), and raphe pallidus nucleus (B1) from the rostral side, while the remaining cell bodies are localized at other sites. Serotonin neurons number between 25-50,000 in rodents and $\geq$150,000 in primates (monkeys), and the B7 group accounts for a large proportion of the serotonin neurons (Nieuwenhuys, 1985; Takeuchi, 1988).

2.2 Serotonin fiber distribution

Axons and dendrites of serotonin neurons are distributed more densely and widely than previously predicted in the central nervous system. The density and area of their distribution exceed those of the noradrenalin neuron system. The distribution of serotonergic axons and dendrites differs markedly between species. This tendency is most remarkable in the neocortex. The distribution pattern of serotonin fibers in the motor, sensory, and visual cortices of rodents is similar, whereas it differs markedly in the neocortex in primates. These distribution patterns of serotonin fibers are presumed to be dependent on the functions of the neocortex. In general, serotonin fibers are distributed more densely in the granular cell layer in the input system of the cortices than in the pyramidal cell layer in the output system, and the density of serotonin distribution is much higher in the primary visual cortex than in the primary motor cortex in primates. On the other hand, noradrenalin fibers are distributed more densely in the pyramidal cell layer in the output system and the distribution is complementary to that of serotonin fibers, i.e. "mutually exclusive distribution pattern" (Nieuwenhuys, 1985; Takeuchi, 1988).

2.3 Projections to other monoamine neuron systems

The serotonin neuron system regulates the activity of other monoamine neurons, in particular, dopamine neurons. For example, serotonin fibers are distributed in a markedly high density in the substantia nigra, and synapses are formed with a number of dopamine neurons in the substantia nigra in all animal species. Thus, the serotonin neuron system controls dopamine neurons in the substantia nigra at the cell body level via the 5-HT2A receptor. Serotonin fibers are distributed not only at sites rich in dopamine fiber endings, such as the striatum and nucleus accumbens, but also in the prefrontal area of the frontal cortex, which is important for the pathogenesis of developmental disorders.

2.4 High sprouting capability

The serotonin neuron system exhibits a markedly high sprouting capability compared with that of other neuron systems. In general, serotonin fibers are sparsely distributed at the striatum. When the substantia nigra is chemically ablated unilaterally during development (by 12 days after birth in mice) using 6-hydroxydopamine, dopamine neurons in the substantia nigra and dopamine fibers in the striatum disappear. In contrast, axon sprouting of serotonin fibers appears in the striatum of the experimental animals and generates a markedly dense distribution. These findings suggest that the serotonin neuron system has a strong influence on the dopamine neuron system (Yamazoe, 2001). This phenomenon, called "heterotypic sprouting", is observed particularly between the serotonin neuron and dopamine neuron systems. Heterotypic sprouting of serotonin fibers during development is thought to be involved in the plasticity of the brain and the modification of symptoms of diseases derived from dysfunction of the dopamine neuron system. Therefore, heterotypic sprouting is particularly intriguing in investigating the pathogenesis of developmental disorders.

2.5 Volume transmission

As mentioned previously, the serotonin neuron system is called the total control system or the diffuse projection system, and plays a major role in central nervous system function. The conventional projection system is called the "wiring transmission" system, and the serotonin neuron system is thought to be responsible for non-functional ("volume transmission") and junctional neurotransmission (Bunin, 1999).

2.6 Development and the serotonin neuron system

During brain development, serotonin has been shown to influence the maturation of target tissues, including dendritic elaboration, synaptogenesis, neurogenesis and organization of the cortex (Lauder, 1978, 1990; Kondoh, 2004). At early stages of development, when the blood–brain barrier is not yet fully formed, serotonin can enter the brain of a developing fetus, and cause a loss of serotonin terminals through negative feedback. This loss of serotonin innervation persists throughout subsequent development and the symptoms of autism appear (Whitaker-Azmitia, 2005).

Serotonin neurons appear at 5 weeks of gestation and their numbers increase dramatically through the tenth week. Raphe nuclei can be detected at 15 weeks gestation. Axons from serotonin neurons grow into the cortex prenatally and expression of the serotonin transporter begins at the end of the first trimester. Serotonin levels in the brain increase postnatally for 2-5 years, then decline to adult levels at approximately 50% of the peak values (Whitaker-Azmitia, 2005).

In rats, serotonin fibers exhibit a temporary control over the primary sensory cortex during fetal days 2-14, which corresponds to the period of synapse formation. Serotonin transporters appear temporarily in glutamatergic thalamocortical afferent nerve fibers within 2 weeks after birth in rats. Serotonin concentrations that are too high or too low influence normal development during this period (Erzurumlu & Kind, 2001). An insufficient concentration of serotonin in the brain during synapse formation interferes with the development of the barrel in the rat somatosensory area. In contrast, excessive serotonin induces superabundant axonal branching in the somatosensory area and obscures the boundary of the barrel, as observed in monoamine oxydase A-knockout mice (Erzurumlu & Kind, 2001). Abnormalities in synaptic connection induced by a reduced or excess intracerebral serotonin concentration greatly reduces the number of dendrites not only in the sensory cortex, but also in the hippocampus. Considering that serotonin fibers are densely distributed in the granular cell layer in the cerebral neocortex, polymorphisms of the serotonin transporter gene reported in autism spectrum disorders patients may influence serotonin regulation in the thalamocortical nerve pathway, which may lead to abnormalities in sensory and cognitive functions in most autism spectrum disorders patients.

Development of the serotonin neuron system is influenced by a variety of environmental factors. In mice, studies have shown that malnutrition and hypoxia during the neonatal period induce irreversible changes in the serotonin neuron system, that vulnerability of the serotonin neuron system is dependent on the region, and that a variety of environmental factors may induce brain injury associated with the serotonin neuron system (Ishimura, 1989). Moreover, the serotonin neuron system is influenced greatly by aging, and S100β in astrocytes is involved in the dynamics of serotonin fibers in the hippocampus. Axons originating from the B7 and B8 groups show different intracerebral distribution and degenerative process associated with aging, and each has specific neurobiological significance (Nishimura, 1995). Considering the pathogenesis of autism spectrum disorders, it is of interest that the serotonergic system is activated through rhythmic movements, such as gait, chewing and respiration, and that adequate physical activity is important for serotonin activation (Kohyama, 2011). Thus, accumulating evidence indicates that serotonergic projections undergo continuous age-related change through early childhood and that the serotonin neuron system is particularly plastic and fragile.

3. Serotonergic neurotransmission in autism spectrum disorders

In this review, autistic disorder (Kanner's 'autism'), childhood disintegrative disorder, pervasive development disorder not otherwise specified ('atypical autism'), and Asperger syndrome are collectively termed autism spectrum disorders (Persico & Bourgeron, 2006). According to previous studies, abnormalities in the brain stem-subcortical structure (caudate nucleus, putamen, and pallidum)-cerebellum network, as well as in the hippocampus, piriform gyrus, and cingulate gyrus, play a central role in the pathogenesis. A number of studies have examined the relationship between autism spectrum disorders and abnormalities in the serotonin neuron system, and these abnormalities are considered to be the core of the pathogenesis of autism spectrum disorders (Johnston & Blue, 2006; Persico & Bourgeron, 2006).

3.1 Hyperserotonemia

Increased blood serotonin concentrations in approximately 30% of autism spectrum disorders patients strongly suggests that serotonin is involved with the onset of autism spectrum disorders. Although an increase in serotonin uptake by platelets elevates serotonin concentrations in the blood, an abnormality in the serotonin neuron system has not been identified in these patients. The relationship of blood serotonin to the behavioral aspects of autism spectrum disorders and the range of blood serotonin concentrations in the autism spectrum disorders group are not clear. Serotonin levels have not been correlated consistently with degree of mental retardation or other symptoms (Anderson, 2005). High serotonin levels in the blood have been suggested to cause a low intracerebral serotonin concentration in autism spectrum disorders, and are considered to be the result of negative feedback on serotonin neurons (Whitaker-Azmitia, 2005). This negative feedback is likely mediated by the 5-HT1A, which is the earliest appearing serotonin receptor and has a prenatal peak both in human and rodents. In some cases, the receptor appears transiently during development in brain regions, such as the cerebellum, that do not express the receptor in the adult.

Although most serotonin-related neurochemical research in autism spectrum disorders has focused on hyperserotonemia, several studies examining serotonin in cerebrospinal fluid have been reported (Anderson, 2005). However, cerebrospinal fluid studies are in general agreement that little or no difference exists between the mean levels of the major serotonin metabolite, 5-hydroxyindoleacetic acid in autism spectrum disorders patients and control groups.

3.2 Serotonin hypothesis based on clinical viewpoints

The concept of autism has changed over time. The main symptoms of autism spectrum disorders include onset in early infancy, male predominance, age-dependent symptomatic development, and symptomatic vulnerability to the influence of environmental factors. The neuronal system responsible for these clinical distinctions should have the following characteristics: development in early infancy, predominant vulnerability in males, involvement in the regulation of the growth and function of the brain by projecting to the subcortical to cortical levels, and susceptibility to environmental factors. These features suggest that a brainstem monoamine neuron system may be the primary contributor to the pathogenesis of autism spectrum disorders. Among them, the serotonin neuron system is considered the most favored candidate since it fulfills all of the characteristics outlined above (Segawa & Nomura 2005). Abnormalities in the serotonin neuron system are thought to cause disturbances in the sleep/arousal rhythm observed from the early stages of autism spectrum disorders, poor social skills, adaptation disorder to a new environment, and impaired cognitive function.

3.3 Positron emission tomography studies

Analysis of serotonin metabolism by positron emission tomography (PET) revealed that serotonin synthesis is disturbed in the frontal lobe, thalamus, and cerebellum, which are thought to be involved in the pathogenesis of autism spectrum disorders. PET using α [C-11] methyl-L-tryptophan showed that the uptake of α [C-11] methyl-L-tryptophan in autism spectrum disorders is different between the right and left sides in the frontal lobe, thalamus, and cerebellar dentate nucleus. When compared with the adult levels of serotonin synthesis in the whole brain, the level of serotonin synthesis in children up to five years old without autism spectrum disorders was ≥200% that of adults, and gradually decreased thereafter to the level in adults. The serotonin synthesis levels decreased earlier in female children than in male children. However, in children with autism spectrum disorders, serotonin synthesis gradually increased from 2 to 15 years old to a level 1.5-fold that of the adult (Chugani, 1999; Chugani, 2002). PET studies using [11C](+)McN5652, which is highly selective for the serotonin transporter, as a tracer showed that serotonin transporter density in Asperger syndrome patients was significantly decreased in a wide area containing the midbrain, basal ganglia, and cerebral cortex. These results need to be verified fully.

3.4 Neuropathologic changes implicating the serotonin neuron system

Most regions of the brain reported to be responsible for autism receive rich projections from the serotonin neuron system pathologically, in that they have dense serotonergic terminals. Thus, abnormalities in serotonin neurotransmission may be involved in the pathogenesis of autism spectrum disorders. To date, potential foci responsible for autism spectrum disorders have been reported, including the superior and inferior olivary nuclei, facial nerve nucleus, dorsal raphe nucleus in the brain stem, the frontal lobe (prefrontal area) and lateral lobe (superior temporal gyrus) in the cerebral cortex, cerebellar vermis and hemispheres, and the cingulate gyrus, septal area, amygdala, and hippocampus in the cerebral limbic system (Amaral, 2008). Neuropathologic foci of autism spectrum disorders are present widely, suggesting that brain injury occurs at an early stage of development.

In regards to neurotransmission, most of the foci reported to be responsible of autism spectrum disorders are strongly projected by the serotonin neuron system, except for the cerebellum. A particularly strong rationale has been developed for involvement of the amygdala and associated areas of the limbic cortex. More specifically, the core social relatedness deficits in autism spectrum disorders serve to focus attention on the rostral limbic system, including the amygdala, septum, medial orbitofrontal cortex, anterior insular cortex, anterior cinglate cortex, and the nucleus accumbens (Anderson, 2005). The various limbic areas are richly innervated with serotonergic projections, and the nucleus accumbens, an area crucial in appetitive and reward processes, has an especially dense serotonergic innervation. Meanwhile, in the cerebellum, where serotonin fiber density is lower than in most other brain areas, the serotonin neuron system is thought to be crucial for cerebellar function. As described before, 5HT-1A receptors have a rich and early expression in the rodent and human cerebellum. Therefore, neuropathological findings strongly suggest that abnormalities in serotonergic neurotransmission may be involved in the pathogenesis of autism spectrum disorders. Furthermore, neuropathological findings in the brain were compared between reeler mice deficient in reelin and autism spectrum disorders patients, and common lesion distribution and microscopic findings between the two conditions were observed (Persico & Bourgeron, 2006). In addition, the serotonin agonist, 5-methoxytryptamine, influences the level of reelin, which shows an abnormal level in the brains of autism spectrum disorders patients.

3.5 Animal models of autism spectrum disorders

In animals, serotonin is involved in mediating behaviors of sleep, arousal, aggression, impulsivity, and affiliation, all of which are relevant to autism. Reduced serotonergic function has been associated with worsened sleep, depressed mood, altered arousal, increased aggression, and increased impulsivity (Anderson, 2005). In general, serotonin plays an inhibitory role in the brain. Its actions are complex and depend greatly on the specific location or distribution of serotonergic terminals and classes of receptors stimulated. In regards to animal models for autism spectrum disorders, a thalidomide-administered model (rat), a valproic acid-administered model (rat), a Borna disease virus infection model (rat), an amygdala injury model (monkey, rat), and a hypothyroidism model (rat) have been constructed. A majority of the animal models show abnormal early development of the serotonin neuron system, although abnormalities in the dopamine and noradrenalin neuron systems have also been suggested. Furthermore, as in the autism spectrum disorders animal models, oxytocin-deficient and oxytocin receptor-deficient mice have been reported.

3.6 Other models related to serotonin

The involvement of the serotonin neuron system in the pathogenesis of autism spectrum disorders has been reported from a variety of fields. (1) A mutation in the tryptophan 2,3-dioxygenase gene that encodes the rate-limiting enzyme in tryptophan-serotonin-kynurenine metabolism was reported in autism spectrum disorders patients. (2) Serotonin depletion during the neonatal period increased the thickness of the cerebral cortex, which is a common finding with an increased cerebral volume in autism spectrum disorders patients. (3) 5-methoxytryptamine has been shown to influence the level of reelin, which shows an abnormal level in the brain in autism spectrum disorders patients (Janusonis, 2004). (4) A selective serotonin reuptake inhibitor was shown to improve clinical symptoms. However, the factors that may critically affect the response to selective serotonin reuptake inhibitor treatment and adverse effects in individuals with autism spectrum disorders need to be clarified. (5) Risperidone, another frequently used medication for patients with autism spectrum disorders, is an atypical neuroleptic that acts as a potent antagonist at the 5-HT2A receptor.

3.7 Oxytocin and autism spectrum disorders

Oxytocin is a peptide hormone produced by neurosecretory cells at the hypothalamus supraoptic nucleus and paraventricular nucleus in mothers and it is secreted from the posterior pituitary lobe. The hormone exhibits a variety of functions in the brain, as well as roles in delivery and galactopoiesis. Oxytocin regulates emotion in the company of somebody and oxytocin concentrations in the blood are low in autism spectrum disorders patients, therefore, the hormone attracts the most attention in investigations of the pathogenesis of autism spectrum disorders (Kirsch. et al; 2005). Oxytocin plays a role in signal transmission between the mother and the fetus so that neurons in the fetus are prepared for delivery. Oxytocin temporarily switches intracerebral GABAergic neurotransmission from the excitatory to inhibitory state in the fetus at delivery and exerts neuroprotective action. Serotonin fiber endings are abundant on oxytocin neurons at the hypothalamus supraoptic nucleus and paraventricular nucleus, and it should be noted that oxytocin neurons are regulated by the serotonin neuron system.

Studies to date show that prenatal treatment with a serotonin agonist, 5-methoxytryptame, results in "autistic-like" behaviors such as decreased social bonding, sensory hyper-responsiveness, seizures and motor changes (Whitaker-Azmitia, 2005). 5-methoxytryptame (serotonin agonist)-treated animals have a loss of oxytocin-containing cells in the hypothalamus, as well as an apparent loss of oxytocin projections towards other brain regions such as the amygdala and the supraoptic nucleus. The amygdala is of interest in autism spectrum disorders because there have been reported changes in volume, cell-packing density and function in this region. In addition, the central nucleus of the amygdala receives an intense serotonergic innervation from the dorsal raphe nucleus, which may modulate behavioral responses to fear (Adolphs, 2002).

3.8 Glutamic acid, neuroligin, and neurexin in autism spectrum disorders

Glutamic acid, as well as serotonin plays an important role in the development of the brain. Glutamic acid is essential for the development and plasticity of the cerebral cortex, and is responsible for the collaborative function with serotonin in the development of the thalamo-cortical pathway. In the cerebellum in autism spectrum disorders patients, up regulation of the glutamate receptor 1 subunit of AMPA mRNA has been reported. In addition, genetic mutations in neuroligin-3 and 4 genes result in changes in glutamatergic synapses. Furthermore, as it has been shown that the balance between excitatory and inhibitory neurotransmission is disturbed by the down-regulation of neuroligin-1, shifting to the excitatory side. Synapse dynamics such as plasticity, production, and pruning is impaired; therefore, the importance of glutamic acid in autism spectrum disorders has been suggested (Johnston & Blue, 2006). Neuroligin and neurexin are cell adhesion molecules located at the postsynaptic and presynaptic region, respectively. They promote synapse formation bidirectionally in the glutamatergic nerve system and GABAergic nerve system, and they have been called "the bridge between molecules and the mind" (Dean & Dresbach, 2006). Insufficient inhibitory neurotransmission is related to impaired cognitive processes, physical control, and the tendency to be complicated with abnormal brain waves and epilepsy in autism spectrum disorders. While mutated proteins of neuroligin 3, neuroligin 4, and neurexin 1 have been reported in autism spectrum disorders patients, the involvement of neuroligin 4 in attention-deficit hyperactivity disorder and learning disorder has also been suggested (Dean & Dresbach, 2006; Betancur, 2009).

3.9 Autism spectrum disorders-related proteins

Autism spectrum disorders-related proteins have a variety of functions including remodeling of chromatin and transcription regulation, the actin cell cytoskeleton, the induction and maintenance of dendritic spines, neurotransmission (neurotransmitters, receptors, and transporters), as a second messenger, apoptosis, cell adhesion, and nerve cell mobilization (Persico & Bourgeron, 2006). SLC6A4, which encodes the serotonin transporter, has been intensively investigated as an autism spectrum disorders-related gene. In addition, genes encoding a GABA receptor subunit (GABAR: GABRA4 and GABRB1), an NMDA receptor subunit (GRIN2A), the oxytocin receptor (OCTR), and the vasopressin receptor (AVPR1) have also been investigated as autism spectrum disorders-related.

4. Serotonergic neurotransmission in attention deficit hyperactivity disorder symptoms in patients with autism spectrum disorders

The majority of the patients with autism spectrum disorders are diagnosed as having attention deficit hyperactivity disorder at an early stage of the disease, suggesting a common neuronal dysfunction between attention deficit hyperactivity disorder and autism spectrum disorders. Attention deficit hyperactivity disorder is a disease that involves a number of genetic factors. It is a heterogeneous disorder with a wide variation in symptoms among individuals, and is influenced by environmental factors. Dysfunction in the catecholaminergic (dopaminergic and noradrenergic) system in the neocortex (prefrontal area)-subcortical structure (the caudate nucleus, putamen, pallidum, and thalamus)-cerebellum pathway is the core of the pathogenesis of attention deficit hyperactivity disorder. The serotonin neuron system is related not only with the abnormality in the catecholaminergic system, but also directly with the pathogenesis of attention deficit hyperactivity disorder (Biederman, 2005).

4.1 Dopaminergic neurotransmission in attention deficit hyperactivity disorder

It is generally accepted that abnormalities in the dopaminergic system are the core of the pathogenesis of attention deficit hyperactivity disorder. Among monoamines, dopamine is located profusely in the brain. The dopaminergic system is also considered a diffuse projection system, however, the distribution is relatively localized compared with the serotonergic and noradrenergic systems. The cell bodies of the dopamine neuron system are distributed in a concentrated manner at the midbrain substantia nigra pars compacta and ventral tegmental area. Dopamine neurons are also localized in the mesocorticolimbic dopaminergic pathway, which has projections to the ventral tegmental area, accumbens nucleus, amygdala, septal area, olfactory tubercule, and prefrontal area, and is deeply involved in functions such as motivation, sustained attention, cognitive function, and reward behavior (Björklund & Dunnett 2007).

A dopamine transporter, localized at the cell membrane, plays an important role in dopamine metabolism by incorporating extracellular dopamine into the cell and decreasing extracellular dopamine concentrations. Single photon emission computed tomography with the administration of altropane, which binds the dopamine transporter specifically, showed that the binding capacity of ^{123}I altropane in the striatum increased by 70% in adults with attention deficit hyperactivity disorder, suggesting the over expression of the dopamine transporter in attention deficit hyperactivity disorder.

After methylphenidate was administered for 4 weeks in the treatment for attention deficit hyperactivity disorder adults, single photon emission computed tomography showed that dopamine transporter activity in the striatum, which was higher than that in the control group before treatment, decreased after treatment and showed no significant difference from controls. The normalization of dopaminergic neurotransmission confirmed the clinical effect of methylphenidate. Furthermore, positron emission tomography with C^{11}altropanea was carried out for a comparative experiment that took into account a history of medication and smoking, and the results showed that the dopamine transporter was significantly increased at the caudate nucleus (on the right) in adult attention deficit hyperactivity disorder cases. Therefore, deregulation of the dopamine transporter resulted in attention deficit hyperactivity disorder (Spencer, 2007). These findings are consistent with the hypothesis that the striatum is involved in the regulation of complicated cognitive function and that the putamen is involved mostly in the control of physical function.

The over-expression of the dopamine transporter in the striatum in humans was confirmed by a number of studies using single photon emission computed tomography and positron emission tomography, but examinations of a bilateral difference of the dopamine transporter in the striatum are not consistent. The distribution of the dopamine transporter is greatly dependent on site, and distribution density is much higher in the striatum than in the frontal lobe, which is known to have the dopaminergic system at the highest density in the cerebral cortex (Stahl, 2008).

As described previously, the monoamine system is responsible for not only conventional neurotransmission via synapses, but also volume transmission without the involvement of synapses. Dopamine carries out neurotransmission in a wide area in the prefrontal area by spreading to the outside of synapses, as well as synaptic clefts, and it is involved in frontal cortex functions, such as selective attention at the dorsal anterior cingulate gyrus, sustained attention in the dorsal lateral prefrontal area, regulation of hyperkinesias in the premotor cortex, and impulsivity control in the orbitofrontal cortex.

4.2 Serotonergic neurotransmission in attention deficit hyperactivity disorder

4.2.1 Interaction between dopaminergic and serotonergic systems

Previous studies of the dopaminergic system in the mesocorticolimbic pathway in attention deficit hyperactivity disorder model rats have demonstrated a loss of function in 6-hydroxydopamine rats, a loss of function and gain of function in spontaneously hypertensive rats, and a gain of function in dopamine transporter knockout/knock-down mice. The anatomical characteristics of the serotonin neuron system suggest a mechanism for disinhibition of the dopaminergic system in association with loss of function in the serotonergic system. Therefore, it is highly likely that abnormalities not only in the dopaminergic system, but also in the serotonergic system are involved in the pathogenesis of attention deficit hyperactivity disorder.

When 6-hydroxydopamine is administered to neonatal rats, they show hyperkinesias for a specific period after birth. In this model, 6-hydroxydopamine destroyed dopaminergic fibers in the prefrontal area and striatum, and dysfunction of the dopaminergic system or the ensuing dopamine receptor hypersensitivity is considered the underlying mechanism for hyperkinesias. Serotonin fiber density is increased remarkably in the striatum of these 6-hydroxydopamine model rats and mice. It has been reported that the excessive control of the serotonergic system is involved in the inhibition of hyperkinesias by a central nervous system stimulant. The delicate balance between the dopaminergic and serotonergic systems is thought to be necessary for normal behavior and monoamines are likely involved in heterogeneity (sub-classification) of attention deficit hyperactivity disorder. In general, when this disorder in the dopaminergic system is predominant, destructive and aggressive behavior is less frequent, and hyperkinesias and attention deficit are the chief complaints of attention deficit hyperactivity disorder. However, when impairment in the serotonergic system is predominant, not only hyperkinesias and attention deficit, but also destructive and aggressive behavior are observed in attention deficit hyperactivity disorder.

4.2.2 Direct involvement of the serotonergic system in attention deficit hyperactivity disorder

Dopamine transporter knockout mice are known as an attention deficit hyperactivity disorder animal model. The administration of methylphenidate alleviates the hyperkinesias in dopamine transporter knockout mice. However, since dopamine transporter, a target of methylphenidate, is deficient in this mouse model, it is reasonable to conclude that the effect is exerted via other neuron systems. When a serotonin transporter inhibitor (a selective serotonin reuptake inhibitor) is administered in this mouse model, hyperkinesias are alleviated despite no change in extracellular dopamine concentrations (Gainetdinov, 1999). On the other hand, when a selective noradrenalin transporter inhibitor (a selective noradrenalin reuptake inhibitor) is administered, hyperkinesias are not alleviated, which suggests that serotonergic neurotransmission is involved in the functional mechanism of methylphenidate for the improvement in attention deficit hyperactivity disorder symptoms. The serotonergic system per se is directly involved in attention deficit hyperactivity disorder and is involved mostly in sustained attention, particularly in the dorsal lateral prefrontal area. Recently, serotonin at the dorsal striatum and prefrontal area has been suggested to play an important role in inhibiting impulsivity in decision-making and aggressiveness (Doya, 2008). Selective serotonin reuptake inhibitors improve attention deficit hyperactivity disorder symptoms via the 5-HT1A, 5-HT2A, and 5-HT2C receptors. Since 5-HT1B receptor-deficient mice show hyperkinesias and aggressiveness, the 5-HT1B receptor has attracted attention as a genetic factor of attention deficit hyperactivity disorder, and the relationship between the subtypes of attention deficit hyperactivity disorder and the attention deficit-predominant type has also been examined (Smoller, 2006).

5. Conclusions

Serotonin is the earliest developing neurotransmitter system in the mammalian brain, and eventually becomes the mostly widely distributed system in the brain, contacting most cells of the cortex. Thus, the serotonin neuron system develops early enough, and is sufficiently widespread that it can influence maturation of many other regions in the brain. The serotonin neuron system that develops at the early stage has a core influence on neural development, morphogenesis, and synapse formation and maintenance. The control of the serotonin neuron system over other monoamine neurons, in particular, dopamine neurons, has been demonstrated. These neuroanatomical findings are important for understanding the pathogenesis of developmental disorders.

In autism spectrum disorders, abnormalities in the brain stem-subcortical structure (caudate nucleus, putamen, and pallidum)-cerebellum network, as well as the hippocampus, piriform gyrus, and cingulate gyrus, plays a central role in the pathogenesis. Since most parts of the brain that have been reported to be responsible for autism spectrum disorders strongly receive projections from the serotonin neuron system, dysfunction of serotonergic neurotransmission is considered to be the core of the pathogenesis of autism spectrum disorders. Abnormalities in the serotonergic system are thought to cause disturbances in the sleep/arousal rhythm observed from the early stages of autism spectrum disorders, poor social skills, and adaptation disorder to a new environment. Dysfunction in the dopaminergic and noradrenergic systems in the neocortex-subcortical structure-cerebellum pathway is the core of the pathogenesis of attention deficit hyperactivity disorder, and the serotonin neuron system is related not only with the abnormality in the catecholaminergic system, but also directly with the pathogenesis of attention deficit hyperactivity disorder.

6. References

Adolphs, R., Baron-Cohen, S., Tranel, J. (2002) Impaired recognition of social emotions following amygdala damage, Cognitive. Neuroscience. Vol. 14 , No. 8, pp. 1264–1274, ISSN 1530-8898

Amaral, D.G., Schumann C.M., Nordahl. C.W. (2008) Neuroanatomy of autism. Trends in Neurosciences. Vol. 31, No.3, pp. 137-143, ISSN 0166-2236

Anderson, M. (2005) Serotonin in autism, In: The neurobiology of autism, Second Edition, Bauman L, and Kemper L, pp. 303-318, The Johns Hopkins University Press, ISBN 0-8018-8046-7, Baltimore, USA.

Betancur, C., Sakurai, T., Buxbaum, J.D. (2009) The emerging role of synaptic cell-adhesion pathways in the pathogenesis of autism spectrum disorders. Trends in Neurosciences. Vol. 32, No. 7, pp. 402-412, ISNS 0166-2236

Biederman, J. (2005) Attention-deficit/hyperactivity disorder: A selective overview. Biological Psychiatry. Vol. 57, No.11, pp. 1215-1220, ISNS 0006-3223

Björklund. A., Dunnett S. (2007) Dopamine neuron systems in the brain: an update. Trends in Neurosciences, Vol. 30, No. 5,194-202, ISSN 0166-2236

Bunin, M.A., Wightman R. (1999) Paracrine neurotransmission in the CNS: involvement of 5-HT. Trends in Neurosciences, Vol. 22, No. 9, pp. 377-382, ISSN 0166-2236

Chugani, D.C. , Muzik, O., Behen, M., R. Rothermel, R., Janisse, J.J. , Lee J. Chugani, H.T. (1999) Developmental changes in brain serotonin synthesis capacity in autistic and nonautistic children, Annals of Neurology. Vol. 45, No. 3, pp. 287–295. ISNS 1531-8249

Chugani, C. (2002) Role of altered brain serotonin mechanisms in autism. Molecular Psychiatry, Vol. 7, S16-17, ISSN 1359-4184

Dean, C., Dresbach, T. (2006) Neuroligins and neurexins: linking cell adhesion, synapse formation and cognitive function. Trends in Neurosciences. Vol. 29, No. 1, pp. 21-29, ISSN 0166-2236

Doya, K. (2008) Modulators of decision making. Nature Neuroscience; Vol. 11, No , pp. 410-416, ISSN 1097-6256

Erzurumlu, R.S., Kind P.C. (2001) Neural activity: sculptor of barrels in the neocortex. Trends in Neurosciences, Vol. 24, No.10, pp. 589-596, ISSN 0166-2236

Gainetdinov, R., (1999) Role of serotonin in the paradoxical calming effect of psychostimurants on hyperactivity. Science, Vol. 15, No. 397-400, ISSN 0036-8075

Ishimura, A., Takeuchi, Y., Fujiwara, K., Sawada, T., Kusunoki, T. (1989) Effects of undernutrition on the serotonin neuron system in the developing brain: an immunohistochemical study. Developmental Brain Research, Vol. 50, No. 2, pp 225-231, ISSN 0165-3806

Janusonis, S., Gluncic, V., Rakic, P. (2004) Early serotonergic projections to Cajal-Retzius cells: relevance for cortical development, Journal of Neuroscience. Vol. 24, No. 7, pp. 1652–1659, ISSN 0270-6474

Johnston, M., Blue M. (2006) Neurobiology of Autism. In: Autism; A neurobiological disorder of early brain development. International Review of Child Neurology Series, T. Tuchman, I. Rapin, (Eds.) 80-92, Mac Keith Press, ISBN: 1-8986683-49-2, London, England.

Kirsch P., Esslinger, C., Chen, Q., Mier, D., Lis, S., Siddhant, S., Gruppe, H., Mattay, V.S., Gallhofer, B., Meyer-Lindenberg, A. (2005) Oxytocin modulates neural circuitry for social cognition and fear in humans. Journal of Neuroscience. Vol. 25, No.49, pp. 11489-11493. ISSN 0270-6474

Kohyama, J. (2011) Sleep health and asynchronization. Brain and Development, Vol. 33, No. 3, pp. 252-259, ISSN 0387-7604

Kondoh, M., Shiga. M., Okado, N. (2004) Regulation of dendrite formation of Purkinje cells by serotonin through serotonin1A and serotonin2A receptors in culture, Neuroscience Research. Vol. 48, No. 1, pp. 101–109, ISSN 0168-0102

Lauder, JM., Krebs, H. (1978) Serotonin as a differentiation signal in early neurogenesis, Developmental Neuroscience Vol. 1, No.1, pp. 15–30. ISSN 0378-5866

Lauder, JM.(1990) Ontogeny of the serotonergic system in the rat: serotonin as a developmental signal, Annals of the New York Academy of Sciences Vol. 600, No.1, pp. 297–313. ISSN 1749-6632

Nieuwenhuys R. (1985) Chemoarchtecture of the brain. Springer-Verlag, ISBN 3-540-15349-7, Berlin

Nishimura, A., Ueda, S., Takeuchi, Y., Sawada, T., Kawata, M. (1995) Age-related decrease of serotonergic fibers and S-100B immunoreactivity in the rat dentate gyrus. NeuroReport. Vol. 6, No.10, pp. 1317-1320. ISSN 0959-4965

Persico, A.M., Bourgeron, T. (2006): Searching for ways out of the autism maze: genetic, epigenetic and environmental clues. Trends in Neurosciences, Vol. 29, No. 7, pp. 349-358, ISSN 0166-2236

Segawa, M., Nomura, Y. (2006): Pathophysiology of Autism. In: Autism; A neurobiological disorder of early brain development. International Review of Child Neurology Series. Tuchman, T., Rapin, I. (eds) 248-264, Mac Keith Press, ISBN 1-8986683-49-2, London.

Smoller, J.W., Biederman, J., Arbeitman, L. (2006) Association between the 5HT1B receptor gene (HTR1B) and the inattentive subtype of ADHD. Biological Psychiatry. Vol. 59, No. 5, pp. 460-467, ISSN 0006-3223

Spencer, T. Biederman, J., Madras B.K. Dougherty, D.D., Bonab, A.A., Livni, E., Meltzer, P.C., Martin,J., Rauch, S., Fischman, A.J. (2007) Further evidence of dopamine transporter dysregulation in ADHD: A controlled PET imaging study using altropane. Biological Psychiatry, Vol. 62, No.1, pp. 1059-1061, ISSN 0006-3223

Stahl, SM. (2008) Stahl's Essential Psychopharmacology, 3rd ed, Cambridge University Press, ISBN 978-0-521-85702-4, New York

Takeuchi, Y. (1988) Distribution of serotonin neurons in the central nervous system. In; Neuronal Serotonin, Osborne NN, Haman M (eds). pp. 25-56, John Wiley & Sons, ISBN 0-471-91154-2, Chichester

Verney, C., Lebrand, C., Gaspar, P. (2002) Changing distribution of monoaminergic markers in the developing humancerebral cortex with special emphasis on the serotonin transporter, Anatomical Record. Vol. 267, No.2, pp. 87–93. ISSN 1932-8494

Whitaker-Azmitia, PM. (2005) Behavioral and cellular consequences of increasing serotonergic activity during brain development: a role in autism? International Journal of Developmental Neuroscience, Vol. 23, No.1. pp. 73-83, ISSN 0736-5748

Yamazoe, Y., Takeuchi, Y., Matsushita, H., Kawano, H., Sawada, T. (2001) Serotonergic heterotypic sprouting in unilaterally dopamine-depleted mouse striatum. Developmental Neuroscience, Vol. 23, No.1, pp 78-83, ISSN 0378-5866

6

Adenosine and Autism - Recent Research and a New Perspective

Susan A. Masino[1,2], Julia Svedova[1], Masahito Kawamura, Jr.[3], Francis D. DiMario, Jr.[4,5] and Inge-Marie Eigsti[6]
[1]Neuroscience, Trinity College, Hartford, CT,
[2]Psychology, Trinity College, Hartford, CT,
[3]Pharmacology, Jikei University, Tokyo,
[4]Pediatric Neurology, Connecticut Children's Medical Center, Hartford, CT,
[5]Pediatrics, University of Connecticut Health Center, Farmington, CT,
[6]Psychology, University of Connecticut, Storrs, CT,
[1,2,4,5,6]USA
[3]Japan

1. Introduction

Autism Spectrum Disorders (ASD) are associated with atypical social, behavioral and physiological characteristics. Here we outline an emerging connection among the increased incidence of epilepsy, disrupted sleep and perseverative behaviors exhibited and sought by persons with autism. Specifically, we propose that persons with autism can benefit from increased levels of adenosine, a powerful inhibitory neuromodulator and the core molecule of adenosine triphosphate (ATP). We review the literature and present recent data obtained via a customized questionnaire administered to parents of children with a confirmed autism diagnosis. This customized questionnaire demonstrates that symptoms of autism are reduced subsequent to stimuli predicted to increase adenosine. In addition, we present evidence from the literature and pilot data from a retrospective study of children with epilepsy or epilepsy and autistic behavior who were treated with a ketogenic diet, a long-established anticonvulsant therapy that recently has been shown to suppress seizures via the adenosine A_1 receptor (A_1R) subtype. Our discussion focuses on the actions of adenosine in the central nervous system, with multiple implications for ASD, and the potential for developing new evidence-based therapies. Taken together, published peer-reviewed research and recent preliminary research suggest that adenosine could help resolve multiple physiological and behavioral symptoms of ASD.

2. Adenosine - role in physiology and behavior

Adenosine is a purine molecule present throughout the body. In the central nervous system it is increasingly appreciated as a homeostatic bioenergetic network regulator (Boison et al., 2011). Fundamental to its homeostatic influences on neuronal networks are adenosine's integral roles in metabolism and cell signaling, and its regulatory influence as an obligate by-product of transmethylation reactions. In terms of metabolism, adenosine is the core of arguably the most important purine, adenosine triphosphate (ATP), the pre-eminent cell energy molecule, and also of related adenine nucleotides adenosine diphosphate (ADP) and adenosine monophosphate (AMP); these purines determine overall cell energy charge (Dunwiddie & Masino, 2001). In terms of signaling, adenosine binds to a family of membrane-bound G protein-coupled cell-surface receptors (A_1, A_{2A}, A_{2B}, A_3) to influence membrane potential, synaptic transmission and second messenger signaling (Ralevic & Burnstock, 1998; Dunwiddie & Masino, 2001). Its ubiquitous presence in every cell, widespread receptor distribution, and tonic levels throughout the extracellular space - sufficient to activate a subset of these receptors - make adenosine a key player in neuron-glia interactions (Fields & Burnstock, 2006), and able to exert a dynamic and broad regulatory influence.

Adenosine A_1 receptors (A_1Rs) and adenosine A_{2A} receptors (A_{2A}Rs) have the highest affinity for adenosine and the most predominant ongoing functional effects in the central nervous system (Fredholm et al., 2005). Generally, A_1Rs serve to decrease synaptic transmission and decrease cAMP, whereas A_{2A}Rs facilitate transmission and increase cAMP (Schulte & Fredholm, 2003). A_1Rs are expressed widely throughout the brain (Goodman et al., 1982), and A_{2A}Rs are highly concentrated in the basal ganglia and olfactory tubercle (Rosin et al., 2003). Thus, many effects of adenosine are location- and subtype-specific. These two receptor subtypes are both targeted by caffeine, a non-selective A_1R/A_2R antagonist and the most widely used psychoactive drug world-wide (Fredholm et al., 1999).

A_1Rs exert a tonic inhibition on excitatory transmission via pre- and postsynaptic actions (Ralevic & Burnstock, 1998; Dunwiddie & Masino, 2001). The inhibitory effects of A_1Rs are well-known to prevent and stop seizures (Dunwiddie & Worth, 1982), reduce anxiety (Florio et al., 1998; Johansson et al., 2001), and promote neuronal survival during times of severe cell stress (Fredholm, 1997). Blocking the ongoing influence of A_1Rs can precipitate or prolong seizures and reduce neuronal survival, and enhancing their actions offers seizure protection (Dunwiddie, 1999) and neuroprotection (Fredholm, 1997). Events such as strokes, seizures, or other metabolic dysfunctions are associated with a net dephosphorylation of ATP and a large increase in extracellular adenosine (Latini & Pedata, 2001) and its inhibitory influence via A_1Rs can reduce excitotoxicity and short-circuit further metabolic demand. Thus adenosine was initially regarded as a "retaliatory metabolite," as it acts to maintain metabolic and network homeostasis during cell stress (Newby, 1984).

More recently we have recognized that the homeostatic impact of adenosine is regulated dynamically under non-pathological conditions, and regulated largely by astrocytes (Halassa et al., 2009). ATP and adenosine are released directly from astrocytes (Pascual et al., 2005), and ATP is dephosphorylated rapidly to adenosine in the extracellular space (Dunwiddie et al., 1997; Cunha et al., 1998); the level of extracellular adenosine is regulated primarily by intracellular astrocyte-based adenosine kinase (Gouder et al., 2004; Boison, 2006).

3. Purines and autism

Adenosine affects general arousal, and increased adenosine at A_1Rs reduces anxiety; conversely, decreased A_1R activation can increase anxiety (Florio et al., 1998; Johansson et al., 2001). Both A_1Rs and A_{2A}Rs may be involved in adenosine's ability to promote sleep

(Rainnie et al., 1994; Porkka-Heiskanen, 1999; Huang et al., 2005). Notably, increased seizure propensity, disturbed sleep and anxiety are all found commonly in persons with ASD (Malow, 2004; Polimeni et al., 2005; Oswald & Sonenklar, 2007), suggesting that an increased influence of adenosine could reduce all three co-morbidities.

In dopamine-containing areas, A_{2A}Rs are expressed at higher levels than in other regions, and A_{2A}Rs have an opposing relationship with dopamine D_2 receptors. Increased adenosine receptor activation reduces D_2 receptor activation (Fuxe et al., 2007), and the two receptors form a functional heteromer (Ferre et al., 2007). This complex also seems to be under the control of A_1Rs, and adenosine agonists have been shown to reduce repetitive behaviors (Poleszak & Malec, 2000; Tanimura et al., 2010). Therefore, in addition to reducing general excitability through increased A_1R activation, adenosine can influence dopamine-related behaviors through multiple adenosine receptor subtypes and reduce behavioral pathologies - such as repetitive behaviors, which are particularly relevant for ASDs - through these interactions. At this time the research evidence linking adenosine to repetitive behaviors is growing, but not as strong as the link between adenosine and epilepsy. It is important to note that dopamine is involved in many aspects of behavior, and increasing adenosine in dopamine-containing areas could impact motivation (Salamone & Correa, 2009); the full complement of positive and negative effects and their interactions in ASD would be difficult to predict.

Abnormalities in purine metabolism have been observed in autism (Nyhan et al., 1969; Page & Coleman, 2000; Bottini et al., 2001; Marie et al., 2002; Page & Moseley, 2002), but have not been linked directly to core symptoms of the disorder or receptor-mediated effects on neuronal function, and purinergic strategies have not been exploited for practical clinical benefits. Whereas some symptoms and behaviors are likely due to aberrant neuroanatomical development and other genetically-determined substrates, increasing the inhibitory influence of adenosine could help significantly with multiple behavioral and physiological sequelae.

3.1 Symptoms of autism and regulation of adenosine

Recent work suggests adenosine dysregulation may be associated with psychiatric disorders such as schizophrenia (Boison et al., 2011). In general it is not surprising that a brain-wide metabolic/neuronal regulator such as adenosine could influence multiple co-morbidities associated with complex psychiatric disorders. The co-existence of neurological/psychiatric disorders is also becoming more appreciated, and observed in adults as well as children and adolescents (Jones et al., 2008). As just a few examples, there are well-documented comorbidities between epilepsy and depression/anxiety (Kanner, 2004, 2008; LaFrance et al., 2008; Ekinci et al., 2009), schizophrenia (Hyde & Weinberger, 1997), and, more broadly, sleep disorders and psychiatric disorders (Reeves et al., 2010), including autism (Spence & Schneider, 2009; Jeste, 2011). As noted above, based on behavioral and physiological characteristics of ASD, including impaired sleep (Malow, 2004; Polimeni et al., 2005), increased seizures (Malow, 2004; Canitano, 2007; Spence & Schneider, 2009), anxiety (Chalfant et al., 2007), and perseverative behaviors (Ridley, 1994; Liss et al., 2006), an insufficient influence of adenosine might underlie some symptoms. These parallels are summarized in **Figure 1**. Alternatively, even if levels are normal, increased adenosine may still be beneficial, offering short term improvements and potentially facilitating more long-term changes.

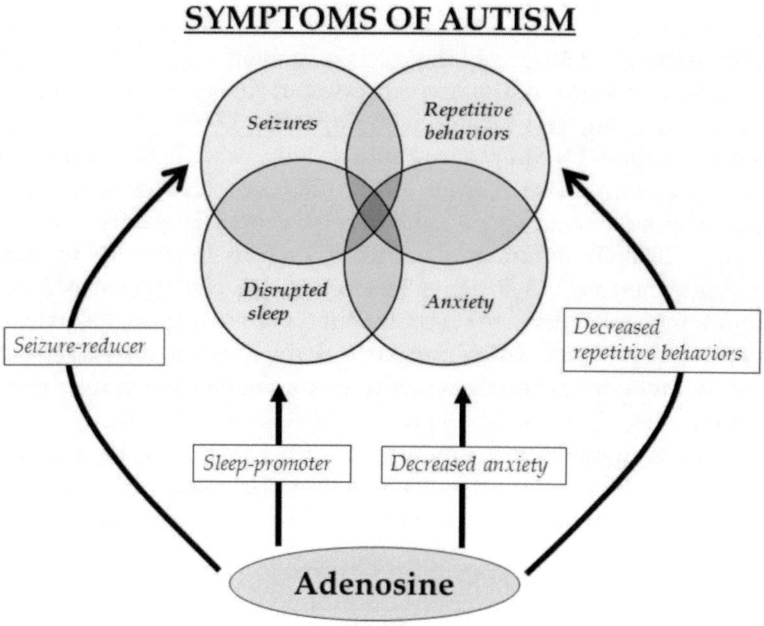

Fig. 1. Key actions of adenosine in the central nervous system can target multiple symptoms of autism. Increased adenosine at the cellular level alters physiology and behavior (promoting sleep, and decreasing seizures, anxiety and repetitive behaviors). This cohort of outcomes can target overlapping co-morbidities and symptoms of autism (sleep disruption and increased seizure propensity, repetitive behaviors and anxiety).

Research has shown that diverse stimuli such as mechanical pressure or sudden physical impact (Franke et al., 2006), seizures (Whitcomb et al., 1990), intense exercise (Dworak et al., 2007), increased temperature (Masino & Dunwiddie, 1999), decreased pH (Dulla et al., 2005; Dulla et al., 2009) and reduced glucose (Kawamura et al., 2010) can all increase brain adenosine directly (or indirectly via ATP dephosphorylation: Dunwiddie et al., 1997; Cunha, 2001; Zhang et al., 2003; Pascual et al., 2005) and thus modulate synaptic transmission within minutes. Most of these effects appear to reverse rapidly; others are relatively long-lasting (up to hours) or related to chronic changes in physiology. For example, acupuncture has been shown to increase local adenosine, perhaps related to the mechanical stimulation (Franke et al., 2006; Goldman et al., 2010). Other recent work suggests that a ketogenic diet, an effective treatment for pediatric epilepsy, might act via A_1Rs (Masino & Geiger, 2008; Kawamura et al., 2010; Masino et al., 2011b). Unfortunately, it is impractical to measure brain adenosine in humans (necessarily an invasive procedure), and adenosine levels outside the central nervous system (e.g., in plasma or urine) are not informative.

In compiling a list of stimuli known to increase adenosine it becomes apparent that many behaviors exhibited by persons with ASDs or shown to improve symptoms could be related to one of these physiological changes – e.g. mechanical pressure, including rocking or spinning and Grandin's "hug machine" (Edelson et al., 1999; Escalona et al., 2001), fever

(Curran et al., 2007), etc. Therefore, some behaviors expressed by persons with autism could be attempts at metabolic and neuronal homeostasis via increased adenosine. Engaging in adenosine-increasing activities could thus lead immediately to decreased symptoms of autism.

Ongoing work testing the relationship between adenosine and autism is proceeding in animal models. However, because no animal model represents the human condition entirely (which itself is a broad spectrum of complex behaviors), we felt it was important to test initially for evidence for a link between adenosine and ASD in humans. As noted, direct tests in humans to assay or manipulate adenosine in a meaningful way would involve highly invasive measurements or manipulations, or, potentially, drugs which are not approved in children. Initial work, described below, and coupled with evidence in animals, will be building blocks for pursuing this further.

3.2 Predictions and outcomes - adenosine and autism spectrum disorders symptomatology

Based on animal research suggesting non-pathological stimuli that could increase adenosine, we explored the possibility of a beneficial relationship between autism and adenosine via a customized behavioral questionnaire focused on behaviors typical of ASD (Masino et al., 2011a). Subjects were recruited with the assistance of the Interactive Autism Network (IAN) Research Database at the Kennedy Krieger Institute and Johns Hopkins Medicine – Baltimore, sponsored by the Autism Speaks Foundation. Participants had received a formal autism diagnosis according to DSM-IV criteria made by a qualified professional (e.g., psychologist, psychiatrist) and were above a threshold score of 12 on the Social Communication Questionnaire (SCQ) parent-report measure (Rutter et al., 2003). Autism diagnosis was confirmed by IAN staff; fully 98% of participants ascertained as on the spectrum according to standard IAN phenotyping procedures were ASD-positive according to clinicians' best estimate (Lee et al., 2010).

Along with basic information such as age, diagnosis and verbal level, we queried behavioral changes following activities identified as expected to increase, decrease or have no effect on adenosine. Our final data set included responses from 155 parents of children with a confirmed ASD diagnosis. Parents were naive to all study hypotheses.

Results from this study revealed a significant relationship between engaging in stimuli pre-assigned as expected to increase adenosine and parental report of decreased severity of ASD symptoms (Masino et al., 2011a). We found a trend for persons diagnosed with Asperger's to show a stronger behavioral effect of adenosine-increasing activities, a finding which deserves additional exploration. In general, the significant effects of adenosine-increasing activities were unrelated to participant age, gender, or verbal level. These results suggest that, if validated, increasing adenosine could have broad applicability to ASD.

Interestingly, we found a significant relationship between reported caffeine intake and reduced symptoms of autism. Because the dose of caffeine was not controlled, and because the effect of caffeine on the nervous system is determined by both the dose and the frequency, it is not possible to make strong interpretations based on this finding. However, akin to the trend for more beneficial effects on persons with Asperger's, the relationship between caffeine and autism deserves further exploration; along these lines, recently others have also posited a relationship between caffeine and autism (Ghanizadeh, 2010). Because of its well-established safety profile and limited side effects, vast epidemiological database, and known actions in the nervous system, caffeine is being considered as a low-cost adjuvant or for its own therapeutic potential in multiple neurological disorders (Arendash & Cao, 2010; Cunha & Agostino, 2010; Prediger, 2010; Chen & Chern, 2011).

4. Metabolic regulation of adenosine - ketogenic diet

The ketogenic diet is a restrictive diet high in fat and low in carbohydrate that significantly reduces the frequency of seizures. Epilepsy is characterized as a disorder of unprovoked spontaneous seizures, which affects approximately 1% to 3% of population (Shneker & Fountain, 2003); rates of epilepsy in ASD are substantially higher – with estimates ranging from 5% up to 40% (Canitano, 2007). The most common treatment for epilepsy is antiepileptic drug therapy; the ketogenic diet is a metabolic treatment whereby limiting carbohydrates restricts available glucose and initiates ketone-based (ketogenic) metabolism. This dietary approach forces the use of ketones for energy, a metabolic shift which also occurs during fasting (Aoki, 1981; Hartman & Vining, 2007). Although the exact neural mechanisms of the ketogenic diet are currently unknown (although under active investigation by a number of laboratories), its efficacy in reduction of seizures has been described extensively. In particular, studies have demonstrated that the ketogenic diet can decrease seizure frequency by more than 50% in one-third to one-half of children, and eliminate seizures in 10-15% of children in 6-12 months (Vining et al., 1998; Keene, 2006).

Despite its high success rate, the ketogenic diet is typically offered very infrequently. Notable exceptions include proper diagnosis of a very few metabolic conditions where it is becoming known as the preferred therapy, and others where it is gaining ground as a potential first-line therapy (e.g. GLUT-1 deficiency syndrome). In parallel, it is specifically contraindicated for a small set of metabolic conditions (Kossoff et al., 2009). When a ketogenic diet is offered, it is typically a "last resort" therapy - after attempting at least two (and often more) antiepileptic drugs. The major problems associated with the diet are the inconvenience - preparation of special meals – and negative side effects in a subset of patients including vomiting, weight loss, increase in serum lipids, acidosis, gastric problems or renal stones (Groesbeck et al., 2006; Keene, 2006; Freeman et al., 2007). Although most of these are rare or short-term, the diet is usually only prescribed to patients with multiple seizures per day, unacceptable drug-related side effects, or seizures that are not sufficiently responsive to anti-epileptic drugs.

Despite a clear link between energy metabolism and brain activity established by historical observations and clinical and basic research, the first randomized, prospective, clinically controlled study using a ketogenic diet has been published only recently (Neal et al., 2008). At last, this much-needed clinical research report proved (at a higher scientific standard – prospective, randomized) what had been reported in many other research studies and observed clinically for many decades: the ketogenic diet is an effective treatment for pediatric epilepsy, and it can be successful even in cases of medically-intractable epilepsy (Kinsman et al., 1992; Vining et al., 1998). In this study, the number of seizures observed in children placed on the diet decreased by 38% in three months. In contrast, the number of seizures observed in a matched group of children who continued their current treatment increased by 37% in 3 months. In some children who started the ketogenic diet seizures stopped entirely, whereas this was never observed in the control group. This report and an accompanying commentary

(Wiznitzer, 2008) note that the critical mechanisms underlying the success of the diet remain unknown, and this knowledge gap has thus far stymied efforts to develop treatments based on the metabolic changes precipitated by the ketogenic diet.

Recent research suggests that the ketogenic diet may act by increasing the influence of adenosine acting at A_1Rs (Masino & Geiger, 2008; Kawamura et al., 2010; Masino et al., 2011b). Other research suggests that the ketogenic diet acts by reducing reactive oxygen species and oxidative stress (Sullivan et al., 2004; Davis et al., 2008), direct actions of ketone bodies on ion channels (Ma et al., 2007) or transporters (Juge et al., 2010), or increased GABA (Yudkoff et al., 2007). Any of these mechanisms could be highly beneficial in ASD, which is associated with disinhibition (Hussman, 2001; Fatemi et al., 2009a,b) and oxidative stress (James et al., 2004; McGinnis, 2005; James et al., 2006). While the link between the efficacy of the ketogenic diet and increased adenosine is still emerging, there are at least 3 strong reasons why - based on well-established evidence - a ketogenic diet could be helpful in autism:

1. The ketogenic diet treats epilepsy. A high percentage of children with autism have epilepsy. Importantly, these numbers might be higher with the routine ability to measure electrographic (non-behavioral) seizures and altered subcortical activity.
2. The ketogenic diet reduces reactive oxygen species and oxidative stress. Autism is associated with increased oxidative stress.
3. Most importantly, an established, medically-supervised diet-based strategy is cost-effective and available to translate immediately into practical benefits.

It is important to note that both adenosine and a ketogenic diet can stop drug-resistant epilepsy, suggesting mechanisms other than those targeted by existing drugs. Both mobilize a broad homeostatic bioenergetic regulation (Boison et al., 2011) and thus may be best suited to address a broad spectrum of symptoms with limited side effects.

Since we published a hypothesis that a key therapeutic aspect of a ketogenic diet is to increase adenosine acting via A_1Rs (Masino & Geiger, 2008) this hypothesis has been tested in three ways: (1) we explored an in vitro model of ketogenic diet electrophysiologically in individual neurons, (2) we administered a ketogenic diet in vivo to three types of transgenic mice, all with spontaneous electrographic seizures due to decreased A_1R signaling, and (3) we tested the ability of a ketogenic diet to reduce pain and inflammation, a prediction of this hypothesis. Respectively, we found that (1) metabolic conditions designed to mimic a ketogenic diet in brain slices mobilized a novel autocrine regulation of neuronal activity via adenosine acting at A_1Rs (Kawamura et al., 2010), (2) a ketogenic diet reduced electrographic seizures in mice with intact A_1Rs, but did not in mice lacking A_1Rs (Masino et al., 2011b), and (3) a ketogenic diet reduced pain and inflammation, as predicted (Ruskin et al., 2009).

While data continue to emerge, and multiple mechanisms could play a role, together these studies suggest that a ketogenic diet reduces seizures by increasing adenosine-mediated inhibition and offer insight into therapies for other clinical conditions where adenosine is known or hypothesized to offer clinical benefits. Using a retrospective analysis, we explored the effects of the ketogenic diet on behavior and temperament of children with epilepsy with or without hallmark symptoms of autism.

4.1 Ketogenic diet - retrospective analysis in children

To evaluate the effects of a ketogenic diet on symptoms of autism, alongside changes in seizure frequency, behavior and mood, we identified children who began dietary therapy because of their epilepsy diagnosis. Based on decades of research and consistent clinical findings, we expected diet therapy to reduce seizures significantly in the majority of these children. At the same time, because of the high incidence of epilepsy in children with autism, a subset of these children with pediatric epilepsy exhibited co-morbid symptoms of ASD. Thus, we were able to assess changes in behavior in children with or without symptoms of ASD in parallel with other physiological and neurological outcomes of diet therapy. In this way we examined the outcomes based on treatment with a standard diet protocol, available currently to any clinician and in use at many centers nationally and globally.

Using a retrospective analysis we examined clinical and behavioral features of children with epilepsy between 18-months and 12-years of age who initiated the ketogenic diet at Connecticut Children's Medical Center between January 2004 and January 2009. All aspects of the study were approved by the Institutional Review Board. Potential participants in this retrospective study were identified using the ketogenic diet database at the Connecticut Children's Medical Center. Because some measures used in this study are only available currently in English, children with non-English speaking parents were excluded.

Parents of selected children were invited to participate in the study with a phone call. After obtaining a verbal consent, questionnaires were mailed to the parents. The participating children, whose parents filled out the questionnaires, were divided into two groups: subjects currently on the ketogenic diet (KD group) and subjects who discontinued the ketogenic diet (non-KD group). Children in the KD group had stayed on the diet for at least 6 months, and children in the non-KD group stopped the diet at least 2 months before the testing. Demographic and clinical data were collected from chart review.

Statistical analyses (Student *t*-test, Pearson test, and chi-square test) were performed to compare the differences between the KD group and the non-KD group. To dissociate differences related to either autism or cognitive impairment, subjects who were assessed as autistic based on their Child Autism Rating Scale score (filled out by parent) and subjects who were evaluated as cognitively impaired by a neurologist were analyzed separately.

In this initial retrospective analysis there was no difference in the number of boys versus girls (10 each) and no substantial difference in the number of children who were still on the ketogenic diet versus those who continued the diet (11 and 9, respectively) or the average age of the KD versus non-KD group (6.4±0.8 and 8.3±1.1, respectively, P=0.18). We expected to find a significant different in seizure frequency, and although there was a trend in the average number of seizures per week in each group (KD group, 16±8 seizures per week; non-KD group, 99±52 seizures per week) there was high variability (and thus the lack of significance) due primarily to two clinically-expected outcomes: (1) some children in the non-KD group became seizure-free after being on the diet, and went off the diet with a current score of zero seizures per week; (2) children who did not comply and stay on the diet, or for whom the diet didn't work, had more than 100 seizures per week. Both of these types of cases were included in the non-KD group. These data are summarized in **Table 1.**

	KD Group	Non-KD group	P value
Number	11 (6 females)	9 (4 females)	1.0
Age	6.4±0.8	8.3±1.1	0.18
Number of Seizures/week	16±8	99±52	0.10

Table 1. Summary statistics of children who were on the ketogenic diet (KD-group) or children who started but were not currently on the ketogenic diet (non-KD group) at Connecticut Children's Medical Center from January 2004-January 2009. See text for additional details.

We also considered whether the children were autistic only, cognitively impaired only, or autistic and cognitively impaired. There was no difference in the number of children who were autistic and cognitively impaired in each group (KD group: 5/11; non-KD group: 4/9); the numbers in other categories were too small for statistical comparison.

Statistical analysis of data from the Child Behavior Checklist showed that subjects who were still on the ketogenic diet (KD group, n=11) had significantly fewer behavioral problems than subjects who were no longer on the ketogenic diet at the time of testing (non-KD group, n=9; χ^2 *(1)*, P<0.05). When analyzing separately cognitively impaired children and autistic-like children, we found that in both of these subgroups, the subjects who were currently on the ketogenic diet (KD group) had better behavioral traits than subjects who had already stopped ketogenic diet treatment (non-KD group); P<0.001 and P<0.01, respectively; (Svedova et al., 2010). Effects of the ketogenic diet on behavior of children will continue in ongoing retrospective and prospective studies.

Although these data are preliminary, they add to positive outcomes discussed below in previous prospective studies (Pulsifer et al., 2001; Evangeliou et al., 2003) and anecdotal reports. Along with evidence that a ketogenic diet increases A_1R activation, evidence that A_1Rs decrease anxiety, and evidence that either a ketogenic diet or A_1R activation decreases seizures, a proven metabolic approach such as a ketogenic diet should be considered for ASDs.

4.2 Ketogenic diet, adenosine and autism

Previous studies have shown that autism is caused by one or more gene defects, which can affect brain development, cell signaling, cell transport or structure (Santangelo & Tsatsanis, 2005; Zhao et al., 2007). It has been suggested that some of these gene defects may result in metabolic dysfunctions that can eventually cause behavioral abnormalities associated with autism, and several metabolic disorders present symptoms similar to autism, underscoring a link between autism and metabolism (Zecavati & Spence, 2009).

Based on these findings, there have been numerous attempts to treat autism with different types of restrictive diets, and dietary therapies remain popular in the autism community. Here we have focused our discussion primarily on the ketogenic diet, based on (1) the proven clinical success of the ketogenic diet for epilepsy, and preliminary positive results with autism, (2) in vivo and in vitro research evidence linking the ketogenic diet to adenosine, and (3) a recent retrospective study of the classic ketogenic diet in children with epilepsy and autism, discussed above.

In general, gastrointestinal problems and metabolic disturbances appear to be closely related to ASD, and there are several known metabolic impairments that are associated with autistic symptoms. These include phenylketonuria, adenylsuccinate lyase deficiency, histidinemia, or hyperuricosuric autism (Page & Coleman, 2000). The prevalence of food allergies (Gurney et al., 2006) and the incidence of at least 25-30% of children with autism suffering from chronic diarrhea, constipation or food issues (Page & Coleman, 2000; Ibrahim et al., 2009), suggest that some types of ASD may be treated with dietary restrictions. In the past few decades, several diets have been described in terms of their effects on symptoms of autism, including gluten-free, casein-free (GFCF) diet, specific carbohydrate diet, body ecology diet, or ketogenic diet (Srinivasan, 2009). A survey by Witwer and Lecavalier (2005) estimated that 15.5% of autistic children or adolescents were on modified diets for treatment of autism. It should be noted that that in several studies, including a double-blind placebo-controlled intervention, the GFCF diet was found to have no benefits for ASD (Elder et al., 2006; Millward et al., 2008), and has been associated with protein malnutrition (Arnold et al., 2003). Thus, the specificity of GFCF approaches to symptoms of ASD should be viewed with caution.

Direct evidence supporting a beneficial relationship between a ketogenic diet and autism comes from one published report - a prospective study - which tested the effects of a ketogenic diet in children with autism (Evangeliou et al., 2003). However, there are multiple unpublished anecdotal reports (Beth Zupec-Kania, personal communication), including reports of dramatic success with a ketogenic diet in autism. In the prospective study, Evangeliou et al. applied a ketogenic diet protocol to children with autism and found that 60% of diet-compliant children aged 4-10 who participated for 6 months showed an improvement in their symptoms. Significant improvement (> 12 points on the Childhood Autism Rating Scale) was recorded in 2/18 patients, average improvement (> 8-12 points) in 8/18 and minor improvement (> 2-8 points) recorded in 8/18. Notably, the diet protocol included 4 weeks on and 2 weeks off the ketogenic diet - a different regimen than administered typically for treatment of epilepsy. Still, these results are promising and provide preliminary data supporting a link between a ketogenic diet and autism, and potentially among autism, adenosine and a ketogenic diet. The biggest improvements were quantified in those patients who showed only mild autistic behavior, similar to our findings for a trend for the biggest benefit for adenosine-increasing activities in individuals with Asperger's (Masino et al., 2011a).

In addition to a reduction of their seizures, children on the ketogenic diet can enjoy other benefits - including reduced drug intake and general improvements in cognition, alertness, development, attention and social life (Kinsman et al., 1992; Sirven et al., 1999; Pulsifer et al., 2001; Hallböök et al., 2007; Hallböök et al., 2007), and - as we found in our retrospective study - fewer behavioral problems (Svedova et al., 2010). However, the number of studies researching the effects of the ketogenic diet on behavior and cognition of children remains surprisingly limited. A single prospective study conducted by Pulsifer et al. (2001) focused on the effects of the ketogenic diet on behavior and development of children with intractable epilepsy. The children were evaluated at the diet initiation and after one year using three parental report measures: Developmental Profile-2nd edition, Child Behavior Checklist, and Parenting Stress Index-Short Form. The results demonstrated an overall significant improvement in developmental functioning. In addition, the Child Behavior Checklist showed a significant improvement in attention and social problems. Interestingly, the authors commented that the developmental and behavioral improvements were not statistically related to better seizure control or to reduced intake of antiepileptic drugs.

A recent study examining the efficacy of writing parental goal letters prior to initiating the ketogenic diet showed that improvements in cognition and alertness were stated as the third most common goal (Farasat et al., 2006). Interestingly, the study suggested that cognition and alertness improvements were more likely to lead to long-term adherence to the diet – more so than improved seizure control or a decreased intake of antiepileptic drugs. From this observation, it is evident that potential improvements in cognition and behavior are crucial for many parents. Unfortunately, as noted above, and despite many decades of clinical use, there is a limited literature exploring the effects of the ketogenic diet on behavior and cognition of children. However, a wealth of resources about the ketogenic diet can be found on-line at the Charlie Foundation (www.charliefoundation.org).

5. Conclusions

Here we propose adenosine as a beneficial neuromodulator for ASD symptoms based on published research, behavioral and physiological symptoms of autism, and recent indirect evidence obtained in children (parent report via a customized adenosine questionnaire and retrospective analysis of children with epilepsy and with or without hallmark symptoms of autism). Obtaining direct evidence (measuring adenosine accurately in humans) is too invasive, and drugs that may increase adenosine have not been specifically established as such, would need to be used "off-label," and are not approved for use in children. Gathering correlated evidence regarding changes in adenosine in animal models can proceed in parallel with the type of "proof-of-principle" work gathered in humans and presented herein.

Beyond clinical endpoints indicating that increased adenosine would be beneficial in autism - e.g., reduced seizures - adenosine would be predicted to be increased by the atypical behaviors sought by persons with autism and offer short term benefits. Similarly, a ketogenic diet would be predicted to reduce symptoms of autism. Our customized questionnaire demonstrated highly significant changes - ASD symptoms were reduced after stimuli predicted to increase adenosine. These initial results should be followed up with more detailed controlled and prospective studies, and include stimuli that have been verified to increase adenosine. In this respect, two conditions that might increase adenosine (in different ways) include acupuncture and a ketogenic diet. Published research (Evangeliou et al., 2003) and our retrospective study, described in more detail here, suggest that a ketogenic diet does indeed offer multiple benefits for children with ASD.

Ketogenic diet is an existing treatment option at many institutions treating pediatric epilepsy, and diet therapy appears to be increasing in popularity. Diet therapy works typically within two weeks and certainly within three months, and thus clinical efficacy is determined relatively rapidly after starting treatment. Even though the ketogenic diet is now established as a clinically effective treatment in pediatric epilepsy, it has not been explored adequately in other developmental disorders - such as autism - that may benefit from seizure control and relief from diverse co-morbid symptoms. A more comprehensive overview of the diverse potential benefits of a relationship between a ketogenic diet and adenosine, including autism, is outlined in Masino et al. (2009).

It is important to remember that ketogenic diet therapy can result in a permanent reduction in epileptic seizures, and some children remain seizure-free even after tapering off the diet. The implication that permanent benefits could be achieved in ASD is incredibly enticing - although there is no evidence at this time. Notably, however, the ketogenic diet is often effective in treating seizures which are medically-refractory. In parallel with ongoing use of the classic ketogenic diet, related dietary strategies such as low glycemic index therapy (LGIT; Pfeifer & Thiele, 2005; Muzykewicz et al., 2009) and the modified Atkins diet (MAD; Kossoff et al., 2003; Kossoff et al., 2008) are actively investigated for efficacy in epilepsy. Because these diets are less restrictive than the standard ketogenic diet, they offer more popular appeal and may represent different "doses" of a ketogenic diet. The classical ketogenic diet is the most strict – highest "dose" and in some cases can offer seizure control in patients who did not achieve sufficient control with the MAD (Kossoff et al., 2010).

Because there is neither a proprietary drug nor drug development required, clinically-proven dietary strategies are translatable immediately to another pediatric population such as autism. Although used less often, the ketogenic diet is effective in adults with epilepsy, and thus could be considered for adults with ASD. There are currently no established effective treatments for ASD in children or adults, and no approved psychiatric drugs for children, making the risks relatively small and the potential benefits quite large. A drug-based strategy would be easier to administer and appeal to a wider patient population; a "ketogenic diet in a pill" remains a hotly pursued therapeutic target (Rho & Sankar, 2008).

In general, a metabolic approach to increase adenosine could offer positive reinforcement, including improved sleep and learning, and reduced seizures and anxiety. The combined evidence of abnormal purine metabolism, stereotyped behaviors, increased seizures and disrupted sleep suggests a general dysregulation of this modulator and thus potential new insight into therapies for ASD. Success with metabolic adenosine-increasing approaches could also provide an impetus for exploring adenosine-enhancing drugs in autism, and establishing their safety and efficacy in children.

This work linking adenosine and autism integrates behavioral and neurobiological data, puts forth a new theory regarding ongoing symptoms of autism, and - if translated to clinical benefits - could address several areas of critical need. First, it targets co-morbidities in the form of anxiety, sleep, and seizure disorders. It is well known that sufficient quality and quantity of sleep is crucial for general mental health and optimizing learning and memory. Second, it proposes a specific and novel neurobiological target for reducing the ongoing symptoms of autism spectrum disorder – increased extracellular adenosine. Third, it targets metabolic mechanisms that alter adenosine, rather than receptor-based strategies. For decades researchers have attempted to develop adenosine-receptor based therapeutics but side effects, particularly due to peripheral actions of adenosine, remain a roadblock (Dunwiddie, 1999; Boison et al., 2011). This physiologic/metabolic approach, which shifts the dynamic level of adenosine itself, is likely to have far fewer side effects when translated into humans. Thus, some stimuli for increasing adenosine levels and decreasing ASD symptoms may be immediately accessible environmental interventions involving changes in diet or behavior, in parallel with pursuing more traditional pharmaceutical interventions.

6. Acknowledgements

Supported by National Institutes of Health NS066392 and NS065957, National Science Foundation IOS-0843585, Connecticut Children's Medical Center, and Trinity College. We acknowledge the support of Interactive Autism Network (IAN), and Dr. Deborah Fein for her time and encouragement.

7. References

Aoki, T.T. (1981). Metabolic adaptations to starvation, semistarvation, and carbohydrate restriction. *Progress in Clinical and Biological Research,* Vol. 67, pp. 161-177, 0361-7742

Arendash, G.W. & Cao, C. (2010). Caffeine and coffee as therapeutics against Alzheimer's disease. *Journal of Alzheimer's Disease,* Vol. 20 Suppl 1, pp. S117-S126, 1387-2877

Arnold, G.L., Hyman, S.L., Mooney, R.A. & Kirby, R.S. (2003). Plasma amino acids profiles in children with autism: potential risk of nutritional deficiencies. *Journal of Autism and Developmental Disorders,* Vol. 33, No. 4, (Aug), pp. 449-454, 0162-3257

Boison, D. (2006). Adenosine kinase, epilepsy and stroke: mechanisms and therapies. *Trends in Pharmacological Sciences,* Vol. 27, No. 12, (Dec), pp. 652-658, 0165-6147

Boison, D., Masino, S.A. & Geiger, J.D. (2011). Homeostatic bioenergetic network regulation: a novel concept to avoid pharmacoresistance in epilepsy. *Expert Opinion on Drug Discovery,* in press, 1746-0441

Boison, D., Singer, P., Shen, H.-Y., Feldon, J. & Yee, B.K. (2011). Adenosine hypothesis of schizophrenia - opportunities for pharmacotherapy. *Neuropharmacology,* in press, 0028-3908

Bottini, N., De Luca, D., Saccucci, P., Fiumara, A., Elia, M., Porfirio, M.C., Lucarelli, P. & Curatolo, P. (2001). Autism: evidence of association with adenosine deaminase genetic polymorphism. *Neurogenetics,* Vol. 3, No. 2, (Mar), pp. 111-113, 1364-6745

Canitano, R. (2007). Epilepsy in autism spectrum disorders. *European Child & Adolescent Psychiatry,* Vol. 16, No. 1, (Feb), pp. 61-66, 1018-8827

Chalfant, A.M., Rapee, R. & Carroll, L. (2007). Treating anxiety disorders in children with high functioning autism spectrum disorders: a controlled trial. *Journal of Autism and Developmental Disorders,* Vol. 37, No. 10, (Nov), pp. 1842-1857, 0162-3257

Chen, J.F. & Chern, Y. (2011). Impacts of methylxanthines and adenosine receptors on neurodegeneration: human and experimental studies. *Handbook of Experimental Pharmacology,* Vol. 200, pp. 267-310, 0171-2004

Cunha, R.A. (2001). Adenosine as a neuromodulator and as a homeostatic regulator in the nervous system: different roles, different sources and different receptors. *Neurochemistry International,* Vol. 38, No. 2, (Feb), pp. 107-125, 0197-0186

Cunha, R.A. & Agostino, P.M. (2010). Chronic caffeine consumption prevents memory disturbance in different animal models of memory decline. *Journal of Alzheimer's Disease,* Vol. 20 Suppl. 1, pp. S95-S116, 1387-2877

Cunha, R.A., Sebastião, A.M. & Ribeiro, J.A. (1998). Inhibition by ATP of hippocampal synaptic transmission requires localized extracellular catabolism by ecto-nucleotidases into adenosine and channeling to adenosine A_1 receptors. *Journal of Neuroscience,* Vol. 18, No. 6, (Mar 15), pp. 1987-1995, 1813-1948

Curran, L.K., Newschaffer, C.J., Lee, L.C., Crawford, S.O., Johnston, M.V. & Zimmerman, A.W. (2007). Behaviors associated with fever in children with autism spectrum disorders. *Pediatrics,* Vol. 120, No. 6, (Dec), pp. e1386-e1392, 1012-232X

Davis, L.M., Rho, J.M. & Sullivan, P.G. (2008). UCP-mediated free fatty acid uncoupling of isolated cortical mitochondria from fasted animals: correlations to dietary modulations. *Epilepsia,* Vol. 49, Suppl. 8, (Nov), pp. 117-119, 0013-9580

Dulla, C.G., Dobelis, P., Pearson, T., Frenguelli, B.G., Staley, K.J. & Masino, S.A. (2005). Adenosine and ATP link P_{CO2} to cortical excitability via pH. *Neuron,* Vol. 48, No. 6, (Dec 22), pp. 1011-1023, 0896-6273

Dulla, C.G., Frenguelli, B.G., Staley, K.J. & Masino, S.A. (2009). Intracellular acidification causes adenosine release during states of hyperexcitability in the hippocampus. *Journal of Neurophysiology,* Vol. 102, No. 3, (Sep), pp. 1984-1993, 0022-3077

Dunwiddie, T.V. (1999). Adenosine and suppression of seizures. *Advances in Neurology,* Vol. 79, pp. 1001-1010, 0091-3952

Dunwiddie, T.V., Diao, L. & Proctor, W.R. (1997). Adenine nucleotides undergo rapid, quantitative conversion to adenosine in the extracellular space in rat hippocampus. *Journal of Neuroscience,* Vol. 17, No. 20, (Oct 15), pp. 7673-7682, 1813-1948

Dunwiddie, T.V. & Masino, S.A. (2001). The role and regulation of adenosine in the central nervous system. *Annual Review of Neuroscience,* Vol. 24, pp. 31-55, 0147-006X

Dunwiddie, T.V. & Worth, T.S. (1982). Sedative and anticonvulsant effects of adenosine analogs in mouse and rat. *Journal of Pharmacology and Experimental Therapeutics,* Vol. 220, No. 1, (Jan), pp. 70-76, 0022-3565

Dworak, M., Diel, P., Voss, S., Hollmann, W. & Strüder, H.K. (2007). Intense exercise increases adenosine concentrations in rat brain: implications for a homeostatic sleep drive. *Neuroscience,* Vol. 150, No. 4, (Dec 19), pp. 789-795, 0306-4522

Edelson, S.M., Edelson, M.G., Kerr, D.C. & Grandin, T. (1999). Behavioral and physiological effects of deep pressure on children with autism: a pilot study evaluating the efficacy of Grandin's Hug Machine. *American Journal of Occupational Therapy,* Vol. 53, No. 2, (Mar-Apr), pp. 145-152, 0272-9490

Ekinci, O., Titus, J.B., Rodopmann, A.A., Berkem, M. & Trevatham, E. (2009). Depression and anxiety in children and adolescents with epilepsy: prevalence, risk factors, and treatment. *Epilepsy & Behavior,* Vol. 14, No. 1, (Jan), pp. 8-18, 1525-5050

Elder, J.H., Shankar, M., Shuster, J., Theriaque, D., Burns, S. & Sherrill, L. (2006). The gluten-free, casein-free diet in autism: results of a preliminary double blind clinical trial. *Journal of Autism and Developmental Disorders,* Vol. 36, No. 3, (Apr), pp. 413-420, 0162-3257

Escalona, A., Field, T., Singer-Strunck, R., Cullen, C. & Hartshorn, K. (2001). Brief report: improvements in the behavior of children with autism following massage therapy. *Journal of Autism and Developmental Disorders,* Vol. 31, No. 2, (Apr), pp. 513-516, 0162-3257

Evangeliou, A., Vlachonikolis, I., Mihailidou, H., Spilioti, M., Skarpalezou, A., Makaronas, N., Prokopiou, A., Christodoulou, P., Liapi-Adamidou, G., Helidonis, E., Sbyrakis, S. & Smeitink, J. (2003). Application of a ketogenic diet in children with autistic behavior: pilot study. *Journal of Child Neurology,* Vol. 18, No. 2, (Feb), pp. 113-118, 0883-0738

Farasat, S., Kossoff, E.H., Pillas, D.J., Rubenstein, J.E., Vining, E.P.G. & Freeman, J.M. (2006). The importance of parental expectations of cognitive improvements for their children with epilepsy prior to starting the ketogenic diet. *Epilepsy & Behavior,* Vol. 8, No. 2, (Mar), pp. 406-410, 1525-5050

Fatemi, S.H., Folsom, T.D., Reutiman, T.J. & Thuras, P.D. (2009a). Expression of $GABA_B$ receptors is altered in brains of subjects with autism. *Cerebellum,* Vol. 8, No. 1, (Mar), pp. 64-69, 1473-4222

Fatemi, S.H., Reutiman, T.J., Folsom, T.D. & Thuras, P.D. (2009b). $GABA_A$ receptor downregulation in brains of subjects with autism. *Journal of Autism and Developmental Disorders,* Vol. 39, No. 2, (Feb), pp. 223-230, 0162-3257

Ferré, S., Diamond, I., Goldberg, S.R., Yao, L., Hourani, S.M.O., Huang, Z.L., Urade, Y. & Kitchen, I. (2007). Adenosine A_{2A} receptors in ventral striatum, hypothalamus and nociceptive circuitry Implications for drug addiction, sleep and pain. *Progress in Neurobiology,* Vol. 83, No. 5, (Dec), pp. 332-347, 0301-0082

Fields, R.D. & Burnstock, G. (2006). Purinergic signalling in neuron-glia interactions. *Nature Reviews. Neuroscience,* Vol. 7, No. 6, (Jun), pp. 423-436, 1471-003X

Florio, C., Prezioso, A., Papaioannou, A. & Vertua, R. (1998). Adenosine A1 receptors modulate anxiety in CD1 mice. *Psychopharmacology,* Vol. 136, No. 4, (Apr), pp. 311-319, 0033-3158

Franke, H., Krügel, U. & Illes, P. (2006). P2 receptors and neuronal injury. *Pflugers Archiv: European Journal of Physiology,* Vol. 452, No. 5, (Aug), pp. 622-644, 0031-6768

Fredholm, B.B. (1997). Adenosine and neuroprotection. *International Review of Neurobiology,* Vol. 40, No., pp. 259-280, 0074-7742

Fredholm, B.B., Bättig, K., Holmén, J., Nehlig, A. & Zvartau, E.E. (1999). Actions of caffeine in the brain with special reference to factors that contribute to its widespread use. *Pharmacological Reviews,* Vol. 51, No. 1, pp. 83-133, 0031-6997

Fredholm, B.B., Chen, J.F., Cunha, R.A., Svenningsson, P. & Vaugeois, J.M. (2005). Adenosine and brain function. *International Review of Neurobiology,* Vol. 63, pp. 191-270, 0074-7742

Freeman, J.M., Kossoff, E.H. & Hartman, A.L. (2007). The ketogenic diet: one decade later. *Pediatrics,* Vol. 119, No. 3, (Mar), pp. 535-543, 0031-4005

Fuxe, K., Ferré, S., Genedani, S., Franco, F. & Agnati, L.F. (2007). Adenosine receptor-dopamine receptor interactions in the basal ganglia and their relevance for brain function. *Physiology & Behavior,* Vol. 92, No. 1-2, (Sep 10), pp. 210-217, 0031-9384

Ghanizadeh, A. (2010). Possible role of caffeine in autism spectrum disorders, a new testable hypothesis. *Journal of Food Science,* Vol. 75, No. 6, (Aug 1), pg. ix, 0022-1147

Goldman, N., Chen, M., Fujita, T., Xu, Q., Peng, W., Liu, W., Jensen, T.K., Pei, Y., Wang, F., Han, X., Chen, J.-F., Schnermann, J., Takano, T., Bekar, L., Tieu, K. & Nedergaard, M. (2010). Adenosine A1 receptors mediate local anti-nociceptive effects of acupuncture. *Nature Neuroscience,* Vol. 13, No. 7, (Jul), pp. 883-888, 1097-6256

Goodman, R.R., Cooper, M.J., Gavish, M. & Snyder, S.H. (1982). Guanine nucleotide and cation regulation of the binding of [^{3}H]cyclohexyladenosine and [^{3}H]diethylphenylxanthine to adenosine A1 receptors in brain membranes. *Molecular Pharmacology,* Vol. 21, No. 2, (Mar), pp. 329-335, 0026-895X

Gouder, N., Scheurer, L., Fritschy, J.-M. & Boison, D. (2004). Overexpression of adenosine kinase in epileptic hippocampus contributes to epileptogenesis. *Journal of Neuroscience,* Vol. 24, No. 3, (Jan 21), pp. 692-701, 1813-1948

Groesbeck, D.K., Bluml, R.M. & Kossoff, E.H. (2006). Long-term use of the ketogenic diet in the treatment of epilepsy. *Developmental Medicine and Child Neurology,* Vol. 48, No. 12, (Dec), pp. 978-981, 0012-1622

Gurney, J.G., McPheeters, M.L. & Davis, M.M. (2006). Parental report of health conditions and health care use among children with and without autism. *Archives of Pediatric & Adolescent Medicine,* Vol. 160, No. 8, (Aug), pp. 825-830, 1072-4710

Halassa, M.M., Florian, C., Fellin, T., Munoz, J.R., Lee, S.-Y., Abel, T., Haydon, P.G. & Frank, M.G. (2009). Astrocytic modulation of sleep homeostasis and cognitive consequences of sleep loss. *Neuron,* Vol. 61, No. 2, (Jan 29), pp. 213-219, 0896-6273

Hallböök, T., Kohler, S., Rosén, I. & Lundgren, J. (2007). Effects of ketogenic diet on epileptiform activity in children with therapy resistant epilepsy. *Epilepsy Research* Vol. 77, No. 2-3, (Dec), pp. 134-140, 0920-1211

Hallböök, T., Lundgren, J. & Rosén, I. (2007). Ketogenic diet improves sleep quality in children with therapy-resistant epilepsy. *Epilepsia,* Vol. 48, No. 1, (Jan), pp. 59-65, 0013-9580

Hartman, A.L. & Vining, E.P. (2007). Clinical aspects of the ketogenic diet. *Epilepsia,* Vol. 48, No. 1, (Jan), pp. 31-42, 0013-9580

Huang, Z.-L., Qu, W.-M., Eguchi, N., Chen, J.-R., Schwartzschild, M.A., Fredholm, B.B., Urade, Y. & Hayaishi, O. (2005). Adenosine A_{2A}, but not A_1, receptors mediate the arousal effect of caffeine. *Nature Neuroscience,* Vol. 8, No. 7, (Jul), pp. 858-859, 1097-6256

Hussman, J.P. (2001). Suppressed GABAergic inhibition as a common factor in suspected etiologies of autism. *Journal of Autism and Developmental Disorders,* Vol. 31, No. 2, (Apr), pp. 247-248, 0162-3257

Hyde, T.M. & Weinberger, D.R. (1997). Seizures and schizophrenia. *Schizophrenia Bulletin,* Vol. 23, No. 4, pp. 611-622, 0586-7614

Ibrahim, S.H., Voigt, R.G., Katusic, S.K., Weaver, A.L. & Barbaresi, W.J. (2009). Incidence of gastrointestinal symptoms in children with autism: a population-based study. *Pediatrics,* Vol. 124, No. 2, (Aug), pp. 680-686, 1012-232X

James, S.J., Cutler, P., Melnyk, S., Jernigan, S., Janak, L., Gaylor, D.W. & Neubrander, J.A. (2004). Metabolic biomarkers of increased oxidative stress and impaired methylation capacity in children with autism. *American Journal of Clinical Nutrition,* Vol. 80, No. 6, (Dec), pp. 1611-1617, 0002-9165

James, S.J., Melnyk, S., Jernigan, S., Cleves, M.A., Halsted, C.H., Wong, D.H., Cutler, P., Bock, K., Boris, M., Bradstreet, J.J., Baker, S.M. & Gaylor, D.W. (2006). Metabolic endophenotype and related genotypes are associated with oxidative stress in children with autism. *American Journal of Medical Genetics, Part B Neuropsychiatric Genetics,* Vol. 141, No. 8, (Dec 5), pp. 947-956, 1552-4841

Jeste, S.S. (2011). The neurology of autism spectrum disorders. *Current Opinion in Neurology,* Vol. 24, No. 2, (Apr), pp. 132-139, 1350-7540

Johansson, B., Halldner, L., Dunwiddie, T.V., Masino, S.A., Poelchen, W., Giménez-Llort, L., Escorihuela, L.M., Fernández-Teruel, A., Wiesenfeld-Hallin, Z., Xu, X.-J., Hårdemark, A., Betscholtz, C., Herlenius, E. & Fredholm, B.B. (2001). Hyperalgesia, anxiety, and decreased hypoxic neuroprotection in mice lacking the adenosine A_1 receptor. *Proceedings of the National Academy of Sciences of the United Statse of America,* Vol. 98, No. 16, (Jul 31), pp. 9407-9412, 0027-8424

Jones, J.E., Austin, J.K., Caplan, R., Dunn, D., Plioplys, S. & Salpekar, J.A. (2008). Psychiatric disorders in children and adolescents who have epilepsy. *Pediatrics in Review,* Vol. 29, No. 2, (Feb), pp. e9-14, 0191-9601

Juge, N., Gray, J.A., Omote, H., Miyaji, T., Inoue, T., Hara, C., Uneyama, H., Edwards, R.H., Nicoll, R.A. & Moriyama, Y. (2010). Metabolic control of vesicular glutamate transport and release. *Neuron,* Vol. 68, No. 1, (Oct 6), pp. 99-112, 0896-6273

Kanner, A.M. (2004). Is major depression a neurologic disorder with psychiatric symptoms? *Epilepsy & Behavior,* Vol. 5, No. 5, (Oct), pp. 636-644, 1525-5050

Kanner, A.M. (2008). Mood disorder and epilepsy: a neurobiologic perspective of their relationship. *Dialogues in Clinical Neuroscience,* Vol. 10, No. 1, pp. 39-45, 1294-8322

Kawamura, M., Jr., Ruskin, D.N. & Masino, S.A. (2010). Metabolic autocrine regulation of neurons involves cooperation among pannexin hemichannels, adenosine receptors and K_{ATP} channels. *Journal of Neuroscience,* Vol. 30, No. 11, (Mar 17), pp. 3886-3895, 1813-1948

Keene, D.L. (2006). A systematic review of the use of the ketogenic diet in childhood epilepsy. *Pediatric Neurology,* Vol. 35, No. 1, (Jul), pp. 1-5, 0887-8994

Kinsman, S.L., Vining, E.P.G., Quaskey, S.A., Mellits, D. & Freeman, J.M. (1992). Efficacy of the ketogenic diet for intractable seizure disorders: review of 58 cases. *Epilepsia,* Vol. 33, No. 6, (Nov-Dec), pp. 1132-1136, 1409-617X

Kossoff, E.H., Bosarge, J.L., Miranda, M.J., Wiemer-Kruel, A., Kang, H.C. & Kim, H.D. (2010). Will seizure control improve by switching from the modified Atkins diet to the traditional ketogenic diet? *Epilepsia,* Vol. 51, No. 12, (Dec), pp. 2496-2499, 0013-9580

Kossoff, E.H., Krauss, G.L., McGrogan, J.R. & Freeman, J.M. (2003). Efficacy of the Atkins diet as therapy for intractable epilepsy. *Neurology,* Vol. 61, No. 12, (Dec 23), pp. 1789-1791, 0028-3878

Kossoff, E.H., Rowley, H., Sinha, S.R. & Vining, E.P.G. (2008). A prospective study of the modified Atkins diet for intractable epilepsy in adults. *Epilepsia,* Vol. 49, No. 2, (Feb), pp. 316-319, 0013-9580

Kossoff, E.H., Zupec-Kania, B.A., Åmark, P.E., Ballaban-Gil, K.R., Bergqvist, A.G.C., Blackford, R., Buchhalter, J.R., Caraballo, R.H., Cross, J.H., Dahlin, M.G., Donner, E.J., Klepper, J., Jehle, R.S., Kim, H.D., Liu, Y.M.C., Nation, J., Nordli, D.R., Jr., Pfeifer, H.H., Rho, J.M., Stafstrom, C.E., Thiele, E.A., Turner, Z., Wirrell, E.C., Wheless, J.W., Veggiotti, P., Vining, E.P.G., the Charlie Foundation, & the Practice Committee of the Child Neurology Society (2009). Optimal clinical management of children receiving the ketogenic diet: Recommendations of the International Ketogenic Diet Study Group. *Epilepsia,* Vol. 50, No. 2, (Feb), pp. 304-317, 0013-9580

LaFrance, W.C., Jr., Kanner, A.M. & Hermann, B. (2008). Psychiatric comorbidities in epilepsy. *International Review of Neurobiology,* Vol. 83, pp. 347-383, 0074-7742

Latini, S. & Pedata, F. (2001). Adenosine in the central nervous system: release mechanisms and extracellular concentrations. *Journal of Neurochemistry,* Vol. 79, No. 3, (Nov), pp. 463-484, 0022-3042

Lee, H., Marvin, A.R., Watson, T., Piggot, J., Law, J.K., Law, P.A., Constantino, J.N. & Nelson, S.F. (2010). Accuracy of phenotyping of autistic children based on internet implemented parent report. *American Journal of Medical Genetics. Part B, Neuropsychiatric Genetics,* Vol. 153B, No. 6, (Sep), pp. 1119-1126, 1552-4841

Liss, M., Saulnier, C., Fein, D. & Kinsbourne, M. (2006). Sensory and attention abnormalities in autistic spectrum disorders. *Autism,* Vol. 10, No. 2, (Mar), pp. 155-172, 1362-3613

Ma, W., Berg, J. & Yellen, G. (2007). Ketogenic diet metabolites reduce firing in central neurons by opening K_{ATP} channels. *Journal of Neuroscience,* Vol. 27, No. 14, (Apr 4), pp. 3618-3625, 1813-1948

Malow, B.A. (2004). Sleep disorders, epilepsy, and autism. *Mental Retardation and Developmental Disabilities Research Reviews,* Vol. 10, No. 2, pp. 122-125, 1080-4013

Marie, S., Race, V., Nassogne, M.C., Vincent, M.F. & Van den Berghe, G. (2002). Mutation of a nuclear respiratory factor 2 binding site in the 5' untranslated region of the ADSL gene in three patients with adenylosuccinate lyase deficiency. *American Journal of Human Genetics,* Vol. 71, No. 1, (Jul), pp. 14-21, 0002-9297

Masino, S.A. & Dunwiddie, T.V. (1999). Temperature-dependent modulation of excitatory transmission in hippocampal slices is mediated by extracellular adenosine. *Journal of Neuroscience,* Vol. 19, No. 6, (Mar 15), pp. 1932-1939, 1813-1948

Masino, S.A. & Geiger, J.D. (2008). Are purines mediators of the anticonvulsant/neuroprotective effects of ketogenic diets? *Trends in Neuroscience,* Vol. 31, No. 6, (Jun), pp. 273-278, 0166-2236

Masino, S.A. & Geiger, J.D. (2009). The ketogenic diet and epilepsy: Is adenosine the missing link? *Epilepsia,* Vol. 50, No. 2, (Feb), pp. 332-333, 0013-9580

Masino, S.A., Kawamura, M., Jr., Plotkin, L.M., Svedova, J., DiMario, F.J., Jr. & Eigsti, I.-E. (2011a). The relationship between the neuromodulator adenosine and behavioral symptoms of autism. *Neuroscience Letters,* Vol. 500, No. 1, pp. 1-5, 0304-3940

Masino, S.A., Kawamura, M., Jr., Wasser, C.D., Pomeroy, L.T. & Ruskin, D.N. (2009). Adenosine, ketogenic diet and epilepsy: the emerging therapeutic relationship between metabolism and brain activity. *Current Neuropharmacology,* Vol. 7, No. 3, (Sep), pp. 257-268, 1570-159X

Masino, S.A., Li, T., Theofilas, P., Ruskin, D.N., Fredholm, B.B., Geiger, J.D., Aronica, E. & Boison, D. (2011b). A ketogenic diet suppresses seizures in mice through adenosine A_1 receptors. *Journal of Clinical Investigation,* Vol. 121, No. 7, pp. 2679-2683, 0021-9738

McGinnis, W.R. (2005). Oxidative stress in autism. *Alternative Therapies in Health and Medicine,* Vol. 11, No. 1, (Jan-Feb), pg. 19, 1078-6791

Millward, C., Ferriter, M., Calver, S. & Connell-Jones, G. (2008). Gluten- and casein-free diets for autistic spectrum disorder. *Cochrane Database of Systematic Reviews,* Vol. (2), (Apr 16), CD003498, 1361-6137

Muzykewicz, D.A., Lyczkowski, D.A., Memon, N., Conant, K.D., Pfeifer, H.H. & Thiele, E.A. (2009). Efficacy, safety, and tolerability of the low glycemic index treatment in pediatric epilepsy. *Epilepsia,* Vol. 50, pp. 1118-1126, 0013-9580

Neal, E.G., Chaffe, H., Schwartz, R.H., Lawson, M.S., Edwards, N., Fitzsimmons, G., Whitney, A. & Cross, J.H. (2008). The ketogenic diet for the treatment of childhood epilepsy: a randomised controlled trial. *Lancet Neurology,* Vol. 7, No. 6, (Jun), pp. 500-506, 1474-4422

Newby, A.C. (1984). Adenosine and the concept of "retaliatory metabolites." *Trends in Biochemical Science,* Vol. 9, No. 2, (Feb), pp. 42-44, 0968-0004

Nyhan, W.L., James, J.A., Teberg, A.J., Sweetman, L. & Nelson, L.G. (1969). A new disorder of purine metabolism with behavioral manifestations. *Journal of Pediatrics,* Vol. 74, No. 1, (Jan), pp. 20-27, 0022-3476

Oswald, D.P. & Sonenklar, N.A. (2007). Medication use among children with autism spectrum disorders. *Journal of Child and Adolescent Psychopharmacology,* Vol. 17, No. 3, (Jun), pp. 348-355, 1044-5463

Page, T. & Coleman, M. (2000). Purine metabolism abnormalities in a hyperuricosuric subclass of autism. *Biochimica et Biophysica Acta,* Vol. 1500, No. 3, (Mar 17), pp. 291-296, 0006-3002

Page, T. & Moseley, C. (2002). Metabolic treatment of hyperuricosuric autism. *Progress in Neuro-psychopharmacology & Biological Psychiatry,* Vol. 26, No. 2, (Feb), pp. 397-400, 0278-5846

Pascual, O., Casper, K.B., Kubera, C., Zhang, J., Revilla-Sanchez, R., Sul, J.-Y., Takano, H., Moss, S.J., McCarthy, K. & Haydon, P.G. (2005). Astrocytic purinergic signaling coordinates synaptic networks. *Science,* Vol. 310, No. 5745, (Oct 7), pp. 113-116, 0036-8075

Pfeifer, H. & Thiele, E. (2005). Low-glycemic-index treatment: a liberalized ketogenic diet for treatment of intractable epilepsy. *Neurology,* Vol. 65, No. 11, (Dec 13), pp. 1810-1812, 0028-3878

Poleszak, E. & Malec, D. (2000). Influence of adenosine receptor agonists and antagonists on ampheramine-induced stereotypy in rats. *Polish Journal of Pharmacology,* Vol. 52, No. 6, (Nov-Dec), pp. 423-429, 1230-6002

Polimeni, M.A., Richdale, A.L. & Francis, A.J. (2005). A survey of sleep problems in autism, Asperger's disorder and typically developing children. *Journal of Intellectual Disability Research,* Vol. 49, No. 4, (Apr), pp. 260-268, 0964-2633

Porkka-Heiskanen, T. (1999). Adenosine in sleep and wakefulness. *Annals of Medicine,* Vol. 31, No. 2, (Apr), pp. 125-129, 0785-3890

Prediger, R.D. (2010). Effects of caffeine in Parkinson's disease: from neuroprotection to the management of motor and non-motor symptoms. *Journal of Alzheimer's Disease,* Vol. 20 Suppl 1, pp. S205-S220, 1387-2877

Pulsifer, M.B., Gordon, J.M., Brandt, J., Vining, E.P.G. & Freeman, J.M. (2001). Effects of ketogenic diet on development and behavior: preliminary report of a prospective study. *Developmental Medicine and Child Neurology,* Vol. 43, No. 5, (May), pp. 301-306, 0012-1622

Rainnie, D.G., Grunze, H.C., McCarley, R.W. & Greene, R.W. (1994). Adenosine inhibition of mesopontine cholinergic neurons: implications for EEG arousal. *Science,* Vol. 263, No. 5147, (Feb 4), pp. 689-692, 0036-8075

Ralevic, V. & Burnstock, G. (1998). Receptors for purines and pyrimidines. *Pharmacological Reviews,* Vol. 50, No. 3, (Sep), pp. 413-492, 0031-6997

Reeves, G., Blaisdell, C., Lapidus, M., Langenberg, P., Ramagopal, M., Cabassa, J., Bollinger, M.B., Nijjar, G.V., Anthony, B., Achenbach, T. & Postolache, T.T. (2010). Sleep architecture and behavioral abnormalities in children and adolescents. *International Journal of Adolescent Medicine and Health,* Vol. 22, No. 4, (Oct-Dec), pp. 535-545, 0334-0139

Rho, J. & Sankar, R. (2008). The ketogenic diet in a pill: is this possible? *Epilepsia,* Vol. 49 Suppl 8, (Nov), pp. 127-133, 0013-9580

Ridley, R.M. (1994). The psychology of perseverative and stereotyped behaviour. *Progress in Neurobiology,* Vol. 44, No. 2, (Oct), pp. 221-231, 0301-0082

Rosin, D.L., Hettinger, B.D., Lee, A. & Linden, J. (2003). Anatomy of adenosine A2A receptors in brain: morphological substrates for integration of striatal function. *Neurology,* Vol. 61, No. 11 Suppl 6, (Dec 9), pp. S12-S18, 0028-3878

Ruskin, D.N., Kawamura, M., Jr. & Masino, S.A. (2009). Reduced pain and inflammation in juvenile and adult rats fed a ketogenic diet. *PLoS One,* Vol. 4, No. 16, (Dec 23), e8349, 1932-6203

Rutter, M., Bailey, A. & Lord, C. (2003). *Social Communication Questionnaire,* Western Psychological Services, Los Angeles, California

Salamone, J.D. & Correa, M. (2009). Dopamine/adenosine interactions involved in effort-related aspects of food motivation. *Appetite,* Vol. 53, No. 3, (Dec), pp. 422-425, 0195-6663

Santangelo, S.L. & Tsatsanis, K. (2005). What is known about autism: genes, brain, and behavior. *American Journal of Pharmacogenomics,* Vol. 5, No. 2, pp. 71-92, 1175-2203

Schulte, E. & Fredholm, B.B. (2003). Signalling from adenosine receptors to mitogen-activated protein kinases. *Cellular Signalling,* Vol. 15, No. 9, (Sep), pp. 813-827, 0898-6568

Shneker, B.F. & Fountain, N.B. (2003). Epilepsy. *Disease-a-Month,* Vol. 49, No. 7, (Jul), pp. 427-478, 0011-5029

Sirven, J., Whedon, B., Caplan, D., Liporace, J., Glosser, D., O'Dwyer, J. & Sperling, M.R. (1999). The ketogenic diet for intractable epilepsy in adults: preliminary results. *Epilepsia,* Vol. 40, No. 12, (Dec), pp. 1721-1726, 0013-9580

Spence, S.J. & Schneider, M.T. (2009). The role of epilepsy and epileptiform EEGs in autism spectrum disorders. *Pediatric Research,* Vol. 65, No. 6, (Jun), pp. 599-606, 0031-3998

Srinivasan, P. (2009). A review of dietary interventions in autism. *Annals of Clinical Psychiatry,* Vol. 21, No. 4, (Oct-Dec), pp. 237-247, 1040-1237

Sullivan, P.G., Rippy, N.A., Dorenbos, K., Concepcion, R.C., Agarwal, A.K. & Rho, J.M. (2004). The ketogenic diet increases mitochondrial uncoupling protein levels and activity. *Annals of Neurology,* Vol. 55, No. 4, (Apr), pp. 576-580, 0364-5134

Svedova, J., Masino, S.A. & DiMario, F.J., Jr. (2010). Retrospective analysis of the effects of the ketogenic diet on behavior and temperament of children with epilepsy, *Proceedings of the Society for Neuroscience,* San Diego, CA, Nov 2010

Tanimura, Y., Vaziri, S. & Lewis, M.H. (2010). Indirect basal ganglia pathway mediation of repetitive behavior: Attenuation by adenosine receptor agonists. *Behavioural Brain Research,* Vol. 210, No. 1, (Jun 26), pp. 116-122, 0166-4328

Vining, E.P.G., Freeman, J.M., Ballaban-Gil, K., Camfield, C.S., Camfield, P.R., Holmes, G.L., Shinnar, S., Shuman, R., Trevathan, E. & Wheless, J.W. (1998). A multicenter study of the efficacy of the ketogenic diet. *Archives of Neurology,* Vol. 55, No. 11, (Nov), pp. 1433-1437, 0003-9942

Whitcomb, K., Lupica, C.R., Rosen, J.B. & Berman, R.F. (1990). Adenosine involvement in postictal events in amygdala-kindled rats. *Epilepsy Research,* Vol. 6, No. 3, (Aug), pp. 171-179, 0920-1211

Witwer, A. & Lecavalier, L. (2005). Treatment incidence and patterns in children and adolescents with autism spectrum disorders. *Journal of Child and Adolescent Psychopharmacology,* Vol. 15, No. 4, (Aug), pp. 671-681, 1044-5463

Wiznitzer, M. (2008). From observations to trials: the ketogenic diet and epilepsy. Lancet Neurology, Vol. 7, No. 6, (Jun), pp. 471-472, 1474-4422

Yudkoff, M., Daikhin, Y., Melø, T.M., Nissim, I., Sonnewald, U. & Nissim, I. (2007). The ketogenic diet and brain metabolism of amino acids: relationship to the anticonvulsant effect. *Annual Review of Nutrition,* Vol. 27, pp. 415-430, 0199-9885

Zecavati, N. & Spence, S.J. (2009). Neurometabolic disorders and dysfunction in autism spectrum disorders. *Current Neurology and Neuroscience Reports,* Vol. 9, No. 2, (Mar), pp. 129-136

Zhang, J.-m., Wang, H.-k., Ye, C.-q., Ge, W., Chen, Y., Jiang, Z.-l., Wu, C.-p., Poo, M.-m. & Duan, S. (2003). ATP released by astrocytes mediates glutamatergic activity-dependent heterosynaptic suppression. *Neuron,* Vol. 40, No. 5, (Dec 4), pp. 971-982, 0896-6273

Zhao, X., Leotta, A., Kustanovich, V., Lajonchere, C., Geschwind, D.H., Law, K., Law, P., Qiu, S., Lord, C., Sebat, J., Ye, K. & Wigler, M. (2007). A unified genetic theory for sporadic and inherited autism. *Proceedings of the National Academy of Sciences of the United States of America,* Vol. 104, No. 31, (Jul 31), pp. 12831-12836, 0027-8424

7

Nicotinic Acetylcholine Receptor Alterations in Autism Spectrum Disorders – Biomarkers and Therapeutic Targets

Rene Anand[1,2], Stephanie A. Amici[1], Gerald Ponath[1], Jordan I. Robson[1], Muhammad Nasir[1] and Susan B. McKay[1]

[1]*Department of Pharmacology*
[2]*Department of Neuroscience,*
The Ohio State University, College of Medicine, Columbus, Ohio, USA

1. Introduction

Autism Spectrum Disorders (ASD) are a set of complex neurodevelopmental disorders defined behaviorally by impaired social interaction, delayed and disordered language, repetitive or stereotypic behavior and a restricted range of interest (Fombonne, 1999). ASD affect nearly 1 in 110 children, and disproportionally affect four times as many boys as girls. Comorbid symptoms often include seizures, sleep problems, gastrointestinal disorders, and metabolic deregulation (Coury, 2010). As such, ASD are an enormous challenge for parents, medical professionals, and educators. Their treatments put a significant financial strain on healthcare systems worldwide. There is no pharmacotherapy proven effective for treating the core deficits in ASD. There is also a paucity of biomarkers for autism. Both genetic and environmental factors are thought to contribute to autism susceptibility (Courchesne, 2007; Geschwind, 2009; Südhof, 2008; Ramocki & Zoghbi, 2008), but because only some of the genetic factors have been identified unequivocally thus far (Cook & Scherer, 2008; Levitt & Campbell, 2009), finding effective treatments that target the underlying causes of ASD remains a major challenge.

Identifying endophenotypes and biomarkers for complex and heterogeneous disorders such as ASD are important not only to elucidate their etiologies, but also to identify suitable biochemical molecules and pathways to target the treatment of core deficits. In this review, we present a rationale that neuronal nicotinic acetylcholine receptor (nAChR) alterations are biomarkers for ASD and that specific nAChRs subtypes are likely to be useful therapeutic targets for the treatment of core deficits. This rationale is based on the synthesis of emerging evidence from multiple types of studies, including our own, using postmortem, genetic, functional, and molecular neurobiological methodologies from two disparate areas of research – autism spectrum disorders and nicotine dependence.

2. Neuronal nicotinic acetylcholine receptors

Neuronal nAChRs are a family of ion channels that are permeable to both monovalent (Na^+ and K^+) and divalent (Ca^{++}) cations and are formed by assembly of different combinations of subunits termed α2 to α10 and β2 to β4. These channels are heteropentamers with the exception of the α7 nAChR, which is usually a homopentamer (Lindstrom, 1996; Lindstrom, 1997; Sargent, 1993). In neurons, nAChRs regulate the release of many different neurotransmitters including acetylcholine, dopamine, γ-aminobutyric acid (GABA), glutamate, and serotonin at presynaptic sites (McGehee & Role, 1996) and mediate fast synaptic transmission at postsynaptic sites (Zhang et al., 1996; Frazier et al., 1998a; Frazier et al., 1998b). These functions have a broad range of physiological effects on reward, analgesia, anxiety, affect, locomotion, attention, mood, learning, memory, and executive function (Miwa et al., 2011). nAChRs can also modulate neurite growth (Pugh & Berg, 1994; Lipton et al., 1988) and cell survival (Pugh & Margiotto, 2000; Messi et al., 1997; Kihara et al., 1997; 1998; 2001). nAChRs have been intensely studied for many decades not only to understand their normal physiological roles, but more importantly to elucidate their pathophysiological role in mediating addiction to nicotine in tobacco, because tobacco use among smokers, in particular, results in greater than 400,000 deaths per year in the U.S. alone. In addition to their role in nicotine addiction, nAChR dysfunctions are also implicated in other disorders, including Alzheimer's disease, Parkinson's disease, schizophrenia, attention deficit-hyperactivity disorder, anxiety disorders, Tourette's syndrome, and depression (Newhouse & Kelton, 2000; Newhouse et al., 2004; Mineur & Picciotto, 2010).

3. Alterations of nAChRs in ASD

3.1 Changes in α4β2 nAChR expression

Examination of postmortem brains of individuals with ASD has identified major nAChR abnormalities in multiple postmortem studies. In the first such study to be undertaken, postmortem tissue from 7 adults with a mean age of 24 years was examined. High-affinity ^{3}[H]epibatidine binding was reported to be significantly reduced in the frontal and parietal cortex of these individuals with ASD compared to age-matched controls. Furthermore, immunohistochemical analyses showed that the loss of ^{3}[H]epibatidine correlated with reduced expression of the α4 and β2 nAChR subunits. Notably, the mRNA for these two nAChR subunits was not significantly decreased, suggesting that the reduction in nAChR subunit levels resulted from an impaired posttranslational mechanism. Also, ^{3}[H]pirenzepine binding to M1 and M2 muscarinic AChRs (mAChRs) was not significantly altered, suggesting that the loss of nAChR expression resulted from deregulation of a posttranslational mechanism that specifically affected nAChRs, but not mAChRs (Perry et al., 2001). In a subsequent study, postmortem tissue from 8 adults with a mean age of 24 years was examined. Again, high-affinity [^{3}H]epibatidine binding was reported to be significantly reduced by greater than 50% in the cerebellar cortex of individuals with ASD. High-resolution analyses of the autoradiographic data indicated that the loss of ^{3}[H]epibatidine binding occurred in the granule cell layer, the Purkinje layer, and the molecular cell layer of the cerebellum of individuals with ASD compared to age-matched controls. Significant reduction in the expression of the α4 nAChR subunit, but not its mRNA (Lee et al., 2002), was also observed and is consistent with the notion that α4β2 nAChR loss results from an impaired posttranslational mechanism regulating it expression. In a third study, immunohistochemical analysis of postmortem brains from 3 adults with ASD of mean age 29 years surprisingly showed no changes in the expression of the α4 nAChR subunit in the thalamus compared to age-matched controls. However, reduction of the β2 nAChR subunit was observed in the paraventricular nucleus and nucleus reuniens of the thalamus (Martin-Ruiz et al., 2004).

3.2 Changes in α7 nAChR expression

In contrast to the loss of ^{3}[H]epibatidine binding and decreased expression of the α4 and β2 nAChR subunits, no significant change in the binding of ^{125}I-α-bungarotoxin to the α7 nAChR or immunohistological detection of the α7 nAChR (Perry et al., 2001) was reported in the frontal and parietal cortex. In the cerebellar cortex, however, binding of α-bungarotoxin to the α7 nAChR and immunohistological detection of the α7 nAChR did show a significant increase in the expression of the α7 nAChR in the granule cell layer, but not in the Purkinje cells or the molecular cell layer. Interestingly, similar to the β2 nAChR subunit, reduction of the α7 nAChR subunit was also observed in the paraventricular nucleus and nucleus reuniens of the thalamus. Thus, alterations in the expression of both the α4β2 nAChR and the α7 nAChR in individuals with ASD appears to show regional specificity (Perry et al., 2001; Lee et al., 2002; Martin-Ruiz et al., 2004), suggesting that these changes are compensatory and result from altered homeostasis of neural networks, rather than the direct effect of a single molecule in a particular molecular pathway.

Two recent studies on rare genomic microdeletions and copy-number variations (CNVs) revealed a possible involvement of the CHRNA7 gene in some cases of autism. The first study investigated segmental duplications at breakpoints (BP4–BP5) of chromosome 15q13.2q13.3 from 1441 individuals with autism from 751 families in the Autism Genetic Resource Exchange (AGRE) repository (Miller et al, 2009). This genomic sequence spans over 1.5 Mb and includes CHRNA7. From this cohort 10 patients were identified with genomic imbalance at chromosome 15q13.2q13.3, including five with BP4–BP5 microdeletions. Among the 1420 parents and 132 unaffected/unknown siblings no cases of BP4–BP5 microdeletion were found. The second study on genomic CNVs explored the genetic contribution to ASD in a large cohort of families (Simons Simplex Collection consisting of 915 families) with a single autistic child and at least one unaffected sibling (Levy et al., 2011). The contribution of the transmission of "ultrarare" variants to ASD, in particular inherited genomic duplications was also estimated. A transmitted duplication within the CHRNA7 gene was observed in 8 autistic children and 3 unaffected siblings within 6 families. A further network-based analysis of genetic associations (NETBAG) of that dataset strengthened the involvement of CHRNA7 as one of the genes affected by rare de novo CNVs in autism (Gilman et al., 2011).

4. nAChRs modulate multiple behaviors deficient in ASD

ASD is defined by three behavioral deficits, impaired social interactions, repetitive behaviors, and delayed language. Multiple studies using animal models implicate a functional role for nAChRs in some of these behavioral deficits in ASD. β2-containing nAChRs regulate executive and social behaviors in studies using β2 nAChR subunit knockout mice (Granon et al., 2003). Knockout β2 nAChR mice show a decrease in slow exploratory behavior - a measure of cognitive function during which animals slowly and precisely explore their environment, a lack of sensitization to novel stimuli, and abnormal social behavior during aggressive confrontations with other mice (Granon et al., 2003). Recovery of the slow exploratory behavior was observed by injecting a lentiviral vector expressing the β2 nAChR subunit into the ventral tegmental area (VTA) in the knockout mice (Maskos et al., 2005). Re-expressing the β2 nAChR subunit in the prefrontal cortex also improves social abnormalities in this knockout mouse. Increased social interaction and decreased novel exploration in a social interaction paradigm with concurrent motivation was ameliorated after stereotaxically injecting the β2 nAChR subunit into the prelimbic area of the prefrontal cortex (PFC) (Avale et al., 2011).

As previously mentioned, nAChR dysfunction is also implicated in several other neurological disorders with repetitive behavior. We suggest here that similarities in behaviors across those neurological conditions, as well as high prevalence of simultaneity suggest a possible shared underlying mechanism. Moreover, there has been a recent push to redefine repetitive behavior in these neuropsychiatric disorders and instead characterize stereotypies into disorder-related endophenotypes rather than separate disorder-specific symptoms (Kas et al., 2007, Langen et al., 2011). Tourette's syndrome (TS), obsessive compulsive disorder (OCD), and attention deficit hyperactivity disorder (ADHD), all involve disordered cortical-basal ganglia circuitry and all can be successfully treated with drugs acting on nAChRs. The basal ganglia and orbitofrontal cortex, both regions highly innervated by nicotinic acetylcholine receptor rich interneurons are hyperactive during PET/SPECT studies of OCD (Baxter et al., 1988) and hypoactive in studies of ADHD (Zametkin et al., 1990) and TS (Braun et al., 1995). The orbitofrontal cortex controls inhibition and disinhibition of behavior, and lesions in this area are sufficient to cause impulsive and inappropriate behavior. Nicotine or an analog alone has demonstrated potential to treat repetitive behaviors in these disorders. A transdermal nicotine patch, administered as therapy for TS, decreases the severity and frequency of tics, a compulsory symptom of TS (Sanberg, 1997). Nicotine gum administered to OCD patients previously resistant to other treatment clinically improved behavior (Carlsson, 2001; Pasquini et al., 2005). Interestingly, clomipramine, an SSRI commonly prescribed for the treatment of OCD, also acts on nAChRs (Lopez-Valdes, 2002). Lastly, (-)-Nicotine and ABT-418, an α4β2 nAChR agonist (Potter et al., 1999), both successfully treat adults with ADHD (Levin and Simon, 1998; Wilens et al., 1999). It is interesting to note that hyperactivity, tics, and obsessive compulsive disorder are all common comorbid disorders seen in patients with ASD with approximately 59% of ASD patients having impulsivity problems, 8-10% having tics, and 37% having OCD (Levy et al., 2009). Although it is clear that similar neurocircuitry is involved in several disorders with repetitive behavior, further research is needed to determine whether the underlying mechanisms causing this dysfunction overlap in TS, OCD, ADHD, and in ASD.

nAChRs also are involved in several other non-core, but frequently occurring symptoms in ASD. The most common comorbid disorders and symptoms associated with ASD are psychiatric (e.g., depression and anxiety), neurological (e.g., epilepsy), sleep, and sensory (e.g., tactile) disorders. Epilepsy occurs in 5-49% of people with autism (Levy et al., 2009). Genetic abnormalities in CHRN4A and CHRNB2, encoding the α and β nAChR subunits respectively, are sufficient to cause autosomal dominant nocturnal frontal lobe epilepsy (ADNFLE) (De Fusco, 2000; Bertrand, 2002; Steinlein, 2002; Hoda, 2009), however ADNFLE is not associated with ASD. 52-73% of patients with ASD experience sleep disruption and 43-84% experience anxiety disorders. Knocking out the α4 nAChR subunit increases anxiety in mice (Ross et al., 2000) and the β2 nAChR knockout animal shows abnormal sleep pattern (Lena et al., 2004). These studies demonstrate that behaviors regulated by nAChRs are disparate and commonly aberrant in ASD and suggest the potential for nAChR-acting drugs in the treatment of ASD.

Lastly, there is accumulating evidence that the immune system is disrupted in individuals with ASD (Careaga et al., 2010). Elevated levels of chemokines have been detected in the brains and cerebrospinal fluid (Chez et al., 2007; Wills et al., 2009) as well as the plasma (Ashwood et al., 2011a) of individuals with ASD, and this elevation correlated with more impaired behavior (Ashwood et al., 2011b). Furthermore, postmortem studies of individuals with autism also detected presence of activated neuroglial cells in their brain (Vargas et al., 2005). In a recent study, activated microglia were detected in the dorsolateral PFC in 5 out of 13 samples, 2 of which were under the ages of 6 years (Morgan et al., 2010). These results suggest that inflammation of the central nervous system, at least in some individuals, may contribute to the neuropathology of ASD. Thus, suppression of neuroinflammation by targeting α7 nAChRs in ASD may be potentially beneficial.

5. Neurexin and neuroligin deficits in ASD

The neurexins are cell adhesion molecules encoded by three genes corresponding to neurexins 1, 2 and 3 (Missler & Südhof, 1998; Lise & El-Husseini, 2006). As a result of transcriptional initiation from two different promoters, each neurexin gene encodes a longer α-neurexin protein and a shorter β-neurexin protein. The proteins are identical from their intracellular C-termini through their transmembrane domains, glycosylation-rich domains and the sixth LNS domain of α-neurexin, which corresponds to the only LNS domain in β-neurexin. They have divergent N-terminal extracellular domains, which allow for interactions with multiple proteins. Additionally, alternative splicing at multiple splice sites within each gene can give rise to more than one thousand different isoforms, which differ only in their extracellular domains. Neurexins recruit N- and P/Q-type calcium channels to active zones of presynaptic terminals through scaffolding proteins, including calmodulin-associated serine/threonine kinase (CASK) (Hata et al., 1996; Missler et al., 2003; Zhang et al., 2005). α-neurexins were reported to specifically induce GABAergic postsynaptic differentiation (Kang et al., 2008). The enormous structural diversity of the neurexins suggests that they are involved in a multitude of physiological functions yet to be elucidated.

Results from a linkage and copy number variation analysis conducted by the Autism Genome Project Consortium (Szatmari et al., 2007) show that neurexin-1 dysfunction is associated with ASD. This conclusion has been corroborated in multiple linkage analysis studies since (Kim et al., 2008; Marshall et al., 2008) and in analysis of structural variants in the α- and β-neurexin genes (Zahir et al., 2008; Feng et al., 2006; Yan et al., 2008; Gai et al., 2011; Gauthier et al., 2011). Neurexin knock-out animals have provided insights into the functions of the neurexin family. Neurexin 1/2/3α- triple knock-out animals die perinatally and have reduced spontaneous and evoked neurotransmission at glutamatergic and GABAergic synapses, demonstrating that α-neurexins are necessary for neurotransmitter release at synapses (Missler et al., 2003). Additionally, mice lacking neurexins have impaired neuroendocrine secretion (Dudanova et al., 2006), which may mirror some children with autism that exhibit dysfunction of the hypothalamic-pituitary-adrenocortical system, possibly due to altered neuroendocrine regulation (Corbett et al., 2006). Similar to the neurexin triple knockout animals, mice lacking neurexin1/2α or neurexin2/3α die within 1 month after birth and have reduced neurotransmission. Analyses of brain morphology in α-neurexin knockouts revealed no major impairments in synapse formation, but minor reductions in dendrite branch length and spine numbers were detected, suggesting they are important in synapse maturation more so than formation (Dudanova et al., 2007). None of the single α-neurexin knock-out animals have dramatic phenotypes, with the neurexin-2α knock-out animals showing the least severe phenotype (Craig & Kang, 2007). In the absence of neurexin-1α, miniature excitatory postsynaptic currents were reduced in recordings from hippocampal slices. Behaviorally, the neurexin-1α–deficient mice were identical to wild-type mice in multiple social interactions, but displayed decreased grooming behavior, impaired nest building, decreased pre-pulse inhibition, and improved motor learning in behavioral studies (Etherton et al., 2009). While the neurexin-1α–deficient mice display behavioral phenotypes similar to what is seen in autism they are not sufficient to explain ASD yet they still provide a useful but limited model of ASD. The β-neurexin and combined α- and β-neurexin knockout animals have not yet been fully evaluated.

The neuroligins are encoded by five differentially spliced genes that encode multiple neuroligin isoforms (Zhang et al., 2005; Boucard et al., 2005). In complementary roles, neuroligins, the postsynaptic binding partners of neurexins, recruit N-methyl-D-aspartate (NMDA) receptors and $GABA_A$ receptors through their interactions with scaffolding proteins such as post-synaptic density 95 (PSD-95) and gephyrin, respectively (Graf et al., 2004; Nam & Chen, 2005; Chih et al., 2006; Poulopoulos et al., 2009). Thus, bi-directional interactions between neurexins and neuroligins appear to serve a critical function in the assembly and maturation of both glutamatergic and GABAergic synapses through recruitment of the requisite presynaptic and postsynaptic components of neurons (Dean & Dreshbach, 2006; Craig & Kang, 2007; Sudhof, 2008).

Neuroligins are strongly implicated in ASD. Chromosomal rearrangements and copy number variations in neuroligin-1 are linked to autism (Konstantareas & Homatidis, 1999; Ylisaukko-oja et al., 2005; Glessner et al., 2009). There is also evidence that mutations in neuroligin-3 and neuroligin-4 are found in patients with ASD (Laumonnier et al., 2004; Jamain et al., 2003). In addition mouse models support a role for neuroligins in ASD. Neuroligin-1 knock-out mice are viable and fertile, but also have synaptic dysfunctions (Chubykin et al., 2007). At the molecular level, the NMDA/AMPA ratio at corticostriatal synapses is reduced, which is associated with repetitive grooming that may mirror some of the repetitive behaviors seen in autistic patients (Blundell et al., 2010). In contrast to neuroligin-1-deficient mice, which show impaired NMDA receptor signaling, neuroligin-2 knock-out animals have deficits in inhibitory synaptic transmission (Chubykin et al., 2007). Behaviorally, neuroligin-2 knock-out mice exhibit increased anxiety, but normal social interactions (Blundell et al., 2009), similar to the neurexin-1α-deficient mice. Mutations in neuroligin-3 and neuroligin-4 lead to intracellular retention of the mutant proteins (Chih et al., 2004; Comoletti et al., 2004). The neuroligin-3 R451C mutation is a gain of function mutation. Mice with this point mutation exhibited impaired social interactions and increased inhibitory synaptic transmission (Tabuchi et al., 2007). Mice lacking neuroligin-4 correspond to loss-of-function mutations in human neuroligin-4 and show deficits in reciprocal social interactions and ultrasonic communication (Jamain et al., 2008). Neuroligin 1/2/3α triple knock-out animals die at birth, but similar to their α-neurexin-deficient counterparts, do not show dramatic reductions in synapse numbers or brain architecture, but do have severely impaired synaptic transmission (Varoqueaux et al., 2006).

The studies of the neurexin and neuroligin functions indicate a role for them in proper synaptic function but not synapse formation. Although it is clear that the deficits of neurexins and neuroligins play a role in ASD, understanding their interactions with receptors will provide additional insight into their functions.

6. Neurexins associate with multiple receptors, including nAChRs

Accumulating evidence indicates that neurexins interact directly with more than the neuroligins. Our laboratory was the first to provide experimental evidence for direct interactions between neurexins and receptors by showing that neurexin-1β coimmunoprecipitates with recombinant α4β2 nAChRs when expressed in heterologous cells (Cheng et al., 2009). Functionally, the neurexin-1β regulates targeting of α4β2 nAChRs to pre-synaptic terminals in neurons (Cheng et al., 2009). Complementary studies report a role for neurexin-1 and neuroligin-1 in recruitment of α3-containing nAChRs to the post-synaptic density (Conroy et al., 2007; Ross & Conroy, 2008). In addition, recent studies show that neurexins interact with multiple receptors. First, neurexin-1β interacts with $GABA_A$ receptors; this interaction modulates the cell surface expression levels of the $GABA_A$ receptors but not its functions *per se* (Zhang et al., 2010). Second, leucine-rich repeat transmembrane protein (LRRTM2) binds trans-synaptically to both neurexin-1α and-1β and induces presynaptic differentiation at excitatory synapses (Ko et al., 2009; de Wit et al., 2009; Siddiqui et al., 2010). Knock-down of LRRTM2 in the rat dentate gyrus shows a large reduction in AMPAR-mediated EPSCs in *in vivo* recordings from granule cells in hippocampal slices. Furthermore, the association between neurexin-1 and LRRTM2 is a functional interaction. When neurexin-1 is knocked-down in hippocampal neurons, LRRTM2 is unable to induce presynaptic differentiation (de Wit et al., 2009). Finally, neurexins associate with GluRδ2 receptors via a soluble protein called cerebellin -1 precursor protein (Cbln1) (Uemura et al., 2010). In the Cbln1 knockout mice, the synaptogenic activity of GluRδ2 receptor is lost. Thus, GluRδ2 mediates cerebellar synapse formation by interacting with presynaptic neurexins via Cbln1.

7. Genetic variants of neurexin-1 are linked to nicotine dependence

A recent high-density genome-wide association study for nicotine dependence linked single nucleotide polymorphisms (SNP) in the neurexin-1 gene to the development of nicotine dependence and thus smoking behavior (Bierut et al., 2007). A second independent study also showed linkage between a variant of the neurexin-1 gene and nicotine dependence in smokers of European and African-American ancestry (Nussbaum et al., 2008). These results, along with the fact that neurexins functionally target α4β2 nAChRs to synapses, implicate neurexins in the etiology of other neurological diseases typically associated with pathophysiological functions of nAChRs. α4β2 nAChRs mediate essential features of nicotine addiction including reward, tolerance, and sensitization (Tapper et al., 2004). Thus, functions are likely to be affected by changes in the expression levels of neurexin-1. The exact mechanism by which neurexin-1α and -1β splicing is regulated to generate the predicted hundreds of neurexin-1 isoforms remains to be elucidated. It is possible that a regulatory SNP in the intron of the neurexin-1 gene could modulate neurexin-1 expression or splicing efficiency and thus influence nAChR functions by regulating their synaptic targeting efficiency. Because there are hundreds of neurexin-1α isoforms, the linkage between neurexin-1 gene variants, α4β2 nAChR synaptic targeting, and nicotine dependence requires additional studies. Nevertheless, the functional linkage between neurexin-1 and α4β2 nAChR and their converging roles in nicotine dependence suggests that α4β2 nAChR activity may regulate neurexin-1 gene expression.

8. nAChR modulate excitation-inhibition balance

There is strong evidence that some forms of ASD are caused by an imbalance of excitatory and inhibitory synaptic transmission in neuronal circuits that are responsible for the establishment of language processing and social behavior during prenatal and postnatal brain development. Increased glutamatergic (excitatory) signaling or suppressed GABAergic (inhibitory) signaling is sufficient to disrupt the excitatory/inhibitory balance in local circuit-plasticity (Rubenstein & Merzenich, 2003). A hyperexcitable cortex is poorly differentiated functionally and therefore inherently unstable and susceptible to epilepsy. This might explain why, in addition to the autistic core symptoms, an average of ~30% of individuals with ASD develop clinically apparent seizures (Gillberg & Billstedt, 2000). In several mouse models of autism this lack of homeostasis of excitatory and inhibitory signaling was observed (Tabuchi et al., 2007; Gogolla et al., 2009). In the frontal cortex, cholinergic transmission can modulate cortical tone establishing a homeostasis of excitatory and inhibitory signals (Aracri et al., 2010). In layer V of the prefrontal cortex, nAChR activation increases the threshold for activating glutamatergic synapses (Couey et al., 2007), whereas GABA release is stimulated in several cortical layers by nAChR activation (Alkondon et al., 2000).

We posit that some of the regulatory effects of balancing inhibitory and excitatory synaptic transmission are mediated by synaptic targeting of nAChRs by neurexins. This results in the change of expression levels of nAChRs in various brain regions of autistic individuals. Therefore allosteric modulators or direct agonists targeting nAChRs by might be useful to restore the imbalance of excitatory and inhibitory synaptic transmission caused by deregulated expression of neurexin-1.

9. Nicotinic receptors as biomarkers for ASD

9.1 Positron Emission Tomography ligands for α4β2 nAChRs

The alterations in nAChRs in ASD may also serve as an early molecular biomarker, detectable by imaging tools such as positron emission tomography (PET), the most advanced modality for non-invasive study of receptors. Monitoring the reversal of the loss of α4β2 nAChR in the frontal, parietal, and cerebellar cortex and the upregulation of α7 nAChR in the cerebellar cortex by PET imaging in the brains of individuals with ASD might provide a clinical tool to complement behavioral tests needed to assess the effectiveness of novel pharmacotherapies for autism.

Three radiotracers, [^{11}C]nicotine, (S)-3- (azetidin-2-ylmethoxy)-2-[^{18}F]fluoropyridine (2-[^{18}F]FA) and (S)-5- (azetidin-2-ylmethoxy)-2-[^{18}F]fluoropyridine (6-[^{18}F]FA), have been used for studying α4β2 nAChRs in the human brain using PET. The PET imaging properties of these radioligands are not perfect however. Poor signal-to-noise ratios and other drawbacks of [^{11}C]nicotine suggest that this radiotracer is not well suited for quantitative imaging in animals and humans. 2-[^{18}F]FA is the only currently available radioligand for quantitative imaging nAChR in humans. The "slow" brain kinetics of 2-[^{18}F]FA and 6-[^{18}F]FA hamper mathematical modeling and reliable kinetic parameter estimation since it takes many hours of PET scanning (5–7 h) for the tracer radioactivity to reach a spatial-temporal steady state (Horti et al., 2010). Another crucial problem with 2-[^{18}F]FA and 6-[^{18}F]FA is relatively low binding potential (BP) in extrathalamic regions (BP $\leq$ 0.6–0.8), including the cortex, which has a lower nAChR density. Altered densities of cortical and striatal nAChRs in neurodegenerative diseases (Pimlott et al., 2004) and schizophrenia (Ochoa & Lasalde-Dominicci, 2007) illustrates the importance of imaging extrathalamic nAChRs.

A variety of radioligands with fast regional brain kinetics have been presented in non-human primates and pigs. Analogs of epibatidine showed "rapid" brain kinetics and improved BP (Gao et al., 2007, 2008). One compound of the series, (-)-2-(6-[^{18}F]fluoro-2,3'-bipyridin-5'-yl)-7-methyl-7-aza-bicyclo[2.2.1]heptane ([^{18}F]JHU87522 or [^{18}F]AZAN) exhibited better imaging properties in animal studies than those of 2[^{18}F]FA and 6-[^{18}F]FA including a greater BP value and faster brain kinetics. In addition, the brain uptake of [^{18}F]AZAN is greater and its acute toxicity is lower. Most available PET and single photon emission computed tomography (SPECT) imaging agents for nAChR are agonists and these nAChR-agonists are toxic when injected at high doses. Unlike 2-FA that is nAChR agonist, AZAN displays properties of functional antagonist of α4β2 nAChR. Currently, AZAN is undergoing toxicological studies that will determine if this radioligand is sufficiently safe for clinical application as a PET radiotracer. If [^{18}F]AZAN is safe for human PET studies, there are strong indications that it could become the radiotracer of choice for PET imaging of nAChR in human brain (Horti et al., 2010).

9.2 Positron Emission Tomography ligands for α7 nAChRs

Several radiotracers were developed for selective imaging of the α7 nAChRs in the human brain for PET and SPECT (Dolle et al., 2001; Pomper et al., 2005; Ogawa et al., 2006). Despite these efforts, there have been no clinical studies using these radioligands for α7 nAChRs in the human brain.

Very recently, 4-[^{11}C]methylphenyl 2,5-diazabicy- clo[3.2.2]nonane-2-carboxylate ([^{11}C]CHIBA-1001) was developed as a novel PET ligand for α7 nAChRs in the conscious monkey brain. An *in vitro* binding study showed that the IC_{50} value of CHIBA-1001 for [^{125}I]α-bungarotoxin binding to the rat brain homogenates was 45.8 nM. [^{11}C]CHIBA-1001 distribution in the monkey brain measured by PET was consistent with the regional distribution of α7 nAChRs. Moreover, brain uptake of [^{11}C]CHIBA-1001 was dose-dependently blocked by pretreatment with the selective α7 nAChR agonist SSR180711, but was not altered by the selective α4β2 nAChR agonist A-85380 (Hashimoto et al., 2008).

In the human brain, [^{11}C]CHIBA-1001 was found widely distributed in all brain regions. The regional distribution pattern of [^{11}C]CHIBA- 1001 is consistent with what is expected *in vitro* (Falk et al., 2003; Court et al., 1999; 2001; Marutle et al., 2001), but different from that of α4β2 nAChRs (Clementi, 2004). However, it is slightly different from the regional distribution in the monkey brain (Hashimoto et al., 2008). In the human brain, remarkable radioactivity accumulation was observed in the cerebellum. These findings suggest that [^{11}C]CHIBA-1001 is a suitable radioligand for imaging α7 nAChRs in the human brain, as it offers acceptable dosimetry and pharmacological safety at the dose required for adequate PET imaging (Toyohara et al., 2009).

These recent advances in the development of new nAChR PET radioligands, like [^{18}F]AZAN for $\alpha4\beta2$ nAChRs and [^{11}C]CHIBA-1001 for $\alpha7$ nAChRs with fast kinetics and low toxicity will provide promising tools for monitoring alterations of brain nAChR especially in young patients with ASD. The principal downside to the use of PET is the unknown risk of using radioactive ligands and sedatives, especially in younger individuals, to perform PET scans.

10. Nicotinic drugs as therapeutic agents for ASD

10.1 Agonists

10.1.1 $\alpha4\beta2$ nAChRs

The extensive loss of $\alpha4\beta2$ nAChRs in some individuals with ASD provide a rationale for exploratory trials of drugs that can upregulate and activate $\alpha4\beta2$ nAChRs and thus compensate for their loss both physically and functionally. The panoply of drugs developed over the last few decades for smoking cessation therapy as well as other disorders with pathophysiological roles for nAChRs (Taly et al., 2009), offers a large selection of drugs that are likely to be specific for $\alpha4\beta2$ nAChRs and capable of upregulating them. Varenicline (Chantix), one such drug that has FDA approval for use in smoking cessation therapy is a partial agonist of the $\alpha4\beta2$ nAChRs (Coe et al; 2005) and of interest for treatment of ASD. Although varenicline is also a full agonist of the $\alpha7$ AChR (Mihalak et al., 2006), its relative specificity for $\alpha4\beta2$ nAChRs is thought to be due to differences in its EC_{50} for activation of $\alpha4\beta2$ nAChRs versus $\alpha7$ nAChRs, as well as a function of the low concentrations at which it is used clinically for anti-smoking therapy (Niaura et al., 2006). Thus it has become one of the most widely used smoking cessation drugs with millions of users worldwide and shows little sympathetic and parasympathetic complications from cross activation of ganglionic nAChRs ($\alpha3\beta4$ nAChRs). Interestingly, much like nicotine, varenicline can upregulate $\alpha4\beta2$ nAChRs *in vitro*. Finally, as a partial agonist, it has the additional benefit of providing chronic low-level activation of $\alpha4\beta2$ nAChRs (Papke et al., 2011) and possibly associated downstream intracellular signaling pathways. Varenicline has been shown to change behaviors in some smokers, and a public health advisory from the FDA includes warnings of increased suicidal thoughts and actions. It is important to note, however, that the increase in suicidal thoughts and actions may occur in only a subpopulation of individuals taking varenicline as there is ample evidence that smoking may be more prevalent in those individuals with comorbid neuropsychiatric conditions, including schizophrenia (Adler et al., 1993; Dalack et al., 1999). This may explain behavioral changes reported among smokers using varenicline if individuals have subclinical neuropsychiatric conditions. This idea has been supported by a recent study reporting that there was no clear evidence that varenicline use in itself was associated with an increased risk for depression or suicidal thoughts (Gunnell et al., 2010). Also, unlike in schizophrenia, the prevalence of smoking in individuals with ASD is low (Bejerot & Nylander, 2003), possibly because the loss of $\alpha4\beta2$ nAChRs occurs early in development - a clinical feature further strengthening the utility of using $\alpha4\beta2$ nAChRs loss as a biomarker for ASD. Nevertheless, any clinical trial of varenicline for individuals with ASD should require close monitoring of possible suicidal ideation given the heterogeneity of causes expected for ASD, some of which may overlap with schizophrenia (Kirov et al., 2009).

10.1.2 α7 nAChRs

It is possible to use α7 nAChR agonists to treat neuroinflammation in ASD. There is strong evidence that activation of the α7 nAChR expressed on monocytes and macrophage, by inhibiting NF-kappaB nuclear translocation, suppresses cytokine release by them (Wang et al., 2003), and that this cholinergic anti-inflammatory pathway that provides a bidirectional link between the nervous and immune system, inhibits the innate immune response (Rosas-Ballina & Tracey, 2009). Hence, a reasonable case can be made for the use of α7 nAChR agonists to treat neuroinflammation in ASD. Individuals could be stratified by monitoring brain inflammation by the uptake of the microglial marker, [^{11}C]PK11195, a PET ligand useful for detecting peripheral benzodiazepine receptors expressed in high amounts in activated microglia (Rojas et al., 2007). However, given that α7 AChR appears to be pathologically upregulated in cerebellum of some individuals with ASD, caution is advocated in the use of α7 AChR agonists to treat ASD. The primary challenge is that the net behavioral benefit from suppressing neuroinflammation mediated by microglia versus over stimulating upregulated α7 AChRs in the granule cell layer, cannot be predicted *a priori*. Two different α7 nAChR agonists have been used to treat schizophrenia; drugs that might be repurposed for use in individuals with ASD and detectable neuroinflammation.

One of these drugs, GTS-21, or 3-(2,4-dimethoxybenzylidene)-anabaseine (DMXB-A) is a partial agonist of α7 nAChRs may have beneficial effects in ASD patients. In healthy control subjects, DMXB-A improves attention, working memory, and episodic memory (Kitagawa et al., 2003). The default network, which has been widely reported to be abnormal in schizophrenia (Garrity et al., 2007), is a functionally connected network of brain regions that includes the posterior cingulate cortex, cuneus/precuneus, medial prefrontal cortex, medial temporal lobe, and inferior parietal cortices (Buckner et al., 2008; Tregellas et al, 2011). Altered default network activity has been shown to be a result of DMXB-A administration to patients with schizophrenia (Tregellas et al., 2011), with decreased expression of α7 nAChRs (Freedman et al., 1995).

A second candidate drug, Tropisetron is a partial agonist of the α7 nAChR. Auditory sensory gating P50 deficits are correlated with neuropsychological deficits in attention, one of the principal cognitive disturbances in schizophrenia. In a clinical trial with 33 schizophrenic patients administration of tropisetron, without placebo, significantly improved auditory sensory gating P50 deficits in non-smoking patients with schizophrenia (Shiina et al., 2010). In mice, the early postnatal period represents a critical time window essential for brain development. The administration of tropisetron from postnatal days 2-12 (P2-P12) in mice did not induce significant cognitive, schizophrenia-like or emotional alterations in tropisetron-treated animals as compared to controls, when tested in multiple behavioral assays (Inta et al., 2011).

10.2 Positive allosteric modulators

Galantamine is an acetylcholinesterase inhibitor that also acts as a positive allosteric modulator at the α4β2 and α7 nAChRs (Dajas-Bailador et al., 2003; Samochocki et al., 2003; Schilström et al., 2007). In two studies with small numbers of subjects it has been reported that galantamine showed potential benefits for attention, memory, and psychomotor speed in schizophrenia (Schubert et al., 2006; Lee et al., 2007). An unpublished study from Johnson and Johnson failed to find an advantage for galantamine on a measure of global cognition (clinicaltrials.gov, trial number: NCT 00077727). In a 12-week open-label trial of galantamine, thirteen children with autism, previously unmedicated, (mean age, 8.8 +/- 3.5 years) showed a significant reduction in parent-rated irritability and social withdrawal on the Aberrant Behavior Checklist (ABC), as well as significant improvements in emotional lability and inattention on the Conners' Parent Rating Scale—Revised (Nicolson et al., 2006). Similarly, clinical ratings showed reductions in the anger subscale of the Children's Psychiatric Rating Scale. Eight of 13 participants were rated as responders on the basis of their improvement scores on the Clinical Global Impressions scale. The allosteric properties of galantamine could directly lead to increased release of acetylcholine and activation of postsynaptic nAChRs (Samochocki et al., 2003) or act indirectly through its effects on the release of other neurotransmitters, especially glutamate and dopamine (Schilström et al., 2007; Wang et al., 2007).

It has been demonstrated that amyloid-β precursor protein (APP) is upregulated in a mouse model for Fragile X mental retardation (FXS) (Westmark et al., 2008) and two clinical studies have reported higher levels of APP in children with autism. In the first study, affected children expressed sAPP at 2 or more times the levels of children without autism and up to 4 times more than children with mild autism (Sokol et al., 2006). In the second study, elevated plasma sAPPα was found in 60% of known autistic children (n = 25) compared to healthy age-matched controls (Bailey et al., 2008). Recent studies showed that galantamine allosterically modulates microglial nAChRs and increases microglial beta-amyloid (Aβ) phagocytosis (Wang et al., 2007; Takata et al., 2010).

Collectively, these studies suggest that positive allosteric modulators of α4β2 nAChRs, when used by themselves or in conjunction with agonists, may be beneficial in correcting deficits in the functions of α4β2 nAChRs and thereby core deficits of ASD.

11. Conclusions

This review presents a reasonable rationale based on synthesis of the literature that nAChRs are suitable biomarkers as well as therapeutic targets for addressing core deficits in ASD. Multiple lines of evidence show that nAChRs can modulate many of the functions deficient in individuals with ASD. Furthermore, neuropathological findings, albeit small in numbers, show significant alterations in both α4β2 nAChRs and α7 nAChRs. In the cerebellum, an anatomical area contributing significantly to the etiology of ASD, α4β2 nAChRs are deficient, and α7 nAChRs are upregulated. These findings suggest that well developed PET ligands for both these nAChR subtypes can be used to monitor changes in their expression in response to treatment, behavioral or pharmacological. A novel functional linkage between neurexin-1 and α4β2 nAChR and their converging roles in nicotine dependence suggests that α4β2 nAChR activity may regulate neurexin-1 gene expression. Additionally, agonists and positive allosteric modulators of the α4β2 AChRs are likely to be therapeutic agents that can help restore α4β2 nAChRs expression levels in the brains of individuals with ASD, based on known effects of these agents. A case can be made for the use of α7 nAChRs to reduce neuroinflammation in the brain in those ASD individuals with such clinical pathology. The ultimate hope is that these agents, when administered early in development, by their presumed ability to modulate a number of different neurotransmitter systems and associated signaling pathways, could help correct core deficits associated with ASD.

12. Acknowledgements

R. A. was a recipient of an Essel Independent Investigator Award from the National Alliance for Research on Schizophrenia and Depression. S. A. A. is a recipient of a Ruth Kirschstein National Research Service Award from the National Institute of Drug Abuse. Support from the National Institutes of Health (NIDA and NIGMS), Autism Speaks, the Ohio State University College of Medicine Medical Research Fund, the Marci and Bill Ingram Comprehensive Center for Autism Spectrum Disorders, and the Gertz family to R. A. is gratefully acknowledged. Thanks to Dr. Eugene Arnold at the OSU Nisonger Center for providing a clinical perspective on ASD. Thanks to Dr. Gregg Wells for editorial comments. Thanks to all the families affected by ASD whose tireless dedication to raising awareness, advocacy and research funds through the annual Columbus Walk Now for Autism Speaks inspire and support the authors' efforts in ASD research.

13. References

Adler, L. E., Hoffer, L. D., Wiser, A. & Freedman, R. (1993). Normalization of auditory physiology by cigarette smoking in schizophrenic patients. *Am J Psychiatry* 150: 1856–1861.

Alkondon, M., Pereira, E. F., Eisenberg, H. M. & Albuquerque, E. X. (2000). Nicotinic receptor activation in human cerebral cortical interneurons: a mechanism for inhibition and disinhibition of neuronal networks. *J Neurosci* 20:66-75.

Aracri, P., Consonni, S., Morini, R., Perrella, M., Rodighiero, S., Amadeo, A. & Becchetti, A. (2010). Tonic modulation of GABA release by nicotinic acetylcholine receptors in layer V of the murine prefrontal cortex. *Cereb Cortex* 20(7):1539-55.

Ashwood, P., Krakowiak, P., Hertz-Picciotto, I., Hansen, R., Pessah, I. N. & Van de Water, J. (2011a). Associations of impaired behaviors with elevated plasma chemokines in autism spectrum disorders. *J Neuroimmunol* 232(1-2):196-9.

Ashwood, P., Krakowiak, P., Hertz-Picciotto, I., Hansen, R., Pessah, I. & Van de Water, J. (2011b). Elevated plasma cytokines in autism spectrum disorders provide evidence of immune dysfunction and are associated with impaired behavioral outcome. *Brain Behav Immun* 25(1):40-5.

Avale, M. E., Chabout, J., Pons, S., Serreau, P., De Chaumont, F., Olivo-Marin, J. C., Bourgeois, J. P., Maskos, U., Changeux, J. P. & Granon, S. (2011). Prefrontal nicotinic receptors control novel social interaction between mice. *FASEB J* Mar 18. [Epub ahead of print]

Bailey, A.R., Giunta, B.N., Obregon D., et al. (2008). Peripheral biomarkers in autism: secreted amyloid precursor protein-alpha as a probable key player in early diagnosis. *Int J Clin Exp Med* 1:338–344.

Baxter, L. R. Jr., Schwartz, J. M., Mazziotta, J. C., Phelps, M. E., Pahl, J. J., Guze, B. H. & Fairbanks, L. (1988). Cerebral glucose metabolic rates in nondepressed patients with obsessive-compulsive disorder. *Am J Psychiatry* 145(12):1560-3.

Bejerot, S. & Nylander, L. (2003). Low prevalence of smoking in patients with autism spectrum disorders. *Psychiatry Res* 119(1-2):177-82.

Bencherif, M., Lippiello, P. M., Lucas, R. & Marrero, M. B. (2011). Alpha7 nicotinic receptors as novel therapeutic targets for inflammation-based diseases. *Cell Mol Life Sci* 68(6):931-49.

Bertrand, D., Picard, F., Le Hellard, S., Weiland, S., Favre, I., Phillips, H., Bertrand, S., Berkovic, S. F., Malafosse, A. & Mulley, J. (2002). How mutations in the nAChRs can cause ADNFLE epilepsy. *Epilepsia* 43 Suppl 5:112-22.

Bierut, L. J., Madden, P. A., Breslau, N., Johnson, E. O., Hatsukami, D., Pomerleau, O. F., Swan, G. E., Rutter, J., Bertelsen, S., Fox, L., Fugman, D., Goate, A. M., Hinrichs, A. L., Konvicka, K., Martin, N. G., Montgomery, G. W., Saccone, N. L., Saccone, S. F., Wang, J. C., Chase, G. A., Rice, J. P. & Ballinger, D. G. (2007). Novel genes identified in a high-density genome wide association study for nicotine dependence. *Hum Mol Genet* 16(1):24-35.

Blundell, J., Tabuchi, K., Bolliger, M. F., Blaiss, C. A., Brose, N., Liu, X., Südhof, T. C. & Powell, C. M. (2009). Increased anxiety-like behavior in mice lacking the inhibitory synapse cell adhesion molecule neuroligin 2. *Genes Brain Behav* 8(1):114-26.

Blundell, J., Blaiss, C. A., Etherton, M. R., Espinosa, F., Tabuchi, K., Walz, C., Bolliger, M. F., Südhof, T. C. & Powell, C. M. (2010). Neuroligin-1 deletion results in impaired spatial memory and increased repetitive behavior. *J Neurosci* 30(6):2115-29.

Boucard, A. A., Chubykin, A. A., Comoletti, D., Taylor, P. & Südhof, T. C. (2005). A splice code for trans-synaptic cell adhesion mediated by binding of neuroligin 1 to alpha- and beta-neurexins. *Neuron* 48(2):229-36.

Braun, A. R., Randolph, C., Stoetter, B., Mohr, E., Cox, C., Vladar, K., Sexton, R., Carson, R. E., Herscovitch, P. & Chase, T. N. (1995). The functional neuroanatomy of Tourette's syndrome: an FDG-PET Study. II: Relationships between regional cerebral metabolism and associated behavioral and cognitive features of the illness. *Neuropsychopharmacology* 13(2):151-68.

Buckner, R., Andrews-Hanna, J. & Schacter, D. (2008). The brain's default network. *Ann N Y Acad Sci* 1124:1–38.

Careaga, M., Van de Water, J. & Ashwood, P. (2010). Immune dysfunction in autism: a pathway to treatment. *Neurotherapeutics* 7(3):283-92.

Carlsson, M. L. (2001). On the role of prefrontal cortex glutamate for the antithetical phenomenology of obsessive compulsive disorder and attention deficit hyperactivity disorder. *Prog Neuropsychopharmacol Biol Psychiatry* 25(1):5-26.

Cheng, S. B., Amici, S. A., Ren, X. Q., McKay, S. B., Treuil, M. W., Lindstrom, J. M., Rao, J. & Anand, R. (2009). Presynaptic targeting of alpha4beta 2 nicotinic acetylcholine receptors is regulated by neurexin-1beta. *J Biol Chem* 284(35):23251-9.

Chez, M. G., Dowling, T., Patel, P. B., Khanna, P. & Kominsky, M. (2007). Elevation of tumor necrosis factor-alpha in cerebrospinal fluid of autistic children. *Pediatr Neurol* 36(6):361-5.

Chih, B., Afridi, S. K., Clark, L., Scheiffele, P. (2004). Disorder-associated mutations lead to functional inactivation of neuroligins. *Hum Mol Genet* 13(14):1471-7.

Chih, B., Gollan, L., Scheiffele, P. (2006). Alternative splicing controls selective trans-synaptic interactions of the neuroligin neurexin complex. *Neuron* 51(2):171-8.

Chubykin, A. A., Atasoy, D., Etherton, M. R., Brose, N., Kavalali, E. T., Gibson, J. R. & Südhof, T. C. (2007). Activity dependent validation of excitatory versus inhibitory synapses by neuroligin-1 versus neuroligin-2. *Neuron* 54(6):919-31.

Clementi, G. C. (2004). Neuronal nicotinic receptors: from structure to pathology. *Prog Neurobiol* 74:363–96.

Coe, J. W., Brooks, P. R., Vetelino, M. G., Wirtz, M. C., Arnold, E. P., Huang, J., Sands, S. B., Davis, T. I., Lebel, L. A., Fox C. B., Shrikhande, A., Heym, J. H., Schaeffer, E., Rollema, H., Lu, Y., Mansbach, R. S., Chambers, L. K, Rovetti, C. C., Schulz, D. W., Tingley, F. D. 3rd & O'Neill, B. T. (2005). Varenicline: an alpha4beta2 nicotinic receptor partial agonist for smoking cessation.*J Med Chem* 48:3474-7.

Comoletti, D., De Jaco, A., Jennings, L. L., Flynn, R. E., Gaietta, G., Tsigelny, I., Ellisman, M. H. & Taylor P. (2004). The Arg451Cys-neuroligin-3 mutation associated with autism reveals a defect in protein processing. *J Neurosci* 24(20):4889-93.

Conroy, W. G., Nai, Q., Ross, B., Naughton, G. & Berg, D. K. (2007). Postsynaptic neuroligin enhances presynaptic inputs at neuronal nicotinic synapses. *Dev Biol* 307(1):79-91.

Cook, E. H. Jr. & Scherer, S. W. (2008). Copy-number variations associated with neuropsychiatric conditions. *Nature* 455(7215):919-23.

Corbett, B. A., Mendoza, S., Abdullah, M., Wegelin, J. A. & Levine, S. (2006). Cortisol circadian rhythms and response to stress in children with autism. *Psychoneuroendocrinology* 31(1):59-68.

Couey, J. J., Meredith, R. M., Spijker, S., Poorthuis, R. B., Smit, A. B., Brussaard, A. B. & Mansvelder, H. D. (2007). Distributed network actions by nicotine increase the threshold for spike-timing-dependent plasticity in prefrontal cortex. *Neuron* 54(1):73-87.

Courchesne, E., Pierce, K., Schumann, C. M., Redcay, E., Buckwalter, J. A., Kennedy, D. P., Morgan, J. (2007). Mapping early brain development in autism. *Neuron* 56(2):399-413.

Court, J., Spurden, D., Lloyd, S., McKeith, I., Ballard, C., Cairns, N., Kerwin, R., Perry, R. & Perry, E. (1999). Neuronal nicotinic receptors in dementia with Lewy bodies and schizophrenia: α-bungarotoxin and nicotinic binding in thalamus. *J Neurochem* 73:1590–7.

Court, J., Martin-Ruiz, C., Piggott, M., Spurden, D., Griffiths, M. & Perry, E. (2001). Nicotinic receptor abnormalities in Alzheimer's disease. *Biol Psychiatry* 49:175–84.

Coury, D. (2010). Medical treatment of autism spectrum disorders. *Curr Opin Neurol* 23(2):131-6.

Craig, A. M. & Kang, Y. (2007). Neurexin-neuroligin signaling in synapse development. *Curr Opin Neurobiol* 17(1):43-52.

Dajas-Bailador, F. A., Heimala, K. & Wonnacott, S. (2003). The allosteric potentiation of nicotinic acetylcholine receptors by galantamine is transduced into cellular responses in neurons: Ca2+ signals and neurotransmitter release. *Mol Pharmacol* 64:1217–1226.

Dalack, G. W., Becks, L., Hill, E., Pomerleau, O. F. & Meador-Woodruff, J. H. (1999). Nicotine withdrawal and psychiatric symptoms in cigarette smokers with schizophrenia. *Neuropsychopharmacology* 21: 195–202.

De Fusco, M., Becchetti, A., Patrignani, A., Annesi, G., Gambardella, A., Quattrone, A., Ballabio, A., Wanke, E. & Casari, G. (2000). The nicotinic receptor beta 2 subunit is mutant in nocturnal frontal lobe epilepsy. *Nat Genet* 26(3):275-6.

de Wit, J., Sylwestrak, E., O'Sullivan, M. L., Otto, S., Tiglio, K., Savas, J. N., Yates, J. R. 3rd, Comoletti, D., Taylor, P. & Ghosh, A. (2009). LRRTM2 interacts with Neurexin1 and regulates excitatory synapse formation. *Neuron* 64(6):799-806.

Dean, C. & Dresbach, T. (2006). Neuroligins and neurexins: linking cell adhesion, synapse formation and cognitive function. *Trends Neurosci* 29(1):21-9.

Dolle, F., Valette, H., Hinnen, F., Vaufrey, F., Demphel, S., Coulon, C., et al. (2001). Synthesis and preliminary evaluation of a carbon-11-labelled agonist of the α7 nicotinic acetylcholine receptor. *J Labelled Cpd Radiopharm* 44:785–95.

Dudanova, I., Sedej, S., Ahmad, M., Masius, H., Sargsyan, V., Zhang, W., Riedel, D., Angenstein, F., Schild, D., Rupnik, M. & Missler, M. (2006). Important contribution of alpha-neurexins to Ca2+-triggered exocytosis of secretory granules. *J Neurosci* 26(41):10599-613.

Dudanova, I., Tabuchi, K., Rohlmann, A., Südhof, T. C. & Missler, M. (2007). Deletion of alpha-neurexins does not cause a major impairment of axonal pathfinding or synapse formation. *J Comp Neurol* 502(2):261-74.

Etherton, M. R., Blaiss, C. A., Powell & C. M., Südhof, T. C. (2009). Mouse neurexin-1alpha deletion causes correlated electrophysiological and behavioral changes consistent with cognitive impairments. *Proc Natl Acad Sci U S A* 106(42):17998-8003.

Falk, L., Nordberg, A., Seiger, A˚., Kjældgaard, A. & Hellstro¨m-Lindahl, E. (2003). Higher expression of α7 nicotinic acetylcholine receptors in human fetal compared to adult brain. *Dev Brain Res* 142: 151–60.

Feng, J., Schroer, R., Yan, J., Song, W., Yang, C., Bockholt, A., Cook, E. H. Jr., Skinner, C., Schwartz, C. E. & Sommer, S. S. (2006). High frequency of neurexin 1beta signal peptide structural variants in patients with autism. *Neurosci Lett* 409(1):10-3.

Fombonne, E. (1999). The epidemiology of autism: a review. *Psychol Med* 29(4):769-86.

Frazier, C. J., Buhler, A. V., Weiner, J. L., & Dunwiddie, T. V. (1998a). Synaptic potentials mediated via alpha-bungarotoxin-sensitive nicotinic acetylcholine receptors in rat hippocampal interneurons. *J Neurosci.* 18:8228-8235.

Frazier, C. J., Rollins, Y. D., Breese, C. R., Leonard, S., Freedman, R. & Dunwiddie, T. V. (1998b). Acetylcholine activates an alpha-bungarotoxin-sensitive nicotinic current in rat hippocampal interneurons, but not pyramidal cells, *J Neurosci* 18:1187-1195.

Freedman, R., Hall, M., Adler, L. E. & Leonard, S. (1995). Evidence in postmortem brain tissue for decreased numbers of hippocampal nicotinic receptors in schizophrenia. *Biol Psychiatry* 38:22–33.

Gai, X., Xie, H. M., Perin, J. C., Takahashi, N., Murphy, K., Wenocur, A. S., D'arcy, M., O'Hara, R. J., Goldmuntz, E., Grice, D. E., Shaikh, T. H., Hakonarson, H., Buxbaum, J. D., Elia, J. & White, P. S. (2011). Rare structural variation of synapse and neurotransmission genes in autism. *Mol Psychiatry* Mar 1. [Epub ahead of print].

Gao, Y., Horti, A. G., Kuwabara, H., Ravert, H. T., Hilton, J., Holt, D. P., et al. (2007). Derivatives of (-)-7-methyl-2-(5-(pyridinyl)pyridin-3- yl)-7-azabicyclo[2.2.1]heptane are potential ligands for positron emission tomography imaging of extrathalamic nicotinic acetylcholine receptors. *J Med Chem* 50(16):3814–3824.

Gao, Y., Kuwabara, H., Spivak, C. E., Xiao, Y., Kellar, K., Ravert, H. T., et al. (2008). Discovery of (-)-7-methyl-2-exo-[3'-(6-[18F] fluoropyridin-2-yl)-5'-pyridinyl]-7-azabicyclo[2.2.1]heptane, a radiolabeled antagonist for cerebral nicotinic acetylcholine receptor (alpha4beta2-nAChR) with optimal positron emission tomography imaging properties. *J Med Chem* 51(15):4751–4764.

Garrity, A. G., Pearlson, G. D., McKiernan, K., Lloyd, D., Kiehl, K. A. & Calhoun, V. D. (2007). Aberrant "default mode" functional connectivity in schizophrenia. *Am J Psychiatry* 164:450 - 457.

Gauthier, J., Siddiqui, T. J., Huashan, P., Yokomaku, D., Hamdan, F. F., Champagne, N., Lapointe, M., Spiegelman, D., Noreau, A., Lafrenière, R. G., Fathalli, F., Joober, R., Krebs, M. O., Delisi, L. E., Mottron, L., Fombonne, E., Michaud, J. L., Drapeau, P., Carbonetto, S., Craig, A. M. & Rouleau, G. A. (2011). Truncating mutations in NRXN2 and NRXN1 in autism spectrum disorders and schizophrenia. *Hum Genet.* 2011 Mar 22. [Epub ahead of print].

Geschwind, D. H. (2009). Advances in autism. *Annu Rev Med* 60:367-80.

Gillberg, C. & Billstedt, E. (2000). Autism and Asperger syndrome: coexistence with other clinical disorders. *Acta Psychiatr Scand* 102, 321–330.

Gilman, S., R., Iossifov, I., Levy, D., Ronemus, M., Wigler, M., Vitkup, D. (2011). Rare De Novo Variants Associated with Autism Implicate a Large Functional Network of Genes Involved in Formation and Function of Synapses. Neuron 70(5): 898-907.

Glessner, J. T., Wang, K., Cai, G., Korvatska, O., Kim, C. E., Wood, S., Zhang, H., Estes, A., Brune C. W., Bradfield, J. P., Imielinski, M., Frackelton, E. C., Reichert, J., Crawford, E. L., Munson, J., Sleiman, P. M., Chiavacci, R., Annaiah, K., Thomas, K., Hou, C., Glaberson, W., et al. (2009). Autism genome-wide copy number variation reveals ubiquitin and neuronal genes. *Nature* 459:569–573.

Gogolla, N., Leblanc, J. J., Quast, K. B., Südhof, T., Fagiolini, M. & Hensch, T. K. (2009). Common circuit defect of excitatory-inhibitory balance in mouse models of autism. *J Neurodev Disord* 1(2):172-181.

Graf, E. R., Zhang, X., Jin, S. X., Linhoff, M. W. & Craig, A. M. (2004). Neurexins induce differentiation of GABA and glutamate postsynaptic specializations via neuroligins. *Cell* 119: 1013–1026.

Granon, S., Faure, P. & Changeux, J. P. (2003). Executive and social behaviors under nicotinic receptor regulation. *Proc Natl Acad Sci U S A* 100(16):9596-601.

Gunnell, D., Irvine, D., Wise, L., Davies, C. & Martin, R. M. (2009). Varenicline and suicidal behaviour: a cohort study based on data from the General Practice Research Database. *BMJ* 339:b3805.

Hashimoto, K., Nishiyama, S., Ohba, H., Matsuo, M., Kobashi, T., Takahagi, M., et al. (2008). [11C]CHIBA-1001 as a novel PET ligand for α7 nicotinic receptors in the brain: a PET study in conscious monkeys. *PLoS ONE* 3:e3231.

Hata, Y., Butz, S. & Südhof, T. C. (1996). CASK: a novel dlg/PSD95 homolog with an N-terminal calmodulin-dependent protein kinase domain identified by interaction with neurexins. *J Neurosci* 16(8):2488-94.

Hoda, J. C., Wanischeck, M., Bertrand, D. & Steinlein, O. K. (2009). Pleiotropic functional effects of the first epilepsy-associated mutation in the human CHRNA2 gene. *FEBS Lett* 583(10):1599-604.

Horti, A. G., Gao, Y., Kuwabara, H. & Dannals, R. F. (2010). Development of radioligands with optimized imaging properties for quantification of nicotinic acetylcholine receptors by positron emission tomography. *Life Sci* 86(15-16):575-84.

Inta, D., Vogt, M. A., Lima-Ojeda, J. M., Pfeiffer, N., Schneider, M. & Gass, P. (2011). Lack of long-term behavioral alterations after early postnatal treatment with tropisetron:

Implications for developmental psychobiology. *Pharmacol Biochem Behav* 99(1):35-41.

Jamain, S., Quach, H., Betancur, C., Råstam, M., Colineaux, C., Gillberg, I. C., Soderstrom, H., Giros, B., Leboyer, M., Gillberg, C., Bourgeron, T.; Paris Autism Research International Sibpair Study. (2003). Mutations of the X-linked genes encoding neuroligins NLGN3 and NLGN4 are associated with autism. *Nat Genet* 34(1):27-9.

Jamain, S., Radyushkin, K., Hammerschmidt, K., Granon, S., Boretius, S., Varoqueaux, F., Ramanantsoa, N., Gallego, J., Ronnenberg, A., Winter, D., Frahm, J., Fischer, J., Bourgeron, T., Ehrenreich, H. & Brose, N. (2008). Reduced social interaction and ultrasonic communication in a mouse model of monogenic heritable autism. *Proc Natl Acad Sci U S A* 105(5):1710-5.

Kang, Y., Zhang, X., Dobie, F., Wu, H. & Craig, A. M. (2008). Induction of GABAergic postsynaptic differentiation by alpha-neurexins. *J Biol Chem* 283(4):2323-34.

Kas, M.J., Fernandes, C., Schalkwyk, L.C., & Collier, D.A. (2007). Genetics of behavioural domains across the neuropsychiatric spectrum; of mice and men. *Mol Psychiatry* 12(4):324-30.

Kihara, T., Shimohama, S., Sawada, H., Kimura, J., Kume, T., Kochiyama, H., Maeda, T., & Akaike, A. (1997). Nicotinic receptor stimulation protects neurons against beta-amyloid toxicity. *Ann Neurol* 42:159-63.

Kihara, T., Shimohama, S., Urushitani, M., Sawada, H., Kimura, J., Kume, T., Maeda, T., and Akaike, A. (1998). Stimulation of alpha4beta2 nicotinic acetylcholine receptors inhibits beta-amyloid toxicity. *Brain Res* 792:331-4.

Kihara, T., Shimohama, S., Sawada, H., Honda, K., Nakamizo, T., Shibasaki, H., Kume, T., & Akaike, A. (2001). alpha 7 Nicotinic Receptor Transduces Signals to Phosphatidylinositol 3-Kinase to Block A beta -Amyloid-induced Neurotoxicity. *J Biol Chem* 276:13541-6.

Kim, H. G., Kishikawa, S., Higgins, A. W., Seong, I. S., Donovan, D. J., Shen, Y., Lally, E., Weiss, L. A., Najm, J., Kutsche, K., Descartes, M., Holt, L., Braddock, S., Troxell, R., Kaplan, L., Volkmar, F., Klin, A., Tsatsanis, K., Harris, D. J., Noens, I., Pauls, D. L., Daly, M. J., MacDonald, M. E., Morton, C. C., Quade, B. J. & Gusella, J. F. (2008). Disruption of neurexin 1 associated with autism spectrum disorder. *Am J Hum Genet* 82(1):199-207.

Kirov, G., Rujescu, D., Ingason, A., Collier, D. A., O'Donovan, M. C. & Owen, M. J. (2009). Neurexin 1 (NRXN1) deletions in schizophrenia. *Schizophr Bull* 35(5):851-4.

Kitagawa, H., Takenouchi, T., Azuma, R., Wesnes, K., Kramer, W., Clody, D. & Burnett, A. L. (2003). Safety, pharmacokinetics, and effects on cognitive function of multiple doses of GTS-21 in healthy, male volunteers. *Neuropsychopharmacology* 28:542–551.

Ko, J., Fuccillo, M. V., Malenka, R. C. & Südhof, T. C. (2009). LRRTM2 functions as a neurexin ligand in promoting excitatory synapse formation. *Neuron* 64(6):791-8.

Konstantareas, M. M. & Homatidis, S. (1999). Chromosomal abnormalities in a series of children with autistic disorder. *J Autism Dev Disord* 29:275–285.

Langen, M., Durston, S., Kas, M.J., van Engeland, H., & Staal, W.G. (2011). The neurobiology of repetitive behavior: … and men. *Neurosci and Biobehavioral Rev* 35:356-65.

Laumonnier, F., Bonnet-Brilhault, F., Gomot, M., Blanc, R., David, A., Moizard, M. P., Raynaud, M., Ronce, N., Lemonnier, E., Calvas, P., Laudier, B., Chelly, J., Fryns, J. P., Ropers, H. H., Hamel, B. C., Andres, C., Barthélémy, C., Moraine, C. & Briault, S.

(2004). X-linked mental retardation and autism are associated with a mutation in the NLGN4 gene, a member of the neuroligin family. *Am J Hum Genet* 74(3):552-7.

Lee, M., Martin-Ruiz, C., Graham, A., Court, J., Jaros, E., Perry, R., Iversen, P., Bauman M. & Perry, E. (2002). Nicotinic receptor abnormalities in the cerebellar cortex in autism. *Brain* 125(Pt 7):1483-95.

Lee, S. W., Lee, J. G., Lee, B. J. & Kim, Y. H. (2007). A 12-week, double-blind, placebo-controlled trial of galantamine adjunctive treatment to conventional antipsychotics for the cognitive impairments in chronic schizophrenia. *Int Clin Psychopharm* 22:63–68.

Léna, C., Popa, D., Grailhe, R., Escourrou, P., Changeux, J. P. & Adrien, J. (2004). Beta2-containing nicotinic receptors contribute to the organization of sleep and regulate putative micro-arousals in mice. *J Neurosci* 24(25):5711-8.

Levin, E. D. & Simon, B. B. (1998). Nicotinic acetylcholine involvement in cognitive function in animals. *Psychopharmacology* 138(3-4):217-30.

Levitt, P. & Campbell, D. B. (2009). The genetic and neurobiologic compass points toward common signaling dysfunctions in autism spectrum disorders. *J Clin Invest* 119(4):747-54.

Levy, D., Ronemus, M., Yamrom, B., Lee, Y., Leotta, A., Kendall, J., et al. (2011). Rare De Novo and Transmitted Copy-Number Variation in Autistic Spectrum Disorders. Neuron 70(5): 886-897.

Levy, S. E., Mandell, D. S. & Schultz, R. T. (2009). Autism. *Lancet* 374(9701):1627-38.

Lipton, S. A., Frosch, M. P., Phillips, M. D., Tauck, D. L. & Aizenman, E. (1998). Nicotinic antagonists enhance process outgrowth by rat retinal ganglion cells in culture. *Science* 239:1293-1296.

Lisé, M. F. & El-Husseini, A. (2006). The neuroligin and neurexin families: from structure to function at the synapse. *Cell Mol Life Sci* 63(16):1833-49.

Lindstrom, J. (1996). Neuronal nicotinic acetylcholine receptors. *Ion Channels* 4:377-450.

Lindstrom, J. (1997). Nicotinic acetylcholine receptors in health and disease. *Mol. Neurobiol* 15:193-222.

López-Valdés, H. E., García-Colunga, J. & Miledi, R. (2002). Effects of clomipramine on neuronal nicotinic acetylcholine receptors. *Eur J Pharmacol* 444(1-2):13-9.

Marshall CR, Noor A, Vincent JB, Lionel AC, Feuk L, Skaug J, Shago M, Moessner R, Pinto D, Ren Y, Thiruvahindrapduram B, Fiebig A, Schreiber S, Friedman J, Ketelaars CE, Vos YJ, Ficicioglu C, Kirkpatrick S, Nicolson R, Sloman L, Summers A, Gibbons CA, Teebi A, Chitayat D, Weksberg R, Thompson A, Vardy C, Crosbie V, Luscombe S, Baatjes R, Zwaigenbaum L, Roberts W, Fernandez B, Szatmari P, Scherer SW. (2008). Structural variation of chromosomes in autism spectrum disorder. *Am J Hum Genet* 82(2):477-88.

Martin-Ruiz, C. M., Lee, M., Perry, R. H., Baumann, M., Court, J. A. & Perry, E.K. (2004). Molecular analysis of nicotinic receptor expression in autism. *Brain Res Mol Brain Res* 123(1-2):81-90.

Marutle, A., Zhang, X., Court, J., Piggott, M., Johnson, M., Perry, R., Perry, E. & Nordberg, A. (2001). Laminar distribution of nicotinic receptor subtypes in cortical regions in schizophrenia. *J Chem Neuroanat* 22:115–26.

Maskos, U., Molles, B. E., Pons, S., Besson, M., Guiard, B. P., Guilloux, J. P., Evrard, A., Cazala, P., Cormier, A., Mameli-Engvall, M., Dufour, N., Cloëz-Tayarani, I.,

Bemelmans, A. P., Mallet, J., Gardier, A. M., David, V., Faure, P., Granon, S. & Changeux, J. P. (2005). Nicotine reinforcement and cognition restored by targeted expression of nicotinic receptors. *Nature* 436(7047):103-7.

McClure, J. B., Swan, G. E., Jack, L., Catz, S. L., Zbikowski, S. M., McAfee, T. A., Deprey, M., Richards, J. & Javitz, H. (2009). Mood, side-effects and smoking outcomes among persons with and without probable lifetime depression taking varenicline. *J Gen Intern Med* 24(5):563-9.

McGehee, D. S. & Role, L. W. (1996). Presynaptic ionotropic receptors. *Curr Opin Neurobiol* 6:342-349.

Merikangas, A. K., Corvin, A. P. & Gallagher, L. (2009) Copy-number variants in neurodevelopmental disorders: promises and challenges. *Trends Genet* 25(12):536-44.

Messi, M. L., Renganathan, M., Grigorenko, E. & Delbono, O. (1997). Activation of alpha7 nicotinic acetylcholine receptor promotes survival of spinal cord motoneurons. *FEBS Lett* 411:32-8.

Mihalak, K.B., Carroll, F. I. & Luetje, C.W. (2006). Varenicline is a partial agonist at alpha4beta2 and a full agonist at alpha7 neuronal nicotinic receptors. *Mol Pharmacol* 70(3):801-5.

Miller, D.,T., Shen, Y., Weiss, L., A., Korn, J., Anselm, I., Bridgemohan, C., et al. (2009). Microdeletion/duplication at 15q13.2q13.3 among individuals with features of autism and other neuropsychiatric disorders. J Med Genet. 46(4):242-8.

Mineur, Y. S. & Picciotto, M.R. (2010). Nicotine receptors and depression: revisiting and revising the cholinergic hypothesis. *Trends Pharmacol Sci* 31(12):580-6.

Missler, M. & Südhof TC. (1998). Neurexins: three genes and 1001 products. *Trends Genet* 14(1):20-6.

Missler, M., Zhang, W., Rohlmann, A., Kattenstroth, G., Hammer, R. E., Gottmann, K. & Südhof, T.C. (2003). Alpha-neurexins couple Ca2+ channels to synaptic vesicle exocytosis. *Nature* 423(6943):939-48.

Miwa, J. M., Freedman, R. & Lester, H. A. (2011). Neural systems governed by nicotinic acetylcholine receptors: emerging hypotheses. *Neuron.* 70(1):20-33.

Morgan, J. T., Chana, G., Pardo, C. A., Achim, C., Semendeferi, K., Buckwalter, J., Courchesne, E. & Everall, I. P. (2010) Microglial activation and increased microglial density observed in the dorsolateral prefrontal cortex in autism. *Biol Psychiatry* 68(4):368-76.

Nam, C. I. & Chen, L. (2005). Postsynaptic assembly induced by neurexin-neuroligin interaction and neurotransmitter. *Proc Natl Acad Sci USA* 102(17):6137-42.

Newhouse, P. A. & Kelton, M. (2000). Nicotinic systems in central nervous systems disease: degenerative disorders and beyond. *Pharm Acta Helv* 74(2-3):91-101.

Newhouse, P., Singh, A. & Potter, A. (2004). Nicotine and nicotinic receptor involvement in neuropsychiatric disorders. *Curr Top Med Chem* 4(3):267-82.

Niaura, R., Jones, C. & Kirkpatrick, P. (2006). Varenicline. *Nat Rev Drug Discov* 5:537-8.

Nicolson, R., Craven-Thuss, B. & Smith, J. (2006). A prospective, open-label trial of galantamine in autistic disorder. *J Child Adolesc Psychopharmacol* 16(5):621-9.

Nussbaum, J., Xu, Q., Payne, T. J., Ma, J. Z., Huang, W., Gelernter, J. & Li, M. D. (2008). Significant association of the neurexin-1 gene (NRXN1) with nicotine dependence in European- and African-American smokers. *Hum Mol Genet* 17(11):1569-77.

Ochoa, E.L. & Lasalde-Dominicci, J. (2007). Cognitive deficits in schizophrenia: Focus on neuronal nicotinic acetylcholine receptors and smoking. *Cell Mol Neurobiol* 27(5):609–639.

Ogawa, M., Tatsumi, R., Fujio, M., Katayama, J. & Magata, Y. (2006). Synthesis and evaluation of [125I]I-TSA as a brain nicotinic acetylcholine receptor $\alpha7$ subtype imaging agent. *Nucl Med Biol* 33:311–36.

Papke, R. L., Trocmé-Thibierge, C., Guendisch, D., Al Rubaiy, S. A. & Bloom, S.A. (2011). Electrophysiological perspectives on the therapeutic use of nicotinic acetylcholine receptor partial agonists. *J Pharmacol Exp Ther* 337(2):367-79.

Pasquini, M., Garavini, A. & Biondi, M. (2005). Nicotine augmentation for refractory obsessive-compulsive disorder. A case report. *Prog Neuropsychopharmacol Biol Psychiatry* 29(1):157-9.

Perry, E. K., Lee, M. L., Martin-Ruiz, C. M., Court, J. A., Volsen, S. G., Merrit, J., Folly, E., Iversen, P. E., Bauman, M. L., Perry, R. H. & Wenk, G. L. (2001). Cholinergic activity in autism: abnormalities in the cerebral cortex and basal forebrain. *Am J Psychiatry* 158(7):1058-66.

Pimlott, S. L., Piggott, M., Owens, J., Greally, E., Court, J. A., Jaros, E., Perry, R. H., Perry, E. K. & Wyper, D. (2004). Nicotinic acetylcholine receptor distribution in Alzheimer's disease, dementia with Lewy bodies, Parkinson's disease, and vascular dementia: In vitro binding study using 5-[125I]-A-85380. *Neuropsychopharmacology* 29(1):108–116.

Pomper, M. G, Phillips, E., Fan, H., McCarthy, D. J., Keith, R. A., Gordon, J. C., Scheffel, U., Dannals, R. F. & Musachio, J. L. (2005). Synthesis and biodistribution of radiolabeled □7 nicotinic acetylcholine receptor ligands. *J Nucl Med* 46: 326–34.

Poulopoulos, A., Aramuni, G., Meyer, G., Soykan, T., Hoon, M., Papadopoulos, T., Zhang, M., Paarmann, I., Fuchs, C., Harvey, K., Jedlicka, P., Schwarzacher, S. W., Betz, H., Harvey, R. J., Brose, N., Zhang, W. & Varoqueaux, F. (2009). Neuroligin 2 drives postsynaptic assembly at perisomatic inhibitory synapses through gephyrin and collybistin. *Neuron* 63(5):628-42.

Potter, A., Corwin, J., Lang, J., Piasecki, M., Lenox, R. & Newhouse, P. A. (1999). Acute effects of the selective cholinergic channel activator (nicotinic agonist) ABT-418 in Alzheimer's disease. *Psychopharmacology* 142(4):334-42.

Pugh, P. C., and Berg, D. K. (1994). Neuronal acetylcholine receptors that bind alpha-bungarotoxin mediate neurite retraction in a calcium-dependent manner, *J Neurosci* 14:889-896.

Pugh, P. C. & Margiotta, J. F. (2000). Nicotinic acetylcholine receptor agonists promote survival and reduce apoptosis of chick ciliary ganglion neurons. *Mol Cell Neurosci* 15:113-22.

Ramocki, M.B. & Zoghbi, H.Y. (2008). Failure of neuronal homeostasis results in common neuropsychiatric phenotypes. *Nature* 455(7215):912-8.

Rojas, S., Martín, A., Arranz, M. J., Pareto, D., Purroy, J., Verdaguer, E., Llop, J., Gómez, V., Gispert, J.D., Millán, O., Chamorro, A. & Planas, A.M. (2007) Imaging brain inflammation with [(11)C]PK11195 by PET and induction of the peripheral-type benzodiazepine receptor after transient focal ischemia in rats. *J Cereb Blood Flow Metab* 27(12):1975-86.

Rosas-Ballina, M. & Tracey, K. J. (2009). The neurology of the immune system: neural reflexes regulate immunity. *Neuron* 64(1):28-32.

Ross, B. S & Conroy, W. G. (2008) Capabilities of neurexins in the chick ciliary ganglion. *Dev Neurobiol.* 68(3):409-19.

Ross, S. A., Wong, J. Y., Clifford, J. J., Kinsella, A., Massalas, J. S., Horne, M. K., Scheffer, I. E., Kola, I., Waddington, J. L., Berkovic, S. F. & Drago, J. (2000) Phenotypic characterization of an alpha 4 neuronal nicotinic acetylcholine receptor subunit knock-out mouse. *J Neurosci.* 20(17):6431-41.

Rubenstein, J. & Merzenich, M. (2003). Model of autism: increased ratio of excitation/inhibition in key neural systems. *Genes, Brain, and Behavior* 2(5):255–267.

Samochocki, M., Höffle, A., Fehrenbacher, A., Jostock, R., Ludwig, J., Christner, C., Radina, M., Zerlin, M., Ullmer, C., Pereira, E. F. R., Lübert, H., Albuquerque, E. X. & Maelicke, A. (2003). Galantamine is an allosterically potentiating ligand of neuronal nicotinic but not of muscarinic acetylcholine receptors. *J Pharmacol Exp Ther* 305:1024–1036.

Sanberg, P. R., Silver, A. A., Shytle, R. D., Philipp, M. K., Cahill, D. W., Fogelson, H. M., McConville, B. J. (1997) Nicotine for the treatment of Tourette's syndrome. *Pharmacol Ther* 74(1):21-5.

Sargent, P. B. (1993). The diversity of neuronal nicotinic acetylcholine receptors. *Annu Rev Neurosci* 16:403-443.

Schilström, B., Ivanov, V. B., Wiker, C. & Svensson, T.H. (2007). Galantamine enhances dopaminergic neurotransmission in vivo via allosteric potentiation of nicotinic acetylcholine receptors. *Neuropsychopharmacology* 32:43–53.

Schubert, M. X., Young, K. A. & Hicks, P. B. (2006). Galantamine improves cognition in schizophrenic patients stabilized on risperidone. *Biol Psychiatry* 60:530–533.

Shiina, A., Shirayama, Y., Niitsu, T., Hashimoto, T., Yoshida, T., Hasegawa, T., Haraguchi, T., Kanahara, N., Shiraishi, T., Fujisaki, M., Fukami, G., Nakazato, M., Iyo, M. & Hashimoto, K. (2010). A randomised, double-blind, placebo-controlled trial of tropisetron in patients with schizophrenia. *Ann Gen Psychiatry* 24(9):27.

Sokol, D. K., Chen, D., Farlow, M. R., Dunn, D. W., Maloney, B., Zimmer, J. A. & Lahiri, D. K. (2006). High levels of Alzheimer beta-amyloid precursor protein (APP) in children with severely autistic behavior and aggression. *J Child Neurol* 21(6):444-9.

Steinlein, O. K. (2002). Nicotinic acetylcholine receptors and epilepsy. *Curr Drug Targets CNS Neurol Disord* 1(4):443-8.

Südhof, T. C. (2008) Neuroligins and neurexins link synaptic function to cognitive disease. *Nature* 455(7215):903-11.

Szatmari P, Paterson AD, Zwaigenbaum L, Roberts W, Brian J, Liu XQ, et al. Autism Genome Project Consortium. (2007). Mapping autism risk loci using genetic linkage and chromosomal rearrangements. *Nat Genet* 39(3):319-28.

Siddiqui, T. J., Pancaroglu, R., Kang, Y., Rooyakkers, A. & Craig, A. M. (2010). LRRTMs and neuroligins bind neurexins with a differential code to cooperate in glutamate synapse development. *J Neurosci* 30:7495–7506.

Tabuchi K, Blundell, J., Etherton, M. R., Hammer, R. E., Liu, X., Powell, C. M. & Sudhof, T. C. (2007). A neuroligin-3 mutation implicated in autism increases inhibitory synaptic transmission in mice. *Science* 318:71–76.

Takata, K., Kitamura, Y., Saeki, M., Terada, M., Kagitani, S., Kitamura, R., Fujikawa, Y., Maelicke, A., Tomimoto, H., Taniguchi, T. & Shimohama, S. (2010). Galantamine-induced amyloid-{beta} clearance mediated via stimulation of microglial nicotinic acetylcholine receptors. *J Biol Chem* 285(51):40180-91.

Taly, A., Corringer, P. J., Guedin, D., Lestage, P. & Changeux, J.P. (2009) Nicotinic receptors: allosteric transitions and therapeutic targets in the nervous system. *Nat Rev Drug Discov* 8(9):733-50.

Tapper, A. R., McKinney, S. L., Nashmi, R., Schwarz, J., Deshpande, P., Labarca, C., Whiteaker, P., Marks, M. J., Collins, A.C. & Lester, H. A. (2004). Nicotine activation of alpha4* receptors: sufficient for reward, tolerance, and sensitization. *Science* 306(5698):1029-32.

Toyohara, J., Sakata, M., Wu, J., Ishikawa, M., Oda, K., Ishii, K., et al. (2009). Preclinical and the first clinical studies on [11C]CHIBA-1001 for mapping alpha7 nicotinic receptors by positron emission tomography. *Ann Nucl Med* 23(3):301-9.

Tregellas, J. R., Tanabe, J., Rojas, D. C., Shatti, S., Olincy, A., Johnson, L., Martin, L. F., Soti, F., Kem, W. R., Leonard, S. & Freedman, R. (2011). Effects of an alpha 7-nicotinic agonist on default network activity in schizophrenia. *Biol Psychiatry* 69(1):7-11.

Uemura,T., Lee, S. J., Yasumura, M., Takeuchi, T., Yoshida, T., Ra, M., Taguchi, R., Sakimura, K. & Mishina, M. (2010). Trans-synaptic interaction of GluRdelta2 and Neurexin through Cbln1 mediates synapse formation in the cerebellum. *Cell* 141(6):1068-79.

Vargas, D. L., Nascimbene, C., Krishnan, C., Zimmerman, A. W. & Pardo, C. A. (2005). Neuroglial activation and neuroinflammation in the brain of patients with autism. *Ann Neurol* 57(1):67-81.

Varoqueaux, F., Aramuni, G., Rawson, R. L., Mohrmann, R., Missler, M., Gottmann, K., Zhang, W., Südhof, T.C. & Brose, N. (2006). Neuroligins determine synapse maturation and function. *Neuron* 51(6):741-54.

Wang, D., Noda, Y., Zhou, Y., Mouri, A., Mizoguchi, H., Nitta, A., Chen, W. & Nabeshima, T. (2007). The allosteric potentiation of nicotinic acetylcholine receptors by galantamine ameliorates the cognitive dysfunction in beta amyloid25–35 icv-injected mice: involvement of dopaminergic systems. *Neuropsychopharmacology* 32:1261–1271.

Wang, H., Yu, M., Ochani, M., Amella, C. A., Tanovic, M., Susarla, S., Li, J. H., Wang, H., Yang, H., Ulloa, L., Al-Abed, Y., Czura, C. J. & Tracey, K. J. (2003). Nicotinic acetylcholine receptor alpha7 subunit is an essential regulator of inflammation. *Nature* 421(6921):384-8.

Westmark, C. J., Westmark, P. R., Beard, A. M., Hildebrandt, S. M. & Malter, J. S. (2008). Seizure susceptibility and mortality in mice that over-express amyloid precursor protein. *Int J Clin Exp Pathol.* 1(2):157-68.

Wilens, T. E., Biederman, J., Spencer, T. J., Bostic, J., Prince, J., Monuteaux, M. C., Soriano, J., Fine, C., Abrams, A., Rater, M. & Polisner, D. (1999) A pilot controlled clinical trial of ABT-418, a cholinergic agonist, in the treatment of adults with attention deficit hyperactivity disorder. *Am J Psychiatry* 156(12):1931-7.

Wills, S., Cabanlit, M., Bennett, J., Ashwood, P., Amaral, D. G. & Van de Water, J. (2009). Detection of autoantibodies to neural cells of the cerebellum in the plasma of subjects with autism spectrum disorders. *Brain Behav Immun* 23(1):64-74.

Yan, J., Noltner, K., Feng, J., Li, W., Schroer, R., Skinner, C, Zeng, W., Schwartz, C. E. & Sommer, S. S. (2008). Neurexin 1α structural variants associated with autism. *Neurosci Lett* 438(3):368-70.

Ylisaukko-oja, T., Rehnström, K., Auranen, M., Vanhala, R., Alen, R., Kempas, E., Ellonen, P., Turunen, J. A., Makkonen, I., Riikonen, R., Nieminen-von Wendt, T., von Wendt, L., Peltonen, L. & Järvelä, I. (2005). Analysis of four neuroligin genes as candidates for autism. *Eur J Hum Genet* 13:285–1292.

Zahir, F. R., Baross,,A., Delaney, A. D., Eydoux, P., Fernandes, N. D., Pugh, T., Marra, M. A., & Friedman, J. M. (2008). A patient with vertebral, cognitive and behavioural abnormalities and a de novo deletion of NRXN1alpha. *J Med Genet* 45(4):239-43.

Zametkin, A. J., Nordahl, T. E., Gross, M., King, A. C., Semple, W. E., Rumsey, J., Hamburger, S. & Cohen, R. M. (1990) Cerebral glucose metabolism in adults with hyperactivity of childhood onset. *N Engl J Med.* 323(20):1361-6.

Zhang, Z. W., Coggan, J. S. & Berg, D. K. (1996). Synaptic currents generated by neuronal acetylcholine receptors sensitive to alpha-bungarotoxin. *Neuron* 17:1231-40.

Zhang, W., Rohlmann, A., Sargsyan, V., Aramuni, G., Hammer, R. E., Südhof, T. C. & Missler, M. (2005). Extracellular domains of alpha-neurexins participate in regulating synaptic transmission by selectively affecting N- and P/Q-type Ca2+ channels. *J Neurosci.* 25(17):4330-42.

Zhang, C., Atasoy, D., Araç, D., Yang, X., Fucillo, M. V., Robison, A.J., Ko, J., Brunger, A.T., & Südhof T.C. (2010). Neurexins physically and functionally interact with GABA(A) receptors. *Neuron* 66(3):403-16.

8

Identifying Variations Within Unstable Regions of the Genome Reveal Autism Associated Patterns

Abdullah Alqallaf and Ali Hajjiah
Electrical Engineering Department, College of Engineering and Petroleum, Kuwait University, Kuwait city, Kuwait

1. Introduction

Autism is a collection of neurodevelopment and abnormal behaviors which can be characterized by social isolation, language deficits and repetitive or stereotyped behaviors. It is a lifelong disorder that starts at early childhood and becomes apparent before three years old up to adulthood that ranging in severity from case to case. Autism spectrum disorder (ASD) has received a great deal of attention in recent years since the apparent prevalence of children with this spectrum of neurological and behavioral deficits is on the rise. It is currently estimated to be approximately 1 in 150 children based on a 14 state survey conducted by the Centers for Disease Control (CDC) in the United States of America (Kuehn, 2007) and it is predominately in males with a ratio of approximately of 4 males to 1 female (Fombonne, 2003).

While it is hotly debated in both the lay and academic communities as to whether ASD incidence is truly increasing and not just a function of increased reporting and changes in diagnostic criteria, it is uncontested that the number of children diagnosed with ASD presents an important pediatric health problem. The social and economical impacts on individuals with ASD and their families as well as the society maybe considerably reduced if early identification and diagnosing can be achieved using simple and accurate approach. Although it is initially described in the 1940s, the exact etiology and pathology of ASD remains rudimentary and challenging. A number of studies have reported links between the development of the ASD and various factors such as genetics, environmental, immunological, nutritional and neurological. It is likely to result from a combination of these factors. Different methodologies have been proposed to identify and diagnose ASD using different criteria. The autism diagnostic observation schedule (ADOS) is a protocol consists of a series of structural tasks that involve social interactions used to diagnose and assess ASD. Others are using functional magnetic resonance imaging (f-MRI) to scan the brain as pattern recognition method of the defected neurons in the autistic individuals. However, these methodologies depend on the interactions between the examiner and the patient. On the other hand, studying the function of the biological system provides alternative way to embrace the complexity of ASD. Although the neurobiological and genetics basis of ASD and related disorders is unclear, multiple lines of evidence have converged on abnormal brain functions. Using previous knowledge of biological processes and protein interactions of neurological disorders related to ASD, there were able to identify several genes and genetic contributors that had been strongly associated to ASD (Sebat et al., 2007 & Abrahams, 2008). Alterations of these contributors have been proposed as a factor involved in the etiology of ASD.

Understanding the biological mechanisms related to ASD at early stage is essential for identifying and diagnosing the disease and will lead to better treatments. Our main objective in this chapter is to understand the molecular and cellular underpinnings of ASD by identifying the genetic contributors to this set of complex disorders. We are also keenly interested in developing DNA-based methods that can serve to improve our diagnostic evaluation of ASD. Accurate and simple diagnostic methods would go a long ways in promoting early and appropriate interventions. Our research is grounded in recent work showing that deletions and duplications of DNA contribute a very significant degree of genetic variation in human populations. Finally, the work presented in this chapter focuses primarily on determining if DNA copy number changes are associated with ASD.

2. High-resolution genetic data

Data on genome structural and functional features for various organisms is being accumulated and analyzed aiming to explore in depth the biological information and to convert data into meaningful biological knowledge. To date, different experimental technologies such as microarray and DNA sequencing had been proposed to generate high-resolution genetic data and to understand the complex dynamic interactions between complex diseases and the biological system components of genes and genes products. These approaches made it possible to enhance our understanding of biological variations in healthy and diseased organisms through computational-based models. However, these technologies contain many sources of errors. Some types of errors are of our interests that have biological origins. Other types of errors are undesirable and need to be eliminated before further analysis. In particular, these technologies produce certain systematic sources of errors due to the experimental design process used in generating the genetic data such as labeling, printing, and scanning the examined samples. Figure 1 illustrates a simple description of generating DCN data using aCGH technology. Identifying the genomic locations and genetic contributors responsible for these variations is a problem of great importance to biologists. Current estimates indicate that DNA sequence differences due to changes in DNA copy number account for 3-4 fold more variation than that provided by single nucleotide polymorphisms, the most widely studied type of variation. It is also apparent that certain segments of the genome are susceptible to copy number alterations on account of particular sequence features, such as low copy repeats (LCRs).

LCRs are relatively large (>1 Kb), highly related elements (>90% identity) that are typically repeated a modest number of times and frequently found on the same chromosome arm. Many regions of genomic instability are known to be involved in genetic syndromes, termed "genomic disorders", where similar, but not identical, copy number changes produce specific developmental syndromes. It is remarkable that many LCR-rich intervals are located within chromosomal regions where rearrangements are known to be associated with neurobehavioral disorders, including autism (Christian et al., 2008; Marshall et al., 2008; Sebat et al., 2007 & Kirov et al., 2008), mental retardation (Sharp et al., 2006, 2008) and schizophrenia (Cantor et al.; Stefansson et al.; Stone et al. & Walsh et al., 2008). To determine

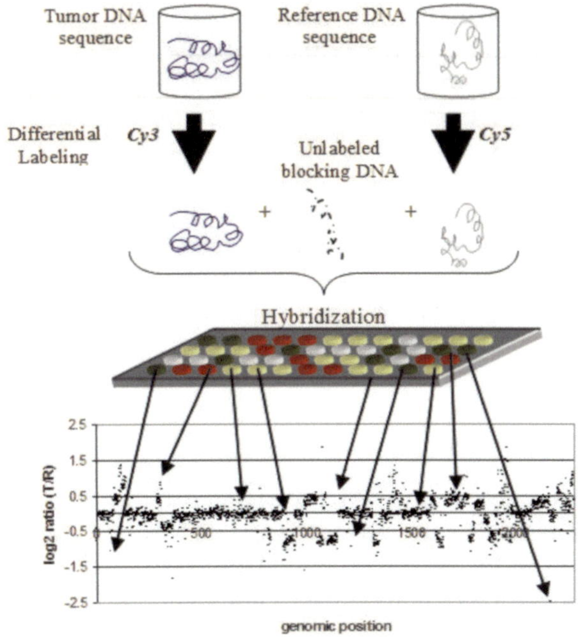

Fig. 1. Illustration of Microarray-based comparative genomic hybridization (array-CGH) process. The tumor and reference DNA are labeled and competitively hybridized to the array together with unlabeled blocking DNA to block repeated sequences. The ratio of the fluorescence intensity for each spot represented as a point in the relative copy number profile.

if copy number variants found within unstable segments of the genome are associated with autism susceptibility, we have conducted a high-resolution array CGH analysis of five genomic intervals that are rich in LCRs and where chromosomal rearrangements are associated with neurodevelopmental disorders. These regions include 7q11 (61-82Mb), 10q22.3-23.31 (77-92Mb), 15q11-13 (18-35 Mb), 17p11 (12-22Mb), and 22q11 (14-26Mb). The 7q11 interval spans the segment involved in Williams-Beuren Syndrome, a contiguous gene syndrome that produces a variety of cognitive and adaptive deficiencies (Greer et al., 1997). The reciprocal duplications of the Williams-Beuren deletion interval are associated with language delay and autism (Somerville et al., 2005; Van der Aa et al., 2009 & Depienne et al., 2007), suggesting that duplications in this genomic region are more closely linked to behavioral deficits that fall within the spectrum of autism disorder. Deletions flanked by segmental duplications are associated with language delay, attention deficit hyperactivity disorder (ADHD), and autism for the 10q22-23 interval (Balciuniene et al., 2007), and a balanced translocation affecting the KCNMA1 gene, which encodes a calcium-activated large conductance potassium channel, on 10q22 has also been reported in a child with autism (Laumonnier et al., 2006). Maternally-derived duplications of the 15q11-13 interval are the most common cytogenetic abnormalities associated with autism (Cook et al., 2001), and maternal as well as paternal-derived deletions are responsible for Angelman and Prader-Willi syndromes, respectively. In addition, deletions in the 15q11-13 interval are associated with mental retardation (Sharp et al., 2008), epilepsy (Sharp et al., 2008 & Helbig et al., 2009), and schizophrenia (Stefansson et al., & Stone et al., 2008). LCR-mediated chromosomal rearrangements within 17p11 result in various nervous system dysfunctions (Lee et al., 2006), including Smith-Magenis and Potocki-Lupski syndromes. Deletions within the 22q11.2 interval are the most frequent interstitial deletions in humans, occurring in approximately 1 in 4000 live births (Papolos et al., 1996). These deletions cause congenital multisystem abnormalities referred to as 22q11 deletion syndrome, and include clinical entities such as Velocardiofacial syndrome, DiGeorge syndrome, and CATCH22 syndrome. Autism spectrum symptoms were reported in 20-50% of patients with 22q11 deletion syndrome; 15-20% of the patients have schizophrenia, and 40% of the patients manifest ADHD (Niklasson et al., 2001; Antonell et al., 2005 & Vorstman et al., 2006). Large deletions within the Velocardiofacial-DiGeorge syndromes critical region of 22q11 are found in patients with schizophrenia at a frequency of less than 1% (Stone et al., 2008). In the next section, we will present a novel methodology for the analysis of genetic data.

3. Method

In this section, we present a framework to evaluate the predictive power of recurrent variations at multiple genomic sites. The section is divided into two main parts. First, as a preprocessing step for feature extraction, a robust methodology based on statistical signal processing techniques is presented to clearly map and detect structural variations in the form of DNA copy number along the genome. Second, as a feature selection method prior to further analysis, a regional evaluation analysis is presented. It includes statistical learning procedures to measure the statistical and biological significance of the predicted variations. Then, classification techniques applied to segregate the tested samples into groups and to provide insight into the complex pattern of the predicted variations as well as discovering the relationship among them. There are three critical elements of our analysis that are novel: 1) we are detecting copy number changes as small as 1000 bp[1] (previous studies provided sensitivity typically hundreds of thousands of bp); this allows us to monitor genetic variants that might contribute incrementally to ASD susceptibility, 2) we are using oligo-arrays as a genotyping tool, performing a case-control association analysis, where copy number changes are the genetic variation being assessed, 3) we are developing algorithms to improve the sensitivity and specificity of array CGH data, assessing false positive and false negative rates.

3.1 Data preprocessing

Microarray data analysis is subject to multiple sources of variation, of which biological sources are of interest whereas most others are due to experimental sources. In other words, the goal of aCGH data analysis is to find the true boundaries of the variant regions (segments) which correspond to chromosomal variations and to remove other variations due to human factors, array printer performance, labeling, and hybridization efficiency (Kallioniemi et al., 1992). It consists of three key steps; 1) data preparation, 2) noise reduction, and 3) variation detection. In the data preparation step, copy number data is generated experimentally through aCGH process and then combined with their genomic positions. The next step is to reduce the experimental errors. This step is generally divided into two parts, data normalization, and data filtering. After normalizing the raw DCN data

[1] base-pairs

and before detecting the variant segments, the necessary step is to filter the normalized data for noise reduction.

3.1.1 Data modeling

According to the data description and properties generated from microarray technologies discussed in the previous section, we approximate a given DCN data sample as a one-dimensional piecewise discrete signal corrupted by additive white Gaussian noise with zero-mean and small variance. A good model for describing DNA copy number data is:

$$y[n] = f[n] + \varepsilon_n, \qquad n=1, 2, \ldots, N. \tag{1}$$

where $y[n]$ and $f[n]$ are the observed and true intensities of the DCN data probe at n^{th} location along the x-axis respectively. Here N is the length of DCN data and ε represents a vector of independent identically distributed (*i.i.d.*) random variables drawn from the Gaussian distribution of zero-mean and small variance (Wang et al., 2007).

3.1.2 Irregular probe position

Most prior works considered the DNA copy number profiles as discrete signals under the assumption that the probes are uniformly distributed along the chromosomes. This assumption may lead to wrong decisions with false positive or/and false negative points. More recent studies (Wang et al., 2007 & Willenbrock et al., 2005) show that considering the nonuniform spacing distance between the probes of the DCN data profiles could be beneficial for detecting and measuring the DNC variations.

Hence, we remodeled the DCN data discussed in the previous section as nonuniformly distributed discrete signals as follows:

$$y[x_n] = f[x_n] + \varepsilon_n, \qquad n=1, 2, \ldots, N. \tag{2}$$

where x_n in this case is the nonuniform distributed probe at n^{th} location along the x-axis. The x_n's are not uniformly distributed and the distance between two adjacent probes x_n and x_{n+1} may vary randomly. The $y[x_n]$ and $f[x_n]$ are the observed and true intensities of the DCN data probe location x_n respectively. The ε_n represent *i.i.d.* random variable from the Gaussian distribution with zero-mean and small variance σ^2.

3.2 Maximum likelihood estimator for genetic variation detection

Generally, Copy Number variations (CNVs) detection techniques fall into two categories: statistical based models and smoothing techniques. In the statistical based models, the noise free signal and noise models are required. Unfortunately, these models are usually unknown or impossible to describe adequately with simple random processes. As a result, the important details (i.e., breakpoints) of the CNVs regions will be included in the segmentation process. In addition, the techniques are computationally costly. Furthermore, most statistical models proposed to analyze array CGH data involve modeling the association between changes in neighboring probes. While this is helpful to find wide changes, it tends to ignore local changes. In the literature, there are various statistical approaches that have been proposed to detect changes in the DCN data.

On the other hand, the smoothing techniques provide alternative methods for processing the DCN data that are characterized by small and long intervals with sharp transitions and singularities at boundaries edges (breakpoints). The techniques are particularly suitable for denoising DCN data as they do not require a parametric model in finding structures in the data. In these methods, local operators are applied to the noisy data. Only those points in a small local neighborhood are involved in the computation. The main advantage of these techniques is their computational efficiency. They can process the data in parallel without waiting for their neighboring points to be processed.

To this end, the proposed smoothing techniques provide efficient run-time speed and they are well suited to predict the variations in the discontinuous nature of such data. However, the smoothing techniques suffer from two main drawbacks. First, the breakpoints of the variation regions are involved in the smoothing process and these techniques exhibit artifacts in the neighborhood of these discontinuities that tend to blur the variation edges. Second, they did not consider the physical distances between the adjacent probes and simply assumed that they were uniformly spaced. This simplification will lead to suboptimal results. In this section, we propose a robust method based on maximum likelihood principle (Alqallaf et al., 2009) to clearly map and detect structural variation in the form of DNA copy number along the human genome. We apply dynamic programming to compute the DNA copy number estimates and reduce the computational complexity. Furthermore, we employ the minimum description length approach to estimate the number of unknown parameters. To evaluate our proposed method, we examine and compare the ability to reliably predict variations using molecular test, quantitative polymerase chain reaction. We take the comparison a step further by conducting two experiments designed specifically to assess the sensitivity and specificity of our proposed methods using high-density oligonucleotide array that have been examined by a number of different platforms and laboratories. Using well-characterized cell lines and custom tiled arrays, we show that the proposed method outperforms other popular commercial software and published algorithms in terms of detection performance and computational complexity.

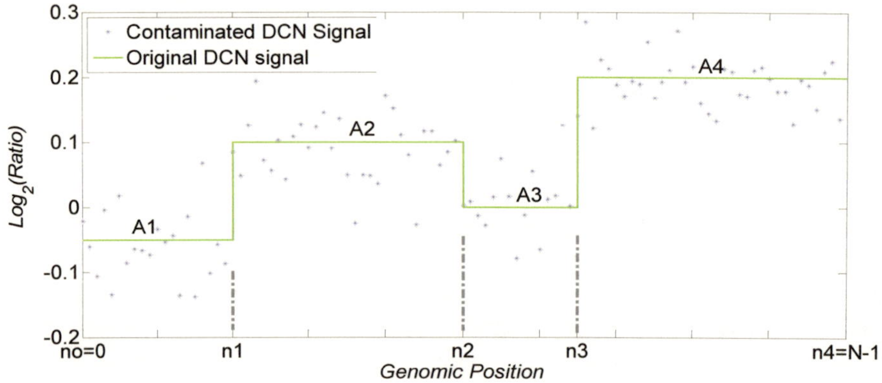

Fig. 2. Illustration of the observed DNA copy number data modeling of (3) with 4 segments.

As described in the previous section, the DNA copy number observations can be modeled as one-dimensional discrete time series with multilevel and jumps at unknown transition times, corrupted by additive white Gaussian noise (AWGN) of zero-mean and small variance σ^2. Figure 2 displays a graphical representation of the observed DNA copy number data modeling with 3 segments. Here we define $f[n]$ is the true piece-wise constant DCN signal to be estimated. Then, we define

$$f[n] = \sum_{i=1}^{M} A_i \left[u[n_{i-1}] - u[n_i]\right] = \begin{cases} A_1 & n = 0,\ 1,\ \cdots,\ n_1 - 1, \\ A_2 & n = n_1,\ n_1 + 1,\ \cdots,\ n_2 - 1, \\ \vdots & \\ A_M & n = n_{M-1},\ n_{M-1} + 1,\ \cdots,\ N - 1. \end{cases} \quad (3)$$

where n_0=0<n_1<n_2< … <n_{M-1}<n_M=N and $u[n]$ is the unit step function. Here A_i and n_i are the intensity level and the length of the ith variant segment, respectively, with a total of M segments.

Based on the data assumption, we wish to design a detector to detect or equivalently estimate the uknown parameters. To do so, we first apply dynamic programming (DP) (Larson & Castie, 1982) to estimate the minimum number of the variant regions M using the minimum description principle (MDL) technique (Rissanen, 1978). Next, we apply the principle of maximum likelihood (ML) to estimate the values of breakpoints locations and intensity levels corresponding to these regions. Assuming that the number of variant regions M is known, then the ith variant region can be characterized by the probability density function (PDF) $p_i([y[n_{i-1}]:y[n_i\text{-}1]];A_i)$, where A_i and n_i are the unknown parameters representing the intensity level and the breakpoint of the ith variant segment, respectively. Moreover, each variant region is assumed to be statistically independent of all other regions. Hence, the PDF of the entire data record can be written as

$$p(\boldsymbol{y};\boldsymbol{A},\boldsymbol{n}) = \prod_{i=1}^{M} p_i(y[n_{i-1}:n_i - 1];A_i) = \frac{1}{(2\pi\sigma^2)^{N/2}} \exp\left[-\frac{1}{2\sigma^2}\sum_{i=1}^{M}\left[\sum_{n=n_{i-1}}^{n_i-1}\left(y[n]-A_i\right)^2\right]\right]. \quad (4)$$

The DP algorithm can also be applied here to reduce the computational complexity to a more manageable level that is linearly proportional with the number of variant regions M.

3.3 Comparison study

In this section, we conduct two experiments to compare our proposed method with recent approaches to improve the sensitivity and specificity of array CGH data (assessing false positive and false negative rates).

3.3.1 Self-self hybridization experiment

In this experiment, we compare the performance of our proposed method MLE (Alqallaf et al., 2009) with Circular Binary Segmentation (CBS) algorithm (Venkatraman et al., 2007) and Copy Number Professional software package (BioDiscovery) Nexus algorithm by direct measurement of false positives. The same DNA sample is used as both the test and reference and hence any copy number variant assigned by an algorithm is incorrect and a false positive. In other words, we compare the DNA sample with itself in the aCGH process to generate the DCN data as described in section 2. In the ideal case, the intensity level, the difference between the tested sample and a known reference measured in $\log_2$ ratio, should equal to zero. However, due to the experimental noise, we expect to detect segments with relatively small intensity level value that are below cut-offs criteria. Otherwise, the detected segments would be considered as false positives. As shown in Table 1, the average number of events detected by CBS algorithm is lower than the events detected by other algorithms. However, the average length of the events detected by our proposed algorithm MLE is relatively shorter than the average length of the events detected by CBS and Nexus.

Array#	Nexus		CBS		MLE	
	E	*L*	*E*	*L*	*E*	*L*
1	10	40336	2	790	6	8481
2	30	391030	4	625	19	27866
3	100	2894494	7	4130039	55	42440
Avg.	47	1108620	5	1377151	27	26262

Table 1. Comparison of the proposed algorithms using number of detected events, *E*, and their length, *L*, in base-pair for the three tested array samples.

3.3.2 Duplicated dye-swap experiments for two HapMap samples

Here we take the comparison a step further by conducting experiments designed specifically to assess our proposed algorithm, MLE, using high-density oligonucleotide array CGH. In this experiment, replicate dye-swap experiments were conducted comparing DNA samples from two hapmap (Redon et al., 2006) subjects that have been examined by a number of different platforms and laboratories, NA15510 and NA10851, for a total of four arrays. The relative intensities differences are measured and reported. It should be noted that the directionality of any detected variant is expected to be opposite when the dyes are swapped. That is, deletions with the first array will appear to be duplications with the second array. This is due to the convention of reporting the $\log_2$ ratios as described in section 2. This experiment allows us to assess the sensitivity of the proposed algorithms. Table 2 shows that the number of CNVs detected by the MLE is considerably higher than those detected using CBS (a range of 4.5% to 36% more for the 4 different array experiments). Our results show that applying the averaging window of 2Kb allow the algorithms to be well suited for detecting variations in high-density oligonucleotide array aCGH.

Array #	CBS	MLE
1	14	20
2	10	20
3	20	21
4	13	21

Table 2. List of the number of events (CNVs) detected by CBS and MLE algorithms.

4. Statistical significance

After filtering multiple DCN datasets of normal control and test samples, we need to apply a statistical analysis to reveal the randomness and to classify the genes or genomic locations that are involved or play roles in the targeted disease, ASD. In this section, we present two statistical approaches to measure the significance of common CNVs across the samples and especially in the complex LCRs regions. First, we measure the relative frequency at each genomic position within the LCRs regions. Second, based on the relative frequency, a regional evaluation scheme is used to measure the significance of the overlapping recurrent CNVs and to classify the tested DCN samples.

4.1 Statistical-based model

In summary, most of the proposed algorithms in the literature did not consider the statistical and biological significance of the analysis of multiple DCN data samples. In particular, they did not address the task of identifying common variations that overlap a set or subset of the study samples to reveal the randomness of the predicted CNVs. Indeed, few studies have addressed class discovery across multiple samples of DCN data (Grant et al., 1999 & Diskin et al., 2006). However, they did not consider denoising the data prior to applying the statistical analysis. Although these are effective methods for searching statistically for common variations across multiple samples, it suffers from two main issues which can be summarized as follows: First, it does not take into considerations that different variation types (gain and loss) may occur within the same genomic locations. They simply discard these locations and indicate them as missing values. This will lead to decreases in the data resolution. Second, it does not differentiate between the intensity levels. This is an important issue for characterizing the variations in the complex areas of low copy repeats (LCR). For this, we propose in our statistical analysis to identify nonrandom gains and losses across multiple samples with the consideration of these issues.

To reveal the randomness and identify the genes or genomic locations that are involved or play roles in the targeted disease, we apply a statistical analysis to measure the significance of recurrent CNVs including those in the complex regions of LCRs. Here, we plot the frequency of the occurrence of the predicted CNVs (deletions and/or duplications) that are overlapped across multiple case samples with respect to control samples. Suppose that a set of M filtered DCN samples each with N probes, then the normalized frequency at the nth position can be measured as

$$G[n] = \frac{\sum_{s \in M} v_{s,n}}{M}. \qquad n = 1,\ 2,\ \ldots,\ N \tag{5}$$

where s represents the sample of the same variation type and $v_{s,n}$ is a binary number which equals to 1 if the variation is present and 0 otherwise. Figure 3 shows the differences in the frequency of occurrence of the gains and losses between 71 normal control and 71 autistic samples of chromosome 7. The differences suggest further analysis to discover the relationship between the predicted CNVs and to classify the tested samples.

4.2 Putative recurrent CNVs classification

Although the predicted variant segments of each aCGH profile have their own importance, finding recurrent copy number variations that overlap and share the same type adds another dimension to link them with the targeted disease. The size of our aCGH profiles is relatively large and many of the variants regions of the same type (deletions/duplications) are found in both cases and controls. We therefore include a filtering step by removing these CNVs to make it easier to find the interesting variations and reduce the number of data points to some subset of concatenated CNVs.

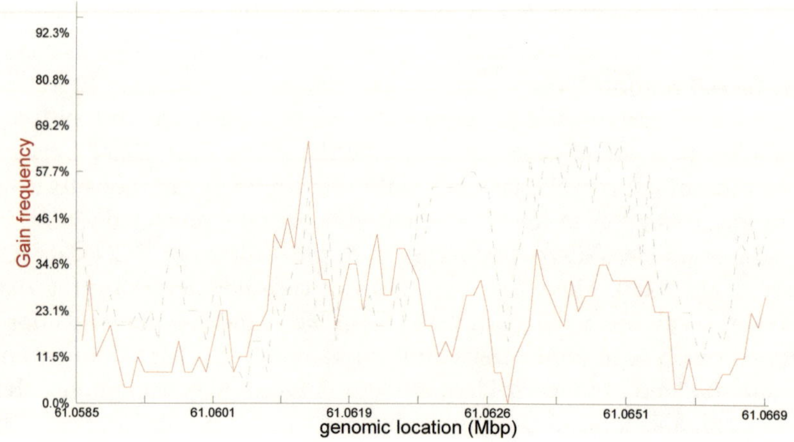

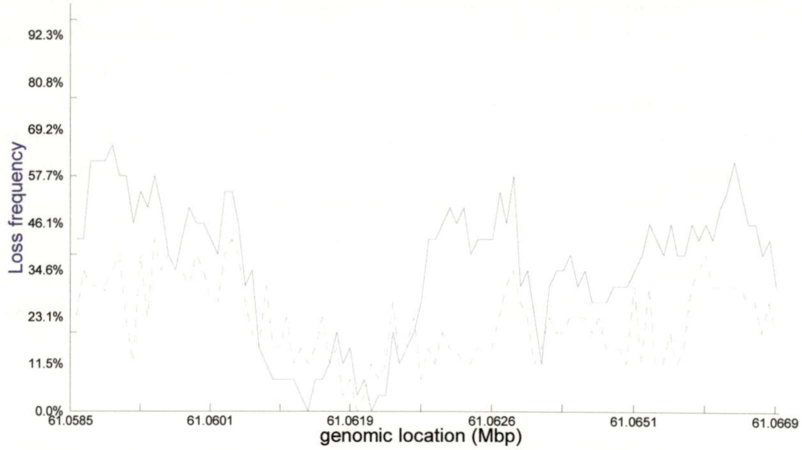

Fig. 3. Frequency plots of the filtered samples. A) The solid red and the dash gray lines represent the gain frequency of the typically developed (TD) and autistic (AU) samples, respectively. B) The solid blue and the dash gray lines represent the loss frequency of the TD and AU samples, respectively.

Before we make a decision on the predicted segments, in this section, we extend our method by imposing cutoff criteria based on regional genetic information as an optimal feature selection. The reasons for performing this procedure are as follows. First, we seek the genetic structure and thus the genetic mechanisms responsible for the progression of the disease. Second, we would like to remove or eliminate the irrelevant features (e.g., CNVs) from the classification and hence, to increase the run-time speed and to improve the accuracy of the classification. After ranking the CNVs, a suitable set is identified and declared as an optimal feature set to be used for classification analysis. Although the feature selection step is a major step attempting to discover and reveal genetic mechanisms, it can not be claimed to discover the true biological relationship without further experimental evaluation. The extension accounts for the minimal number of probes within in each

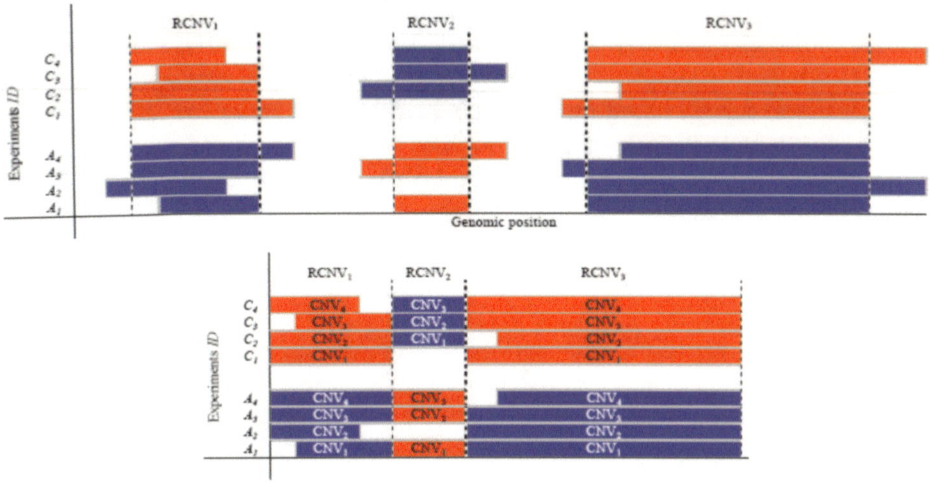

Fig. 4. Schematic representation of 3 recurrent copy number variations (RCNVs) with different lengths (Top) and concatenation vectors of the predicted combinations of RCNVs (Bottom). The x-axis represents the genomic position and the y-axis represents the experiments indices, C_i for normal control samples and A_i for autistic samples, respectively. The vertical dashed lines represent the RCNVs boundaries. The red and blue bars represent duplication and deletion for the corresponding positions.

segment, the intensity level represented by the $\log_2$ ratio value, and the repeat content of the region where the CNV is located.

Each segment that met biological and statistical cut-off criteria is considered a CNV and assembled into a segmentation table for further biological analysis. Figure 4 is an illustration of three RCVNs with different sizes of filtered DCN data for multiple samples of normal control, C_i's, and autistic, A_i's, individuals, respectively.

With this setup, we apply the traditional clustering algorithms (Fuzzy c-means and k-means) to the concatenation vectors of the predicted combinations of RCNVs to classify the DCN data samples and to provide insight into the pattern of the variations using the concatenated recurrent CNVs that are statistically significant.

In the next section, we will investigate the classification performance using the predicted combinations of multiple RCNV sites of different chromosomes produced by the regional evaluation method presented in this section that may have direct role in the targeted disease, ASD.

5. Visualization and pattern recognition

To visualize the microarray data, we apply agglomerative hierarchical clustering algorithm to decide the level or scale of clustering that is most appropriate for our clustering analysis. It provides a graphical representation of the samples to explore the number of ways to look for relationships between the samples and to provide insight into the pattern of the recurrent CNVs. The algorithm groups the data samples based on the defined measure of the distances between the samples elements using similarities functions to create the clusters. It starts from each single sample as a cluster and it merge the samples into clusters (groups or subgroups) based on the updated similarity measures (linkage), where clusters at one level are joined as clusters at the next level. The definition of the similarity measures depends on the clustering algorithm and the biological meaning of similarity. For example, a correlation distance, $d_p(\boldsymbol{x},\boldsymbol{y})$, based on Pearson's correlation (6) may bring together samples whose probes intensity levels are different, but have a similar behavior, and which would be considered different by the Euclidean distance $d_e(\boldsymbol{x},\boldsymbol{y})$ (7) which is suitable for discovering the common CNVs. Specifically,

$$d_p(\mathbf{x},\mathbf{y}) = \frac{\sum_{i=1}^{N}(x_i - \bar{x})(y_i - \bar{y})}{\sqrt{\sum_{i=1}^{N}(x_i - \bar{x})^2 \sum_{i=1}^{N}(y_i - \bar{y})^2}}, \tag{6}$$

$$d_e(\mathbf{x},\mathbf{y}) = \sqrt{\sum_{i=1}^{N}(x_i - y_i)^2}, \tag{7}$$

where $\bar{x}$ and $\bar{y}$ are the sample mean values of the two data vectors $\boldsymbol{x}$ and $\boldsymbol{y}$ with N data points, respectively.

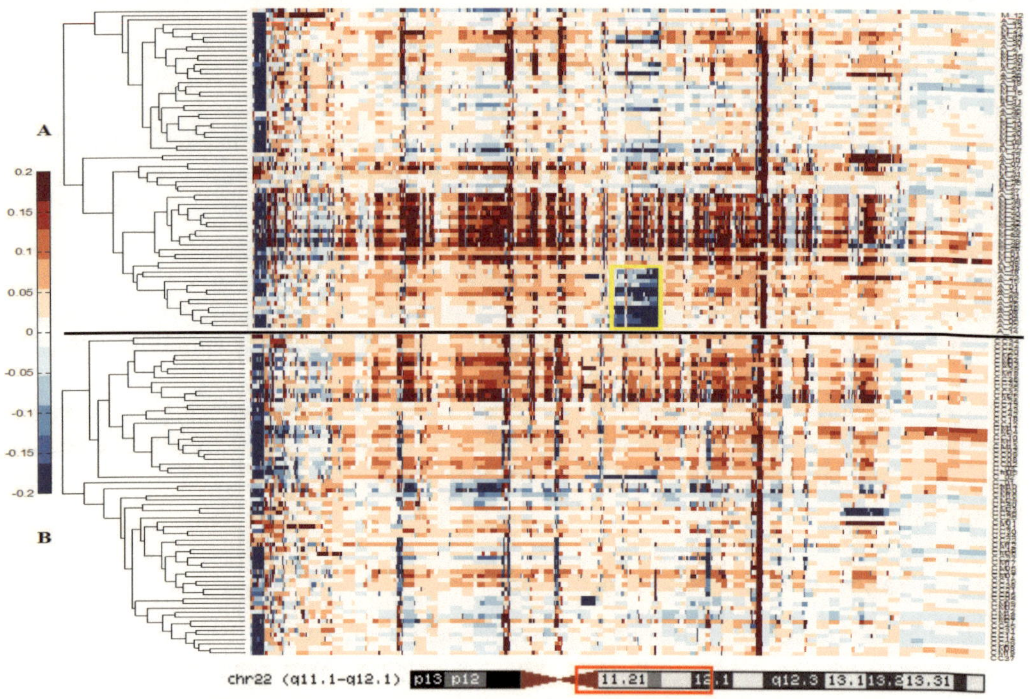

Fig. 5. Hierarchical clustering of chromosome 22 using 142 samples. A) 71 autistic (AU) samples. B) 71 typically developed (TD) samples. Dark red represents duplications and dark blue represents deletions. The solid black line used to separate the AU and TD samples. Yellow square represents deletion region within the AU subgroup.

To explore the dataset before imposing the cut-off criteria, we perform unsupervised hierarchical clustering with the Euclidean distance as a distance metric to calculate the pair-wise distances between the tested samples and centroid linkage method to create linkage between the clusters tree. The heat map is used to represent thousands of $\log_2$ ratios (intensity values) of the probes of each sample it uses two-color of a matrix of colored cells, red for duplication, where the $\log_2$ ratio is positive, and blue for deletion, where the $\log_2$ ratio is negative. The rows represent the tested samples and the columns represent the probes positions, and the brightness of the cells is proportional to their intensity levels. For this analysis, our case-control population of study consists of 71 individuals with a diagnosis of autism compared to 71 typically developing controls matched for gender and ethnicity. Figure 5 shows an example of one of the five chromosomal regions used in this study. By simple comparison between the recurrent CNVs detected in the entire or subset of the autistic samples (Figure 5. A) and those detected in the typically developed samples (Figure 5. B), we can detect patterns of variations that are exclusively or selectively represented in one or the other group (see for example the deletions noted with yellow box). The yellow square show long deletion region within the AU subgroup compared to the other members in the AU individuals and TD controls.

6. Conclusion

In this chapter, we presented an overview for the analysis of genetic variations in the form of DNA copy number changes and their association with the targeted disease, autism spectrum disorder. Our study shows that our proposed algorithm, MLE, is computationally efficient and it can achieve even better detection capabilities by considering the effect of the nonuniform genomic spacing distance between the biomarkers. Moreover, to enhance our algorithm's ability to map and identify regions of variation across multiple samples, we preformed statistical analysis on the filtered samples searching for common variations. The potential impact of the statistical analysis is to provide insight into the patterns of the variations by characterizing and classifying the samples that are involved in the targeted diseases. Indeed, the high frequency of variants (duplications and/or deletions) detected in these regions across the samples allowed the assembly of a copy number map of both typically developed and diseased individuals. The mapping approach reveals patterns of copy number change along these chromosomal intervals that are not currently represented in the assembly of genomic variants compiled from relatively low-resolution genome-wide platforms. Our findings indicate that Low copy repeat-rich intervals, known to be relatively susceptible to copy number changes and sequence rearrangement, show a greater degree of copy number alteration in diseased compared to typically developed individuals. A larger contribution of variations detected (duplications and/or deletions) in the total copy number burden differences have been reported to be associated with different genetic diseases. Our findings also show ethnicity is an important consideration that should be integrated into case-control study design. The findings suggest that autism is associated with an increased amount of copy number alteration in unstable segments of the genome. The experimental results also show that using high-resolution custom-tiled oligonucleotide array comparative genomic hybridization samples, improve the accuracy of the proposed methods to detect the true amount of structural variations of the human genome including previously reported variations with known biological and clinical relevance and new variations that warrant further investigated. To explore the idea that patterns of relatively common copy number variations can increase the power of discrimination between autistic and typically developing patients, a set of recurrent variants that are statistically differed between the two groups is identified and presented. The findings suggest that combinations of copy number variations could provide the basis for discriminating autistic and typically developing groups and potentially identifying distinct subgroups within the phenotypic heterogeneity of autism spectrum disorder. Finally, the analysis presented in this chapter is broadly applicable to case-control studies of genetic diseases beyond the targeted disease, autism.

7. References

Freeman, JL.; Perry, GH.; Feuk, L.; Redon, R.; McCarroll, SA.; Altshuler, DM.; Aburatani, H.; Jones, KW.; Tyler-Smith, C. & Hurles, ME. (2006). Copy number variation: new insights in genome diversity. *Genome Research*, Vol.16, No.8, pp. 949-961.

Lee, JA. & Lupski, JR. (2006). Genomic rearrangements and gene copy-number alterations as a cause of nervous system disorders. *Neuron*, Vol.52, No.1, pp. 103-121.

Stankiewicz, P. & Lupski, JR. (2002). Molecular-evolutionary mechanisms for genomic disorders. *Current Opinion in Genetics & Development on ScienceDirect*, Vol.12, No.3, pp. 312-319.

Christian, SL.; Brune, CW.; Sudi, J.; Kumar, RA.; Liu, S.; Karamohamed, S.; Badner, JA.; Matsui, S.; Conroy, J. & McQuaid, D. (2008). Novel submicroscopic chromosomal abnormalities detected in autism spectrum disorder. *Biological Psychiatry*, Vol.63, No.12, pp. 1111-1117.

Marshall, CR.; Noor, A.; Vincent, JB.; Lionel, AC.; Feuk, L.; Skaug, J.; Shago, M.; Moessner, R.; Pinto, D. & Ren, Y. (2008). Structural variation of chromosomes in autism spectrum disorder. *American Journal of Human Genetics*, Vol.82, No.2, pp. 477-488.

Sebat, J.; Lakshmi, B.; Malhotra, D.; Troge, J.; Lese-Martin, C.; Walsh, T.; Yamrom, B.; Yoon, S.; Krasnitz, A. & Kendall, J.; (2007) Strong association of de novo copy number mutations with autism. *Science*, Vol.316, No.5823, pp. 445-449.

Kirov, G.; Gumus, D.; Chen, W.; Norton, N.; Georgieva, L.; Sari, M.; O'Donovan, MC.; Erdogan, F.; Owen, MJ. & Ropers, HH. (2008). Comparative genome hybridization suggests a role for NRXN1 and APBA2 in schizophrenia. *Human Molecular Genetics*, Vol.17, No.3 pp. 458-465.

Sharp, AJ.; Hansen, S.; Selzer, RR.; Cheng, Z.; Regan, R.; Hurst, JA.; Stewart, H.; Price, SM.; Blair, E. & Hennekam, RC. (2006). Discovery of previously unidentified genomic disorders from the duplication architecture of the human genome. *Nature Genetics*, Vol.38, No.9, pp. 1038-1042.

Sharp, AJ.; Mefford, HC.; Li, K.; Baker, C.; Skinner, C.; Stevenson, RE.; Schroer, RJ.; Novara, F.; DeGregori, M. & Ciccone, R. (2008). A recurrent 15q13.3 microdeletion syndrome associated with mental retardation and seizures. Nature Genetics, Vol.40, No.3, pp. 322-328.

Cantor, RM. & Geschwind, DH. (2008). Schizophrenia: genome, interrupted. *Neuron*, Vol.58, No.2, pp. 165-167.

Stefansson, H.; Rujescu, D.; Cichon, S.; Pietilainen, OP.; Ingason, A.; Steinberg, S.; Fossdal, R.; Sigurdsson, E.; Sigmundsson, T. & Buizer-Voskamp, JE. (2008). Large recurrent microdeletions associated with schizophrenia. Nature, Vol.455, No.7210, pp. 232-236.

Stone, JL.; O'Donovan, MC.; Gurling, H.; Kirov, GK.; Blackwood, DH.; Corvin, A.; Craddock, NJ.; Gill, M.; Hultman, CM. & Lichtenstein, P. (2008). Rare chromosomal deletions and duplications increase risk of schizophrenia. Nature, Vol.455, No.7210, pp. 237-241.

Walsh, T.; McClellan, JM.; McCarthy, SE.; Addington, AM.; Pierce, SB.; Cooper, GM.; Nord, AS.; Kusenda, M.; Malhotra, D. & Bhandari, A. (2008). Rare structural variants disrupt multiple genes in neurodevelopmental pathways in schizophrenia. Science, Vol.320, No.5875, pp. 539-543.

Estivill, X. & Armengol, L. (2007). Copy number variants and common disorders: filling the gaps and exploring complexity in genome-wide association studies. *PLoS Genetics*, Vol.3, No.10, pp. 1787-1799.

Greer, MK.; Brown, FR.; Pai, GS.; Choudry, SH. & Klein, AJ. (1997). Cognitive, adaptive, and behavioral characteristics of Williams syndrome. *American Journal of Human Genetics*, Vol.74, No.5, pp. 521-525.

Somerville, MJ.; Mervis, CB.; Young, EJ.; Seo, EJ.; del Campo, M.; Bamforth, S.; Peregrine, E.; Loo, W.; Lilley, M. & Perez-Jurado, LA. (2005). Severe expressive-language delay related to duplication of the Williams-Beuren locus. *The New England Journal of Medicine*, Vol.353, No.16, pp. 1694-1701.

Van der Aa, N.; Rooms, L.; Vandeweyer, G.; van den Ende, J.; Reyniers, E.; Fichera, M.; Romano, C.; Delle Chiaie, B.; Mortier, G. & Menten, B. (2009). Fourteen new cases contribute to the characterization of the 7q11.23 microduplication syndrome. *European Journal of Medical Genetics*, Vol.52, No.2-3, pp. 94-100.

Depienne, C.; Heron, D.; Betancur, C.; Benyahia, B.; Trouillard, O.; Bouteiller, D.; Verloes, A.; LeGuern, E.; Leboyer, M. & Brice, A. (2007). Autism, language delay and mental retardation in a patient with 7q11 duplication. *Journal of Medical Genetics*, Vol.44, No.7, pp. 452-458.

Balciuniene, J.; Feng, N.; Iyadurai, K.; Hirsch, B.; Charnas, L.; Bill, BR.; Easterday, MC.; Staaf, J.; Oseth, L. & Czapansky-Beilman, D. (2007). Recurrent 10q22-q23 deletions: a genomic disorder on 10q associated with cognitive and behavioral abnormalities. *American Journal of Human Genetics*, Vol.80, No.5, pp. 938-947.

Laumonnier, F.; Roger, S.; Guerin, P.; Molinari, F.; M'Rad, R.; Cahard, D.; Belhadj, A.; Halayem, M.; Persico, AM. & Elia, M. (2006). Association of a functional deficit of the BKCa channel, a synaptic regulator of neuronal excitability, with autism and mental retardation. *American Journal of Psychiatry*, Vol.163, No.9, pp. 1622-1629.

Cook, EH. Jr. (2001). Genetics of autism. *Child and Adolescent Psychiatric Clinics of North America*, Vol.10, No.2, pp. 333-350.

Helbig, I.; Mefford, HC.; Sharp, AJ.; Guipponi, M.; Fichera, M.; Franke, A.; Muhle, H.; de Kovel, C.; Baker, C. & von Spiczak, S. (2009). 15q13.3 microdeletions increase risk of idiopathic generalized epilepsy. *Nature Genetics*, Vol.41, No.2, pp. 160-162.

Papolos, DF.; Faedda, GL.; Veit, S.; Goldberg, R.; Morrow, B.; Kucherlapati, R. & Shprintzen, RJ. (1996). Bipolar spectrum disorders in patients diagnosed with velo-cardio-facial syndrome: does a hemizygous deletion of chromosome 22q11 result in bipolar affective disorder? *American Journal of Psychiatry*, Vol.153, No.12, pp. 1541-1547.

Niklasson, L.; Rasmussen, P.; Oskarsdottir, S. & Gillberg, C. (2001) Neuropsychiatric disorders in the 22q11 deletion syndrome. *Genetics in Medicine*, Vol.3, No.1, pp. 79-84.

Antonell, A.; de Luis, O.; Domingo-Roura, X. & Perez-Jurado, LA. (2005). Evolutionary mechanisms shaping the genomic structure of the Williams-Beuren syndrome chromosomal region at human 7q11.23. *Genome Research*, Vol.15, No.9, pp. 1179-1188.

Vorstman, JA.; Morcus, ME.; Duijff, SN.; Klaassen, PW.; Heineman-de Boer, JA.; Beemer, FA.; Swaab, H.; Kahn RS. & van Engeland, H. (2006). The 22q11.2 deletion in children: high rate of autistic disorders and early onset of psychotic symptoms. *Journal of the American Academy of Child & Adolescent Psychiatry*, Vol.45, No.9, pp. 1104-1113.

Hertz-Picciotto, I.; Croen, LA.; Hansen, R.; Jones, CR.; van de Water, J. & Pessah, IN. (2006). The CHARGE study: an epidemiologic investigation of genetic and environmental factors contributing to autism. *Environmental Health Perspectives*, Vol.114, No.7, pp. 1119-1125.

Redon, R.; Ishikawa, S.; Fitch.; KR.; Feuk, L.; Perry, GH.; Andrews, TD.; Fiegler, H.; Shapero, MH.; Carson, AR. & Chen, W. (2006). Global variation in copy number in the human genome. *Nature*, Vol.444, No.7118, pp. 444-454.

Shen F, Huang J, Fitch KR, Truong VB, Kirby A, Chen W, Zhang J, Liu G, McCarroll SA, Jones KW (2008). Improved detection of global copy number variation using high density, non-polymorphic oligonucleotide probes. *BMC Genetics*, Vol.9 No.27.

Conrad, DF.; Pinto, D.; Redon, R.; Feuk, L.; Gokcumen, O.; Zhang, Y.; Aerts, J.; Andrews, TD.; Barnes, C. & Campbell, P. (2010). Origins and functional impact of copy number variation in the human genome. *Nature*, Vol.464, No.7289, pp. 704-712.

Perry, GH.; Ben-Dor, A.; Tsalenko, A.; Sampas, N.; Rodriguez-Revenga, L.; Tran, CW.; Scheffer, A.; Steinfeld, I.; Tsang, P. & Yamada, NA. (2008). The fine-scale and complex architecture of human copy-number variation. *American Journal of Human Genetics*, Vol.82, No.3, pp 685-695.

Berg, JS.; Brunetti-Pierri, N.; Peters, SU.; Kang, SH.; Fong, CT.; Salamone, J.; Freedenberg, D.; Hannig, VL.; Prock, LA. & Miller, DT. (2007). Speech delay and autism spectrum behaviors are frequently associated with duplication of the 7q11.23 Williams-Beuren syndrome region. *Genetics in Medicine*, Vol.9, No.7, pp. 427-441.

Potocki, L.; Bi, W.; Treadwell-Deering, D.; Carvalho, CM.; Eifert, A.; Friedman, EM.; Glaze, D.; Krull, K.; Lee, JA. & Lewis, RA.; (2007). Characterization of Potocki-Lupski syndrome (dup(17)(p11.2p11.2)) and delineation of a dosage-sensitive critical interval that can convey an autism phenotype. *American Journal of Human Genetics*, Vol.80, No.4, pp. 633-649.

Carvalho, CM. & Lupski, JR. (2008). Copy number variation at the breakpoint region of isochromosome 17q. *Genome Research*, Vol.18, No.11, pp. 1724-1732.

Mukaddes, NM. & Herguner, S. (2007). Autistic disorder and 22q11.2 duplication. *World Journal of Biological Psychiatry*, Vol.8, No.2, pp. 127-130.

Barabash, A.; Marcos, A.; Ancin, I.; Vazquez-Alvarez, B.; de Ugarte, C.; Gil, P.; Fernandez, C.; Encinas, M.; Lopez-Ibor, JJ. & Cabranes, JA. (2009) APOE, ACT and CHRNA7 genes in the conversion from amnestic mild cognitive impairment to Alzheimer's disease. *Neurobiol Aging*, Vol.30, No.8, pp. 1254-1264.

Miller, DT.; Shen, Y.; Weiss, LA.; Korn, J.; Anselm, I.; Bridgemohan, C.; Cox, GF.; Dickinson, H.; Gentile, J. & Harris, DJ. (2009). Microdeletion/duplication at 15q13.2q13.3 among individuals with features of autism and other neuropsychiatric disorders. *Journal of Medical Genetics*, Vol.46, No.4 pp. 242-248.

Shinawi, M.; Schaaf, CP.; Bhatt, SS.; Xia, Z.; Patel, A.; Cheung, SW.; Lanpher, B.; Nagl, S.; Herding, HS. & Nevinny-Stickel, C. (2009). A small recurrent deletion within 15q13.3 is associated with a range of neurodevelopmental phenotypes. *Nature Genetics*, Vol.41, No.12, pp. 1269-1271.

Tsuchiya, KD.; Wiesner, G.; Cassidy, SB.; Limwongse, C.; Boyle, JT. & Schwartz, S. (1998). Deletion 10q23.2-q23.33 in a patient with gastrointestinal juvenile polyposis and other features of a Cowden-like syndrome. *Genes, Chromosomes & Cancer*, Vol.21, No.2, pp. 113-118.

Zhou, XP.; Woodford-Richens, K.; Lehtonen, R.; Kurose, K.; Aldred, M.; Hampel, H.; Launonen, V.; Virta, S.; Pilarski, R. & Salovaara, R. (2001). Germline mutations in BMPR1A/ALK3 cause a subset of cases of juvenile polyposis syndrome and of Cowden and Bannayan-Riley-Ruvalcaba syndromes. *American Journal of Human Genetics*, Vol.69, No.4, pp. 704-711.

Fombonne, E. (2003). Epidemiological Surveys of Autism and Other Pervasive Developmental Disorders: An Update. *Journal of Autism and Developmental Disorders*, Vol.33, No.4, pp. 365–382.

Kuehn, B. (2007). CDC: Autism Spectrum Disorders Common. *Journal of the American Medical Association*, Vol.297, No.940.

Abrahams, B. & Geschwind, D. (2008). Advances in autism genetics: on the threshold of a new neurobiology. *Nature Reviews Genetics*. Vol.9, No.5, pp. 341–355.

Alqallaf, A.; Tewfik, A. & Selleck, S. (2009). Genetic variation detection using maximum likelihood estimator. *IEEE International Workshop on Genomic Signal Processing and Statistics*. ISBN: 9781424447619, Minnesota, USA.

Kallioniemi, A.; Kallioniemi, O.; Sudar, D.; Rutovitz, D.; Gray, J.; Waldman, F. & Pinkel, D. (1992). Comparative genomic hybridization for molecular cytogenetic analysis of solid tumors. Science, Vol.258, No.5083, pp. 818-821.

Wang, Y. & Wang, S. (2007). A novel stationary wavelet denoising algorithm for array-based DNA copy number data. International Journal of Bioinformatics Research and Applications, Vol.3, No.2, pp. 206–222.

Willenbrock, H. & Fridlyand, J. (2005). A comparison study: applying segmentation to array CGH data for downstream analyses. *Bioinformatics*, Vol.21, No.22, pp. 4084–4091.

Rissanen J. (1978). Modeling by Shortest Data Description. *Automatica*, Vol.14, pp. 465-471.

Larson, R. & Castie, J. (1982). Principles of Dynamic Programming. Vol.1–2, *Marcel Dekker* Inc., NY.

Nexus: Copy Number Professional software package (BioDiscovery), Inc., El Segundo, CA) BioDiscovery Inc. available from: http://www.biodiscovery.com.

Venkatraman, E. & Olshen, A. (2007). A faster circular binary segmentation algorithm for the analysis of array CGH data. Bioinformatics, Vol.23, No.6 pp. 657-663.

Grant, G.; Manduchi, E.; Cheung, V. & Ewens, W. (1999). Significance testing for direct identity-by-descent mapping. Annals of Human Genetics, Vol.63, No.5, pp. 441-454.

Diskin, S.; Eck, T.; Greshock, J.; Mosse, Y.; Naylor, T.; Stoeckert, C.; Weber, B.;Maris, J. & Grant, G. (2006). STAC: A method for testing the significance of DNA copy-number aberrations across multiple array-CGH experiments. Genome Research, Vol.16, No.9, pp. 1149-1158.

9

Genome-Wide Association Studies of Copy Number Variation in Autism Spectrum Disorder

Hae-Jin Hu and Yeun-Jun Chung
Integrated Research Center for Genome Polymorphism, The Catholic University of Korea, College of Medicine, Seoul Korea

1. Introduction

Autism is a syndrome with a broad spectrum of phenotypes characterized by deficits in social interaction and communication, repetitive or stereotyped behaviors, and restricted interests (Rutter, 2005). Autism spectrum disorder (ASD) manifests mostly before 3 years of age (Klauck, 2006). ASDs include two related diagnoses; pervasive developmental disorder (PDD) including atypical autism, impairment in the same areas, but not meeting criteria for autism; Asperger's syndrome, which is milder than PDD, showing similar impairments in social interaction, behaviors and interests, but no significant delay in linguistic and cognitive development (Weiss, 2009). The prevalence rate of ASDs is ~0.6% and ASDs are approximately four times more common in males than in females (Veenstra-VanderWeele, 2004). Many studies have been performed to elucidate the pathogenesis of ASDs, but identified risk factors do not explain a significant proportion of the disease prevalence.

Genetic epidemiological data have been suggesting that ASDs are heritable both in autism families and in the general population (Freitag, 2007). The concordance rates of autism in monozygotic twins were reported to be significantly higher (~ 60–90%) than those in dizygotic twins (~ 10%) and the recurrence rates are known to be approximately 10-20 times higher in siblings than in normal population (Folstein & Rosen-Sheidley, 2001; Cohen et al., 2005; Bailey et al., 1995; Lauritsen et al., 2005). ASD is not a single-gene disorder with Mendelian inheritance but rather a component of various genetic disorders with apparent cytogenetic abnormalities (Eapen, 2011). Cytogenetic alterations were detected in 7.4% of ASD (Vorstman et al., 2006), and some of them have been suggested as causative factors of neurodevelopmental disorders (Merikangas et al., 2009). However, discrepancies in study results and diverse modes of inheritance have hindered the discovery of common genetic susceptibility factors to ASDs. For these reasons, despite the growing evidence supporting the genetic susceptibility to ASD development (Folstein & Rosen-Sheidley, 2001; Veenstra-VanderWeele & Cook, 2004), the genetic mechanisms of ASD is still largely unknown.

Recent technical advance in microarray-based whole-genome analysis has enabled identification of common and rare genetic alterations associated with ASDs. Several recent studies have suggested that ASDs are associated with genetic variations including single nucleotide polymorphisms (SNPs) and copy number variations (CNVs), and that these genetic variations may work together (Veenstra-VanderWeele & Cook, 2004). For example, de novo CNVs were found in ~7% of idiopathic ASD families via oligoarray-comparative genomic hybridization and whole-genome SNP array analysis (Abrahams & Geschwind, 2008; Psychiatric GWAS Consortium Coordinating Committee et al., 2009). In addition to rare de novo variations, common genetic variations such as SNPs on 5p14.1 were found to be associated with ASDs and this finding was replicated in two independent studies (Wang et al., 2009). Recently introduced next-generation sequencing (NGS) will further accelerate mining of genetic variations linked with ASDs (Ropers, 2010). Graphical overview of the reported ASD-associated CNVs and SNP are illustrated in the Figure 1. In this chapter, we will review the recent results of CNV and SNP genome-wide association studies (GWAS) on ASDs and discuss the perspectives of the genetic susceptibility study of ASDs.

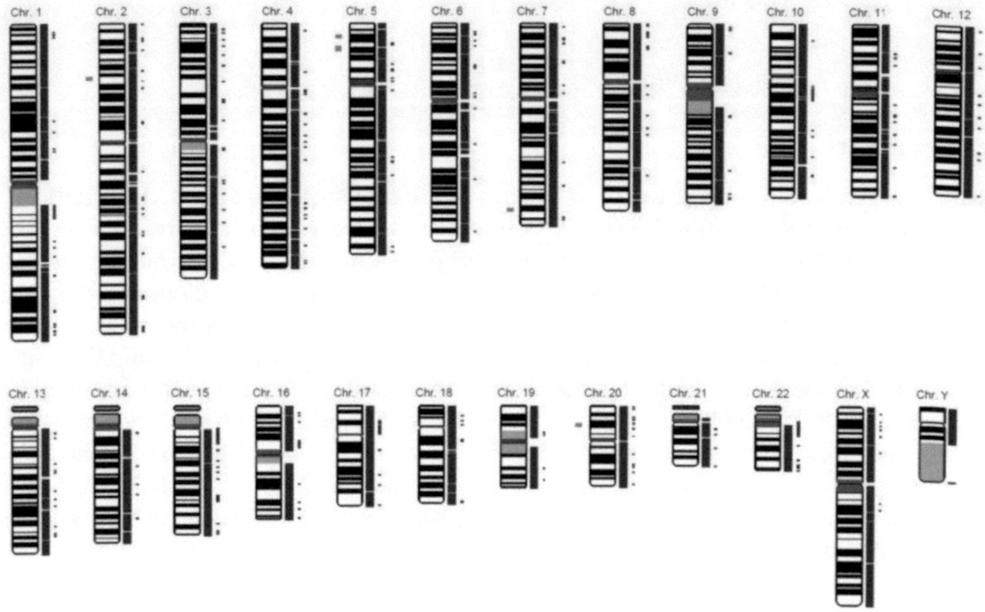

Fig. 1. Genomic map of CNVs and SNPs associated with ASDs identified by GWAS. Green and red bars on the left and right side of the karyograms indicate chromosomal locations of SNPs and CNVs, respectively. Blue bars on the right side of the karyograms present the locations of known genes. This figure was drawn by IdeogramBrowser (Müller et al, 2007).

2. CNVs associated with ASD

2.1 What is CNV?

Using array-CGH, a combination of microarray and comparative genomic hybridization (CGH) technologies, two pioneering groups of scientists have identified wide-spread CNVs in apparently healthy, normal individuals in 2004 (Iafrate et al., 2004; Sebat et al., 2004). CNV is defined as any type of genetic variant that alters the chromosomal structure, including duplications and deletions (Iafrate et al., 2004; Sebat et al., 2004; Redon et al., 2006) and now known to be one of the most prevalent types of genetic variations in the human genome (Feuk et al., 2006; Hurles et al., 2008; Carter, 2007; Estivill & Armengol, 2007). In addition to SNPs, CNVs in normal individuals have been widening our understanding of genetic heterogeneity (Iafrate et al., 2004; Sebat et al., 2004; Redon et al., 2006). Commonly used working definition of CNV was a copy number change involving a DNA segment sized 1 kilobases (kb) or larger (Freeman et al., 2006; Feuk et al., 2006). Nowadays, definition of CNVs includes any DNA structural variants including duplications, deletions and inversions (Hurles et al., 2008). When the frequency of CNV is common (>1%) in the population, CNV is also called copy number polymorphism (CNP). However, due to lack of standardized technologies to define CNV, the size and frequency of CNV have not been well defined in human populations. Since the two pioneering studies discovered the evidence of the existence of CNVs (Iafrate et al., 2004; Sebat et al., 2004), more than 66,000 CNVs and 34,000 InDels have been identified in various populations (Redon et al., 2006; Simon-Sanchez et al., 2007; de Smith et al., 2007; Perry et al., 2008; Díaz de Ståhl et al., 2008; Yim et al., 2010; Conrad et al., 2010; Park et al., 2010) and catalogued in the public database, Database of Genomic Variants (http://projects.tcag.ca/variation/) (Feuk et al, 2006). More CNVs have been uncovered using the NGS analysis (Mills et al., 2011; Kidd et al., 2010; Kim et al., 2009).

CNVs can affect gene functions in several ways and have a potential to affect gene expression levels presumably larger than that of SNPs. Deletion or duplication may disrupt the genes located inside those regions, resulting in changes in the gene structure, which can affect the gene expression. Alternatively, disruption of the transcription regulatory regions and the enhancers can also affect the gene expression. During the recombination which is thought to be an important mechanism of CNV development, novel fusion products may be generated, which may exert positive or negative effects on gene expression and epigenetic regulations (Feuk et al, 2006; Zhang et al., 2009; Hampton et al., 2009; Przybytkowski et al., 2011; Reymond et al., 2007). Taken together, structural variations are likely to be responsible for the phenotypic variation of human beings and comprehensive mapping of CNVs can facilitate the understanding of inter-individual phenotypic differences including disease susceptibility and responsiveness to drugs (Feuk et al, 2006; Estivill & Armengol, 2007). Indeed, CNVs have been found to be associated with various types of Mendelian traits and also a substantial number of complex diseases including neurodevelopmental disorders (Buchanan & Scherer, 2008; Lee & Lupski, 2006).

2.2 CNVs in ASD

To assess the role of CNV in ASD, several different whole-genome microarray platforms based on oligonucleotides, SNPs and BAC clones have been used for ASD family studies or case-control studies (Abrahams & Geschwind, 2008; Cook & Scherer, 2008). As a result, lines of evidence have been accumulated that multiple rare de novo CNVs contribute to the susceptibility to ASD. For example, duplications and/or deletions on chromosome 15q11–q13 confer increased risk of ASD (15q11–q13 duplication syndrome, Prader-Willi syndrome and Angelman syndrome). Approximately one fourth of the individuals who have a 22q11.2 deletion and over 90% of individuals with duplication of 17p11.2 show characteristics of ASD (Cohen et al., 2005; Abrahams & Geschwind, 2008; Fernández et al., 2009). Significant associations have been reported between ASD and CNV of various genes, such as *NRXN1* (2p16.3), *NLGN3* (Xq13.1), *NLGN4* (Xp22.23) and *SHANK3* (22q13.3). There have been many reports on CNVs associated with ASDs, but, due to technical limitations and lack of standardized methods for defining the CNVs and CNV regions (CNVRs), there are inconsistencies among studies which should be removed by further GWAS. Table 1 summarizes the major CNVs identified by GWAS in ASD.

Discovery Sample		Replication Sample		Study design	Platform	CNV Detection method	Number of CNVs identified	Strong candidate loci	CN change	Reference
Case	Control	Case	Control							
1496 families with 7,917 subjects	Unaffected family members	-	-	Family-based	Affymetrix 10K	dChip	254	NRXN1 1q21 17p12 22q11.2	del	Szatmari et al., 2007
165 families	99 unaffected families	-	-	Family-based	Agilent 244K 390K ROMA	HMM	17	SLC4A10, FHIT FHIT FLJ16237 A2BP1	del del dup del del	Sebat et al., 2007
180	372	532	465	Family-based and Case-control	Array-CGH	NimbleGen	1	16p11.2	microdel	Kumar et al., 2008
751 multiplex families with 1441 cases	1420 (AGRE parents) 2814 (bipolar disporder or NIMH controls)	512 (CHB) 299 (deCODE)	434 (CHB) 18,834 (decode)	Family-based	Affymetrix 5.0 (AGRE) Affymetrix 500K (controls)	COPPER/ Birdseye (AGRE) ADM-2(CHB) HMM(deCODE)	47	16p11.2	del/dup	Weiss et al., 2008
397	372	-	-	Family-based	19K BAC Microarray	-	51	15q11-q13 22q11 16p11.2	dup dup microdel	Christian et al., 2008
427 families	500	-	1,152 matched controls	Case-control	Affymetrix 500K	dChip, CNAG, GEMCA	277	16p11.2 SHANK3- NLGN4- NRXN1-PSD DPP6- DPP10- PCDH9 ANKRD11 DPYD PTCHD1 15q24	del/dup	Marshall et al., 2008
859	1409	1,336	1,110	Case-control	Illumina HumanHap550	PennCNV	78,490	15q11-13 22q11.21 NRXN1 CNTN4 PARK2 RFWD2 AK123120 UNQ3037 GRID1 NLGN1 GYPELOC44	dup dup del del/dup del dup dup del del dup dup	Glessner et al., 2009
912 multiplex families	1,488 (CHOP) 542 (NINDS)	859	1,051	Case-control	Illumina HumanHap550	PennCNV	> 150	NRXN1 UBE3A 15q11-q13 BZRAP1 MDGA2	del dup del/dup del/dup del	Bucan et al, 2009

Discovery Sample		Replication Sample		Study design	Platform	CNV Detection method	Number of CNVs identified	Strong candidate loci	CN change	Reference
Case	Control	Case	Control							
28 children	62 Adults	-	-	Case-control	Array-CGH	Array-CyGHt	38	8p23.1 17p11.2	del del	Cho et al, 2009
996	1,287	-	3,677	Case-control and Family-based	Illumina 1M	QuantiSNP iPattern	5,478	SHANK2 SYNGAP1 DLGAP2 CSNK1D/SLC16A3 NRXN1 22q11.21 DDX53/PTCHD1	del del dup dup/del dup/del del del	Pinto rt al., 2010

ACC:Autism Case-Control cohort
ADM : aberration detection method
AGP: Autism Genome Project
AGRE: Autism Genetic Resource Exchange
CHB: Children's Hospital Boston
CHOP: Children's Hospital of Philadelphia
CNAG: Copy Number Analysis for GeneChip
COPPER: copy-number polymorphism evaluation routine
dChip: DNA Chip Analyzer
GEMCA: Genotyping Microarray based CNV Analysis
HMM: hidden Markov model
NIMH: National Institute of Mental Health
NINDS: National Institute of Neurological Disorders and Stroke

Table 1. Genome-wide CNV association studies of autism

In 2007, two pioneering studies demonstrated the association of CNVs with ASD. The Autism Genome Project Consortium performed linkage and CNV analyses using Affymetrix 10K SNP array for 1,181 ASD families with at least two affected individuals (Autism Genome Project Consortium et al., 2007). Of the 254 highly significant CNVs, the investigators emphasized four CNVs and the most interesting finding was a 300-kb sized CNV loss on chromosome 2p16 identified recurrently in two families. The deletion of this region disrupted the coding exons of the neurexin 1 gene (*NRXN1*), which interacts with neurologins and involves in synaptogenesis. Therefore, deterioration of the neurexin 1 function by deletion may affect susceptibility to ASD or its phenotypes. The structural variation in the *NRXN1* gene was reported from the previous autism studies (Chubykin et al., 2005; Feng et al., 2006). The other three interesting CNVs were 1.1-Mb sized CNV gain on chromosome 1q21, 933-kb sized de novo duplication on 17p12, and duplication on 22q11.2. The duplication on 17p12 is known to cause Charcot-Marie-Tooth 1A (CMT1A) disease (Houlden et al., 2006). In addition, other micro-duplications of the same chromosomal region have been reported in individuals with mental retardation, linguistic delay, autism and related phenotypes (Moog et al., 2004).

Sebat and his colleagues performed array-CGH analysis with 264 families and explored the association of de novo CNVs with ASD, which are not present in their respective parents (Sebat et al., 2007). The authors identified 17 de novo CNVs in 16 subjects. According to their result, the frequency of spontaneous mutation was 10% in the sporadic cases and 3% in the multiplex families, while 1% in unaffected individuals. One of the de novo CNV loci was a 4.3-Mb sized deletion at 22q13.31-q13.33, where *SHANK3* gene is located. Recurrent deletion of this region has been previously reported in ASD (Manning et al., 2004). Durand et al. reported that mutations in *SHANK3* gene were associated with ASD and abnormal gene dosage of *SHANK3* was associated with severe cognitive deficits, linguistic delay and ASD (Durand et al., 2007). SHANK3 is a scaffolding protein found in excitatory synapses directly opposite to the presynaptic active zone. This gene has been suggested to be associated with the neurobehavioral symptoms observed in individuals with 22q13 deletions.

In 2008, four independent studies consistently reported the association of the CNV on 16p11.2 locus with autism. Weiss et al. adopted Affymetrix 5.0 SNP array to find CNVs in 751 multiplex families from the Autism Genetic Resource Exchange (AGRE) (Weiss et al., 2008). They identified 32 high- and 15 low- confidence regions. Among the candidate loci, microdeletion and microduplication on 16p11.2 were validated to be associated with ASD. This association was further confirmed in clinical testing data from Children's Hospital Boston and in a large population data from Iceland (deCODE genetics data). Kumar et al. screened 180 ASD cases and 372 controls using a 19K whole-genome tiling bacterial artificial chromosome (BAC) array to identify submicroscopic copy number changes specific to autism (Kumar et al., 2008). They observed ~500-kb sized recurrent microdeletion on 16p11.2 in two cases with autism but not in the controls. When they assessed the frequency of this putative autism-associated genomic disorder, 0.6% of the ASD cases showed the alterations while none in controls. The variation was confirmed by FISH, microsatellite analyses and array-CGH. Christian et al. also used the same 19K whole-genome tiling BAC array to identify ASD-associated CNVs in the 397 cases and 372 control set (Christian et al., 2008). Among the 51 candidate CNVs, recurrent CNVs were identified in the loci including 15q11-q13, 22q11, and 16p11.2. They were confirmed by FISH, microsatellite analysis, or quantitative polymerase chain reaction (PCR) analysis. Marshall et al. performed whole-genome screening for 427 ASD cases and 500 controls using Affymetrix 500K SNP arrays (Marshall et al., 2008). Of the 277 CNVs identified only in the cases, the CNVs on 16p11.2 locus appeared in around 1% of the ASD cases, which included both duplications and deletions. There exist SHANK3-NLGN4-NRXN1 postsynaptic density genes, DPP6-DPP10-PCDH9 (synapse complex), ANKRD11, DPYD and PTCHD1 in other associated CNVs.

New CNVs in addition to the known ones have been suggested to be associated with ASD in the subsequent studies. Glessner et al. performed a whole-genome CNV analysis with 859 cases and 1,409 controls using Illumina HumanHap550 BeadChip (Glessner et al., 2009). They generated 78,490 CNV calls and the positive findings were further evaluated in an independent cohort of 1,336 ASD cases and 1,110 controls. Through this approach, they identified several known ASD-associated genes as well as novel candidate CNVs. For example, they identified the CNVs in the loci including 15q11–q13, 22q11.21, *NRXN1* and *CNTN4*, which were previously reported to be associated with autism (Kim et al., 2009; Roohi et al., 2009; Fernandez et al., 2008). However, some of the genes or loci previously known to be associated with ASD such as *AUTS2* (Kalscheuer et al., 2007), *NLGN3* (Jamain et al., 2003), *SHANK3* (Moessner et al., 2007) and 16p11.2(Weiss et al., 2008) were not replicated in their study. Especially 16p11.2, a locus consistently reported to be associated in four previous independent studies, did not show a significant association in this study. Several new susceptibility genes such as *NLGN1* and *ASTN2* were identified in this study. Both genes encode neuronal cell-adhesion molecules. In Chubykin et al's report, mutations in neuroligin superfamily members were identified in the individuals with ASD (Chubykin et al., 2005). ASTN1 is a neuronal protein receptor integral in the process of glial-guided granule cell migration during development (Zheng et al., 1996). Furthermore, CNVs of the genes involved in the ubiquitin pathways, such as *UBE3A*, *PARK2*, *RFWD2* and *FBXO40*, were observed in the ASD cases but not in the controls. Bucan et al. conducted high -density genotyping of 912 multiplex families from the AGRE collection and 1,488 controls using Illumina HumanHap550 BeadChip (Bucan et al., 2009). They identified more than 150 loci harboring rare variants in multiple unrelated patients and the positive findings were further validated in an independent cohort of 859 ASD cases and 1,051 controls by genomic quantitative PCR. Among the candidate loci, there are previously reported ones such as *NRXN1*(Marshall et al., 2008), *UBE3A* (Glessner et al., 2009), and 15q11-q13 (Christian et al., 2008) and novel ones such as *BZRAP1* and *MDGA2*.

In 2009, Cho et al. reported the ASD associated CNVs in east-Asians. They performed whole-genome BAC array-CGH with 28 ASD cases and with 62 controls and identified 38 CNVs including those harboring two significant loci, 8p23.1 and 17p11.2 (Cho et al., 2009). *DEFENSIN* gene family are located in the 8p23.1 CNV locus and often showed copy number polymorphisms in earlier studies (Linzmeier & Ganz, 2005). Although there have been no direct clues to connect the copy number loss of *DEFENSIN* gene and ASD, immunological dysfunction has been suggested to be associated with autism (Rutter, 2005).

Most recently, Pinto et al. analyzed the genome-wide features of rare CNVs in autism using Illumina 1M SNP arrays (Pinto et al., 2010). Based on 996 cases and 1,287 controls, they identified 5,478 rare CNVs. By examining parent-child transmission, the authors found the 226 de novo and inherited CNVs which were not present in controls. As a whole, ASD cases were found to carry a higher number of de novo CNVs than controls (1.69 fold, P=3.4X10^{-4}). A number of novel genes such as *SHANK2*, *SYNGAP1*, *DLGAP2* and the *DDX53–PTCHD1* in the CNVs were found to be associated with ASD in this study. Also, through gene set enrichment analysis, cellular proliferation, projection and motility, and GTPase/Ras signaling were found to be affected by the CNVs identified in their study. This approach demonstrated the new paradigm of autism research based on functional pathway and cross-talk.

3. SNPs in autism

Before the establishment of GWAS, the genome-wide linkage analysis has been used for the discovery of the mutations in diverse diseases (OMIM http://www.ncbi.nlm.nih.gov/omim). Location of the disease genes were successfully narrowed down by linkage disequilibrium mapping studies, but linkage approach was not always successful especially for complex diseases. In many cases, the significant linkage loci were not replicated. One potential reason is that the effect of a single risk variant on the pathogenesis of complex disease might be too small to be detected. Small genetic effects could be detected with greater power by association analyses such as GWAS with large case-control population (Risch & Merikangas, 1996). In other words, to identify common risk alleles in the common complex diseases, population-wide analysis with more common and dense variants is required. SNP-GWAS can be an ideal approach for unbiased screening and also be adopted for high-density linkage analysis. SNP-GWAS became a matured technology for exploring novel associations between genetic variants and complex diseases because over 12 million of SNAs have been catalogued and high density array fabrication/analysis technologies have been developed. In neuropsychiatric disorders with unknown etiology such as ASD, SNP-GWAS have been actively adopted to explore the genetic background of the diseases (Table 2). In this chapter, we will review the major SNP-GWAS results for ASDs.

<table>
<tr><th>Initial Sample Size</th><th>Replication Sample Size</th><th>Region</th><th>Reported Gene(s)</th><th>Strongest SNP</th><th>p-value of discovery set (OR*)</th><th>p-value of replication set (OR*)</th><th>Platform</th><th>Reference</th></tr>
<tr><td>72 families (148 cases)</td><td>1,295 trios</td><td>7q35</td><td>CNTNAP2</td><td>rs7794745</td><td>2.14E-05</td><td><0.005</td><td>Affymetrix 500K</td><td>Arking et al., 2008</td></tr>
<tr><td>1,031 families (1,553 cases)</td><td>2,073 trios</td><td>5p15</td><td>SEMA5A, TAS2R1</td><td>rs10513025</td><td>1.7E-06 (0.55)</td><td>6.1E-03 (0.76)</td><td>Affymetrix 5.0 /500K</td><td>Weiss et al., 2009</td></tr>
<tr><td>438 families</td><td>487 families</td><td>5p14.1</td><td>Intergenic</td><td>rs10038113</td><td>2.75E-05 (0.67)</td><td>3.28E-03</td><td>Illumina Human 1Mv1 beadchip</td><td>Ma et al., 2009</td></tr>
<tr><td>780 families (1,299 cases)</td><td>447 families</td><td rowspan="2">5p14.1</td><td rowspan="2">CDH10,CD H9</td><td rowspan="2">rs4307059</td><td>1.1E-05</td><td>1.2E-2</td><td>550K/1M Illumina</td><td rowspan="2">Wang et al., 2009</td></tr>
<tr><td>1,204 cases 6,491 controls</td><td>108 cases 540 controls</td><td>2.2E-04 (1.19)</td><td>1.6E-2</td><td>550K/300K Illumina</td></tr>
<tr><td>745 boys in social group 870 boys in nonsocial group</td><td>1,400 boys</td><td>2p21</td><td>Intergenic (social traits)</td><td>rs11894053 (social traits)</td><td>0.02</td><td>-</td><td>Affymetrix 500K</td><td>Ronald et al., 2010</td></tr>
<tr><td>1,558 families</td><td>2,179 families</td><td>20p12.1</td><td>MACROD2 (Str | Eur)**</td><td>rs4141463 (Str | Eur)</td><td>2.1E-08 (0.56)</td><td>4.7E-08 (0.65)</td><td>1M Illumina</td><td>Anney et al., 2010</td></tr>
</table>

*OR: Odds Ratio
**Str | Eur : strict diagnosis and European ancestry

Table 2. Genome-wide SNP association studies of autism

3.1 Common SNPs associated with ASD

Arking et al. performed a two-stage study on ASD using genome-wide linkage and family-based association mapping by whole-genome SNP genotyping (Arking et al., 2008). For stage I, they selected 72 multiplex ASD families and genotyped the samples using Affymetrix 500K arrays. In this approach, they could not find any significant SNPs or haplotypes. However, through the genome-wide linkage analysis, they identified 2 significant loci associated with ASD, 7q35 and 10p13-14. In the most significant locus (7q35), they identified that a polymorphism in contactin-associated protein-like 2 (*CNTNAP2*) gene, a member of the neurexin superfamily, is associated with ASD. In the second stage, they validated the significant findings of the stage I by examining 145 multiplex families and confirmed that *CNTNAP2* was an autism-susceptibility gene. This result was the first evidence that a common genetic variant in the neurexin superfamily member increases risk of autism.

In 2009, Weiss et al. explored a linkage and SNP association analysis with 1,031 multiplex autism families using Affymetrix 5.0 SNP array (Weiss et al., 2009). They found that a SNP on 5p15 locus between *SEMA5A* and *TAS2R1* gene was significantly associated with autism. In addition, the expression of SEMA5A was found to decrease in brains of autistic patients. Taken together the authors suggested a possibility of *SEMA5A* as an autism risk gene. Wang et al. used higher density SNA array and larger study populations to identify common genetic risk factors underlying ASDs (Wang et al., 2009). They used two different sets of study subjects in discovery stage using Illumina Human 1M beadchip. First set was 780 families with 1,299 affected children and the second set was 1,204 patients and 6,491 controls. They identified six significant SNPs located between cadherin 10 (*CDH10*) and cadherin 9 (*CDH9*) genes strongly associated with ASD. These two genes encode neuronal cell-adhesion molecules. Among the 6 SNPs, the most significant one was rs4307059 ($P = 3.4 \times 10^{-8}$, odds ratio = 1.19). The SNP was replicated in two independent datasets of 447 families and 108 case-540 control sets. Combined analysis using all four datasets showed that all six SNPs are associated with autism (P values ranging from 7.4×10^{-8} to 2.1×10^{-10}). Interestingly, 5p14.1 was consistently suggested as a novel risk locus in an independent study in the same year. Ma et al. performed GWAS with 438 Caucasian autistic families using Illumina Human 1M beadchip (Ma et al., 2009). They found that 96 SNPs were strongly associated with autism ($P < 0.0001$). They validated all 96 significant associations in independent samples of 487 families using 550K Illumina BeadChip, which was the same array platform to Wang et al's. A novel locus on 5p14.1 was found to be significantly associated with autism both in the discovery and validation dataset.

The Autism Genome Project (AGP) Consortium performed high-resolution genotyping with 1,558 families to identify significant SNPs (Anney et al., 2010). For primary analysis, they partitioned the dataset along axes of diagnosis and ancestry; spectrum versus strict; European versus all ancestries. Based on these partitioned data, they conducted four GWAS. They observed the strongest association for SNP rs4141463 in one of the four primary association analyses. Located within MACROD2, this marker crossed the GWA significance threshold of $P < 5 \times 10^{-8}$. They are performing analysis of combining data to validate the results of the primary analysis.

Despite the expected advantages of large-scale GWAS analysis, none of the candidate associations have been replicated so far, which may underscore the genetic and phenotypic heterogeneity of ASD and indicate the fact that the effect size of common alleles contributing to common disorders is much smaller than expectation (Eapen, 2011).

3.2 Rare SNPs associated with ASD

Definition of a rare variant is a variant with frequency <1%. The most deleterious variants might be naturally eliminated during evolution, but some remain as rare variants. In 'common disease–common variant' model, most of the rare SNP associations have been missed by current GWAS concept. However, rapid development of NGS will facilitate the discovery of rare variants. Rare SNP associations are more likely to be detected by re-sequencing of relevant regions in hundreds or thousands of individuals. It is anticipated that advances in re-sequencing technologies will make it feasible to search systematically for rare variant effects.

4. Conclusion

Human Genome Project has provided insight into a complete sequence of the haploid human genome and we also have got new insight of the human genetic variations. Based on this new insight, conventional target gene oriented and hypothesis-driven research design has been quickly moved to a new paradigm, hypothesis-free mining of novel disease associated genes. Indeed, over hundreds of genetic variants which may affect the susceptibility of pathogenesis of complex disease have been identified by the GWAS. The GWAS have been actively adopted in studying the causative factors of neurodevelopmental disorders including ASD. Through GWAS approach, several robust ASD-associated variants in the genes such as *NRXN1, SHANK3, NLGN4* and *CNTNAP2* were uncovered. However, it is too early to say that GWAS have brought reliable-enough insight into ASD. Many of the significant CNVs identified in one study were not consistent or not successfully replicated in the following studies. New improved algorithms for CNV and CNVR will be needed for defining the CNVs more robustly. To sort out the platform to platform variation of CNV call, which is one of the obstacles for the meta-analysis, more reliable experimental methods should be developed. Re-sequencing large number of individuals without CNVs will help to discover the new rare variants. Considering the speed of technological innovations including algorithm, high-throughput analysis and NGS, we anticipate that current obstacles of GWAS in autism research will be removed soon. However, GWAS result itself will not be enough to get clinically applicable insight about the pathophysiology of ASD. Integration of GWAS data with other resources such as improved bio-imaging, personal whole-genome sequencing, gene-environmental interaction and metagenome analysis data about gastrointestinal commensal bacteria will enable us to get a more comprehensive insight in designing future personalized care of autism.

5. Useful website for ASD related data

- Psychiatric GWAS Consortium(PGC): https://pgc.unc.edu/index.php
- National Institute of Mental Health Center for Collaborative Genetic Studies on Mental Disorders : http://nimhgenetics.org/
- Autism Genetic Resource Exchange (AGRE): http://www.agre.org/
- Autism Genome Project (AGP): http://www.well.ox.ac.uk/monaco/autism/AGP.shtml
- The International Schizophrenia Consortium (ISC): http://pngu.mgh.harvard.edu/isc/
- Genetic Association Information Network(GAIN): http://www.genome.gov/19518664#al-4
- CNV project at the Children's Hospital of Philadelphia: http://cnv.chop.edu
- The SGENE project: http://www.sgene.eu/Summary.php
- Database of Genomic Variants (DGV): http://projects.tcag.ca/variation
- A Catalog of Published Genome-Wide Association Studies: http://www.genome.gov/gwastudies/index.cfm?#searchForm
- DECIPHER: https://decipher.sanger.ac.uk
- CNV project: http://www.sanger.ac.uk/humgen/cnv/
- GEN2PHEN: http://www.gen2phen.org/

6. Acknowledgements

Work in the authors' laboratory is supported by the Korea Health 21 R&D Project (A040002), Ministry of Health and Welfare, Republic of Korea.

7. References

[1] Abrahams BS, Geschwind DH. (2008). Advances in autism genetics: on the threshold of a new neurobiology. *Nat Rev Genet*. 9:341-355.

[2] Anney R, Klei L, Pinto D, Regan R, Conroy J, Magalhaes TR, Correia C, Abrahams BS, Sykes N, Pagnamenta AT, Almeida J, Bacchelli E, Bailey AJ, Baird G, Battaglia A,

Berney T, Bolshakova N, Bölte S, Bolton PF, Bourgeron T, Brennan S, Brian J, Carson AR, Casallo G, Casey J, Chu SH, Cochrane L, Corsello C, Crawford EL, Crossett A, Dawson G, de Jonge M, Delorme R, Drmic I, Duketis E, Duque F, Estes A, Farrar P, Fernandez BA, Folstein SE, Fombonne E, Freitag CM, Gilbert J, Gillberg C, Glessner JT, Goldberg J, Green J, Guter SJ, Hakonarson H, Heron EA, Hill M, Holt R, Howe JL, Hughes G, Hus V, Igliozzi R, Kim C, Klauck SM, Kolevzon A, Korvatska O, Kustanovich V, Lajonchere CM, Lamb JA, Laskawiec M, Leboyer M, Le Couteur A, Leventhal BL, Lionel AC, Liu XQ, Lord C, Lotspeich L, Lund SC, Maestrini E, Mahoney W, Mantoulan C, Marshall CR, McConachie H, McDougle CJ, McGrath J, McMahon WM, Melhem NM, Merikangas A, Migita O, Minshew NJ, Mirza GK, Munson J, Nelson SF, Noakes C, Noor A, Nygren G, Oliveira G, Papanikolaou K, Parr JR, Parrini B, Paton T, Pickles A, Piven J, Posey DJ, Poustka A, Poustka F, Prasad A, Ragoussis J, Renshaw K, Rickaby J, Roberts W, Roeder K, Roge B, Rutter ML, Bierut LJ, Rice JP, Salt J, Sansom K, Sato D, Segurado R, Senman L, Shah N, Sheffield VC, Soorya L, Sousa I, Stoppioni V, Strawbridge C, Tancredi R, Tansey K, Thiruvahindrapduram B, Thompson AP, Thomson S, Tryfon A, Tsiantis J, Van Engeland H, Vincent JB, Volkmar F, Wallace S, Wang K, Wang Z, Wassink TH, Wing K, Wittemeyer K, Wood S, Yaspan BL, Zurawiecki D, Zwaigenbaum L, Betancur C, Buxbaum JD, Cantor RM, Cook EH, Coon H, Cuccaro ML, Gallagher L, Geschwind DH, Gill M, Haines JL, Miller J, Monaco AP, Nurnberger JI Jr, Paterson AD, Pericak-Vance MA, Schellenberg GD, Scherer SW, Sutcliffe JS, Szatmari P, Vicente AM, Vieland VJ, Wijsman EM, Devlin B, Ennis S, Hallmayer J. (2010). A genome-wide scan for common alleles affecting risk for autism. *Hum Mol Genet.* 19:4072-4082.

[3] Arking DE, Cutler DJ, Brune CW, Teslovich TM, West K, Ikeda M, Rea A, Guy M, Lin S, Cook EH, Chakravarti A. (2008). A common genetic variant in the neurexin superfamily member CNTNAP2 increases familial risk of autism. *Am J Hum Genet.* 82:160-164.

[4] Autism Genome Project Consortium, Szatmari P, Paterson AD, Zwaigenbaum L, Roberts W, Brian J, Liu XQ, Vincent JB, Skaug JL, Thompson AP, Senman L, Feuk L, Qian C, Bryson SE, Jones MB, Marshall CR, Scherer SW, Vieland VJ, Bartlett C, Mangin LV, Goedken R, Segre A, Pericak-Vance MA, Cuccaro ML, Gilbert JR, Wright HH, Abramson RK, Betancur C, Bourgeron T, Gillberg C, Leboyer M, Buxbaum JD, Davis KL, Hollander E, Silverman JM, Hallmayer J, Lotspeich L, Sutcliffe JS, Haines JL, Folstein SE, Piven J, Wassink TH, Sheffield V, Geschwind DH, Bucan M, Brown WT, Cantor RM, Constantino JN, Gilliam TC, Herbert M, Lajonchere C, Ledbetter DH, Lese-Martin C, Miller J, Nelson S, Samango-Sprouse CA, Spence S, State M, Tanzi RE, Coon H, Dawson G, Devlin B, Estes A, Flodman P, Klei L, McMahon WM, Minshew N, Munson J, Korvatska E, Rodier PM, Schellenberg GD, Smith M, Spence MA, Stodgell C, Tepper PG, Wijsman EM, Yu CE, Rogé B, Mantoulan C, Wittemeyer K, Poustka A, Felder B, Klauck SM, Schuster C, Poustka F, Bölte S, Feineis-Matthews S, Herbrecht E, Schmötzer G, Tsiantis J, Papanikolaou K, Maestrini E, Bacchelli E, Blasi F, Carone S, Toma C, Van Engeland H, de Jonge M, Kemner C, Koop F, Langemeijer M, Hijmans C, Staal WG, Baird G, Bolton PF, Rutter ML, Weisblatt E, Green J, Aldred C, Wilkinson JA, Pickles A, Le Couteur A, Berney T, McConachie H, Bailey AJ, Francis K, Honeyman G, Hutchinson A, Parr

JR, Wallace S, Monaco AP, Barnby G, Kobayashi K, Lamb JA, Sousa I, Sykes N, Cook EH, Guter SJ, Leventhal BL, Salt J, Lord C, Corsello C, Hus V, Weeks DE, Volkmar F, Tauber M, Fombonne E, Shih A, Meyer KJ. (2007). Mapping autism risk loci using genetic linkage and chromosomal rearrangements. *Nat Genet* 39:319-328.

[5] Bailey A, Le Couteur A, Gottesman I, Bolton P, Simonoff E, Yuzda E, Rutter M. (1995). Autism as a strongly genetic disorder: evidence from a British twin study. *Psychol. Med* 25:63–77.

[6] Bucan M, Abrahams BS, Wang K, Glessner JT, Herman EI, Sonnenblick LI, Alvarez Retuerto AI, Imielinski M, Hadley D, Bradfield JP, Kim C, Gidaya NB, Lindquist I, Hutman T, Sigman M, Kustanovich V, Lajonchere CM, Singleton A, Kim J, Wassink TH, McMahon WM, Owley T, Sweeney JA, Coon H, Nurnberger JI, Li M, Cantor RM, Minshew NJ, Sutcliffe JS, Cook EH, Dawson G, Buxbaum JD, Grant SF, Schellenberg GD, Geschwind DH, Hakonarson H. (2009). Genome-wide analyses of exonic copy number variants in a family-based study point to novel autism susceptibility genes. *PLoS Genet*. 5:e1000536.

[7] Buchanan, J. A. & Scherer, S. W. (2008). Contemplating effects of genomic structural variation. *Genet. Med*. 10:639–647.

[8] Carter, N.P. (2007). Methods and strategies for analyzing copy-number variation using DNA microarrays. *Nat. Genet*. 39: S16–21

[9] Cho SC, Yim SH, Yoo HK, Kim MY, Jung GY, Shin GW, Kim BN, Hwang JW, Kang JJ, Kim TM, Chung YJ. (2009). Copy number variations associated with idiopathic autism identified by whole-genome microarray-based comparative genomic hybridization. *Psychiatr Genet*. 19:177-185.

[10] Christian SL, Brune CW, Sudi J, Kumar RA, Liu S, Karamohamed S, Badner JA, Matsui S, Conroy J, McQuaid D, Gergel J, Hatchwell E, Gilliam TC, Gershon ES, Nowak NJ, Dobyns WB, Cook EH Jr. (2008). Novel submicroscopic chromosomal abnormalities detected in autism spectrum disorder. *Biol Psychiatry*. 63:1111-1117.

[11] Chubykin AA, Liu X, Comoletti D, Tsigelny I, Taylor P, Sudhof TC. (2005). Dissection of synapse induction by neuroligins: effect of a neuroligin mutation associated with autism. *J Biol Chem*. 280:22365-22374

[12] Cohen D., Pichard, N., Tordjman, S., Baumann, C., Burglen, L., Excoffier, E., Lazar, G., Mazet, P., Pinquier, C., Verloes, A., Héron, D. (2005). Specific genetic disorders and autism: clinical contribution towards their identification. *J Autism Dev Disord* 35:103–116.

[13] Conrad DF, Pinto D, Redon R, Feuk L, Gokcumen O, Zhang Y, Aerts J, Andrews TD, Barnes C, Campbell P, Fitzgerald T, Hu M, Ihm CH, Kristiansson K, Macarthur DG, Macdonald JR, Onyiah I, Pang AW, Robson S, Stirrups K, Valsesia A, Walter K, Wei J; Wellcome Trust Case Control Consortium, Tyler-Smith C, Carter NP, Lee C, Scherer SW, Hurles ME. (2010). Origins and functional impact of copy number variation in the human genome. *Nature*. 464:704-712.

[14] Cook EH Jr, Scherer SW. (2008). Copy-number variations associated with neuropsychiatric conditions. *Nature*. 455:919-923

[15] de Smith AJ, Tsalenko A, Sampas N, Scheffer A, Yamada NA, Tsang P, Ben-Dor A, Yakhini Z, Ellis RJ, Bruhn L, Laderman S, Froguel P, Blakemore AI. (2007). Array CGH analysis of copy number variation identifies 1284 new genes variant in

healthy white males: implications for association studies of complex diseases. Hum. Mol. Genet. 16:2783–2794.

[16] Díaz de Ståhl T, Sandgren J, Piotrowski A, Nord H, Andersson R, Menzel U, Bogdan A, Thuresson AC, Poplawski A, von Tell D, Hansson CM, Elshafie AI, Elghazali G, Imreh S, Nordenskjöld M, Upadhyaya M, Komorowski J, Bruder CE, Dumanski JP. (2008). Profiling of copy number variations (CNVs) in healthy individuals from three ethnic groups using a human genome 32 K BAC-clone-based array. *Hum. Mutat*. 29:398–408.

[17] Durand CM, Betancur C, Boeckers TM, Bockmann J, Chaste P, Fauchereau F, Nygren G, Rastam M, Gillberg IC, Anckarsäter H, Sponheim E, Goubran-Botros H, Delorme R, Chabane N, Mouren-Simeoni MC, de Mas P, Bieth E, Rogé B, Héron D, Burglen L, Gillberg C, Leboyer M, Bourgeron T. (2007). Mutations in the gene encoding the synaptic scaffolding protein SHANK3 are associated with autism spectrum disorders. *Nat Genet*. 39:25-27.

[18] Eapen, V. (2011). Genetic basis of autism: is there a way forward? *Curr Opin Psychiatry*. 24:226-236.

[19] Estivill, X. and Armengol, L. (2007). Copy number variants and common disorders: filling the gaps and exploring complexity in genome-wide association studies. *PLoS Genet*. 3:1787–1799.

[20] Feng J, Schroer R, Yan J, Song W, Yang C, Bockholt A, Cook EH Jr, Skinner C, Schwartz CE, Sommer SS. (2006). High frequency of neurexin 1 [beta] signal peptide structural variants in patients with autism. *Neuroscience letters*. 409:10-13.

[21] Fernández L, Nevado J, Santos F, Heine-Suñer D, Martinez-Glez V, García-Miñaur S, Palomo R, Delicado A, Pajares IL, Palomares M, García-Guereta L, Valverde E, Hawkins F, Lapunzina P. (2009). A deletion and a duplication in distal 22q11.2 deletion syndrome region. Clinical implications and review. *BMC Med Genet*. 10:48.

[22] Fernandez T, Morgan T, Davis N, Klin A, Morris A, Farhi A, Lifton RP, State MW. (2008). Disruption of Contactin 4 (CNTN4) results in developmental delay and other features of 3p deletion syndrome. *Am J Hum Genet*. 82:1385.

[23] Feuk, L., Carson, A.R., and Scherer, S.W. (2006). Structural variation in the human genome. *Nat. Rev. Genet*. 7: 85–97.

[24] Feuk, L., Marshall, C.R., Wintle, R.F. and Scherer, S.W. (2006). Structural variants: changing the landscape of chromosomes and design of disease studies. *Hum. Mol. Genet*. 15: R57–R66.

[25] Folstein, S.E., Rosen-Sheidley, B. (2001). Genetics of autism: complex aetiology for a heterogeneous disorder. *Nat Rev Genet* 2:943–955.

[26] Freeman JL, Perry GH, Feuk L, Redon R, McCarroll SA, Altshuler DM, Aburatani H, Jones KW, Tyler-Smith C, Hurles ME, Carter NP, Scherer SW, Lee C. (2006). Copy number variation: new insights in genome diversity. *Genome Res*. 16:949-961.

[27] Freitag, C.M. (2007). The genetics of autistic disorders and its clinical relevance: a review of the literature. *Mol Psychiatry*. 12:2-22.

[28] Glessner JT, Wang K, Cai G, Korvatska O, Kim CE, Wood S, Zhang H, Estes A, Brune CW, Bradfield JP, Imielinski M, Frackelton EC, Reichert J, Crawford EL, Munson J, Sleiman PM, Chiavacci R, Annaiah K, Thomas K, Hou C, Glaberson W, Flory J, Otieno F, Garris M, Soorya L, Klei L, Piven J, Meyer KJ, Anagnostou E, Sakurai T, Game RM, Rudd DS, Zurawiecki D, McDougle CJ, Davis LK, Miller J, Posey DJ,

Michaels S, Kolevzon A, Silverman JM, Bernier R, Levy SE, Schultz RT, Dawson G, Owley T, McMahon WM, Wassink TH, Sweeney JA, Nurnberger JI, Coon H, Sutcliffe JS, Minshew NJ, Grant SF, Bucan M, Cook EH, Buxbaum JD, Devlin B, Schellenberg GD, Hakonarson H. (2009). Autism genome-wide copy number variation reveals ubiquitin and neuronal genes. *Nature*. 459:569-573.

[29] Hampton OA, Den Hollander P, Miller CA, Delgado DA, Li J, Coarfa C, Harris RA, Richards S, Scherer SE, Muzny DM, Gibbs RA, Lee AV, Milosavljevic A. (2009). A sequence-level map of chromosomal breakpoints in the MCF-7 breast cancer cell line yields insights into the evolution of a cancer genome. *Genome Res*. 19:167-177.

[30] Houlden, H. & Reilly, M.M. (2006). Molecular genetics of autosomal-dominant demyelinating Charcot-Marie-Tooth disease. *Neuromolecular Med*. 8: 43–62

[31] Hurles ME, Dermitzakis ET, Tyler-Smith C. (2008). The functional impact of structural variation in humans. *Trends Genet*. 24: 238–245

[32] Iafrate AJ, Feuk L, Rivera MN, Listewnik ML, Donahoe PK, Qi Y, Scherer SW, Lee C. (2004). Detection of large-scale variation in the human genome. *Nat. Genet*. 36:949–951

[33] Jamain S, Quach H, Betancur C, Råstam M, Colineaux C, Gillberg IC, Soderstrom H, Giros B, Leboyer M, Gillberg C, Bourgeron T; Paris Autism Research International Sibpair Study. (2003). Mutations of the X-linked genes encoding neuroligins NLGN3 and NLGN4 are associated with autism. *Nat Genet*. 34:27-29.

[34] Kalscheuer VM, FitzPatrick D, Tommerup N, Bugge M, Niebuhr E, Neumann LM, Tzschach A, Shoichet SA, Menzel C, Erdogan F, Arkesteijn G, Ropers HH, Ullmann R. (2007). Mutations in autism susceptibility candidate 2 (AUTS2) in patients with mental retardation. *Hum Genet*. 121:501-509.

[35] Kidd JM, Sampas N, Antonacci F, Graves T, Fulton R, Hayden HS, Alkan C, Malig M, Ventura M, Giannuzzi G, Kallicki J, Anderson P, Tsalenko A, Yamada NA, Tsang P, Kaul R, Wilson RK, Bruhn L, Eichler EE. (2010). Characterization of missing human genome sequences and copy-number polymorphic insertions. *Nat Methods*. 7:365-371.

[36] Kim HG, Kishikawa S, Higgins AW, Seong IS, Donovan DJ, Shen Y, Lally E, Weiss LA, Najm J, Kutsche K, Descartes M, Holt L, Braddock S, Troxell R, Kaplan L, Volkmar F, Klin A, Tsatsanis K, Harris DJ, Noens I, Pauls DL, Daly MJ, MacDonald ME, Morton CC, Quade BJ, Gusella JF. (2008). Disruption of neurexin 1 associated with autism spectrum disorder. *Am J Hum Genet*. 82:199-207.

[37] Kim JI, Ju YS, Park H, Kim S, Lee S, Yi JH, Mudge J, Miller NA, Hong D, Bell CJ, Kim HS, Chung IS, Lee WC, Lee JS, Seo SH, Yun JY, Woo HN, Lee H, Suh D, Lee S, Kim HJ, Yavartanoo M, Kwak M, Zheng Y, Lee MK, Park H, Kim JY, Gokcumen O, Mills RE, Zaranek AW, Thakuria J, Wu X, Kim RW, Huntley JJ, Luo S, Schroth GP, Wu TD, Kim H, Yang KS, Park WY, Kim H, Church GM, Lee C, Kingsmore SF, Seo JS. (2009). A highly annotated whole-genome sequence of a Korean individual. *Nature*. 460:1011-1015.

[38] Klauck, S.M. (2006). Genetics of autism spectrum disorder. *Eur. J. Hum. Genet*. 14: 714-720.

[39] Kumar RA, KaraMohamed S, Sudi J, Conrad DF, Brune C, Badner JA, Gilliam TC, Nowak NJ, Cook EH Jr, Dobyns WB, Christian SL. (2008). Recurrent 16p11.2 microdeletions in autism. *Hum Mol Genet*. 17:628-638.

[40] Lauritsen MB, Pedersen CB, Mortensen PB. (2005). Effects of familial risk factors and place of birth on the risk of autism: a nationwide register-based study. *J. Child Psychol. Psychiatry* 46:963–971.

[41] Lee, J. A. & Lupski, J. R. (2006). Genomic rearrangements and gene copy-number alterations as a cause of nervous system disorders. *Neuron* 52:103–121.

[42] Linzmeier RM, Ganz T. (2005). Human defensin gene copy number polymorphisms: comprehensive analysis of independent variation in alpha- and beta-defensin regions at 8p22-p23. *Genomics* 86:423–430.

[43] Ma D, Salyakina D, Jaworski JM, Konidari I, Whitehead PL, Andersen AN, Hoffman JD, Slifer SH, Hedges DJ, Cukier HN, Griswold AJ, McCauley JL, Beecham GW, Wright HH, Abramson RK, Martin ER, Hussman JP, Gilbert JR, Cuccaro ML, Haines JL, Pericak-Vance MA. (2009). A genome-wide association study of autism reveals a common novel risk locus at 5p14.1. *Ann Hum Genet.* 73:263-273.

[44] Manning MA, Cassidy SB, Clericuzio C, Cherry AM, Schwartz S, Hudgins L, Enns GM, Hoyme HE. (2004). Terminal 22q deletion syndrome: a newly recognized cause of speech and language disability in the autism spectrum. *Pediatrics.* 114:451-457.

[45] Marshall CR, Noor A, Vincent JB, Lionel AC, Feuk L, Skaug J, Shago M, Moessner R, Pinto D, Ren Y, Thiruvahindrapduram B, Fiebig A, Schreiber S, Friedman J, Ketelaars CE, Vos YJ, Ficicioglu C, Kirkpatrick S, Nicolson R, Sloman L, Summers A, Gibbons CA, Teebi A, Chitayat D, Weksberg R, Thompson A, Vardy C, Crosbie V, Luscombe S, Baatjes R, Zwaigenbaum L, Roberts W, Fernandez B, Szatmari P, Scherer SW. (2008). Structural variation of chromosomes in autism spectrum disorder. *Am J Hum Genet.* 82:477-488.

[46] Merikangas AK, Corvin AP, Gallagher L. (2009). Copy-number variants in neurodevelopmental disorders: promises and challenges. *Trends Genet.* 25:536-544.

[47] Mills RE, Walter K, Stewart C, Handsaker RE, Chen K, Alkan C, Abyzov A, Yoon SC, Ye K, Cheetham RK, Chinwalla A, Conrad DF, Fu Y, Grubert F, Hajirasouliha I, Hormozdiari F, Iakoucheva LM, Iqbal Z, Kang S, Kidd JM, Konkel MK, Korn J, Khurana E, Kural D, Lam HY, Leng J, Li R, Li Y, Lin CY, Luo R, Mu XJ, Nemesh J, Peckham HE, Rausch T, Scally A, Shi X, Stromberg MP, Stütz AM, Urban AE, Walker JA, Wu J, Zhang Y, Zhang ZD, Batzer MA, Ding L, Marth GT, McVean G, Sebat J, Snyder M, Wang J, Ye K, Eichler EE, Gerstein MB, Hurles ME, Lee C, McCarroll SA, Korbel JO; 1000 Genomes Project. (2011). Mapping copy number variation by population-scale genome sequencing. *Nature.* 470:59-65.

[48] Moessner R, Marshall CR, Sutcliffe JS, Skaug J, Pinto D, Vincent J, Zwaigenbaum L, Fernandez B, Roberts W, Szatmari P, Scherer SW. (2007). Contribution of SHANK3 mutations to autism spectrum disorder. *Am J Hum Genet.* 81:1289-1297.

[49] Moog U, Engelen JJ, Weber BW, Van Gelderen M, Steyaert J, Baas F, Sijstermans HM, Fryns JP. (2004). Hereditary motor and sensory neuropathy (HMSN) IA, developmental delay and autism related disorder in a boy with duplication (17)(p11.2p12). *Genet. Couns.* 15: 73–80

[50] Müller A, Holzmann K, Kestler HA. Visualization of genomic aberrations using Affymetrix SNP arrays. *Bioinformatics* 23:496-497.

[51] Park H, Kim JI, Ju YS, Gokcumen O, Mills RE, Kim S, Lee S, Suh D, Hong D, Kang HP, Yoo YJ, Shin JY, Kim HJ, Yavartanoo M, Chang YW, Ha JS, Chong W, Hwang GR, Darvishi K, Kim H, Yang SJ, Yang KS, Kim H, Hurles ME, Scherer SW, Carter NP,

Tyler-Smith C, Lee C, Seo JS. (2010). Discovery of common Asian copy number variants using integrated high-resolution array CGH and massively parallel DNA sequencing. *Nat Genet*. 42:400-405.

[52] Perry GH, Ben-Dor A, Tsalenko A, Sampas N, Rodriguez-Revenga L, Tran CW, Scheffer A, Steinfeld I, Tsang P, Yamada NA, Park HS, Kim JI, Seo JS, Yakhini Z, Laderman S, Bruhn L, Lee C. (2008) The fine-scale and complex architecture of human copy-number variation. *Am. J. Hum. Genet*. 82: 685–695.

[53] Pinto D, Pagnamenta AT, Klei L, Anney R, Merico D, Regan R, Conroy J, Magalhaes TR, Correia C, Abrahams BS, Almeida J, Bacchelli E, Bader GD, Bailey AJ, Baird G, Battaglia A, Berney T, Bolshakova N, Bölte S, Bolton PF, Bourgeron T, Brennan S, Brian J, Bryson SE, Carson AR, Casallo G, Casey J, Chung BH, Cochrane L, Corsello C, Crawford EL, Crossett A, Cytrynbaum C, Dawson G, de Jonge M, Delorme R, Drmic I, Duketis E, Duque F, Estes A, Farrar P, Fernandez BA, Folstein SE, Fombonne E, Freitag CM, Gilbert J, Gillberg C, Glessner JT, Goldberg J, Green A, Green J, Guter SJ, Hakonarson H, Heron EA, Hill M, Holt R, Howe JL, Hughes G, Hus V, Igliozzi R, Kim C, Klauck SM, Kolevzon A, Korvatska O, Kustanovich V, Lajonchere CM, Lamb JA, Laskawiec M, Leboyer M, Le Couteur A, Leventhal BL, Lionel AC, Liu XQ, Lord C, Lotspeich L, Lund SC, Maestrini E, Mahoney W, Mantoulan C, Marshall CR, McConachie H, McDougle CJ, McGrath J, McMahon WM, Merikangas A, Migita O, Minshew NJ, Mirza GK, Munson J, Nelson SF, Noakes C, Noor A, Nygren G, Oliveira G, Papanikolaou K, Parr JR, Parrini B, Paton T, Pickles A, Pilorge M, Piven J, Ponting CP, Posey DJ, Poustka A, Poustka F, Prasad A, Ragoussis J, Renshaw K, Rickaby J, Roberts W, Roeder K, Roge B, Rutter ML, Bierut LJ, Rice JP, Salt J, Sansom K, Sato D, Segurado R, Sequeira AF, Senman L, Shah N, Sheffield VC, Soorya L, Sousa I, Stein O, Sykes N, Stoppioni V, Strawbridge C, Tancredi R, Tansey K, Thiruvahindrapduram B, Thompson AP, Thomson S, Tryfon A, Tsiantis J, Van Engeland H, Vincent JB, Volkmar F, Wallace S, Wang K, Wang Z, Wassink TH, Webber C, Weksberg R, Wing K, Wittemeyer K, Wood S, Wu J, Yaspan BL, Zurawiecki D, Zwaigenbaum L, Buxbaum JD, Cantor RM, Cook EH, Coon H, Cuccaro ML, Devlin B, Ennis S, Gallagher L, Geschwind DH, Gill M, Haines JL, Hallmayer J, Miller J, Monaco AP, Nurnberger JI Jr, Paterson AD, Pericak-Vance MA, Schellenberg GD, Szatmari P, Vicente AM, Vieland VJ, Wijsman EM, Scherer SW, Sutcliffe JS, Betancur C. (2010). Functional impact of global rare copy number variation in autism spectrum disorders. *Nature*. 466:368-372.

[54] Przybytkowski E, Ferrario C, Basik M. (2011). The use of ultra-dense array CGH analysis for the discovery of micro-copy number alterations and gene fusions in the cancer genome. *BMC Med Genomics*. 4:16.

[55] Psychiatric GWAS Consortium Coordinating Committee, Cichon S, Craddock N, Daly M, Faraone SV, Gejman PV, Kelsoe J, Lehner T, Levinson DF, Moran A, Sklar P, Sullivan PF. (2009). Genomewide association studies: history, rationale, and prospects for psychiatric disorders. *Am J Psychiatry*. 166:540-556.

[56] Redon R, Ishikawa S, Fitch KR, Feuk L, Perry GH, Andrews TD, Fiegler H, Shapero MH, Carson AR, Chen W, Cho EK, Dallaire S, Freeman JL, González JR, Gratacòs M, Huang J, Kalaitzopoulos D, Komura D, MacDonald JR, Marshall CR, Mei R, Montgomery L, Nishimura K, Okamura K, Shen F, Somerville MJ, Tchinda J,

Valsesia A, Woodwark C, Yang F, Zhang J, Zerjal T, Zhang J, Armengol L, Conrad DF, Estivill X, Tyler-Smith C, Carter NP, Aburatani H, Lee C, Jones KW, Scherer SW, Hurles ME. (2006). Global variation in copy number in the human genome. *Nature* 444: 444–454

[57] Reymond A, Henrichsen CN, Harewood L, Merla G. (2007). Side effects of genome structural changes. *Curr Opin Genet Dev.* 17:381-386

[58] Risch N, Merikangas K (1996). The future of genetic studies of complex human diseases. *Science* 273:1516–1517

[59] Ropers HH. (2010). Genetics of early onset cognitive impairment. *Annu Rev Genomics Hum Genet.* 11:161-187

[60] Roohi J, Montagna C, Tegay DH, Palmer LE, DeVincent C, Pomeroy JC, Christian SL, Nowak N, Hatchwell E. (2009). Disruption of contactin 4 in three subjects with autism spectrum disorder. *J Med Genet.* 46:176-182.

[61] Rutter, M. (2005). Autism research: lessons from the past and prospects for the future. *J Autism Dev Disord* 35:241–257.

[62] Sebat J, Lakshmi B, Troge J, Alexander J, Young J, Lundin P, Månér S, Massa H, Walker M, Chi M, Navin N, Lucito R, Healy J, Hicks J, Ye K, Reiner A, Gilliam TC, Trask B, Patterson N, Zetterberg A, Wigler M. (2004). Large-scale copy-number polymorphism in the human genome. *Science* 305:525–528.

[63] Sebat J, Lakshmi B, Malhotra D, Troge J, Lese-Martin C, Walsh T, Yamrom B, Yoon S, Krasnitz A, Kendall J, Leotta A, Pai D, Zhang R, Lee YH, Hicks J, Spence SJ, Lee AT, Puura K, Lehtimäki T, Ledbetter D, Gregersen PK, Bregman J, Sutcliffe JS, Jobanputra V, Chung W, Warburton D, King MC, Skuse D, Geschwind DH, Gilliam TC, Ye K, Wigler M. (2007). Strong association of de novo copy number mutations with autism. *Science.* 316:445-449.

[64] Simon-Sanchez J, Scholz S, Fung HC, Matarin M, Hernandez D, Gibbs JR, Britton A, de Vrieze FW, Peckham E, Gwinn-Hardy K, Crawley A, Keen JC, Nash J, Borgaonkar D, Hardy J, Singleton A. (2007). Genome-wide SNP assay reveals structural genomic variation, extended homozygosity and cell-line induced alterations in normal individuals. *Hum. Mol. Genet.* 16: 1–14.

[65] Veenstra-VanderWeele, J., Christian, S.L., Cook, E.H. Jr. (2004). Autism as a paradigmatic complex genetic disorder. *Annu Rev Genomics Hum Genet* 5: 379–405.

[66] Veenstra-VanderWeele J, Cook EH Jr. (2004). Molecular genetics of autism spectrum disorder. *Mol Psychiatry* 9:819–832.

[67] Vorstman, J.A., Staal, W.G., van Daalen, E., van Engeland, H., Hochstenbach, P.F., Franke, L. (2006). Identification of novel autism candidate regions through analysis of reported cytogenetic abnormalities associated with autism. *Mol. Psychiatry* 11:18-28.

[68] Wang K, Zhang H, Ma D, Bucan M, Glessner JT, Abrahams BS, Salyakina D, Imielinski M, Bradfield JP, Sleiman PM, Kim CE, Hou C, Frackelton E, Chiavacci R, Takahashi N, Sakurai T, Rappaport E, Lajonchere CM, Munson J, Estes A, Korvatska O, Piven J, Sonnenblick LI, Alvarez Retuerto AI, Herman EI, Dong H, Hutman T, Sigman M, Ozonoff S, Klin A, Owley T, Sweeney JA, Brune CW, Cantor RM, Bernier R, Gilbert JR, Cuccaro ML, McMahon WM, Miller J, State MW, Wassink TH, Coon H, Levy SE, Schultz RT, Nurnberger JI, Haines JL, Sutcliffe JS, Cook EH, Minshew NJ, Buxbaum JD, Dawson G, Grant SF, Geschwind DH, Pericak-Vance MA,

Schellenberg GD, Hakonarson H. (2009). Common genetic variants on 5p14.1 associate with autism spectrum disorders. *Nature*. 459:528-533.

[69] Weiss, L.A. (2009). Autism genetics: emerging data from genome-wide copy-number and single nucleotide polymorphism scans. *Expert Rev Mol Diagn*. 9:795-803.

[70] Weiss LA, Shen Y, Korn JM, Arking DE, Miller DT, Fossdal R, Saemundsen E, Stefansson H, Ferreira MA, Green T, Platt OS, Ruderfer DM, Walsh CA, Altshuler D, Chakravarti A, Tanzi RE, Stefansson K, Santangelo SL, Gusella JF, Sklar P, Wu BL, Daly MJ; Autism Consortium. (2008). Association between microdeletion and microduplication at 16p11.2 and autism. *N Engl J Med*. 358:667-675.

[71] Weiss LA, Arking DE, Daly MJ, Chakravarti A. (2009). A genome-wide linkage and association scan reveals novel loci for autism. *Nature*. 461:802-808.

[72] Yim SH, Kim TM, Hu HJ, Kim JH, Kim BJ, Lee JY, Han BG, Shin SH, Jung SH, Chung YJ. (2010). Copy number variations in East-Asian population and their evolutionary and functional implications. *Hum Mol Genet*. 19:1001-1008.

[73] Zhang F, Gu W, Hurles ME, Lupski JR. (2009). Copy number variation in human health, disease, and evolution. *Annu Rev Genomics Hum Genet*. 10:451-481.

[74] Zheng, C.,Heintz, N., Hatten, M.E. (1996). CNS gene encoding astrotactin, which supports neuronal migration along glial fibers. *Science* 272: 417–419

10

A Missense Mutation in *CD38* Associated with Autism Spectrum Disorder in Three Pedigrees

Haruhiro Higashida[1,2], Toshio Munesue[2], Shigeru Yokoyama[1], Minako Hashii[1], Keita Koizumi[2] and Akihiro Matsushima[3]
[1]*Department of Biophysical Genetics, Kanazawa University Graduate School of Medicine, Kanazawa,*
[2]*Research Center for Child Mental Development, Kanazawa University, Kanazawa,*
[3]*Nanao National Hospital, Nanao,*
Japan

1. Introduction

Autism spectrum disorder (ASD) or pervasive developmental disorder (PDD) is a neurodevelopmental disease, beginning in childhood but extending through to adulthood. ASD is characterised by impairments in reciprocal social interaction and communication, and by restricted or stereotyped patterns of interests and activities. This disorder has received much scientific and social attention[1,2]. ASD is more common than previously supposed with a frequency of 0.6-3 out of 100 births[2-6] and occurs either sporadically or in a familial pattern, and far more commonly in males[7-9]. The etiology remains largely unknown[10]. Previously we demonstrated that CD38 acts as a 'niceness' protein for mouse social behavior, by regulating release of oxytocin (OT)[11], which seems to be essential for mutual recognition and trust[12,13]. Therefore, here, we describe our results on single nucleotide polymorphisms (SNPs) of CD38 in ASD patients and control subjects[14]. In addition, we report our experience of treatment of one ASD patient with a CD38 SNP by nasal OT administration.

2. Results

Figures 1 and 2 show human CD38 expression in the frontal cortex, cerebellum, hypothalamus and amygdala, by RT-PCR with human brain RNA samples which were used for synthesizing cDNAs. CD38 mRNA was highly expressed in the hypothalamus in the human brain, suggesting that CD38 has an important role on human social behavior, as in the mouse[11].

Armed with this new information about CD38 in the human brain, we set out to examine the human *CD38* gene. The mRNA for CD38 is transcribed from human chromosome 4p15[15,16]. The *CD38* gene consists of 8 exons, spanning a genomic stretch of 70.51 kb (mRNA: 1227 bases) (http://www.broad.mit.edu/mpg/haploview; Figure 3a). SNP screening in 8 exons and their flanking introns was performed by direct sequencing in 29 unrelated

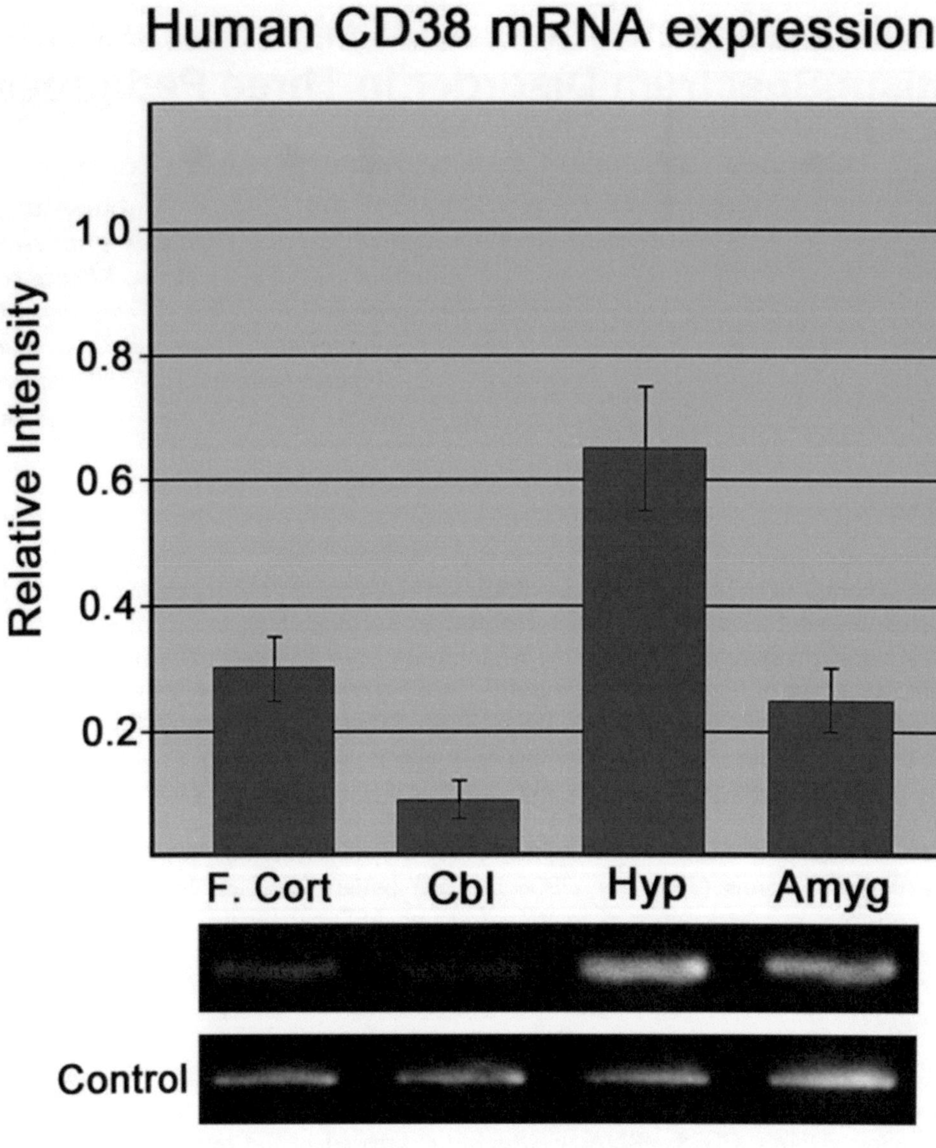

Fig. 1. **Semi-quantitative RT-PCR confesses human CD38 expression in the frontal cortex (F. Cort), the cerebellum (Cbl), the hypothalamus (Hyp) and the amygdala (Amyg).** Human brain RNA samples provided commercially (Ambion) were used for synthesizing cDNAs. Relative intensity was calculated by comparing with β-actin expression as a control.

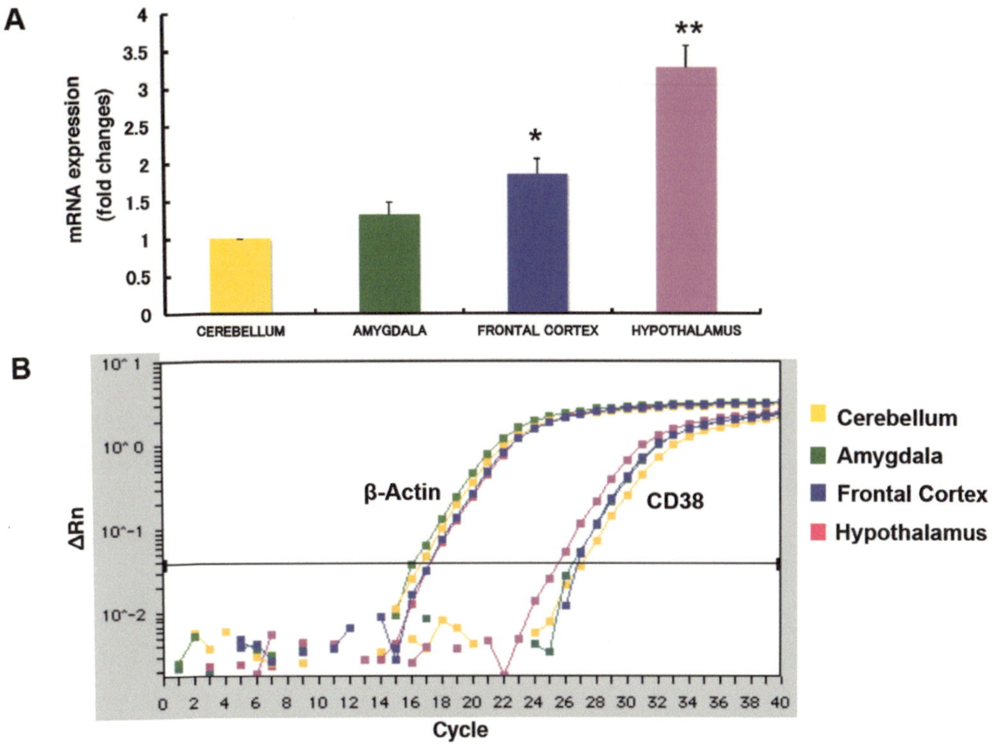

Fig. 2. **mRNA expression levels of CD38 gene in various human brain tssues.** (a) The relatve expression level of human CD38 gene was determined in RNA samples from frontal cortex, cerebellum, hypothalamus and amygdala using real-tme quantitatve PCR. Housekeeping normalized units (threshold cycle) for each gene obtained in the PCR analysis were used to determine the fold-change among samples. Bars represent fold-changes of the mRNA level of CD38 when comparing cerebellum with other tssues. Data are expressed as the mean ± S.D., performed in duplicate and repeated 3 times. *p<0.01, **p<0.01, Significantly increased from cerebellum value (P<0.01 and P<0.001). (b) Each plot represents the baseline-subtracted fluorescence intensity (ΔRn) that reflects mRNA levels of CD38 or β-actin genes. Horizontal lines indicate threshold lines set in the exponentially increasing area calculated by using SDS software.

subjects (the sample set A in Table 1) fulfilling *Diagnostic and Statistical Manual of Mental Disorders*, 4th edition (DSM-IV, American Psychiatric Association, 1994), criteria for ASD, and in 201 non-clinical control subjects (sample set E), in the Kanazawa area in Japan. We detected 12 SNPs that had already been reported, plus 3 novel mutations (Figure 3a). Allelic and genotypic frequencies in these samples are summarized in Tables 2 and 3. Among them, as shown in Figure 3b, we detected the C4693T mutation in exon 3 (SNP13; rs1800561) that leads to an arginine (R)-to-tryptophan (W) substitution at amino acid 140, R140W.

In the following experiments, we focused mainly on this mutation, because of functional abnormality in R140W-substituted-CD38: (1) The R140 is relatively well conserved among multiple species except for the rodent (Figure 3c). R140 is located in the flexible loop (137-141) at the midpoint of the N- and C-ternimus domains between two helical domains (αa4

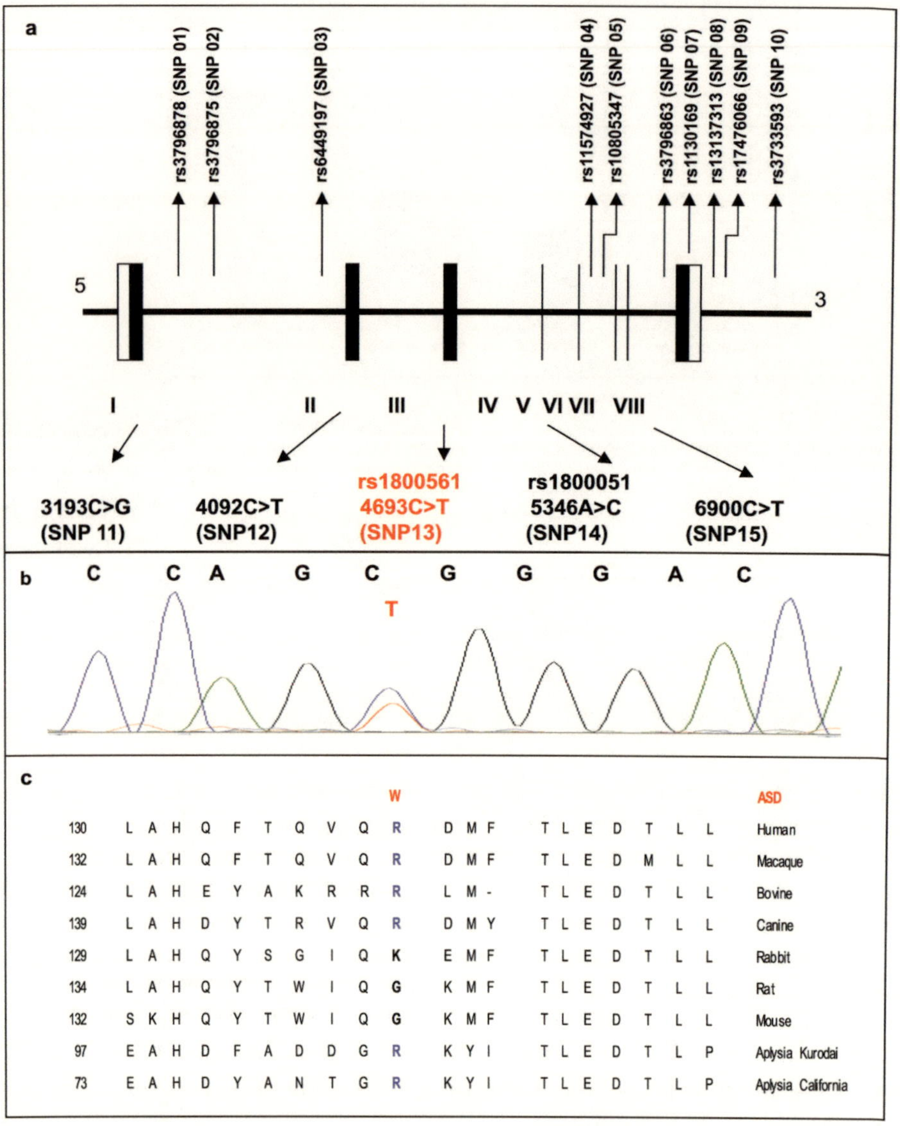

Fig. 3. **Genome and molecular structures and mutations of CD38.**
(**a**) Genomic structure of *CD38* and locations of SNPs in introns (upper) and in exons (lower). Exons are indicated by boxes, with translated regions in closed boxes, and untranslated regions in open boxes. Mutations at amino acids at positions of 47, 116, 140, 168 and 264 of CD38 are indicated. Numbering of the nucleotides starts at the A of ATG encoding the initial methionine and refers to Genbank accession number D84284.
(**b**) Sequence trace was derived from a blood DNA sample of 4693C/T heterozygote.
(**c**) Conservation of R at the 140th amino acid among different species, except for the rodent. Sequences were obtained through the accession numbers of NM001775, AY555148, NM175798, AF117714, AF272974, NM013127, NM007646, D30048, and M85206/M37644 for the indicated species, respectively. It is noted that mutant structure has more open-form conformation than wild-type and its variation is slightly larger.

and αa5) and is the pivot of the hinge region connecting two regions of L-shaped molecule[17]. Therefore, the mutation (W140) causes severe perturbations of the predicted protein structure, if compared with the human (R140), rabbit (K140) or mouse (G140) CD38 (see Figure 7 in ref.14). (2) Indeed, the mutant W140-CD38 showed one third of ADP-ribosyl cyclase activity of wild-type CD38 expressed in the CHO cells[15]. (3) Social amnesia was not rescued by local expression of W140-CD38 in the hypothalamus in *Cd38* knockout mice[11].

Sample set	Subject number	Male/Female	Age range	Age average	Description	Country	W140 allele frequency	Reference
A	29	(23/6)	12 to 44	22.8+/-7.6	Unrelated ASD	Japan*	0.052	
B	3	(3/0)	21 to 44	30.0+/- 7.1	3 probands in A	Japan*	1	
C	25	(15/10)	21 to 84	53.0+/-4.6	3 families in B	Japan*	0.32	
D	252	(252/0)			ASD	USA**	0	20, 21
E	201	(106/95)	22 to 64	32.5+/-0.9	Unscreened control	Japan*	0.007	

*In the Kanazawa area **AGRE samples

Table 1. **Sample sets in this experiment**

SNP*			Control N	Allele counts	Frquency	ASD N	Allele counts	Frequency
SNP01	rs3796878	G>A	400	4	0.01	58	0	0
SNP02	rs3796875	A>G	398	100	0.25	56	18	0.321
SNP03	rs6449197	C>T	392	81	0.207	58	16	0.275
SNP04	rs11574927	A>G	402	78	0.195	58	6	0.103
SNP05	rs10805347	A>G	400	180	0.452	58	22	0.379
SNP06	rs3796863	C>A	398	148	0.371	58	20	0.345
SNP07	rs1130169	C>T	398	36	0.341	56	15	0.268
SNP08	rs13137313	A>G	398	176	0.442	58	28	0.483
SNP09	rs17476066	T>C	392	46	0.117	58	6	0.103
SNP10	rs3733593	C>T	398	128	0.32	56	17	0.305
SNP11		C>G	402	1	0.002	58	0	0
SNP12		C>T	402	0	0	58	1	0.017
SNP13	rs1800561	C>T	402	3	0.008	58	3	0.052
SNP14	rs1800051	A>C	402	59	0.146	58	13	0.22
SNP15		C>T	402	6	0.015	58	0	0

*SNPs are denoted as major allele>minor allele

Table 2. **Alleleic frequency of SNPs in control and ASD subjects**

The 140R/W (C4693T) heterozygotes were found in 3 male subjects (sample set B; two autistic and one Asperger) out of 29 ASD patients examined (23 males and 6 females with the mean age = 22.8 ± 7.6; prevalence of 10.3% of ASD samples). We examined whether or not the W140 allele seems to be co-segregated with ASD and ASD-related traits in 3 probands' families. Twenty five members of the 3 kindred families (sample set C) were available for detailed clinical and genetic analyses (Figure 4). The 4693C-to-T change was found in all probands' fathers in the 3 families and brothers in the 2 families (Family #1 and #3). The mutation is present in the grandmother of the father's side in the Family #1 (1-I-4) and is also predicted to be transmitted from the late grandmother of the father's side in the

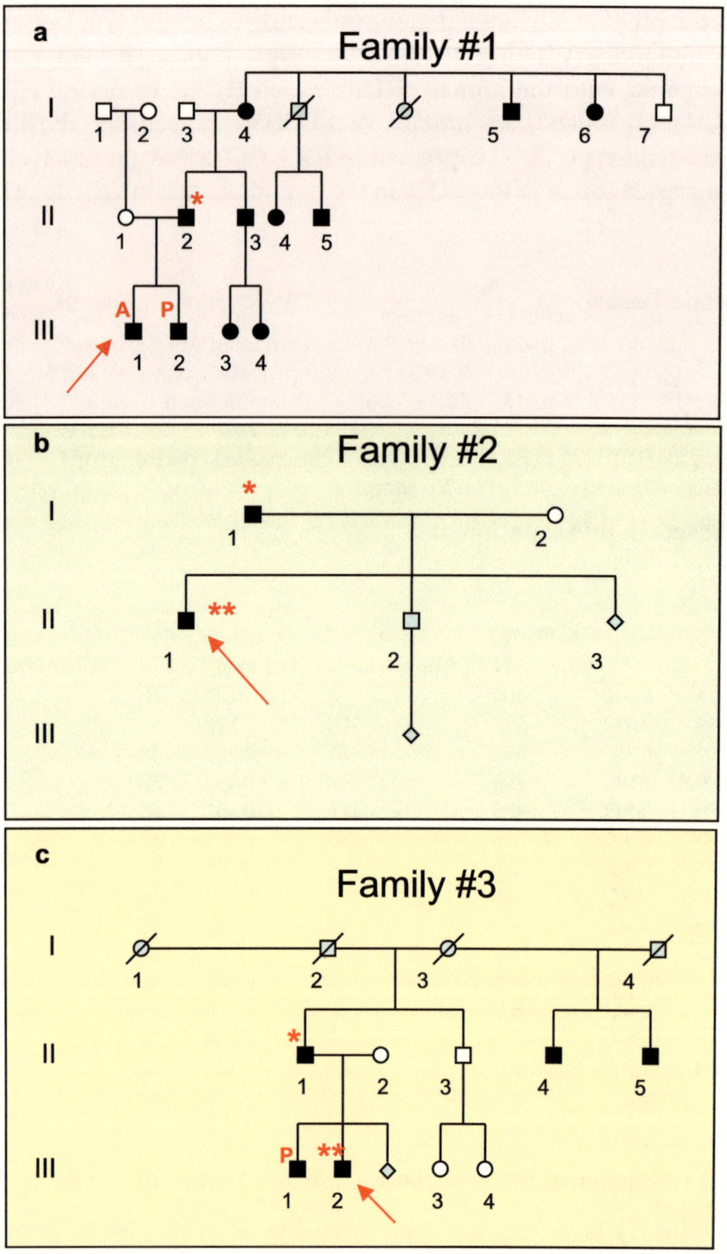

Fig. 4. **Pedigrees for three families of ASD probands carrying the W140.** Affected males (probands) are indicated with red arrows. Affected or carrier males or females are indicated by filled squares or circles, respectively. Empty symbols denote individuals with no mutation. Gray symbols indicate undetermined with no DNA available for analysis. Sexes are hidden by diamond symbols upon request. The subjects are indicated by progressive Arabic numbers according to the three generations. The current study was approved by the ethical committee of Kanazawa University Graduate School of Medicine. Subjects marked with symbols are: **, denotes ASD; *, ASD traits; P, PDD-NOS; A, Asperger disorder.

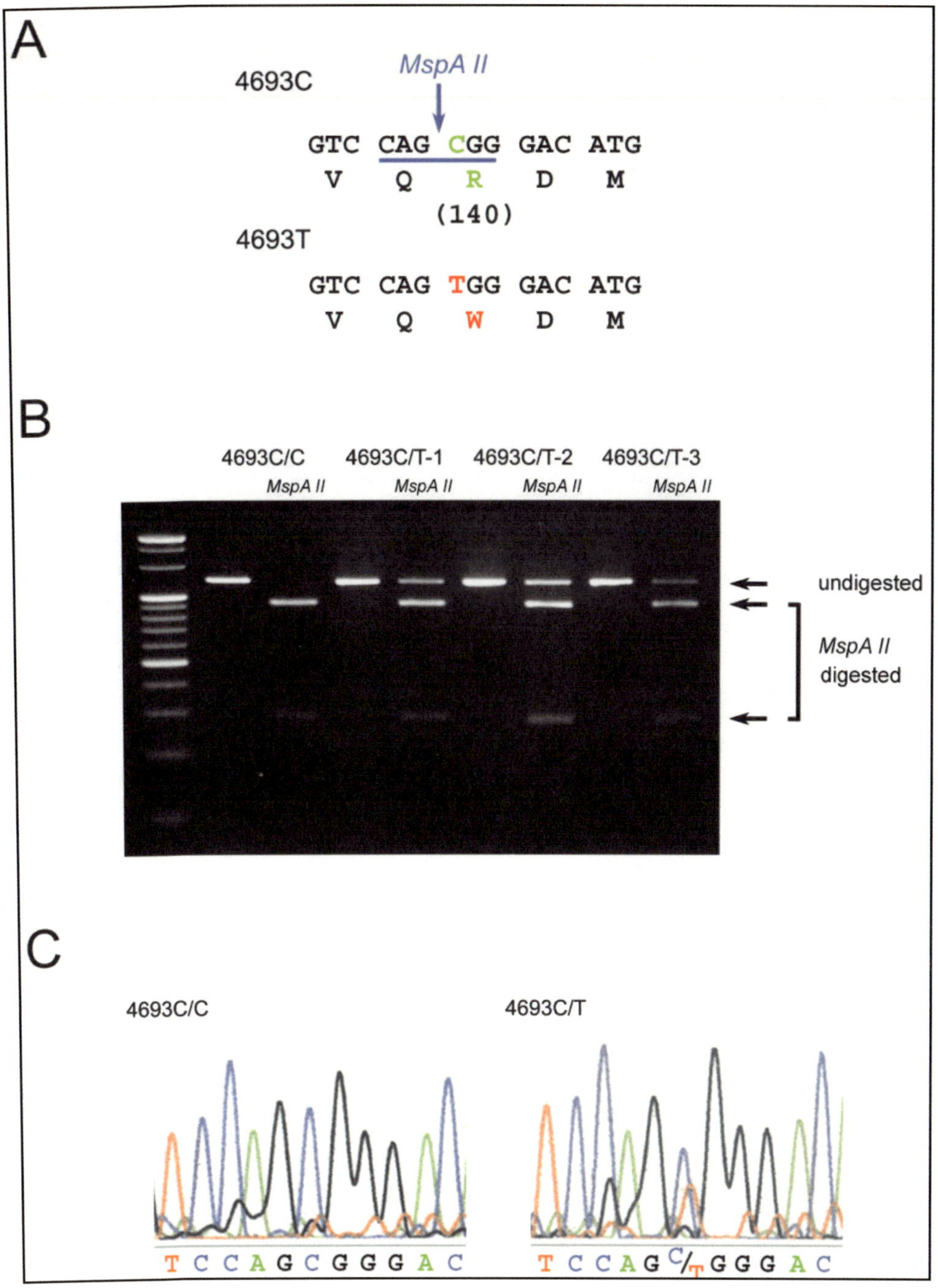

Fig. 5. **Expression of the minor allele of rs1800561**
(A) cDNA with 4693C has *Msp*AII site. (B) RT-PCR products from blood RNA samples were digested by *Msp*AII. An RT-PCT product from a homozygous 4693C/C subject gives two bands, while a heterozygous 4693C/T subject giving three bands. (C) Sequencing of RT-PCR products confirms the SNP.

Family #3 (3-I-3), in an apparently autosomal dominant fashion. We found a total of 18 carriers in 28 family members cooperative (prevalence=64%). In all cases the mutation was heterozygous (allelic frequency=0.32). The mutant allele was indeed transcribed in the subjects tested (Figure 5).

The kindred were clinically evaluated by interviewing. The probands' young (1-III-2 in the Family #1) and old (3-III-1 in the family #3) brothers showed clinical features conforming to PDD-NOS (PDD not otherwise specified) or Asperger. Two fathers (1-II-2 and 3-II-1) in their 50s and another father (2-I-1) in his 70s were all diagnosed as having with ASD traits. Most other adults over 50 years old in these pedigrees had not been clinically diagnosed with ASD or other psychiatric diseases, though some showed personal traits such as eccentricity, resulting in 8 ASD subjects out of 13 male carriers (62%). Interestingly, four young female cousins with (1-III-3 and 1-III-4 in the family #1) and without (3-III-3 and 3-III-4 in the family #3) the mutation, had no clinical ASD phenotype.

We also evaluated them from the score of the Autism-Spectrum Quotient (AQ)[18,19], in which older subjects esteemed themselve by recalling behaviours at their life period of 20s. AQ scores in two young male carriers in the family #1 (1-III-1 and -2) fulfilled the criteria (cut-off point of 28) of ASD, though this score was not obtained from two other ASD probands (in the families #2 and #3), because of low intelligence (Figure 6). Some carriers' scores were above the standard deviation of average values in noncarrier family members who showed normal control scores, indicating that such carriers may be considered to manifest ASD traits, even though not affected at the clinical level (Figure 6a). Statistically there is no difference between three different age groups (young, middle and old generations), but the males' score was significantly higher than that of females (p<0.05; Figure 6b). These clinical and self-describing evaluations suggest that this gene polymorphism is important to determine the ASD or ASD trait phenotype.

Given these results, we obtained serum samples from the kindreds to further study the connection between the human *CD38* mutation and plasma OT or arginine vasopressin (AVP) levels, since we previously showed that a null mutation of *Cd38* resulted in the selective decrease of plasma OT levels in mice[11], and low levels of OT have been reported in autistic children[20]. The plasma OT levels in the carriers (161.3 ± 26.5 pg/ml, n=12) were lower than those of kindred non-carriers (345.8 ± 61.3 pg/ml, n=10; p<0.01), as shown in Figure 7. The differences seem to be found in the younger generation but not so in older subjects (Figure 7c). The OT levels of three probands and two young carriers were compared with ASD patients without the W140 mutation (in the sample set A): the levels of five W140 carriers (79.2 ± 16.6 pg/ml; n=5) were lower than those without the mutation (147.7 ± 15.0 pg/ml; p<0.01, n=26). Furthermore, the OT level of the carrier ASD probands was significantly lower when compared with that in 101 adult control (198.2 ± 24.7 pg/ml; p<0.01). As expected, there is no difference in AVP levels between *CD38*-mutation-carriers and –noncarriers in the pedigrees (Figure 7b and d). Low plasma hormone levels were frequently observed in subjects with high AQ scores in carriers in the pedigrees (Figure 7e and f).

We also analysed the R140W mutation in 252 ASD subjects (excluding Hispanic and Asian peoples) recruited to the Autism Genetic Resource Exchange (AGRE; http://www.agre.org)[21] in USA (sample set D[22]). No mutation was found, suggesting ethnicity-dependent frequency differences. Finally, from 201 healthy unscreened control subjects (sample set E), 2 females and 1 male were positive for the mutation, representing allelic frequency of 0.007 (Tables 2 and 3). This frequency is 7.4-fold lower than those (0.052) in ASD patient group in the same residential area (p<0.028; Table 3).

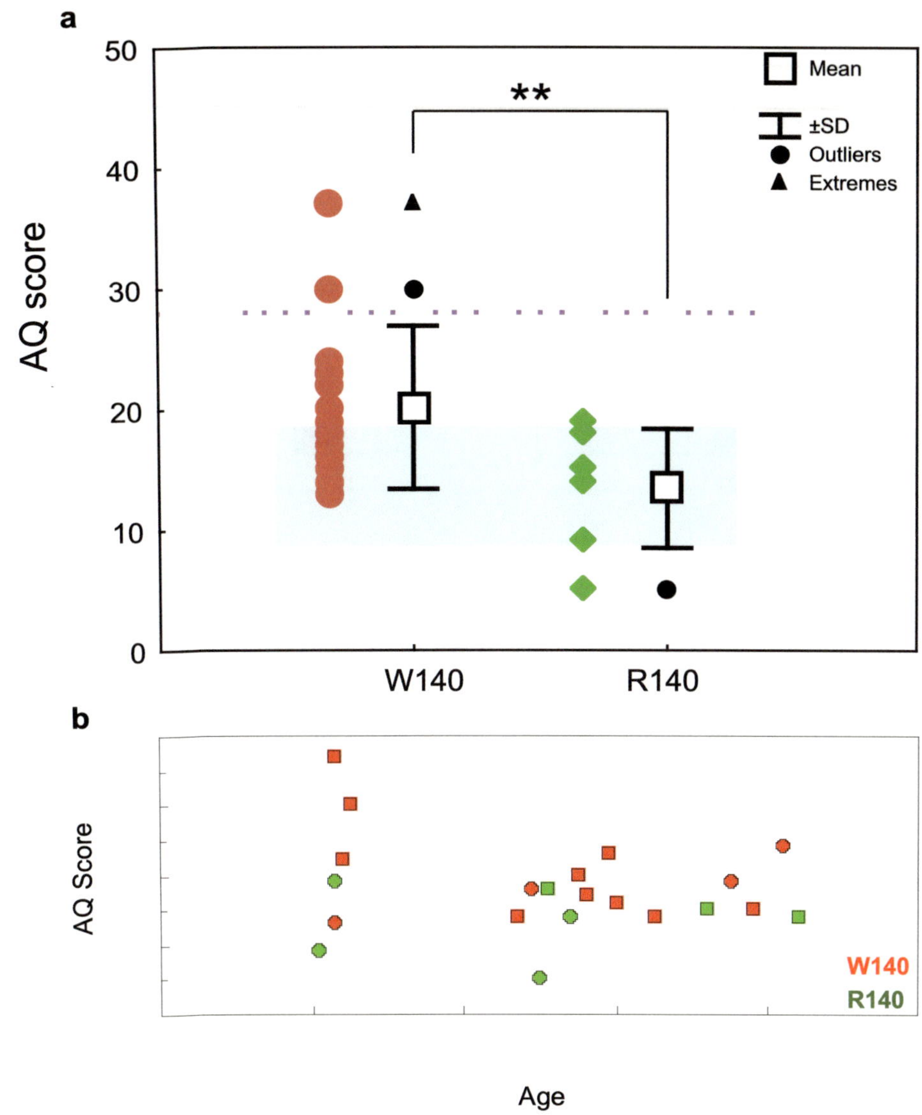

Fig. 6. **AQ score in family members in the three pedigrees.** (**a**) Assessment groups were: W140: n=14 family members with the monoallelic R140W mutation; R140: n=7 persons without the mutation in the families. Horizontal dashed bar indicates the critical score of 28 obtained from clinical ASD group in a separate experiment[18]. The mean and standard deviation range of AQ scores were illustrated. Note that 5 individuals show the intermediate score above the control range but below the ASD score. The score with R140W is higher than that without (one-way ANOVA, $p<0.05$). (**b**) A plot of AQ scores of each individual of family members with or without the R140W mutation according to age. No significance was found between three generations (20<age<40, 40<age<60 and 60<age) by two-way ANOVA. The scores of males are significantly higher than those of females ($p<0.05$). Circles indicate female, and squares, male.

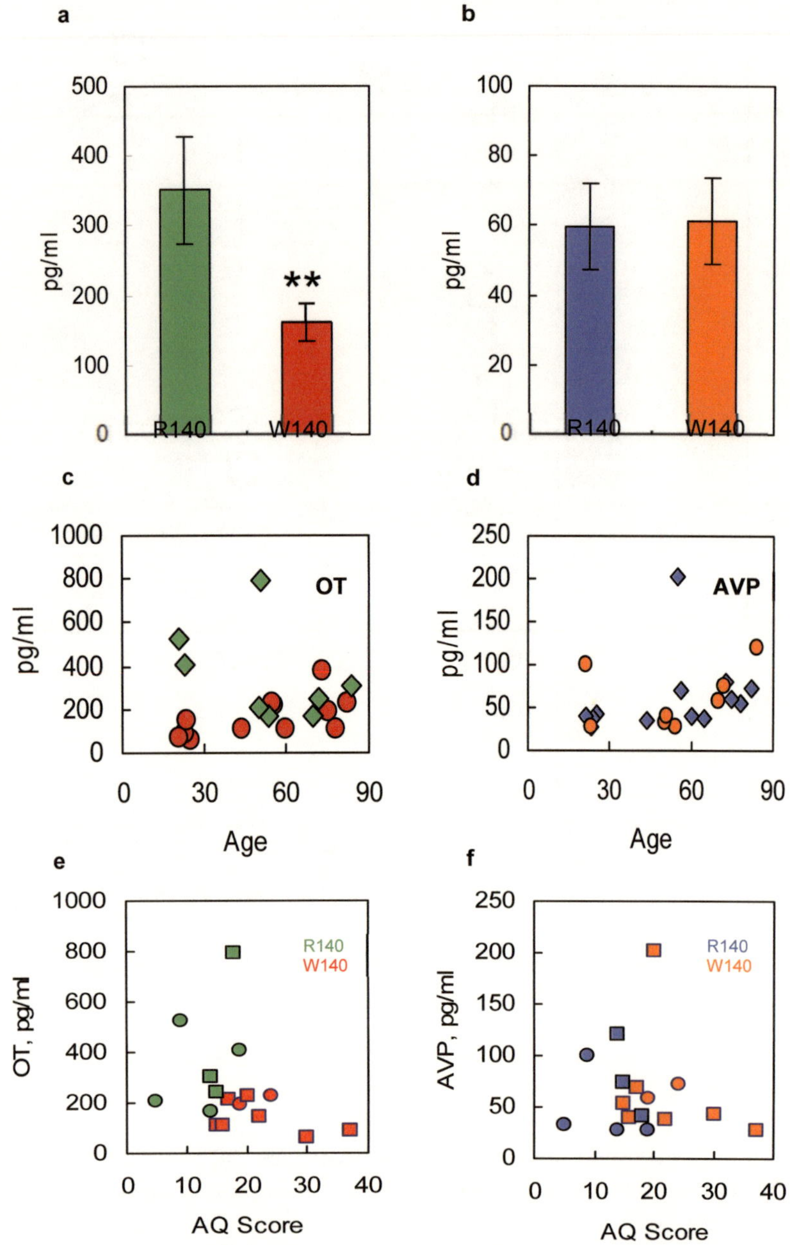

Fig. 7. **Plasma oxytocin and vasopressin levels in family members.** Plasma concentrations of OT (**a**) and AVP (**b**) levels in family members with (W140; n=12, red or orange bar) or without (R140; n=10, green or blue bar) heterozygous R140W allele. Mean ± s.e.m. **, p<0.01 (one-way ANOVA). OT (**c**) or AVP (**d**) levels in the three kindred according to age. OT (**e**) or AVP (**f**) levels as a function of AQ scores. Red and orange symbols or green and blue symbols indicate levels from persons with or without the R140W mutation, respectively. Circles denote female and diamonds or squares, male.

SNP	Sample	HWE	Genotype (No. & Frequency)						*p-value	Allele (No. & Frequency)				*p-value
SNP01			A/A		A/G		G/G			A		G		
	Control	0.89	0	0.00	4	0.02	196	0.98	1	4	0.01	396	0.99	1
	ASD	-	0	0.00	0	0.00	29	1.00		0	0.00	58	1.00	
SNP02			A/A		A/G		G/G			A		G		
	Control	0.83	111	0.56	76	0.38	12	0.06	0.602	298	0.75	100	0.25	0.416
	ASD	0.6	13	0.46	13	0.46	2	0.07		39	0.70	17	0.30	
SNP03			T/T		T/C		C/C			T		C		
	Control	0.55	7	0.04	67	0.34	122	0.62	0.236	81	0.21	311	0.79	0.234
	ASD	0.26	1	0.03	14	0.48	14	0.48		16	0.28	42	0.72	
SNP04			A/A		A/G		G/G			A		G		
	Control	0.78	129	0.65	64	0.32	7	0.04	0.089	322	0.81	78	0.20	0.104
	ASD	0.17	24	0.83	4	0.14	1	0.03		52	0.90	6	0.10	
SNP05			A/A		A/G		G/G			A		G		
	Control	0.12	55	0.28	110	0.55	35	0.18	0.363	220	0.55	180	0.45	0.672
	ASD	0.43	11	0.38	12	0.41	6	0.21		34	0.59	24	0.41	
SNP06			A/A		A/C		C/C			A		C		
	Control	0.17	23	0.12	102	0.51	74	0.37	0.959	148	0.37	250	0.63	0.771
	ASD	0.71	3	0.10	14	0.48	12	0.41		20	0.34	38	0.66	
SNP07			T/T		T/C		C/C			T		C		
	Control	0.18	19	0.10	98	0.49	82	0.41	0.226	136	0.34	262	0.66	0.293
	ASD	0.34	3	0.11	9	0.32	16	0.57		15	0.27	41	0.73	
SNP08			A/A		A/G		G/G			A		G		
	Control	0.75	63	0.32	96	0.48	40	0.20	0.594	222	0.56	176	0.44	0.574
	ASD	0.36	9	0.31	12	0.41	8	0.28		30	0.52	28	0.48	
SNP09			T/T		T/C		C/C			T		C		
	Control	0.83	153	0.78	40	0.20	3	0.02	0.261	346	0.88	46	0.12	0.657
	ASD	0.06	25	0.86	3	0.10	1	0.03		53	0.91	5	0.09	
SNP10			T/T		T/C		C/C			T		C		
	Control	0.85	20	0.10	88	0.44	91	0.46	0.373	128	0.32	270	0.68	1
	ASD	0.1	1	0.04	16	0.57	11	0.39		18	0.32	38	0.68	
SNP12			T/T		T/C		C/C			T		C		
	Control	-	0	0.00	0	0.00	201	1.00	0.126	0	0.00	402	1.00	0.126
	ASD	0.92	0	0.00	1	0.03	28	0.97		1	0.02	57	0.98	
SNP13			T/T		T/C		C/C			T		C		
	Control	0.91	0	0.00	3	0.01	198	0.99	**0.028**	3	0.01	399	0.99	**0.029**
	ASD	0.77	0	0.00	3	0.10	26	0.90		3	0.05	55	0.95	

P-value from Fisher's exact test. SNPs11, 14 and 15 were not analysed.

Table 3. **Statistical analysis of SNPs in 29 ASD and 201 control subjects in the Kanazawa area**

One proband receiving intranasal OT for 6 months showed improvements in the area of social approach, eye-contact and communication behavior without serious adverse effects. These results suggest that the *CD38* W140 allele could be a possible risk factor for one form of ASD by abrogating OT function and the carriers become candidates for the OT treatment[14].

3. Discussion

ASD is heterogeneous and forms a continuum, and thus is likely to involve many genes[7-10,23-26]. *De novo* mutations related to ASD are rarely inherited[7], but some ASD also can be inherited from pre-existing genetic variants in parents, for both of which a unified mechanism was proposed[9]. Our results shed light on genetic mutations of *CD38* in relatively mature ASD patients and the inheritable ASD subgroups of three independent families in which the CD38-R140W mutation showed relative co-segregation with diverse phenotypes of ASD. In such multiplex families, the risk of autism or ASD traits in young male offsprings is nearly ~100%, representing dominant transmission of a mutation with high penetrance. However, we cannot completely exclude as yet unidentified contributing factors.

The heterozygous 4693C/T carriers were identified in Japanese and Italian general populations[16,27]. The presence of this mutation in Japanese has already been reported in the HapMap site (http://www.hapmap.org/) and by Yagui *et al.* in 1998[15]. Surprisingly, the carriers were also identified in females. Therefore, it is highly possible that even clinically-unaffected female carriers with this mutant allele could in turn transmit the mutation to their offspring with high penetrance in males. Currently, we are intensively collecting blood samples for mutation analysis from many ASD and control subjects in different cities and countries to support our finding, which so far has been carried out in a relatively small group in a restricted area.

A noteworthy role for OT in social recognition has been shown in rodent and human studies[12,13]. Recently, a 1.1-Mb deletion of 20p13 including the OT gene (copy number decrease) has been detected in a child with Asperger disorder[8]. An association of one or two intronic SNPs in the OT receptor gene with autism has also been reported[25,28,29], suggesting that defects in OT signaling confer genetic vulnerability to ASD. Though the R140W mutation was not found in 252 American AGRE samples, the association study with tagSNPs showed one SNP (SNP06; rs3796863) that is positively related ($p<0.004$) with American high functioning autism[14].

In conclusion, *CD38* mutations provide one genetic basis for those instances of ASD that arises from disruption of the OT signaling. Thus, our finding provides the first theoretical background for the evidence-based treatment by OT infusion for a subgroup of ASD patients with the lower plasma OT level[14], which has already been tried for non-selected ASD subjects[30,31].

4. Methods summary

Clinical and genetic studies were carried out according to institutional guidelines after ethical approval of participating institutions and informed consent was obtained from all participating patients. A total of 29 unrelated affected individuals out of 96 in the Kanazawa area in Japan (Sample set A) were admitted the Kanazawa University Hospital diagnosed with DSM-IV in accordance with clinical criteria. Blood and platelet biochemical analyses were performed in 29 ASD probands and their parents and family members who agreed to supply. Genotyping for the association study and mutation screening were performed by direct sequencing or TaqMan technology. PCR products were sequenced with the BigDye Terminator Cycle Sequencing Kit (V3.1, Applied Biosystems, Foster City, CA, USA). Samples were then subjected to electrophoresis, using an ABI PRISM genetic analyzer (Applied Biosystems). Absence of genotyping errors was controlled by sequencing the PCR product with the opposite primer in a subset of patients.

Statistics: Data are expressed as mean ± s.d. or s.e.m. Statistical analysis was performed using one-way or two-way ANOVA. The criterion for significance in all cases was $p<0.05$.

5. References

[1] Singh J, Hallmayer J & Illes J. Interacting and paradoxical forces in neuroscience and society. *Nature Rev. Neurosci.* 8, 153-60 (2007).

[2] Editorial. New guidance on autism. *Lancet* 370, 1590 (2007).

[3] Williams J G, Higgins J P & Brayne C E. Systematic review of prevalence studies of autism spectrum disorders. *Arch. Dis. Child.* 91, 8-15 (2006).

[4] Baird, G. *et al*. Prevalence of disorders of the autism spectrum in a population cohort of children in South Thames: the Special Needs and Autism Project (SNAP). *Lancet* 368, 210-215 (2006).

[5] Honda H, Shimizu Y, Imai M & Nitto Y. Cumulative incidence of childhood autism: a total population study of better accuracy and precision. *Dev. Med. Child. Neurol.* 47, 10-18 (2005).

[6] Sumi S, Taniai H, Miyachi T & Tanemura M. Sibling risk of pervasive developmental disorder estimated by means of an epidemiologic survey in Nagoya. *J. Hum. Genet.* 51, 518-522 (2006).

[7] Beaudet AI. Autism: highly heritable but not inherited. *Nature Med.* 13, 534-536 (2007).
[8] Sebat J, Lakshmi B, Malhotra D, Troge J, Lese-Martin C, Walsh T, Yamrom B, Yoon S, Krasnitz A, Kendall J, Leotta A, Pai D, Zhang R, Lee YH, Hicks J, Spence SJ, Lee AT, Puura K, Lehtimäki T, Ledbetter D, Gregersen PK, Bregman J, Sutcliffe JS, Jobanputra V, Chung W, Warburton D, King MC, Skuse D, Geschwind DH, Gilliam TC, Ye K & Wigler M. Strong association of de novo copy number mutations with autism. *Science* 316, 445-449 (2007).
[9] Zhao X, Leotta A, Kustanovich V, Lajonchere C, Geschwind DH, Law K, Law P, Qiu S, Lord C, Sebat J, Ye K & Wigler M. A unified genetic theory for sporadic and inherited autism. *Proc. Natl. Acad. Sci. USA* 104, 12831-12836 (2007).
[10] Happé F, Ronald A & Plomin R. Time to give up on a single explanation for autism. *Nature Neurosci.* 9, 1218-12120 (2006).
[11] Jin D, Liu HX, Hirai H, Torashima T, Nagai T, Lopatina O, Shnayder NA, Yamada K, Noda M, Seike T, Fujita K, Takasawa S, Yokoyama S, Koizumi K, Shiraishi Y, Tanaka S, Hashii M, Yoshihara T, Higashida K, Islam MS, Yamada N, Hayashi K, Noguchi N, Kato I, Okamoto H, Matsushima A, Salmina A, Munesue T, Shimizu N, Mochida S, Asano M & Higashida H. CD38 is critical for social behaviour by regulating oxytocin secretion. *Nature* 446, 41-45 (2007).
[12] Insel TR. The challenge of translation in social neuroscience: a review of oxytocin, vasopressin, and affiliative behavior. *Neuron* 65, 768-779 (2010) .
[13] Kosfeld M, Heinrichs M, Zak P J, Fischbacher U & Fehr E. Oxytocin increases trust in humans. *Nature* 435, 673-676 (2005).
[14] Munesue T, Yokoyama S, Nakamura K, Anitha A, Yamada K, Hayashi K, Asaka T, Liu HX, Jin D, Koizumi K, Islam MS, Huang JJ, Ma WJ, Kim UH, Kim SJ, Park K, Kim D, Kikuchi M, Ono Y, Nakatani H, Suda S, Miyachi T, Hirai H, Salmina A, Pichugina YA, Soumarokov AA, Takei N, Mori N, Tsujii M, Sugiyama T, Yagi K, Yamagishi M, Sasaki T, Yamasue H, Kato N, Hashimoto R, Taniike M, Hayashi Y, Hamada J, Suzuki S, Ooi A, Noda M, Kamiyama Y, Kido MA, Lopatina O, Hashii M, Amina S, Malavasi F, Huang EJ, Zhang J, Shimizu N, Yoshikawa T, Matsushima A, Minabe Y & Higashida H. *Neurosci. Res.* 67, 181-91 (2010).
[15] Yagui K, Shimada F, Mimura M, Hashimoto N, Suzuki Y, Tokuyama Y, Nata K, Tohgo A, Ikehata F, Takasawa S, Okamoto H, Makino H, Saito Y & Kanatsuka A. A missense mutation in the CD38 gene, a novel factor for insulin secretion: association with Type II diabetes mellitus in Japanese subjects and evidence of abnormal function when expressed in vitro. *Diabetologia* 41, 1024-1028 (1998).
[16] Malavasi F, Deaglio S, Funaro A, Ferrero E, Horenstein AL, Ortolan E, Vaisitti T & Aydin S. Evolution and function of the ADP ribosyl cyclase/CD38 gene family in physiology and pathology. *Physiol. Rev.* 88, 841-886 (2008).
[17] Liu Q, Kriksunov IA, Graeff R, Lee H C & Hao Q. Structural basis for formation and hydrolysis of the calcium messenger cyclic ADP-ribose by human CD38. *J. Biol. Chem.* 282, 5853-5861 (2007).
[18] Baron-Cohen S, Hoekstra RH,, Knickmeyer R & Wheelwright S. The Autism-Spectrum Quotient (AQ)-adolescent version. *J. Autism Dev. Disord.* 36, 343-350 (2006).
[19] Munesue T, Ono Y, Mutoh K, Shimoda K, Nakatani H & Koshino Y. Autonomic and motor responses to Utena's simple psychiatric function test in autism spectrum disorder (in Japanese). *Seishin Igaku* 49, 599-606 (2007).

[20] Modahl C, Green L, Fein D, Morris M, Waterhouse L, Feinstein C & Levin H. Plasma oxytocin levels in autistic childen. *Biol. Psychiatry* 43, 270-277 (1998).

[21] Geschwind DH, Sowinski J, Lord C, Iversen P, Shestack J, Jones P, Ducat L, Spence SJ; AGRE Steering Committee. The autism genetic resource exchange: a resource for the study of autism and related neuropsychiatric conditions. *Am. J. Hum. Genet.* 69, 463-466 (2001).

[22] Toyoda T, Nakamura K, Yamada K, Thanseem I, Anitha A, Suda S, Tsujii M, Iwayama Y, Hattori E, Toyota T, Miyachi T, Iwata Y, Suzuki K, Matsuzaki H, Kawai M, Sekine Y, Tsuchiya K, Sugihara G, Ouchi Y, Sugiyama T, Takei N, Yoshikawa T & Mori N. SNP analyses of growth factor genes EGF, TGFbeta-1, and HGF reveal haplotypic association of EGF with autism. *Biochem. Biophys. Res. Commun.* 360, 715-720 (2007).

[23] Pinto D *et al.* Functional impact of global rare copy number variation in autism spectrum disorders. *Nature* 466, 368-372 (2010).

[24] Szatmari, P. *et al.* Mapping autism risk loci using genetic linkage and chromosomal rearrangements. *Nature Genet.* 39, 319-328 (2007).

[25] Weiss LA, Arking DE. Gene Discovery Project of Johns Hopkins & the Autism Consortium, Daly MJ, Chakravarti A. A genome-wide linkage and association scan reveals novel loci for autism. *Nature 461*, 802-808 (2009).

[26] Skuse, GH. Rethinking the nature of genetic vulnerability to autistic spectrum disorders. *Trends Genet.* 23, 387-395 (2007).

[27] Mallone R, Ortolan E, Baj G, Funaro A, Giunti S, Lillaz E, Saccucci F, Cassader M, Cavallo-Perin P & Malavasi F. Autoantibody response to CD38 in Caucasian patients with type 1 and type 2 diabetes: immunological and genetic characterization. *Diabetes* 50, 752-762 (2001).

[28] Wu S, Jia M, Ruan Y, Liu J, Guo Y, Shuang M, Gong X, Zhang Y, Yang X & Zhang D. Positive association of the oxytocin receptor gene (OXTR) with autism in the Chinese Han population. *Biol. Psychiatry* 58, 74-77 (2005).

[29] Wermter AK, Kamp-Becker I, Hesse P, Schulte-Körne G, Strauch K & Remschmidt H. Evidence for the involvement of genetic variation in the oxytocin receptor gene (OXTR) in the etiology of autistic disorders on high-functioning level. *Am. J. Med. Genet. B Neuropsychiatr. Genet.* 153B, 629-39 (2010).

[30] Hollander E, Novotny S, Hanratty M, Yaffe R, deCaria C & Aronowitz B & Mosovich S. Oxytocin infusion reduces repetitive behaviors in adults with autistic and Asperger's disorders. *Neuropsychopharmacol.* 28, 193-198 (2003).

[31] Hollander E, Bartz J, Chaplin W, Phillips A, Sumner J, Soorya L, Anagnostou E & Wasserman S. Oxytocin increases retention of social cognition in autism. *Biol. Psychiatry* 61, 498-503 (2007).

11

Brain Oxytocin is a Main Regulator of Prosocial Behaviour - Link to Psychopathology

Lukas Michael and Neumann Inga D
University of Regensburg
Germany

1. Introduction

The neuropeptide oxytocin (OT) becomes increasingly attractive not only for neurobiologists, psychologists and psychiatrists, but also for sociologists and economists promoted by the discovery of amazing behavioural functions it regulates, especially in the context of social interactions. The discovery of multiple pro-social as well as anti-stress effects of OT, which have been discovered in recent years, makes the OT system of the brain a promising target for psychotherapeutic intervention and treatment of numerous psychiatric illnesses, for example, anxiety disorders, social phobia, depression or autism (Slattery & Neumann, 2010). The endogenous OT system of the brain can be found in different activity forms. Consequently, neuronal OT synthesis, OT release within distinct regions of the brain, OT receptor binding and the intensity of OT behavioural effects can vary in dependence on the physiological (or pathophysiological) activity state, which will be outlined in more detail below. It is generally assumed that psychopathologies associated with impaired social interactions, such as autism, are accompanied by an impaired activity of the brain OT system, which may affect at least one of the above mentioned parameters.

1.1 The oxytocin system

Chemically, OT consists of nine amino acids. Two cystein residues form a disulfid bridge creating the circular structure of the nonapeptide discovered almost 60 years ago (Du Vigneaud et al., 1953). OT and the structural related neuropeptide arginine vasopressin (AVP) belong to the arginine vasotocin family (Acher et al., 1972). Neuropeptides of this family are ubiquitous within vertebrates and evolutionary highly conserved, both in structure and functions (Hoyle, 1999).

Together with AVP, OT is an essential part of the hypothalamo-neurohypophysial system. OT and AVP are mainly synthesized in a well-defined arrangement of magnocellular neurons located within the hypothalamic supraoptic (SON) and paraventriular nuclei (PVN) of the hypothalamus at the base of the brain. Via axonal projections OT and AVP reach the neurohypophysis where they are released into the blood stream.

Originally, OT has been reported as a hormonal key regulator of female reproductive functions in all mammalian species when secreted into blood. Thus, OT accelerates the delivery process as it promotes uterine contractions, and is essential for milk ejection during

lactation. However, starting with the fundamental discoveries of David DeWied and Cort Pedersen (De Wied, 1965, Pedersen & Prange, 1979), OT (and also AVP) emerged as a neuromodulator of the brain regulating a broad variety of behaviours. Among the various behavioural effects of OT are prosocial actions such as the promotion of maternal care and aggression, pair-bonding, sexual behaviour in males and females, social cognition, social memory and social support (for review see Bielsky & Young, 2004, Donaldson & Young, 2008, Neumann, 2009). Moreover, OT has the potential to reduce anxiety and to inhibit physiological stress responses which is likely to accompany these prosocial actions (Neumann, 2002, 2009).
In the context of these multiple behavioural effects, neuronal release of OT within the brain is of main relevance and has recently attracted the attention of many neurobiologists. The locally restricted intracerebral release of OT or AVP can be monitored using microdialysis in combination with sensitive radioimmunoassays, even in freely behaving animals (for review see Landgraf & Neumann, 2004, Veenema & Neumann, 2008). Monitoring of central release patterns of OT during ongoing behavioural performance is challenging, but possible. Thus, even during mating, mother-offspring interactions and suckling, or the display of aggressive or defensive behaviours, dialysates can be sampled without interference with the behaviour of the animal. Moreover, using this technique, the dynamic changes in the concentration of OT in the extracellular fluid surrounding oxytocinergic neuronal structures can be correlated with ongoing behavioural performance.
The release of OT and AVP within the brain should occur from dendrites or perikarya of magnocellular neurons described within the hypothalamus (Ludwig & Leng, 2006), but also from axonal or collateral projections of parvo- or magnocellular neurons targeting, for example, regions of the limbic brain (Buijs et al., 1983). OT and AVP systems served as a suitable model arrangement not only for the discovery of important molecular and cellular mechanisms of neuropeptide synthesis, precursor processing, and cellular trafficking, but also for the stimuli and neuronal mechanisms of intracerebral neuropeptide release within distinct brain regions (Landgraf & Neumann, 2004, Ludwig & Leng, 2006, Ludwig & Pittman, 2003, Neumann, 2007). In the context of this chapter it is important to mention that a variety of social stimuli trigger the activation of OT neurons and, thus, local OT release within the brain (see below), in some, but not all, instances accompanying peripheral secretion into blood. We hypothesize that, in humans, lack of OT activation in a social context might be associated with social dys-functions as seen in a variety of psychopathologies.

1.2 Autism and social behaviors

Various psychiatric disorders are associated not only with emotional disturbances, but also with dysfunctions and deficiencies in social interactions (Kohn & Asnis, 2003, Neumann et al., 2010). Thus, impaired social functions such as social withdrawal, social phobia, aggression and violence, or impaired social cognition are core symptoms for, for example, major depression, anxiety disorder, posttraumatic stress disorder (PTSD), borderline syndrome, schizophrenia, and autism spectrum disorders (ASD) including the Asperger's Syndrome. Deficits in sociability seen in ASD become apparent during standard nonverbal social interactions, e.g. eye contact or affective expression. Reduced empathy the inability to share blissful and tearful emotions with others, and a lack of social and emotional reciprocity further demonstrate reduced sociability. Moreover, individuals with ASD fail to recognize faces and to integrate facial expressions of emotions caused by impaired social cognitive abilities (Harony & Wagner, 2010).

Due to its profound pro-social actions discovered in animal studies, OT effects on human social behaviour started to be in the focus of interest of psychologists and psychiatrists, among others. Consequently, a potential involvement of the brain OT system in maladaptations during these diseases and the potential of synthetic OT as therapeutic strategy has been suggested (Harony & Wagner, 2010).

This chapter aims at summarizing various activity states of the OT system and provides evidence from both animal and human studies for its role as a key regulator of complex social behaviors.

2. The brain OT system and rodent social behaviour

There are various physiological conditions associated with an altered activity state of the endogenous OT system. These include, for example, the peripartum period in females, sexual activity in both males and females, social interactions between conspecifics, both offensive and friendly, or pairbonding in the female monogamous prairie vole. Under these conditions of a naturally occurring high-activity state, the functional relevance of the endogenous brain OT system can be nicely investigated using various pharmacological tools such as selective receptor antagonists. Moreover, studying the dynamic release patterns of OT within a relevant brain region during such an activity state tells us a lot about the neurobiological activity of the endogenous brain OT system, which is important for the final aim to either use OT as a possible therapeutics or, alternatively, to target the brain OT system therapeutically. Most of our knowledge regarding the intracerebral release of OT during such high-activity states as well as the functional significance for pro-social behaviours arrives from rodent studies.

2.1 Intracerebral OT release in the context of female social behaviours peripartum

Physiological and behavioural changes have been extensively described in the mammalian maternal brain occurring in the peripartum period (Neumann, 2001, Slattery & Neumann, 2008). They continue in lactation as a direct result of close social interactions between mother and offspring, for example during suckling, maternal care and protection. The OT system is highly activated peripartum as it plays a predominant role in female reproduction for speeding up birth and facilitating milk ejection. The general activation of the OT system is reflected by increased OT synthesis in neurons of the hypothalamus, OT secretion into blood, and local OT release and OT receptor binding in several brain regions (for review see Neumann, 2003, Numan & Insel, 2003, Slattery & Neumann, 2008).

Using intracerebral microdialysis or push-pull perfusion techniques, local release of OT has been shown, for example, within the hypothalamic SON and PVN, the septum, hippocampus, and the olfactory bulb during parturition and suckling (Kendrick et al., 1988, Neumann et al., 1993, for review see Neumann, 2009). Such centrally released OT has been shown to be involved in several neuroendocrine and behavioural functions: both during birth and in response to suckling, brain OT regulates the pulsatile release of OT into the blood in a positive feedforward mechanism (Moos & Richard, 1989, Neumann et al., 1994). Further, centrally released OT simultaneously promotes the fine-tuned maintenance of mother-offspring interactions, such as offspring recognition (sheep), maternal care, but also maternal defence behaviour (Bosch, 2011, Bosch & Neumann, 2008, Kendrick et al., 1988, Lubin et al., 2003, Pedersen, 1997). Together these physiological functions are important prerequisites for the survivial of the offspring. In addition, OT at this high activity state

inhibits emotional, e.g. anxiety-related, and neuroendocrine responses to acute stressful stimuli (Neumann et al., 2000).
Moreover, we could monitor increased release of OT within the hypothalamic PVN and the central amygdala during the defence of the pups against a virgin intruder rat in lactating rats. Intracerebral microdialysis allowed simultaneous monitoring of locally restricted release patterns as well as of various maternal aggressive and non-aggressive aspects of social behaviour of the dam (Bosch et al., 2005). Interestingly, central OT release was found to be correlated with the intensity of maternal aggression, indicating a direct link between local neuropeptide release and behavioural performance (Bosch et al., 2005). In order to reveal the functional significance of locally released OT, we blocked local OT receptors using a specific OT receptor antagonist before behavioural testing. Within the PVN and the central amygdala, bilateral application of the OT antagonist reduced aggressive behaviour towards the virgin intruder (Bosch et al., 2005), but pup-directed maternal behaviour was not altered. Thus, in lactation, OT released within the hypothalamus and amygdala and acting at local receptors fulfils an important role in promoting maternal defence behaviour and offspring protection. Within the amygdala, OT directly regulates neuronal activity (Huber et al., 2005, Neumann, 2002). Moreover, both within the PVN and the amygdala OT exerts anxiolytic effects in female and male rats (Bale et al., 2001, Blume et al., 2008, Neumann, 2002). Therefore, it is tempting to suggest a functional link between locally released OT in response to a social challenge, i.e. exposure to the intruder, the reduction in anxiety and the display of maternal aggressive behaviour. Indeed, different levels of maternal behaviour and maternal aggression have been found in rats selectively bred for high versus low trait anxiety (Bosch, 2011, Bosch & Neumann, 2008).
In addition to the link between the high activity of the brain OT system and respective pro-social and defensive behaviours of the mother, the high activity state of the OT system peripartum has further been associated with a state of anxiolysis and general attenuation of physiological stress responses (Heinrichs et al., 2001, Neumann et al., 2000, Torner et al., 2002, Windle et al., 1997) (for review see Neumann, 2009, Slattery & Neumann, 2008).

2.2 Intracerebral OT release during the display of male sexual behaviour

OT is also released within the male brain, in response to social and non-social stimuli (for review see Engelmann et al., 2004, Landgraf & Neumann, 2004, Neumann, 2007). As sexual interaction is the most intense social interaction found in males and increased OT secretion into blood has been found during orgasm (Carmichael et al., 1987, Stoneham et al., 1985) we and others have studied the activation of the brain OT system during sexual activity and its role in the regulation of sexual behaviour.
In response to sexual stimuli, there is increased Fos-expression within the hypothalamus reflecting increased neuronal activity in OT neurons of the PVN (Witt & Insel, 1994). This is likely to reflect both increased secretion of OT into blood, but also intracerebral neuropeptide release. Indeed, we could recently demonstrate using intracerebral microdialysis in freely moving male rats that successful mating triggers local release of OT within the PVN (Waldherr & Neumann, 2007). Interestingly, OT release already started to rise during the presence of the primed female behind a perforated wall, which allowed olfactory and visual, but not physical, contact or mating. As males clearly displayed signs of behavioural arousal under these conditions, OT activation may already be induced by the presence of a receptive female even without mating (Waldherr & Neumann, 2007). Within

the hypothalamic PVN OT has been shown to play an important role in the regulation of male sexual behaviour (Argiolas & Gessa, 1991), but also of anxiety (Blume et al., 2008). Specifically, during mating, such locally released OT could be shown to exert an anxiolytic effect in male rats (Waldherr & Neumann, 2007).

We therefore conclude that the release of OT within the brain during sexual activity has far reaching behavioural consequences and beneficial effects for the male rat, i.e. reducing the level of anxiety and stress responses for several hours. In humans, there is anecdotal and experimental evidence of a link between sexual activity, and sedation, increased relaxation and calmness in the post-coital period (Brody, 2006, Krüger et al., 2002). Our data show that these effects are mediated, at least in part, by an activated brain OT system. As OT was shown to exert reinforcing and rewarding actions (Liberzon et al., 1997), the possibility further exists that enforced and reinforced trust to the sexual partner also involves brain OT (Kosfeld et al., 2005), although this is still highly speculative.

In summary, the central release of OT during close social interaction, such as suckling the offspring in lactating mammals, or sexual activity in males, is likely to be involved not only in the regulation of the associated particular social behaviours, but also in the beneficial effects of these pro-social interactions. Positive effects such as anxiolysis, attenuated stress responses, increased calmness and sedation (Carter et al., 2001, Heinrichs et al., 2003, Neumann, 2002, Waldherr & Neumann, 2007) are likely to be rewarding and to further promote social interactions (Neumann 2009).

2.3 Central OT and social cognition

Social interactions, especially long-lasting social bonds, require different forms of social memory and social recognition. As shown in both male and female rodents social recognition largely depends on an intact brain OT system (for review see Bielsky & Young, 2004)

Centrally applied OT facilitates social memory in a dose-dependent manner as shown in male rats. In contrast, infusion of the OT receptor antagonist blocked this effect but was not successful to impair their social memory per se (Benelli et al., 1995). Literatur on male mice seems more straight forward. In male mice lacking the OT gene (Ferguson et al., 2000) or in mice with deficient OT release (Jin et al., 2007) impaired social cognition and social memory skills were found. Importantly, OT bilaterally infused into the amygdala was able to restore the cognitive deficits seen in OT knockout mice, whereas OT receptor antagonist infusions impaired social memory in male wildtype mice (Ferguson et al., 2001). Other regions responsive to synthetic or endogenous OT in the context of social recognition are the lateral septum (Popik et al., 1992), the olfactory bulb (Dluzen et al., 1998, Larrazolo-Lopez et al., 2008), the medial preoptic area (Popik & van Ree, 1991), and the ventral hippocampus (van Wimersma Greidanus & Maigret, 1996).

Also, in female rats brain OT seems to be important for social discrimination of two juvenile rats (Engelmann et al., 1998). The medial amygdala (Choleris et al., 2007) and the olfactory bulb (Larrazolo-Lopez et al., 2008) could be identified as sites of action using microdialysis and local pharmacological blockade or downregulation of OT receptors.

OT seems to be an important factor in female social cognition in a more complex context thus promoting long-lasting bonds such as mother-infant bonding or pair bonding. In ewes, lamb recognition and bonding could clearly be related to the release of OT, for example within the olfactory bulb, during birth and in response to suckling (Kendrick et al., 1988, Lévy et al., 1995). Moreover, in the monogamous prairie vole, social recognition of the mate is a prerequisite for monogamous behaviour and the ability to form a selective pair-bond. Similar to the offspring bonding in ewes, OT plays a critical role in pair-bonding (see below) (Insel & Hulihan, 1995, Young & Wang, 2004). Thus, parturition- and mating-induced stimulation of OT release within distinct brain regions seems to be a promoting factor for social cognition and a requirement for the formation of lasting social bonds.

In female OT knockout mice, the essential role of OT in social memory has also been demonstrated in the context of the Bruce effect. The Bruce effect refers to the ability of a female mouse to discriminate between her mate and a novel mate. Contact with a novel male consequently leads to an interruption of pregnancy. OT knockout females failed to remain pregnant, if re-exposed to either their mate or a novel male. Only females that were allowed to remain with their mate maintained pregnancy (Wersinger et al., 2008). This inability to distinguish between the mate and a novel male in females with deficits in the OT systems further demonstrates the importance of OT in long-term social memory as well as short-term social recognition especially in females.

2.4 Central OT and female pair-bonding

There is a large body of literature concerning the facilitating effects of the neuropeptides OT and AVP in pair-bonding of monogamous voles (for review see Young & Wang, 2004). Whereas AVP is well established to be involved in pair-bonding and partner preference in male prarie voles, OT seems to play a major role in female pair-bonding (Cho et al., 1999, Cushing & Carter, 2000). This was shown by the facilitating effect of centrally applied OT on the development of partner preference even without prior mating (Williams et al., 1994), and receptor binding studies demonstrated an increased OT receptor binding in the nucleus accumbens and caudate putamen of female monogamous prairie voles compared with non-monogamous vole species (Insel & Shapiro, 1992). The involvement of OT within several brain regions in female pair-bonding was further demonstrated by blockade of mating-induced female pair-bonding following infusion of an OT receptor antagonist into the prefrontal cortex or the nucleus accumbens, but not in the caudate putamen (Young et al., 2001). Recently, we succeeded in demonstrating OT release within the nucleus accumbens during mating in female prarie voles (Ross et al., 2009). Such locally released OT was shown to originate most likely within the hypothalamic SON and PVN. Consequently, activation of magnocellular OT neurons during mating and OT secretion into blood may, simultaneously, result in locally restricted release of OT from neuronal collaterals in dependence on gender and species.

In this context, it is important to mention that the nucleus accumbens is part of the mesolimbic dopamine reward system (Wise, 2002). As mating is a rewarding stimulus, and was shown to induce pair-bonding in female voles, endogenous OT release triggered by sexual stimuli may potentially mediate its facilitating effects on partner formation and pair-bonding via these circuitries (Neumann, 2009, Wang & Aragona, 2004).

2.5 Central OT release during social stress

The OT system is highly responsive to all kind of stressors, both being non-social and social in nature (Engelmann et al., 2004, Landgraf & Neumann, 2004, Neumann, 2007) For example, exposure to forced swimming or to a larger and aggressive conspecifics during the social defeat are both stimuli for OT release in selected brain target regions (Ebner et al., 2000, Engelmann et al., 1999, Wigger & Neumann, 2002, Wotjak et al., 1998).

Exposure of male rats to social defeat selectively stimulates OT release within the hypothalamic SON (Engelmann et al., 1999) and the septal area (Ebner et al., 2000). In contrast, OT secretion into blood remains unchanged in response to this social stressor indicating independent release patterns into blood and within the brain.

Also, in virgin female rats, social defeat is a stimulus for the brain OT system and triggers OT release within the hypothalamic PVN, but not the amygdala or the lateral septum (Bosch et al., 2004). In females, social defeat can be achieved by exposure to a lactating dam in the presence of her litter (Neumann et al., 2001). During lactation, dams are highly aggressive protecting their offspring. Maternal defence behaviour is also stressful for the lactating dam resulting in elevated stress responses and an increased release of OT within the PVN and amygdala, especially in dams displaying highly aggressive behaviour (Bosch et al., 2005, Neumann et al., 2001). Interestingly, the amount of OT locally released within these regions was found to be directly correlated with the total amount of aggression displayed by the dam during the maternal defence test (Bosch et al., 2005).

Thus, both in males and females, aversive social interactions as seen during social defeat and maternal defence, respectively, are strong stimuli for the brain OT system. In lactation, brain OT strongly promotes maternal aggression (Bosch et al., 2005), but we have to keep in mind that maternal aggression is a defensive strategy important for the protection of the offspring. The neurobiological mechanisms of maternal defensive aggression and intermale aggression, for example, are almost completely independently regulated. However, to which extend central OT regulates aggressive behaviour in males is rather unclear (Ebner et al., 2000); so far, we could not reveal a clear effect of OT on aggression in male rats (unpublished observation).

3. Behavioural effects of OT in humans

In humans, intranasal or intravenous application of OT was reported to improve a broad variety of complex social behaviours (for reviews see Heinrichs et al., 2009, MacDonald & MacDonald, 2010, Meyer-Lindenberg, 2008).

Specifically, intranasal OT increased trust in healthy men (Kosfeld et al., 2005) and even prevented betrayal-triggered decrease in trust (Baumgartner et al., 2008). In this context, OT increased ratings for trustworthiness and attractiveness of unfamiliar faces (Theodoridou et al., 2009). Moreover, OT-treated subjects were significantly more generous than placebo-treated men during a generosity game (Zak et al., 2007). In general, OT seems to improve the interpretation of social cues (Domes et al., 2007b, Kosfeld et al., 2005), especially the recognition of fear (Fischer-Shofty el al., 2010). . OT also facilitates the recognition of faces (Rimmele et al., 2009) most effectively when they express positive emotion (Guastella et al., 2008b, Savaskan et al., 2008). In the context of ASD associated with avoidance of eye contact it is important to mention that OT promotes a gaze-shift towards the eye region of presented faces (Guastella et al., 2008a) also independent of their valence as this normally occurs during presentation of fearful faces (Gamer & Buchel, 2009). In the context of OT promoting social bondings it is of interest to mention initial studies demonstrating that OT enhanced attachment security (Buchheim et al., 2009)

On a more neurophysiological level, human functional imaging studies (fMRI) indicated that OT reduces amygdala responses to threatening, non-social scenes and to angry and fearful faces (Kirsch et al., 2005). More specifically, it could be shown that OT promotes the activity in amygdala regions involved in the processing of positive social stimuli (Gamer et al., 2010), an effect that was shown to generalize to facial expressions, irrespective of their valence (Domes et al., 2007a) .

These studies show that OT promontes trust and generosity, improves "mind-reading" and facilitates the ability to recognize faces and facial expressions of social cues. Thus, there is substantial evidence for complex behavioural effects of OT, as a result of acute intranasal application, on social competence. These prosocial effects give hope for OT as a potential treatment option under condition of social dysfunctions as seen in ASD.

4. Autism spectrum disorders - involvement of brain OT

Various animal and human studies strongly support the hypothesis of an involvement of OT in complex social interactions, both under healthy and pathological conditions associated with social dysfunctions. Therefore the following paragraph will summarize human data concerning OT actions within the amygdala in the context of ASD.

4.1 Amygdala and autism

The amygdala as part of the limbic brain has been implicated in the neurobiology of autism as seen from morphometric data (Dalton et al., 2007, Dalton et al., 2005, Dziobek et al., 2006). The subnuclei of the amygdala are key areas regulating arousal and vigilance to emotionally relevant stimuli (Davis & Whalen, 2001, Fitzgerald et al., 2006, Hsu et al., 2005, Yang et al., 2002), and spontaneous social cognition by processing the rapid evaluation of social stimuli. Thus, it is very likely that the amygdala is significantly involved in the processing of both social as well as non-social, e.g. stressful and emotional stimuli, both being of high relevance for the individual. However, the amygdala is also likely to be involved in complex emotional and cognitive processes such as empathy and affiliation.

Various neurotransmitters and neuromodulators are involved in the neuronal processing within the amygdala, among them the neuropeptides OT and AVP. OT receptors were localized within the central and medial amygdala (Lukas et al., 2010, Tribollet et al., 1988), OT is locally released (Bosch et al., 2005, Ebner et al., 2005) and exerts local neuronal effects regulating the electrophysiological activity of central amygdala neurons (Huber et al., 2005). In this way, OT might be involved in the complex regulation of social behaviour and emotional responses to various social and stressful cues both under healthy and pathological conditions also in humans.

4.2 OT system activity alterations and therapeutical implications in autism

Indeed, dysfunctions of the brain OT system have been intensively discussed to contribute to the development of social deficits in autism (Carter, 2007, Hammock & Young, 2006, Harony & Wagner, 2010). For example, plasma OT concentrations, although reflecting central OT system activity only to a certain degree, were found to be attenuated in individuals with ASD (Green et al., 2001, Lane, 2009). More and more studies indicate the potential use of OT applied intravenously or intranasally in the treatment of ASD. Thus, OT reduced repetitive behaviors in patients with Asperger or autism (Hollander et al., 2002) and promoted prosocial behaviors in high-functioning autism (Andari et al., 2010). Moreover, OT improved affective speech comprehension in adults (Hollander et al., 2007) and emotion recognition in youths with ASD (Guastella et al., 2010). Furthermore, OT was used as an adjunct to exposure therapy (Guastella et al., 2009) and was shown to attenuate amygdala reactivity to fear (Labuschagne et al., 2010) in social anxiety, another core symptom of ASD. Association studies on several ethnic groups linked polymorphisms in the OT receptor gene with ASD (Jacob et al., 2007, Lerer et al., 2007, Wu et al., 2005). Furthermore another study was able to detect epigenetic modifications within the OT receptor promotor region of ASD patients indicating altered levels of OT receptor expression (Gregory et al., 2009).

5. Conclusion

Our chapter summarizes our knowledge about different activity states of the brain OT system in the context of varying social interactions. Increased OT system activity triggered by social interactions is linked to high levels of OT release within distinct brain regions, which, in turn, is involved in the promotion of these social interactions, as seen during maternal and sexual behavior, pair-bonding or social memory functions. Moreover, high OT activity and release of OT, for example within regions of the reward circuitry, further promote social interactions as they are experienced as being rewarding. Indeed, a high activity state of the OT system is associated with calmness, anxiolysis and attenuated stress responsiveness, as seen during lactation and after sexual activity, further contributing to feeling of wellness. In contrast, although experimental evidence is extremely limited, the present data give support to the hypothesis that social dysfunctions are associated with lack of brain OT system activation, either under basal conditions or in reposnse to social stimuli, or both. Thus, appropriate brain OT system activation seems a prerequisite for healthy and normal sociability. We also discuss that translation of these findings mainly from rodents to human studies is possible. Effects of intranasal OT could be directly related to the promotion of complex human social behaviors, also under conditions of disease like autism. Consequently, targeting the brain OT system or application of synthetic OT to compensate for deficits in endogenous OT system activity in combimnation with psychotherapy, appears to be a promising treatment option promoting pro-social behaviors in humans especially under pathological conditions of impaired sociability.

6. References

Acher, R., Chauvet, J. & Chauvet, M. T., (1972). Phylogeny of the neurohypophysial hormones. Two new active peptides isolated from a cartilaginous fish, squalus acanthias. *Eur J Biochem.* **29:** 12-9.

Andari, E., Duhamel, J. R., Zalla, T., Herbrecht, E., Leboyer, M. & Sirigu, A., (2010). Promoting social behavior with oxytocin in high-functioning autism spectrum disorders. *Proc Natl Acad Sci U S A.* 107: 4389-94.

Argiolas, A. & Gessa, G. L., (1991). Central functions of oxytocin. *Neuroscience & Biobehavioral Reviews.* 15: 217-231.

Bale, T. L., Davis, A. M., Auger, A. P., Dorsa, D. M. & McCarthy, M. M., (2001). Cns region-specific oxytocin receptor expression: Importance in regulation of anxiety and sex behavior. *J. Neurosci.* **21:** 2546-2552.

Baumgartner, T., Heinrichs, M., Vonlanthen, A., Fischbacher, U. & Fehr, E., (2008). Oxytocin shapes the neural circuitry of trust and trust adaptation in humans. *Neuron*. 58: 639-50.

Benelli, A., Bertolini, A., Poggioli, R., Menozzi, B., Basaglia, R. & Arletti, R., (1995). Polymodal dose-response curve for oxytocin in the social recognition test. *Neuropeptides*. 28: 251-5.

Bielsky, I. F. & Young, L. J., (2004). Oxytocin, vasopressin, and social recognition in mammals. *Peptides*. 25: 1565-74.

Blume, A., Bosch, O. J., Miklos, S., Torner, L., Wales, L., Waldherr, M. & Neumann, I. D., (2008). Oxytocin reduces anxiety via erk1/2 activation: Local effect within the rat hypothalamic paraventricular nucleus. *European Journal of Neuroscience*. 27: 1947-1956.

Bosch, O. J., (2011). Maternal nurturing is dependent on her innate anxiety: The behavioral roles of brain oxytocin and vasopressin. *Hormones and Behavior*. 59: 202-212.

Bosch, O. J., Krömer, S. A., Brunton, P. J. & Neumann, I. D., (2004). Release of oxytocin in the hypothalamic paraventricular nucleus, but not central amygdala or lateral septum in lactating residents and virgin intruders during maternal defence. *Neuroscience*. 124: 439-448.

Bosch, O. J., Meddle, S. L., Beiderbeck, D. I., Douglas, A. J. & Neumann, I. D., (2005). Brain oxytocin correlates with maternal aggression: Link to anxiety. *J Neurosci*. 25: 6807-15.

Bosch, O. J. & Neumann, I. D., (2008). Brain vasopressin is an important regulator of maternal behavior independent of dams' trait anxiety. *Proceedings of the National Academy of Sciences*. 105: 17139-17144.

Brody, S., (2006). Blood pressure reactivity to stress is better for people who recently had penile-vaginal intercourse than for people who had other or no sexual activity. *Biological Psychology*. 71: 214-222.

Buchheim, A., Heinrichs, M., George, C., Pokorny, D., Koops, E., Henningsen, P., O'Connor, M. F. & Gundel, H., (2009). Oxytocin enhances the experience of attachment security. *Psychoneuroendocrinology*. 34: 1417-22.

Buijs, R. M., De Vries, G. J., Van Leeuwen, F. W. & Swaab, D. F., (1983). Vasopressin and oxytocin: Distribution and putative functions in the brain. *Prog Brain Res*. 60: 115-22.

Carmichael, M. S., Humbert, R., Dixen, J., Palmisano, G., Greenleaf, W. & Davidson, J. M., (1987). Plasma oxytocin increases in the human sexual response. *J Clin Endocrinol Metab*. 64: 27-31.

Carter, C. S., (2007). Sex differences in oxytocin and vasopressin: Implications for autism spectrum disorders? *Behavioural Brain Research*. 176: 170-186.

Carter, C. S., Altemus, M. & Chrousos, G. P., (2001). Neuroendocrine and emotional changes in the post-partum period. *Prog Brain Res*. 133: 241-9.

Cho, M. M., DeVries, A. C., Williams, J. R. & Carter, C. S., (1999). The effects of oxytocin and vasopressin on partner preferences in male and female prairie voles (microtus ochrogaster). *Behav Neurosci*. 113: 1071-9.

Choleris, E., Little, S. R., Mong, J. A., Puram, S. V., Langer, R. & Pfaff, D. W., (2007). Microparticle-based delivery of oxytocin receptor antisense DNA in the medial amygdala blocks social recognition in female mice. *Proceedings of the National Academy of Sciences*. 104: 4670-4675.

Cushing, B. S. & Carter, C. S., (2000). Peripheral pulses of oxytocin increase partner preferences in female, but not male, prairie voles. *Horm Behav*. 37: 49-56.

Dalton, K. M., Nacewicz, B. M., Alexander, A. L. & Davidson, R. J., (2007). Gaze-fixation, brain activation, and amygdala volume in unaffected siblings of individuals with autism. *Biol Psychiatry*. 61: 512-20.

Dalton, K. M., Nacewicz, B. M., Johnstone, T., Schaefer, H. S., Gernsbacher, M. A., Goldsmith, H. H., Alexander, A. L. & Davidson, R. J., (2005). Gaze fixation and the neural circuitry of face processing in autism. *Nat Neurosci*. 8: 519-26.

Davis, M. & Whalen, P. J., (2001). The amygdala: Vigilance and emotion. *Mol Psychiatry*. 6: 13-34.

De Wied, D., (1965). The influence of the posterior and intermediate lobe of the pituitary and pituitary peptides on the maintenance of a conditioned avoidance reponse in rats. *Int. J. Neuropharmacol*. 4: 157-67.

Dluzen, D. E., Muraoka, S., Engelmann, M. & Landgraf, R., (1998). The effects of infusion of arginine vasopressin, oxytocin, or their antagonists into the olfactory bulb upon social recognition responses in male rats. *Peptides*. 19: 999-1005.

Domes, G., Heinrichs, M., Gläscher, J., Büchel, C., Braus, D. F. & Herpertz, S. C., (2007a). Oxytocin attenuates amygdala responses to emotional faces regardless of valence. *Biological Psychiatry*. 62: 1187-1190.

Domes, G., Heinrichs, M., Michel, A., Berger, C. & Herpertz, S. C., (2007b). Oxytocin improves "Mind-reading" In humans. *Biol Psychiatry*. 61: 731-3.

Donaldson, Z.R. & Young, L.J., (2008) Oxytocin, Vasopressin, and the Neurogenetics of Sociality. *Science*. 322 900-904

Du Vigneaud, V., Ressler, C. & Trippett, S., (1953). The sequence of amino acids in oxytocin, with a proposal for the structure of oxytocin. *J Biol Chem*. 205: 949-57.

Dziobek, I., Fleck, S., Rogers, K., Wolf, O. T. & Convit, A., (2006). The [`]amygdala theory of autism' revisited: Linking structure to behavior. *Neuropsychologia*. 44: 1891-1899.

Ebner, K., Bosch, O. J., Kromer, S. A., Singewald, N. & Neumann, I. D., (2005). Release of oxytocin in the rat central amygdala modulates stress-coping behavior and the release of excitatory amino acids. *Neuropsychopharmacology*. 30: 223-30.

Ebner, K., Wotjak, C. T., Landgraf, R. & Engelmann, M., (2000). A single social defeat experience selectively stimulates the release of oxytocin, but not vasopressin, within the septal brain area of male rats. *Brain Res*. 872: 87-92.

Engelmann, M., Ebner, K., Landgraf, R., Holsboer, F. & Wotjak, C. T., (1999). Emotional stress triggers intrahypothalamic but not peripheral release of oxytocin in male rats. *J Neuroendocrinol*. 11: 867-72.

Engelmann, M., Ebner, K., Wotjak, C. T. & Landgraf, R., (1998). Endogenous oxytocin is involved in short-term olfactory memory in female rats. *Behav Brain Res*. 90: 89-94.

Engelmann, M., Landgraf, R. & Wotjak, C. T., (2004). The hypothalamic-neurohypophysial system regulates the hypothalamic-pituitary-adrenal axis under stress: An old concept revisited. *Front Neuroendocrinol*. 25: 132-49.

Ferguson, J. N., Aldag, J. M., Insel, T. R. & Young, L. J., (2001). Oxytocin in the medial amygdala is essential for social recognition in the mouse. *J. Neurosci*. 21: 8278-8285.

Ferguson, J. N., Young, L. J., Hearn, E. F., Matzuk, M. M., Insel, T. R. & Winslow, J. T., (2000). Social amnesia in mice lacking the oxytocin gene. *Nat Genet*. 25: 284-8.

Fischer-Shofty, M., Shamay-Tsoory, S. G., Harari, H. & Levkovitz, Y., (2010) The effect of intranasal administration of oxytocin on fear recognition. *Neuropsychologia*. 48: 179-84.

Fitzgerald, D. A., Angstadt, M., Jelsone, L. M., Nathan, P. J. & Phan, K. L., (2006). Beyond threat: Amygdala reactivity across multiple expressions of facial affect. *Neuroimage*. 30: 1441-8.

Gamer, M., Zurowski, B. & Buechel, C., (2010). Different amygdala subregions mediate valence-related and attentional effects of oxytocin in humans. *Proceedings of the National Academy of Sciences*. 107: 9400-9405.

Green, L., Fein, D., Modahl, C., Feinstein, C., Waterhouse, L. & Morris, M., (2001). Oxytocin and autistic disorder: Alterations in peptide forms. *Biol Psychiatry*. 50: 609-13.

Gregory, S., Connelly, J., Towers, A., Johnson, J., Biscocho, D., Markunas, C., Lintas, C., Abramson, R., Wright, H., Ellis, P., Langford, C., Worley, G., Delong, G. R., Murphy, S., Cuccaro, M., Persico, A. & Pericak-Vance, M., (2009). Genomic and epigenetic evidence for oxytocin receptor deficiency in autism. *BMC Medicine*. 7: 62.

Guastella, A. J., Einfeld, S. L., Gray, K. M., Rinehart, N. J., Tonge, B. J., Lambert, T. J. & Hickie, I. B., (2010). Intranasal oxytocin improves emotion recognition for youth with autism spectrum disorders. *Biological Psychiatry*. 67: 692-694.

Guastella, A. J., Howard, A. L., Dadds, M. R., Mitchell, P. & Carson, D. S., (2009). A randomized controlled trial of intranasal oxytocin as an adjunct to exposure therapy for social anxiety disorder. *Psychoneuroendocrinology*. 34: 917-923.

Guastella, A. J., Mitchell, P. B. & Dadds, M. R., (2008a). Oxytocin increases gaze to the eye region of human faces. *Biol Psychiatry*. 63: 3-5.

Guastella, A. J., Mitchell, P. B. & Mathews, F., (2008b). Oxytocin enhances the encoding of positive social memories in humans. *Biol Psychiatry*.

Hammock, E. A. & Young, L. J., (2006). Oxytocin, vasopressin and pair bonding: Implications for autism. *Philos Trans R Soc Lond B Biol Sci*. 361: 2187-98.

Harony, H. & Wagner, S., (2010). The contribution of oxytocin and vasopressin to mammalian social behavior: Potential role in autism spectrum disorder. *Neurosignals*. 18: 82-97.

Heinrichs, M., Baumgartner, T., Kirschbaum, C. & Ehlert, U., (2003). Social support and oxytocin interact to suppress cortisol and subjective responses to psychosocial stress. *Biol Psychiatry*. 54: 1389-98.

Heinrichs, M., Meinlschmidt, G., Neumann, I., Wagner, S., Kirschbaum, C., Ehlert, U. & Hellhammer, D. H., (2001). Effects of suckling on hypothalamic-pituitary-adrenal axis responses to psychosocial stress in postpartum lactating women. *J Clin Endocrinol Metab*. 86: 4798-4804.

Heinrichs, M., von Dawans, B. & Domes, G., (2009). Oxytocin, vasopressin, and human social behavior. *Frontiers in Neuroendocrinology*. 30: 548-557.

Hollander, E., Bartz, J., Chaplin, W., Phillips, A., Sumner, J., Soorya, L., Anagnostou, E. & Wasserman, S., (2007). Oxytocin increases retention of social cognition in autism. *Biological Psychiatry*. 61: 498-503.

Hollander, E., Novotny, S., Hanratty, M., Yaffe, R., DeCaria, C. M., Aronowitz, B. R. & Mosovich, S., (2002). Oxytocin infusion reduces repetitive behaviors in adults with autistic and asperger's disorders. *Neuropsychopharmacology*. 28: 193-198.

Hoyle, C. H., (1999). Neuropeptide families and their receptors: Evolutionary perspectives. *Brain Res*. 848: 1-25.

Hsu, M., Bhatt, M., Adolphs, R., Tranel, D. & Camerer, C. F., (2005). Neural systems responding to degrees of uncertainty in human decision-making. *Science*. 310: 1680-3.

Huber, D., Veinante, P. & Stoop, R., (2005). Vasopressin and oxytocin excite distinct neuronal populations in the central amygdala. *Science*. 308: 245-248.

Insel, T. R. & Hulihan, T. J., (1995). A gender-specific mechanism for pair bonding: Oxytocin and partner preference formation in monogamous voles. *Behav Neurosci*. 109: 782-9.

Insel, T. R. & Shapiro, L. E., (1992). Oxytocin receptor distribution reflects social organization in monogamous and polygamous voles. *Proc Natl Acad Sci U S A*. 89: 5981-5.

Jacob, S., Brune, C. W., Carter, C. S., Leventhal, B. L., Lord, C. & Cook Jr, E. H., (2007). Association of the oxytocin receptor gene (oxtr) in caucasian children and adolescents with autism. *Neuroscience Letters*. 417: 6-9.

Jin, D., Liu, H.-X., Hirai, H., Torashima, T., Nagai, T., Lopatina, O., Shnayder, N. A., Yamada, K., Noda, M., Seike, T., Fujita, K., Takasawa, S., Yokoyama, S., Koizumi, K., Shiraishi, Y., Tanaka, S., Hashii, M., Yoshihara, T., Higashida, K., Islam, M. S., Yamada, N., Hayashi, K., Noguchi, N., Kato, I., Okamoto, H., Matsushima, A., Salmina, A., Munesue, T., Shimizu, N., Mochida, S., Asano, M. & Higashida, H., (2007). Cd38 is critical for social behaviour by regulating oxytocin secretion. *Nature*. 446: 41-45.

Kendrick, K. M., Keverne, E. B., Chapman, C. & Baldwin, B. A., (1988). Intracranial dialysis measurement of oxytocin, monoamine and uric acid release from the olfactory bulb and substantia nigra of sheep during parturition, suckling, separation from lambs and eating. *Brain Res*. 439: 1-10.

Kirsch, P., Esslinger, C., Chen, Q., Mier, D., Lis, S., Siddhanti, S., Gruppe, H., Mattay, V. S., Gallhofer, B. & Meyer-Lindenberg, A., (2005). Oxytocin modulates neural circuitry for social cognition and fear in humans. *J. Neurosci*. 25: 11489-11493.

Kohn, S.R., & Asnis, G.M., (2003). "Aggression in psychatric disorders" in *Neurobiology of Agression: Understanding and preventing Violence*, ed. M. Mattson (Totowa, NJ: Humana Press Inc.)

Kosfeld, M., Heinrichs, M., Zak, P. J., Fischbacher, U. & Fehr, E., (2005). Oxytocin increases trust in humans. *Nature*. 435: 673-6.

Krüger, T. H. C., Haake, P., Hartmann, U., Schedlowski, M. & Exton, M. S., (2002). Orgasm-induced prolactin secretion: Feedback control of sexual drive? *Neuroscience & Biobehavioral Reviews*. 26: 31-44.

Labuschagne, I., Phan, K. L., Wood, A., Angstadt, M., Chua, P., Heinrichs, M., Stout, J. C. & Nathan, P. J., (2010). Oxytocin attenuates amygdala reactivity to fear in generalized social anxiety disorder. *Neuropsychopharmacology*. 35: 2403-2413.

Landgraf, R. & Neumann, I. D., (2004). Vasopressin and oxytocin release within the brain: A dynamic concept of multiple and variable modes of neuropeptide communication. *Front Neuroendocrinol*. 25: 150-76.

Larrazolo-Lopez, A., Kendrick, K. M., Aburto-Arciniega, M., Arriaga-Avila, V., Morimoto, S., Frias, M. & Guevara-Guzman, R., (2008). Vaginocervical stimulation enhances social recognition memory in rats via oxytocin release in the olfactory bulb. *Neuroscience*. 152: 585-93.

Lerer, E., Levi, S., Salomon, S., Darvasi, A., Yirmiya, N. & Ebstein, R. P., (2007). Association between the oxytocin receptor (oxtr) gene and autism: Relationship to vineland adaptive behavior scales and cognition. *Mol Psychiatry*. 13: 980-988.

Lévy, F., Kendrick, K. M., Goode, J. A., Guevara-Guzman, R. & Keverne, E. B., (1995). Oxytocin and vasopressin release in the olfactory bulb of parturient ewes: Changes with maternal experience and effects on acetylcholine, [gamma]-aminobutyric acid, glutamate and noradrenaline release. *Brain Research*. 669: 197-206.

Liberzon, I., Trujillo, K. A., Akil, H. & Young, E. A., (1997). Motivational properties of oxytocin in the conditioned place preference paradigm. *Neuropsychopharmacology*. 17: 353-9.

Lubin, D. A., Elliott, J. C., Black, M. C. & Johns, J. M., (2003). An oxytocin antagonist infused into the central nucleus of the amygdala increases maternal aggressive behavior. *Behav Neurosci*. 117: 195-201.

Ludwig, M. & Leng, G., (2006). Dendritic peptide release and peptide-dependent behaviours. *Nat Rev Neurosci*. 7: 126-36.

Ludwig, M. & Pittman, Q. J., (2003). Talking back: Dendritic neurotransmitter release. *Trends Neurosci*. 26: 255-61.

Lukas, M., Bredewold, R., Neumann, I. D. & Veenema, A. H., (2010). Maternal separation interferes with developmental changes in brain vasopressin and oxytocin receptor binding in male rats. *Neuropharmacology*. 58: 78-87.

MacDonald, K. & MacDonald, T. M., (2010). The peptide that binds: A systematic review of oxytocin and its prosocial effects in humans. *Harvard Review of Psychiatry*. 18: 1-21.

Meyer-Lindenberg, A., (2008). Impact of prosocial neuropeptides on human brain function. *Prog Brain Res*. Volume 170: 463-470.

Moos, F. & Richard, P., (1989). Paraventricular and supraoptic bursting oxytocin cells in rat are locally regulated by oxytocin and functionally related. *The Journal of Physiology*. 408: 1-18.

Neumann, I., Koehler, E., Landgraf, R. & Summy-Long, J., (1994). An oxytocin receptor antagonist infused into the supraoptic nucleus attenuates intranuclear and peripheral release of oxytocin during suckling in conscious rats. *Endocrinology*. 134: 141-148.

Neumann, I., Russell, J. A. & Landgraf, R., (1993). Oxytocin and vasopressin release within the supraoptic and paraventricular nuclei of pregnant, parturient and lactating rats: A microdialysis study. *Neuroscience*. 53: 65-75.

Neumann, I. D., (2001). Alterations in behavioral and neuroendocrine stress coping strategies in pregnant, parturient and lactating rats. *Prog Brain Res*. 133: 143-52.

Neumann, I. D., (2002). Involvement of the brain oxytocin system in stress coping: Interactions with the hypothalamo-pituitary-adrenal axis. *Prog Brain Res*. 139: 147-62.

Neumann, I. D., (2003). Brain mechanisms underlying emotional alterations in the peripartum period in rats. *Depression and Anxiety*. 17: 111-121.

Neumann, I. D., (2007). Stimuli and consequences of dendritic release of oxytocin within the brain. *Biochem Soc Trans*. 35: 1252-7.

Neumann, I. D., (2009). The advantage of social living: Brain neuropeptides mediate the beneficial consequences of sex and motherhood. *Front Neuroendocrinol*. 30: 483-96.

Neumann, I. D., Torner, L. & Wigger, A., (2000). Brain oxytocin: Differential inhibition of neuroendocrine stress responses and anxiety-related behaviour in virgin, pregnant and lactating rats. *Neuroscience*. 95: 567-75.

Neumann, I. D., Toschi, N., Ohl, F., Torner, L. & Kromer, S. A., (2001). Maternal defence as an emotional stressor in female rats: Correlation of neuroendocrine and behavioural parameters and involvement of brain oxytocin. *Eur J Neurosci*. 13: 1016-24.

Neumann, I. D., Veenema, A. H. & Beiderbeck, D. I., (2010). Aggression and anxiety: Social context and neurobiological links. *Frontiers in Behavioral Neuroscience*. 5: 12.

Pedersen, C. A., (1997). Oxytocin control of maternal behavior regulation by sex steroids and offspring stimulia. *Annals of the New York Academy of Sciences*. 807: 126-145.

Pedersen, C. A. & Prange, A. J., Jr., (1979). Induction of maternal behavior in virgin rats after intracerebroventricular administration of oxytocin. *Proc Natl Acad Sci U S A*. 76: 6661-5.

Popik, P. & van Ree, J. M., (1991). Oxytocin but not vasopressin facilitates social recognition following injection into the medial preoptic area of the rat brain. *Eur Neuropsychopharmacol*. 1: 555-60.

Popik, P., Vos, P. E. & Van Ree, J. M., (1992). Neurohypophyseal hormone receptors in the septum are implicated in social recognition in the rat. *Behav Pharmacol*. 3: 351-358.

Rimmele, U., Hediger, K., Heinrichs, M. & Klaver, P., (2009). Oxytocin makes a face in memory familiar. *J. Neurosci*. 29: 38-42.

Ross, H. E., Cole, C. D., Smith, Y., Neumann, I. D., Landgraf, R., Murphy, A. Z. & Young, L. J., (2009). Characterization of the oxytocin system regulating affiliative behavior in female prairie voles. *Neuroscience*. 162: 892-903.

Savaskan, E., Ehrhardt, R., Schulz, A., Walter, M. & Schächinger, H., (2008). Post-learning intranasal oxytocin modulates human memory for facial identity. *Psychoneuroendocrinology*. 33: 368-374.

Slattery, D. A. & Neumann, I. D., (2008). No stress please! Mechanisms of stress hyporesponsiveness of the maternal brain. *J Physiol*. 586: 377-85.

Slattery, D. A. & Neumann, I. D., (2010). Chronic icv oxytocin attenuates the pathological high anxiety state of selectively bred wistar rats. *Neuropharmacology*. 58: 56-61.

Stoneham, M. D., Everitt, B. J., Hansen, S., Lightman, S. L. & Todd, K., (1985). Oxytocin and sexual behaviour in the male rat and rabbit. *J Endocrinol*. 107: 97-106.

Theodoridou, A., Rowe, A. C., Penton-Voak, I. S. & Rogers, P. J., (2009). Oxytocin and social perception: Oxytocin increases perceived facial trustworthiness and attractiveness. *Horm Behav*. 56: 128-32.

Torner, L., Toschi, N., Nava, G., Clapp, C. & Neumann, I. D., (2002). Increased hypothalamic expression of prolactin in lactation: Involvement in behavioural and neuroendocrine stress responses. *Eur J Neurosci*. 15: 1381-9.

Tribollet, E., Barberis, C., Jard, S., Dubois-Dauphin, M. & Dreifuss, J. J., (1988). Localization and pharmacological characterization of high affinity binding sites for vasopressin and oxytocin in the rat brain by light microscopic autoradiography. *Brain Res.* 442: 105-18.

van Wimersma Greidanus, T. B. & Maigret, C., (1996). The role of limbic vasopressin and oxytocin in social recognition. *Brain Res.* 713: 153-9.

Veenema, A. H. & Neumann, I. D., (2008). Central vasopressin and oxytocin release: Regulation of complex social behaviours. *Prog Brain Res.* 170: 261-76.

Waldherr, M. & Neumann, I. D., (2007). Centrally released oxytocin mediates mating-induced anxiolysis in male rats. *Proc Natl Acad Sci U S A.* 104: 16681-4.

Wang, Z. & Aragona, B. J., (2004). Neurochemical regulation of pair bonding in male prairie voles. *Physiology & Behavior.* 83: 319-328.

Wersinger, S. R., Temple, J. L., Caldwell, H. K. & Young, W. S., 3rd, (2008). Inactivation of the oxytocin and the vasopressin (avp) 1b receptor genes, but not the avp 1a receptor gene, differentially impairs the bruce effect in laboratory mice (mus musculus). *Endocrinology.* 149: 116-21.

Wigger, A. & Neumann, I. D., (2002). Endogenous opioid regulation of stress-induced oxytocin release within the hypothalamic paraventricular nucleus is reversed in late pregnancy: A microdialysis study. *Neuroscience.* 112: 121-129.

Williams, J. R., Insel, T. R., Harbaugh, C. R. & Carter, C. S., (1994). Oxytocin administered centrally facilitates formation of a partner preference in female prairie voles (microtus ochrogaster). *J Neuroendocrinol.* 6: 247-50.

Windle, R. J., Shanks, N., Lightman, S. L. & Ingram, C. D., (1997). Central oxytocin administration reduces stress-induced corticosterone release and anxiety behavior in rats. *Endocrinology.* 138: 2829-34.

Wise, R. A., (2002). Brain reward circuitry: Insights from unsensed incentives. *Neuron.* 36: 229-40.

Witt, D. M. & Insel, T. R., (1994). Increased fos expression in oxytocin neurons following masculine sexual behavior. *J Neuroendocrinol.* 6: 13-8.

Wotjak, C. T., Ganster, J., Kohl, G., Holsboer, F., Landgraf, R. & Engelmann, M., (1998). Dissociated central and peripheral release of vasopressin, but not oxytocin, in response to repeated swim stress: New insights into the secretory capacities of peptidergic neurons. *Neuroscience.* 85: 1209-1222.

Wu, S., Jia, M., Ruan, Y., Liu, J., Guo, Y., Shuang, M., Gong, X., Zhang, Y., Yang, X. & Zhang, D., (2005). Positive association of the oxytocin receptor gene (oxtr) with autism in the chinese han population. *Biol Psychiatry.* 58: 74 - 77.

Yang, T. T., Menon, V., Eliez, S., Blasey, C., White, C. D., Reid, A. J., Gotlib, I. H. & Reiss, A. L., (2002). Amygdalar activation associated with positive and negative facial expressions. *Neuroreport.* 13: 1737-41.

Young, L. J., Lim, M. M., Gingrich, B. & Insel, T. R., (2001). Cellular mechanisms of social attachment. *Horm Behav.* 40: 133-8.

Young, L. J. & Wang, Z., (2004). The neurobiology of pair bonding. *Nat Neurosci.* 7: 1048-54.

Zak, P. J., Stanton, A. A. & Ahmadi, S., (2007). Oxytocin increases generosity in humans. *PLoS ONE.* 2: e1128.

12

Environmental Factors in the Aetiology of Autism – Lessons from Animals Prenatally Exposed to Valproic Acid

Tomasz Schneider[1] and Ryszard Przewłocki[2,3]
[1]*University of Oxford,*
[2]*Institute of Pharmacology PAS*
[3]*Jagiellonian University,*
[1]*United Kingdom*
[2,3]*Poland*

1. Introduction

1.1 Diagnosis

Autism was first described by a child psychologist Leo Kanner as 'inability to relate themselves in the ordinary way to people and situations from the beginning of life' (Kanner, 1943). For many years autism was thought to be a consequence of bad parenting; fortunately, in the late 60-ties it was shown that there was no difference in parenting style between the parents of autistic and non-autistic children and a neurobiological basis of autism was suggested (Rutter et al., 1967; Rutter, 1968). Nowadays, autism is classified within the broad domain of pervasive developmental disorders (PDD) which also includes Rett syndrome, childhood disintegrative disorder, Asperger syndrome, and PDD not otherwise specified (American Psychiatric Association, 1994). There is a high phenotypic heterogeneity within this class of disorders and the debate regarding their clinical boundaries is ongoing. Instead of a classical categorical approach, a more useful description of this group of disorders as "autistic spectrum disorders" (ASD), not including Rett syndrome, has been proposed (reviewed in Willemsen-Swinkels & Buitelaar, 2002). ASD is characterized behaviourally by impairments in three core domains: social interaction, verbal and nonverbal communication, and restricted, repetitive, and stereotyped patterns of behaviour, interests, and activities appearing before the age of three (American Psychiatric Association, 1994). Clinical pattern and severity of impairment vary along these dimensions from mild to severe including a complete absence of interest in interacting with others but also subtle dysfunctions in managing social interactions, a complete absence of spoken language but also mild impairment and idiosyncratic vocabulary or even hyper-linguism in some of the Asperger cases, simple motor stereotypies and a preference for sameness but also complex rituals accompanied by distress and aggression when they cannot be fulfilled. The core symptoms are frequently accompanied by a spectrum of neurobehavioral derangements, including: hyperactivity, aberrant sensitivity to sensory stimulation, reduced joint attention, and anxiety (Ayres & Tickle, 1980; Bodfish et al., 2000; Dawson et al., 2004; Militerni et al., 2000; Muris et al., 1998; Pierce & Courchesne, 2001). Additional commonly associated characteristics include large head circumference, mental retardation, seizures, self injury, sleep disturbances, upset to change in routine, and lack of theory of mind (Baron-Cohen et al., 1985; Dawson et al., 2002; Richler et al., 2007; Volkmar & Nelson, 1990; Woodhouse et al., 1996). Similarly to core symptoms, intellectual capabilities also vary across the entire IQ spectrum with the majority of autistic individuals displaying low IQs (Charman et al., 2011) and the high functioning Asperger patients at the high end of the IQ spectrum (Pring, 2005). Prevalence estimates suggest a rate of 0.1-0.2% for autism and 0.6% for a broader autistic phenotype (Fombonne, 2009). Studies based on both clinical and epidemiological samples find a higher incidence of autism in boys than in girls, with reported ratios averaging around 4 to 1 (Newschaffer et al., 2007).

1.2 Intervention

The past 50 years have seen a myriad of interventions targeted at people with autism. Attempts to develop drugs that specifically improve social and communicative functioning have failed. However, medications such as atypical antipsychotics (e.g., risperidone, olanzapine, ziprasidone, quetiapine, aripiprazole), psychostimulants (methylphenidate), presynaptic noradrenergic blocking agents (clonidine and guanfacine) and serotonin reuptake inhibitors (clomipramine, fluoxetine, sertraline) have been shown to reduce symptoms of affective instability, irritability, hyperactivity and inattentiveness, aggression, self-injury and stereotypies (reviewed by (Myers, 2007). Several reports have also suggested efficacy of early intensive behavioural therapy in attenuation or reversal of the core autistic symptoms (Lovaas, 1987; Ozonoff & Cathcart, 1998). The gains of the behavioural therapy appear to sustain over time (McEachin et al., 1993).

1.3 Pathology

Pathological studies have revealed an association between autistic symptoms and certain pathological changes in brain structure and cellularity (reviewed in (Bauman & Kemper, 2005). Postmortem autopsy of autistic brains revealed alterations in neuronal anatomy within frontal (Bailey et al., 1998), temporal (Bachevalier, 1994), parietal (Courchesne et al., 1993), limbic (Bauman, 1991), brainstem and cerebellar (Kemper & Bauman, 1998) regions. The few cross-sectional studies that examined age-related changes revealed a complex pattern of growth abnormalities in the cerebellum, cortex, amygdala, and hippocampus (Courchesne, 2004; Hashimoto et al., 1995; Schumann et al., 2004; Schumann et al., 2010). It has been suggested that early dysfunction of the amygdala and frontal cortex might be the substrate for the severe socio-emotional deficits seen in autism (Bachevalier, 1994; Baron-Cohen et al., 2000). However, the most consistent findings have been found in the cerebellum: marked reduction in Purkinje cell number in the cerebellar hemispheres, a reduction in granule cell numbers, and decreased size of the deep cerebellar nuclei. The globose and emboliform nuclei were most severely affected, while the dentate nucleus was spared. The abnormalities noted in the deep nuclei were age dependent: younger individuals had large cells in the deep nuclei relative to controls, but in older cases the cell were reduced in size (Courchesne, 2004; Fatemi et al., 2002). Loss of Purkinje cells may play an important role in the aetiopathogenesis of the disorder (Kern, 2003).

1.4 Neurotransmitter systems

Many neurotransmitters have been implicated in aberrations observed in autism including serotonin (5-HT), dopamine, norepinephrine, acetylcholine (Ach), glutamate, gamma-aminobutyric acid (GABA), endogenous opiods, oxytocin, and cortisol (reviewed in Lam et al., 2006). The most consistent finding was hyperserotoninemia. Serotonin blood levels are highly elevated in a significant number of autistic children (Betancur et al., 2002; Singh et al., 1997). Direct in vivo measurements using positron emission tomography demonstrated asymmetries of 5-HT synthesis in the frontal cortex, thalamus and cerebellum in autistic boys (Chugani et al., 1997). Also, while 5-HT synthesis is usually high in young children and then gradually declines, in autistic children 5-HT levels are persistently high (Chugani et al., 1999). Because metabolism, catabolism and transport mechanisms for 5-HT do not seem to be affected in autism (Cook & Leventhal, 1996; Persico et al., 2002), it has been suggested that the elevated 5-HT level is a consequence of excessive stimulation of synthesis and release. While it is tempting to view this elevated 5-HT level as a fundamental cause of ASDs, treatments that increase 5-HT levels seem to improve some symptoms of autism, such as perseverations and social relatedness (Fatemi et al., 1998), while depletion of tryptophan, a serotonin precursor, exacerbate autistic symptoms especially stereotypies, self-injury and anxiety (McDougle et al., 1996). The role of 5-HT in ASDs therefore remains unclear.

1.5 Genetic background

The aetiology of autism is not known but it has a strong genetic component revealed by up to 60% concordance between monozygotic twins and almost 90% heritability (Wassink et al., 2004). Sibling risk is around 2–7%, which is much higher than in the general population (0.01–0.08%) (Orsmond & Seltzer, 2007). This might suggest that autism may lay at the extreme of a continuum of some autistic traits. In line with such an assumption, relatives of individuals with autism show a plethora of mild autistic traits, including personality, social-behavioural and language aberrations (Bolton et al., 1998; Piven et al., 1997). Despite its high heritability, only approximately 10% of autism cases can be traced to a known genetic aberration (Barton & Volkmar, 1998). Several large linkage analyses have identified genome regions of significant linkage for autism harbouring multiple candidate genes, suggesting a complex multigenic cause of this disorder, which might involve anything from 2 to a dozen genes with complex gene–gene and/or gene–environment interactions (Persico & Bourgeron, 2006). Genetic studies have suggested several molecular pathways with the potential to disrupt neurodevelopmental trajectories *in utero* that might be involved in the pathogenesis of autism, including signalling molecules such as neurotrophin, Reelin, and hepatocyte growth factor, and synaptic proteins such as neurexin and neuroligin (reviewed in Pardo & Eberhart, 2007). Despite considerable effort, these underlying risk alleles have been remarkably elusive, with the exception of a few large effect genes and several single gene disorders associated with an increased risk for autism (Freitag, 2007). The obstacles encountered in mapping the risk alleles have led to new models of inheritance including contributions of *de novo* mutations and/or epigenetic mechanisms in the underlying genetic susceptibility to autism (Persico & Bourgeron, 2006).

1.5.1 Epigenetic factors in autism

Epigenetics refers to the reversible regulation of various genomic functions mediated through partially stable modifications of DNA and chromatin histones (Henikoff & Matzke,

1997). Epigenetic processes are essential for normal cellular development and differentiation, and allow the regulation of gene function through non-mutagenic mechanisms. Systems that initiate and sustain the epigenetic state of DNA, and hence an epigenotype, include histone modifications, RNA- associated silencing, and DNA methylation (Petronis, 2010). Involvement of epigenetic factors in autism spectrum disorders is demonstrated by the central role of epigenetic mechanisms in the pathogenesis of Rett syndrome and fragile X syndrome (FXS), single gene disorders commonly associated with autism (Hagerman, 2006; Samaco et al., 2005). Rett syndrome arises from mutation in the gene encoding the methyl-CpG-binding protein 2 (MeCP2), one of the key mediators of epigenetic regulation of gene expression (Amir et al., 1999). FXS arises through silencing of FMR1 (fragile X mental retardation 1) gene (Hagerman et al., 2005). Importantly, brain tissues from patients with autism have reduced expression of MeCP2 (Samaco et al., 2005) and about 5% of patients with autism have duplications of the imprinted region of chromosome 15q11-q13 (Cook, Jr. et al., 1997). Additionally, autism has been observed in conjunction with an incidence of prenatal exposure to the histone deacetylase inhibitor, valproic acid (Rasalam et al., 2005). Children with autism have also a lower level of methylation intermediates in their plasma, possibly causing an impaired methylation capacity (James et al., 2004). A possible role of epigenetic factors in autism might be one of the reasons for the problems encountered in mapping risk alleles for autism; disruption of gene expression via epigenetic mechanisms is not reflected in the primary nucleotide sequence and may evade detection by standard mapping strategies. The epigenome is most vulnerable to the effect of environmental factors during embryogenesis when the rate of DNA synthesis is high and the epigenetic marks needed for normal tissue differentiation and development are being established (Dolinoy, 2007). Therefore, epigenetic adaptations in response to *in utero* environmental exposure may play an important role in development and disease susceptibility after birth (Dolinoy et al., 2007; Persico & Bourgeron, 2006). Indeed, evidence has accumulated that exposure to toxic substances during early embryogenesis and/or very soon after birth can trigger the onset of autism (Arndt et al., 2005).

2. Environmental risk factors for autism

Converging lines of evidence suggest that autism spectrum disorders (ASDs) have their origins in early prenatal life (for in depth reviews see Arndt et al., 2005; Miller et al., 2005). This assumption is based on reports showing that (1) many of minor malformations that occur frequently in people with autism as well as anomalies reported from histological studies of the autistic brains are known to arise during embryogenesis; (2) congenital syndromes with high rates of autism include somatic aberrations originating in the first trimester; (3) the environmental factors known to increase the risk of autism have critical periods of action in the first 3 months of pregnancy. In this section we will present evidence showing that the first trimester is a critical time point for environmental/drug insults triggering autism.

2.1 Onset of symptoms vs. time of exposure to insults

Autism spectrum disorders are commonly diagnosed at the age of two or three when typically developing children become increasingly social and communicative; however, most, if not all, children who will be diagnosed with autism show symptoms long before that age. This was first noticed in the late 70ties by Ornitz and colleagues (1977). Using parental questionnaires they showed that children who were later diagnosed with autism exhibited developmental delays even during the first year of life. This was confirmed by studies using home videos from the first two years of life of children later diagnosed with psychoses. Video recordings from children's natural environment showed fewer age appropriate behaviours in this group (Rosenthal et al., 1980) especially aberrant social attention (e.g., looking at people) and social behaviour (e.g., smiling at people and vocalizing to people), while most measures related to objects (e.g., looking at objects and smiling at objects) did not differ from control subjects in the first six months of life (Maestro et al., 2001). Better controlled studies of movies from the first birthday parties showed that most children later diagnosed with autism can be distinguished from controls at one year of age by such anomalies as failure to point to objects and failure to respond to their name (Osterling & Dawson, 1994). These results were confirmed by more recent studies (e.g., (Zwaigenbaum et al., 2005). Even at the biochemical level neonates who would later be diagnosed with either autism or mental retardation were distinguished by increased blood concentrations of neuron-related products: vasoactive intestinal peptide (VIP), calcitonin gene-related peptide (CGRP), brain-derived neurotrophic factor (BDNF) and neurotrophin 4/5 (NT4/5) (Nelson et al., 2001; Nelson et al., 2006). The two clinical groups did not differ from each other. These results clearly suggest that children with autism are already different from typically developing children at the time of birth.

2.2 Evidence for early injury in autism

2.2.1 Dysmorphic features

Dysmorphic facial features and increased occipitofrontal circumference reported in most cases of idiopathic autism (not co-morbid with the known genetic disorder) can not be induced postnatally, and most of the reported minor anomalies must have originated from the first eight weeks post-conception (e.g., (Miles & Hillman, 2000; Rodier et al., 1997a; Walker, 1977). For example, Rodier and colleagues (1997a) described minor physical malformations linked to autism in a population in Nova Scotia, Canada. Posterior rotation of the external ears was the most characteristic feature of children with autism compared to their unaffected siblings and non-autistic children with developmental delay. Importantly, these types of ear abnormalities have also been found in children with autism following exposure to thalidomide or valproic acid (Miller et al., 1998; Moore et al., 2000; Stromland et al., 1994). Dysmorphic features observed in idiopathic autism are strongly correlated with the extend of brain abnormalities revealed by imaging techniques and with increased male to female ratio (Miles et al., 2005; Miles & Hillman, 2000). Although, all of the physical anomalies reported in autism are also seen in children with other developmental disabilities, and in the population of typically-developing children, they might be useful in specifying critical periods for neurodevelopmental injuries in autism.

2.2.2 Neuroanatomical abnormalities

There are several neuroanatomical and histological arguments in favour of very early alterations of development as part of the aetiology of ASDs. For example, autopsies reported by Bailey and colleagues (1998) showed extra tracts running through the pontine tegmentum in one of the autistic brains studied and two brains with smaller and not fully separated pyramidal tracts. Because the basic tracts of the neuroaxis are established before parturition, the described anomalies were clearly of prenatal origins. Further, histological studies of one of the most consistent findings in pathology of autism, a marked reduction in Purkinje cell number in the cerebellar hemispheres, showed that the lose of these cells must have taken place during first weeks of neurodevelopment: (1) Purkinje cells body are normally wrapped by neighbouring basket cells and the late loss of Purkinje cells is characterized by the presence of "empty baskets", which were not observed in autistic brains (Bailey et al., 1998); (2) if reduction in Purkinje cell numbers appears after the 30th week *in utero* this results in massive retrograde loss of neurons in the inferior olive (Sakai et al., 1994; Takashima, 1982); while abnormalities of the inferior olive have been reported in autistic brains (Bailey et al., 1998; Bauman & Kemper, 1985), the nucleus is not affected to the extend that would be expect if cells were degenerating in response to a late loss of Purkinje cells. Similarly, a brain from a patient with autism, studied by Rodier and colleagues (1996), had a massive reduction in the number of motor neurons in the facial nucleus (400 vs. 9000). In the normal brain used for comparison, this nucleus was outlined by a capsule created by passing fibres, but in the brain of autistic patient no capsule was present, which suggest that the neurons forming the facial nucleus were not in position when passing fibres made their tracts. If the facial neurons had been lost in late gestation or postnatal life, the capsule would still be there. Thus, again the obvious conclusion is that the nucleus failed to form early during neurodevelopment. Each of these findings supports the conclusion that the neuroanatomy of people with autism is altered in a very early gestation.

2.2.3 Syndromes with high rates of autism

There are several congenital conditions originating from disruption of very early development with highly increased rate of autism including (1) Moebius sequence (Briegel et al., 2009); (2) the CHARGE association (Hartshorne et al., 2005); (3) Goldenhar syndrome (Johansson et al., 2007); (4) Duane syndrome (Miller et al., 2009); (5) Joubert syndrome (Ozonoff et al., 1999); (6) Cornelia de Lange syndrome (Moss et al., 2008); and (7) Smith-Limli-Opitz syndrome (Sikora et al., 2006). These syndromes are different in many ways, but they all involve abnormal development in the embryonic period. For example, (1) Möbius syndrome involve a variety of functional anomalies linked to congenital aberrations of the sixth and seventh cranial nerve, limb defects, and craniofacial anomalies involving the tongue and lip; (2) the CHARGE association is characterized by a co-occurrence of congenital malformations including colobomas, heart defects, choanal atresia, retarded growth or development, genital anomalies, and ear abnormalities and/or hearing loss; (3) Goldenhar syndrome is a combination of epibulbar dermoids, lipodermoids, and preauricular skin tags and fistula, upper lid coloboma, facial, and vertebral anomalies; (4) Duane syndrome is characterized by absent or hypoplastic abducens nucleus and nerve, and innervation of the lateral rectus eye muscle by a branch of the oculomotor nerve; (5) Joubert syndrome is an extremely rare recessively inherited disorder characterized by breathing difficulties, hypotonia, ataxia, eye movement anomalies, failure of development of the cerebellar vermis and the cranial nerve motor nuclei, and failure of the superior cerebellar peduncles to cross; (6) the phenotype of Cornelia de Lange syndrome includes epicanthal folds, ptosis, broad nasal bridge, short nose, long upper lip, micrognathia, anomalies of the limbs, heart, and gastrointestinal tract, and growth reduction; (7) Smith-Limli-Opitz syndrome (SLOS) is characterised by facial features including a broad, high forehead, hypertelorism, ptosis, epicanthal folds, broad nasal bridge, short nose with reverted nares, and micrognathia, low set and small ears, sometimes cleft palate, syndactyly, and genital anomalies. The dysplasias of the brain stem nuclei and cerebellar vermis observed in these syndromes suggest that their neurodevelopment have been disturbed in the fourth or fifth week postconception, when those structures are forming. Autism has been reported in a very high rate in all these syndromes, 25-68% for full autism and 50-100% for autistic features (Johansson et al., 2010; Moss et al., 2008; Ozonoff et al., 1999; Sikora et al., 2006). Why autism exists in a significant number in these syndromes is still a mystery and we do not know how early-onset insult affecting multiple brainstem structures can lead to malfunction of the higher centres not yet formed at the time of initial injury and most commonly linked to autism.

2.2.4 Critical period of exposure to teratogens increasing the risk of autism

A number of teratogenic substances have been identified in epidemiological studies as the key triggers of autism including maternal rubella infection (Chess, 1971), ethanol (Nanson, 1992), misoprostol (Bandim et al., 2003), thalidomide (Stromland et al., 1994), and valproic acid (Moore et al., 2000; Rasalam et al., 2005). The critical period for exposure to these teratogens appears to be during the first trimester.

2.2.4.1 Thalidomide exposure

A study of patients in the Swedish thalidomide registry revealed that about 30% of children exposed to thalidomide on the 20–24th day of gestation became autistic, which was 250 times the rate in the general population at this time (Stromland et al., 1994). The original aim of this study was to describe the ocular motility dysfunctions (strabismus) and other eye anomalies or visual disturbances in thalidomide victims, but this was fortunately accompanied by psychiatric evaluation. The neurological abnormalities observed in the five thalidomide-autistic cases included the following: three had Duane syndrome (failure of the abducens cranial nerve to innervate the lateral rectus muscle by the eye with subsequent re-innervation of the muscle by the oculomotor cranial nerve); one patient had face paresis (oculomotor palsy); four had Möbius syndrome (failure of the facial cranial nerve to innervate the facial muscles); two had abnormal lacrimation (due to a failure of the neurons of the superior salivatory nucleus to innervate the lacrimal apparatus). All five patients had ear malformations and hearing deficits. Importantly, ear malformations (Walker, 1977), eye motility problems (Scharre & Creedon, 1992), and Möbius syndrome (Gillberg & Steffenburg, 1989) had previously been associated with autism. In fact, external ear malformations are the most common physical abnormality observed in autism and the ones which best distinguish between autism and mental retardation (Rodier et al., 1997a; Walker, 1977). What makes the thalidomide cases so informative is that the external signs of thalidomide teratogeny allow accurate dating of the stages of development when exposure to the teratogen took place (Miller, 1991). On the bases of that timetable of teratogenic effects of thalidomide it was concluded that the ophthalmologic and cranial nerve dysfunction involving ocular structures observed in thalidomide-autistic cases occurred from teratogen intake during the 4th week post-conception (Stromland et al., 1994); i.e., at the time of neural tube closure and development of the first neurons, which form the motor nuclei of the cranial nerves (Altman & Bayer, 1982). That was the first study related to autism in which we had a known cause, a set of physical, neurological, and psychiatric autistic symptoms, and identified stage of development when the triggering insult occurred.

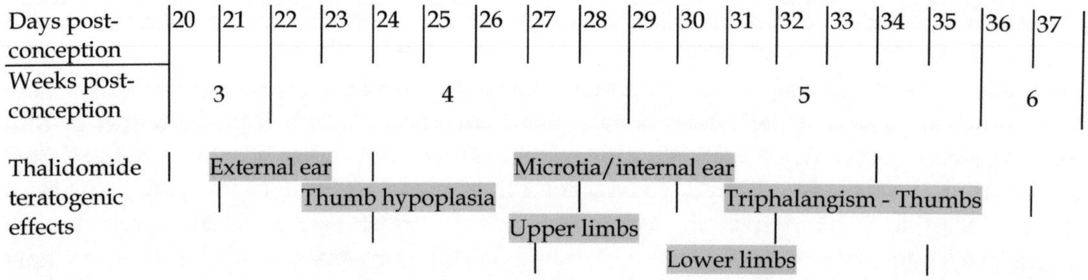

Fig. 1. Timing for teratogenic effects of thalidomide in humans (Lenz & Knapp, 1962; Miller, 1991).

2.2.4.2 Valproic acid

An antiepileptic drug increasing the risk for autism, valproic acid (Moore et al., 2000; Rasalam et al., 2005), is typically taken throughout the entire pregnancy, thus the timing for its teratogenic effect cannot be estimated directly. However, the timing of injury to the developing nervous system can be estimated from accompanying somatic features in exposed children. Children prenatally exposed to VPA exhibit similar patterns of physical malformations as those exposed to thalidomide, but with a decreased severity of symptoms (Ardinger et al., 1988; DiLiberti et al., 1984; Kozma, 2001). These include dysmorphic features indicative of injury around the time of neural tube closure, e.g., neural tube defects, congenital heart disease, craniofacial abnormalities, and abnormally shaped or posteriorly rotated ears, which are common in idiopathic autism (e.g., (Miles & Hillman, 2000; Rodier et al., 1997a) and were also reported in thalidomide-autism cases with the known exposure time to teratogen during the 4th week postconception. Based on this pattern of abnormalities we can estimate that VPA and thalidomide injured the developing nervous system at a similar time during embryogenesis.

2.2.4.3 Misoprostol

Misoprostol (Cytotec®), a prostaglandin type E analogue, was used as a self-administered abortion drug, especially in Brazil. Misoprostol is typically used in the first trimester of pregnancy. Self-induced, but failed abortions lead to several cases of infants born with malformations involving limbs and cranial nerves (Gonzalez et al., 1998; Shepard, 1995). In some of these cases, the children had Möbius syndrome (Shepard, 1995) and a rate of autism was highly increased in this population (Bandim et al., 2003).

2.2.4.4 Rubella

Epidemiological studies showed an increased risk for autism after rubella infection and noted that all the children in rubella-induced autism had multiple symptoms of rubella injury (Chess, 1977; Chess et al., 1978). Previous studies found that cases with multiple symptoms after rubella exposure came mainly from mothers exposed within the first eight weeks postconception, with mothers of children suffering from severe mental retardation showing an onset of rash even earlier at the 2nd to 5th week postconception (Ueda et al., 1979). This suggests that exposure to rubella virus in autistic cases took place during the first weeks of development.

The coincidence of critical periods for the described environmental risk factors for autism is a strong evidence that autism arises very early in development; however, the fact that each of these teratogens appears to act during the embryonic period (the first eight weeks of life) does not obviously rule out the possibility that autism could be initiated at other stages of development. Again, we do not know how early-onset insults affecting multiple brainstem structures can lead to malfunction of the higher centres not yet formed at the time of initial injury and most commonly linked to autism. It might be that a group of unidentified progenitor cells for higher brain centres get injured during early insults or that failure of axonal guidance under conditions in which important landmarks are absent or misplaced in the lower brain structures disturb normal development of higher brain centres. These are hypothesis that might be studied in animal models.

3. Animal models of autism

Because the primary diagnostic criteria of autism are abnormal behaviours, rather than biochemical or neuroanatomical factors, the use of animal models in the study of autism is unavoidable, as this is the only way to study behavioural defects in context of a whole organism. Rodent models in particular allow an extensive multi-omics approach to autism with a spectrum of non-invasive and invasive approaches at the genetic, molecular, cellular, synaptic, and behavioural levels. Currently used animal models of autism can be classified into three categories: (1) models created by neonatal lesions of brain areas shown to be abnormal in autism, e.g., cerebellum, the amygdala, hippocampus, or the medial prefrontal cortex; (2) genetically modified animals, e.g., targeted mutations in genes associated with autism or localized in chromosomal regions identified by linkage analyses; and (3) models mimicking environmental factors that increase the risk for autism in humans, e.g., prenatal exposure to valproic acid (VPA, 2-propyl-pentanoic acid) or pre- and neonatal immunological challenges. The first two approaches were extensively reviewed elsewhere (Ey et al., 2011; Moy & Nadler, 2008; Tordjman et al., 2007). In the present chapter we will focus on two animal models of autism induced by environmental factors: neonatal exposure to Borna disease virus and prenatal exposure to valproic acid. We will use these model to show and explain how human data can inform creation of animal models of autism and how animal model can be used to extend our understanding of autism and potentially lead to new therapeutic interventions for this still incurable disorder; but first a short introduction to *in vivo* modelling.

3.1 Concepts for *in vivo* modelling

There is an ongoing discussion about the criteria to be used in the evaluation of animal models. It has been suggested that the only meaningful initial evaluating criterion for an animal model is its ability to lead to accurate predictions about the disorder in question (predictive validity). Other authors stressed that the demonstration of construct validity (similarity to the underlying causes of the disease) represents the most important and necessary component in validation of an animal model. Others claim that the modelling of symptoms (face validity) must remain the primary goal of animal models of psychiatric disorders (for review see Fernando & Robbins, 2011). The resolution of this puzzling controversy depends on the desired purpose of the model that one wishes to validate. Although, regarding complex psychiatric disorders such as autism, a multidimensional approach should be used as an optimal strategy.

3.2 Challenges for animal models of autism

In recent years, several rodent models of autism have been developed that reflect some behavioural, genetic, and neuroanatomical alterations associated with this disorder. One of the main problems for the development of a relevant model is to define markers of autism, addressing its complexity and diversity. An ideal rodent model of autism should display symptoms of aberrant social interaction and communication, as well as repetitive behaviours. Tasks that could examine these behavioural symptoms in rodents have been already developed (for reviews see Crawley, 2004; Crawley, 2007) and are summarized in Table 1. However, an animal model of autism based solely on behavioural assays would be incomplete. Animal model should also address a combination of the neuropathological, biochemical and genetic factors implicated in autism. Another challenge lies in the fact that many clinical hallmarks of autism are difficult or almost impossible to replicate in rodents, e.g., theory of mind (ability to intuit the feelings and intentions of others) or speech deficits. It is also important to realize that assumed "core" psychopathological phenomena observed in autism are present as common clinical features in schizophrenia, depression, obsessive-compulsive disorder and other medical and psychiatric illnesses. What is specific for autism is a pattern of socio-behavioural aberrations and their appearance before the age of three, and animal models should try to address this fact.

3.3 Behavioural tests for autistic-like aberrations in rodents

Impairment in social interaction is a critical component for animal models of autism. Rodents are highly social animals displaying a plethora of different social behaviours. The propensity of animals to spend time with conspecific rather than non-social novel objects can be used as one of the measures. This can be best done in an automated three-chambered apparatus in which social interactions, social recognition, and social memory can also be scored (Moy et al., 2004; Nadler et al., 2004). Measures of the level of social approach can be accompanied by more specific analyses of reciprocal social interactions, including, nose-to-nose contacts, anogenital inspections, aggression, escape behaviour, nesting patterns, juvenile rough and tumble play, etc. (Moy et al., 2008; Moy et al., 2009). Impairments in social communication may be measured in rodents using olfactory and auditory communication tasks. Different kinds of ultrasonic vocalizations can be elicited in rodents, starting from separation calls in pups isolated from their mothers, to frequency modulated vocalizations in social situations in adult animals, treated, at least in rats, as indicators of positive and negative affective states (Portfors, 2007; Scattoni et al., 2009). Both frequency and time structures of ultrasonic vocalizations can be analyzed. Olfactory social signals, including deposition of pheromones, are another form of rodent communication (Arakawa et al., 2008). Rodents' chemical signals play a particularly important role in determining social dominance and intersexual relationships. Finally, repetitive behaviour encompasses both motor stereotypy and self-injury, can be scored using standardized scales, and behaviours reflecting general cognitive rigidity, such as ability to change, resistance to change, and responses to the change in routine can be investigated by exploratory choices and reversal tasks using spatially contingent reinforcers, e.g., reversal learning in T-maze or water maze (Crawley, 2004; Silverman et al., 2010). Because autism is accompanied by a plethora of neurobehavioral aberrations, it is important to include a range of additional behavioural tasks assessing, e.g., anxiety, seizure susceptibility, sleep patterns, sensitivity to sensory stimuli, learning and memory, and maturation and development. Finally, both pathological features of autism (e.g., decreased number of Purkinje neurons) and biological findings (e.g., increased serotonin levels) should be addressed.

Tests analogous to core autistic symptoms	Tests analogues to associated symptoms
1. *Impairment in social interactions* • reciprocal social interactions (partner grooming, nose-to-nose contacts, anogenital inspections, aggression, escape behaviour, juvenile rough and tumble play etc.) • propensity to spend time with conspecific vs. novel objects • conditioned place preference to conspecifics • preference for social novelty • aggression (resident-intruder test) • social recognition • nesting patterns in the home cage 2. *Impairments in social communication* • behavioural responses to social olfactory cues from conspecifics • deposition of social olfactory pheromones • vocalizations emitted during social interactions • responses to vocalizations from conspecifics • parental retrieval of separated pups • ultrasonic vocalizations by separated pups 3. *Repetitive, stereotyped patterns of behavior, interests, and activities* • motor stereotypies • extinction of a learned response in an operant chamber • reversal of a position habit in an appetitive T-maze task, aversive Y-maze task or the Morris water maze • spontaneous responses to errors during reversal tasks	1. *Anxiety* elevated plus maze; light-dark box exploration; Vogel conflict licking test; marble burying; 2. *Theory of Mind deficits* location of buried food following observation of conspecifics; social transmission of food preference; avoidance of aggressive encounters; 3. *Mental retardation* acquisition of Morris water maze task; acquisition of T-maze tasks; contextual and cued fear conditioning; operant learning tasks; attentional measures in the five choice serial reaction time task; 4. *Seizure susceptibility* spontaneous seizure activity; sensitivity to audiogenic seizures; sensitivity to drug-induced seizures; 5. *Motor clumsiness* balance beam foot slips; rotarod motor coordination and balance; gait analysis; 6. *Sleep disturbances* circadian running wheels; videotaped observations of home cage sleep and activity patterns; 7. *Idiosyncratic responses to sensory stimuli* acoustic startle; tactile startle; hot plate; Von Frey filaments; unresponsiveness to sensory attentional cues (failure to disengage attention); discriminative eyeblink conditioning and reversal; 8. *Developmental milestones and progression* brain weight and volume; size of structures and pathways; repeated testing of all relevant behaviours at juvenile and adult ages;

Table 1. Rodent Behavioural Tasks Relevant to Autism

3.4. Models created by environmental and immune factors

3.4.1 Neonatal Borna Disease Virus (BDV) infection in rat

Disturbed functioning of the immune system, especially pre- and/or neonatal immunological challenge has been long implicated in the pathogenesis of autism (Chess, 1977; Singh et al., 1991; Warren et al., 1986; Warren et al., 1990; Warren et al., 1996). The best characterized model of autism utilizing this approach is induced by neonatal Borna disease virus (BDV) infection in rat (Pletnikov et al., 2002a). Borna disease virus, an atypical, neurotropic, noncytolytic, negative strand RNA virus, shows affinity for limbic and cerebellar circuitry (de la Torre et al., 1990; de la Torre, 2002). Infection causes a spectrum of behavioural deficits depending on the age, immune status, central nervous system maturity, and genetics of the host (de la Torre, 2002; Pletnikov et al., 2001). Intracranial injection of the BDV in a newborn rat pup within the first 24–48 h after birth is the most common way of inducing neonatal BDV infection in rats (Pletnikov et al., 2002a). Neonatally infected Lewis rats have only transient inflammation (Hornig et al., 1999) but nonetheless have abnormalities of hippocampal and cerebellar development (Hornig et al., 1999; Pletnikov et al., 2003), growth (Bautista et al., 1994), spatial and aversive learning (Rubin et al., 1999), locomotor activity (Hornig et al., 1999), emotional reactivity (Hornig et al., 1999; Pletnikov et al., 1999a), and play behaviour (Pletnikov et al., 1999b), although duration of non-play social investigation (e.g. sniffing, approach, and follow) was higher in BDV-infected rats (Pletnikov et al., 1999b). These abnormalities mimic the impaired social interaction and atypical responses to sensory and emotional stimuli characteristic in autism. Injury to the cerebellum is one of the most salient morphological features of neonatal infection. BDV infection induces a prominent loss of Purkinje cells during the first seven months of life (up tp 75%). A loss of Purkinje cells and their dendrites in the molecular layer might be responsible for markedly reduced cerebellum size in this model (Pletnikov et al., 2003). In addition to injury of the cerebellum, neonatal BDV infection affects the postnatal maturation of the hippocampus and leads to continuing loss and eventual complete disappearance of dentate gyrus neurons by 45–55 postnatal days and their replacement by reactive glial cells (Gonzalez-Dunia et al., 2000; Pletnikov et al., 2002b). Neonatal BDV infection also induces cortical shrinkage. It has been shown that up to 30% of cortical neurons are lost in BDV infected rats by postnatal day 45 (Pletnikov et al., 2002b). At the cellular level diminished immunoreactivity for GAP-43 and synaptophysin is observed both in the neocortex and the hippocampus of neonatally BDV-infected rats (Gonzalez-Dunia et al., 2000). This model bears obvious similarities to behavioural and anatomical aberrations observed in autism; however, there is little serological evidence that suggests BDV infects humans (Chalmers et al., 2005), and its role in psychiatric disorders remains controversial. The disadvantage of infection models is that they lead to a persistent infection of the brain precluding more precise characterization of the time course and structural specificity of the created neuronal aberrations.

3.4.2 Animal model of autism induced by prenatal exposure to valproic acid

3.4.2.1 VPA exposure in humans

Clinically, VPA was first introduced in the 60-ties as an anticonvulsant and mood-stabilizing drug. Currently VPA (Depakene, Valproate, Valrelease) is used for treatment of epilepsy, bipolar disorder, and migraine prophylaxis (Rosenberg, 2007). Although there is clear evidence for teratogenic effects of VPA (Jentink et al., 2010), it is still used during pregnancy when the benefits to the mother outweigh the risks to the embryo or foetus (category "D" classification by the FDA). These include situations where the mother's health would be at risk without taking the drug and for which safer drugs are not available.

3.4.2.2 Studies implicating VPA in autism

Many children exposed in utero to VPA exhibit fetal valproate syndrome (FVS), a syndrome characterized by a constellation of malformations, developmental delays and behavioural aberrations (Ardinger et al., 1988; DiLiberti et al., 1984; Kozma, 2001). Common facial features of FVS include epicanthal folds, broad nasal bridge, short nose with inverted nostrils, long upper lip, and low set, posteriorly rotated ears. The link between prenatal VPA exposure and autism was based on seven case studies of children with FVS diagnosed with autism (Christianson et al., 1994; Williams et al., 2001; Williams & Hersh, 1997). The first population study on 57 children with fetal anticonvulsant syndromes (a syndrome caused by a variety of anticonvulsant drugs) conducted by Moore and colleagues (2000) showed highly increased prevalence of autistic symptoms in this group of children. The children were exposed to either VPA alone (60%), VPA in combination with another anticonvulsant drug (21%), or to non-VPA anticonvulsant drugs (19%). Moore reported 46 (81%) kids with speech delays and 34 (60%) kids with two or more autistic features, of which 6 (11%) had a diagnosis of ASD. Furthermore, 46 (81%) had behavioural problems, 22 (39%) displayed hyper-activity and poor concentration, of which 4 (7%) had a diagnosis of attention deficit/hyper-activity disorder. Forty-four (77%) kids had learning difficulties, 34 (60%) had gross motor delay, and 24 (42%) had minor motor delay. A more recent long-term study of the effects of prenatal exposure to antiepileptic drugs in 260 children (122 males, 138 females) showed a very similar pattern of behavioural aberration: 26 (16 males) children were reported by parents to have social or behavioural difficulties, 11 children (6 males, 5 females) fulfilled the DSM-IV criteria for autistic disorder and one (female) fulfilled the DSM-IV criteria for ASDs (Rasalam et al., 2005). These children comprised 4.6% of the exposed group studied, and 1.9% of all exposed children born during the study period. Other children from this group (26 in total) had difficulties in areas of speech and language development and social communication but did not meet the criteria for autism spectrum disorders. Sodium valproate was the drug most commonly associated with ASDs, five of 56 (8.9%) children exposed to sodium valproate alone had either autistic disorder or ASDs. Thus, the rate of autism in humans prenatally exposed to VPA is much higher than in the general population (approximately 0.1-0.16%, (Fombonne, 2005).

3.4.2.3 Valproic acid mechanisms of action

The mechanism of action of valproic acid is not fully understood. Effects of the drug may be related, at least in part, to increased brain concentrations of the inhibitory neurotransmitter GABA. Animal studies have shown that valproic acid inhibits GABA transferase and succinic aldehyde dehydrogenase, enzymes which are important for GABA catabolism (Johannessen, 2000); however, current hypotheses for the mechanism for VPA action are mostly focused on its epigenetic properties. VPA is an inhibitor of histone deacetylases (HDACs) (Gottlicher, 2004; Phiel et al., 2001) and its teratogenic effects are mediated specifically by inhibition of HDACs (Gurvivh et al., 2005). HDACs regulate chromatin structure by removing acetyl groups from lysines in the amino-terminal tails of core histones in the nucleosome regulating chromatin structure and gene expression (Dokmanovic & Marks, 2005). This process leads to expression of a small number of normally silent genes.

3.4.2.4 VPA-exposure as an animal model of autism

A study of patients in the Swedish thalidomide registry revealed that about 30% of children exposed to thalidomide on the 20–24th day of gestation became autistic, which was 250 times the rate in the general population at this time (Stromland et al., 1994). The 20–24th day of gestation is the time of neural tube closure and development of the first neurons, which form the motor nuclei of the cranial nerves. Since thalidomide does not have the same teratogenic effect in rodents as in primates (Schumacher et al., 1972), valproic acid was used to injure rats' brainstems in utero (Rodier et al., 1996). VPA exposure induces similar patterns of abnormal development across species with skeletal and cranial nerves abnormalities reported in mice (Nau et al., 1991), rats (Vorhees, 1987), and monkeys (Mast et al., 1986). Most importantly, prenatal VPA exposure leads to several-fold increase in the rate of autism (Bromley et al., 2008; Moore et al., 2000; Rasalam et al., 2005).

3.4.2.4.1 Neuroanatomical similarities to autism

In rats, the neural tube closes on day 11; by the 12th day of gestation production of the motor nuclei of trigeminal, abducens, and hypoglossal nerves is completed (Altman & Bayer, 1980). Malfunctions of the targets of these neurons were reported in autism (reviewed in Arndt et al., 2005; Miller et al., 2005). Offspring of female rats injected with VPA during neural tube closure show several brain abnormalities, resembling those found at autopsy and in brain-imaging studies of autistic patients: abnormalities of the cranial nerve motor nuclei, hypoplasia of brainstem structures, reduced volume of posterior parts of cerebellar vermis and hemispheres, a loss of Purkinje cells, and injury to deep nuclei of the cerebellum (Ingram et al., 2000; Rodier et al., 1996; Rodier et al., 1997b). Changes in the timing of exposure to VPA were used to produce different injuries. In brief, a single intraperitoneal NaVPA injection on PND 11.5 at a dose of 350 mg/kg VPA resulted in a significant reduction in the trigeminal and hypoglossal nuclei. Exposure on day E12 resulted in abnormalities of the abducens, trigeminal, and hypoglossal nuclei, and exposure on day E12.5 resulted in reductions of neurons in the oculomotor, abducens, trigeminal, and hypoglossal nuclei. Insults affecting these neurons are associated with abnormalities in facial features that are common in idiopathic autism and were observed in all of the five autistic thalidomide cases (Rodier et al., 1997a; Stromland et al., 1994). Cerebellar abnormalities consistent with human cases of autism were found following exposure of 600 mg/kg sodium valproate on day E12.5. Purkinje cell numbers in posterior lobules (VI–VIII and X) of the vermis were reduced, but were normal in the anterior lobes (IV–V) (Ingram et al., 2000), which resembles human MRI studies that have shown decreased size of the posterior cerebellar vermis in autism (Courchesne et al., 1994a; Courchesne et al., 1994b; Hashimoto et al., 1995).Further, the interpositus nucleus, corresponding to the globose and emboliform nuclei in humans, but not dentate nucleus was significantly reduced in VPA rats (Rodier et al., 1997b). In human cases, the globose and emboliform nuclei are much more severely affected than the dentate nucleus (reviewed in Kemper & Bauman, 1998). The overall brain volume in the treated animals was reduced by 18% (using brain weight).

3.4.2.4.2 Behavioural similarities to autism

There is a vast body of evidence that administration of VPA on day 12 of gestation has long-term effects on postnatal behaviours in male but not female rats, which include (1) lower sensitivity to pain and higher sensitivity to nonpainful stimuli, (2) diminished acoustic prepulse inhibition, (3) locomotor and repetitive/stereotypic-like hyperactivity combined with lower exploratory activity, (4) decreased number of social behaviours and increased latency to social behaviours, (5) decreased seizure threshold; (6) and higher anxiety (Markram et al., 2008a; Narita et al., 2010; Schneider et al., 2001; Schneider et al., 2007; Schneider et al., 2008; Schneider & Przewlocki, 2005; Schneider & Przewlocki, 2007). In addition, VPA rats showed delayed maturation, lower body weight, delayed motor development, and attenuated integration of a coordinated series of reflexes, and delayed nest-seeking response mediated by olfactory system (Schneider & Przewlocki, 2005). Interestingly, all behavioural aberrations described in those animals appear in adolescent animals, which could distinguish the VPA rat model of autism from other animal models of neurodevelopmental disorders, especially rodent models of schizophrenia. Similar behavioural abnormalities and developmental delays have been observed in autism. Indeed, autistic patients express: social interaction deficits (Kaufmann et al., 2004); hyperactivity with decreased exploratory activity (Pierce & Courchesne, 2001); motor repetitive/stereotypic behaviours (Militerni et al., 2002); lowered sensitivity to pain (Militerni et al., 2000); deficits of information processing and attention with impaired sensorimotor gating (Allen & Courchesne, 2001; McAlonan et al., 2002), higher anxiety and phobias (Evans et al., 2005; Gillott et al., 2001); a greater risk for developing seizure disorder (Volkmar & Nelson, 1990); and delayed sensorimotor development (Losche, 1990). VPA-treated offspring also exhibited greatly amplified conditioned cued and contextual fear responses when tested up to 3 months after conditioning. Fear memories were not only amplified, but also more generalized to other stimuli configurations and more resistant to extinction than in control animals (Markram et al., 2008a). Again, autistic children are known to have impairments in extinction learning and to display strong perseverations (Coldren & Halloran, 2003; Mullins & Rincover, 1985). Those results confirm existence of similarities between the observed pattern of aberrations in VPA rats and features of disturbed behaviour in autistic patients. Noteworthy, VPA rats express a very specific pattern of aberration in eye-blink conditioning test (Murawski et al., 2009; Stanton et al., 2007), which is similar to that reported by Sears and colleagues (1994) in autistic patients, i.e., normal basic sensory and motor function, but exaggerated amplitude of conditioned blinks, and altered timing of conditioned blinks. One difference between children with autism and the VPA-exposed rats was that the animals did not display a higher rate of acquisition of conditioned responses. Almost all studies using eye-blink conditioning in human disorders (e.g., schizophrenia, (Bolbecker et al., 2009); mental retardation, (Hogg et al., 1979); Huntington disease, (Woodruff-Pak & Papka, 1996); Fragile X Syndrome, (Tobia & Woodruff-Pak, 2009) show impairment in conditioning ability compared to control subjects. It is therefore surprising that children with autism and rats exposed to VPA *in utero* display an enhancement of eye-blink conditioning, which might actually lead to the first 'autism-specific' test for between species comparisons.

3.4.2.4.3 Biochemical similarities to autism

Prenatal exposure to VPA in Sprague–Dawley rats on embryonic day 9 (neural plate stage) leads to increased serotonin levels in the hippocampus, increased dopamine in the frontal cortex, and hyperserotonemia (Miyazaki et al., 2005; Narita et al., 2002; Tsujino et al., 2007). VPA administration also alters serotonergic neuronal differentiation and migration in the dorsal raphé nucleus (Miyazaki et al., 2005; Tsujino et al., 2007). These results are strikingly similar to the data obtained on the serotonergic system in human autism (Anderson et al.,

1990; Lam et al., 2006). Serotonin is not only a neurotransmitter, but also a regulator of development of several brain areas, such as the neocortex, hippocampus, and cerebellum. It has been previously shown in other models that depletion of serotonin results in a significant delay in maturation of cortical structures (Bennett-Clarke et al., 1995; Vitalis et al., 2007) and excessive serotonin during early development results in hyper-innervation and expansion of cortical architecture (Vitalis et al., 1998). Thus, increased serotonin level in VPA rats may be one of the factors triggering altered developmental patterns observed in this model. Indeed, a more complex dendritic arborization in apical dendrites of pyramidal cells in motor cortex had been shown in VPA-exposed animals (Snow et al., 2008), suggesting disturbed pruning process, which is consistent with theories related to abnormal human brain development in autism. VPA rats also express altered functioning of opioidergic (Schneider et al., 2007) and glutamatergic (Rinaldi et al., 2007) systems. Recently, the same model has been replicated in mice. VPA-mice share behavioural phenotype described previously in rats but also had decreased NLGN3 mRNA expression in hippocampus and somatosensory cortex (Kolozsi et al., 2009; Roullet et al., 2010), which again resembles human literature showing that mutations in neuroligin3 gene may predispose to autism (Bourgeron, 2009; Jamain et al., 2003).

3.4.2.4.4 Immunological similarities to autism

Prenatal VPA exposure in rats led also to immunological aberrations resembling decreased cellular immunity observed in autism (reviewed in Cohly & Panja, 2005). Decreased weight of the thymus, decreased splenocytes proliferative response to concanavaline A, lower IFN-γ/IL-10 ratio, and increased production of NO by peritoneal macrophages were described in male VPA rats, whereas females exhibited only decreased IFN-γ/IL-10 ratio (Schneider et al., 2008). Some of these might have been mediated by increased basal level of corticosterone (Schneider et al., 2008). The neuroimmune network is involved in the adaptation to stressful stimuli and an inadequate response to environmental stressors has been linked to its dysfunction (reviewed in (Merlot et al., 2008; Petrovsky, 2001). Previous studies have shown that prenatal exposure to VPA on day 8 of gestation leads to lesions in thymus (Gossrau & Graf, 1989), even though colonization of the thymus by pluripotent stem cells occurs in rat around day 13 of gestation (Dietert et al., 2000), which is a typical exposure time in VPA models. Described changes were transient, but in VPA rats thymus atrophy was persistent even in adult rats, which might suggest a potential role of increased susceptibility to stress and of increased level of corticosterone observed in VPA male rats, as corticosteroids are known to cause thymic atrophy (Gorski et al., 1988). There is also mounting evidence that stress may induce a shift in the type 1/type 2 cytokine balance toward a type 2 cytokine response (Agarwal & Marshall, Jr., 1998; Elenkov & Chrousos, 1999), which is consistent with decreased IFN-γ/IL-10 ratio observed in VPA rats. The ratio of IFN-γ to IL-10 in culture supernatants is of critical importance in determining their pro- or anti-inflammatory capacity, i.e., either activation (IFN-γ) or inhibition (IL-10) of monocytic and lymphocytic function (Katsikis et al., 1995). Similar imbalance in Th1/Th2 response has been reported in autism (e.g., Gupta et al., 1998). Thus, the mechanism of the immunomodulatory effect of prenatal VPA exposure may be related to its indirect impact exerted through activation of autonomic nervous system and/or the hypothalamic–pituitary–adrenal axis.

3.4.2.4.5 Cellular level similarities to autism

On the cellular level VPA induced overexpression of NR2A and NR2B subunits of NMDA receptors and kinase calcium/calmodulin-dependent protein kinase I (CAMKI1), which is a key signalling enzyme associated with NMDA receptor-mediated synaptic plasticity (Rinaldi et al., 2007). In contrast, AMPA receptor subunits GluR1, GluR2, and GluR3 and the obligatory subunit of the NMDA receptor, NR1, extracellular signal-regulated kinase (ERK) and cAMP response element binding protein (CREB), some phosphorylated forms of signaling proteins (pCREB-S133, pCaMKII-T286/287, GluR1-S831, pGluR1-S845, pNR1-S896, pNR1-S897, pNR2B-S1303), as well as the main metabotropic glutamate receptor subunits (mGluR1, mGluR5, mGluR4, mGluR6/7) and the kainite receptor subunits (GluR6/7), were not affected in the VPA-treated neocortex. This indicated a highly selective abnormality within the glutamatergic system in VPA-induced model of autism (Rinaldi et al., 2007).

Reactivity of microcircuits in VPA rats' somatosensory cortex, prefrontal cortex, and amygdala measured by a multi-electrode array (MEA) stimulator showed dramatic increase (x2) in reactivity to electrical stimulation and boosted synaptic plasticity as well as a deficit in inhibition in amygdala (Markram et al., 2008b; Rinaldi et al., 2007; Rinaldi et al., 2008a; Rinaldi et al., 2008b; Silva et al., 2009). Taking into account that paired recordings of excitatory AMPA-mediated synaptic responses were weaker in the VPA rats and the numbers of synapses per synaptic connection smaller, it was suggested that an excessive recurrent circuitry may play a crucial role in the observed hyperreactivity to electrical stimulation in brain microcircuits. Further, synaptic plasticity experiments between pairs of pyramidal neurons revealed increased postsynaptic long-term potentiation in the VPA treated slices (Rinaldi et al., 2007), which suggested hyperplasticity of glutamatergic synapses. At the same time, the pyramidal neurons required much more current to drive their voltage to spiking threshold and the number of spikes generated for a series of current stimuli was much lower than in control slices. There were no differences in the passive conductance. This hypo-excitability of pyramidal neurons may be an attempt to counter the hyper-reactivity as a compensatory mechanism (Rinaldi et al., 2007; Rinaldi et al., 2008b). Morphological examination of 3D reconstructions of pyramidal neurons did not show any significant differences between VPA-treated tissue and control. Hyper-reactivity of the neocortical microcircuitry is therefore not caused by larger or more elaborate neurons, more excitable neurons, an increase in neuron numbers, stronger synaptic connections, or by a loss of inhibition. Indeed, changes in these parameters seem to act in the opposite direction, perhaps part of a compensatory strategy (Rinaldi et al., 2007; Rinaldi et al., 2008a; Rinaldi et al., 2008b). The described hyper-connectivity was found only in neurons forming the typical dimensions of a neocortical minicolumn (~50µm somatic distance), but not for pairs of pyramidal neurons (100–200µm apart). The minicolumn is thought to be the smallest computational circuit in the brain (Lucke & Malsburg C., 2004). It consists of a core line of vertically ascending pyramidal and inhibitory neurons, their connections and input/output axons. Hyper-connectivity in microcircuits can lead to exaggerated recruitment of neurons when presented with a stimulus and could account for the hyper-reactivity found in these local circuits after VPA treatment. An important aspect of hyper-connectivity induced by VPA exposure is that pyramidal neurons target more neurons even at the expense of using less synapses per connection. Thus, the form of hyper-connectivity observed in this model can be seen as a hypertrophy of connectivity between neurons. Local hyperconnectivity may render cortical modules more sensitive to stimulation and once activated, more autonomous and more difficult to command (Rinaldi et al., 2007; Rinaldi et al., 2008b). Studies on minicolumnar arrangements in the frontal and temporal lobes showed altered neuronal anatomy and circuitry in autism with minicolumns abnormally narrow, both in the column core and the neurophil containing inhibitory interneurons (Casanova et al., 2002). This suggests that the autistic brain may exhibits an increased number of minicolumns, i.e., more processing units, and more excitable/less inhibit microcircuitry.

3.4.2.4.6 Gender dimorphism in VPA model

Most of behavioural and immunological aberrations induced by prenatal exposure to VPA were observed in male but not female rats (Schneider et al., 2008). It is unclear how gender *in utero* can affect VPA-induced behavioural, endocrine, and immunological effects. Observed gender differences in functional outcome of prenatal exposure to VPA cannot be explained by differences in direct VPA effect on foetuses as morphological studies in VPA rats have not reported any differences between males and females in a kind of or extension of brain injuries induced by VPA (Ingram et al., 2000; Rodier et al., 1996; Rodier et al., 1997b). Gender differences in the model cannot be either explained by VPA influence on estrogenic receptors, which has recently been shown in vitro (Fortunati et al., 2008; Reid et al., 2005), as estrogen receptor system of the rat brain is not detected before day 16 of gestation (Miranda et al., 1994; Miranda & Toran-Allerand, 1992). Thus, attenuation of behavioural and immunological alterations observed in female VPA rats is probably not related to differences in direct teratogenic action of VPA. We would rather suggest that protective effects of estrogen and progesterone and sex-related differences in neurotransmitters systems development and/or functioning during consecutive developmental stages may play the crucial role in the observed attenuation of the VPA-induced aberrations in female rats. Estrogen and progesterone might reduce the consequences of brain injuries by enhancing anti-oxidant mechanisms, decreasing excitotoxicity (altering glutamate receptor activity, reducing immune inflammation, providing neurotrophic support, stimulating axonal remyelinization), and enhancing synaptogenesis and dendritic arborization (Roof & Hall, 2000a; Roof & Hall, 2000b; Stein, 2001). Importantly, it has been recently suggested that sex hormone action may be mediated via gene-specific epigenetic modifications of DNA and histones (Kaminsky et al., 2006). Hormone-induced DNA methylation and histone modifications at specific gene regulatory regions may modify the risk of a disease and lead to disproportion in male to female ratio in autism as well as to sex-specific phenotypes in VPA rats.

3.4.2.4.7 Beneficial effects of environmental enrichment in VPA rats

The availability of a valid animal model of autism opened the door to rigorous evaluation of the effects of environmental manipulations on the behavioural expression of neuropathological deficits in VPA exposed animals. Our own experiments have shown that environmental enrichment reverses almost all autistic-like behavioural aberrations observed in VPA rats (Schneider et al., 2006). VPA rats subjected to environmental enrichment compared to a VPA non-enriched group exhibited higher sensitivity to pain and lower sensitivity to nonpainful stimuli; stronger acoustic prepulse inhibition; lower locomotor, repetitive/stereotypic-like activity, and enhanced exploratory activity; decreased anxiety; increased number of social behaviours, and shorter latency to social explorations, and when compared to control non-enriched animals, increased number of pinnings in adolescence and social explorations in adulthood, and more numerous entries to open arms and longer time spent in the open arms of the elevated plus-maze, which suggest decreased anxiety. Mechanisms of this remarkable behavioural improvement are unknown. However, converging lines of evidence suggest that environmental enrichment in rodents leads to better performance in various learning tasks (Bruel-Jungerman et al., 2005; Rampon et al.,

2000), enhanced social play behaviour (Morley-Fletcher et al., 2003; Schneider et al., 2006), and lower anxiety (Gortz et al., 2008; Sztainberg et al., 2010). Improvements in behavioural performances were accompanied by changes in various neurochemical and anatomical features in rodent brains, e.g., increased dendritic spine density and branching in the cerebral cortex (Diamond et al., 1972; Greer et al., 1981), hippocampus (Bruel-Jungerman et al., 2005; Rampon et al., 2000), striatum (Turner et al., 2003), and cerebellum (Angelucci et al., 2009); reduced apoptosis (Young et al., 1999), and enhanced neurogenesis (Levi & Michaelson, 2007). Enrichment also causes a significant change in the expression of genes whose products are involved in neuronal structure, plasticity, and neurotransmission (Rampon et al., 2000). Finally, environmental enrichment reverses behavioural, cognitive and molecular aberrations resulting from prenatal or early postnatal factors such as maternal separation (Bredy et al., 2003) or developmental Pb^{2+} exposure (Guilarte et al., 2003), and has been shown to attenuate behavioural and morphological phenotype in mouse genetic models of Rett Syndrome (Lonetti et al., 2010) and the fragile X syndrome (Restivo et al., 2005), both commonly associated with autism.

4. Conclusions

The advantage of animal models of autism is to study developmental and behavioural deficits in context of a whole organism. We can use such models to clarify complex relationships between genetic, behavioural and environmental variables to better understand and potentially cure autism. Rodent models described in this chapter allow an extensive multi-omics approach to autism with a spectrum of non-invasive and invasive approaches at the genetic, molecular, cellular, synaptic, and behavioural levels, which has greatly extended our knowledge about autism and mechanism underlying both autistic-like and normal behaviours in animals and humans. What we need now is to combine these different approaches into multidisciplinary studies determining the consequences of environmental factors (e.g., stress, teratogenic substances, enriched environment) on the development of autistic-like behavioural changes in genetically modified animals, and vice versa, to determine the genetic basis of negative and positive effects of environmental factors in animal models. Among currently used animal models of autism described in this chapter, the one induced by prenatal exposure to valproic acid seems to fulfil criteria for construct, face and predictive validity and may be used to further elucidate the neurobiological mechanisms underlying the functional effects of genetic and environmental factors relevant to autism. This model could become a common experimental platform to validate new pharmacological and behavioural interventions with potential relevance to autism.

4.1 Hypothesis based on VPA model and their implications for autism

4.1.1 Environmental enrichment

Of the vast number of animal studies that yield results of interest to human research, studies on the impact of an enriched environment on brain development and behaviour can be of enormous interest. What we need now is the development of a sound theory of the effects of specific environmental experiences on neurobehavioural development and a rationale for the external and internal mechanisms that mediate these effects. Studies on the influence of an enriched environment are one of the major attempts to understand the interaction between the environment and the genome in the regulation of the phenotype. At the very least, study of the beneficial effects of environmental enrichment in the VPA model indicates that there are many opportunities for enhancing brain activity and behaviour, and that they can have pronounced therapeutic effects on behavioural alteration in the animal model of autism induced by prenatal exposure to VPA (Schneider et al., 2006). This leads us to a consideration of the relevance of this model to brain damage rehabilitation and behavioural-cognitive therapy attenuation of autistic features in humans. Having such a model we can start asking questions about genetic/epigenetic, molecular, biochemical and structural changes in the brain induced by environmental enrichment and how these changes can be related and used to improve therapeutic interventions in autism. For example, we can assume that the combination of enriched experience with pharmacological treatments may further strengthen these beneficial effects and improve their therapeutic effectiveness, and this can be tested first in VPA model. Thus, what we need now is both to identify the key external factors and understand the internal mechanisms that mediate the beneficial effects of environmental enrichment on neurobehavioural development of VPA rodents and to use this knowledge to help people suffering from autism.

4.1.2 Gender dimorphism

Gender dimorphism is another interesting phenomenon observed both in the VPA model and autism. Studies based on both clinical and epidemiological samples find a higher incidence of autism in boys than in girls, with reported ratios averaging around 4 to 1 (Fombonne, 2003; Newschaffer et al., 2007; Volkmar et al., 1993). The reason for this gender discrepancy is unknown. Accordingly, most of behavioural and immunological aberrations induced by prenatal exposure to VPA were observed in males but not females (Schneider et al., 2008). Gender differences in functional outcome of prenatal exposure to VPA cannot be explained by differences in direct VPA effect on foetuses as morphological studies in VPA rats have not reported any differences between males and females in the extend of brain injuries induced by VPA (Ingram et al., 2000; Rodier et al., 1996; Rodier et al., 1997b). Similarly, there are no structural or functional imaging data suggesting gender differences among autistic patients. We would rather suggest that protective effects of estrogen and progesterone and sex-related differences in neurotransmitters systems development and/or functioning during consecutive developmental stages may play the crucial role in the observed attenuation of the VPA-induced aberrations in female rats and skewed sex ratio in autism. As mentioned above, estrogen and progesterone might reduce the consequences of brain injuries by enhancing anti-oxidant mechanisms, decreasing excitotoxicity (altering glutamate receptor activity, reducing immune inflammation, providing neurotrophic support, stimulating axonal remyelinization), and enhancing synaptogenesis and dendritic arborization (Roof & Hall, 2000a; Roof & Hall, 2000b; Stein, 2001). Some of these effects might be mediated by epigenetic mechanisms (Kaminsky et al., 2006) suspected to play an important role in autism. This might be a very fruitful direction for future studies in neurotherapeutics for autism.

4.1.3 Intense World Syndrome as a unifying theory of autism

Based on results obtained in the VPA-induced animal model of autism a new unifying theory has been recently proposed describing autism as an 'intense world syndrome' (Markram et al., 2007; Markram & Markram, 2010), in which hyper-reactivity and hyper-plasticity of the brains' microcircuits causes excessive neuronal information processing and storage in the brain. Such excessive information overload is proposed to produce hyper-perception, hyper-attention, and hyper-memory, which may lead to exaggerated perception/hyper-focusing of fragments of a sensory world instead of creating holistic, multimodal representations, and can diminish an ability to shift one's attention to new stimuli due to the difficulty for top-down mechanisms to coordinate the overly autonomous microcircuits. Inability to disengage attentional capacities has been frequently reported in autism (e.g., Landry & Bryson, 2004). Hyper-plasticity may also lead to exaggerated memorization of non-related stimuli and thus to over-generalization, which has been observed in fear conditioning test in VPA rats and might be related to decreased inhibition and hyper-reactivity in VPA treated amygdala. Therefore an autistic person may perceive the world not only as overwhelmingly intense due to hyper-reactivity of primary sensory areas, but also as aversive and highly stressful due to a hyper-reactive amygdala, which due to overgeneralization can make fear associations with usually neutral stimuli. The autistic person may well try to cope with the intense and aversive world by withdrawal. Thus, impaired social interactions in autism might be the result of an intense, overwhelming, and fragmented perception, which escapes any holistic interpretation. This theory is consistent with high anxiety levels as well as the hypertrophy of the amygdala in autism (Gillott et al., 2001; Sparks et al., 2002). Increased anxiety and phobias are common in autistic patients (Evans et al., 2005; Gillott et al., 2001) and their relatives (Micali et al., 2004), and frequency of autistic-like symptoms is highly increased in children with mood and anxiety disorders (Towbin et al., 2005). In line with this hypothesis, decreased amygdala activation has been linked to genetic hyper-sociability in Williams syndrome (Martens et al., 2009; Meyer-Lindenberg et al., 2005), whereas increased activation is observed in social phobia (Stein et al., 2002; Stein et al., 2007). Moreover, autistic children exhibit increased autonomic responses, indicative of enhanced amygdala activity (Corbett et al., 2006; Hirstein et al., 2001; Tordjman et al., 1997), and increased corticosterone blood level was observed in VPA rats (Schneider et al., 2008). Although this is still a preliminary proposal and needs further clarification it is well based on both human and animal data described in this chapter and may be used as an alternative unifying theory for previous theoretical proposals such as the 'weak central coherence theory of autism' (Happe & Frith, 2006), the 'executive function theory of autism' (Hughes et al., 1994; Russell & Hill, 2001) or the "theory of mind" conception of autism (Baron-Cohen, 1991; Frith & Happe, 1994). In fact, if this is hyper-reactivity that makes autistic individuals withdrawn from the world, a completely new approach to pharmacological intervention in autism should be considered. For example, while most of the commonly prescribed medications try to increase neuronal and cognitive functioning in autism, the autistic brain might rather need to be 'calmed down' and cognitive functions diminished in order to re-instate balance and functionality.

5. Acknowledgment

RP was supported by statutory funds of the Institute of Pharmacology PAS, Krakow, Poland.

6. Reference

Agarwal, S. K. & Marshall, G. D., Jr. (1998). Glucocorticoid-induced type 1/type 2 cytokine alterations in humans: a model for stress-related immune dysfunction. *J Interferon Cytokine Res*, Vol.18, No.12, pp. 1059-1068

Allen, G. & Courchesne, E. (2001). Attention function and dysfunction in autism. *Front Biosci*, Vol.6, pp. D105-D119

Altman, J. & Bayer, S. A. (1980). Development of the brain stem in the rat. IV. Thymidine-radiographic study of the time of origin of neurons in the pontine region. *J Comp Neurol*, Vol.194, No.4, pp. 905-929

Altman, J. & Bayer, S. A. (1982). Development of the cranial nerve ganglia and related nuclei in the rat. *Adv Anat Embryol Cell Biol*, Vol.74, pp. 1-90

American Psychiatric Association (1994). Diagnostic and Statistical Manual of Mental Disorders. Fourth Edition. Vol.4, pp.

Amir, R. E., Van den Veyver, I. B., Wan, M., Tran, C. Q., Francke, U., & Zoghbi, H. Y. (1999). Rett syndrome is caused by mutations in X-linked MECP2, encoding methyl-CpG-binding protein 2. *Nat Genet*, Vol.23, No.2, pp. 185-188

Anderson, G. M., Horne, W. C., Chatterjee, D., & Cohen, D. J. (1990). The hyperserotonemia of autism. *Ann N Y Acad Sci*, Vol.600, pp. 331-340

Angelucci, F., De, B. P., Gelfo, F., Foti, F., Cutuli, D., Bossu, P., Caltagirone, C., & Petrosini, L. (2009). Increased concentrations of nerve growth factor and brain-derived neurotrophic factor in the rat cerebellum after exposure to environmental enrichment. *Cerebellum*, Vol.8, No.4, pp. 499-506

Arakawa, H., Blanchard, D. C., Arakawa, K., Dunlap, C., & Blanchard, R. J. (2008). Scent marking behavior as an odorant communication in mice. *Neurosci Biobehav Rev*, Vol.32, No.7, pp. 1236-1248

Ardinger, H. H., Atkin, J. F., Blackston, R. D., Elsas, L. J., Clarren, S. K., Livingstone, S., Flannery, D. B., Pellock, J. M., Harrod, M. J., Lammer, E. J., & . (1988). Verification of the fetal valproate syndrome phenotype. *Am J Med Genet*, Vol.29, No.1, pp. 171-185

Arndt, T. L., Stodgell, C. J., & Rodier, P. M. (2005). The teratology of autism. *Int J Dev Neurosci*, Vol.23, No.2-3, pp. 189-199

Ayres, A. J. & Tickle, L. S. (1980). Hyper-responsivity to touch and vestibular stimuli as a predictor of positive response to sensory integration procedures by autistic children. *Am J Occup Ther*, Vol.34, No.6, pp. 375-381

Bachevalier, J. (1994). Medial temporal lobe structures and autism: a review of clinical and experimental findings. *Neuropsychologia*, Vol.32, No.6, pp. 627-648

Bailey, A., Luthert, P., Dean, A., Harding, B., Janota, I., Montgomery, M., Rutter, M., & Lantos, P. (1998). A clinicopathological study of autism. *Brain*, Vol.121 (Pt 5), pp. 889-905

Bandim, J. M., Ventura, L. O., Miller, M. T., Almeida, H. C., & Costa, A. E. (2003). Autism and Mobius sequence: an exploratory study of children in northeastern Brazil. *Arq Neuropsiquiatr*, Vol.61, No.2A, pp. 181-185

Baron-Cohen, S. (1991). The development of a theory of mind in autism: deviance and delay? *Psychiatr Clin North Am*, Vol.14, No.1, pp. 33-51

Baron-Cohen, S., Leslie, A. M., & Frith, U. (1985). Does the autistic child have a "theory of mind"? *Cognition*, Vol.21, No.1, pp. 37-46

Baron-Cohen, S., Ring, H. A., Bullmore, E. T., Wheelwright, S., Ashwin, C., & Williams, S. C. (2000). The amygdala theory of autism. *Neurosci Biobehav Rev*, Vol.24, No.3, pp. 355-364

Barton, M. & Volkmar, F. (1998). How commonly are known medical conditions associated with autism? *J Autism Dev Disord*, Vol.28, No.4, pp. 273-278

Bauman, M. & Kemper, T. L. (1985). Histoanatomic observations of the brain in early infantile autism. *Neurology*, Vol.35, No.6, pp. 866-874

Bauman, M. L. (1991). Microscopic neuroanatomic abnormalities in autism. *Pediatrics*, Vol.87, No.5 Pt 2, pp. 791-796

Bauman, M. L. & Kemper, T. L. (2005). Neuroanatomic observations of the brain in autism: a review and future directions. *Int J Dev Neurosci*, Vol.23, No.2-3, pp. 183-187

Bautista, J. R., Schwartz, G. J., de la Torre, J. C., Moran, T. H., & Carbone, K. M. (1994). Early and persistent abnormalities in rats with neonatally acquired Borna disease virus infection. *Brain Res Bull*, Vol.34, No.1, pp. 31-40

Bennett-Clarke, C. A., Lane, R. D., & Rhoades, R. W. (1995). Fenfluramine depletes serotonin from the developing cortex and alters thalamocortical organization. *Brain Res*, Vol.702, No.1-2, pp. 255-260

Betancur, C., Corbex, M., Spielewoy, C., Philippe, A., Laplanche, J. L., Launay, J. M., Gillberg, C., Mouren-Simeoni, M. C., Hamon, M., Giros, B., Nosten-Bertrand, M., & Leboyer, M. (2002). Serotonin transporter gene polymorphisms and hyperserotonemia in autistic disorder. *Mol Psychiatry*, Vol.7, No.1, pp. 67-71

Bodfish, J. W., Symons, F. J., Parker, D. E., & Lewis, M. H. (2000). Varieties of repetitive behavior in autism: comparisons to mental retardation. *J Autism Dev Disord*, Vol.30, No.3, pp. 237-243

Bolbecker, A. R., Mehta, C. S., Edwards, C. R., Steinmetz, J. E., O'Donnell, B. F., & Hetrick, W. P. (2009). Eye-blink conditioning deficits indicate temporal processing abnormalities in schizophrenia. *Schizophr Res*, Vol.111, No.1-3, pp. 182-191

Bolton, P. F., Pickles, A., Murphy, M., & Rutter, M. (1998). Autism, affective and other psychiatric disorders: patterns of familial aggregation. *Psychol Med*, Vol.28, No.2, pp. 385-395

Bourgeron, T. (2009). A synaptic trek to autism. *Curr Opin Neurobiol*, Vol.19, No.2, pp. 231-234

Bredy, T. W., Humpartzoomian, R. A., Cain, D. P., & Meaney, M. J. (2003). Partial reversal of the effect of maternal care on cognitive function through environmental enrichment. *Neuroscience*, Vol.118, No.2, pp. 571-576

Briegel, W., Schimek, M., Kamp-Becker, I., Hofmann, C., & Schwab, K. O. (2009). Autism spectrum disorders in children and adolescents with Moebius sequence. *Eur Child Adolesc Psychiatry*, Vol.18, No.8, pp. 515-519

Bromley, R. L., Mawer, G., Clayton-Smith, J., & Baker, G. A. (2008). Autism spectrum disorders following in utero exposure to antiepileptic drugs. *Neurology*, Vol.71, No.23, pp. 1923-1924

Bruel-Jungerman, E., Laroche, S., & Rampon, C. (2005). New neurons in the dentate gyrus are involved in the expression of enhanced long-term memory following environmental enrichment. *Eur J Neurosci*, Vol.21, No.2, pp. 513-521

Casanova, M. F., Buxhoeveden, D. P., Switala, A. E., & Roy, E. (2002). Minicolumnar pathology in autism. *Neurology*, Vol.58, No.3, pp. 428-432

Chalmers, R. M., Thomas, D. R., & Salmon, R. L. (2005). Borna disease virus and the evidence for human pathogenicity: a systematic review. *QJM*, Vol.98, No.4, pp. 255-274

Charman, T., Pickles, A., Simonoff, E., Chandler, S., Loucas, T., & Baird, G. (2011). IQ in children with autism spectrum disorders: data from the Special Needs and Autism Project (SNAP). *Psychol Med*, Vol.41, No.3, pp. 619-627

Chess, S. (1971). Autism in children with congenital rubella. *J Autism Child Schizophr*, Vol.1, No.1, pp. 33-47

Chess, S. (1977). Follow-up report on autism in congenital rubella. *J Autism Child Schizophr*, Vol.7, No.1, pp. 69-81

Chess, S., Fernandez, P., & Korn, S. (1978). Behavioral consequences of congenital rubella. *J Pediatr*, Vol.93, No.4, pp. 699-703

Christianson, A. L., Chesler, N., & Kromberg, J. G. (1994). Fetal valproate syndrome: clinical and neuro-developmental features in two sibling pairs. *Dev Med Child Neurol*, Vol.36, No.4, pp. 361-369

Chugani, D. C., Muzik, O., Behen, M., Rothermel, R., Janisse, J. J., Lee, J., & Chugani, H. T. (1999). Developmental changes in brain serotonin synthesis capacity in autistic and nonautistic children. *Ann Neurol*, Vol.45, No.3, pp. 287-295

Chugani, D. C., Muzik, O., Rothermel, R., Behen, M., Chakraborty, P., Mangner, T., da Silva, E. A., & Chugani, H. T. (1997). Altered serotonin synthesis in the dentatothalamocortical pathway in autistic boys. *Ann Neurol*, Vol.42, No.4, pp. 666-669

Cohly, H. H. & Panja, A. (2005). Immunological findings in autism. *Int Rev Neurobiol*, Vol.71, pp. 317-341

Coldren, J. T. & Halloran, C. (2003). Spatial reversal as a measure of executive functioning in children with autism. *J Genet Psychol*, Vol.164, No.1, pp. 29-41

Cook, E. H. & Leventhal, B. L. (1996). The serotonin system in autism. *Curr Opin Pediatr*, Vol.8, No.4, pp. 348-354

Cook, E. H., Jr., Lindgren, V., Leventhal, B. L., Courchesne, R., Lincoln, A., Shulman, C., Lord, C., & Courchesne, E. (1997). Autism or atypical autism in maternally but not paternally derived proximal 15q duplication. *Am J Hum Genet*, Vol.60, No.4, pp. 928-934

Corbett, B. A., Mendoza, S., Abdullah, M., Wegelin, J. A., & Levine, S. (2006). Cortisol circadian rhythms and response to stress in children with autism. *Psychoneuroendocrinology*, Vol.31, No.1, pp. 59-68

Courchesne, E. (2004). Brain development in autism: early overgrowth followed by premature arrest of growth. *Ment Retard Dev Disabil Res Rev*, Vol.10, No.2, pp. 106-111

Courchesne, E., Press, G. A., & Yeung-Courchesne, R. (1993). Parietal lobe abnormalities detected with MR in patients with infantile autism. *AJR Am J Roentgenol*, Vol.160, No.2, pp. 387-393

Courchesne, E., Saitoh, O., Townsend, J. P., Yeung-Courchesne, R., Press, G. A., Lincoln, A. J., Haas, R. H., & Schriebman, L. (1994a). Cerebellar hypoplasia and hyperplasia in infantile autism. *Lancet*, Vol.343, No.8888, pp. 63-64

Courchesne, E., Saitoh, O., Yeung-Courchesne, R., Press, G. A., Lincoln, A. J., Haas, R. H., & Schreibman, L. (1994b). Abnormality of cerebellar vermian lobules VI and VII in patients with infantile autism: identification of hypoplastic and hyperplastic subgroups with MR imaging. *AJR Am J Roentgenol*, Vol.162, No.1, pp. 123-130

Crawley, J. N. (2004). Designing mouse behavioral tasks relevant to autistic-like behaviors. *Ment Retard Dev Disabil Res Rev*, Vol.10, No.4, pp. 248-258

Crawley, J. N. (2007). Mouse behavioral assays relevant to the symptoms of autism. *Brain Pathol*, Vol.17, No.4, pp. 448-459

Dawson, G., Munson, J., Estes, A., Osterling, J., McPartland, J., Toth, K., Carver, L., & Abbott, R. (2002). Neurocognitive function and joint attention ability in young children with autism spectrum disorder versus developmental delay. *Child Dev*, Vol.73, No.2, pp. 345-358

Dawson, G., Toth, K., Abbott, R., Osterling, J., Munson, J., Estes, A., & Liaw, J. (2004). Early social attention impairments in autism: social orienting, joint attention, and attention to distress. *Dev Psychol*, Vol.40, No.2, pp. 271-283

de la Torre, J. C. (2002). Bornavirus and the brain. *J Infect Dis*, Vol.186 Suppl 2, pp. S241-S247

de la Torre, J. C., Carbone, K. M., & Lipkin, W. I. (1990). Molecular characterization of the Borna disease agent. *Virology*, Vol.179, No.2, pp. 853-856

Diamond, M. C., Rosenzweig, M. R., Bennett, E. L., Lindner, B., & Lyon, L. (1972). Effects of environmental enrichment and impoverishment on rat cerebral cortex. *J Neurobiol*, Vol.3, No.1, pp. 47-64

Dietert, R. R., Etzel, R. A., Chen, D., Halonen, M., Holladay, S. D., Jarabek, A. M., Landreth, K., Peden, D. B., Pinkerton, K., Smialowicz, R. J., & Zoetis, T. (2000). Workshop to identify critical windows of exposure for children's health: immune and respiratory systems work group summary. *Environ Health Perspect*, Vol.108 Suppl 3, pp. 483-490

DiLiberti, J. H., Farndon, P. A., Dennis, N. R., & Curry, C. J. (1984). The fetal valproate syndrome. *Am J Med Genet*, Vol.19, No.3, pp. 473-481

Dokmanovic, M. & Marks, P. A. (2005). Prospects: histone deacetylase inhibitors. *J Cell Biochem*, Vol.96, No.2, pp. 293-304

Dolinoy, D. C. (2007). Epigenetic gene regulation: early environmental exposures. *Pharmacogenomics*, Vol.8, No.1, pp. 5-10

Dolinoy, D. C., Weidman, J. R., & Jirtle, R. L. (2007). Epigenetic gene regulation: linking early developmental environment to adult disease. *Reprod Toxicol*, Vol.23, No.3, pp. 297-307

Elenkov, I. J. & Chrousos, G. P. (1999). Stress Hormones, Th1/Th2 patterns, Pro/Anti-inflammatory Cytokines and Susceptibility to Disease. *Trends Endocrinol Metab*, Vol.10, No.9, pp. 359-368

Evans, D. W., Canavera, K., Kleinpeter, F. L., Maccubbin, E., & Taga, K. (2005). The fears, phobias and anxieties of children with autism spectrum disorders and Down syndrome: comparisons with developmentally and chronologically age matched children. *Child Psychiatry Hum Dev*, Vol.36, No.1, pp. 3-26

Ey, E., Leblond, C. S., & Bourgeron, T. (2011). Behavioral profiles of mouse models for autism spectrum disorders. *Autism Res*, Vol.4, No.1, pp. 5-16

Fatemi, S. H., Halt, A. R., Realmuto, G., Earle, J., Kist, D. A., Thuras, P., & Merz, A. (2002). Purkinje cell size is reduced in cerebellum of patients with autism. *Cell Mol Neurobiol*, Vol.22, No.2, pp. 171-175

Fatemi, S. H., Realmuto, G. M., Khan, L., & Thuras, P. (1998). Fluoxetine in treatment of adolescent patients with autism: a longitudinal open trial. *J Autism Dev Disord*, Vol.28, No.4, pp. 303-307

Fernando, A. B. & Robbins, T. W. (2011). Animal models of neuropsychiatric disorders. *Annu Rev Clin Psychol*, Vol.7, pp. 39-61

Fombonne, E. (2003). The prevalence of autism. *JAMA*, Vol.289, No.1, pp. 87-89

Fombonne, E. (2005). Epidemiology of autistic disorder and other pervasive developmental disorders. *J Clin Psychiatry*, Vol.66 Suppl 10, pp. 3-8

Fombonne, E. (2009). Epidemiology of pervasive developmental disorders. *Pediatr Res*, Vol.65, No.6, pp. 591-598

Fortunati, N., Bertino, S., Costantino, L., Bosco, O., Vercellinatto, I., Catalano, M. G., & Boccuzzi, G. (2008). Valproic acid is a selective antiproliferative agent in estrogen-sensitive breast cancer cells. *Cancer Lett*, Vol.259, No.2, pp. 156-164

Freitag, C. M. (2007). The genetics of autistic disorders and its clinical relevance: a review of the literature. *Mol Psychiatry*, Vol.12, No.1, pp. 2-22

Frith, U. & Happe, F. (1994). Autism: beyond "theory of mind". *Cognition*, Vol.50, No.1-3, pp. 115-132

Gillberg, C. & Steffenburg, S. (1989). Autistic behaviour in Moebius syndrome. *Acta Paediatr Scand*, Vol.78, No.2, pp. 314-316

Gillott, A., Furniss, F., & Walter, A. (2001). Anxiety in high-functioning children with autism. *Autism*, Vol.5, No.3, pp. 277-286

Gonzalez, C. H., Marques-Dias, M. J., Kim, C. A., Sugayama, S. M., Da Paz, J. A., Huson, S. M., & Holmes, L. B. (1998). Congenital abnormalities in Brazilian children associated with misoprostol misuse in first trimester of pregnancy. *Lancet*, Vol.351, No.9116, pp. 1624-1627

Gonzalez-Dunia, D., Watanabe, M., Syan, S., Mallory, M., Masliah, E., & de la Torre, J. C. (2000). Synaptic pathology in Borna disease virus persistent infection. *J Virol*, Vol.74, No.8, pp. 3441-3448

Gorski, J. R., Muzi, G., Weber, L. W., Pereira, D. W., Iatropoulos, M. J., & Rozman, K. (1988). Elevated plasma corticosterone levels and histopathology of the adrenals and thymuses in 2,3,7,8-tetrachlorodibenzo-p-dioxin-treated rats. *Toxicology*, Vol.53, No.1, pp. 19-32

Gortz, N., Lewejohann, L., Tomm, M., Ambree, O., Keyvani, K., Paulus, W., & Sachser, N. (2008). Effects of environmental enrichment on exploration, anxiety, and memory in female TgCRND8 Alzheimer mice. *Behav Brain Res*, Vol.191, No.1, pp. 43-48

Gossrau, R. & Graf, R. (1989). Lesions and repair of cells of maternal mice after valproic acid (VPA) treatment on day 8 of pregnancy: an enzyme histochemical analysis. *Acta Histochem*, Vol.86, No.1, pp. 23-32

Gottlicher, M. (2004). Valproic acid: an old drug newly discovered as inhibitor of histone deacetylases. *Ann Hematol*, Vol.83 Suppl 1, pp. S91-S92

Greer, E. R., Diamond, M. C., & Tang, J. M. (1981). Increase in thickness of cerebral cortex in response to environmental enrichment in Brattleboro rats deficient in vasopressin. *Exp Neurol*, Vol.72, No.2, pp. 366-378

Guilarte, T. R., Toscano, C. D., McGlothan, J. L., & Weaver, S. A. (2003). Environmental enrichment reverses cognitive and molecular deficits induced by developmental lead exposure. *Ann Neurol*, Vol.53, No.1, pp. 50-56

Gupta, S., Aggarwal, S., Rashanravan, B., & Lee, T. (1998). Th1- and Th2-like cytokines in CD4+ and CD8+ T cells in autism. *J Neuroimmunol*, Vol.85, No.1, pp. 106-109

Hagerman, R. J. (2006). Lessons from fragile X regarding neurobiology, autism, and neurodegeneration. *J Dev Behav Pediatr*, Vol.27, No.1, pp. 63-74

Hagerman, R. J., Ono, M. Y., & Hagerman, P. J. (2005). Recent advances in fragile X: a model for autism and neurodegeneration. *Curr Opin Psychiatry*, Vol.18, No.5, pp. 490-496

Happe, F. & Frith, U. (2006). The weak coherence account: detail-focused cognitive style in autism spectrum disorders. *J Autism Dev Disord*, Vol.36, No.1, pp. 5-25

Hartshorne, T. S., Grialou, T. L., & Parker, K. R. (2005). Autistic-like behavior in CHARGE syndrome. *Am J Med Genet A*, Vol.133A, No.3, pp. 257-261

Hashimoto, T., Tayama, M., Murakawa, K., Yoshimoto, T., Miyazaki, M., Harada, M., & Kuroda, Y. (1995). Development of the brainstem and cerebellum in autistic patients. *J Autism Dev Disord*, Vol.25, No.1, pp. 1-18

Henikoff, S. & Matzke, M. A. (1997). Exploring and explaining epigenetic effects. *Trends Genet*, Vol.13, No.8, pp. 293-295

Hirstein, W., Iversen, P., & Ramachandran, V. S. (2001). Autonomic responses of autistic children to people and objects. *Proc Biol Sci*, Vol.268, No.1479, pp. 1883-1888

Hogg, J., Remington, R. E., & Foxen, T. H. (1979). Classical conditioning of profoundly retarded, multiply handicapped children. *Dev Med Child Neurol*, Vol.21, No.6, pp. 779-786

Hornig, M., Weissenbock, H., Horscroft, N., & Lipkin, W. I. (1999). An infection-based model of neurodevelopmental damage. *Proc Natl Acad Sci U S A*, Vol.96, No.21, pp. 12102-12107

Hughes, C., Russell, J., & Robbins, T. W. (1994). Evidence for executive dysfunction in autism. *Neuropsychologia*, Vol.32, No.4, pp. 477-492

Ingram, J. L., Peckham, S. M., Tisdale, B., & Rodier, P. M. (2000). Prenatal exposure of rats to valproic acid reproduces the cerebellar anomalies associated with autism. *Neurotoxicol Teratol*, Vol.22, No.3, pp. 319-324

Jamain, S., Quach, H., Betancur, C., Rastam, M., Colineaux, C., Gillberg, I. C., Soderstrom, H., Giros, B., Leboyer, M., Gillberg, C., & Bourgeron, T. (2003). Mutations of the X-linked genes encoding neuroligins NLGN3 and NLGN4 are associated with autism. *Nat Genet*, Vol.34, No.1, pp. 27-29

James, S. J., Cutler, P., Melnyk, S., Jernigan, S., Janak, L., Gaylor, D. W., & Neubrander, J. A. (2004). Metabolic biomarkers of increased oxidative stress and impaired methylation capacity in children with autism. *Am J Clin Nutr*, Vol.80, No.6, pp. 1611-1617

Jentink, J., Loane, M. A., Dolk, H., Barisic, I., Garne, E., Morris, J. K., & de Jong-van den Berg LT (2010). Valproic acid monotherapy in pregnancy and major congenital malformations. *N Engl J Med*, Vol.362, No.23, pp. 2185-2193

Johannessen, C. U. (2000). Mechanisms of action of valproate: a commentatory. *Neurochem Int*, Vol.37, No.2-3, pp. 103-110

Johansson, M., Billstedt, E., Danielsson, S., Stromland, K., Miller, M., Granstrom, G., Flodmark, O., Rastam, M., & Gillberg, C. (2007). Autism spectrum disorder and underlying brain mechanism in the oculoauriculovertebral spectrum. *Dev Med Child Neurol*, Vol.49, No.4, pp. 280-288

Johansson, M., Gillberg, C., & Rastam, M. (2010). Autism spectrum conditions in individuals with Mobius sequence, CHARGE syndrome and oculo-auriculo-vertebral spectrum: diagnostic aspects. *Res Dev Disabil*, Vol.31, No.1, pp. 9-24

Kaminsky, Z., Wang, S. C., & Petronis, A. (2006). Complex disease, gender and epigenetics. *Ann Med*, Vol.38, No.8, pp. 530-544

Kanner, L. (1943). Autistic Disturbances of Affective Contact. *Nervous Child*, No.2, pp. 217-250

Katsikis, P. D., Cohen, S. B., Londei, M., & Feldmann, M. (1995). Are CD4+ Th1 cells pro-inflammatory or anti-inflammatory? The ratio of IL-10 to IFN-gamma or IL-2 determines their function. *Int Immunol*, Vol.7, No.8, pp. 1287-1294

Kaufmann, W. E., Cortell, R., Kau, A. S., Bukelis, I., Tierney, E., Gray, R. M., Cox, C., Capone, G. T., & Stanard, P. (2004). Autism spectrum disorder in fragile X syndrome: communication, social interaction, and specific behaviors. *Am J Med Genet A*, Vol.129A, No.3, pp. 225-234

Kemper, T. L. & Bauman, M. (1998). Neuropathology of infantile autism. *J Neuropathol Exp Neurol*, Vol.57, No.7, pp. 645-652

Kern, J. K. (2003). Purkinje cell vulnerability and autism: a possible etiological connection. *Brain Dev*, Vol.25, No.6, pp. 377-382

Kolozsi, E., Mackenzie, R. N., Roullet, F. I., Decatanzaro, D., & Foster, J. A. (2009). Prenatal exposure to valproic acid leads to reduced expression of synaptic adhesion molecule neuroligin 3 in mice. *Neuroscience*, Vol.163, No.4, pp. 1201-1210

Kozma, C. (2001). Valproic acid embryopathy: report of two siblings with further expansion of the phenotypic abnormalities and a review of the literature. *Am J Med Genet*, Vol.98, No.2, pp. 168-175

Lam, K. S., Aman, M. G., & Arnold, L. E. (2006). Neurochemical correlates of autistic disorder: a review of the literature. *Res Dev Disabil*, Vol.27, No.3, pp. 254-289

Landry, R. & Bryson, S. E. (2004). Impaired disengagement of attention in young children with autism. *J Child Psychol Psychiatry*, Vol.45, No.6, pp. 1115-1122

Lenz, W. & Knapp, K. (1962). Thalidomide embryopathy. *Arch Environ Health*, Vol.5, pp. 100-105

Levi, O. & Michaelson, D. M. (2007). Environmental enrichment stimulates neurogenesis in apolipoprotein E3 and neuronal apoptosis in apolipoprotein E4 transgenic mice. *J Neurochem*, Vol.100, No.1, pp. 202-210

Lonetti, G., Angelucci, A., Morando, L., Boggio, E. M., Giustetto, M., & Pizzorusso, T. (2010). Early environmental enrichment moderates the behavioral and synaptic phenotype of MeCP2 null mice. *Biol Psychiatry*, Vol.67, No.7, pp. 657-665

Losche, G. (1990). Sensorimotor and action development in autistic children from infancy to early childhood. *J Child Psychol Psychiatry*, Vol.31, No.5, pp. 749-761

Lovaas, O. I. (1987). Behavioral treatment and normal educational and intellectual functioning in young autistic children. *J Consult Clin Psychol*, Vol.55, No.1, pp. 3-9

Lucke, J. & Malsburg C. (2004). Rapid processing and unsupervised learning in a model of the cortical macrocolumn. *Neural Comput*, Vol.16, No.3, pp. 501-533

Maestro, S., Muratori, F., Barbieri, F., Casella, C., Cattaneo, V., Cavallaro, M. C., Cesari, A., Milone, A., Rizzo, L., Viglione, V., Stern, D. D., & Palacio-Espasa, F. (2001). Early behavioral development in autistic children: the first 2 years of life through home movies. *Psychopathology*, Vol.34, No.3, pp. 147-152

Markram, H., Rinaldi, T., & Markram, K. (2007). The intense world syndrome--an alternative hypothesis for autism. *Front Neurosci*, Vol.1, No.1, pp. 77-96

Markram, K. & Markram, H. (2010). The intense world theory - a unifying theory of the neurobiology of autism. *Front Hum Neurosci*, Vol.4, pp. 224

Markram, K., Rinaldi, T., La, M. D., Sandi, C., & Markram, H. (2008a). Abnormal fear conditioning and amygdala processing in an animal model of autism. *Neuropsychopharmacology*, Vol.33, No.4, pp. 901-912

Markram, K., Rinaldi, T., La, M. D., Sandi, C., & Markram, H. (2008b). Abnormal fear conditioning and amygdala processing in an animal model of autism. *Neuropsychopharmacology*, Vol.33, No.4, pp. 901-912

Martens, M. A., Wilson, S. J., Dudgeon, P., & Reutens, D. C. (2009). Approachability and the amygdala: insights from Williams syndrome. *Neuropsychologia*, Vol.47, No.12, pp. 2446-2453

Mast, T. J., Cukierski, M. A., Nau, H., & Hendrickx, A. G. (1986). Predicting the human teratogenic potential of the anticonvulsant, valproic acid, from a non-human primate model. *Toxicology*, Vol.39, No.2, pp. 111-119

McAlonan, G. M., Daly, E., Kumari, V., Critchley, H. D., van, A. T., Suckling, J., Simmons, A., Sigmundsson, T., Greenwood, K., Russell, A., Schmitz, N., Happe, F., Howlin, P., & Murphy, D. G. (2002). Brain anatomy and sensorimotor gating in Asperger's syndrome. *Brain*, Vol.125, No.Pt 7, pp. 1594-1606

McDougle, C. J., Naylor, S. T., Cohen, D. J., Aghajanian, G. K., Heninger, G. R., & Price, L. H. (1996). Effects of tryptophan depletion in drug-free adults with autistic disorder. *Arch Gen Psychiatry*, Vol.53, No.11, pp. 993-1000

McEachin, J. J., Smith, T., & Lovaas, O. I. (1993). Long-term outcome for children with autism who received early intensive behavioral treatment. *Am J Ment Retard*, Vol.97, No.4, pp. 359-372

Merlot, E., Couret, D., & Otten, W. (2008). Prenatal stress, fetal imprinting and immunity. *Brain Behav Immun*, Vol.22, No.1, pp. 42-51

Meyer-Lindenberg, A., Hariri, A. R., Munoz, K. E., Mervis, C. B., Mattay, V. S., Morris, C. A., & Berman, K. F. (2005). Neural correlates of genetically abnormal social cognition in Williams syndrome. *Nat Neurosci*, Vol.8, No.8, pp. 991-993

Micali, N., Chakrabarti, S., & Fombonne, E. (2004). The broad autism phenotype: findings from an epidemiological survey. *Autism*, Vol.8, No.1, pp. 21-37

Miles, J. H. & Hillman, R. E. (2000). Value of a clinical morphology examination in autism. *Am J Med Genet*, Vol.91, No.4, pp. 245-253

Miles, J. H., Takahashi, T. N., Bagby, S., Sahota, P. K., Vaslow, D. F., Wang, C. H., Hillman, R. E., & Farmer, J. E. (2005). Essential versus complex autism: definition of fundamental prognostic subtypes. *Am J Med Genet A*, Vol.135, No.2, pp. 171-180

Militerni, R., Bravaccio, C., Falco, C., Fico, C., & Palermo, M. T. (2002). Repetitive behaviors in autistic disorder. *Eur Child Adolesc Psychiatry*, Vol.11, No.5, pp. 210-218

Militerni, R., Bravaccio, C., Falco, C., Puglisi-Allegra, S., Pascucci, T., & Fico, C. (2000). Pain reactivity in children with autistic disorder. *J Headache Pain*, No.1, pp. 53-56

Miller, M. T. (1991). Thalidomide embryopathy: a model for the study of congenital incomitant horizontal strabismus. *Trans Am Ophthalmol Soc*, Vol.89, pp. 623-674

Miller, M. T., Stromland, K., Gillberg, C., Johansson, M., & Nilsson, E. W. (1998). The puzzle of autism: an ophthalmologic contribution. *Trans Am Ophthalmol Soc*, Vol.96, pp. 369-385

Miller, M. T., Stromland, K., Ventura, L., Johansson, M., Bandim, J. M., & Gillberg, C. (2005). Autism associated with conditions characterized by developmental errors in early embryogenesis: a mini review. *Int J Dev Neurosci*, Vol.23, No.2-3, pp. 201-219

Miller, M. T., Ventura, L., & Stromland, K. (2009). Thalidomide and misoprostol: Ophthalmologic manifestations and associations both expected and unexpected. *Birth Defects Res A Clin Mol Teratol*, Vol.85, No.8, pp. 667-676

Miranda, R. C., Sohrabji, F., & Toran-Allerand, D. (1994). Interactions of estrogen with the neurotrophins and their receptors during neural development. *Horm Behav*, Vol.28, No.4, pp. 367-375

Miranda, R. C. & Toran-Allerand, C. D. (1992). Developmental expression of estrogen receptor mRNA in the rat cerebral cortex: a nonisotopic in situ hybridization histochemistry study. *Cereb Cortex*, Vol.2, No.1, pp. 1-15

Miyazaki, K., Narita, N., & Narita, M. (2005). Maternal administration of thalidomide or valproic acid causes abnormal serotonergic neurons in the offspring: implication for pathogenesis of autism. *Int J Dev Neurosci*, Vol.23, No.2-3, pp. 287-297

Moore, S. J., Turnpenny, P., Quinn, A., Glover, S., Lloyd, D. J., Montgomery, T., & Dean, J. C. (2000). A clinical study of 57 children with fetal anticonvulsant syndromes. *J Med Genet*, Vol.37, No.7, pp. 489-497

Morley-Fletcher, S., Rea, M., Maccari, S., & Laviola, G. (2003). Environmental enrichment during adolescence reverses the effects of prenatal stress on play behaviour and HPA axis reactivity in rats. *Eur J Neurosci*, Vol.18, No.12, pp. 3367-3374

Moss, J. F., Oliver, C., Berg, K., Kaur, G., Jephcott, L., & Cornish, K. (2008). Prevalence of autism spectrum phenomenology in Cornelia de Lange and Cri du Chat syndromes. *Am J Ment Retard*, Vol.113, No.4, pp. 278-291

Moy, S. S., Nadler, J. J., Perez, A., Barbaro, R. P., Johns, J. M., Magnuson, T. R., Piven, J., & Crawley, J. N. (2004). Sociability and preference for social novelty in five inbred strains: an approach to assess autistic-like behavior in mice. *Genes Brain Behav*, Vol.3, No.5, pp. 287-302

Moy, S. S., Nadler, J. J., Young, N. B., Nonneman, R. J., Grossman, A. W., Murphy, D. L., D'Ercole, A. J., Crawley, J. N., Magnuson, T. R., & Lauder, J. M. (2009). Social approach in genetically engineered mouse lines relevant to autism. *Genes Brain Behav*, Vol.8, No.2, pp. 129-142

Moy, S. S., Nadler, J. J., Young, N. B., Nonneman, R. J., Segall, S. K., Andrade, G. M., Crawley, J. N., & Magnuson, T. R. (2008). Social approach and repetitive behavior in eleven inbred mouse strains. *Behav Brain Res*, Vol.191, No.1, pp. 118-129

Moy, S. S. & Nadler, J. J. (2008). Advances in behavioral genetics: mouse models of autism. *Mol Psychiatry*, Vol.13, No.1, pp. 4-26

Mullins, M. & Rincover, A. (1985). Comparing autistic and normal children along the dimensions of reinforcement maximization, stimulus sampling, and responsiveness to extinction. *J Exp Child Psychol*, Vol.40, No.2, pp. 350-374

Murawski, N. J., Brown, K. L., & Stanton, M. E. (2009). Interstimulus interval (ISI) discrimination of the conditioned eyeblink response in a rodent model of autism. *Behav Brain Res*, Vol.196, No.2, pp. 297-303

Muris, P., Steerneman, P., Merckelbach, H., Holdrinet, I., & Meesters, C. (1998). Comorbid anxiety symptoms in children with pervasive developmental disorders. *J Anxiety Disord*, Vol.12, No.4, pp. 387-393

Myers, S. M. (2007). The status of pharmacotherapy for autism spectrum disorders. *Expert Opin Pharmacother*, Vol.8, No.11, pp. 1579-1603

Nadler, J. J., Moy, S. S., Dold, G., Trang, D., Simmons, N., Perez, A., Young, N. B., Barbaro, R. P., Piven, J., Magnuson, T. R., & Crawley, J. N. (2004). Automated apparatus for quantitation of social approach behaviors in mice. *Genes Brain Behav*, Vol.3, No.5, pp. 303-314

Nanson, J. L. (1992). Autism in fetal alcohol syndrome: a report of six cases. *Alcohol Clin Exp Res*, Vol.16, No.3, pp. 558-565

Narita, M., Oyabu, A., Imura, Y., Kamada, N., Yokoyama, T., Tano, K., Uchida, A., & Narita, N. (2010). Nonexploratory movement and behavioral alterations in a thalidomide or valproic acid-induced autism model rat. *Neurosci Res*, Vol.66, No.1, pp. 2-6

Narita, N., Kato, M., Tazoe, M., Miyazaki, K., Narita, M., & Okado, N. (2002). Increased monoamine concentration in the brain and blood of fetal thalidomide- and valproic acid-exposed rat: putative animal models for autism. *Pediatr Res*, Vol.52, No.4, pp. 576-579

Nau, H., Hauck, R. S., & Ehlers, K. (1991). Valproic acid-induced neural tube defects in mouse and human: aspects of chirality, alternative drug development, pharmacokinetics and possible mechanisms. *Pharmacol Toxicol*, Vol.69, No.5, pp. 310-321

Nelson, K. B., Grether, J. K., Croen, L. A., Dambrosia, J. M., Dickens, B. F., Jelliffe, L. L., Hansen, R. L., & Phillips, T. M. (2001). Neuropeptides and neurotrophins in neonatal blood of children with autism or mental retardation. *Ann Neurol*, Vol.49, No.5, pp. 597-606

Nelson, P. G., Kuddo, T., Song, E. Y., Dambrosia, J. M., Kohler, S., Satyanarayana, G., Vandunk, C., Grether, J. K., & Nelson, K. B. (2006). Selected neurotrophins, neuropeptides, and cytokines: developmental trajectory and concentrations in neonatal blood of children with autism or Down syndrome. *Int J Dev Neurosci*, Vol.24, No.1, pp. 73-80

Newschaffer, C. J., Croen, L. A., Daniels, J., Giarelli, E., Grether, J. K., Levy, S. E., Mandell, D. S., Miller, L. A., Pinto-Martin, J., Reaven, J., Reynolds, A. M., Rice, C. E., Schendel, D., & Windham, G. C. (2007). The epidemiology of autism spectrum disorders. *Annu Rev Public Health*, Vol.28, pp. 235-258

Ornitz, E. M., Guthrie, D., & Farley, A. H. (1977). The early development of autistic children. *J Autism Child Schizophr*, Vol.7, No.3, pp. 207-229

Orsmond, G. I. & Seltzer, M. M. (2007). Siblings of individuals with autism spectrum disorders across the life course. *Ment Retard Dev Disabil Res Rev*, Vol.13, No.4, pp. 313-320

Osterling, J. & Dawson, G. (1994). Early recognition of children with autism: a study of first birthday home videotapes. *J Autism Dev Disord*, Vol.24, No.3, pp. 247-257

Ozonoff, S. & Cathcart, K. (1998). Effectiveness of a home program intervention for young children with autism. *J Autism Dev Disord*, Vol.28, No.1, pp. 25-32

Ozonoff, S., Williams, B. J., Gale, S., & Miller, J. N. (1999). Autism and autistic behavior in Joubert syndrome. *J Child Neurol*, Vol.14, No.10, pp. 636-641

Pardo, C. A. & Eberhart, C. G. (2007). The neurobiology of autism. *Brain Pathol*, Vol.17, No.4, pp. 434-447

Persico, A. M. & Bourgeron, T. (2006). Searching for ways out of the autism maze: genetic, epigenetic and environmental clues. *Trends Neurosci*, Vol.29, No.7, pp. 349-358

Persico, A. M., Pascucci, T., Puglisi-Allegra, S., Militerni, R., Bravaccio, C., Schneider, C., Melmed, R., Trillo, S., Montecchi, F., Palermo, M., Rabinowitz, D., Reichelt, K. L., Conciatori, M., Marino, R., & Keller, F. (2002). Serotonin transporter gene promoter variants do not explain the hyperserotoninemia in autistic children. *Mol Psychiatry*, Vol.7, No.7, pp. 795-800

Petronis, A. (2010). Epigenetics as a unifying principle in the aetiology of complex traits and diseases. *Nature*, Vol.465, No.7299, pp. 721-727

Petrovsky, N. (2001). Towards a unified model of neuroendocrine-immune interaction. *Immunol Cell Biol*, Vol.79, No.4, pp. 350-357

Phiel, C. J., Zhang, F., Huang, E. Y., Guenther, M. G., Lazar, M. A., & Klein, P. S. (2001). Histone deacetylase is a direct target of valproic acid, a potent anticonvulsant, mood stabilizer, and teratogen. *J Biol Chem*, Vol.276, No.39, pp. 36734-36741

Pierce, K. & Courchesne, E. (2001). Evidence for a cerebellar role in reduced exploration and stereotyped behavior in autism. *Biol Psychiatry*, Vol.49, No.8, pp. 655-664

Piven, J., Palmer, P., Landa, R., Santangelo, S., Jacobi, D., & Childress, D. (1997). Personality and language characteristics in parents from multiple-incidence autism families. *Am J Med Genet*, Vol.74, No.4, pp. 398-411

Pletnikov, M. V., Jones, M. L., Rubin, S. A., Moran, T. H., & Carbone, K. M. (2001). Rat model of autism spectrum disorders. Genetic background effects on Borna disease virus-induced developmental brain damage. *Ann N Y Acad Sci*, Vol.939, pp. 318-319

Pletnikov, M. V., Moran, T. H., & Carbone, K. M. (2002a). Borna disease virus infection of the neonatal rat: developmental brain injury model of autism spectrum disorders. *Front Biosci*, Vol.7, pp. d593-d607

Pletnikov, M. V., Rubin, S. A., Moran, T. H., & Carbone, K. M. (2003). Exploring the cerebellum with a new tool: neonatal Borna disease virus (BDV) infection of the rat's brain. *Cerebellum*, Vol.2, No.1, pp. 62-70

Pletnikov, M. V., Rubin, S. A., Schwartz, G. J., Moran, T. H., Sobotka, T. J., & Carbone, K. M. (1999a). Persistent neonatal Borna disease virus (BDV) infection of the brain causes chronic emotional abnormalities in adult rats. *Physiol Behav*, Vol.66, No.5, pp. 823-831

Pletnikov, M. V., Rubin, S. A., Vasudevan, K., Moran, T. H., & Carbone, K. M. (1999b). Developmental brain injury associated with abnormal play behavior in neonatally Borna disease virus-infected Lewis rats: a model of autism. *Behav Brain Res*, Vol.100, No.1-2, pp. 43-50

Pletnikov, M. V., Rubin, S. A., Vogel, M. W., Moran, T. H., & Carbone, K. M. (2002b). Effects of genetic background on neonatal Borna disease virus infection-induced neurodevelopmental damage. I. Brain pathology and behavioral deficits. *Brain Res*, Vol.944, No.1-2, pp. 97-107

Portfors, C. V. (2007). Types and functions of ultrasonic vocalizations in laboratory rats and mice. *J Am Assoc Lab Anim Sci*, Vol.46, No.1, pp. 28-34

Pring, L. (2005). Savant talent. *Dev Med Child Neurol*, Vol.47, No.7, pp. 500-503

Rampon, C., Jiang, C. H., Dong, H., Tang, Y. P., Lockhart, D. J., Schultz, P. G., Tsien, J. Z., & Hu, Y. (2000). Effects of environmental enrichment on gene expression in the brain. *Proc Natl Acad Sci U S A*, Vol.97, No.23, pp. 12880-12884

Rasalam, A. D., Hailey, H., Williams, J. H., Moore, S. J., Turnpenny, P. D., Lloyd, D. J., & Dean, J. C. (2005). Characteristics of fetal anticonvulsant syndrome associated autistic disorder. *Dev Med Child Neurol*, Vol.47, No.8, pp. 551-555

Reid, G., Metivier, R., Lin, C. Y., Denger, S., Ibberson, D., Ivacevic, T., Brand, H., Benes, V., Liu, E. T., & Gannon, F. (2005). Multiple mechanisms induce transcriptional silencing of a subset of genes, including oestrogen receptor alpha, in response to deacetylase inhibition by valproic acid and trichostatin A. *Oncogene*, Vol.24, No.31, pp. 4894-4907

Restivo, L., Ferrari, F., Passino, E., Sgobio, C., Bock, J., Oostra, B. A., Bagni, C., & Ammassari-Teule, M. (2005). Enriched environment promotes behavioral and morphological recovery in a mouse model for the fragile X syndrome. *Proc Natl Acad Sci U S A*, Vol.102, No.32, pp. 11557-11562

Richler, J., Bishop, S. L., Kleinke, J. R., & Lord, C. (2007). Restricted and repetitive behaviors in young children with autism spectrum disorders. *J Autism Dev Disord*, Vol.37, No.1, pp. 73-85

Rinaldi, T., Kulangara, K., Antoniello, K., & Markram, H. (2007). Elevated NMDA receptor levels and enhanced postsynaptic long-term potentiation induced by prenatal exposure to valproic acid. *Proc Natl Acad Sci U S A*, Vol.104, No.33, pp. 13501-13506

Rinaldi, T., Perrodin, C., & Markram, H. (2008a). Hyper-connectivity and hyper-plasticity in the medial prefrontal cortex in the valproic Acid animal model of autism. *Front Neural Circuits*, Vol.2, pp. 4

Rinaldi, T., Silberberg, G., & Markram, H. (2008b). Hyperconnectivity of local neocortical microcircuitry induced by prenatal exposure to valproic acid. *Cereb Cortex*, Vol.18, No.4, pp. 763-770

Rodier, P. M., Bryson, S. E., & Welch, J. P. (1997a). Minor malformations and physical measurements in autism: data from Nova Scotia. *Teratology*, Vol.55, No.5, pp. 319-325

Rodier, P. M., Ingram, J. L., Tisdale, B., & Croog, V. J. (1997b). Linking etiologies in humans and animal models: studies of autism. *Reprod Toxicol*, Vol.11, No.2-3, pp. 417-422

Rodier, P. M., Ingram, J. L., Tisdale, B., Nelson, S., & Romano, J. (1996). Embryological origin for autism: developmental anomalies of the cranial nerve motor nuclei. *J Comp Neurol*, Vol.370, No.2, pp. 247-261

Roof, R. L. & Hall, E. D. (2000a). Estrogen-related gender difference in survival rate and cortical blood flow after impact-acceleration head injury in rats. *J Neurotrauma*, Vol.17, No.12, pp. 1155-1169

Roof, R. L. & Hall, E. D. (2000b). Gender differences in acute CNS trauma and stroke: neuroprotective effects of estrogen and progesterone. *J Neurotrauma*, Vol.17, No.5, pp. 367-388

Rosenberg, G. (2007). The mechanisms of action of valproate in neuropsychiatric disorders: can we see the forest for the trees? *Cell Mol Life Sci*, Vol.64, No.16, pp. 2090-2103

Rosenthal, J., Massie, H., & Wulff, K. (1980). A comparison of cognitive development in normal and psychotic children in the first two years of life from home movies. *J Autism Dev Disord*, Vol.10, No.4, pp. 433-444

Roullet, F. I., Wollaston, L., Decatanzaro, D., & Foster, J. A. (2010). Behavioral and molecular changes in the mouse in response to prenatal exposure to the anti-epileptic drug valproic acid. *Neuroscience*, Vol.170, No.2, pp. 514-522

Rubin, S. A., Sylves, P., Vogel, M., Pletnikov, M., Moran, T. H., Schwartz, G. J., & Carbone, K. M. (1999). Borna disease virus-induced hippocampal dentate gyrus damage is associated with spatial learning and memory deficits. *Brain Res Bull*, Vol.48, No.1, pp. 23-30

Russell, J. & Hill, E. L. (2001). Action-monitoring and intention reporting in children with autism. *J Child Psychol Psychiatry*, Vol.42, No.3, pp. 317-328

Rutter, M. (1968). Concepts of autism: a review of research. *J Child Psychol Psychiatry*, Vol.9, No.1, pp. 1-25

Rutter, M., Greenfeld, D., & Lockyer, L. (1967). A five to fifteen year follow-up study of infantile psychosis. II. Social and behavioural outcome. *Br J Psychiatry*, Vol.113, No.504, pp. 1183-1199

Sakai, T., Matsuda, H., Watanabe, N., Kamei, A., & Takashima, S. (1994). Olivocerebellar retrograde trans-synaptic degeneration from the lateral cerebellar hemisphere to the medial inferior olivary nucleus in an infant. *Brain Dev*, Vol.16, No.3, pp. 229-232

Samaco, R. C., Hogart, A., & Lasalle, J. M. (2005). Epigenetic overlap in autism-spectrum neurodevelopmental disorders: MECP2 deficiency causes reduced expression of UBE3A and GABRB3. *Hum Mol Genet*, Vol.14, No.4, pp. 483-492

Scattoni, M. L., Crawley, J., & Ricceri, L. (2009). Ultrasonic vocalizations: a tool for behavioural phenotyping of mouse models of neurodevelopmental disorders. *Neurosci Biobehav Rev*, Vol.33, No.4, pp. 508-515

Scharre, J. E. & Creedon, M. P. (1992). Assessment of visual function in autistic children. *Optom Vis Sci*, Vol.69, No.6, pp. 433-439

Schneider, T., Labuz, D., & Przewlocki, R. (2001). Nociceptive changes in rats after prenatal exposure to valproic acid. *Pol J Pharmacol*, Vol.53, No.5, pp. 531-534

Schneider, T. & Przewlocki, R. (2005). Behavioral alterations in rats prenatally exposed to valproic acid: animal model of autism. *Neuropsychopharmacology*, Vol.30, No.1, pp. 80-89

Schneider, T. & Przewlocki, R. (2007). Increased sensitivity to pentylenetetrazole induced but not audiogenic seizures in an animal model of autism created by prenatal exposure to valproic acid. *Neural Plasticity*, pp. 79

Schneider, T., Roman, A., Basta-Kaim, A., Kubera, M., Budziszewska, B., Schneider, K., & Przewlocki, R. (2008). Gender-specific behavioral and immunological alterations in an animal model of autism induced by prenatal exposure to valproic acid. *Psychoneuroendocrinology*, Vol.33, No.6, pp. 728-740

Schneider, T., Turczak, J., & Przewlocki, R. (2006). Environmental enrichment reverses behavioral alterations in rats prenatally exposed to valproic acid: issues for a therapeutic approach in autism. *Neuropsychopharmacology*, Vol.31, No.1, pp. 36-46

Schneider, T., Ziolkowska, B., Gieryk, A., Tyminska, A., & Przewlocki, R. (2007). Prenatal exposure to valproic acid disturbs the enkephalinergic system functioning, basal hedonic tone, and emotional responses in an animal model of autism. *Psychopharmacology (Berl)*, Vol.193, No.4, pp. 547-555

Schumacher, H. J., Terapane, J., Jordan, R. L., & Wilson, J. G. (1972). The teratogenic activity of a thalidomide analogue, EM 12 in rabbits, rats, and monkeys. *Teratology*, Vol.5, No.2, pp. 233-240

Schumann, C. M., Bloss, C. S., Barnes, C. C., Wideman, G. M., Carper, R. A., Akshoomoff, N., Pierce, K., Hagler, D., Schork, N., Lord, C., & Courchesne, E. (2010). Longitudinal magnetic resonance imaging study of cortical development through early childhood in autism. *J Neurosci*, Vol.30, No.12, pp. 4419-4427

Schumann, C. M., Hamstra, J., Goodlin-Jones, B. L., Lotspeich, L. J., Kwon, H., Buonocore, M. H., Lammers, C. R., Reiss, A. L., & Amaral, D. G. (2004). The amygdala is enlarged in children but not adolescents with autism; the hippocampus is enlarged at all ages. *J Neurosci*, Vol.24, No.28, pp. 6392-6401

Sears, L. L., Finn, P. R., & Steinmetz, J. E. (1994). Abnormal classical eye-blink conditioning in autism. *J Autism Dev Disord*, Vol.24, No.6, pp. 737-751

Shepard, T. H. (1995). Mobius syndrome after misoprostol: a possible teratogenic mechanism. *Lancet*, Vol.346, No.8977, pp. 780

Sikora, D. M., Pettit-Kekel, K., Penfield, J., Merkens, L. S., & Steiner, R. D. (2006). The near universal presence of autism spectrum disorders in children with Smith-Lemli-Opitz syndrome. *Am J Med Genet A*, Vol.140, No.14, pp. 1511-1518

Silva, G. T., Le Be, J. V., Riachi, I., Rinaldi, T., Markram, K., & Markram, H. (2009). Enhanced long-term microcircuit plasticity in the valproic Acid animal model of autism. *Front Synaptic Neurosci*, Vol.1, pp. 1

Silverman, J. L., Yang, M., Lord, C., & Crawley, J. N. (2010). Behavioural phenotyping assays for mouse models of autism. *Nat Rev Neurosci*, Vol.11, No.7, pp. 490-502

Singh, V. K., Warren, R. P., Odell, J. D., & Cole, P. (1991). Changes of soluble interleukin-2, interleukin-2 receptor, T8 antigen, and interleukin-1 in the serum of autistic children. *Clin Immunol Immunopathol*, Vol.61, No.3, pp. 448-455

Singh, V. K., Singh, E. A., & Warren, R. P. (1997). Hyperserotoninemia and serotonin receptor antibodies in children with autism but not mental retardation. *Biol Psychiatry*, Vol.41, No.6, pp. 753-755

Snow, W. M., Hartle, K., & Ivanco, T. L. (2008). Altered morphology of motor cortex neurons in the VPA rat model of autism. *Dev Psychobiol*, Vol.50, No.7, pp. 633-639

Sparks, B. F., Friedman, S. D., Shaw, D. W., Aylward, E. H., Echelard, D., Artru, A. A., Maravilla, K. R., Giedd, J. N., Munson, J., Dawson, G., & Dager, S. R. (2002). Brain structural abnormalities in young children with autism spectrum disorder. *Neurology*, Vol.59, No.2, pp. 184-192

Stanton, M. E., Peloso, E., Brown, K. L., & Rodier, P. (2007). Discrimination learning and reversal of the conditioned eyeblink reflex in a rodent model of autism. *Behav Brain Res*, Vol.176, No.1, pp. 133-140

Stein, D. G. (2001). Brain damage, sex hormones and recovery: a new role for progesterone and estrogen? *Trends Neurosci*, Vol.24, No.7, pp. 386-391

Stein, M. B., Goldin, P. R., Sareen, J., Zorrilla, L. T., & Brown, G. G. (2002). Increased amygdala activation to angry and contemptuous faces in generalized social phobia. *Arch Gen Psychiatry*, Vol.59, No.11, pp. 1027-1034

Stein, M. B., Simmons, A. N., Feinstein, J. S., & Paulus, M. P. (2007). Increased amygdala and insula activation during emotion processing in anxiety-prone subjects. *Am J Psychiatry*, Vol.164, No.2, pp. 318-327

Stromland, K., Nordin, V., Miller, M., Akerstrom, B., & Gillberg, C. (1994). Autism in thalidomide embryopathy: a population study. *Dev Med Child Neurol*, Vol.36, No.4, pp. 351-356

Sztainberg, Y., Kuperman, Y., Tsoory, M., Lebow, M., & Chen, A. (2010). The anxiolytic effect of environmental enrichment is mediated via amygdalar CRF receptor type 1. *Mol Psychiatry*, Vol.15, No.9, pp. 905-917

Takashima, S. (1982). Olivocerebellar lesions in infants born prematurely. *Brain Dev*, Vol.4, No.5, pp. 361-366

Tobia, M. J. & Woodruff-Pak, D. S. (2009). Delay eyeblink classical conditioning is impaired in Fragile X syndrome. *Behav Neurosci*, Vol.123, No.3, pp. 665-676

Tordjman, S., Anderson, G. M., McBride, P. A., Hertzig, M. E., Snow, M. E., Hall, L. M., Thompson, S. M., Ferrari, P., & Cohen, D. J. (1997). Plasma beta-endorphin, adrenocorticotropin hormone, and cortisol in autism. *J Child Psychol Psychiatry*, Vol.38, No.6, pp. 705-715

Tordjman, S., Drapier, D., Bonnot, O., Graignic, R., Fortes, S., Cohen, D., Millet, B., Laurent, C., & Roubertoux, P. L. (2007). Animal models relevant to schizophrenia and autism: validity and limitations. *Behav Genet*, Vol.37, No.1, pp. 61-78

Towbin, K. E., Pradella, A., Gorrindo, T., Pine, D. S., & Leibenluft, E. (2005). Autism spectrum traits in children with mood and anxiety disorders. *J Child Adolesc Psychopharmacol*, Vol.15, No.3, pp. 452-464

Tsujino, N., Nakatani, Y., Seki, Y., Nakasato, A., Nakamura, M., Sugawara, M., & Arita, H. (2007). Abnormality of circadian rhythm accompanied by an increase in frontal cortex serotonin in animal model of autism. *Neurosci Res*, Vol.57, No.2, pp. 289-295

Turner, C. A., Lewis, M. H., & King, M. A. (2003). Environmental enrichment: effects on stereotyped behavior and dendritic morphology. *Dev Psychobiol*, Vol.43, No.1, pp. 20-27

Ueda, K., Nishida, Y., Oshima, K., & Shepard, T. H. (1979). Congenital rubella syndrome: correlation of gestational age at time of maternal rubella with type of defect. *J Pediatr*, Vol.94, No.5, pp. 763-765

Vitalis, T., Cases, O., Callebert, J., Launay, J. M., Price, D. J., Seif, I., & Gaspar, P. (1998). Effects of monoamine oxidase A inhibition on barrel formation in the mouse somatosensory cortex: determination of a sensitive developmental period. *J Comp Neurol*, Vol.393, No.2, pp. 169-184

Vitalis, T., Cases, O., Passemard, S., Callebert, J., & Parnavelas, J. G. (2007). Embryonic depletion of serotonin affects cortical development. *Eur J Neurosci*, Vol.26, No.2, pp. 331-344

Volkmar, F. R. & Nelson, D. S. (1990). Seizure disorders in autism. *J Am Acad Child Adolesc Psychiatry*, Vol.29, No.1, pp. 127-129

Volkmar, F. R., Szatmari, P., & Sparrow, S. S. (1993). Sex differences in pervasive developmental disorders. *J Autism Dev Disord*, Vol.23, No.4, pp. 579-591

Vorhees, C. V. (1987). Teratogenicity and developmental toxicity of valproic acid in rats. *Teratology*, Vol.35, No.2, pp. 195-202

Walker, H. A. (1977). Incidence of minor physical anomaly in autism. *J Autism Child Schizophr*, Vol.7, No.2, pp. 165-176

Warren, R. P., Margaretten, N. C., Pace, N. C., & Foster, A. (1986). Immune abnormalities in patients with autism. *J Autism Dev Disord*, Vol.16, No.2, pp. 189-197

Warren, R. P., Singh, V. K., Averett, R. E., Odell, J. D., Maciulis, A., Burger, R. A., Daniels, W. W., & Warren, W. L. (1996). Immunogenetic studies in autism and related disorders. *Mol Chem Neuropathol*, Vol.28, No.1-3, pp. 77-81

Warren, R. P., Yonk, L. J., Burger, R. A., Cole, P., Odell, J. D., Warren, W. L., White, E., & Singh, V. K. (1990). Deficiency of suppressor-inducer (CD4+CD45RA+) T cells in autism. *Immunol Invest*, Vol.19, No.3, pp. 245-251

Wassink, T. H., Brzustowicz, L. M., Bartlett, C. W., & Szatmari, P. (2004). The search for autism disease genes. *Ment Retard Dev Disabil Res Rev*, Vol.10, No.4, pp. 272-283

Willemsen-Swinkels, S. H. & Buitelaar, J. K. (2002). The autistic spectrum: subgroups, boundaries, and treatment. *Psychiatr Clin North Am*, Vol.25, No.4, pp. 811-836

Williams, G., King, J., Cunningham, M., Stephan, M., Kerr, B., & Hersh, J. H. (2001). Fetal valproate syndrome and autism: additional evidence of an association. *Dev Med Child Neurol*, Vol.43, No.3, pp. 202-206

Williams, P. G. & Hersh, J. H. (1997). A male with fetal valproate syndrome and autism. *Dev Med Child Neurol*, Vol.39, No.9, pp. 632-634

Woodhouse, W., Bailey, A., Rutter, M., Bolton, P., Baird, G., & Le, C. A. (1996). Head circumference in autism and other pervasive developmental disorders. *J Child Psychol Psychiatry*, Vol.37, No.6, pp. 665-671

Woodruff-Pak, D. S. & Papka, M. (1996). Huntington's disease and eyeblink classical conditioning: normal learning but abnormal timing. *J Int Neuropsychol Soc*, Vol.2, No.4, pp. 323-334

Young, D., Lawlor, P. A., Leone, P., Dragunow, M., & During, M. J. (1999). Environmental enrichment inhibits spontaneous apoptosis, prevents seizures and is neuroprotective. *Nat Med*, Vol.5, No.4, pp. 448-453

Zwaigenbaum, L., Bryson, S., Rogers, T., Roberts, W., Brian, J., & Szatmari, P. (2005). Behavioral manifestations of autism in the first year of life. *Int J Dev Neurosci*, Vol.23, No.2-3, pp. 143-152

13

Environmentally Induced Oxidative Stress and Disruption of Brain Thyroid Hormone Homeostasis in Autism Spectrum Disorders

Elizabeth M. Sajdel-Sulkowska
Department of Psychiatry, Harvard Medical School and Brigham and Women's Hospital, Boston, MA, USA

In memory of my son and a budding neuroscientist, Zachary L. Sulkowski, BS

1. Introduction

Autism Spectrum Disorder (ASD) is a group of neuropsychiatric disorders characterized by impairments of language, social interactions, and movements and involves neurochemical, morphological, and neuroanatomical changes in specific brain regions including several cortical regions: the cerebellum, corpus callosum, basal ganglia and the limbic system (Sajdel-Sulkowska et al., 2010). It affects 1% of the births and its incidence is on the rise. Although its causation is assumed to have a strong genetic component, most of the known genetic risks have been associated with copy number variants (CNVs). Even an international genome-wide scan (AGP; Anney et al., 2010) failed to discover "the critical autism loci". Furthermore, ASD concordance for monozygotic twins aged 18 years and younger, is less than 90 percent (Rosenberg et al., 2009). Thus nongenetic, environmental triggers of ASD pathology are gaining recognition as likely causal factors although the mechanisms involved in the environmental impact are not fully understood.

This chapter focuses on the developmental impact of environmental pollutants that interfere with the thyroid hormone (TH), a key hormone involved in the regulation of brain development (Oppenheimer and Schwartz, 1997), as a possible factor contributing to autistic pathology. Many environmental toxicants, such as herbicides, polychlorinated biphenyls (PCBs), bisphenol A (BPA) and organic mercury compounds are potent disruptors of the endocrine system including TH. TH plays a critical role in brain development by virtue of regulating cellular metabolism, growth, differentiation and maturation and is indispensable for the proper development of the central nervous system (CNS). TH deficiency during CNS development results in disorders such as cretinism and a spectrum of psychoneurological disorders including both neurological and cognitive deficiencies (Vermiglio et al., 1995). It has been suggested that maternal hypothyroxinemia during critical periods may disrupt the developmental processes and produce morphological brain changes leading to autism (Roman, 2007). Hypothyroidism during pregnancy has been proposed as one of the twelve autism risk factors (King, 2011). Yet studies addressing TH plasma levels, both 3′,3,5-triiodothyronine (T3) and 3,5,3′,5′-tetraiodothyronine (thyroxine,T4), and thyroid stimulating hormone (TSH) failed to show major abnormalities in autism; thus TH involvement in autistic pathology has been ruled out. What has been overlooked in dismissing a TH-autism relationship is the fact that the majority of active TH hormone (T3) in the brain does not come from circulation, but is converted from the prohormone (T4) locally in the brain by deiodinase activity through the removal of iodine. Thus, while plasma TH levels may be within the normal range, its levels in the brain may be inadequate to support normal developmental processes. *In vivo* animal studies suggest that environmental toxicants can affect brain deiodinase activity and are supported by *in vitro* studies suggesting a direct inhibition of deiodinase enzymes by environmental triggers of oxidative stress (Mori et al., 1996; Lamirand et al., 2008). This chapter focuses on the developmental impact of environmental pollutants that trigger oxidative stress and disrupt brain homeostasis of TH, a key hormone involved in the regulation of brain development (Oppenheimer and Schwartz, 1997), as a plausible factor contributing to autistic pathology.

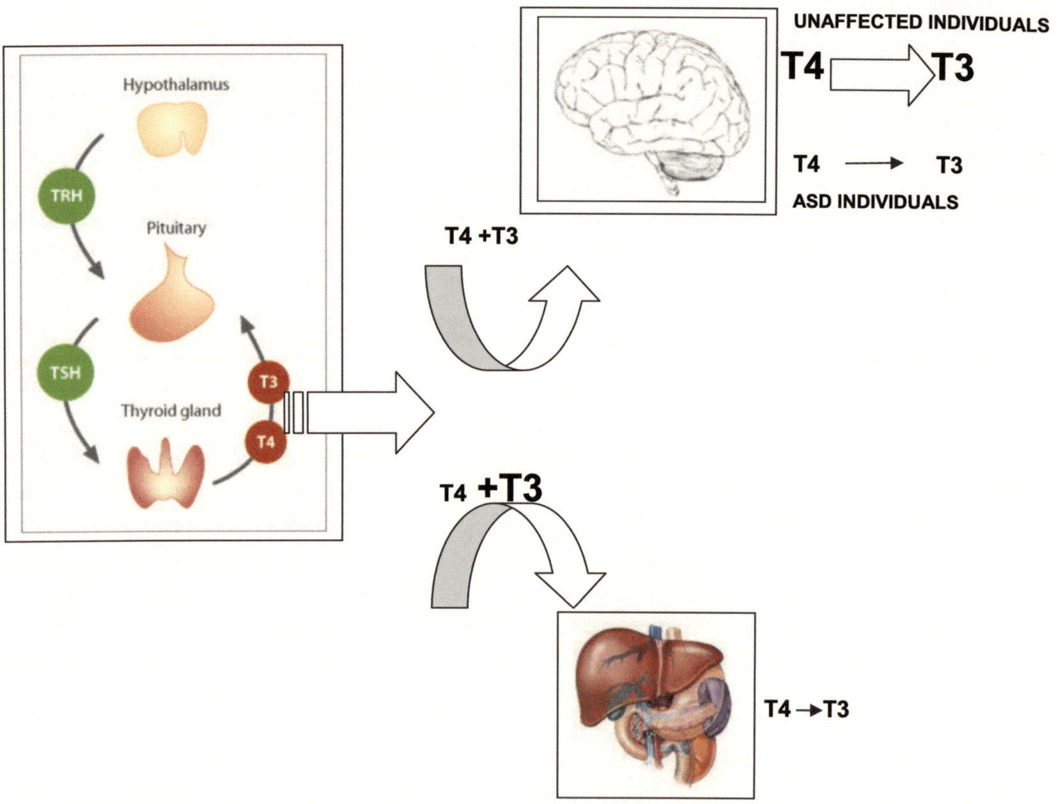

Fig. 1. Relative contribution of plasma-originated vs. tissue originated T3 in the CNS and the peripheral tissues.

2. Brain TH homeostasis: contribution of circulating vs. locally produced active forms of TH, T3

THs, both T3 and T4, are produced in the thyroid gland. The ratio of T3 to T4 released into the blood is 1:20. Both T3 and T4 then reach the individual body organs, where the prohormone T4 is converted to the biologically active hormone T3. The organ/tissue levels of T3 are regulated locally primarily by the activity of two different selenoenzymes, deiodinases type 2 (D2) and type 3 (D3), (Leonard, 1992; Silva et al., 1982; Bianco et al., 2002), although deiodinase type 1 is also involved (D1; Bates et al., 1999). In the CNS, approximately 70-80% of T3 originates from intracerebral T4 to T3 conversion, while the plasma contribution amounts to 20-30 % (Leonard, 1992); Bianco et al., 2002), and D2 is responsible for most of the T3 supply within the brain (Crantz et al., 1982). Mice with a globally targeted disruption of the *Dio2* gene (D2KO mice) have ~50% less T3 content in their cerebral cortex, cerebellum, and hypothalamus (Galton et al., 2007). The extent of the local brain T4 to T3 conversion is in contrast to the peripheral tissues where T3 comes mostly from plasma. In the brain T3 exerts its major effect by binding to the nuclear TH receptor (TR), a ligand-regulated transcription factor, and regulates T3-dependent gene transcription. TR-mediated transcription may be modulated by various substances. The nuclear hormone receptor superfamily contains more than 40 transcriptional factors and most of these receptors are present in the brain. (Koibuchi et al., 2003).

Excess T4 and T3 are then converted to inactive metabolites rT3 and 3,3′-diiodothyronine (T2) by D3. D2 is localized mainly in the glial cells (Guadano-Ferraz et al., 1997), but the Purkinje cell localization has been observed during specific developmental periods (Verhoelst et al., 2005). D3 is localized mainly in neurons including the Purkinje cells (Verhoelst et al., 2002). D3 activity increases in hyperthyroidism, and decreases in hypothyroidism (Dratman et al., 1983) and is thought to protect neurons from excessive T3 levels. D2 and D3 activity balance has been shown to be critical for the regulation of the intraneuronal level of the active form of T3 (Leonard, 1982; Bianco et al., 2002). Both D2 and D3 activities have been demonstrated in the human brain (Campos-Barros et al., 1996).

While the majority of brain T3 is derived through the conversion of T4 to T3 by D2 coded by the Dio2 gene, some T3 is transported from plasma through the blood-brain barrier, a process mediated in part by the monocarboxylate transporter 8 (Mct8/MCT8). Using mice with inactivated Mct8 (Slc16a2) and Dio2 genes it has been shown that T3 from plasma and intracerebrally generated T3 play a distinct role in the brain and specifically in the regulation of TH-dependent gene expression. Inactivation of the Mct8 gene (Mct8KO) was without effect on the expression of 31 of these genes, but Dio2 inactivation selectively affected the expression of negatively regulated genes (Morte et al., 2010). In our recent study, thimerosal (TM) exposure resulted in decreased cerebellar D2 activity and overexpression of genes negatively regulated by TH (Sulkowski et al., accepted).

3. Systemic changes in TH in autism

Several clinical studies, to date, have shown no evidence of TH abnormalities in autism. The study of a small group of patients between the ages of 7 and 21 years showed no clinical evidence of hypothyroidism with reported levels of plasma T3, T4, and TSH all within the normal range (Abbassi et al., 1978). Similarly, a study of a larger population of autistic children showed normal levels of T3, T4 and TSH (Cohen et al., 1980). On the other hand, others reported significantly lower TSH levels in autistic boys as compared to mentally retarded or control groups (Hashimoto et al., 1991) and marginal changes in diurnal rhythms of serum TSH (Nir et al., 1995). Thus, while the evidence for the involvement of TH in autistic pathology is not compelling, there appears to be a tendency for TSH abnormalities in autism. Based on these findings further research for TH involvement in autism has been abandoned. However, others have tested the theory of a mild neonatal hypothyroidism in autism in animal models (Sadamatsu et al., 2006)

4. Altered deiodinase activities and brain TH homeostasis in other pathologies

Considering that most of the brain T3 is generated by the activity of D2, it is surprising that no studies of the deiodinase activity in autism have been reported. Interestingly, in Alzheimer's disease, where there is also no evidence of systemic TH abnormalities is also missing, as assessed by serum TH levels (McKhann et al., 1984), there is evidence of localized intra-brain hypothyroidism. Direct measures of T3 and T4 in the postmortem AD brains indicated no changes in T4 levels, but significantly lower T3 levels in advanced stages of the disease (Davis et al., 2008), suggesting decreased conversion of T4 to T3, possibly due to decreased D2 activity. Furthermore, both the level of rT3 and the rT3:T4 ratio in the cerebrospinal fluid (CSF) are significantly increased, suggesting an abnormal intracerebellar TH metabolism most likely due to an increase in D3 activity (Sampaolo et al., 2005). An increase in the CSF rT3 concentration has been found in other disorders involving the CNS. The CSF levels of T4 and free rT3 were increased during endogenous depression as compared to levels after recovery suggesting increased production of rT3 from T4 in the brain (Kirkegaard and Faber, 1991). These observations lend further support to the concept of local intra-brain regulation of TH homeostasis and its relevance to various pathological conditions.

5. Disruption of brain TH homeostasis by environmental toxicants

Considering the absence of systemic TH abnormalities in autism and postulating the impact of environmental toxicants on brain TH homeostasis, we will examine some of their neurotoxic properties. Many environmental pollutants, including BPA, PCBs, organochlorine (dicofol, endosulfan) and organophosphate (Diazinon) pesticides, as well as metals such as lead, mercury and cadmium (Schantz and Windholm (2001) are considered to be endocrine disruptors. While most of them have been classified as endocrine disruptors, some of them, like PCBs (Venkataraman et al., 2007) and PBDE (Messer et al., 2010) and perchlorates (Brar et al., 2010), are also classified as TH disruptors. PCBs and PBDEs compete with T3 by virtue of having a similar chemical structure (Koibuchi et al, 2003). Table 1 summarizes the data on environmental toxicants implicated in ASD pathology.

TH plays a critical role in brain development, and thus toxicants that affect TH homeostasis are most likely to interfere with brain development. It has been proposed that transient maternal hypothyroxinemia induced by environmental antithyroid agents such as PCBs, perchlorates, mercury and coal derivatives, could contribute to autistic pathology (Roman, 2009). This hypothesis was based on a leading ecological study in Texas that correlated higher levels of autism with the environmental release of mercury from industrial sources (Palmer et al., 2006). A potential association between autism and metal concentrations, including mercury, has been reported in the San Francisco Bay area (Windham et al., 2006). A similar relationship has been postulated between autism and polybrominated diphenyl esters (PBDEs), potent thyroid hormone mimetics, used in home furnishings and electronics (Messer et al., 2010). Some of the toxicants interfere directly with TH synthesis and alter plasma TH levels, others bind plasma to the TH transport protein, transthyretin (TTR), resulting in a lower rate of T4 transport to the fetal brain (Schroder-van der Elst et al., 1998). However, others like mercury do not produce changes in the circulating TH (Watanabe et al., 1999), and yet they disrupt TH actions.

TOXICANT	SOURCE	ASSOCIATION WITH AUTISM	ENDOCRINE DISRUPTOR	TH DISRUPTOR	PLASMA TH (T3,T4, TSH)	BRAIN T3/T4 D2, D3	OXIDATIVE STRESS IN BRAIN
BPA	Plastics	Brown, 2009	Aydogan et al., 2008		↑ TH Zoeller et al., 2005; no change: Nieminen et al., 2002; Kobayashi et al., 2002	?	Jain et al., 2011
DICOFOL	Pesticides	Roberts et al., 2007			↓ T4:Van den Berg et al., 1991	?	?
ENDOSUL-FAN	Pesticides	Roberts et al., 2007	Schoeters et al., 2008		?	?	Hinkal et al., 1995
METHYL-MERCURY/ ETHYL-MERCURY	Industrial byproducts/ Pharmaceutical Air, food	Tan et al., 2009; Windham et al., 2006	Sringari et al., 2008		No change: Watanabe et al., 1999; lower: Tan et al., 2009	(Sulkowski et al., submitted)	Stringari et al., 2008
PCBs	Industrial byproducts, food	Roman, 2009; Kimura-Kuroda et al., 2007	Venkataraman et al., 2007	Brar et al., 2008	↓ Total and free T4: Morse et al., 1996		Vendkataraman et al., 2007; Hassoun et al., 2010
PBDE	Flame retardants	Messer et al., 2010	Messer et al, 2010	TH mimetic: Messer et al., 2010			Giordano et al., 2008; Zhang et al., 2010
PERCHLOR ATES	Drinking water	Roman et al, 2009	Roman, 2009	Bekkedal et al., 2004	Liuet al., 2008		Liu et al., 2008

Table 1. Environmental toxicants associated with autistic pathology.

As discussed above, the major source of the biologically active hormone T3 in the brain is the local intra-brain conversion of T4 to T3, while a small fraction comes from circulating T3. Thus it is possible that a direct action on some of the endocrine disruptors on brain deiodinases affects brain TH homeostasis. Indeed, we have observed the inhibition of the brain deiodinase D2 following perinatal exposure to TM (Sulkowski et al., accepted).

Most of the toxicants implicated in ASD pathology are also potent triggers of oxidative stress (Table 1). As evidence derived from in vitro studies suggests, in response to oxidative stress D3 increases while D2 decreases (Lamirand et al., 2008; Freitas et al., 2010). Thus it is likely that the effect of many of these toxicants on brain deiodinases is mediated via mechanisms involving oxidative stress (Sulkowski et al., accepted).

Many of the toxicants, including heavy metals, (Bokara et al., 2008) and specifically mercury (Hg; Windham et al., 2006; Palmer et al., 2009), have been identified as factors exerting a range of harmful neurological and cognitive effects in humans and experimental animals, and have been implicated in the etiology of a number of neuropsychiatric disorders including Alzheimer's disease (Gerhardsson et al., 2008), Parkinson's disease (Monnet-Tschudi et al., 2006) and autism (Windham et al., 2006; Palmer et al., 2009). A specifically strong association has been observed between Hg exposure and autism; we will thus consider the Hg effect in relation to brain TH homeostasis in greater detail. The major environmental organic compounds of mercury include methylmercury (Met-Hg) and ethylmercury (Et-Hg). Met-Hg can be found in contaminated fish; Et-Hg is a metabolite of TM used in the United States in some maternal flu vaccines and in infant vaccines in the developing countries (Sulkowski et al., accepted). Hg compounds accumulate significantly in the pituitary and thyroid glands in both animals (Nishida et al., 1986) and humans (Kosta et al., 1975), and interfere with the hypothalamic-pituitary-thyroid (HTP) axis. Exposure to Met-Hg can produce hypothyroid conditions (Nishida et al., 1989), although changes in TH plasma levels based on both animal and human studies are inconsistent (Tan et al., 2009).

Met-Hg has been shown to cross the placenta (Nordenhall et al., 1995) and Hg also enters the milk (Morgan et al., 2006) and is taken up by suckling pups (Oskarsson et al., 1995). Hg accumulates in both fetal and neonatal brains (Linares et al., 2004; Orct et al., 2006; Zareba et al., 2007) potentially affecting neurodevelopment (Orct et al., 2006). In rats, postnatal exposure (P1-P30) results in impairments in motor coordination and learning (Sakamoto et al., 2004). Perinatal TM exposure in rats results in the impairment of auditory functions and motor learning (Sulkowski et al., accepted). In humans, Met-Hg exposure in expectant mothers due to fish consumption is associated with `increased mercury accumulation in the infant brains accompanied by behavioral abnormalities, which include deficits in motor, attention, and verbal performance that are more pronounced in males (Gao et al., 2007), while the postnatal Met-Hg exposure in humans appears to have no recognizable effects (Debes et al., 2006). Hg compounds in general are potent endocrine disruptors (Heath et al., 2005;Windham et al., 2006; Palmer et al., 2009; Tan et al., 2009) and are also specifically TH disruptors (Stingari et al., 2008).

Organic Hg compounds are also potent triggers of oxidative stress. Exposure to Met-Hg or Et-Hg *in vivo* or *in vitro* (Linares et al., 2004; Kaur et al., 2006; Rush et al., 2009; Glaser et al., 2010; Yin et al., 2011), induces oxidative stress that leads to a cascade of other changes including decreased neurogenesis, increased neuronal apoptosis and impaired synaptic plasticity in the neonatal brain. Results of one of our recently completed studies indicate that perinatal TM exposure increases cerebellar 3-nitrotyrosine (3-NT; Sulkowski et al., accepted), a well accepted marker of oxidative stress found in over fifty different pathologies including autism (Sajdel-Sulkowska, 2010).

Further, Met-Hg is not only a potent trigger of oxidative stress, but also a disruptor of antioxidant defense systems (Chang and Tsai, 2008; Barcelos et al., 2011). Gestational exposure to Met-Hg in mice results in increased lipid peroxidation via interference in brain GSH levels (Stringari et al., 2008), while gestational exposure (G12-G14) in rats to Met-Hg (5 mg/kg) induces oxidative stress and reduces the antioxidant enzyme superoxide dismutase (SOD) in the hippocampus (Vincente et al., 2004).

Hg compounds have been shown to target tissue deiodinases (Sulkowski et al., accepted). Our data derived from *in vivo* experiments in rats, supports results of earlier *in vitro* studies (Lamirand et al., 2008). Other *in vitro* studies indicated that the exposure of neuronal cells to

Met-Hg (Kim et al., 2005) or neuroblastoma cells to TM (James et al., 2005) results in a depletion of GSH which is both an antioxidant and a cofactor of deiodinases (Goswani and Rosenberg, 1988; Bhat et al., 1989; Croteau et al., 1998; Goemann et al., 2010). Thus, cerebellar D2 activity might be impaired due to a lack of the reducing cofactor. In primary astrocyte culture, GSH counteracts the impact of oxidative stress, and decreases D3 activity but increases D2 activity (Lamirannd et al., 2008). It is of interest that T3 regulates GSH levels in the developing brain and treatment of astrocyte cultures with TH results in increased GSH levels and improved antioxidative defense, suggesting that TH plays a positive role in maintaining GSH homeostasis and protecting the brain from oxidative stress (Dasgupta et al., 2007). Thus is it is also possible that a decrease in D2 activity could further amplify the effects of oxidative stress.

As discussed above, tissue levels of T3 are regulated by D2 and D3, which are selenoproteins and are consequently sensitive to selenium availability. Selenium is not only a cofactor of deiodinases but also a potent antioxidant. Thus, environmental contaminants that sequester selenium or induce oxidative stress are likely to affect deiodinase activity. Met-Hg has been shown to interact with selenium (Soldin et al., 2008) and can inhibit the function of selenoproteins such as the deiodinases (Watanabe et al., 2001). We have also shown that TM exposure increases levels of oxidative stress (Sulkowski et al., accepted), which has been found previously to decrease expression of the *Dio2* gene (Lamirannd et al., 2008).

6. Sexually dimorphic responses to environmental endocrine disruptors and sex ratio in autism

When discussing the impact of environmental factors on CNS, it is critical to recognize the sexual dimorphism of their effects (Nguon et al., 2005a). Sex-dependent responses to a number of environmental pollutants including organophosphate pesticides (Dam et al., 2000), have been previously reported. Our earlier studies on the perinatal exposure to PCBs in rats demonstrated sex-dependent effects on cerebellar and motor functions with males being more sensitive (Nguon et al., 2005b). Even at low concentrations, different PCB congeners interfere with TH status in a sex-dependent manner (Abdelouahab et al., 2008). Our recently completed study on the perinatal exposure to TM revealed not only sex- but also strain-dependent effects on motor learning and cerebellar oxidative stress and D2 activity (Sulkowski et al., accepted). Specifically, in the Spontaneously Hypertensive Rats (SHR), a strain more sensitive to inflammation (Ballerio et al., 2007), perinatal exposure to TM resulted in decreased cerebellar D2 activity in male, but not in female neonates, and this decrease was correlated with a disruption of T3-dependent gene expression (Sulkowski et al, accepted). Our findings are in agreement with earlier observations both in humans (Gao et al., 2007) and in animals (Sobutskii et al., 2007) showing that the developing males appear to be more sensitive to Hg exposure. Furthermore, gene profiling studies in the rat cerebellum following perinatal exposure to a number of toxicants including PCBs, pesticides and methylmercury, showed differential sex-dependent effects of on gene expression (Padhi et al., 2008). Although the precise mechanism involved in this dimorphism is not known, in the cerebellum, developmentally-timed progesterone synthesis in the Purkinje cells (Sakamoto et al., 2003), differential regulation of progesterone-receptors by estradiol (Quadros et al., 2002; Guerra-Araiza et al., 2002), and the formation of estradiol from testosterone in the Purkinje cells (Sakamoto et al., 2003), have been implicated in these differential effects. It is thus interesting that the Purkinje cells express D2 at specific times during development (Verhoelst et al., 2005). Therefore it is possible that environmental toxicants interfere with TH homeostasis by acting on the Purkinje cells.

7. Could localized, intra-brain TH deficiency contribute to the pathology of ASD and present new venues for the diagnosis and treatment of autism

It is clear from the preceding discussion that there are no systemic TH changes in autism, that some environmental factors disrupt TH regulation without any effect on systemic TH status, and that it is the local intra-brain T4 to T3 conversion rather than circulating T3 levels that are responsible for the majority of brain T3.

Furthermore, the T3 generated locally in the brain by D2 controls the expression of genes negatively regulated by TH, while plasma T3 controls the expression of the positively regulated genes (Morte et al., 2010). Thus, systemic hypothyroidism that is known to interfere with normal brain development may regulate the expression of genes distinct from those that are regulated by the locally generated T3, and is thus likely to result in a different set of morphological and functional abnormalities. Animal studies have indicated that in the developing rat cerebellum, systemic TH deficiency affects cerebellar granule cell migration. Also, Purkinje cell migration requires reelin (Miyata et al., 2010). Reelin is one of the genes whose abnormal expression is implicated in autism (Fatemi et al., 2005) and is also regulated by T3 produced locally in the fetal brain from T4 by deiodinase activity mostly in astrocytes but also in Purkinje cells (Verhoelst et al., 2005). It is possible that the aberrant Purkinje cell migration in ASD contributes to the decrease in Purkinje cells in ASD (Courchesne, 1991). Furthermore, in ASD, the lower intra-brain T3 levels occur in the absence of a systemic T3 deficiency (Davis et al., 2008), most likely due to the increased activity of D3 (Sampaolo et al., 2005). Similar studies involving postmortem ASD brains are now being initiated in our laboratory.

Although none of the studies so far provide direct evidence for the disruption of brain TH metabolism in autism, there is a sufficient amount of indirect data to warrant pursuing the hypothesis that environmentally induced oxidative stress and local brain hypothyroidism contributes to ASD pathology.

According to this hypothesis, brain region-specific oxidative stress in autism may be associated with increased D3 and decreased D2 activity resulting in a region-specific T3 deficiency in the brain. Future human studies utilizing the CSF of living ASD individuals or postmortem brain tissue of ASD donors will support its validity. Such findings would have several significant implications. They may result in methods of early ASD diagnosis; detection of high brain D3 levels in postmortem human brains may suggest the benefits of measuring the levels of its product (rT3) in the CSF of living patients to assess the risks, monitor the disease progression and efficacy of ongoing treatment. Furthermore, several tissue-specific and TH receptor (TR)-specific thyromimetics have been developed as potential treatment for atherosclerosis, obesity and Type 2 diabetes and might be able to correct local brain TH deficiency without systemic thyrotoxicity (Baxter and Webb, 2009) and may thus be considered as potential therapeutic agents. Finally, confirmation that autism may be associated with increased D3 and decreased D2 activity resulting in a region-specific T3 deficiency in the brain could lead to or reinforce dietary treatments, because D2 activity can be modulated not only by selenium but also by xenobiotic compounds (da-Silva et al., 2007). In conclusion, TH abnormalities in autism warrant a second look.

8. Acknowledgments

The author is indebted to Dr. Ann Marie Zavacki for introducing her to the concept of local brain TH regulation. The author also expresses her gratitude to her friend, Patricia McCann, and son, Professor Adam J. Sulkowski, for their expert editorial work.

9. References

Abbassi V, Linscheid T, Coleman M. Triiodothyronine (T3) concentration and therapy in autistic children. J Autism Child Schizophr. 1978;8:383-7.

Abdelouahab N, Mergler D, Takser L, Vanier C, St-jean M, Baldwin M, Spear PA, Chan HM. Gender differences in the effects of organochlorines, mercury and lead on thyroid hormone levels in lakeside communities of Quebeck (Canada). Environ Res. 2008;107:380-92.

Anney R, Klei L, Pinto D, Regan R, Conroy J, Magalhaes TR. A genome-wide scan for common alleles affecting risk for autism. Hum Mol Genet. 2010;19:4072-82.

Aydoğan M, Korkmaz A, Barlas N, Kolankaya D. The effect of vitamin C on bisphenol A, nonylphenol and octylphenol induced brain damages of male rats. Toxicology. 2008;249:35-9.

Ballerio R, Gianazza E, Mussoni L, Miller I, Gelosa P, Guerrini U, Eberini I, Gemeiner M, Belcredito S, Tremoli E, Sironi L. Gender differences in endothelial function and inflammatory markers along the occurrence of pathological events in stroke-prone rats. Exp Mol Pathol. 2007;82:33-41.

Barcelos GR, Grotto D, Serpeloni JM, Angeli JP, Rocha BA, de Oliveira Souza VC, Vicentini JT, Emanuelli T, Bastos JK, Antunes LM, Knasmuller S, Barbosa F Jr. Protective properties of quercetin against DNA damage and oxidative stress induced by methylmercury in rats. Arch Toxicol. 2011; PMID:21286687.

Bates JM, St. Germain DL, Galton VA. Expression profiles of the three iodothyronine deiodinases, D1, D2, and D3, in the developing brain. Endocrinology 140:844-851, 1999.

Baxter JD, Webb P. Thyroid hormone mimetics: potential applications in atherosclerosis, obesity and type 2 diabetes. Nat Rev Drug Discov. 2009;8:308-20.

Bekkedal MY, Arfsten D, Mattie D. An evaluation of neurobehavioral tests used to assess the neurodevelopmental effects of early ammonium perchlorate exposure. J Toxicol Environ Health 2004;67:835-44.

Bellinger FP, Bedi KS, Wilson P, Wilce PA. Ethanol exposure during the third trimester equivalent results in long-lasting decreased synaptic efficacy but not plasticity in the CA1 region of the rat hippocampus. Synapse 1999;31:51-58.

Bhat GB, Iwase K, Hummel BC, Walfish PG. Kinetic characteristics of a thioredoxin-activated rat hepatic and renal low-Km iodothyronine 5'-deiodinase. Biochem J. 1989;258:785-92.

Bianco AC, Salvatore D, Gereben B, Berry MJ, Larsen PR. Biochemistry, cellular and molecular biology, and physiological roles of the iodothyronine selenodeiodinases. Endocr Rev. 2002;23:38-89.

Bokara KK, Brown E, McCormick R, Yallapragada PR, Rajanna S, Bettaiya R. Lead-induced increase in antioxidant enzymes and lipid peroxidation products in developing rat brain. Biometals 2008; 21:9-16.

Brar NK, Waggoner C, Reyes JA, Fairey R, Kelley KM. Evidence for thyroid endocrine disruption in wild fish in San Francisco Bay, California, USA. Relationships to contaminant exposures. Aquat Toxicol. 2010;96:203-15.

Brown JS Jr. Effects of bisphenol-A and other endocrine disruptors compared with abnormalities of schizophrenia: an endocrine-disruption theory of schizophrenia. Schizophr Bull. 2009;35:256-78.

Campos-Barros A, Hoell T, Musa A, Sampaolo S, Stoltenburg G, Pinna G, Eravci M, Meinhold H, Baumgartner A. Phenolic and tyrosyl ring iodothyronine deiodination and thyroid hormone concentrations in the human central nervous system. J Clin Endocrinol Metab. 1996;81:2179-85.

Chang JY, Tsai PF. Prevention of methylmercury-induced mitochondrial depolarization, glutathione depletion and cell death by 15-deoxy-delta-12,14-prostaglandin J(2). Neurotoxicology 2008;29:1054-1061.

Chen J, Berry MJ. Selenium and selenoproteins in the brain and brain diseases. J Neurochem. 2003;86:1-12.

Cohen DJ, Young JG, Lowe TL, Harcherik D. Thyroid hormone in autistic children. J Autism Dev Disord. 1980;10:445-50.

Crantz FR, Silva JE, Larsen PR. An analysis of the sources and quantity of 3,5,3'-triiodothyronine specifically bound to nuclear receptors in rat cerebral cortex and cerebellum. Endocrinology 1982;110:367-375.

Croteau W, Bodwell JE, Richardson JM, St Germain DL. Conserved cysteines in the type 1 deiodinase selenoprotein are not essential for catalytic activity. J Biol Chem. 1998 ;273:25230-25236.

Dam K, Seidler FJ, Slotkin TA. Chlorpyrifos exposure during a critical neonatal period elicits gender-selective deficits in the development of coordination skills and locomotor activity. Brain Res Dev Brain Res. 2000;121:179-87.

Dasgupta A, Das S, Sarkar PK. Thyroid hormone promotes glutathione synthesis in astrocytes by up regulation of glutamate cysteine ligase through differential stimulation of its catalytic and modulator subunit mRNAs. Free Radic Biol Med. 2007;42:617-626.

da-Silva WS, Harney JW, Kim BW, Li J, Bianco SD, Crescenzi A, Christoffolete MA, Huang SA, Bianco AC. The small polyphenolic molecule kaempferol increases cellular energy expenditure and thyroid hormone activation. Diabetes. 2007;56:767-76.

Davis JD, Podolanczuk A, Donahue JE, Stopa E, Hennessey JV, Luo LG, Lim YP, Stern RA. Thyroid hormone levels in the prefrontal cortex of post-mortem brains of Alzheimer's disease patients. Curr Aging Sci. 2008;1:175-81.

Debes F, Budtz-Jørgensen E, Weihe P, White RF, Grandjean P. Impact of prenatal methylmercury exposure on neurobehavioral function at age 14 years. Neurotoxicol Teratol. 2006;28:363-75.

Dratman MB, Crutchfield FL, Gordon JT, Jennings AS. Iodothyronine homeostasis in rat brain during hypo- and hyperthyroidism. Am J Physiol. 1983;245:E185-93.

Fatemi SH, Snow AV, Stary JM, Araghi-Niknam M, Reutiman TJ, Lee S, Brooks AI, Pearce DA. Reelin signaling is impaired in autism. Biol Psychiatry. 2005;57:777-87.

Freitas BC, Gereben B, Castillo M, Kalló I, Zeöld A, Egri P, Liposits Z, Zavacki AM, Maciel RM, Jo S, Singru P, Sanchez E, Lechan RM, Bianco AC. Paracrine signaling by glial cell-derived triiodothyronine activates neuronal gene expression in the rodent brain and human cells. J Clin Invest. 2010;120:2206-17.

Galton VA, Wood ET, St Germain EA, Withrow CA, Aldrich G, St Germain GM, Clark AS, St Germain DL. Thyroid hormone homeostasis and action in the type 2 deiodinase-deficient rodent brain during development. Endocrinology 2007;148:3080-3088.

Gao Y, Yan CH, Tian Y, Xie HF, Zhou X, Yu XD, Yu XG, Tong S, Zhou QX, Shen XM. Prenatal exposure to mercury and neurobehavioral development of neonates in Zhoushan City, China. Environ Res. 2007;105:390-399.

Gereben B, Zavacki AM, Ribich S, Kim BW, Huang SA, Simonides WS, Zeold A, Bianco AC. Cellular and molecular basis of deiodinase-regulated thyroid hormone signaling. Endocr Rev 2008;29:898-938.

Gerhardsson L, Lundh T, Minthon L, Londos E. Metal concentrations in plasma and cerebrospinal fluid in patients with Alzheimer's disease. Dement Geriatr Cogn Disord. 2008;25:508-15.

Giordano G, Kavanagh TJ, Costa LG. Neurotoxicity of a polybrominated diphenyl ether mixture (DE-71) in mouse neurons and astrocytes is modulated by intracellular glutathione levels. Toxicol Appl Pharmacol. 2008;232:161-8.

Glaser V, Nazari EM, Muller YM, Feksa L, Wannmacher CM, Rocha JB, de Bem AF, Farina M, Latini A. Effects of inorganic selenium administration in methylmercury-induced neurotoxicity in mouse cerebral cortex. Int J Dev Neurosci. 2010;28:631-637.

Goemann IM, Gereben B, Harney JW, Zhu B, Maia AL, Larsen PR. Substitution of serine for proline in the active center of type 2 iodothyronine deiodinase substantially alters its in vitro biochemical properties with dithiothreitol but not its function in intact cells. Endocrinology 2010;151:821-829.

Goswami A, Rosenberg IN. Effects of glutathione on iodothyronine 5'-deiodinase activity. Endocrinology 1988;123:192-202.

Guadaño-Ferraz A, Obregón MJ, St Germain DL, Bernal J. The type 2 iodothyronine deiodinase is expressed primarily in glial cells in the neonatal rat brain. Proc Natl Acad Sci U S A. 1997;94:10391-6.

Guerra-Araiza C, Coyoy-Salgado A, Camacho-Arroyo I. Sex differences in the regulation of progesterone receptor isoforms expression in the rat brain. Brain Res Bull. 2002 Oct 30;59(2):105-9.

Hashimoto T, Aihara R, Tayama M, Miyazaki M, Shirakawa Y, Kuroda Y. Reduced thyroid-stimulating hormone response to thyrotropin-releasing hormone in autistic boys. Dev Med Child Neurol. 1991;33:313-9.

Hassoun EA, Periandri-Steinberg S. Assessment of the roles of antioxidant enzymes and glutathione in 3,3',4,4',5-Pentachlorobiphenyl (PCB 126)-induced oxidative stress in the brain tissues of rats after subchronic exposure. Toxicol Environ Chem. 2010;92:301.

Heath JA, Frederick PC. Relationship among mercury concentrations, hormones, and nesting effort of white Ibises (Eudocimus albus) in the Florida Everglades. Auk 2005;122:255-267.

Hincal F, Gürbay A, Giray B. Induction of lipid peroxidation and alteration of glutathione redox status by endosulfan. Biol Trace Elem Res. 1995;47:321-6.

Jain S, Mahendra Kumar CH, Suranagi UD, Mediratta PK. Protective effect of N-acetylcysteine on bisphenol A-induced cognitive dysfunction and oxidative stress in rats. Food Chem Toxicol. 2011 Mar 30. [Epub ahead of print]

James SJ, Slikker W 3rd, Melnyk S, New E, Pogribna M, Jernigan S. Thimerosal neurotoxicity is associated with glutathione depletion: protection with glutathione precursors. Neurotoxicology. 2005;26:1-8.

Kaur P, Aschner M, Syversen T. Glutathione modulation influences methyl mercury induced neurotoxicity in primary cell cultures of neurons and astrocytes. Neurotoxicology 2006;27:492-500.

Kim CY, Watanabe C, Satoh H. Effects of buthionine sulfoximine (BSO) on mercury distribution after Hg(o) exposure. Toxicology. 1995;98:67-72.

Kimura-Kuroda J, Nagata I, Negishi-Kato M, Kuroda Y. Thyroid hormone-dependent development of mouse cerebellar Purkinje cells in vitro. Develop Brain Res. 2002;137:55-65.

Kimura-Kuroda J, Nagata I, Kuroda Y. Disrupting effect of hydroxyl-polychlorinated biphenyl (PCB) congeners on neuronal development of cerebellar Purkinje cells: a possible causal factor for developmental brain disorders? Chemosphere. 2007;67:S412-20.

King CR. A novel embryological theory of autism causation involving endogenous biochemicals capable of initiating cellular gene transcription: A possible link between twelve autism risk factors and the autism 'epidemic' Med Hypotheses. 2011 Mar 7. (Epub ahead of print)

Kirkegaard C, Faber J. Free thyroxine and 3,3′,5′-triiodothyronine levels in cerebrospinal fluid in patients with endogenous depression. Acta Endocrinol (Copenh). 1991;124:166-72.

Kobayashi K, Miyagawa M, Wang RS, Suda M, Sekiguchi S, Honma T. Effects of in utero and lactational exposure to bisphenol A on thyroid status in F1 rat offspring. Ind Health. 2005;43:685-90.

Koibuchi N, Jingu H, Iwasaki T, Chin WW. Current perspectives on the role of thyroid hormone in growth and development of cerebellum. Cerebellum. 2003;2:279-89.

Lamirand A, Pallud-Mothré S, Ramaugé M, Pierre M, Courtin F. Oxidative stress regulates type 3 deiodinase and type 2 deiodinase in cultured rat astrocytes. Endocrinology. 2008;149:3713-3721.

Leonard JL. Regulation of T3 production in the brain. Acta Med Austriaca. 1992;19:5-8.

Linares AF, Loikkanen J, Jorge MF, Soria RB, Novoa AV. Antioxidant and neuroprotective activity of the extract from the seaweed, Halimeda incrassata (Ellis) Lamouroux, against in vitro and in vivo toxicity induced by methyl-mercury. Vet Hum Toxicol. 2004;46:1-5.

Liu F, Gentles A, Theodorakis CW. Arsenate and perchlorate toxicity, growth effects, and thyroid histopathology in hypothyroid zebrafish Danio rerio. Chemosphere. 2008;71:1369-76.

Malin JP,Ködding R, Fuhrmann H, von zur Mühlen A. T4, T3 and rT3 levels in serum and cerebrospinal fluid of patients with amyotrophic lateral sclerosis. J Neurol. 1989;236:57-9.

McKhann G, Drachman D, Folstein M, Katzman R, Price D, Stadlan EM. Clinical diagnosis of Alzheimer's disease: report of the NINCDS-ADRDA Work Group under the auspices of Department of Health and Human Services Task Force on Alzheimer's Disease. Neurology. 1984;34:939-44.

Messer A. Mini-review: polybrominated diphenyl ether (PBDE) flame retardants as potential autism risk factors. Physiol behave. 2010;100:245-9.

Miyata T, Ono Y, Okamoto M, Masaoka M, Sakakibara A, Kawaguchi A, Hashimoto M, Ogawa M. Migration, early axonogenesis, and Reelin-dependent layer-forming behavior of early/posterior-born Purkinje cells in the developing mouse lateral cerebellum. Neural Dev. 2010;5:23.

Monnet-Tschudi F, Zurich MG, Boschat C, Corbaz A, Honegger P. Involvement of environmental mercury and lead in the etiology of neurodegenerative diseases. Rev Environ Health. 2006;21:105-17.

Mori K, Stone S, Braverman LE, Devito WJ. Involvement of tyrosine phosphorylation in the regulation of 5"-deiodinase in FRTL-5 rat thyroid cells and rat astrocytes. Endocrinology. 1996;137:1313-8.

Morse DC, Wehler EK, Wesseling W, Koeman JH, Brouwer A. Alterations in rat brain thyroid hormone status following pre- and postnatal exposure to polychlorinated biphenyls (Aroclor 1254).Toxicol Appl Pharmacol. 1996;136:269-79.

Morte B, Ceballos A, Diez D, Grijota-Martinez C, Dumitrescu AM, Di Cosmo C, Galton VA, Refetoff S, Bernal J. Thyroid hormone-regulated mouse cerebral cortex genes are differentially dependent on the source of the hormone: a study in monocarboxylate transporter-8- and deiodinase-2-deficient mice. Endocrinology 2010;151:2381-7.

Nguon K, Ladd B, Baxter MG, Sajdel-Sulkowska EM. Sexual dimorphism in cerebellar structure, function, and response to environmental perturbations.Prog Brain Res. 2005;148:341-51.

Nguon K, Baxter MG, Sajdel-Sulkowska EM. Perinatal exposure to polychlorinated biphenyls differentially affects cerebellar development and motor functions in male and female rat neonates. The Cerebellum 2005b;4:112-122.

Nieminen P, Lindström-Seppä P, Mustonen AM, Mussalo-Rauhamaa H, Kukkonen JV. Bisphenol A affects endocrine physiology and biotransformation enzyme activities of the field vole (Microtus agrestis).Gen Comp Endocrinol. 2002;126:183-9.

Nir I, Meir D, Zilber N, Knobler H, Hadjez J, Lerner Y. Brief report: circadian melatonin, thyroid-stimulating hormone, prolactin, and cortisol levels in serum of young adults with autism. J Autism Dev Disoerd. 1995;25:641-54.

Nordenhäll K, Dock L, Vahter M. Transplacental and lactational exposure to mercury in hamster pups after maternal administration of methyl mercury in late gestation. Pharmacol Toxicol. 1995;77:130-135.

Oppenheimer JH, Schwartz HL. Molecular basis of thyroid hormone-dependent brain development. Endocr Rev. 1997;18:462-75.

Orct T, Blanusa M, Lazarus M, Varnai VM, Kostial K. Comparison of organic and inorganic mercury distribution in suckling rat. J Appl Toxicol. 2006;26:536-539.

Oskarsson A, Palminger Hallén I, Sundberg J. Exposure to toxic elements via breast milk. Analyst. 1995;120:765-770.

Padhi BK, Pelletier G, Williams A, Berndt-Weis L, Yauk C, Bowers WJ, Chu I. Gene expression profiling in rat cerebellum following in utero and lactational exposure to mixtures of methylmercury, polychlorinated biphenyls and organochlorine pesticides. Toxicol Lett. 2008;176:93-103.

Palmer RF, Blanchard S, Wood R. Proximity to point sources of environmental mercury release as a predictor of autism prevalence. Health Place 2009; 15:18-24.

Quadros PS, Lopez V, De Vries GJ, Chung WC, Wagner CK. Progesterone receptors and the sexual differentiation of the medial preoptic nucleus.J Neurobiol. 2002;51:24-32.

Roberts EM, English PB, Grether JK, Windham GC, Somberg L, Wolff C. Maternal residence near agricultural pesticide applications and autism spectrum disorders among children in the California Central Valley. Environ Health Perspect. 2007;115:1482-9.

Roman GC. Autism: transient in utero hypothyroxinemia related to maternal flavonoid ingestion during pregnancy and to other environmental antithyroid agents. J Neurol Sci. 2007;262:15-26.

Rosenberg RE, Law JK, Yenokyan G, McGready J, Kaufmann WE, Law PA. Characteristics and concordance of autism spectrum disorders among 277 twin pairs. Arch Pediatr Adolesc Med. 2009;163:907-14.

Rush T, Hjelmhaug J, Lobner D. Effects of chelators on mercury, iron, and lead neurotoxicity in cortical culture. Neurotoxicology 2009;30:47-51.

Sadamatsu M, Kanai H, Xu X, Liu Y, Kato N. Review of animal models for autism: implication of thyroid hormone. Congenit Anom (Kyoto). 2006;46:1-9.

Sajdel-Sulkowska EM, Lipinski B, Windom H, Audhya T, McGinnis W. Oxidative stress in autism: cerebellar 3-nitrotyrosine levels. Am. J. Biochem Biotechnol 2008;4:73-84.

Sajdel-Sulkowska EM. Oxidative stress and neurotrophin signaling in autism. In: Autism: Oxidative stress, inflammation and immune abnormalities. CRC Press (A Chauhan, V. Chauhan, W.T. Brown, edts) pp. 47-60, 2010.

Sajdel-Sulkowska EM, Xu M, McGinnis W, Koibuchi N. Brain region-specific changes in oxidative stress and neurotrophin levels in autism spectrum disorders (ASD). Cerebellum. 2011;10:43-8.

Sakamoto M, Kakita A, de Oliveira RB, Sheng Pan H, Takahashi H. Dose-dependent effects of methylmercury administered during neonatal brain spurt in rats. Brain Res Dev Brain Res. 2004;152:171-176.

Sampaolo S, Campos-Barros A, Mazziotti G, Carlomagno S, Sannino V, Amato G, Carella C, Di Iorio G. Increased cerebrospinal fluid levels of 3,3′,5′-triiodothyronine in patients with Alzheimer's disease. J Clin Endocrinol Metab. 2005;90:198-202.

Schantz SL, Widholm JJ. Cognitive effects of endocrine-disrupting chemicals in animals. Environ Health Perspect. 2001;109:1197-206.

Schoeters G, Den Hond E, Dhooge W, van Larebeke N, Leijs M. Endocrine disruptors and abnormalities of pubertal development. Basic Clin Pharmacol Toxicol. 2008;102:168-75.

Silva JE, Larsen PR. Comparison of iodothyronine 5'-deiodinase and other thyroid-hormone-dependent enzyme activities in the cerebral cortex of hypothyroid neonatal rat. Evidence for adaptation to hypothyroidism. J Clin Invest. 1982;70:1110-23.

Sobutskii MP, Kovan'ko EG, Liutinskii SI, Ivanov SD. Effect of age and gender on genotoxic and biochemical indexes in animal blood after low doses of radiation-mercury exposure. Adv Gerontol. 2007;20:91-96.

Soldin OP, O'Mara DM, Aschner M. Thyroid hormones and methylmercury toxicity. Biol Trace Elem Res. 2008;126:1-12.

Stringari J, Nunes AK, Franco JL, Bohrer D, Garcia SC, Dafre AL, Milatovic D, Souza DO, Rocha JB, Aschner M, Farina M. Prenatal methylmercury exposure hampers glutathione antioxidant system ontogenesis and causes long-lasting oxidative stress in the mouse brain. Toxicol Appl Pharmacol. 2008;227:147-154.

Sulkowski ZL, Chen T, Midha S, Zavacki AM, Sajdel-Sulkowska EM. Perinatal exposure to thimerosal results in a sex- and strain-dependent neurodevelopmental abnormalities. Cerebellum (accepted)

Tan SW, Meiller JC, Mahaffey KR. The endocrine effects of mercury in humans and wildlife. Crit Rev Toxicol. 2009; 39:228-269.

Van den Berg KJ, van Raaij JA, Bragt PC, Notten WR. Interactions of halogenated industrial chemicals with transthyretin and effects on thyroid hormone levels in vivo. Arch Toxicol. 1991;65:15-9.

Venkataraman P, Muthuvel R, Krishnamoorthy G, Arunkumar A, Sridhar M, Srinivasan N, Balasubramanian K, Aruldhas MM, Arunakaran J. PCB (Aroclor 1254) enhances oxidative damage in rat brain regions: protective role of ascorbic acid. Neurotoxicology. 2007;28:490-8.

Verhoelst CH, Vandenborne K, Severi T, Bakker O, Zandieh Doulabi B, Leonard JL, Kuhn ER, van der Geyten S, Darras VM. Specific detection of type III iodothyronine deiodinase protein in chicken cerebellar purkinje cells. Endocrinology. 2002;143:2700-7.

Verhoelst CH, Roelens SA, Darras VM. Role of spatiotemporal expression of iodothyronine deiodinase proteins in cerebellar cell organization. Brain Res Bull. 2005;67:196-202.

Vermiglio F, Lo Presti VP, Scaffidi Argentina G, Finocchiaro MD, Gullo D, Squatrito S, Trimarchi F. Maternal hypothyroxinaemia during the first half of gestation in an iodine deficient area with endemic cretinism and related disorders. Clin Endocrinol (Oxf). 1995;42:409-15.

Vicente E, Boer M, Netto C, Fochesatto C, Dalmaz C, Rodrigues Siqueira I, Gonçalves CA. Hippocampal antioxidant system in neonates from methylmercury-intoxicated rats. Neurotoxicol Teratol. 2004;26:817-823.

Watanabe C. Selenium deficiency and brain functions: the significance for methylmercury toxicity. Nippon Eiseigaku Zasshi. 2001;55:581-589.

Watanabe C, Yoshida K, Kasanuma Y, Kun Y, Satoh H. In utero methylmercury exposure differentially affects the activities of selenoenzymes in the fetal mouse brain. Environ Res. 1999;80:208-14.

Windham GC, Zhang L, Gunier R, Croen LA., Grether JK. Autism spectrum disorders in relation to distribution of hazardous air pollutants in the San Francisco bay area. Environ Health Perspect. 2006; 114:1438-1444.

Yin ZZ, Lee E, Ni M, Jiang H, Milatovic D, Rongzhu L, Farina M, Rocha JB, Aschner M. Methylmercury-induced alterations in astrocyte function are attenuated by ebselen. Neurotoxicology 2011; PMID:21300091.

Zareba G, Cernichiari E, Hojo R, Nitt SM, Weiss B, Mumtaz MM, Jones DE, Clarkson TW. Thimerosal distribution and metabolism in neonatal mice: comparison with methyl mercury. J Appl Toxicol. 2007;27:511-518.

Zavacki AM, Ying H, Christoffolete MA, Aerts G, So E, Harney JW, Cheng SY, Larsen PR, Bianco AC. Type 1 iodothyronine deiodinase is a sensitive marker of peripheral thyroid status in the mouse. Endocrinology 2005;146:1568-75.

Zhang ZX, Wang XH, Zou LW, Ding SS, Zhai JX. [Oxidative stress of decabromodiphenylether in mice brain tissue]. Zhonghua Lao Dong Wei Sheng Zhi Ye Bing Za Zhi. 2010;28:900-3.

Zoeller RT, Bansal R, Parris C. Bisphenol-A, an environmental contaminant that acts as a thyroid hormone receptor antagonist in vitro, increases serum thyroxine, and alters RC3/neurogranin expression in the developing rat brain. Endocrinology. 2005;146:607-12.

14

Perinatal Immune Activation and Risk of Autism

Theoharis Theoharides[1,2,3], Asimenia Angelidou[1,3,5], Konstantinos-Dionysios Alysandratos[1,3,5], Shahrzad Asadi[1], Konstantinos Francis[3,4], Lefteris Lykouras[3,4] and Dimitrios Kalogeromitros[3,4]†
[1]*Tufts University School of Medicine,*
[2]*Tufts Medical Center,*
[3]*University of Athens Medical School,*
[4]*Attikon General Hospital,*
[5]*University of Texas Southwestern, Children's Medical Center*
[1,2,5]*USA*
[3,4]*Greece*

1. Introduction

He could fit in the palms of both hands, seemed to look at you beseechingly while you rushed to thread a vein and snake a tube down a tiny nostril of this 24 week preemie. Already exposed to the prenatal stress that culminated in premature delivery, he has been further exposed to the stress associated with the separation from his mother and multiple medical interventions. Like other premature babies, he is more vulnerable to invasive infections from bacteria and viruses; moreover, the delayed development of his gut-blood-brain barriers could expose him to potential neurotoxins.

Such infants are up to 4 times more likely to develop autism. If their mothers had allergies, mastocytosis or an autoimmune disease, this risk almost doubles.

Autism Spectrum Disorders (ASD) are pervasive developmental disorders that include Autistic Disorder and Asperger's Disorder, although Pervasive Developmental Disorder-Not Otherwise Specified (PDD-NOS) is frequently included (Johnson & Myers, 2007). ASD are characterized by variable deficits in social skills, stereotypic behaviors, and a wide range of behavioral and learning problems. ASD manifest during early childhood and at least 30% present with sudden clinical regression of development around 3 years of age (Matson J.L. & Kozlowski A.M., 2010; Zappella, 2010). Over the last 20 years, there has been an impressive rise in ASD with current prevalence estimates of 1/100 children (Fombonne, 2009; Kogan et al., 2009).

In the majority of cases, the cause of ASD is unknown (Levy et al., 2009). Some autism susceptibility genes have been identified (Weiss et al., 2009), but gene interactions with environmental factors are increasingly suspected (Deth et al., 2008; Herbert, 2010). Recent reviews have focused mostly on genomic screens that suggest there are multiple gene interactions in autism; however no gene abnormality alone can explain the apparent increase in ASD prevalence (Durkin et al., 2010; Herbert, 2010; Miles, 2011). Increasing evidence suggests that there are different ASD endophenotypes, even within the ASD spectrum (Palmieri & Persico, 2010).

---

† *Deceased*

Intrauterine conditions in combination with external environmental triggers or specific genotypes, may lead to developmental disturbances. Indeed, an epidemiologic study, nested within a cohort of 698 autistic children in Denmark, concluded that prenatal environmental factors and parental psychopathology are associated with the risk of autism and these factors seem to act independently (Larsson et al., 2005). It is also recognized that tuberous sclerosis, a neurocutaneous disorder, involves autistic symptoms in approximately 40-45% of cases (Smalley, 1998). It has been proposed that this partial penetrance may be the result of an interaction between gene mutations and environmental factors, such as gestational immune activation (Ehninger et al., 2010). An early environmental insult, such as prenatal infection could cause long-term changes in neural function by altering epigenetic programming. Studies on rodents suggest that during early development, environmental signals can activate intracellular pathways, leading to epigenetic changes and consequently changes in neural function (Zhang & Meaney, 2010). In fact, variations in early maternal care affect stress responses in the offspring by altering the methylation status of the glucocorticoid receptor gene promoter (Weaver et al., 2004).

2. The role of prematurity

Premature births (delivery prior to 37 weeks gestation) currently account for 12.7% of all births in the United States, a rise of approximately 20% in the past two decades (MacDorman et al., 2010). Although infants less than 28 weeks gestation are at the highest risk for long-term neurologic problems, infants born between 32 and 36 weeks are increasingly recognized to be at risk for neurologic injury, such as leukomalacia and grey matter damage (Adams-Chapman, 2006; Argyropoulou, 2010; Okumura et al., 2010; Volpe, 2009). Clinical, epidemiological and experimental studies have revealed that key factors, such as inflammation and oxidative stress contribute considerably to white- and grey-matter injury in premature infants, whose brains are particularly susceptible to damage (Kaindl et al., 2009). In infants surviving premature birth, cerebellar hemorrhagic injury is also associated with a high prevalence of neurodevelopmental disabilities (Limperopoulos et al., 2007). The resulting long-term neurologic complications may include learning difficulties, behavioral and socio-emotional concerns, and poor general health outcomes. These vulnerable near-term (late preterm) infants make up the greatest number of premature births and account for the significant increase in the rate of prematurity in the recent years (Martin, 2011). Although the etiology of premature delivery is often unknown, chronic *in utero* inflammation or infection are well-described conditions that lead to preterm labor and birth (Dubicke et al., 2010; Snegovskikh et al., 2009; Thaxton et al., 2010). Inflammation itself has been strongly associated with adverse neurodevelopmental outcomes in premature infants (Lin et al., 2010; Rovira et al., 2011). An additional 5-8% of deliveries are complicated by pre-eclampsia or gestational diabetes, which may lead to placental insufficiency, abnormal growth, and postnatal metabolic imbalance. Excessive production of cortocotropin releasing hormone (CRH) has also been linked to preterm labour (Campbell et al., 1987; Warren et al., 1992). A number of cytokines are known to cause in vitro secretion of CRH from cultured placental trophoblasts, including IL-1 and IL-6 (Petraglia et al., 1990). In turn, CRH stimulates release of IL-6 from peripheral blood mononuclear cells, which infiltrate the fetal membranes and the placenta in increasing numbers during intrauterine infection (Angioni et al., 1993).

Recent reports suggest a potential association between preterm birth and autism. In particular, one retrospective study investigated preterm children born in Atlanta, GA (1981-93) who survived to three years of age, and identified rates of autism through the

Metropolitan Atlanta Developmental Disabilities Surveillance Program. Preterm birth prior to 33 weeks gestation was associated with a two-fold higher risk of autism in all infants (Limperopoulos et al., 2008). Interestingly, this study reported a gender-specific, four-fold increased risk for autism accompanied by mental retardation in preterm girls with low birth weight (LBW) (<2,500 g at birth). Another study performed a prospective follow-up assessment on 91 ex-preterm very low birth weight (VLBW) infants (<1,500 g at birth) at the mean age of 22 months and found 26% of these children to have a positive Modified Checklist for Autism in Toddlers (M-CHAT) test (Kinney et al., 2008). A more recent study found that 21% of infants (212/988) born before 28 weeks of gestation screened positive using M-CHAT (Kuban et al., 2009) as compared to 5.7% of healthy children 16-30 months old (Kleinman et al., 2008). Much higher rates of positive testing on M-CHAT were found in premature children with motor or sensory impairment (Kuban et al., 2009). It should be noted, however, that a positive CHAT test must be confirmed with more specific diagnostic tools, such as the Autism Diagnostic Observation Schedule-Generic (ADOS-G, a patient observational tool) or the Autism Diagnostic Interview-Revised (ADI-R, a parent interview tool) which provide a more reliable diagnosis for ASD. A prospective study of all births less than 26 weeks gestation in 1995 in the United Kingdom and Ireland concluded that ex-preterm children are at increased risk for ASD in middle childhood, compared with their term-born classmates, after psychiatric, clinical, IQ and SCQ (Social Communication Questionnaire) evaluations (Johnson et al., 2010).

A cohort of 164 families with autistic children (Brimacombe et al., 2007) concluded that the increased risk of autistic disorders related to prematurity is primarily attributed to perinatal complications that occur more commonly among preterm infants, results also confirmed in a Swedish population-based case-control study (Buchmayer et al., 2009). A meta-analysis on prenatal risk factors for autism argued that evidence is insufficient to implicate individual prenatal factors in autism etiology, because many of the studies examined all available prenatal data using designs with methodological weaknesses, so that significant associations may have been observed by chance after multiple testing (Gardener et al., 2009). Findings from population-based studies suggest that suboptimal birth conditions are not independent risk factors for infantile autism, but rather clusters of them increase the risk of ASD (Maimburg & Vaeth, 2006). Finally, a more recent cohort study on infants born in Canada between 1990-2002 concluded that perinatal risk factors including prenatal, obstetrical and neonatal complications, have a lesser overall effect on autistic outcomes among the genetically susceptible pediatric population, compared to children with low genetic susceptibility (Dodds et al., 2010). Reviews of studies evaluating neurobehavioral outcomes following preterm birth reveal a "preterm behavioral phenotype" characterized by symptoms of inattention, anxiety and social difficulties (Johnson & Marlow, 2011; Limperopoulos, 2009).

3. Maternal autoimmune diseases

The relationship between ASD and familial autoimmunity has long been recognized (Money et al., 1971), and has been supported by at least three large population-based studies discussed below; these studies utilized medical records and physician data to determine autoimmunity in families of ASD and typically-developing children.

One case-control study nested within a cohort of infants born in California between 1995-1999, examined the association of "immune-related conditions" with ASD using health records and reported that prevalence of maternal psoriasis, asthma, hay fever and atopic dermatitis during the second trimester of pregnancy correlated with over two-fold elevated risk of ASD in their children (Croen et al., 2005). The second cohort consisted of the pediatric population born in Denmark from 1993 through 2004 (n=689,196), in which 3,325 children were diagnosed with ASD including 1,089 cases of infantile autism. The study confirmed an association between family history of type 1 diabetes and infantile autism, as well as rheumatoid arthritis and ASD; it was also the first to show a significant association between maternal celiac disease and ASD (Atladottir et al., 2009). A significant association between parental rheumatic fever and ASD, as well as several significant correlations between maternal autoimmune diseases and ASD were investigated across 3 Swedish registries by means of a case-control study (n=1,227 ASD cases matched with 25 controls each) (Keil et al., 2010). Auto-antibodies against brain proteins have been reported in a number of mothers with children who developed autism (Croen et al., 2008). A possible explanation for the link between maternal immune dysregulation and ASD would be the transfer of maternal autoantibodies to the developing fetus during pregnancy (Braunschweig et al., 2008; Croen et al., 2008; Singer et al., 2008; Zimmerman et al., 2007), resulting in abnormal neurodevelopment. This phenomenon was manifested in the offsprings of pregnant mice after their transfection with human systemic lupus erythematosus autoantibodies (Lee et al., 2009). A preliminary report also indicated that mothers with a diagnosis of mastocytosis during pregnancy had a high chance of having one or more children with autism (Theoharides, 2009).

Results from animal modeling studies clearly indicate that maternal immune activation (MIA) can cause both acute and lasting changes in behavior and CNS structure and function in the offspring (Boksa, 2010). Administration of bacterial lipopolysaccharide (LPS), a cell wall component from Gram-negative bacteria, activates Toll-like receptor-4 (TLR-4) on immune cells leading to synthesis and release of TNF (Varadaradjalou et al., 2003), IL-1 and IL-6 (Supajatura et al., 2002). It was recently shown that IL-1 receptor antagonism prevented neurodevelopmental anomalies in pregnant rats after systemic end-of-gestation exposure to LPS (Girard et al., 2010). In addition to its direct detrimental effect on the placenta and fetal brain tissue, IL-1 induces selective release of IL-6 from mast cells (Kandere-Grzybowska et al., 2003). IL-6 appears critical for fetal brain development and social behavior development, as demonstrated in a poly(I:C) mouse model for MIA, where co-administration of anti-IL-6 antibody prevented the social deficits and associated gene expression changes in the brain of the offspring (Smith et al., 2007).

However, human studies investigating the role of prenatal infection in the pathogenesis of autism are limited, and mostly focused on viral infections (Chess, 1977; Libbey et al., 2005; Wilkerson et al., 2002). A recent nationwide study in Denmark including children born from 1980 through 2005 points to an increased risk for ASD after maternal viral infection in the first trimester of pregnancy (adjusted hazard ratio = 2.98; CI: 1.29-7.15) or maternal bacterial infection in the second trimester of pregnancy (adjusted hazard ratio = 1.42; CI: 1.08-1.87) (Atladottir et al., 2010). Moreover, a number of rotaviruses have been isolated from asymptomatic neonates (Dunn et al., 1993). Viral double-stranded RNA like poly(I:C) induces release of TNF and IL-6 without degranulation from mast cells through viral TLR-3 (Kulka et al., 2004).

4. Autoimmunity in ASD children

A recent study implied the presence of an endophenotype with complex immune dysfunction both in autistic children and their non-autistic siblings (Saresella et al., 2009). Brain-specific auto-antibodies are present in the plasma of many ASD individuals (Cabanlit et al., 2007; Singh et al., 1997; Singh & Rivas, 2004). Such auto-antibodies suggest a loss of self-tolerance to neural antigens during early neurodevelopment, but their precise role in autism remains unknown (Enstrom et al., 2009; Mostafa et al., 2008; Mostafa & Kitchener, 2009; Wills et al., 2007). They may indicate disruption of the blood-brain barrier (BBB), at least in a subgroup of patients. The presence of an auto-inflammatory response is also supported by the detection of certain inflammation markers. For instance, TNF was high in the cerebrospinal fluid (CSF) (Chez et al., 2007), and IL-6 gene expression was increased in the brain (Li et al., 2009) of autistic patients. CSF and microglia of ASD patients had high levels of macrophage chemoattractant protein-1 (MCP-1) (Vargas et al., 2005), which is also a potent chemoattractant for mast cells (Conti et al., 1997). In contrast, ASD plasma levels of transforming growth factor-beta1 (TGF-β1) were low (Ashwood et al., 2008), which is important in view of the fact that TGF-β1 inhibits mast cell function (Gebhardt et al., 2005). Many children with ASD also report gastrointestinal symptoms (Buie et al., 2010). In a few studies, examination of intestinal biopsies from children with regressive autism reveals features of an autoimmune mucosal pathology, that is not seen in other conditions or inflammatory bowel diseases (Ashwood et al., 2003; Torrente et al., 2002).

Many of the epidemiologic, biochemical and pathologic findings could be explained through activation of mast cells, immune cells important in both innate and acquired immunity (Galli et al., 2005), as well as in inflammation (Theoharides & Kalogeromitros, 2006). Mast cells are well-known for their leading role in allergic reactions, during which they are stimulated by IgE binding to high-affinity receptors (FcεRI), aggregation of which leads to degranulation and secretion of numerous pre-stored and newly-synthesized mediators, including IL-6 and TNF (Blank & Rivera, 2004; Kraft & Kinet, 2007; Schroeder et al., 1995; Schwartz, 1987; Serafin & Austen, 1987; Stone et al., 2010; Torigoe et al., 1997). In addition to IgE, many substances originating in the environment, the intestine or the brain can trigger mast cell activation (Theoharides et al., 2011). These include non-allergic environmental, infectious, neurohormonal and oxidative stress-related triggers, involving release of mediators selectively, without degranulation (Theoharides et al., 2007b). For instance, LPS activates TLR-4 on mast cells and induces selective release of TNF (Varadaradjalou et al., 2003), while IL-1 induces selective release of IL-6 (Kandere-Grzybowska et al., 2003).

Environmental toxins have been implicated in developmental neurotoxicity (Grandjean & Landrigan, 2006) and also in mast cell activation. In particular, polychlorinated biphenyl (PCB) (Hertz-Picciotto et al., 2008) and mercury (Young et al., 2008) have been associated with ASD, and both also activate mast cells (Asadi et al., 2010; Kempuraj et al., 2010; Kwon et al., 2002). Other mast cell triggers include bacterial and viral antigens, as well as peptides such as neurotensin (NT), which we reported to be increased in serum of young children with autism (Angelidou et al., 2010), and to induce mast cell release of extracellular mitochondrial DNA, which was also increased in the serum of these ASD patients (Zhang et al., 2010). This finding may be in addition to mitochondrial dysfunction reported in ASD (Giulivi et al., 2010; Palmieri & Persico, 2010). Given the fact that NT activates mast cells (Theoharides et al., 2004) and abundant NT is located in the gut (Castagliuolo et al., 1996), its elevated levels might lead to gut dysfunction in a cohort of autistic patients. NT also augments the action of CRH, which stimulates selective release of vascular endothelial growth factor (VEGF) (Cao et al., 2005). In fact, CRH acts synergistically with NT to increase vascular permeability (Donelan et al., 2006).

Mast cell-derived cytokines can also increase BBB permeability (Abbott, 2000; Theoharides & Konstantinidou, 2007). BBB disruption has been documented in brain inflammatory diseases, such as multiple sclerosis, where it *precedes* any pathological or clinical symptoms (Minagar & Alexander, 2003; Soon et al., 2007; Stone et al., 1995). We speculate that perinatal mast cell activation, in response to allergic or non-immune triggers, could disrupt the gut-blood-brain barriers (Theoharides & Doyle, 2008) and permit neurotoxic molecules to enter the brain and result in brain inflammation, thus contributing to ASD pathogenesis (Fig. 1) (Theoharides et al., 2008). It is intriguing that mast cell-derived IL-9 induces intestinal permeability and predisposes to oral antigen hypersensitivity in children (Forbes et al., 2008), while it also exacerbates newborn brain toxic lesions (Dommergues et al., 2000). Moreover, IL-33 can synergize with SP (Theoharides et al., 2010) and SCF (Drube et al., 2010) in stimulating mast cell TNF release, while it also activates glial cells to secrete pro-inflammatory cytokines (Yasuoka et al., 2011). The possible involvement of mast cells (Theoharides et al., 2011) in ASD is also supported by the fact that many children with ASD report "allergic-like" symptoms (Angelidou et al., 2011).

5. Perinatal stress

Mast cells have been implicated in inflammatory conditions that worsen by stress (Theoharides & Cochrane, 2004) and in regulating BBB permeability (Esposito et al., 2002). It was shown that early life stress due to maternal separation resulted in an altered brain-gut axis; it was sufficient to cause an increase in the blood concentrations of pro-inflammatory cytokines after a challenge with LPS, and also an increase in plasma corticosterone (O'Mahony et al., 2009). The timing of prenatal stressors was investigated in a case-control study, recording mothers' reports on exposure to stressors during each 4-week block of pregnancy. A higher incidence of stressors at 21-32 weeks gestation was found in autism, in consistency with the embryological age at which pathological cerebellar changes in autism are seen (Beversdorf et al., 2005).

We propose that prenatal or perinatal stress may contribute to the development of ASD through excessive release of CRH. CRH is typically secreted from the hypothalamus, but it can also be secreted from the skin (Slominski et al., 2006) and nerve endings (Skofitsch et al., 1985), where it exerts pro-inflammatory effects (Chrousos, 1995; Slominski et al., 2001; Theoharides et al., 2008). CRH can also be released from immune cells (Karalis et al., 1997) and mast cells (Kempuraj et al., 2004). In fact, CRH released from hair follicles can trigger mast cell proliferation (Ito et al., 2010). This form of tissue CRH sometimes called "immune CRH" may have an immunomodulatory role as an autocrine/paracrine mediator of inflammation during reproduction (Kalantaridou et al., 2007). One of the early effects of immune CRH is the activation of mast cells and the release of several pro-inflammatory cytokines (Theoharides et al., 2004). CRH was increased in the serum of mothers who delivered preterm babies and correlated with their level of anxiety during that period of pregnancy (Makrigiannakis et al., 2007). Maternal serum CRH can cross the placenta, and potentially high amounts of CRH could be produced by the placenta itself in response to external or intrauterine stress (Grammatopoulos, 2008; Torricelli et al., 2011). CRH can then disrupt the BBB (Theoharides & Konstantinidou, 2007), which appears to be compromised in ASD patients, as indicated by the presence of autoantibodies against encephalogenic peptides (Cabanlit et al., 2007; Goines & Van de Water J., 2010; Singer et al., 2006; Vojdani et al., 2002; Wills et al., 2008). BBB disruption due to stress is dependent on both CRH

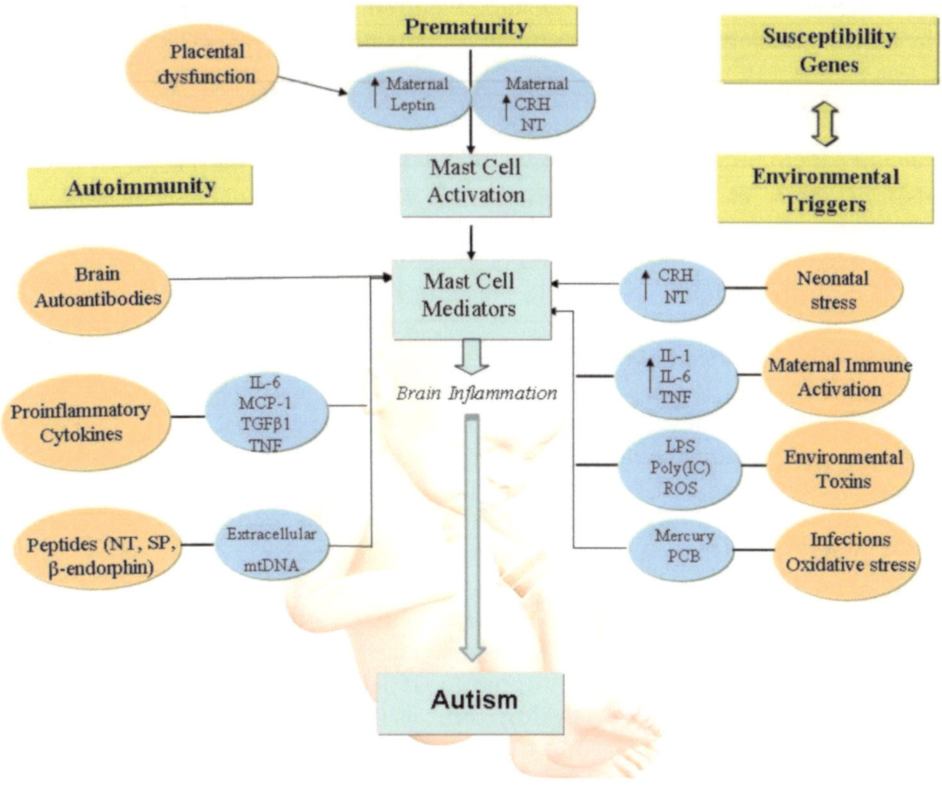

Fig. 1. Diagrammatic depiction of how perinatal immune activation may contribute to brain inflammation and autism. CRH, corticotropin-releasing hormone; IL, interleukin; LPS, lipopolysaccharide; MCP-1, macrophage chemoattractant protein-1; mtDNA, mitochondrial DNA; NT, neurotensin; PCB, polychlorinated biphenyl; poly(IC),polyinosinic:polycytidylic acid; ROS, reactive oxygen species; SP, substance P; TGFβ1, transforming growth factor-beta1; TNF, tumor necrosis factor

(Esposito et al., 2002) and mast cells (Esposito et al., 2001) and is associated with high serum IL-6 that is also mast cell-dependent (Theoharides & Konstantinidou, 2007).

The effect of CRH may be relevant to the behavioral manifestations of ASD. ASD patients had high anxiety levels and were unable to handle stress appropriately (Gillott & Standen, 2007). Evening cortisol levels positively correlated to daily stressors in children with autism (Corbett et al., 2009). Moreover, increase in age of autistic children correlated with increased cortisol levels during social interaction stress (Corbett et al., 2010). CRH has also been shown to increase intestinal permeability of human colonic biopsies, while maternal separation stress and CRH are associated with a dysfunctional mucosal barrier in rodents (Soderholm et al., 2002). A short period of restraint (Chandler et al., 2002) or maternal deprivation stress (Teunis et al., 2002) also increased the severity of experimental autoimmune encephalomyelitis.

The presence of maternal stress may explain why children born within a year of the first child that developed autism had a much higher chance of developing autism than if they were born two years later (Cheslack-Postava et al., 2011). Increased circulating CRH, directly or through immune cell release of other inflammatory and vasoactive molecules, can disrupt the gut-blood-brain barriers during gestation and/or infancy and permit absorption of intestinal-derived inflammatory and neurosensitizing mediators.

6. Other perinatal risk factors

High weight gain in pregnancy has been considered an independent risk factor for autism in the offspring (Stein et al., 2006). Even though a clear mechanism is lacking, leptin is suspected to play a major role. Obese subjects have higher leptin levels than normal weight subjects (Considine et al., 1996; Dardeno et al., 2010). Additionally, it has been suggested that elevated plasma leptin levels during pregnancy are indicative of placental dysfunction (Hauguel-de et al., 2006). Placental insufficiency, possibly associated with anoxia, may also contribute to autism in the offspring (Glasson et al., 2004).

Elevated plasma leptin levels were reported in children with regressive autism (n=37), compared with typically-developing controls (n=50) (Ashwood et al., 2007). One group also measured the variation of circulating leptin at baseline and after one year of follow-up in 35 patients with "classic autism" (according to DSM-IV criteria) aged 14.1±5.4 years old. The authors reported significantly higher leptin values in the patients versus the controls (24.1±17.8 ng/ml vs 9.8±3.2 ng/ml at baseline; 33.5±21.4 ng/ml vs 11.1±3.9 ng/ml at one year), and leptin concentrations were not associated with obesity or pubertal status (Blardi et al., 2010). It is still unclear, however, if the alteration of leptin levels in autism is a primary event or a finding attributable to the disease. Plasma levels of leptin have also been investigated in patients with Rett syndrome (n=16), where it was found that leptin was significantly increased compared to healthy controls (n=16), but also did not correlate with obesity (Blardi et al., 2008), suggesting that there may be more to the actions of leptin in neurodevelopment than weight balance.

The immunomodulatory properties of leptin were first reported on a mouse model, in which obese mice were found to have impaired cell-mediated and humoral immunity, attributed to possible lack of leptin (Chandra, 1980). The epigenetic status in adulthood was shown to be directionally dependent on prenatal nutritional status (Gluckman et al., 2007) and neonatal leptin administration late in the phase of developmental plasticity was able to reverse the developmental programming in rats (Vickers et al., 2005). Mast cells also express leptin and leptin receptors, a finding implicating paracrine or autocrine immunomodulatory effects of leptin on mast cells (Taildeman et al., 2009). Locally released leptin from T lymphocytes doesn't seem to play a major role in immunoregulation in mouse models of intestinal inflammation, suggesting other sources of leptin as critical in modulation of the inflammatory response (Fantuzzi et al., 2005). Despite evidence supporting the role of leptin in immune processes (Lago et al., 2008; Matarese et al., 2005), its precise role in inflammation remains incompletely understood.

7. Oxidative stress and prematurity

A variety of events associated with poor fetal growth or preterm birth are also associated with oxidative stress. These include maternal infection and inflammation that lead to increased lipid peroxidation, but more importantly to alterations in the expression of many genes associated with adverse perinatal outcomes (Ingelfinger, 2007).

Considerable evidence indicates that oxidative stress may be increased in patients with ASD, possibly due to their decreased ability to neutralize free radicals. An earlier study showed increased levels of plasma malondialdehyde, a marker of oxidative stress, ($p<0.05$) in the blood of mothers who delivered preterm and in the cord blood of their preterm neonates, compared to the levels in samples from term deliveries (Joshi et al., 2008). Preterm birth is associated with increased generation of reactive oxygen species (ROS), which places these infants in high risk for injury (Davis & Auten, 2010). In fact, a recent study identified an increase in the oxidative stress marker non-protein bound iron (NPBI) in the cord blood of 168 preterm newborns of gestational age 24-32 weeks (Perrone et al., 2010), suggesting that early identification of neonates at-risk is possible.

The impact of environmental oxidants in the etiology of autism is associated with brain region-specific changes in oxidative stress markers, such as 3-nitrotyrosine (3-NT) and neurotrophin-3 (NT-3), in ASD (Sajdel-Sulkowska et al., 2008; Sajdel-Sulkowska et al., 2011). Deficiencies in anti-oxidant enzymes might, in certain cases, be associated with mercury toxicity, which was shown to be tightly bound to and inactivate thioredoxin (Carvalho et al., 2008). In fact, cytosolic and mitochondrial redox imbalance was found in lymphoblastoid cells of ASD children compared to controls, an event exaggerated by exposure to thimerosal (James et al., 2009).

Several studies have suggested a link between oxidative stress and the immune response (Viora et al., 2001). Because immune cell functions are specially linked to ROS generation, their normal functioning is largely dependent on the oxidant/antioxidant balance. A strong association between oxidative stress and autoimmunity was shown in a group of 44 Egyptian autistic children, 88.64% of whom had elevated plasma F2-isoprostane (a marker of lipid peroxidation) and/or reduced glutathione peroxidase (an anti-oxidant enzyme), compared to 44 age-matched controls. Anti-neuronal antibodies were found in 54.5% of the same cohort, implying immunomodulation (Mostafa et al., 2010). Several groups have hypothesized that oxidative stress is the mechanism by which prenatal LPS affects offspring neurodevelopment (Lante et al., 2008; Paintlia et al., 2008). Potential therapies for oxidative stress and ROS-induced morbidities in the preterm infant include both enzymatic and nonenzymatic antioxidant preparations (Lee & Davis, 2011), such as the naturally-occuring flavonoids quercetin and luteolin (Cotelle, 2001; Middleton, Jr. et al., 2000). Quercetin inhibits mast cells (Kandere-Grzybowska et al., 2006), and also inhibited and reversed acute stress-induced autistic-like behavior and the associated reduced brain glutathione levels in mice (Kumar & Goyal, 2008). Luteolin can also block mast cell activation and superstimulation of activated T-cells (Kempuraj et al., 2008; Theoharides et al., 2007a). Luteolin and its structural analog diosmin prevented MIA-induced behavioral deficits in the mouse offspring by blocking the IL-6-induced JAK2/STAT3 (Janus tyrosine kinase-2/signal transducer and activator of transcription-3) signaling pathway, both *in vivo* and *in vitro* (Parker-Athill et al., 2009). A luteolin-containing dietary supplement, Neuroprotek® was recently made available to help the body reduce brain inflammation.

8. Conclusion

A number of findings suggest the presence of different biological endophenotypes in ASD (Persico et al., 2008), as well as between early-onset and regressive autism, with the latter being associated with poorer outcomes in social reciprocity, verbal IQ and more gastrointestinal symptoms, according to caregiver interviews (Richler et al., 2006).

Increasing evidence indicates that perinatal immune activation, in the mother and/or the fetus, could adversely affect neurodevelopment. Moreover, mast cell activation during this period by environmental, infectious, neurohormonal and immune triggers appears to be involved in gut-blood-brain barrier disruption and subsequent brain inflammation. Reduction of stress during gestation and infancy, potential use of specific CRH receptor antagonists, as well as drugs that could prevent BBB disruption, or block brain inflammation may prove useful in at least a subgroup of infants at high risk for developing autism. These goals may be at least partly achieved by the use of mast cell blockers. Unfortunately, there are no clinically effective mast cell inhibitors, and the available disodium cromoglycate (cromolyn) has proven a weak inhibitor of human mast cells (Theoharides & Kalogeromitros, 2006). Instead, flavonoids such as luteolin are anti-inflammatory and neuroprotective (Dirscherl et al., 2010), and can inhibit mast cell activation (Asadi et al., 2010).

9. Acknowledgements - disclosures

Aspects of research mentioned here were funded by Safe Minds, the National Autism Association, the Autism Research Collaborative, as well as Theta Biomedical Consulting and Development Co., Inc. (Brookline, MA). Asimenia Angelidou and Konstantinos-Dionysios Alysandratos are recipients of scholarships for postgraduate studies from the Hellenic State Scholarships Foundation (Athens, Greece). The authors declare that they have no competing interests. TCT is the inventor of patent application US 12/534,571; US 12/861,152; US 13/009,282 covering the diagnosis and treatment of ASD.

10. References

Abbott, N. J. 2000. Inflammatory mediators and modulation of blood-brain barrier permeability. *Cell Mol Neurobiol*, 20, 131-147.

Adams-Chapman, I. 2006. Neurodevelopmental outcome of the late preterm infant. *Clin Perinatol.*, 33, 4, pp. 947-964.

Angelidou A, Francis K, Vasiadi M, Alysandratos K-D, Zhang B, Theoharides A., Lykouras L, Kalogeromitros D, & Theoharides TC. 2010. Neurotensin is increased in serum of young children with autistic disorder. *J Neuroinflammation*, 7, 48.

Angelidou, A., Alysandratos, K. D., Asadi, S., Zhang, B., Francis, K., Vasiadi, M., Kalogeromitros, D., & Theoharides, T. C. 2011. Brief Report: "Allergic Symptoms" in Children with Autism Spectrum Disorders. More than Meets the Eye? *J Autism Dev.Disord.*, (Jan 6) [Ahead of print].

Angioni, S., Petraglia, F., Gallinelli, A., Cossarizza, A., Franceschi, C., Muscettola, M., Genazzani, A. D., Surico, N., & Genazzani, A. R. 1993. Corticotropin-releasing hormone modulates cytokines release in cultured human peripheral blood mononuclear cells. *Life Sci*, 53, 1735-1742.

Argyropoulou, M. I. 2010. Brain lesions in preterm infants: initial diagnosis and follow-up. *Pediatr Radiol.*, 40, 6, pp. 811-818.

Asadi, S., Zhang, B., Weng, Z., Angelidou, A., Kempuraj, D., Alysandratos, K. D., & Theoharides, T. C. 2010. Luteolin and thiosalicylate inhibit HgCl2 and thimerosal-induced VEGF release from human mast cells. *Int J Immunopathol.Pharmacol*, 23, 4, pp. 1015-1020.

Ashwood, P., Anthony, A., Pellicer, A. A., Torrente, F., Walker-Smith, J. A., & Wakefield, A. J. 2003. Intestinal lymphocyte populations in children with regressive autism:

evidence for extensive mucosal immunopathology. *J Clin Immunol.*, 23, 6, pp. 504-517.

Ashwood, P., Enstrom, A., Krakowiak, P., Hertz-Picciotto, I., Hansen, R. L., Croen, L. A., Ozonoff, S., Pessah, I. N., & de Water, J. V. 2008. Decreased transforming growth factor beta1 in autism: A potential link between immune dysregulation and impairment in clinical behavioral outcomes. *J Neuroimmunol.*, 204, 149-153.

Ashwood, P., Kwong, C., Hansen, R., Hertz-Picciotto, I., Croen, L., Krakowiak, P., Walker, W., Pessah, I. N., & Van de, Water J. 2007. Brief Report: Plasma leptin levels are elevated in autism: Association with early onset phenotype? *J.Autism Dev.Disord.*, 38, 1, pp. 169-175.

Atladottir, H. O., Pedersen, M. G., Thorsen, P., Mortensen, P. B., Deleuran, B., Eaton, W. W., & Parner, E. T. 2009. Association of family history of autoimmune diseases and autism spectrum disorders. *Pediatrics*, 124, 2, pp. 687-694.

Atladottir, H. O., Thorsen, P., Ostergaard, L., Schendel, D. E., Lemcke, S., Abdallah, M., & Parner, E. T. 2010. Maternal infection requiring hospitalization during pregnancy and autism spectrum disorders. *J Autism Dev.Disord.*, 40, 12, pp. 1423-1430.

Beversdorf, D. Q., Manning, S. E., Hillier, A., Anderson, S. L., Nordgren, R. E., Walters, S. E., Nagaraja, H. N., Cooley, W. C., Gaelic, S. E., & Bauman, M. L. 2005. Timing of prenatal stressors and autism. *J Autism Dev.Disord.*, 35, 4, pp. 471-478.

Blank, U.,Rivera, J. 2004. The ins and outs of IgE-dependent mast-cell exocytosis. *Trends Immunol*, 25, 266-273.

Blardi, P., de, Lalla A., Ceccatelli, L., Vanessa, G., Auteri, A., & Hayek, J. 2010. Variations of plasma leptin and adiponectin levels in autistic patients. *Neurosci.Lett.*, 479, 1, pp. 54-57.

Blardi, P., de, Lalla A., D'Ambrogio, T., G V, L C, A A, & J H. 2008. Long-term plasma levels of leptin and adiponectin in Rett Syndrome. *Clin Endocrinol.(Oxf)*, 70, 5, pp. 706-709.

Boksa, P. 2010. Effects of prenatal infection on brain development and behavior: a review of findings from animal models. *Brain Behav.Immun.*, 24, 6, pp. 881-897.

Braunschweig, D., Ashwood, P., Krakowiak, P., Hertz-Picciotto, I., Hansen, R., Croen, L. A., Pessah, I. N., & Van de Water J. 2008. Autism: maternally derived antibodies specific for fetal brain proteins. *NeuroToxicology*, 29, 2, pp. 226-231.

Brimacombe, M., Ming, X., & Lamendola, M. 2007. Prenatal and birth complications in autism. *Matern.Child Health J*, 11, 1, pp. 73-79.

Buchmayer, S., Johansson, S., Johansson, A., Hultman, C. M., Sparen, P., & Cnattingius, S. 2009. Can association between preterm birth and autism be explained by maternal or neonatal morbidity? *Pediatrics*, 124, 5, pp. e817-e825.

Buie, T., Campbell, D. B., Fuchs, G. J., III, Furuta, G. T., Levy, J., Vandewater, J., Whitaker, A. H., Atkins, D., Bauman, M. L., Beaudet, A. L., Carr, E. G., Gershon, M. D., Hyman, S. L., Jirapinyo, P., Jyonouchi, H., Kooros, K., Kushak, R., Levitt, P., Levy, S. E., Lewis, J. D., Murray, K. F., Natowicz, M. R., Sabra, A., Wershil, B. K., Weston, S. C., Zeltzer, L., & Winter, H. 2010. Evaluation, diagnosis, and treatment of gastrointestinal disorders in individuals with ASDs: a consensus report. *Pediatrics*, 125, Suppl 1:S1-18.

Cabanlit, M., Wills, S., Goines, P., Ashwood, P., & Van de Water, J. 2007. Brain-specific autoantibodies in the plasma of subjects with autistic spectrum disorder. *Ann.N Y.Acad.Sci.*, 1107, 92-103.

Campbell, E. A., Linton, E. A., Wolfe, C. D., Scraggs, P. R., Jones, M. T., & Lowry, P. J. 1987. Plasma corticotropin-releasing hormone concentrations during pregnancy and parturition. *J Clin Endocrinol.Metab*, 64, 5, pp. 1054-1059.

Cao, J., Papadopoulou, N., Kempuraj, D., Boucher, W. S., Sugimoto, K, Cetrulo, C. L., & Theoharides, T. C. 2005. Human mast cells express corticotropin-releasing hormone (CRH) receptors and CRH leads to selective secretion of vascular endothelial growth factor. *J Immunol*, 174, 7665-7675.

Carvalho, C. M., Chew, E. H., Hashemy, S. I., Lu, J., & Holmgren, A. 2008. Inhibition of the human thioredoxin system. A molecular mechanism of mercury toxicity. *J Biol.Chem.*, 283, 18, pp. 11913-11923.

Castagliuolo, I., Leeman, S. E., Bartolac-Suki, E., Nikulasson, S., Qiu, B., Carraway, R. E., & Pothoulakis, C. 1996. A neurotensin antagonist, SR 48692, inhibits colonic responses to immobilization stress in rats. *Proc Natl Acad Sci USA*, 93, 12611-12615.

Chandler, N., Jacobson, S., Connolly, R., Esposito, P., & Theoharides, T. C. 2002. Acute stress shortens the time of onset of experimental allergic encephalomyelitis (EAE) in SJL/J mice. *Brain Behav Immun*, 16, 757-763.

Chandra, R. K. 1980. Cell-mediated immunity in genetically obese C57BL/6J ob/ob) mice. *Am.J Clin Nutr.*, 33, 1, pp. 13-16.

Cheslack-Postava, K., Liu, K., & Bearman, P. S. 2011. Closely Spaced Pregnancies Are Associated With Increased Odds of Autism in California Sibling Births. *Pediatrics*,

Chess, S. 1977. Follow-up report on autism in congenital rubella. *J Autism Child Schizophr.*, 127, 2, pp.246-253.

Chez, M. G., Dowling, T., Patel, P. B., Khanna, P., & Kominsky, M. 2007. Elevation of tumor necrosis factor-alpha in cerebrospinal fluid of autistic children. *Pediatr.Neurol.*, 36, 6, pp. 361-365.

Chrousos, G. P. 1995. The hypothalamic-pituitary-adrenal axis and immune-mediated inflammation. *N Engl J Med*, 332, 1351-1362.

Considine, R. V., Sinha, M. K., Heiman, M. L., Kriauciunas, A., Stephens, T. W., Nyce, M. R., Ohannesian, J. P., Marco, C. C., McKee, L. J., Bauer, T. L., & . 1996. Serum immunoreactive-leptin concentrations in normal-weight and obese humans. *N.Engl.J Med.*, 334, 5, pp. 292-295.

Conti, P., Pang, X., Boucher, W., Letourneau, R., Reale, M., Barbacane, R. C., Thibault, J., & Theoharides, T. C. 1997. Impact of Rantes and MCP-1 chemokines on *in vivo* basophilic mast cell recruitment in rat skin injection model and their role in modifying the protein and mRNA levels for histidine decarboxylase. *Blood*, 89, 4120-4127.

Corbett, B. A., Schupp, C. W., Levine, S., & Mendoza, S. 2009. Comparing cortisol, stress, and sensory sensitivity in children with autism. *Autism Res.*, 2, 1, pp. 39-49.

Corbett, B. A., Schupp, C. W., Simon, D., Ryan, N., & Mendoza, S. 2010. Elevated cortisol during play is associated with age and social engagement in children with autism. *Mol.Autism*, 1, 1, pp. 13.

Cotelle, N. 2001. Role of flavonoids in oxidative stress. *Curr.Top.Med.Chem.*, 1, 6, pp. 569-590.

Croen, L. A., Braunschweig, D., Haapanen, L., Yoshida, C. K., Fireman, B., Grether, J. K., Kharrazi, M., Hansen, R. L., Ashwood, P., & Van de, Water J. 2008. Maternal mid-pregnancy autoantibodies to fetal brain protein: the early markers for autism study. *Biological Psychiatry*, 64, 7, pp. 583-588.

Croen, L. A., Grether, J. K., Yoshida, C. K., Odouli, R., & Van de, Water J. 2005. Maternal autoimmune diseases, asthma and allergies, and childhood autism spectrum disorders: a case-control study. *Arch.Pediatr.Adolesc.Med.*, 159, 2, pp. 151-157.

Dardeno, T. A., Chou, S. H., Moon, H. S., Chamberland, J. P., Fiorenza, C. G., & Mantzoros, C. S. 2010. Leptin in human physiology and therapeutics. *Front Neuroendocrinol.*, 31, 3, pp. 377-393.

Davis, J. M.,Auten, R. L. 2010. Maturation of the antioxidant system and the effects on preterm birth. *Semin.Fetal Neonatal Med.*, 15, 4, pp. 191-195.

Deth, R., Muratore, C., Benzecry, J., Power-Charnitsky, V. A., & Waly, M. 2008. How environmental and genetic factors combine to cause autism: A redox/methylation hypothesis. *NeuroToxicology*, 29, 1, pp. 190-201.

Dirscherl, K., Karlstetter, M., Ebert, S., Kraus, D., Hlawatsch, J., Walczak, Y., Moehle, C., Fuchshofer, R., & Langmann, T. 2010. Luteolin triggers global changes in the microglial transcriptome leading to a unique anti-inflammatory and neuroprotective phenotype. *J Neuroinflammation.*, 7, 1, pp. 3.

Dodds, L., Fell, D. B., Shea, S., Armson, B. A., Allen, A. C., & Bryson, S. 2011. The Role of Prenatal, Obstetric and Neonatal Factors in the Development of Autism. *J Autism Dev.Disord.*, 41, 7, pp. 891-902

Dommergues, M. A., Patkai, J., Renauld, J. C., Evrard, P., & Gressens, P. 2000. Proinflammatory cytokines and interleukin-9 exacerbate excitotoxic lesions of the newborn murine neopallium. *Ann.Neurol.*, 47, 1, pp. 54-63.

Donelan, J., Boucher, W., Papadopoulou, N., Lytinas, M., Papaliodis, D., & Theoharides, T. C. 2006. Corticotropin-releasing hormone induces skin vascular permeability through a neurotensin-dependent process. *Proc Natl Acad Sci USA*, 103, 7759-7764.

Drube, S., Heink, S., Walter, S., Lohn, T., Grusser, M., Gerbaulet, A., Berod, L., Schons, J., Dudeck, A., Freitag, J., Grotha, S., Reich, D., Rudeschko, O., Norgauer, J., Hartmann, K., Roers, A., & Kamradt, T. 2010. The receptor tyrosine kinase c-Kit controls IL-33 receptor signaling in mast cells. *Blood*, 115, 19, pp. 3899-3906.

Dubicke, A., Fransson, E., Centini, G., Andersson, E., Bystrom, B., Malmstrom, A., Petraglia, F., Sverremark-Ekstrom, E., & Ekman-Ordeberg, G. 2010. Pro-inflammatory and anti-inflammatory cytokines in human preterm and term cervical ripening. *J Reprod.Immunol*, 84, 2, pp. 176-185.

Dunn, S. J., Greenberg, H. B., Ward, R. L., Nakagomi, O., Burns, J. W., Vo, P. T., Pax, K. A., Das, M., Gowda, K., & Rao, C. D. 1993. Serotypic and genotypic characterization of human serotype 10 rotaviruses from asymptomatic neonates. *J Clin Microbiol.*, 31, 1, pp. 165-169.

Durkin, M. S., Maenner, M. J., Meaney, F. J., Levy, S. E., DiGuiseppi, C., Nicholas, J. S., Kirby, R. S., Pinto-Martin, J. A., & Schieve, L. A. 2010. Socioeconomic inequality in the prevalence of autism spectrum disorder: evidence from a U.S. cross-sectional study. *PLoS.ONE.*, 5, 7, pp. e11551.

Ehninger, D., Sano, Y., de Vries, P. J., Dies, K., Franz, D., Geschwind, D. H., Kaur, M., Lee, Y. S., Li, W., Lowe, J. K., Nakagawa, J. A., Sahin, M., Smith, K., Whittemore, V., & Silva, A. J. 2010. Gestational immune activation and Tsc2 haploinsufficiency cooperate to disrupt fetal survival and may perturb social behavior in adult mice. *Mol.Psychiatry*, (Nov 16) [Ahead of print].

Enstrom, A. M., Van de Water, J. A., & Ashwood, P. 2009. Autoimmunity in autism. *Curr.Opin.Investig.Drugs*, 10, 5, pp. 463-473.

Esposito, P., Chandler, N., Kandere-Grzybowska, K, Basu, S., Jacobson, S., Connolly, R., Tutor, D., & Theoharides, T. C. 2002. Corticotropin-releasing hormone (CRH) and brain mast cells regulate blood-brain-barrier permeability induced by acute stress. *J Pharmacol Exp Ther*, 303, 1061-1066.

Esposito, P., Gheorghe, D., Kandere, K., Pang, X., Conally, R., Jacobson, S., & Theoharides, T. C. 2001. Acute stress increases permeability of the blood-brain-barrier through activation of brain mast cells. *Brain Res*, 888, 117-127.

Fantuzzi, G., Sennello, J. A., Batra, A., Fedke, I., Lehr, H. A., Zeitz, M., & Siegmund, B. 2005. Defining the role of T cell-derived leptin in the modulation of hepatic or intestinal inflammation in mice. *Clin Exp.Immunol*, 142, 1, pp. 31-38.

Fombonne, E. 2009. Epidemiology of pervasive developmental disorders. *Pediatric Research*, 65, 6, pp. 591-598.

Forbes, E. E., Groschwitz, K., Abonia, J. P., Brandt, E. B., Cohen, E., Blanchard, C., Ahrens, R., Seidu, L., McKenzie, A., Strait, R., Finkelman, F. D., Foster, P. S., Matthaei, K. I., Rothenberg, M. E., & Hogan, S. P. 2008. IL-9- and mast cell-mediated intestinal permeability predisposes to oral antigen hypersensitivity. *Journal of Experimental Medicine*, 205, 4, pp. 897-913.

Galli, S. J., Kalesnikoff, J., Grimbaldeston, M. A., Piliponsky, A. M., Williams, C. M., & Tsai, M. 2005. Mast cells as "tunable" effector and immunoregulatory cells: recent advances. *Annu Rev Immunol*, 23, 749-786.

Gardener, H., Spiegelman, D., & Buka, S. L. 2009. Prenatal risk factors for autism: comprehensive meta-analysis. *Br.J Psychiatry*, 195, 1, pp. 7-14.

Gebhardt, T., Lorentz, A., Detmer, F., Trautwein, C., Bektas, H., Manns, M. P., & Bischoff, S. C. 2005. Growth, phenotype, and function of human intestinal mast cells are tightly regulated by transforming growth factor beta1. *Gut*, 54, 7, pp. 928-934.

Gillott, A.,Standen, P. J. 2007. Levels of anxiety and sources of stress in adults with autism. *J.Intellect.Disabil.*, 11, 4, pp. 359-370.

Girard, S., Tremblay, L., Lepage, M., & Sebire, G. 2010. IL-1 receptor antagonist protects against placental and neurodevelopmental defects induced by maternal inflammation. *J Immunol*, 184, 7, pp. 3997-4005.

Giulivi, C., Zhang, Y. F., Omanska-Klusek, A., Ross-Inta, C., Wong, S., Hertz-Picciotto, I., Tassone, F., & Pessah, I. N. 2010. Mitochondrial dysfunction in autism. *JAMA*, 304, 21, pp. 2389-2396.

Glasson, E. J., Bower, C., Petterson, B., de, Klerk N., Chaney, G., & Hallmayer, J. F. 2004. Perinatal factors and the development of autism: a population study. *Arch.Gen.Psychiatry*, 61, 6, pp. 618-627.

Gluckman, P. D., Lillycrop, K. A., Vickers, M. H., Pleasants, A. B., Phillips, E. S., Beedle, A. S., Burdge, G. C., & Hanson, M. A. 2007. Metabolic plasticity during mammalian development is directionally dependent on early nutritional status. *Proc Natl Acad Sci U S A*, 104, 31, pp. 12796-12800.

Goines, P.,Van de Water J. 2010. The immune system's role in the biology of autism. *Curr.Opin.Neurol.*, 23, 2, pp. 111-117.

Grammatopoulos, D. K. 2008. Placental corticotrophin-releasing hormone and its receptors in human pregnancy and labour: still a scientific enigma. *Journal of Neuroendocrinology*, 20, 4, pp. 432-438.

Grandjean, P.,Landrigan, P. J. 2006. Developmental neurotoxicity of industrial chemicals. *Lancet*, 368, 9553, pp. 2167-2178.

Hauguel-de, Mouzon S., Lepercq, J., & Catalano, P. 2006. The known and unknown of leptin in pregnancy. *Am.J Obstet.Gynecol.*, 194, 6, pp. 1537-1545.

Herbert, M. R. 2010. Contributions of the environment and environmentally vulnerable physiology to autism spectrum disorders. *Curr.Opin.Neurol.*, 23, 2, pp. 103-110.

Hertz-Picciotto, I., Park, H. Y., Dostal, M., Kocan, A., Trnovec, T., & Sram, R. 2008. Prenatal exposures to persistent and non-persistent organic compounds and effects on immune system development. *Basic Clin.Pharmacol.Toxicol.*, 102, 2, pp. 146-154.

Ingelfinger, J. R. 2007. Prematurity and the legacy of intrauterine stress. *N.Engl.J Med.*, 356, 20, pp. 2093-2095.

Ito, N., Sugawara, K., Bodo, E., Takigawa, M., van, Beek N., Ito, T., & Paus, R. 2010. Corticotropin-releasing hormone stimulates the in situ generation of mast cells from precursors in the human hair follicle mesenchyme. *J Invest Dermatol.*, 130, 4, pp. 995-1004.

James, S. J., Rose.S., Melnyk, S., Jernigan, S., Blossom, S., Pavliv, O., & Gsylor, D. W. 2009. Cellular and mitochondrial glutathione redox imbalance in lymphoblastoid cells derived from children with autism. *FASEB*, 23(8) pp. 2374-2383.

Johnson, C. P.,Myers, S. M. 2007. Identification and evaluation of children with autism spectrum disorders. *Pediatrics*, 120, 5, pp. 1183-1215.

Johnson, S., Hollis, C., Kochhar, P., Hennessy, E., Wolke, D., & Marlow, N. 2010. Autism spectrum disorders in extremely preterm children. *J Pediatr.*, 156, 4, pp. 525-531.

Johnson, S.,Marlow, N. 2011. Preterm Birth and Childhood Psychiatric Disorders. *Pediatric Research*, (May) [Ahead of print]. , 69, 5 Pt 2, pp. 11R-8R

Joshi, S. R., Mehendale, S. S., Dangat, K. D., Kilari, A. S., Yadav, H. R., & Taralekar, V. S. 2008. High maternal plasma antioxidant concentrations associated with preterm delivery. *Ann.Nutr.Metab*, 53, 3-4, pp. 276-282.

Kaindl, A. M., Favrais, G., & Gressens, P. 2009. Molecular mechanisms involved in injury to the preterm brain. *J Child Neurol.*, 24, 9, pp. 1112-1118.

Kalantaridou, S., Makrigiannakis, A., Zoumakis, E., & Chrousos, G. P. 2007. Peripheral corticotropin-releasing hormone is produced in the immune and reproductive systems: actions, potential roles and clinical implications. *Front Biosci.*, 12, 572-580.

Kandere-Grzybowska, K, Letourneau, R., Kempuraj, D., Donelan, J., Poplawski, S., Boucher, W., Athanassiou, A., & Theoharides, T. C. 2003. IL-1 induces vesicular secretion of IL-6 without degranulation from human mast cells. *J Immunol*, 171, 9, pp. 4830-4836.

Kandere-Grzybowska, K., Kempuraj, D., Cao, J., Cetrulo, C. L., & Theoharides, T. C. 2006. Regulation of IL-1-induced selective IL-6 release from human mast cells and inhibition by quercetin. *Br J Pharmacol*, 148, 208-215.

Karalis, K., Louis, J. M., Bae, D., Hilderbrand, H., & Majzoub, J. A. 1997. CRH and the immune system. *J Neuroimmunol*, 72, 131-136.

Keil, A., Daniels, J. L., Forssen, U., Hultman, C., Cnattingius, S., Soderberg, K. C., Feychting, M., & Sparen, P. 2010. Parental autoimmune diseases associated with autism spectrum disorders in offspring. *Epidemiology*, 21, 6, pp. 805-808.

Kempuraj, D., Asadi, S., Zhang, B., Manola, A., Hogan, J., Peterson, E., & Theoharides, T. C. 2010. Mercury induces inflammatory mediator release from human mast cells. *J Neuroinflammation*, 7, 1, pp. 20.

Kempuraj, D., Papadopoulou, N. G., Lytinas, M., Huang, M., Kandere-Grzybowska, K., Madhappan, B., Boucher, W., Christodoulou, S., Athanassiou, A., & Theoharides, T. C. 2004. Corticotropin-releasing hormone and its structurally related urocortin are synthesized and secreted by human mast cells. *Endocrinology*, 145, 43-48.

Kempuraj, D., Tagen, M., Iliopoulou, B. P., Clemons, A., Vasiadi, M., Boucher, W., House, M., Wolferg, A., & Theoharides, T. C. 2008. Luteolin inhibits myelin basic protein-induced human mast cell activation and mast cell dependent stimulation of Jurkat T cells. *Br J Pharmacol*, 155, 1076-1084.

Kinney, D. K., Munir, K. M., Crowley, D. J., & Miller, A. M. 2008. Prenatal stress and risk for autism. *Neurosci.Biobehav.Rev.*, 32, 8, pp. 1519-1532.

Kleinman, J. M., Robins, D. L., Ventola, P. E., Pandey, J., Boorstein, H. C., Esser, E. L., Wilson, L. B., Rosenthal, M. A., Sutera, S., Verbalis, A. D., Barton, M., Hodgson, S., Green, J., Dumont-Mathieu, T., Volkmar, F., Chawarska, K., Klin, A., & Fein, D. 2008. The modified checklist for autism in toddlers: a follow-up study investigating the early detection of autism spectrum disorders. *J Autism Dev.Disord.*, 38, 5, pp. 827-839.

Kogan, M. D., Blumberg, S. J., Schieve, L. A., Boyle, C. A., Perrin, J. M., Ghandour, R. M., Singh, G. K., Strickland, B. B., Trevathan, E., & van Dyck, P. C. 2009. Prevalence of parent-reported diagnosis of autism spectrum disorder among children in the US, 2007. *Pediatrics*, 5, 124, pp. 1395-1403.

Kraft, S.,Kinet, J. P. 2007. New developments in FcepsilonRI regulation, function and inhibition. *Nat Rev.Immunol*, 7, 5, pp. 365-378.

Kuban, K. C., O'Shea, T. M., Allred, E. N., Tager-Flusberg, H., Goldstein, D. J., & Leviton, A. 2009. Positive screening on the Modified Checklist for Autism in Toddlers (M-CHAT) in extremely low gestational age newborns. *Journal of Pediatrics*, 154, 4, pp. 535-540.

Kulka, M., Alexopoulou, L., Flavell, R. A., & Metcalfe, D. D. 2004. Activation of mast cells by double-stranded RNA: evidence for activation through Toll-like receptor 3. *J Allergy Clin.Immunol*, 114, 1, pp. 174-182.

Kumar, A.,Goyal, R. 2008. Quercetin protects against acute immobilization stress-induced behaviors and biochemical alterations in mice. *J Med Food*, 11, 3, pp. 469-473.

Kwon, O., Lee, E., Moon, T. C., Jung, H., Lin, C. X., Nam, K. S., Baek, S. H., Min, H. K., & Chang, H. W. 2002. Expression of cyclooxygenase-2 and pro-inflammatory cytokines induced by 2,2',4,4',5,5'-hexachlorobiphenyl (PCB 153) in human mast cells requires NF-kappa B activation. *Biol.Pharm.Bull.*, 25, 9, pp. 1165-1168.

Lago, R., Gomez, R., Lago, F., Gomez-Reino, J., & Gualillo, O. 2008. Leptin beyond body weight regulation--current concepts concerning its role in immune function and inflammation. *Cell Immunol*, 252, 1-2, pp. 139-145.

Lante, F., Meunier, J., Guiramand, J., De Jesus Ferreira, M. C., Cambonie, G., Aimar, R., Cohen-Solal, C., Maurice, T., Vignes, M., & Barbanel, G. 2008. Late N-acetylcysteine treatment prevents the deficits induced in the offspring of dams exposed to an immune stress during gestation. *Hippocampus*, 18, 6, pp. 602-609.

Larsson, H. J., Eaton, W. W., Madsen, K. M., Vestergaard, M., Olesen, A. V., Agerbo, E., Schendel, D., Thorsen, P., & Mortensen, P. B. 2005. Risk factors for autism: perinatal factors, parental psychiatric history, and socioeconomic status. *Am.J Epidemiol.*, 161, 10, pp. 916-925.

Lee, J. W.,Davis, J. M. 2011. Future applications of antioxidants in premature infants. *Curr.Opin.Pediatr.*, 23, 2, pp. 161-166.

Lee, J. Y., Huerta, P. T., Zhang, J., Kowal, C., Bertini, E., Volpe, B. T., & Diamond, B. 2009. Neurotoxic autoantibodies mediate congenital cortical impairment of offspring in maternal lupus. *Nat.Med.*, 15, 1, pp. 91-96.

Levy, S. E., Mandell, D. S., & Schultz, R. T. 2009. Autism. *Lancet*, 374, 9701, pp. 1627-1638.

Li, X., Chauhan, A., Sheikh, A. M., Patil, S., Chauhan, V., Li, X. M., Ji, L., Brown, T., & Malik, M. 2009. Elevated immune response in the brain of autistic patients. *J.Neuroimmunol.*, 207, 1-2, pp. 111-116.

Libbey, J. E., Sweeten, T. L., McMahon, W. M., & Fujinami, R. S. 2005. Autistic disorder and viral infections. *J Neurovirol.*, 11, 1, pp. 1-10.

Limperopoulos, C. 2009. Autism spectrum disorders in survivors of extreme prematurity. *Clin Perinatol.*, 36, 4, pp. 791-805, vi.

Limperopoulos, C., Bassan, H., Gauvreau, K., Robertson, R. L., Jr., Sullivan, N. R., Benson, C. B., Avery, L., Stewart, J., Soul, J. S., Ringer, S. A., Volpe, J. J., & duPlessis, A. J. 2007. Does cerebellar injury in premature infants contribute to the high prevalence of long-term cognitive, learning, and behavioral disability in survivors? *Pediatrics*, 120, 3, pp. 584-593.

Limperopoulos, C., Bassan, H., Sullivan, N. R., Soul, J. S., Robertson, R. L., Jr., Moore, M., Ringer, S. A., Volpe, J. J., & du Plessis, A. J. 2008. Positive screening for autism in ex-preterm infants: prevalence and risk factors. *Pediatrics*, 121, 4, pp. 758-765.

Lin, C. Y., Chang, Y. C., Wang, S. T., Lee, T. Y., Lin, C. F., & Huang, C. C. 2010. Altered inflammatory responses in preterm children with cerebral palsy. *Ann.Neurol.*, 68, 2, pp. 204-212.

MacDorman, M. F., Declercq, E., & Zhang, J. 2010. Obstetrical intervention and the singleton preterm birth rate in the United States from 1991-2006. *Am J Public Health*, 100, 11, pp. 2241-2247.

Maimburg, R. D.,Vaeth, M. 2006. Perinatal risk factors and infantile autism. *Acta Psychiatr.Scand.*, 114, 4, pp. 257-264.

Makrigiannakis, A., Semmler, M., Briese, V., Eckerle, H., Minas, V., Mylonas, I., Friese, K., & Jeschke, U. 2007. Maternal serum corticotropin-releasing hormone and ACTH levels as predictive markers of premature labor. *Int.J.Gynaecol.Obstet.*, 97, 2, pp. 115-119.

Martin, J. A. 2011. Preterm Births --- United States, 2007. *MMWR Surveill Summ.*, 60, SU-1, pp. 78-79.

Matarese, G., Moschos, S., & Mantzoros, C. S. 2005. Leptin in immunology. *J Immunol*, 174, 6, pp. 3137-3142.

Matson J.L.,Kozlowski A.M. 2010. Autistic regression. *Res Autism Spectr Disord*, 4, 340-345.

Middleton, E, Jr., Kandaswami, C., & Theoharides, T. C. 2000. The effects of plant flavonoids on mammalian cells: implications for inflammation, heart disease and cancer. *Pharmacol Rev*, 52, 673-751.

Miles, J. H. 2011. Autism spectrum disorders-A genetics review. *Genet.Med*, 13(4):278-294.

Minagar, A.,Alexander, J. S. 2003. Blood-brain barrier disruption in multiple sclerosis. *Mult.Scler.*, 9, 6, pp. 540-549.

Money, J., Bobrow, N. A., & Clarke, F. C. 1971. Autism and autoimmune disease: a family study. *J Autism Child Schizophr.*, 1, 2, pp. 146-160.

Mostafa, G. A., El-Hadidi, E. S., Hewedi, D. H., & Abdou, M. M. 2010. Oxidative stress in Egyptian children with autism: relation to autoimmunity. *J Neuroimmunol.*, 219, 1-2, pp. 114-118.

Mostafa, G. A., El-Sayed, Z. A., El-Aziz, M. M., & El-Sayed, M. F. 2008. Serum anti-myelin-associated glycoprotein antibodies in Egyptian autistic children. *J Child Neurol*, 23, 12, pp. 1413-1418.

Mostafa, G. A.,Kitchener, N. 2009. Serum anti-nuclear antibodies as a marker of autoimmunity in Egyptian autistic children. *Pediatr Neurol*, 40, 2, pp. 107-112.

O'Mahony, S. M., Marchesi, J. R., Scully, P., Codling, C., Ceolho, A. M., Quigley, E. M., Cryan, J. F., & Dinan, T. G. 2009. Early life stress alters behavior, immunity, and microbiota in rats: implications for irritable bowel syndrome and psychiatric illnesses. *Biological Psychiatry*, 65, 3, pp. 263-267.

Okumura, A., Yamamoto, T., Kidokoro, H., Kato, T., Kubota, T., Shoji, H., Sato, H., Shimojima, K., & Shimizu, T. 2010. Altered gene expression in umbilical cord mononuclear cells in preterm infants with periventricular leukomalacia. *Early Hum.Dev.*, 86, 10, pp. 665-667.

Paintlia, M. K., Paintlia, A. S., Contreras, M. A., Singh, I., & Singh, A. K. 2008. Lipopolysaccharide-induced peroxisomal dysfunction exacerbates cerebral white matter injury: attenuation by N-acetyl cysteine. *Exp.Neurol.*, 210, 2, pp. 560-576.

Palmieri, L.,Persico, A. M. 2010. Mitochondrial dysfunction in autism spectrum disorders: Cause or effect? *Biochim.Biophys.Acta*, 1797, 6-7, pp. 1130-1137.

Parker-Athill, E., Luo, D., Bailey, A., Giunta, B., Tian, J., Shytle, R. D., Murphy, T., Legradi, G., & Tan, J. 2009. Flavonoids, a prenatal prophylaxis via targeting JAK2/STAT3 signaling to oppose IL-6/MIA associated autism. *J Neuroimmunol.*, 217, 1-2, pp. 20-27.

Perrone, S., Tataranno, M. L., Negro, S., Longini, M., Marzocchi, B., Proietti, F., Iacoponi, F., Capitani, S., & Buonocore, G. 2010. Early identification of the risk for free radical-related diseases in preterm newborns. *Early Hum.Dev.*, 86, 4, pp. 241-244.

Persico, A. M., Sacco, R., Curatolo, P., Manzi, B., Lenti, C., Saccani, M., Militerni, R., Bravaccio, C., & Elia, A. 2008. Isolation of principal components in autistic disorder symptomatology and their association with biological endophenotypes. *Proc.Society for Neuroscience, Washington DC*, Abstract #446.20.

Petraglia, F., Garuti, G. C., De, Ramundo B., Angioni, S., Genazzani, A. R., & Bilezikjian, L. M. 1990. Mechanism of action of interleukin-1 beta in increasing corticotropin-releasing factor and adrenocorticotropin hormone release from cultured human placental cells. *Am.J Obstet.Gynecol.*, 163, 4 Pt 1, pp. 1307-1312.

Richler, J., Luyster, R., Risi, S., Hsu, W. L., Dawson, G., Bernier, R., Dunn, M., Hepburn, S., Hyman, S. L., McMahon, W. M., Goudie-Nice, J., Minshew, N., Rogers, S., Sigman, M., Spence, M. A., Goldberg, W. A., Tager-Flusberg, H., Volkmar, F. R., & Lord, C. 2006. Is there a 'regressive phenotype' of Autism Spectrum Disorder associated with the measles-mumps-rubella vaccine? A CPEA Study. *J Autism Dev.Disord.*, 36, 3, pp. 299-316.

Rovira, N., Alarcon, A., Iriondo, M., Ibanez, M., Poo, P., Cusi, V., Agut, T., Pertierra, A., & Krauel, X. 2011. Impact of histological chorioamnionitis, funisitis and clinical chorioamnionitis on neurodevelopmental outcome of preterm infants. *Early Hum.Dev*, 8, 4, 253-257.

Sajdel-Sulkowska, E. M., Lipinski, B., Windom, H., Audhya, T., & McGinnis, W. 2008. Oxidative stress in autism: elevated cerebellar 3-nitrotyrosine levels. *AM J Biochem Biotechnol*, 4, 73-84.

Sajdel-Sulkowska, E. M., Xu, M., McGinnis, W., & Koibuchi, N. 2011. Brain region-specific changes in oxidative stress and neurotrophin levels in autism spectrum disorders (ASD). *Cerebellum*, 10, 1, pp. 43-48.

Saresella, M., Marventano, I., Guerini, F. R., Mancuso, R., Ceresa, L., Zanzottera, M., Rusconi, B., Maggioni, E., Tinelli, C., & Clerici, M. 2009. An autistic endophenotype results in complex immune dysfunction in healthy siblings of autistic children. *Biological Psychiatry*, 66, 10, pp. 978-984.

Schroeder, J. T., Kagey-Sobotka, A., MacGlashan, D. W., & Lichtenstein, L. M. 1995. The interaction of cytokines with human basophils and mast cells. *International Archives of Allergy and Immunology*, 107, 79-81.

Schwartz, L. B. 1987. Mediators of human mast cells and human mast cell subsets. *Ann Allergy*, 58, 226-235.

Serafin, W. E.,Austen, K. F. 1987. Mediators of immediate hypersensitivity reactions. *N Engl J Med*, 317, 30-34.

Singer, H. S., Morris, C. M., Gause, C. D., Gillin, P. K., Crawford, S., & Zimmerman, A. W. 2008. Antibodies against fetal brain in sera of mothers with autistic children. *J Neuroimmunol.*, 194, 1-2, pp. 165-172.

Singer, H. S., Morris, C. M., Williams, P. N., Yoon, D. Y., Hong, J. J., & Zimmerman, A. W. 2006. Antibrain antibodies in children with autism and their unaffected siblings. *J Neuroimmunol.*, 178, 1-2, pp. 149-155.

Singh, V. K.,Rivas, W. H. 2004. Prevalence of serum antibodies to caudate nucleus in autistic children. *Neurosci.Lett.*, 355, 1-2, pp. 53-56.

Singh, V. K., Warren, R., Averett, R., & Ghaziuddin, M. 1997. Circulating autoantibodies to neuronal and glial filament proteins in autism. *Pediatr.Neurol.*, 17, 1, pp. 88-90.

Skofitsch, G., Zamir, N., Helke, C. J., Savitt, J. M., & Jacobowitz, D. M. 1985. Corticotropin-releasing factor-like immunoreactivity in sensory ganglia and capsaicin sensitive neurons of the rat central nervous system: colocalization with other neuropeptides. *Peptides*, 6, 307-318.

Slominski, A., Wortsman, J., Pisarchik, A., Zbytek, B., Linton, E. A., Mazurkiewicz, J. E., & Wei, E. T. 2001. Cutaneous expression of corticotropin-releasing hormone (CRH), urocortin, and CRH receptors. *FASEB J*, 15, 1678-1693.

Slominski, A., Zbytek, B., Zmijewski, M., Slominski, R. M., Kauser, S., Wortsman, J., & Tobin, D. J. 2006. Corticotropin releasing hormone and the skin. *Front Biosci*, 11, 2230-2248.

Smalley, S. L. 1998. Autism and tuberous sclerosis. *J Autism Dev.Disord.*, 28, 5, pp. 407-414.

Smith, S. E., Li, J., Garbett, K., Mirnics, K., & Patterson, P. H. 2007. Maternal immune activation alters fetal brain development through interleukin-6. *Journal of Neuroscience*, 27, 40, pp. 10695-10702.

Snegovskikh, V. V., Schatz, F., Arcuri, F., Toti, P., Kayisli, U. A., Murk, W., Guoyang, Luo, Lockwood, C. J., & Norwitz, E. R. 2009. Intra-amniotic infection upregulates decidual cell vascular endothelial growth factor (VEGF) and neuropilin-1 and -2 expression: implications for infection-related preterm birth. *Reprod.Sci.*, 16, 8, pp. 767-780.

Soderholm, J. D., Yates, D. A., Gareau, M. G., Yang, P. C., MacQueen, G., & Perdue, M. H. 2002. Neonatal maternal separation predisposes adult rats to colonic barrier dysfunction in response to mild stress. *Am.J Physiol Gastrointest.Liver Physiol*, 283, 6, pp. G1257-G1263.

Soon, D., Altmann, D. R., Fernando, K. T., Giovannoni, G., Barkhof, F., Polman, C. H., O'Connor, P., Gray, B., Panzara, M., & Miller, D. H. 2007. A study of subtle blood brain barrier disruption in a placebo-controlled trial of natalizumab in relapsing remitting multiple sclerosis. *J Neurol.*, 254, 3, pp. 306-314.

Stein, D., Weizman, A., Ring, A., & Barak, Y. 2006. Obstetric complications in individuals diagnosed with autism and in healthy controls. *Comprehensive Psychiatry*, 47, 1, pp. 69-75.

Stone, K. D., Prussin, C., & Metcalfe, D. D. 2010. IgE, mast cells, basophils, and eosinophils. *The Journal of Allergy and Clinical Immunology*, 125, 2 Suppl 2, pp. S73-S80.

Stone, L. A., Smith, M. E., Albert, P. S., Bash, C. N., Maloni, H., Frank, J. A., & McFarland, H. F. 1995. Blood-brain barrier disruption on contrast-enhanced MRI in patients with mild relapsing-remitting multiple sclerosis: Relationship to course, gender, and age. *Neurology*, 45, 1122-1126.

Supajatura, V., Ushio, H., Nakao, A., Akira, S., Okumura, K., Ra, C., & Ogawa, H. 2002. Differential responses of mast cell Toll-like receptors 2 and 4 in allergy and innate immunity. *J Clin Invest*, 109, 1351-1359.

Taildeman, J., Perez-Novo, C. A., Rottiers, I., Ferdinande, L., Waeytens, A., De, Colvenaer, V, Bachert, C., Demetter, P., Waelput, W., Braet, K., & Cuvelier, C. A. 2009. Human mast cells express leptin and leptin receptors. *Histochem.Cell Biol.*, 131, 6, pp. 703-711.

Teunis, M. A., Heijnen, C. J., Sluyter, F., Bakker, J. M., Van Dam, A. M., Hof, M., Cools, A. R., & Kavelaars, A. 2002. Maternal deprivation of rat pups increases clinical symptoms of experimental autoimmune encephalomyelitis at adult age. *J.Neuroimmunol.*, 133, 1-2, pp. 30-38.

Thaxton, J. E., Nevers, T. A., & Sharma, S. 2010. TLR-mediated preterm birth in response to pathogenic agents. *Infect.Dis.Obstet.Gynecol.*, 2010, pii: 378472.

Theoharides, T. C. 2009. Autism spectrum disorders and mastocytosis. *Int J Immunopathol Pharmacol*, 22, 4, pp. 859-865.

Theoharides, T. C., Angelidou, A., Alysandratos, K. D., Zhang, B., Asadi, S., Francis, K., Toniato, E., & Kalogeromitros, D. 2011. Mast cell activation and autism. *Biochim.Biophys.Acta*, (Dec 23) [Ahead of print].

Theoharides, T. C.,Cochrane, D. E. 2004. Critical role of mast cells in inflammatory diseases and the effect of acute stress. *J Neuroimmunol*, 146, 1-12.

Theoharides, T. C., Donelan, J. M., Papadopoulou, N., Cao, J., Kempuraj, D., & Conti, P. 2004. Mast cells as targets of corticotropin-releasing factor and related peptides. *Trends Pharmacol Sci*, 25, 563-568.

Theoharides, T. C.,Doyle, R. 2008. Autism, gut-blood-brain barrier and mast cells. *J Clin Psychopharmacol*, 28, 5, pp. 479-483.

Theoharides, T. C., Doyle, R., Francis, K., Conti, P., & Kalogeromitros, D. 2008. Novel therapeutic targets for autism. *Trends Pharmacol Sci.*, 29, 8, pp. 375-382.

Theoharides, T. C.,Kalogeromitros, D. 2006. The critical role of mast cell in allergy and inflammation. *Ann NY Acad Sci*, 1088, 78-99.

Theoharides, T. C., Kempuraj, D., & Iliopoulou, B. P. 2007a. Mast cells, T cells, and inhibition by luteolin: implications for the pathogenesis and treatment of multiple sclerosis. *Advances in Experimental Medicine and Biology*, 601, 423-430.

Theoharides, T. C., Kempuraj, D., Tagen, M., Conti, P., & Kalogeromitros, D. 2007b. Differential release of mast cell mediators and the pathogenesis of inflammation. *Immunol Rev*, 217, 65-78.

Theoharides, T. C.,Konstantinidou, A. 2007. Corticotropin-releasing hormone and the blood-brain-barrier. *Front Biosci*, 12, 1615-1628.

Theoharides, T. C., Zhang, B., Kempuraj, D., Tagen, M., Vasiadi, M., Angelidou, A., Alysandratos, K. D., Kalogeromitros, D., Asadi, S., Stavrianeas, N., Peterson, E., Leeman, S., & Conti, P. 2010. IL-33 augments substance P-induced VEGFsecretion from human mast cells and is increased in psoriatic skin. *Proc Natl Acad Sci U S A*, 107, 9, pp. 4448-4453.

Torigoe, C., Goldstein, B., Wofsy, C., & Metzger, H. 1997. Shuttling of initiating kinase between discrete aggregates of the high affinity receptor for IgE regulates the cellular response. *Proc.Natl.Acad.Sci.U.S.A*, 94, 4, pp. 1372-1377.

Torrente, F., Ashwood, P., Day, R., Machado, N., Furlano, R. I., Anthony, A., Davies, S. E., Wakefield, A. J., Thomson, M. A., Walker-Smith, J. A., & Murch, S. H. 2002. Small intestinal enteropathy with epithelial IgG and complement deposition in children with regressive autism. *Mol.Psychiatry*, 7, 4, pp. 375-82, 334.

Torricelli, M., Novembri, R., Bloise, E., De, Bonis M., Challis, J. R., & Petraglia, F. 2011. Changes in placental CRH, urocortins, and CRH-receptor mRNA expression associated with preterm delivery and chorioamnionitis. *J Clin Endocrinol.Metab*, 96, 2, pp. 534-540.

Varadaradjalou, S., Feger, F., Thieblemont, N., Hamouda, N. B., Pleau, J. M., Dy, M., & Arock, M. 2003. Toll-like receptor 2 (TLR2) and TLR4 differentially activate human mast cells. *Eur J Immunol*, 33, 899-906.

Vargas, D. L., Nascimbene, C., Krishnan, C., Zimmerman, A. W., & Pardo, C. A. 2005. Neuroglial activation and neuroinflammation in the brain of patients with autism. *Ann.Neurol.*, 57, 1, pp. 67-81.

Vickers, M. H., Gluckman, P. D., Coveny, A. H., Hofman, P. L., Cutfield, W. S., Gertler, A., Breier, B. H., & Harris, M. 2005. Neonatal leptin treatment reverses developmental programming. *Endocrinology*, 146, 10, pp. 4211-4216.

Viora, M., Quaranta, M. G., Straface, E., Vari, R., Masella, R., & Malorni, W. 2001. Redox imbalance and immune functions: opposite effects of oxidized low-density lipoproteins and N-acetylcysteine. *Immunology*, 104, 4, pp. 431-438.

Vojdani, A., Campbell, A. W., Anyanwu, E., Kashanian, A., Bock, K., & Vojdani, E. 2002. Antibodies to neuron-specific antigens in children with autism: possible cross-reaction with encephalitogenic proteins from milk, Chlamydia pneumoniae and Streptococcus group A. *J Neuroimmunol.*, 129, 1-2, pp. 168-177.

Volpe, J. J. 2009. The encephalopathy of prematurity--brain injury and impaired brain development inextricably intertwined. *Semin.Pediatr Neurol*, 16, 4, pp. 167-178.

Warren, W. B., Patrick, S. L., & Goland, R. S. 1992. Elevated maternal plasma corticotropin-releasing hormone levels in pregnancies complicated by preterm labor. *Am J Obstet Gynecol*, 166, 1198-204.

Weaver, I. C., Cervoni, N., Champagne, F. A., D'Alessio, A. C., Sharma, S., Seckl, J. R., Dymov, S., Szyf, M., & Meaney, M. J. 2004. Epigenetic programming by maternal behavior. *Nat.Neurosci.*, 7, 8, pp. 847-854.

Weiss, L. A., Arking, D. E., Daly, M. J., & Chakravarti, A. 2009. A genome-wide linkage and association scan reveals novel loci for autism. *Nature*, 461, 7265, pp. 802-808.

Wilkerson, D. S., Volpe, A. G., Dean, R. S., & Titus, J. B. 2002. Perinatal complications as predictors of infantile autism. *Int J Neurosci.*, 112, 9, pp. 1085-1098.

Wills, S., Cabanlit, M., Bennett, J., Ashwood, P., Amaral, D., & Van de Water, J. 2007. Autoantibodies in autism spectrum disorders (ASD). *Ann.N.Y.Acad.Sci.*, 1107, 79-91.

Wills, S., Cabanlit, M., Bennett, J., Ashwood, P., Amaral, D. G., & Van de Water, J. 2008. Detection of autoantibodies to neural cells of the cerebellum in the plasma of subjects with autism spectrum disorders. *Brain Behav.Immun.*, 23, 64-74.

Yasuoka, S., Kawanokuchi, J., Parajuli, B., Jin, S., Doi, Y., Noda, M., Sonobe, Y., Takeuchi, H., Mizuno, T., & Suzumura, A. 2011. Production and functions of IL-33 in the central nervous system. *Brain Res*, 1385, pp. 8-17

Young, H. A., Geier, D. A., & Geier, M. R. 2008. Thimerosal exposure in infants and neurodevelopmental disorders: an assessment of computerized medical records in the Vaccine Safety Datalink. *J Neurol Sci*, 271, 1-2, pp. 110-118.

Zappella, M. 2010. Autistic regression with and without EEG abnormalities followed by favourable outcome. *Brain Dev.*, 32, 9, pp. 739-745.

Zhang B, Angelidou A, Alysandratos KD, Vasiadi M, Francis K, Asadi S, Theoharides A., Sideri, k., Lykouras L, Kalogeromitros D, & Theoharides TC. 2010. Mitochondrial DNA and anti-mitochondrial antibodies in serum of autistic children. *J Neuroinflammation*, 7, 1, pp. 80.

Zhang, T. Y.,Meaney, M. J. 2010. Epigenetics and the environmental regulation of the genome and its function. *Annu.Rev Psychol.*, 61, 439-3.

Zimmerman, A. W., Connors, S. L., Matteson, K. J., Lee, L. C., Singer, H. S., Castaneda, J. A., & Pearce, D. A. 2007. Maternal antibrain antibodies in autism. *Brain Behav.Immun.*, 21, 3, pp. 351-357.

15

Immune System and Neurotrophic Factors in Autism

Elvan Iseri and Esra Guney
Gazi University Faculty of Medicine
Turkey

1. Introduction

Evidences related to the fact that there is a immunological functional disorder in autism are gradually increasing. Since immunological mechanisms play a role in the neuronal cell migration, axonal development and formation of synapses, it is mentioned that there is an immune - mediated inflammatory period exists (Bradstreet et al. 2007).

Central nervous system and immune system are systems which are in interaction with each other and which are complex and highly developed (Ashwood and Water 2004). Immune cells and molecules play an important role in forming the brain functions by affecting the cognitive and emotional processes. These molecules have various effects such as infection, inflammation and injury on neural tissues, the modulation of central nervous system response and systematic response. It is indicated that proinflammatory cytokines such as interleukin (IL) – 1, IL -6, IL – 12, interferon (IFN) gamma and tumor necrosis factor (TNF)-alpha in particular directly affect the neural function and development by playing a role in the neurodevelopmental process. Applying the cytokines such as IFN – alpha, IL – 2, TNF-alpha at therapeutical doses lead to the symptoms such as depression, sleep disorder, defects in cognitive functions, motivational change and decline in behavior of making research. It is reported that cytokines applied systematically may lead to increase in hypothalamus, hippocampus and nucleus accumbance and in noradrenergic, dopaminergic and serotenergic metabolism, and it may change the synaptic plasticity and therefore may affect the memory and learning processes (Ashwood et al. 2006). Findings which states that response of maternal immune to the infection, affect the fetal brain development through the increase in the cytokine level on circulation and which are collected from animal models support the role of immune system in the etiology of autism (Yamashita et al. 2003, Patterson 2002). It is mentioned that there is "abnormal" immune findings exists in 15 – 60 % of the autistic children (Pardo et al 2005).

Two basic immune fuctional disorders in autism are associated with immune formation that includes autoimmunity and proinflammatory cytokines. The decline in the number of the CD4 lymphocyte numbers, the change of balance of T helper cell type 1(Th1)/ T helper cell type 2 (Th2) in the direction of Th2 and the response corruption of Th1 (corruption in cell - mediated immunity), the decline in the T lymphocyte response against mitogens, the decline in the natural killer's functions and corruption shown by declined IgA level in humoral immunity are among immune system abnormalities indicated in autism (Cohly and Panja 2005, Careaga et al. 2010).

It is put forward that infectious agents such as lead and measles virus are two basic environmental triggers and the environmental exposure to the lead affects the immune balance in a negative way. Although viruses trigger the process related to the autism, the basic devastating factor is the cytokines that are activated following the infection. In autism, inflammatory agents are generally related to the astrocytes and microglial cell activation. The existence of chronic microglial activation is indicated on the brains of autistic individuals. In autism, there is an increase in the proinflammatory chemokines and modulator cytokines (Cohly and Panja 2005). Detected increase in the level of proinflammatory cytokines such as IL - 6, TNF - alpha and monocyte chemoatractan protein 1 (MCP - 1) in the brain samples and cerebrospinal fluid indicates that active neural inflammatory process is continuing in autism (Careaga et al. 2010). TNF- alpha is a strong immunomodulatory cytokine that is produced by macrophages and partly by active T cells. As well as its important function in the development of brain, it may have a role in the arrangement of glutamatergic transmission. It is put forward that TNF - alpha levels may be the indicator of possible inflammatory damage in autistic disorder (Chez et al. 2007). In the study performed by Jyonouchi and his collegues (2001) an increase in the serum of TNF - alpha, IL-1β and IL-6 levels in autistic individuals is indicated and the existence of improper hereditary immune response in autism is mentioned. However, there is various study results in which no change is observed in the levels of serum of TNF - alpha in autistic disorder (Güney et al. 2008). In different study, increased serum IL - 12 and IFN - gamma levels are determined and no significant change is reported on IFN-alpha, IL-6, TNF-alpha serum levels (Singh 1996). It is indicated that there is a decline occur on anti - inflammatory cytokine levels such as IL - 10 and transforming growth factor beta (TGF-β) on autistic individuals. However, in all studies made in this field no consistent results in terms of cytokine patterns could be provided (Careaga et al. 2010).

Studies in which an increase on autoantibody unique to the brain is indicated support the autoimmune mechanism. In the serum of autistic individuals, autoantibodies against the proteins (myelin basic protein, neuron axon, filament protein, glial fibrillary acidic protein) limited by nervous system are observed (Schmitz and Rezaie 2008). In a study carried out by Singh and Rivas (2004), antibodies are determined against caudat nucleus (49%) cerebral cortex (18%) and cerebellum (9%). It is put forward that these autoantibodies lead to symptoms observed in autism by corrupting neurological functions. Furthermore autoantibodies against receptor of serotonin are observed on autistic individuals (Singh et al. 1997).

In particular for HLA (human leukocyte antigen) gens, it is thought that many gens have relation with risk increase in the formation of certain autoimmune diseases. Different HLA haplotypes are also associated with neural developmental disorders like autism. Many HLA haplotype especially HLA - DR 4 is more frequently observed on autistic children than the general society. HLA genes are within a large genomic region called MHC (major histo - compatibility complex). This region includes a vast of genes. One of these genes is the complement protein C4 gene which is important for natural immunity (Careaga et al. 2010). The relation of complement 4B null allele with autism related to various autoimmune diseases was examined by Mostafa and Shehab (2010). It is indicated that in Egyptian children C4B null allele frequency is significantly higher. The relation of other gens such as macrophage migration inhibitor factor (MIF) gen and serine and threonine kinase C gen, PRKCB with autism is indicated (Careaga et al. 2010).

In many studies, a meaningful relation between familial autoimmunity and autism is found. It is associated with autistic characteristics such as the existence of the familial history of allergic/ autoimmune disease, regression and such as increased head circumference (Croen et al 2005, Molloy et al. 2006, Sacco et al. 2007). It is indicated that in the family of autistic individuals the number of autoimmune disease is much more (Comi et al. 1999). When family members with autoimmune disease goes up from 1 to 3, autism risk will increase and probability ratio goes up from 1.9 to 5.5. The most common autoimmune diseases in these families are diabetes mellitus type 1, adult rheumatoid arthritis, hypothyroidism, celiac disease and systemic lupus erythematous. Yet in a study performed by Croen and his collegues (2005) the relation between the existence of autoimmune disease in mother (auto-immune diseased mother) and autism was supported. It is claimed that rather than a specific disorder in mother, a general immune function disorder will have relation with the increase in autism risk. (Careaga et al. 2010).

The fact that autoantibodies against the critical neural components of fetal brain structure of some mothers of autistic children is detected, shows the possible potential mechanism in the form of transmissing the antibodies to the fetus developing in pregnancy of the mother. It is thought that anti - fetal brain autoantibodies that are available nearly on 12 % of the mothers of autistic children are connected to the neuronal targets during the formation and therefore they affect the neuronal formation. The view stating that antibodies isolated from the mother by animal models may affect the neural formation in the early period and therefore it may lead to behavioral changes (Careaga et al. 2010).

In line with the studies made in order to understand the neurobiology of autism, it is indicated that immunological factors also play a role in the etiopathogenesis of the disorder. Various findings such as the existence of antibodies against neural antigens and autoimmune disease history in the family support the possible role of immunological factors in autism. Although so far achieved symptoms make contribution to data that are related to the neurobiology of autism, they don't affect the diagnosis and treatment of autism. In the future, it is aimed to use these data in the separation of autism from other neurodevelopmental disorders and during the diagnosis and treatment process.

2. Neurotrophic factors

Several lines of evidence suggest that cytokines, growth factors and neurotrophic factors play particular signaling roles with in the brain to produce neurochemical, neuroendocrinological and behavioral changes (Hashimoto et al. 2003). Neurotrophic factors play an important role in the development of peripheral and central nervous system. Neurotrophic factors have an important function in the neurodevelopmental process for survival and differentiation of neurons, and regulation of their functions. Last findings showed us that the pathophysiology of autism is associated with neurotrophic factors. It is suggested that in some of the abnormality of neurotrophic factors may be implicated in mechanism associated with neural dysfunction in autism (Nelson et al 2006, Suzuki et al. 2007, Toyoda et al. 2007).

Neurotrophic factors comprise a range of different protein superfamilies. The well-known group among neurothrophic factors are neurotrophins. Nerve growth factor (NGF), brain-derived growth factor (BDNF), neurotrophin (NT)-3, NT-4, NT-5 ve NT-6 are members of neurotrophin family. Other than the neurotrophins, neurotrophic superfamilies contain too many protein families such as neurokines (for example ciliary NTF (CNTF) and leukaemia inhibitory factor (LIF)), insulin-like growth factors (for example, IGF-1 and IGF-2), fibroblast growth factors (for example FGF-1, FGF-2), as well as transforming growth factor-β (TGF-β) super family and epidermal growth factor superfamily (Barde 1990, Götz et al. 1994, Barbacid 1995, Nickl-Jockschat and Michel 2011).

The most important member of neurotrophin family having too many functions such as neuronal survival, target innervations and synaptogenesis in development of peripheral and central nerveus system is BDNF. All the neurotrophins bind with lower affinity to the p75 receptor and with high affinity to receptors of the tyrosine kinase family: NGF binds to TrkA, BDNF and NT-4 bind TrkB, NT-3 binds to TrkC (McKay et al. 1999). Neurotrophins perform their biological functions through such receptors. In neuron differentiation process, BDNF has too many roles such as neuronal survival, activity-dependent dendritic and axonal outhgrowth/branching, synapse formation and neuronal plasticity (reviewed in Polleux and Lauder 2004, Shieh ve Ghosh 1997). In various studies, the relationship of BDNF gene with autism is examined. Nishimura and his collegues (2007) determined increase of BDNF expression in autistic individuals. In different studies that followed, the probable role of BDNF gene mutation in autism pathogenesis is supported (Cheng et al. 2009). In a recent study by which significant increase of BDNF serum levels are determined in terms of autistic children, it is found that genetic variations of BDNF gene has no significant effect on risk of autism. However, a significant relationship is reported between neurotrophic thyrosine kinase, receptor, type 2=NTRK2 and autism (Correira et al. 2010).

In various studies, it is examined, whether there is any difference in the level of neurotrophic factor in serum and cerebrospinal fluid (CSF) of autistic children or not. In a study performed by Nelson and his collegues (2001), neonatal blood levels of neurotrophic factors are examined in autistic children and non autistic children with mental retardation. When compared with control group, in both groups, an important increase was observed in BDNF and neurotrophin 4/5 neonatal blood levels. In autism and a heterogenic group with cognitive functional defect, this result is very important in terms of indicating some neurotrophins' excessive express in peripheral blood taken in the early days of the life. However it is unknown whether these substances are also at high levels in prenatal life or not, or how long they remain at high levels in postnatal period. In their comprehensive study, Nelson and his collegues determined when compared with control group, that there is significant increase serum neurotrophin -4 and BDNF (the two TrkB ligands) in autism spectrum disorders and mental retardation without autism, but there were no changes with respect to NGF (trkA ligands) and neurotrophin-3 (TrkC ligands) levels. In line with these findings, it is suggested that trkB ligands are expressed or secreted at higher levels in the central nerveus system of children having autism or mental retardation during early infancy (reviewed in Polleux and Lauder 2004). It is suggested that the effect of BDNF and nörotrofin-4 on activity dependent dendritic outgrowth and branching (McAllister et al 1996, 1997), is related to the early and transient brain growth observed in autistic babies (Courchesne et al. 2003, reviewed in Polleux and Lauder 2004).

This increased level of BDNF expression and/or secretion is thought to have a relationship with the role of Methyl-CpG-binding protein 2 (MeCP2) in the control of BDNF transcription (Chen et al. 2003) Mutation in MeCP2 gene is responsible for Rett Syndrome. It is shown that MeCP2 binds selectively to BDNF promoter III and performs the function of repressing the expression of BDNF gene (Shahbazian et al. 2002). The MeCP2 gene mutations, by having effect on BDNF expression and potentially dendritic differentiation in cortex, may become a risk factor for autism (reviewed in Polleux and Lauder 2004). For that purpose, in a study by which MeCP2 gene mutation in autistic individuals is researched, de novo mutation is found in two girls (Carney et al. 2003).

In a different study on children with autism spectrum disorders, the existence of autoantibodies against BDNF in the serum is shown together with the increase on the level of BDNF, and it is put forward that there is an interaction between BDNF and immune system (Connolly and et al. 2006). Morrison and Mason (1998) reported that the levels of increased BDNF and neurotrophin-4 resulted in the decline in the purkinje cell number in the existence of granular cells. Although it is suggested that changes in the levels of neurotrophin on autistic individuals may be the primary factor in the formation of autism, it is still not certain whether these changes are primary etiopathagenic mechanism or secondary reaction against cortical changes (Pardo ve Eberhart 2007).

By taking the changes determined in serum BDNF levels in autism spectrum disorders, Croen and his colleagues (2008) studied the usability of BDNF as an early biological marker for autism. In their evaluation of BDNF concentrations in mid-pregnancy and neonatal blood specimens, they found no differences between subjects with autism compared with subjects with mental retardation or with general population controls. It is reported that in critical period of early development, concentration of BDNF would not be a useful clinic biomarker for autism (Croen et al. 2008).

In a different study, the age-related changes of serum BDNF levels were studied and it was indicated that there were disturbances of BDNF level trajectories in autism cases. In healthy controls, the serum BDNF concentration was found to be increased over the first several years and decreased slightly after reaching the adult level. In the autism cases, mean levels were significantly lower in children between 0-9 ages compared to teenagers or adults, or to age-matched healthy controls, indicating a delayed BDNF increase with development. In the same study, circadian changes, but not seasonal changes, were found in serum levels of BDNF (Katoh-Semba et al. 2007).

There is a little number of studies that concentrated on CSF levels of neurotrophic factors. Riikonen and Vanhala (1999) investigated CSF NGF levels in children with infantile autism and children with Rett syndrome. It was indicated that mainly normal CSF NGF levels were found in autism, whereas there were low to negligible values in Rett Syndrome.

TGF-β signaling, is related with too many biological processes that cover cell growth, differentiation and morphogenesis. By keeping in mind the key role of TGF-β in mental development and in controlling immune responses, it was hyphothesized that TGF-β1 could have a relationship with pathophysiology of autism. In various studies regarding autistic children and adults, it was determined that when compared with control group, the serum TNF-β1 levels were significantly lower in the autistic patients (Okada et al.2007, Ashwood et al. 2008). It is suggested that this finding is in relationship with abnormalities in mental development and immune process regulation, which have been observed in autistic patients.

IGFs are peptides that regulate growth and differentiation as well as being synthesized in most tissues including the developing central nervous system (Bach and Rechler 1995, Jones and Clemmons 1995). It was shown that IGF 1 levels in CSF of autistic children were lower and this might play a role in the pathogenesis of autism (Vanhala et al. 2001). Because IGF 1 is important for the survival of Purkinje cells of the cerebellum, it is put forward that low concentrations of CSF IGF1 at an early stage might be linked with the pathogenesis in autism (Riikonen et al. 2006).

EGF is a member of growth factors which play an important role in the cellular proliferation and the differentiation of the central and peripheral nervous system (Wong and Guillaud 2004). EGF has been suggested to be involved especially in the growth and survival of midbrain dopaminergic neurons (Alexi and Hefti 1993; Casper et al. 1991; Casper and Blum 1995). The most effective drugs in autism treatment are antipsychotics which block dopamin receptors; according to these findings it's suggested that there might be dopaminergic hyperactivity in autism pathophysiology. In a recent study, considering the role of EGF in mental development, it's relation with pathophysiology of autism was investigated and it was demonstrated that levels of EGF were significantly lower in adults with high functioning autism (Suzuki et al. 2007). Contrary to this finding, in a different study carried out on a pediatric autistic group, serum EGF levels in children with autism were significantly higher than those of age-matched normal healthy control subjects. This difference determined among and between adult and pediatric age groups in terms of serum EGF levels might be associated with possible developmental abnormality in the regulation of EGF expression. It can be expected that EGF levels may increase in early ages and begin to decrease as the autistic individual gets older (Işeri et al. 2011).
Due to the fact that behavioral symptoms and functional impairment degree are variable, autism is defined as the heterogeneous symptom cluster that has different etiology and pathology. Although phenotypic heterogeneity is the biggest obstacle in front of the research carried out that are oriented to defining the etiology of the autism, the fact that research techniques related to the biological factors are increased in the recent period and that research made on such field has become widespread, is promising. Due to their key roles during the neurodevelopment, neurotrophic factors have become one of the issues focused on research carried out that are oriented to determining the etiopathogenesis of autism. In most of the studies, significant differences were detected on the levels of serum neurotrophic factor. Studies performed up to now showed that neurotrophic factors could play a role in the pathophysiology of autism. However, it is necessary to integrate the findings of those studies above with new studies investigating the anatomical and functional changes on central nervous system of autistic patients in the future.

3. References

Alexi T, Hefti F (1993) Trophic actions of transforming growth factor alpha on mesencephalic dopaminergic neurons developing in culture. Neuroscience, 55 (4): 903–918.

Ashwood P, Water JV (2004) Is autism an autoimmune disease? Autoimmunity Reviews, 57:557-562.

Ashwood P, Wills S, Water JV (2006) The immune response in autism: a new frontier for autism research. Journal of Leukocyte Biology, 80: 1-15.

Ashwood P, Enstrom A, Krakowiak P, Hertz-Picciotto I, Hansen RL, Croen LA et al. (2008)Decreased transforming growth factor beta 1 in autism: a potential link between immune dysregulation and impairment in clinical behavioral outcomes. Neuroimmunol, 204:149-153.

Bach LA, Rechler MM (1995) Insulin like growth factor binding proteins. Diabete Rev, 3: 38-61.

Barbacid M (1995) Neurotrophic factors and their receptors. Curr Opin Cell Biol, 7:148-155.

Barde YA (1990) The nerve growth factor family. Prog Growth Factor Res, 2: 237-248

Bradstreet JJ, Smith S, Granpeesheh D, El-Dahr JM, Rossignol (2007) Spironolactone might be a desirable immunologic and hormonal intervention in autism spectrum disorders. Med Hypotheses, 68(5): 979-87.

Careaga M, Water JV, Ashwood P (2010) Immune dysfunction in autism: A pathway to treatment. Neurotherapeutics: The Journal of American Society for Experimental NeuroTherapeutics, 7: 283-292.

Carney RM, Wolpert CM, Ravan SA, Shahbazian M, Ashley-Koch A, Cuccaro ML et al (2003) Identification of MeCP2 mutations in a series of females with autistic disorder. Pediatr Neurol, 28: 205-11.

Casper D, Blum M (1995) Epidermal growth factor and basic fibroblast growth factor protect dopaminergic neurons from glutamate toxicity in culture. J Neurochem, 65 (3): 1016–1026.

Casper D, Mytilineou C, Blum M (1991) EGF enhances the survival of dopamine neurons in rat embryonic mesencephalon primary cell culture. J. Neurosci. Res, 30 (2): 372–381.

Chen WG, Chang Q, Lin Y, Meissner A, West AE, Griffith EC et al. (2003) Derepression of BDNF transcription involves calcium-dependent phosphorylation of MeCP2. Science, 302: 885-889.

Cheng L, Ge Q, Xiao P, Sun B, Ke X, Bai Y ve ark (2009) Association study between BDNF gene polymorphism and autism by three-dimensional gel-based microarray. Int J Mol Sci, 10(6): 2487-2500.

Chez MG, Dowling T, Patel PB, Khanna P, Kominsky M (2007) Elavation of Tumor Necrosis Factor-alpha in Cerebrospinal Fluid of Autistic Children. Pediatr Neurol, 36: 361-365.

Cohly HH, Panja A (2005) Immunological dfindings in autism. Int Rev Neurobiol, 71: 317-341.

Comi AM, Zimmerman AW, Frye VH, Law PA, Peeden JN (1999) Familial clustering of autoimmune disorders and evaluation of medical risk factors in autism. J Child Neurol, 14: 388–394.

Connolly AM, Chez M, Streif EM, Keeling RM, Golumbek PT, Kwon JM et al (2006) Brain-derived neurotrophic factor and autoantibodies to neural antigens in sera of children with autistic spectrum disorders, Landau-Kleffner syndrome, and epilepsy. Biol Psychiarty, 59:354-63.

Correira CT, Coutinho AM, Sequeira AF, Sousa IG, Lourenço Venda L, Almeida JP et al. (2010) Increased BDNF levels and NTRK2 gene association suggest a disruption of BDNF/TrkB signaling in autism. Genes, Brain and behavior, 9(7): 841-8.

Courchesne E, Carper R, Akshoomoff N (2003) Evidence of brain overgrowth in the first year of life in autism. JAMA, 290:337-344.

Croen LA, Grether JK, Yoshida CK, Odouli R, Van de Water J (2005) Maternal autoimmune diseases, asthma and allergies, and childhood autism spectrum disorders: A case-control study. Archives of Pediatrics and Adolescent Medicine, 159: 151–157.

Croen LA, Goines P, Braunschweig D, Yolken R, Yoshida CK, Grether JK et al. (2008) Brain-derived neurotrophic factor and autism: Maternal and infant peripheral blood levels in the early markers for autism (EMA) study. Autism Research, 1:130-137

Götz R, Köster R, Winkler C, Raulf F, Lottspeich F, Schartl M et al (1994) Neurotrophin-6 is a new member of the nerve growth factor family. Nature, 372:266-269.

Güney E, İşeri E, Atak A, Aral A, Bodur Ş, Şener Ş (2008) Otistik bozukluğu olan çocuklarda TNF-alfa düzeyi. Türkiye'de Psikiyatri, 10(2): 52-56.

Hashimoto K, Shimizu E, Komatsu N, Nakazato M, Okamura N, Watanabe H, et al. (2003) Increased levels of serum basic fibroblast growth factor in schizophrenia. Psychiatry Research, 120:211-218.

Işeri E, Güney E, Ceylan MF, Yücel A, Aral A, Bodur S et al. (2011) Increased serum levels of epidermal growth factor in children with autism. J Autism Dev Disord, 41(2): 237-41.

Jones JI, Clemmons DR (1995) Insulin like growth factors and their binding protein: biological actions. Endocr Reviews 16: 3-34.

Jyonouchi H, Sun S, Le S (2001) Proinflammatory and regulatory cytokine production associated with innate and adaptive immune responses in children with autism spectrum disorders and developmental regression. J Neuroimmunol, 120: 170-9.

Katoh-Semba R, Wakako R, Komori T, Shigemi H, Miyazaki N, Ito H et al. (2007) Age-related changes in BDNF protein levels in human serum: differences between autism cases and normal controls. Int J Dev Neurosci, 25:367-372.

McAllister AK, Katz LC, Lo DC (1996) Neurotrophin regulation of cortical dendritic growth requires activity . Neuron, 17: 1057-1064.

McAllister AK, Katz LC, Lo DC (1997) Opposing roles for endogenous BDNF and NT-3 in regulating cortical dendritic growth. Neuron, 18:767-778.

McKay SE, Purcell AL, Carew TJ (1999) Regulation of synaptic function by neurotrophic factors in vertebrates and intervertebrates:Implications for development and learning.Learn Mem, 6:193-215.

Molloy CA, Morrow AL, Meinzen-Derr J, Dawson G, Bernier R, Dunn M et al (2006) Familial autoimmune thyroid disease as a risk factor for regression in children with autişm spectrum disorder: ACPEA study. J Autism Dev Disord, 36: 317–324.

Morrison ME, Mason CA (1998) Granule neuron regulation of purkinje cell development: Striking a balance between neurotrophin and glutamate signaling. Journal of neuroscience, 18; 3563-3573.

Mostafa GA, Shehab AA (2010) The link of C4B null allele to autism and to a family histoy of autoimmunity in Egyptian autistic children. J Neuroimmunol, 223 (1-2): 115-9.

Nelson PG, Kuddo T, Song EY, Dambrosia JM, Kohler S, Satyanarayana G et al. (2006) Selected neurotrophins, neuropeptides, and cytokines: developmental trajectory and concentrations in neonatal blood of children with autism or Down syndrome. Int J Dev Neurosci, 24(1):73-80.

Nelson KB, Grether JK, Croen LA, Dambrosia JM, Dickens BF, Jelliffe LL et al. (2001) Neuropeptides and neurotrophins in neonatal blood of children with autism or mental retardation. Ann neurol, 49: 597-606.

Nickl-Jockschat T, Michel TM (2011) The role of neurotrophic factors in autism. Mol Psychiatry, 16: 478-490.

Nishimura K, Nakamura K, Anitha A, Yamada K, Tsujii M, Iwayama Y ve ark (2007) Genetic analyses of the brain -derived neurotrophic factor (BDNF) gene in autism. Biochem Biophys Res Commun, 356 (1):200-6.

Okada K, Hashimoto K, Iwata Y, Nakamura K, Tsujii M, Tsuchiya KJ et al. (2007) Decreased serum levels of transforming growth factor β1 in patients with autism. Prog Neuropsychopharmacol Biol Psychiatry, 31(1): 187-190.

Pardo CA, Eberhart CG (2007) The neurobiology of autism. Brain Pathol, 17: 434-447.

Pardo CA, Vargas DL, Zimmerman AW (2005) Immunity, neuroglia and neuroinflammation in autism. International Review of Psychiatry, 17 (6): 485-495.

Patterson PH (2002) Maternal infection: window on neuroimmuneinteractions in fetal brain development and mental ilness. Curr Opin Neurobiol, 12: 115-118.

Polleux F, Lauder JM (2004) Toward a developmental neurobiology of autism. Mental Retardation and Developmental Disabilities Research Review, 10: 303-317.

Riikonen R, Vanhala R (1999) Levels of cerebrospinal fluid nerve-growth factor differ in infantile autism and Rett Syndrome. Dev Med Child Neurol, 41(3): 148-52.

Riikonen R, Makkonen I, Vanhala R, Turpeinen U, Kuikka J, Kokki H (2006) Cerebrospinal fluid insulin-like growth factors IGF1 and IGF2 in infantile autism. Dev Med Child Neurol 48(9): 751-5.

Sacco R, Militerni R, Frolli A, Bravaccio C, Gritti A, Elia M et al (2007) Clinical, morphological, and biochemical correlates of head circumference in autism. Biol Psychiatry, 62(9): 1038-47.

Schmitz C, Rezaie P (2008) The neuropathology of autism: where do we stand? Neuropathology and Applied Neurobiology, 34: 4-11.

Shahbazian M, Young J, Yuva-Paylor L, Spencer C, Antalffy B, Noebels J et al. (2002) Mice with truncated MeCP2 recapitulate many Rett syndrome features and display hyperacetylation of histone H3. Neuron, 35: 243-254.

Shieh PB, Ghosh A (1997) Neurotrophins: new roles for a seasoned cast. Current Biology, 7: 627-630.

Singh VK (1996) Plasma increase of interleukin-12 and interferon-gamma.Pathological significance in autism. J Neuroimmunol, 66:143-145.

Singh VK, Rivas WH (2004) Prevalence of serum antibodies to caudate nucleus in autistic children. Neuroscience Letters, 355: 53-56.

Singh VK, Singh EA, Warren RP (1997) Hyperserotoninemia and serotonin receptor antibodies in children with autism but not mental retardation. Biol Psychiatry, 41: 753-755.

Suzuki K, Hashimato K, Iwata Y, Nakamura K, Tsujii M, Tsuchiya KJ et al.(2007) Decreased serum levels of epidermal growth factor in adult subjects with high-functioning autism. Biol Psychiatry, 62:267-269.

Toyoda T, Nakamura K, Yamada K, Thanseem I, Anitha A, Suda S, et al. (2007) SNP analyses of growth factor genes EGF, TGFβ-1, and HGF reveal haplotypic association of EGF with autism. Biochemical and Biophysical Research Communications, 360(4): 715-720.

Vanhala R, Turpeinen U, Riikonen R (2001) Low levels of insulin-like growth factor-I in cerebrospinal fluid in children with autism. Dev Med Child Neurol, 43: 614-616.

Wong RW, Guillaud L (2004) The role of epidermal growth factor and its receptors in mammalian CNS. Cytokine Growth Factor Rev, 15:147-156.

Yamashita Y, Fujimoto C, Nakajima E, Isagai T, Matsuishi T (2003) Possible association between congenital cytomegalovirus infection and autistic disorder. J Autism Dev Disord, 33: 455-459.

16

Autoimmunity in Autism Spectrum Disorders

Laila Y. AL-Ayadhi
Autism Research & Treatment Center (ART Center), AL-Amodi Autism Research Chair, Department of Physiology, Faculty of Medicine, King Saud University, Riyadh, Saudi Arabia

1. Introduction

Autism spectrum disorders (ASD) are part of a broad spectrum of neurodevelopmental heterogeneous disorders known as pervasive developmental disorders (PDD), which include autism, Asperger's syndrome, Rett's disorder, and childhood disintegrative disintegrative disorder. By description, ASD are characterized by impairments in verbal and nonverbal communication and social interaction (Association, A. P., 1994), with onset usually around the first 36 months of childhood. Repetitive, stereotyped, purposeless behaviors as well as attention and sensory dysfunctions are common findings in patients with ASD. Over the last few years, the prevalence of ASD has increased dramatically, and this increase, cannot be attributed entirely to improved diagnostic techniques and increased awareness only (Fombonne, 2003). Latest reports estimate that ASD affects approximately one per one hundred persons, with a male-to-female ratio of four to one (4:1) (Fombonne, 2003). Despite that the fact there is increase in ASD research worldwide, the exact etiology of autism and ASD remains largely unknown. Over the last few years, a scientific interest has occurred in the close relationship of the immune system to the central nervous system leading to considerable expansion in the field of psychoneuroimmunology. And currently it is widely accepted that environmental factors can compromise the immune system. A multi directional scientific approach has been adopted by many scientists in their research journey as it is likely to result from a complex combination of environmental, neurological, immunological, and genetic factors. There is emerging evidence and growing concern that a dysregulated or abnormal immune responses play an important role in some forms of ASD. In general, the associations between the immune and neurological systems are becoming more evident in many neurological disorders. Behaviors such as mood and sleep can be altered by cytokines and other products of immune activation, due to widespread effects on neurons. Aberrant immune activity during the early critical periods of brain and neuronal development could potentially play a role in neuronal dysfunction. Several efforts have attempted to link dysfunctional immune activity and ASD, such as maternal immune abnormalities during early pregnancy, increased incidence of familial autoimmunity, and childhood vaccinations. Several lines of research have shown abnormalities of the immune response in autism, including abnormal generation of antibodies, cytokines, and immune cells. The following chapter is a review of current research linking immune dysfunction to ASD.

2. Neurological abnormalities in autism spectrum disorders

Neural development takes place during the early years of life. During this period of time, neuronal differentiation, migration, axonal extension, and synapse formation takes place in an accurately designed sequence of events. In ASD, many neurological abnormalities have been found, which suggest that normal neurodevelopment was disturbed during a critical time of development, which include proliferation, migration, differentiation, synaptogenesis, gliogenesis, myelination, and apoptosis of neurons (Rice & Barone Jr., 2000). This critical time period extend from embryonic stage up until adolescence, with period of overlap at some stages. Quarter of autistic patients undergo a period of autistic regression, around 18th–24th months of age, during which they experience loss of previously acquired language and behavioral skills (Fombonne, 2003). Postmortem and neuroimaging magnetic resonance imaging (MRI) studies played an important role in the discovery of many neurological abnormalities in autistic patient's brains. These studies have suggested that many major brain structures may be affected in autism, including cerebellum, cerebral cortex, amygdala, hippocampus, corpus collosum, basal ganglia, and brain stem (Akshoomoff et al.,2002; Acosta et al., 2004; Courchesne et al., 2001). These abnormalities suggest multiple periods of prenatal onset. Furthermore, brain regions implicated in ASD tend to develop more slowly and are more vulnerable to disturbances. Cerebellar abnormalities have been observed as the most consistent finding in ASD, targeting in particular Purkinje and granular cells (Courchesne, 2002). Another important area which was found to be abnormal in ASD is the limbic system. The limbic system, whose components include the amygdala, hippocampus, cingulate gyrus, and septal nuclei, consists of a group of nuclei unified by a common function. The limbic system controls emotional behavior and any changes in body state that accompany this behavior, such as heart rate blood pressure, and respiration rate. Due to its role in emotion, the limbic system is of major interest in ASD patients; so far, the abnormal findings include increased cell packing and small neuronal size, indicative of cellular, maturational arrest (Akshoomoff et al., 2002; Palmen et al., 2004). Other neurological abnormalities described in ASD are abnormal EEG findings: around third of children with ASD develop epilepsy by adolescence (Volkmar et al., 1999), and an additional, significant minority has subclinical epilepsy, as measured by epileptiform encephalogram, especially during sleep (Tuchman et al., 1991). These findings clearly indicate that there are neurological involvements in ASD that affect the development and differentiation of neurons in the brain.

2.1 Structural magnetic resonance imaging findings

Structural magnetic resonance imaging (SMRI) studies played a major role in highlighting brain changes in ASD. SMRI confirmed the increase in total brain volume in autism, which present as increase in head circumference, starting around 2 to 4 years of age (Hazlett et al., 2005, Courchesne et al., 2001). This increase was attributed mainly to increase in total cerebral white matter and total cortical gray matter. The inner zone of white matter, especially the corpus callosum and internal capsule, showed no volume increase. The volume of the outer radiate white matter was increased in all cerebral lobes but with a frontal predominance. Collectively, these findings were interpreted as evidence of overgrowth of short- and medium-range intrahemispheric corticocortical connections with no detectable involvement of interhemispheric connections or connections between cortex and subcortical structures. The onset of brain overgrowth coincided with the onset of the signs and symptoms of autism, signifying that the overgrowth was part of a pathologic process that disrupted the development of normal brain structure and function in autism. A recent study, by Jieun et. al., (2010), recruiting a narrow age range of children with ASD and age-matched typically developing (TD) children, evaluating alterations in subregional amygdalar morphology. The group showed a bilateral enlargement of laterobasal subregions of the amygdala in 6- to 7-year-old children with ASD and that subregional alterations are associated with deficits in social and communicative behavior (Jieun et al., 2010)

2.2 Functional magnetic resonance imaging studies findings

Further understanding of autism was made from functional magnetic resonance imaging (fMRI) studies. During cognitive processing, subjects with autism use the same cortical areas, compared to aged matched control. Important remarkable variations have been found in the patterns of activation and in the timing or synchronization of the activation across the cortical network recruited to perform different tasks. fMRI study of written sentence comprehension has indicated that high-functioning adults with autism has relatively higher levels of activation in the left posterior superior temporal gyrus (Wernicke) and relatively lower levels of activation in the left frontal inferior gyrus (Broca) compared with age- and IQ-matched controls (Just et al., 2004). In addition, a reduction in functional connectivity, that is the correlation of the time series of the activation among cortical regions participating in performance of higher order tasks, was noted. Lower functional connectivity relative to the control group among participating cortical regions has been found in fMRI studies involving language (Just et al., 2004; Kana et al., 2006) working memory (Koshino et al., 2005), problem solving (Cherkassky et al., 2007), and social cognition (Castelli et al., 2002) providing further evidence of a general problem with functional under connectivity, within and between neocortical systems in autism. Functional imaging findings in autism have been consistent with a cognitive style favoring the use of visuospatial coding strategies, evident in increased reliance on extra striate and parietal regions (Manjaly et al., 2007). This could reflect a disruption in fronto-striatal and fronto-parietal functional connectivity (Just et al., 2007), abnormal activation within frontal and temporal regions has been related to the linguistic difficulties in this population (Groen et al., 2008).

2.3 Cortical connectivity in autism spectrum disorders

Cortical connectivity was examined in autism spectrum disorders by comparing gyral and sulcal thickness as indices of short- and longer-distance cortical connections (Hardan et al., 2006). The results showed an overall increase in cortical thickness in 8 to 12year old boys with autism compared to control. Furthermore, the study demonstrated that cortical thickness in sulci (long connections) was greater (analogous to increased volume of outer radiate white matter) than in gyri (short vertical connections), which is comparable to the findings of Herbert and colleagues for white matter (Herbert et al., 2004). Another significant finding was abnormalities in minicolumns structure in brain of autistic children. Minicolumns are composed of radically oriented arrays of pyramidal neurons (layers II-VI), interneuron's (layers I-VI), axons, and dendrites. Minicolumns assemble into macrocolumns, which form receptive fields. Minicolumns have been hypothesized to be the smallest radial unit of information processing in the cortex, but this function has not been confirmed. In autism spectrum disorders, reports indicate an increase in miniclomns number but narrower in width, with reduced neuronal space, with smaller neuron cell bodies and nucleoli

(Casanova et al., 2006). These abnormalities have been observed bilaterally in cortical areas 3, 4, 9, 17, 21, and 22. The description of these cortical abnormalities provides a critical counterbalance to the numerous reports of increased white matter volume, which might otherwise have led to a white matter model of autism. Findings of atypical patterns in both functional and anatomical connectivity in autism have established that autism is a not a localized neurological disorder, but affects many parts of the brain in many types of cognitive tasks. fMRI studies repeatedly find evidence of decreased coordination between frontal and posterior brain regions in autism, as measured by functional connectivity. In addition, neuroimaging studies have also revealed evidence of an atypical pattern of frontal white matter development in autism. These findings indicate that limitations of brain connectivity give rise to the varied behavioral deficits found in autism. As research continues to investigate these biological mechanisms, new intervention methods may develop to help improve brain connectivity and overcome the behavioral impairments of autism.

3. Autoimmunity in autism spectrum disorders

3.1 General immunological findings in autism spectrum disorders

Autoimmunity develops when the immune system is inappropriately directed to recognize and exert an exaggerated response to self components. The exact mechanism of autoimmunity in autoimmune diseases is not identical, but they all have autoreactive antibodies and T cells. The presence of antibodies directed against components of the CNS in the sera of autistic children is indicative of an autoimmune process that may be involved in the pathology of some cases of ASD. Almost with all autoimmune diseases, genetic, immune and environmental factors, such as, diet, toxic chemicals and infections, play critical roles in the development of the disease (Buehler, 2011; Vojdani et al., 2002; Kiberstis et al., 2002). Casein, casomorphins, gluten (GLU) and gluteomorphins, the opioid peptides, which is present in a range of food sources, are all implicated in the etiology of autism spectrum disorders. These peptides can stimulate T-cells, induce peptide specific T cell responses, and can lead to abnormal levels of cytokine production, which may result in inflammation, autoimmune reactions and disruption of neuroimmune communications (Jyonouchi et al., 2001). In autism spectrum disorders majority of children have wheat and milk protein intolerance. And accordingly, removal of these peptides from the diets significantly improves their conditions (Vojdani et al., 2002). Immunoglobulines, such as, IgG, IgM and IgA, were detected, against nine specific neuron-specific antigens in the sera of children with autism (Vojdani et al., 2002). These antibodies were found to bind with different central nervous system molecules that have sequence homologies to a milk protein.

Long exposure to toxic, environmental or occupational chemicals, have been shown to stimulate the production of autoantibodies to nervous system antigens. Titers of antibodies against neurofilaments and myelin basic protein (MBP) correlated significantly with urinary mercury and blood lead levels, the standard indicators of toxic exposure. In addition, levels of these antibodies proved to correlate with sensorimotor deficits. Gut-associated lymphoid tissues (GALT) can interact with toxins, chemicals and pollutants. If covalent reactions are formed between the drugs or other chemical compounds and the GALT, this can lead to immune responses and chemically-induced Type I- Type IV allergic reactions (Salama et al., 1989). Many infectious agents including measles, Rubella virus and Cytomegalovirus vaccines have long been suggested as etiologic factors in autism (Chess et al., 1978.; Wakefield et al., 1998; Ivarsson et al., 1990).

A complex communication system does exist between the nervous and the immune system, during normal and pathological conditions. Alteration in brain function can result from immune cells and molecules, such as cytokines and chemokines. This might affect cognition and emotions. Furthermore, immune cells and immune molecules can result in neuronal modulation of systemic CNS responses to infection, injury, and inflammation. The cytokines have been shown to directly affect neural tissue function and development, especially the proinflammatory cytokines such as interleukin (IL)-1, IL-6, IL-12, interferon-(IFN), and tumor necrosis factor (TNF) (Jarskog et al., 1997; AL-Ayadhi, 2005a). Inflammatory cytokine IL-6 can induce sleep, and TNF can provoke anorexia (Steinman et al., 2003; Tracey et al., 1988). Cytokines have been suggested to be responsible for many common features of autism, such as mood and sleep disturbances. In turn, neuropeptides, derived from the central and peripheral nervous system, have profound effects on the immune system, including the chemotaxis and recruitment of innate immune cells (Tracey et al., 1988). A dysregulated immune function has been suggested by many scientists. Systemic, immunologic abnormalities in autism have been related to autoimmunity, leading to the generation of antibodies that are reactive to CNS proteins and have the potential for neuronal tissue destruction, or leading to an inappropriate or ineffective immune response to pathogen assault (Korvatska et al., 2002). Several immune abnormalities, suggestive of, dysregulated immune response reported in autistic children include incomplete or partial T cell activation evidenced by increased numbers of T cells without the expression of the -2 receptor (IL-2R) (Warren et al., 1986; Plioplys et al., 1994), dysregulated apoptosis mechanisms (Korvatska et al., 2002), decreased peripheral lymphocyte numbers [30], decreased response to T cell mitogens (Warren et al., 1986; Stubbs et al., 1977) and the imbalance of serum Ig levels [30, 57]. Furthermore, immune-based genes including class II HLA-DRB1 alleles, class III complement C4 alleles, and HLA-extended haplotypes have been linked to autism spectrum disorders (Odell et al., 2005; Torres et al., 2001).

Animal models have also contributed to strengthening the dysregulated immune system hypothesis in the etiology of autism spectrum disorders which indicate that the maternal immune response to infection can influence fetal brain development via increased levels of circulating cytokines (Yamashita et al., 2003; Patterson et al., 2002). Mouse models of maternal influenza virus infection at mid-gestation have similar neuropathological and behavioral abnormalities in the offspring, which are consistent with those seen in autism and were again suggestive of a strong immune component (Patterson, 2002; Shi et al., 2003). Furthermore, infection of neonatal rats with Borna disease virus (BDV) leads to neuronal death in the hippocampus, cerebellum, and neocortex and a behavioral syndrome that has similarities to autism (Hornig, 2002). These abnormalities are correlated with major alterations of cytokine expression in various brain regions, indicating a likely role of cytokines as mediators of CNS injury in this model (Buehler, 2011; Plata-Salaman et al., 1999; Sauder & de la Torre, 1999).

Autoimmunity was first linked to autism in a study of an autistic child, with strong family history of autoimmune diseases (Money et al., 1971). This study suggested that an inherited risk of autoimmunity could increase the risk of developing autism. Another study, investigated the frequency of autoimmune disorders in family members of 61 ASD children and 46 typically developing normal controls, and showed the mean number of autoimmune disorders to be greater in families with autism (Comi et al., 1999). In most of these cases, the individual with the autoimmune disorder was a first-degree relative (i.e., a sibling or a parent) of the autism child (Comi et al., 1999). A variety of anti-brain antibodies have been found in autistic patients, including autoantibodies to serotonin receptor , myelin basic protein (MBP), neuron axon filament protein, cerebellar neurofilaments, nerve growth factor, 2-adrenergic, adrenergicbinding sites, anti-brain endothelial cell proteins, and antibodies directed against an as-yet unknown brain protein (Rosse et al., 2011; Buehler, 2011; AL-Ayadhi, 2005b; Singh et al., 1993; Singh et al., 1997; Todd et al., 1988; Connolly et al., 1999; Silva et al., 2004; Todd & Ciaranello, 1985). However it is still unclear as to the pathophysiological significance of these antibodies reported in children with ASD. For instance, increased autoantibodies is suggestive of increased neuronal damage, as is the case in multiple sclerosis, and other autoimmune diseases, where following demyelination, MBP is unmasked, and there is a subsequent generation of antibodies. Nevertheless, evidence of demyelination in autism has remained indefinable (Rumsey & Ernst, 2000). In one study, Glial fibrillary acidic protein (GFAP) measured in the CSF of 47 ASD children, was significantly elevated compared with 10 age-matched control children, suggesting that gliosis and unspecific brain damage may occur in autism (Ahlsen, et al., 1993). However, as GFAP correlates strongly with age, this is likely to be the result of age-dependent expansion of fibrillary astrocytes, and hence the data needs to be interpreted cautiously (Delneste, et al., 1999). Nevertheless, the importance of the presence of serum antibodies to brain tissues, regardless of the absence of neuronal damage is to be acknowledged. Singh and Rivas (2004), demonstrated antibodies directed to the rat caudate nucleus (the portion of the brain responsible for assembly of peripheral information) in 49% of the autism patients evaluated, but in none of the control cases. The observations of elevated anti-CNS antibodies in autism are at best unconfirmed and in some cases, such as serotonin receptors and MBP, are contradictory. It is important to keep in mind the difficulty in determining whether the autoantibodies present in the plasma of patients with autism contribute to the development of the disorder or if they are a consequence of the disease. Also, findings of autoimmunity in families of children with ASD suggest that in some patients, autoantibodies that target the CNS may be a pathological or an exacerbating factor in the neuronal development of children with ASD. However, increased autoimmunity may be limited to a subset of autistic patients (Buehler, 2011; Singh & Rivas, 2004). An important finding was reached by Wills research group (2009), they demonstrated that 21% the positive autoantibodies samples, reacted intensely with GABAergic Golgi neurons of the cerebellum while no samples from non-sibling, typically developing children showed similar staining (Wills et al., 2009). Further more, Rossi et al., (2011), demonstrated that 42% of controls and subjects with ASD were positively immunoreactive to some neural element, such as, cerebellar Golgi, interneurons, molecular layer of the dentate gyrus, and neuronal nuclei. Interestingly, children whose plasma reacted to brain tissue had scores on the Child Behavior Checklist (CBCL) that indicated increased behavioral and emotional problems. Children whose plasma was immunoreactive with neuronal cell bodies scored higher on multiple CBCL scales (Rossi et. a., 2011).

It is quite interesting to mention the results of a large cohort study consisted of all of the children born in Denmark from 1993 through 2004 (689 196 children). The study concluded the following: associations between family history of type 1 diabetes, infantile autism and maternal history of rheumatoid arthritis and ASDs were confirmed from previous studies. A significant association between maternal history of celiac disease and ASDs was observed for the first time. The observed associations between familial autoimmunity and ASDs/infantile autism are probably attributable to a combination of a common genetic background and a possible prenatal antibody exposure or alteration in fetal environment during pregnancy (Nancy et al., 2011).

3.2 Maternal immune system status findings

Maternal immune abnormalities such as autoimmune diseases, asthma, and allergies during pregnancy were investigated for a link to autism by Croen and colleagues (Croen, et al., 2005). They found no strong evidence linking maternal autoimmune diseases and autism. However, it was found that mothers diagnosed with asthma or allergies during their second trimester were more than twice as likely to have a child with autism (Croen, et al., 2005). To date, no studies have demonstrated that ASD children have an increased frequency of other autoimmune disease, the exception being whether ASD itself could be considered an autoimmune disease. Furthermore, due to the fact that the manifestation of autoimmune disease occur around the age of 30 and upward, and as the ASD cases studied are pediatric cases, it is worth following up the ASD cases to determine whether more autoimmune diseases will be observed as they mature. Serum from a mother with an autistic child was found to bind to Purkinje cells and other neurons, when injected into gestating mice. Furthermore, a behavioral change in mice was observed in the offspring, including altered exploration, motor coordination, and changes in cerebrallar magnetic resonance spectroscopy. On the other hand, mice injected with sera from mothers with typically developing children showed no behavioral changes (Dalton et al., 2003). This study supports the suggestion that maternal antibodies may influence neurodevelopmental processes in a subset of autism cases.

Interlukin 1 (IL-1) plays a key role in mediating severe placental damage and neurodevelopmental anomalies in offspring, as revealed by Girard et al. (2010). This group demonstrated that at the end of gestation, exposure of pregnant rats to systemic microbial product (LPS) triggers placental inflammation and massive cell death, fetal mortality, and both forebrain white matter and motor behavioral alterations in the offspring. All these effects are alleviated by the coadministration of IL-1 receptor antagonist, suggesting a possible protective treatment against human placental and fetal brain damage (Gerard et al., 2010)

3.3 Cytokines role in autism spectrum disorders

Cytokines (Chemokines) are a family of small proteins secreted by immune cells. They have the ability to induce directed chemotaxis in nearby responsive cells. Some chemokines are considered pro-inflammatory and can be induced during an immune response to recruit cells of the immune system to a site of infection. These proteins exert their biological effects by interacting with G protein-linked transmembrane receptors called chemokine receptors found on the surfaces of their target cell. Several studies have demonstrated elevated plasma levels of IL-12 and IFN- in autistic children compared with controls, with no changes for IL-6, TNF- and IFN- (Singh, 1996) plasma levels, suggesting a potential TH1 shift. On the other hand, another study demonstrated, higher plasma IFN- in 10 autistic children compared with adult control subjects (Jyonouchi et al., 2005). Moreover, an increased plasma IFN- levels were observed in 29 autistic children; with a positive correlation with the generation of the intercellular CNS messenger and marker of oxidative stress, nitric oxide (NO) (Sweeten et al., 2004). The same research group observed that the macrophage product neopterin, was higher in serum samples from individuals with ASD compared with controls, which may reflect increased cell-mediated immune activation and IFN- production (Sweeten et al., 2003). Higher IFN- and neopterin levels correlated significantly with elevated, circulating numbers of monocytes observed in autistic children (Sweeten et al., 2003) with elevated urine levels of neopterin and biopterin (Messahel et al., 1998).

In cell culture experiments, in which intracellular cytokine production was examined in 20 autistic patients, compared with 20 aged-matched controls, intracellular production of IL-4 was increased, with a reduction in IFN- and IL-2 in CD4 and CD8 lymphocytes following stimulation (Gupta et al., 1998), suggestive of a TH2 bias. In vitro studies of peripheral blood mononuclear cells stimulated with lipopolysaccharide (LPS), showed an inappropriate innate immune response evidenced by amplified production of proinflammatory cytokines TNF- and IL-1 in ASD patients compared with controls (Jyonouchi et al., 2001). This immune dysregulation of increased TNF- was also found in primary sibling family members of patients with ASD, indicating a possible similar genetic susceptibility in the patients studied. This emphasizes the importance of carefully controlled, age-matched studies in the field of ASD. Moreover, the diversity of the findings reinforces the idea that ASD consists of many different phenotypes, which share the same behavioral commonalities. Cytokines can activate and exert trophic effects on glial cells, which can in turn produce cytokines and chemokines upon such activation. Cell culture studies have shown that neuropoietic cytokines such as IL-6 can have direct effects on neurons and glia, including changes in proliferation, survival, death, neurite outgrowth, and gene expression (Gadient. & Patterson, 1999; Mehler & Kessler, 1998). As the CNS is populated largely by astroglia and microglial cells, these cytokine-cell interactions are important for neuronal cell functioning and development. Immune activation in postmortem brain specimens and CSF from subjects with autism have found neuroinflammation in the cerebral cortex and cerebellum of brain tissue in autism was found. This inflammatory process was characterized by a marked cellular activation of microglial and astroglial cells and the presence of an altered cytokine pattern. In addition, there was an accumulation of perivascular macrophages and monocytes but an absence of lymphocytes and antibody from the brain specimens, suggestive of an innate immune activation. In addition, an enhanced proinflammatory cytokine profile was observed in their CSF. Abnormal immune responses in the neuroglia of autistic patients was suggested, which in turn may influence neural function and neural development, and an aberrant immune response may contribute to the development of autism. In general, the brain and CNS are considered to be protected and isolated from potentially harmful pathogens or agents within the blood, including inflammatory immune cells and proteins, by the blood brain barrier (BBB). Cytokines however, can gain entry into the brain through active transport mechanisms or at circumventricular regions, where the barrier is less controlling (Wilson et al., 2002). Impairment of the BBB function may happen as a result of binding of cytokines and inflammatory mediators to receptors on the endothelial cells directly. In addition, cytokines can migrate into the brain from the blood via the CSF to the choroid plexus or from the blood to either the subarachnoid space or parenchymalperivascular space, resulting in alteration in immune responses and production of cytokines (Ransohoff et al., 2003).

Peripheral cytokines can directly affect afferent neurons and their functions (Dantzer et al., 1998). Immune organs such as bone marrow, thymus, spleen, and lymph nodes play an important role in immune system development. Additionally, immune response is capable of changing expression and distribution of neural receptors in these organs (Mignini et al., 2003). Thus the relationship is reciprocal between immune system and neural receptors. Cytokines can affect many behaviors such as, sleep, mood, appetite and nutritional uptake, exploratory behavior, and, social interactions and communication. Systemic cytokine administration at therapeutic doses of IFN-, IL-2, and TNF can result in mood changes, sleep disorder, decreased exploratory behavior, impaired cognitive function, and changes in enthusiasm and motivation (Larson , 2002; Licinio et al., 1998). Systemic administration of cytokines can produce an increased noradrenergic, dopaminergic, and serotonergic metabolism in the hypothalamus, hippocampus, and nucleus accumbens and modulate synaptic plasticity and thereby alter memory and learning (Zhao, B., & Schwartz, J. P., 1998). Many studies have demonstrated abnormal levels of blood lymphocytes in autism. Significantly decreased CD4 T cells have been observed in ASD (Warren et al., 1990; Denney et al., 1996; Ferrante et al., 2003; Yonk et al., 1990). In animal models, systemic T cell deficiency in mice have been shown to result in learning and memory impairment, which is reversible by T cell replacement (Kipnis et al., 2004). Furthermore, an incomplete or partial activation of T cells following stimulation, with an increased expression of HLA-DR but not the IL 2R chain (CD25), was observed in ASD (Engstrom et al., 2003; Plioplys et al., 1994).

NK cells are an important cytotoxic cell subset of the innate immune system and a major producer of cytokine. Reduced levels of circulating numbers of NK cells number and activity was observed in children with ASD and other related neurodevelopmental disorders, such as, Rett syndrome, compared with controls (Fiumara et al., 1999; Warren et al., 1987). In turn this reflects on its ability to eradicate or prevent viral infections in these children, which could potentially be damaging to neural tissues during critical windows of CNS development. Abnormal concentrations of plasma Ig classes have been observed in some ASD children with increased IgG2, IgG4, IgM and IgG (Ashwood et al., 2003; Croonenberghs et al., 2002; Trajkovski et al., 2004), highly indicative of an underlying autoimmune disorder and/or an atypical susceptibility to infections.

Cytokines activity in the brain tissue of ASD and matching normal subject was examined by Xiaohong et al., (2009). Results showed that proinflammatory cytokines (TNF- α, IL-6 and GM-CSF), Th1 cytokine (IFN-γ) and chemokine (IL-8) were significantly increased in the brains of ASD patients compared with the controls. However the Th2 cytokines (IL-4, IL-5 and IL-10) showed no significant difference. The Th1/Th2 ratio was also significantly increased in ASD patients. They concluded that ASD patients displayed an increased innate and adaptive immune response through the Th1 pathway, suggestive of a localized brain inflammation and autoimmune disorder involvement in the pathogenesis of ASD (Xiaohong et al., 2009). Flow cytometric analysis of NK cells demonstrated increased production of perforin, granzyme B, and interferon gamma (IFNγ) under resting conditions in children with ASD. Following NK cell stimulation in the presence of K562 target cells (cells used to assess NK cell cytotoxicity), the cytotoxicity of NK cells was significantly reduced in ASD compared with controls. Furthermore, under similar stimulation conditions the presence of perforin, granzyme B, and IFNγ in NK cells from ASD children was significantly lower compared with controls. Suggestive of possible dysfunction of NK cells, predisposing to the development of autoimmunity and/or adverse neuroimmune interactions during critical periods of development (Enstrom et al., 2009).

3.4 Neurokines role in autism spectrum disorders

Neurokines (neuropoietic cytokines), are neuronal related cytokines, that regulate cell numbers in the nervous system and influence functional activities of neurons. They are important mediators within the neuroimmune network. These factors are not produced exclusively by brain cells and their activities are not restricted to neurons only. Factors with neuropoietic activity include CNTF (ciliary neuronotrophic factor), CDF (cholinergic differentiation factor, LIF (leukemia inhibitory factor), Oncostatin M but also IL6, (BDNF, NGF, NT-3, NT-4, NT-4, NT-6, GDNF) and factors produced in, or acting upon, the nervous system such as bFGF, Interleukins, or TGF-beta.

There is continual communication between the immune and nervous systems with many peptides playing a role in both. It has been proposed frequently that abnormalities in the levels and actions of neurotransmitters or neuroactive compounds during early critical windows of neurodevelopment may lead to the onset of autism. Neurotransmitters and neuropeptides not only have important key roles in the development and organization of neural tissue but also influence almost all body functions including the immune system. Numerous transmitter systems, including acetylcholine, serotonin (5-HT), dopamine, epinephrine, norepinephrine, oxytocin, vasopressin, glutamate, and _-aminobutryic acid (GABA), have been studied in ASD (Lam et al., 2006). For example, in postmortem brain specimens obtained from patients with ASD, there was a 48– 61% decrease in glutamic acid decarboxylase, an enzyme that converts glutamate into GABA, in the parietal and cerebellar regions of the brain compared with controls (Fatemi et al., 2002). In ASD, this may cause suppression of the GABA-ergic system, resulting in heightened stimulation of the glutamate system, which has been associated with seizures. A positive intens autoantibodies reaction with GABAergic Golgi neurons of the cerebellum in 21% of children with ASD, were demonstrated, while no samples from non-sibling, typically developing children showed similar staining reaction (Wills et al., 2009), which in favor of the autoimmunity theory. Second, excitotoxic damage of neurons, possibly resulting from glutamate hyperactivity, may result in abnormal, structural development of the brain (Bittigau & Ikonomidou, 1997). The neurotransmitter serotonin has a wide range of affects on normal physiological functions including circadian rhythms, appetite, mood, sleep, anxiety, motor activity, and cognition. Serotonin is detected, not only in neurons of the nervous system but also in platelets and lymphocytes of the immune system, where it can exert dose-dependent, suppressive, or proliferative effects. In normal development, serotonin levels are high in the brain up to the age of five and then decrease dramatically (Muzik et al., 1999). Serotonin levels increase in the hypothalamus, hippocampus, and cortex in response to various cytokines, such as IL-1, IFN (Zhao & Schwartz, 1998; Simmons, & Broderick, 2005). Moreover, enzymes that control the conversion of tryptophan into serotonin are under the influence of IFN- and IL-1 (Wirleitner et al., 2003). Increased serotonin levels in peripheral blood platelets have been described in approximately one-third of patients with autism (Anderson et al., 1990). It is interesting that selective serotonin (5-HT) reuptake inhibitors (SSRIs) have been shown to be beneficial in treating obsessional and repetitive behaviors in some ASD patients sometimes (McDougle et al., 1996). The reason for the difference in serotonin levels is unknown; potentially, it may be a result of the presence of inflammatory cytokines or more likely, to alterations in the platelets themselves, which could modify serotonin uptake (Cook et al., 1996). In spite of the fact that imaging studies demonstrated a reduction in brain serotonin system. However, sometimes, treatment with SSRIs, produce a worsening of the symptoms. And accordingly, Azmitia et al., (2011), examined 5-HT axons that were immunoreactive to a serotonin transporter (5-HTT) antibody in a number of postmortem brains from autistic patients and controls with no known diagnosis who ranged in age from 2 to 29 years. Results from this study, demonstrated, a fine, highly branched, and thick straight fibers were found in forebrain pathways, such as, medial forebrain bundle, stria terminalis and ansa lenticularis. Many immunoreactive varicose fine fibers were also seen in target areas, for example, globus pallidus, amygdala and temporal cortex.

Morphometric analysis of the stained axons at all ages studied indicated that the number of serotonin axons was increased in both pathways and terminal regions in cortex from autism donors. Their findings, provide morphological evidence to warrant caution when using serotonin enhancing drugs (e.g. SSRIs and receptor agonist) to treat autistic children (Azmitia et al., 2011)

In addition, proinflammatory cytokines IL-1 and TNF are capable of affecting the activity of the serotonin transporter gene, a potential susceptibility gene in ASD (Coutinho et al., 2004). Cytokines and chemokines play a major role in many stages of development of the CNS and are known to induce the secretion of many neurotransmitters and neuropeptides (Biber et al., 2002). In turn, neuropeptides play an important role in all phases of immune system development, often acting as trophic factors, which has led to the hypothesis that neurotrophins (NTs) should be considered as neurokines, as they act in a cytokine like manner, influencing the development and function of the immune system (Levi-Montalcini et al., 1996). Several NTs with potent immunomodulatory actions, including neuropeptide Y, substance P, calcitonin gene-related peptide (CGRP), vasoactive intestinal peptide (VIP), BDNF, and NT-4/5, which have multiple affects on neurodevelopment and neuron maintenance, have been implicated in ASD. Analysis of neonatal blood spots by recycling immunoaffinity chromatography found that BDNF, VIP, CGRP, and NT-4/5 were elevated in ASD compared with typically developing control children but could not be distinguished from those with mental retardation (Nelson et al., 2001). Brain-derived neurotrophic factor is a major player in neurodevelopment known to regulate neuronal cell survival, growth, plasticity and differentiation, and is now considered to be a growth factor with a wide spectrum of functions outside the nervous system, including modulation and regulation of immune function (Vega et al., 2003; Nockher & Renz, 2003).

Based on animal studies, two structurally related neuropeptides, oxytocin and vasopressin, are believed to play a critical role in the formation of social bonding and recognition and in the processing of social cues (Young et al., 2002). Prairie voles are highly social animals, which form long-lasting pair bonds; in contrast, montane voles are asocial or solitary and do not form pair bonds (Wang et al., 1998). Central infusion of oxytocin in female or vasopressin in male prairie voles helps establish partner-bonding; this phenomenon can be blocked using specific antagonists (Williams et al., 1994; Young et al., 2001). Furthermore, oxytocin knockout mice have normal, cognitive abilities but diminished social recognition, suggesting a key role of oxytocin in social interactions. Other studies (Ferguson et al., 2000; Modahl et al., 1998) have found significantly lower levels of plasma oxytocin when compared with age-matched, normal subjects. Moreover, this decrease in oxytocin levels may be a result of a reduction in the processing of oxytocin, as increased levels of the pro-hormone form of oxytocin were found in autism patients (AL-Ayadhi 2005c; Green et al., 2001). In prairie voles, oxytocin and vasopressin receptors are located in the ventral forebrain, whereas the pattern of expression of oxytocin receptors differs in montane voles (Young et al., 2002). It would seem that not only the concentration of neuropeptides but also the pattern of receptor distribution may be important in the establishment of socially rewarding interactions. So far, signature patterns of neuropeptides and neurotransmitters and their respective receptors have yet to be established in ASD. Further studies that address this issue in ASD may provide clues into the development of impaired social interactions that are present in ASD. It is interesting that Dunzendorfer et al. (Dunzendorfer et al., 2001) have suggested a novel role for neuropeptides in the regulation of dendritic cell (DC) migration. They investigated locomotion of mononuclear cell-derived DCs at different maturation stages toward gradients of sensory neuropeptides in vitro. Calcitonin gene-related peptide, VIP, secretin, and secretoneurin induced immature DC chemotaxis comparable with the potency of the chemokine regulated on activation, normal T expressed and secreted (RANTES), whereas substance P and macrophage-inflammatory protein-3 (MIP-3) stimulated immature cell migration only slightly (Dunzendorfer et al., 2001). Moreover, the neuropeptide VIP synergized with cytokines such as TNF-in the induction of DC maturation (Delneste et al., 1999). In the CNS, DCs have been found in normal meninges, the choroid plexus, and CSF and are actively recruited during inflammation, where they may play equal roles in the defense against infections and contribute to the break-down of tolerance to CNS autoantigens (Pashenkov et al., 2003). These findings suggest a central role for DC- and neuropeptide-mediated chemotaxis in the control of CNS inflammation and the generation of T cell reactivity against CNS antigens, and present an intriguing concept in the context of autism.

4. Conclusion

Autism spectrum disorder is a complex, neurodevelopmental, heterogeneous condition, with multiple phenotypes and subgroups that share behavioral characters. This natural complexity of the disorder has made the pathophysiology and consequently the etiology, exceptionally difficult. There is a considerable controversy in the literature regarding an immune-based factor in the search for pathophysiological cause in ASD. Nevertheless, with increasing reports of immune dysfunction in autism, there is a growing notion and concern that immune dysfunction may play a role in a subgroup of patients with ASD. Attempts have been made to link dysfunctional immune activity and ASD, such as maternal immune abnormalities during early pregnancy, increased incidence of familial autoimmunity, childhood vaccinations, and the generation of autism animal models based on immune parameters. Starting from the embryonic stage of life, to postnatal, and to adult hood, both neurological and immune systems are intertwined and abnormalities in one of them will be reflected on the other with dysregulation altering levels of cytokines, chemokines, neurotransmitters, neuropeptides, as well as hormones. Each of these substances may influence the course of development in the nervous and/or immune systems primarily or through secondary action. Moreover, while the extent to which many of the observations discussed herein are involved in the pathogenesis of autism is unknown, it cannot be discounted that immune dysfunction is an epiphenomenon or a consequence of the disease. Comprehensive extensive studies of autism and age-matched control individuals and their families are mandatory for more conclusive results.

5. References

Acosta, M. T., Pearl, P. L. (2004) Imaging data in autism: from structure to malfunction. *Semin. Pediatr. Neurol.* 11, 205–213.

Ahlsen, G., Rosengren, L., Belfrage, M., Palm, A., Haglid, K., Hamberger, A., Gillberg, C. (1993). Glial fibrillary acidic protein in the cerebrospinal fluid of children with autism and other neuropsychiatric disorders. *Biol. Psychiatry* 33, 734–743

Akshoomoff, N., Pierce, K., Courchesne, E. (2002). The neurobiological basis of autism from a developmental perspective. *Dev. Psychopathol.* 14, 613–634.

AL-Ayadhi L. (2005). Altered Oxytocin and Vasopressin levels in Autistic children in Riyadh Area. *Neurosciences*, Vol. 10 (1):47-50

AL-Ayadhi L. (2005). Pro-inflammatory Cytokines in Autistic children in Riyadh Area, Saudi Arabia. *Neurosciences*, Vol. 10 (2):155-158

AL-Ayadhi L. (2005). The autoimmune Connection of Autism in Riyadh area. *Neurosciences* 2005; Vol. 10 (4): 265-267.

Anderson, G. M., Horne, W. C., Chatterjee, D., Cohen, D. J. (1990). The hyperserotonemia of autism. *Ann. N. Y. Acad. Sci.*, 600, 331–340.

Ashwood, P., Anthony, A., Pellicer, A. A., Torrente, F., Walker-Smith, J. A., Wakefield, A. J. (2003). Intestinal lymphocyte populations in children with regressive autism: evidence for extensive mucosal immunopathology. *J. Clin. Immunol.*, 23, 504–517

Association, A. P. (1994). Diagnostic and Statistical Manual of Mental Disorders, Fourth Edition, Washington, DC, American Psychiatric Association.

Atladóttir HO, Pedersen MG, Thorsen P, Mortensen PO. Bent Deleuran, Eaton. W. (2011). Association of Family History of Autoimmune Diseases and Autism Spectrum. *Pediatrics*, 2009;124;687-694

Azmitia EC, Singh JS, Whitaker-Azmitia PM, (2011). Increased serotonin axons (immunoreactive to 5-HT transporter) in postmortem brains from young autism donors. *Neuropharmacology.*, 2011 Feb 15

Biber, K., Zuurman, M. W., Dijkstra, I. M., Boddeke, H. W. G. M. (2002). Chemokines in the brain: neuroimmunology and beyond. *Curr. Opin. Pharmacol.*, 2, 63–68.

Bittigau, P., Ikonomidou, C. (1997) Glutamate in neurologic diseases. J. Child Neurol. 12, 471–485.

Buehler MR. (2011). A proposed mechanism for autism: an aberrant neuroimmune response manifested as a psychiatric disorder. *Med Hypotheses.*, Mar 19

Casanova MF, van Kooten IAJ, Switala AE, (2006). Minicolumnar abnormalities in autism. *Acta Neuropathol (Berl)*, 112(3):287–303.

Castelli F, Frith C, Happé F, Frith U. (2002). Autism, asperger syndrome and brain mechanisms for the attribution of mental states to animated shapes. *Brain* , 125 (pt 8):1839–1849.

Castelli F, Frith C, Happé F, Frith U. (2002). Autism, asperger syndrome and brain mechanisms for the attribution of mental states to animated shapes. *Brain*, 125(pt 8):1839–1849.

causes brain regional changes of mRNAs for cytokines, cytokine receptor components and neuropeptides. *Brain Res. Bull.* 49, 441–451.

Cherkassky VL, Keller TA, Kana RK, Minshew NJ. (2007). Functional and anatomical cortical underconnectivity in autism: evidence from an fMRI study of an executive function task and corpus callosum morphometry. *Cerb Cortex*,17(4):951–961.

Chess S., P. Fernandez and S. Korn. (1978). Behavioral consequences of congenital rubella. *J. Pediatr.* 93:669.

Chugani, D. C., Muzik, O., Behen, M., Rothermel, R., Janisse, J. J., Lee, J., Chugani, H. T. (1999). Developmental changes in brain serotonin synthesis capacity in autistic and nonautistic children. *Ann. Neurol.* 45, 287–295.

Comi, A. M., Zimmerman, A. W., Frye, V. H., Law, P. A., Peeden, J. N. (1999). Familial clustering of autoimmune disorders and evaluation of medical risk factors in autism. *J. Child Neurol.* 14, 388–394

Connolly, A. M., Chez, M. G., Pestronk, A., Arnold, S. T., Mehta, S., Deuel, R. K. (1999). Serum autoantibodies to brain in Landau-Kleffner variant, autism, and other neurologic disorders. *J. Pediatr.* 134, 607–

Conroy, J., Meally, E., Kearney, G., Fitzgerald, M., Gill, M., Gallagher, L. (2004). Serotonin transporter gene and autism: a haplotype analysis in an Irish autistic population. *Mol. Psychiatry* 9, 587–593.

Cook, E. H., Leventhal, B. L. (1996). The serotonin system in autism. *Curr. Opin. Pediatr.*, 8, 348–354.

Courchesne E, Karns CM, Davis HR, , Ziccardi, R., Carper, R. A., Tigue, Z. D., Chisum, H. J., Moses, P., Pierce, K., Lord, C., (2001). Unusual brain growth patterns in early life in patients with autistic disorder: an MRI study. *Neurology*, 57(2):245–254.

Courchesne E, Karns C, Davis H, Ziccardi R, Carper R, Tigue Z, Chisum HJ, Moses P, Pierce K, Lord C, Lincoln A, Pizzo S, Schreibman L, Haas R, Akshoomoff N, Courchesne R. Unusual brain growth patterns in early life in patients with autistic disorder: an MRI study. *Neurology* 2011 Jun 14;76(24):2111

Coutinho, A. M., Oliveira, G., Morgadinho, T., Fesel, C., Macedo, T. R., Bento, C., Marques, C., Ataide, A., Miguel, T., Borges, L., Vicente, A. M. (2004). Variants of the serotonin transporter gene (SLC6A4) significantly contribute to hyperserotonemia in autism. *Mol. Psychiatry* 9, 264–271.

Croen, L. A., Grether, J. K., Yoshida, C. K., Odouli, R., Van de Water, J. (2005). Maternal autoimmune diseases, asthma and allergies, and childhood autism spectrum disorders: a case-control study. *Arch. Pediatr. Adolesc. Med.* 159, 151–157.

Croonenberghs, J., Wauters, A., Devreese, K., Verkerk, R., Scharpe, S., Bosmans, E., Egyed, B., Deboutte, D., Maes, M. (2002) Increased serum albumin, _ globulin, immunoglobulin IgG, and IgG2 and IgG4 in autism. *Psychol. Med.* 32, 1457–1463.

Dalton, P., Deacon, R., Blamire, A., Pike, M., McKinlay, I., Stein, J., Styles, P., Vincent, A. (2003). Maternal antibodies associated with autism and language disorder. *Ann. Neurol.* 53, 533–537

Dantzer, R., Bluthe, R. M., Laye, S., Bret-Dibat, J. L., Parnet, P., Kelley, K. W. (1998). Cytokines and sickness behavior. *Ann. N. Y. Acad. Sci.* 840, 586–590.

Delneste, Y., Herbault, N., Galea, B., Magistrelli, G., Bazin, I., Bonnefoy, J. Y., Jeannin, P. (1999). Vasoactive intestinal peptide synergizes with TNF-in inducing human dendritic cell maturation. *J. Immunol.*, 163, 3071–3075.

Denney, D. R., Frei, B. W., Gaffney, G. R. (1996). Lymphocyte subsetsand interleukin-2 receptors in autistic children. *J. Autism Dev. Disord.*, 26, 87–97.

Dunzendorfer, S., Kaser, A., Meierhofer, C., Tilg, H., Wiedermann, C. J. (2001). Cutting edge: peripheral neuropeptides attract immature and arrest mature blood-derived dendritic cells. *J. Immunol.* 166, 2167–2172.

Engstrom, A. H., Ohlson, S., Stubbs, E. G., Maciulus, A., Caldwell, V., Odell, J. D., Torres, A. R. (2003). Decreased expression of CD95 (FAS/ APO-1) on CD4_ T-lymphocytes from participants with autism. *J. Dev. Phys. Disabil.* 15, 155–163.

Enstrom A M, Lit L, OnoreCE, Gregg JP, Hansen R, Pessah IN, Hertz-Picciotto I. (2009). Altered Gene Expression and Function of Peripheral Blood Natural Killer Cells in Children with Autism. *Brain Behav Immun.* 2009 January ; 23(1): 124–133

Fatemi, S. H., Halt, A. R., Stary, J. M., Kanodia, R., Schulz, S. C., Realmuto, G. R. (2002). Glutamic acid decarboxylase 65 and 67 kDa proteins are reduced in autistic parietal and cerebellar cortices. *Biol. Psychiatry*, 52, 805–810.

Ferguson, J. N., Young, L. J., Hearn, E. F., Matzuk, M. M., Insel, T. R., Winslow, J. T. (2000). Social amnesia in mice lacking the oxytocin gene. *Nat. Genet.*, 25, 284–288.

Ferrante, P., Saresella, M., Guerini, F. R., Marzorati, M., Musetti, M. C., Cazzullo, A. G. (2003). Significant association of HLA A2-DR11 with CD4 naive decrease in autistic children. *Biomed. Pharmacother.*, 57, 372–374.

Fiumara, A., Sciotto, A., Barone, R., D'Asero, G., Munda, S., Parano, E., Pavone, L. (1999). Peripheral lymphocyte subsets and other immune aspects in Rett syndrome. *Pediatr. Neurol,.* 21, 619–621.

Fombonne, E. (2003). The prevalence of autism. *JAMA* 289, 87–89.

Gadient, R. A., Patterson, P. H. (1999). Leukemia inhibitory factor, interleukin 6, and other cytokines using the GP130 transducing receptor: roles in inflammation and injury. *Stem Cells*, 17, 127–137.

Girard S, Tremblay L, Lepage M, Sébire G., (2010). IL-1 receptor antagonist protects against placental and neurodevelopmental defects induced by maternal inflammation. *J Immunol.*, Apr 1;184(7):3997-4005

Green, L., Fein, D., Modahl, C., Feinstein, C., Waterhouse, L., Morris, M. (2001). Oxytocin and autistic disorder: alterations in peptide forms. *Biol. Psychiatry*, 50, 609–613.

Groen WB, Zwiers MP, van der Gaag RJ, Buitelaar JK. (2008). The phenotype and neural correlates of language in autism: An integrative review. *Neuroscience and Biobehavioral Reviews*, 32(8):1416–1425.

Gupta, S., Aggarwal, S., Rashanravan, B., Lee, T. (1998). Th1- and Th2-like cytokines in CD4_ and CD8_ T cells in autism. *J. Neuroimmunol.* 85, 106-109.

Hardan AY, Muddasani S, Vemulapalli M, Keshavan M, Minshew NJ. (2006). An MRI study of increased cortical thickness in autism. *Am J Psychiatry*, 163(7):1290–1292.

Hazlett HC, Poe M, Gerig G, (2005). Magnetic resonance imaging and head circumference study of brain size in autism: birth through age 2 years. *Arch Gen Psychiatry*, 62(12):1366–1376.

Herbert MR, Ziegler DA, Makris N, (2004). Localization of white matter volume increase in autism and developmental language disorder. *Ann Neurol*, 55 (4):530–540.

Hornig, M., Solbrig, M., Horscroft, N., Weissenbock, H., Lipkin, W. I. (2001). Borna disease virus infection of adult and neonatal rats: models for neuropsychiatric disease. *Curr. Top. Microbiol. Immunol.* 253, 157–17

Ivarsson S.A., L. Bjerre, P. Vegfors and K. Ahlfors (1990). Autism as one of several abnormalities in two children with congenital cytomegalovirus infection. *Neuropediatrics*, 21:102

Jarskog, L. F., Xiao, H., Wilkie, M. B., Lauder, J. M., Gilmore, J. H. (1997). Cytokine regulation of embryonic rat dopamine and serotonin neuronal survival in vitro. *Int. J. Dev. Neurosci.* 15, 711–716

Jieun E. Kim, In Kyoon Lyoo, Annette M., Perry F. Renshaw, Dennis W. Shaw, Seth D. et. al. (2010). Laterobasal Amygdalar Enlargement in 6- to 7-Year-Old Children With Autism Spectrum Disorder. *Arch Gen Psychiatry.*, 2010;67(11):1187-1197

Just MA, Cherkassky VL, Keller TA, Kana RK, Minshew NJ. (2007). Functional and anatomical cortical underconnectivity in autism: Evidence from an fMRI study of an executive function task and corpus callosum morphometry. *Cerebral Cortex,* 17(4):951–961

Just MA, Cherkassky VL, Keller TA, Minshew NJ. (2004). Cortical activation and synchronization during sentence comprehension in high-functioning autism: evidence of underconnectivity. *Brain,* 127 (pt 8):1811–1821.

Jyonouchi H., S.N. Sun and H. Le. (2001). Proinflammatory and regulatory cytokine production associated with innate and adaptive immune responses in children with autism spectrum disorders and developmental regression. *J. Neuroimmunol.,* 120:170

Jyonouchi, H., Geng, L., Ruby, A., Reddy, C., Zimmerman-Bier, B. (2005). Evaluation of an association between gastrointestinal symptoms and cytokine production against common dietary proteins in children with autism spectrum disorders. *J. Pediatr.,* 146, 605–610.

Jyonouchi, H., Sun, S., Le, H. (2001). Proinflammatory and regulatory cytokine production associated with innate and adaptive immune responses in children with autism spectrum disorders and developmental regression. *J. Neuroimmunol.,* 120, 170–179.

Kana RK, Keller TA, Cherkassky VL, Minshew NJ, Just MA. (2006). Sentence comprehension in autism: thinking in pictures with decreased functional connectivity. *Brain,* 129(pt 9):2484–2493.

Kiberstis P. and L. Roberts (2002). It's not just the genes. *Science,* 296:685.

Kipnis, J., Cohen, H., Cardon, M., Ziv, Y., Schwartz, M. (2004). T cell deficiency leads to cognitive dysfunction: implications for therapeutic vaccination for schizophrenia and other psychiatric conditions. *Proc. Natl. Acad. Sci.,* USA 101, 8180–8185.

Korvatska, E., Van de Water, J., Anders, T. F., Gershwin, M. E. (2002). Genetic and immunologic considerations in autism. Neurobiol. Dis., 9, 107–125.

Koshino H, Carpenter PA, Minshew NJ, Cherkassky VL, Keller TA, Just MA. (2005). Functional connectivity in an fMRI working memory task in high-functioning autism. *Neuroimage,* 24(3):810–821.

Lam, K. S., Aman, M. G., Arnold, L. E. (2006). Neurochemical correlates of autistic disorder: a review of the literature. *Res. Dev. Disabil.,* 27, 254–289.

Larson, S. J. (2002). Behavioral and motivational effects of immunesystem ctivation. *J. Gen. Psychol.,* 129, 401–414.

Levi-Montalcini, R., Skaper, S. D., Dal Toso, R., Petrelli, L., Leon, A. (1996). Nerve growth factor: from neurotrophin to neurokine. *Trends Neurosci.* 19, 514–520.

Licinio, J., Kling, M. A., Hauser, P. (1998). Cytokines and brain function: relevance to interferon-_-induced mood and cognitive changes. *Semin. Oncol.,* 25, 30–38

Manjaly ZM, Bruning N, Neufang S, Stephan KE, Brieber S, Marshall JC, et al. (2007). Neurophysiologicalcorrelates of relatively enhanced local visual search in autistic adolescents. *NeuroImage,* 35(1): 283–291.

McDougle, C. J., Naylor, S. T., Cohen, D. J., Volkmar, F. R., Heninger, G. R., Price, L. H. (1996). A double-blind, placebo-controlled study of fluvoxamine in adults with autistic disorder. *Arch. Gen. Psychiatry*, 53, 1001–1008.

Mehler, M. F., Kessler, J. A. (1998). Cytokines in brain development and function. *Adv. Protein Chem.*, 52, 223–251.

Messahel, S., Pheasant, A. E., Pall, H., Ahmed-Choudhury, J., Sungum-Paliwal, R. S., Vostanis, P. (1998). Urinary levels of neopterin and biopterin in autism. *Neurosci. Lett.* 241, 17–20

Mignini, F., Streccioni, V., Amenta, F. (2003). Autonomic innervation of immune organs and neuroimmune modulation. Auton. *Autacoid Pharmacol.* 23, 1–25.

Modahl, C., Green, L., Fein, D., Morris, M., Waterhouse, L., Feinstein, C., Levin, H. (1998). Plasma oxytocin levels in autistic children. *Biol. Psychiatry*, 43, 270–277.

Money, J., Bobrow, N. A., Clarke, F. C. (1971). Autism and autoimmune disease: a family study. *J. Autism Child. Schizophr.*, 1, 146–160.

Nelson, K. B., Grether, J. K., Croen, L. A., Dambrosia, J. M., Dickens, B. F., Jelliffe, L. L., Hansen, R. L., Phillips, T. M. (2001). Neuropeptidesand neurotrophins in neonatal blood of children with autism or mental retardation. *Ann. Neurol.*, 49, 597–606.

Nockher, W. A., Renz, H. (2003). Neurotrophins in inflammatory lung disease; modulators of cell differentiation and neuroimmune interactions. *Cytokine Growth Factor Rev.*, 14, 559–578.

Odell, D., Maciulis, A., Cutler, A., Warren, L., McMahon, W. M., Coon, H., Stubbs, G., Henley, K., Torres, A. (2005). Confirmation of the association of the C4B null allelle in autism. *Hum. Immunol.*, 66, 140–145.

Palmen, S. J., van Engeland, H., Hof, P. R., Schmitz, C. (2004). Neuropathological findings in autism. *Brain*, 127, 2572–2583

Pashenkov, M., Teleshova, N., Link, H. (2003). Inflammation in the central nervous system: the role for dendritic cells. *Brain Pathol.* 13, 23–33

Patterson, P. H. (2002). Maternal infection: window on neuroimmune interactions in fetal brain development and mental illness. *Curr. Opin. Neurobiol.*, 12, 115–118

Plata-Salaman, C. R., Ilyin, S. E., Gayle, D., Romanovitch, A., Carbone, K. M. (1999). Persistent Borna disease virus infection of neonatal rats causes brain regional changes of mRNAs for cytokines, cytokine receptor components and neuropeptides. *Brain Res. Bull.* 49, 441–451

Plioplys, A. V., Greaves, A., Kazemi, K., Silverman, E. (1994). Lymphocyte function in autism and Rett syndrome. *Neuropsychobiology*, 29, 12–16.

Ransohoff, R. M., Kivisakk, P., Kidd, G. (2003). Three or more routes for leukocyte migration into the central nervous system. *Nat. Rev. Immunol.* 3, 569–581

Rice, D., Barone Jr., S. (*2000*). Critical periods of vulnerability for the developing nervous system: evidence from humans and animal models. Environ. *Health Perspect.*, 108 (Suppl. 3), 511–533

Rossi CC, Van de Water J, Rogers SJ, Amaral DG. (2011). Detection of Plasma Autoantibodies to Brain Tissue in Young Children with and without Autism Spectrum Disorders. *Brain Behav Immun.*, Mar 17.

Rumsey, J. M., Ernst, M. (2000). Functional neuroimaging of autistic disorders. Ment. Retard. *Dev. Disabil. Res. Rev.* 6, 171–179

Salama A., B. Schutz, V. Kietel, H. Breithaupt and C. Mueller-Eckhardt. (1989). Immune-mediated agranulocytosis related to drugs and their metabolites: Mode of sensitization and heterogeneity of antibodies. *Br. J. Haematol.*, 72:127

Sauder, C., de la Torre, J. C. (1999). Cytokine expression in the rat central nervous system following perinatal Borna disease virus infection. *J. Neuroimmunol.*, 96, 29–45.

Shi, L., Fatemi, S. H., Sidwell, R. W., Patterson, P. H. (2003). Maternal influenza infection causes marked behavioral and pharmacological changes in the offspring. *J. Neurosci.*, 23, 297–302

Silva, S. C., Correia, C., Fesel, C., Barreto, M., Coutinho, A. M., Marques, C., Miguel, T. S., Ataide, A., Bento, C., Borges, L., Oliveira, G., Vicente, A. M. (2004). Autoantibody repertoires to brain tissue in autism nuclearfamilies. *J. Neuroimmunol.*, 152, 176–182.

Simmons, D. A., Broderick, P. A. (2005). Cytokines, stressors, and clinical depression: augmented adaptation responses underlie depression pathogenesis. Prog. Neuropsychopharmacol. *Biol. Psychiatry*, 29, 793– 807.

Singh, V. K. (1996) Plasma increase of interleukin-12 and interferon_.Pathological significance in autism. *J. Neuroimmunol.*, 66, 143–145

Singh, V. K., Rivas, W. H. (2004). Prevalence of serum antibodies to caudate nucleus in autistic children. *Neurosci. Lett.* 355, 53–56.

Singh, V. K., Warren, R. P., Odell, J. D., Warren, W. L., Cole, P. (1993). Antibodies to myelin basic protein in children with autistic behavior. *Brain Behav. Immun.* 7, 97–103. 50.

Singh, V. K., Warren, R., Averett, R., Ghaziuddin, M. (1997). Circulating autoantibodies to neuronal and glial filament proteins in autism. *Pediatr. eurol.* 17, 88–90.

Steinman, L., Conlon, P., Maki, R., Foster, A. (2003). The intricate interplay among body weight, stress, and the immune response to friend or foe. *J. Clin. Invest.*, 111, 183–185. 24.

Stubbs, E. G., Crawford, M. L. (1977). Depressed lymphocyte responsiveness in autistic children. *J. Autism Child. Schizophr.* 7, 49–55.

Sweeten, T. L., Posey, D. J., McDougle, C. J. (2003). High blood monocyte counts and neopterin levels in children with autistic disorder. *Am. J. Psychiatry*, 160, 1691–1693.

Sweeten, T. L., Posey, D. J., Shanker, S., McDougle, C. J. (2004). High nitric oxide production in autistic disorder: possible role for interferon. *Biol. Psychiatry*, 55, 434–437.

synthesis capacity in autistic and nonautistic children. Ann. *Neurol.*, 45, 287–295.

Todd, R. D., Ciaranello, R. D. (1985). Demonstration of inter- and intraspecies differences in serotonin binding sites by antibodies from an autistic child. *Proc. Natl. Acad. Sci.*, USA 82, 612–616

Todd, R. D., Hickok, J. M., Anderson, G. M., Cohen, D. J. (1988). Antibrain antibodies in infantile autism. *Biol. Psychiatry*, 23, 644–647.

Torres, A. R., Maciulis, A., Odell, D. (2001). The association of MHC genes with autism. Front. *Biosci.* 6, D936–D943.

Tracey, K. J., Wei, H., Manogue, K. R., Fong, Y., Hesse, D. G., Nguyen, H. T., Kuo, G. C., Beutler, B., Cotran, R. S., Cerami, A., et al. (1988). Cachectin/tumor necrosis factor induces cachexia, anemia, and inflammation. *J. Exp. Med.* 167, 1211–1227

Trajkovski, V., Ajdinski, L., Spiroski, M. (2004). Plasma concentration of immunoglobulin classes and subclasses in children with autism in the Republic of Macedonia: retrospective study. *Croat. Med. J.* 45, 746–749.

Tuchman, R. F., Rapin, I., Shinnar, S. (1991). Autistic and dysphasic children. I: clinical characteristics. *Pediatrics* 88, 1211–1218.

Vega, J. A., Garcia-Suarez, O., Hannestad, J., Perez-Perez, M., Gennana, A. (2003). Neurotrophins and the immune system. *J. Anat.* 203, 1 19.

Villalobos, R., T. R., Jayakar, P., Yaylali, I. (1996). Prolonged EEG, monitoring findings in children with pervasive developmental disorder and regression. *Ann. Neurol.*, 40, 300.

Vojdani A., A.W. Campbell, E. Anyanwu, A. Kashanian, K. Bock and E. Vojdani. (2002). Antibodies to neuronspecific antigens in children with autism: possible crossreaction with encephalitogenic proteins from milk, Chlamydia pneumoniae and streptococcus group A. *J. Neuroimmunol.*, 129:168.

Volkmar, F., Cook Jr., E. H., Pomeroy, J., Realmuto, G., Tanguay, P. (1999). Practice parameters for the assessment and treatment of children, adolescents, and adults with autism and other pervasive developmental disorders. American Academy of Child and Adolescent Psychiatry Working Group on Quality Issues. *J. Am. Acad. Child Adolesc. Psychiatry*, 38, 32S–54S.

Wakefield A.J., S.H. Murch, A. Anthony, J. Linnell, D.M. Casson, M. Malik, et al., (1998). Ileal-lymphoidnodular hyperplasia, non-specific colitis, and pervasive developmental disorder in children. *Lancet* 351:637.

Warren, R. P., Foster, A., Margaretten, N. C. (1987). Reduced natural killer cell activity in autism. *J. Am. Acad. Child Adolesc. Psychiatry*, 26, 333–335.

Warren, R. P., Margaretten, N. C., Pace, N. C., Foster, A. (1986). Immune abnormalities in patients with autism. *J. Autism Dev. Disord.* 16, 189- 197.

Warren, R. P., Yonk, L. J., Burger, R. A., Cole, P., Odell, J. D., Warren, W. L., White, E., Singh, V. K. (1990). Deficiency of suppressor-inducer (CD4_CD45RA_) T cells in autism. *Immunol. Invest.* 19, 245–251.

Williams, J. R., Insel, T. R., Harbaugh, C. R., Carter, C. S. (1994). Oxytocin administered centrally facilitates formation of a partner preference in female prairie voles (Microtus ochrogaster). *J. Neuroendocrinol.* 6, 247–250.

Wilson, C. J., Finch, C. E., Cohen, H. J. (2002). Cytokines and cognition the case for a head-to-toe inflammatory paradigm. *J. Am. Geriatr. Soc.*, 50, 2041–2056.

Wirleitner, B., Neurauter, G., Schrocksnadel, K., Frick, B., Fuchs, D. (2003). Interferon induced conversion of tryptophan: immunologic and neuropsychiatric aspects. *Curr. Med. Chem.*, 10, 1581–1591.

Xiaohong L., Abha Chauhna, Ashfaq M. S., Sangita P., Ved Chauhna, Xiu-Min, et. al. (2009). Elevated Immune Response in the Brain of Autistic Patients. *J Neuroimmunol.* 2009 February 15; 207(1-2): 111–116.

Yamashita, Y., Fujimoto, C., Nakajima, E., Isagai, T., Matsuishi, T. (2003). Possible association between congenital cytomegalovirus infection and autistic disorder. *J. Autism Dev. Disord.*, 33, 455–459.

Yonk, L. J., Warren, R. P., Burger, R. A., Cole, P., Odell, J. D., Warren, W. L., White, E., Singh, V. K. (1990). CD4_ helper T cell depression in autism. *Immunol. Lett.* 25, 341–345.

Young, L. J., Lim, M. M., Gingrich, B., Insel, T. R. (2001). Cellular mechanisms of social attachment. *Horm. Behav.* 40, 133–138.

Young, L. J., Pitkow, L. J., Ferguson, J. N. (2002). Neuropeptides and social behavior: animal models relevant to autism. *Mol. Psychiatry*, 7 (Suppl. 2), S38–S39.

Young, L. J., Wang, Z., Insel, T. R. (1998). Neuroendocrine bases of monogamy. *Trends Neurosci.* 21, 71–75.

Zhao, B., Schwartz, J. P. (1998). Involvement of cytokines in normal CNS evelopment and neurological diseases: recent progress and perspectives. *J. Neurosci. Res.*, 52, 7–16.

17

Vaccines and Autism – An Unlikely Connection

Archana Chatterjee and Kody Moffatt
Creighton University School of Medicine
USA

1. Introduction

Autism spectrum disorders (ASDs) encompass a group of chronic developmental disorders characterized by repetitive or stereotypic behaviors, interests and activities, along with marked impairments in a child's ability to socialize and communicate. These debilitating conditions impact every aspect of the life of a child and his/her family. Modern advances in science and technology have provided successful explanations and interventions for many previously life-threatening conditions such as bacterial meningitis and extreme prematurity. However, a scientific cause or definitive treatment for ASDs remains elusive. This lack of evidence regarding the biological causes of ASDs and successful, standardized treatment modalities challenges both parents/caregivers and health care providers in their understanding of these conditions, and effectively addressing the needs of this pediatric population. In some instances, the lack of evidence has fueled the development of hypotheses and possible associations based on the publication of case reports and small cohort studies.

The prevalence of ASDs has increased over the past several decades, but it is unclear whether this is due to a true increase, increasing awareness, or differences in the methods used to diagnose these conditions and assess their prevalence. Given the irrefutable increase in the prevalence of ASDs, there has been interest in both genetic influences and environmental exposures that may have led to this increase over the past several decades. Although a small proportion of ASDs are associated with known congenital conditions, and several genes involved in ASDs have been identified, in most cases the etiology of ASDs is unknown. Some of the environmental triggers for ASDs that have been postulated include lack of breastfeeding, supplemental feeding with infant formulas that do not contain docosahexaenoic acid and arachidonic acid supplementation, childhood vaccinations, the use of acetaminophen and other analgesics, certain viral infections, and sundry other environmental exposures. Among these exposures, vaccinations have received the most widespread interest and attention by both the lay public as well as the medical and scientific communities. Young children are receiving more vaccines than ever, with multiple vaccines given at each visit, to provide protection against a plethora of infectious diseases. ASDs are often diagnosed in children at about the same chronologic age as the peak time for vaccine delivery. Unfortunately, a small, but vocal minority of people have attributed the rise in rates of ASDs to the increase in childhood vaccinations, despite the lack of rigorous scientific evidence to support this contention. The question about vaccines and ASDs continues to cause conflict between public health authorities and worried parent groups. This chapter will provide further details of the arguments on both sides and an analysis of the scientific evidence that supports the view that ASDs and vaccines are unlikely to be linked.

2. Vaccines - victims of their own success?

Vaccination is among the greatest achievements of modern medicine, leading to the eradication of naturally occurring smallpox and the near elimination of polio [1]. Most of the lay public as well as many scientists and physicians do not realize that the first vaccines against smallpox and rabies proved their effectiveness even before the identification of viruses as infectious agents [2]. Vaccination has a short history in medicine and public health when measured against the centuries during which human beings have fought desperately to prevent and treat various plagues and pestilences. Routine vaccination of large populations is a phenomenon of the 20th century [3]. Despite its relatively recent entry into the field of medicine and public health, vaccination has helped in the world-wide eradication/control of 12 major infectious diseases, including smallpox, diphtheria, tetanus, yellow fever, pertussis, *Haemophilus influenzae* type b disease, poliomyelitis, measles, mumps, rubella, typhoid and rabies [3]. In the United States, vaccination has contributed to the significant decline in morbidity from nine vaccine-preventable diseases and their complications between 1900 and 1999 (Table 1) [4]. Vaccines have been described as the single most life-saving accomplishment of the 20th century [5].

Disease	Baseline 20th century annual morbidity	1998 Provisional morbidity	% Decrease
Smallpox	48,164	0	100%
Diphtheria	175,885	1	100%
Pertussis	147,271	6,279	95.7%
Tetanus	1,314	34	97.4%
Poliomyelitis (paralytic)	16,316	0	100%
Measles	503,282	89	100%
Mumps	152,209	606	99.6%
Rubella	47,745	345	99.3%
Congenital rubella syndrome	823	5	99.4%
Haemophilus influenzae type b	20,000	54	99.7%

Table 1. Baseline 20th century annual morbidity and 1998 provisional morbidity from nine diseases with vaccines recommended before 1990 for universal use in children - United States

Parents and many health care providers of the 21st century, particularly in more developed areas of the world such as the United States and Western Europe, have limited or no experience with the devastating effects of these diseases. In the United States public health officials now recommend 28 to 31 vaccine doses before the age of 18 years, many of which are administered together to provide protection early in life, for the convenience of families and health care providers, and to decrease distress to the infant. Public health experts recommend that 95% of the population be vaccinated to provide herd immunity and minimize the possibility of resurgence of these deadly infections. However, parents in developed countries who have not seen these diseases or their disastrous consequences sometimes feel that they are being pressured into immunizing their children involuntarily for public good rather than personal benefit [6]. Some parents even perceive a greater risk to their children from vaccination than from the diseases themselves, not recognizing that the threat from these diseases is reduced simply because we do have effective vaccines to prevent them. Vaccination has thus regrettably become a polarized issue with some parents stressing their own child's well-being at the one extreme and health experts advocating for public health outcomes on the other extreme.

3. Genesis of the "vaccines cause autism" theory

One of the first claims that vaccines might cause autism was made in a book entitled "A Shot in the Dark" by Harris L. Coulter and Barbara Loe Fisher [7]. In it the authors wrote, "With the increasing number of vaccinations American babies have been required to use has come increasing numbers of reports of chronic immune and neurologic disorders ... including ... autism." At the time, little attention was paid to this assertion. The hypothesis received far greater support after a British physician and researcher Dr. Andrew Wakefield along with 12 co-authors published an article describing abnormal gastrointestinal features among 12 children who had been referred to their university pediatric gastroenterology clinic [8]. All of the children were reported to have some type of developmental disorder, and 9 of them had been diagnosed with autism. In 6 of these 9 children, either the parent or a physician had linked the onset of developmental regression with the receipt of the MMR vaccine. In this paper, Wakefield *et al.* proposed an elaborate sequence of events: that measles virus from the live-attenuated MMR vaccine caused intestinal inflammation, the inflamed intestines became "leaky", allowing undefined harmful proteins to enter the bloodstream, travel to the brain and cause autism. In 2000, Wakefield and colleagues published a second paper in which white blood cells in the same 9 autistic children (with what was now referred to as "autistic enterocolitis") were examined for the presence of measles virus [9]. Using polymerase chain reaction, the authors reported that measles virus RNA fragments were found in 3 out of the 9 children, but in none of 22 controls, lending credence to the "leaky-gut" theory [9].

Additional theories of the putative association between vaccines and ASDs include:

1. Concern about the mercury-containing preservative thimerosal (which was used in childhood vaccines for many years) and its potential toxic effects on the developing central nervous system in children;
2. Worry that a combination of MMR and thimerosal-containing vaccines produces additive or synergistic toxic insults on children's brains;
3. Apprehension related to the simultaneous administration of multiple vaccines which might "overwhelm" or "weaken" the relatively immature immune system in young children.

These theories will be explored later in this chapter, but let us first further discuss the most well-known controversy surrounding vaccines and ASDs.

4. Impact of the "MMR causes autism" scare

As a consequence of the publications by Wakefield and his colleagues, many parents anxious about the risk of autism, particularly in the UK, began to refuse the MMR vaccine for their children. After the controversy began, the MMR vaccination compliance dropped in the UK from 92% in 1996 to 82% in 2002 [10]. In some parts of London, it was as low as 62% in 2003, far below the rate needed to avoid an epidemic of measles [10]. By 2006, coverage for MMR for children at 24 months of age in the UK was 85%, significantly lower than the 94% coverage rate for other vaccines [11]. Predictably, the fall in vaccination rates for MMR vaccine was followed by an increase in the incidence in the UK of two of the three diseases that are prevented by it. In 1998 there were 56 confirmed cases of measles in the UK. By the first five months of 2006, there were 449 cases of measles reported in the UK, with the first death since 1992. As expected, the cases occurred in inadequately vaccinated or unvaccinated children [12].

Mumps cases also began rising in 1999 after many years, and by 2005 the UK was in the midst of a mumps epidemic with almost 5000 reports in the first month of 2005 alone [13]. A total of 56,390 notified cases of mumps were reported in England and Wales that year [14]. Interestingly, most patients were aged between 15 and 24 years, too old to have received the routine MMR vaccine around the time the paper by Wakefield *et al.* was published, and too young to have contracted natural mumps as a child. With the decline in mumps that followed the introduction of the MMR vaccine in the UK, these individuals had not been exposed to the disease, and therefore had no immunity, either natural or vaccine-induced. Once immunization rates began to decline following the controversy and the disease re-emerged, they were susceptible to infection [14].

Measles and mumps cases continued in 2006, at incidence rates 13 and 37 times greater than their respective 1998 levels [15]. Two children were severely and permanently injured by measles encephalitis in London [16]. Measles outbreaks also resulted in casualties in nearby countries. Three deaths and 1,500 cases of measles were reported in an outbreak in Ireland, which occurred as a direct result of decreased vaccination rates following the MMR scare [16]. Another study reported the hospitalization of 111 cases of measles mostly with pneumonia, tracheitis or dehydration, with 13 of them requiring ICU admission and 7 of the children needing mechanical ventilation [17]. One editorial has described this as the "fallout" of the paper published by Wakefield et al. [18]. In 2008, for the first time in 14 years, measles was declared to be endemic again in the UK. This was caused by the preceding decade's low MMR vaccination rates, which in turn created a population of susceptible children who could spread the disease [15]. MMR vaccination rates for English children remained at 85% in 2007–08, unchanged from the year before and at too low a level to prevent serious measles outbreaks [19]. In May 2008, a British 17-year-old with an underlying immunodeficiency died of measles [15]. In 2008, measles epidemics were reported from Austria, Italy, and Switzerland [15].

In a study conducted in the US, selective MMR nonreceipt, occurring in as few as 0.77% of children in the 1995 cohort, rose to 2.1% according to the 2000 National Immunization Survey [20]. Children included in the 2000 National Immunization Survey were born at around the time that the putative link between MMR and autism surfaced in the medical literature. Sporadic importations of measles into the US had occurred since the disease was declared eliminated from the US in 2000. However, in 2008, a measles outbreak occurred in the US involving 16 states [21]. Of the individuals affected, 94% were US residents, 93% were unvaccinated and 86% of the cases were imported (69% from Europe).

5. Lack of evidence to support the "MMR causes autism" theory

The scientific limitations of the paper published by Wakefield *et al.* [8] were pointed out soon after it first appeared [22]. It was noted that the paper reported on a small series of cases with no controls, linked three common clinical conditions, and relied on the recall and beliefs of parents [23]. Several large population- and ecologic-based studies were conducted over the following decade that consistently found no evidence of a link between the MMR vaccine and autism and failed to provide any support for Wakefield's theory [24-27]. In fact, the lack of an association between MMR vaccination and autism in children is supported by 19 additional scientific studies performed by diverse groups of investigators using various research methodologies involving disparate groups of patients over more than a decade [28-46]. Several of these studies have been discussed in detail in 4 review articles [47-50]. In other words, despite significant efforts by numerous groups of investigators, the findings of Wakefield *et al.* [8] could not be replicated or confirmed. Interestingly, in a case-control study conducted in Poland, where the MMR vaccine was introduced later than in most other European countries, the investigators report that the risk of autism was lower in children who received the MMR vaccine than in those who did not [44]. The authors do not claim a "protective" effect of the vaccine, but correctly recognize that the decreased risk of autism among vaccinated children may have been due to other confounding factors in their health status such as, healthcare workers or parents who may have noticed signs of developmental delay or disease before the actual autism diagnosis and for this reason have avoided vaccination [44]. This type of critical and honest analysis is missing from studies that support the contention that the MMR vaccine is associated with ASDs [51-53].

In 2004, 10 of the 12 coauthors of Wakefield's acknowledged that "no causal link was established between MMR vaccine and autism as the data were insufficient" in their original paper and asked to "formally retract the interpretation" of their findings [54]. Moreover, an investigation by D'Souza *et al.* using a larger sample size than Wakefield and his colleagues' original study [9], failed to reveal persistence of measles virus RNA in the peripheral blood of children with ASDs [55]. Two additional studies reported no detectable measles virus genome sequence in the blood of autistic children who had received MMR vaccination [56, 57]. Further, in a case-control study conducted by Hornig *et al.*, ileal and cecal tissues from 25 children in the US with autism and gastrointestinal (GI) disturbances and 13 children with GI disturbances alone (controls) undergoing clinically-indicated ileocolonoscopy, were evaluated by real-time reverse transcription (RT)-PCR for presence of measles virus RNA in three laboratories blinded to diagnosis, including one wherein the original findings suggesting a link between measles virus and ASDs were reported [58]. The authors reported no differences between case and control groups in the presence of measles viral RNA in the ileum and cecum [58].

Despite the scientific difficulty with proving a negative, the Institute of Medicine (IOM) in a report on vaccine safety has stated conclusively that there is no causal relationship between the administration of the MMR vaccine and the onset of ASDs [59]. This eighth and final report of the Immunization Safety Review Committee examined the hypothesis that vaccines, specifically the MMR vaccine and thimerosal-containing vaccines, are causally associated with autism. The committee reviewed the extant published and unpublished epidemiological studies regarding causality and studies of potential biologic mechanisms by which immunizations might cause autism and concluded that the body of epidemiological evidence favored *rejection* of a causal relationship between the MMR vaccine and autism

[59]. The committee further found that potential biological mechanisms for vaccine-induced autism that have been generated to date are theoretical only [59]. Thus, the MMR vaccine continues to be safe, efficacious and recommended by the public health authorities and supported by the vast majority of medical professionals. Nevertheless, despite the large body of scientific evidence to the contrary, Barbara Loe Fisher continues to support her initial theory that vaccines cause autism [60].

6. MMR and autism - honest error or deliberate fraud?

The Office of Research Integrity in the United States defines fraud as fabrication, falsification, or plagiarism in proposing, performing, or reviewing research, or in reporting research results [61]. They further explain that:

a. Fabrication is making up data or results and recording or reporting them.
b. Falsification is manipulating research materials, equipment, or processes, or changing or omitting data or results such that the research is not accurately represented in the research record.
c. Plagiarism is the appropriation of another person's ideas, processes, results, or words without giving appropriate credit.
d. Research misconduct does not include honest error or differences of opinion.

Editors at the BMJ claim that it has taken the diligent skepticism of one man, Brian Deer, a journalist standing outside the realms of medicine and science, to show that the initial paper by Wakefield *et al.* [8] was in fact an elaborate fraud [62]. In a series of articles published this year, Deer reports on how Wakefield altered numerous facts about his patients' medical histories in order to support his claim to have identified a new syndrome [63]; how his institution, the Royal Free Hospital and Medical School in London, supported him as he sought to exploit the ensuing MMR scare for financial gain [64]; and how key players failed to investigate thoroughly in the public interest when Deer first raised his concerns [65]. Deer published his first investigation into Wakefield's paper in 2004 [66]. This uncovered the possibility of research fraud, unethical treatment of children, and Wakefield's conflict of interest through his involvement with a lawsuit against manufacturers of the MMR vaccine [62]. Building on these findings, the General Medical Council (GMC) of the UK launched proceedings that focused on whether the research conducted by Wakefield *et al.* [8] was ethical.

While the disciplinary panel was examining the children's medical records in public, Deer compared them with what was published in the *Lancet* article. His focus was not on whether the research was ethical, but whether it was factual. Through interviews, documents, and data made public at the GMC hearings as well as his investigations spanning several years, Deer has unearthed clear evidence of falsification in Wakefield *et al.*'s [8] paper. He found that in every one of the 12 cases reported by Wakefield *et al.* [8], there was misrepresentation or undisclosed alteration, and that in no single case could the children's medical records be fully reconciled with the descriptions, diagnoses, or histories published in the article. The editors of the BMJ have questioned the origins of the falsified data and lay the blame squarely upon Andrew Wakefield [62]. They question whether it is possible that he was wrong, but not dishonest: that he was so incompetent that he was unable to fairly describe the project, or to report even one of the 12 children's cases accurately, and conclude that the article resulted not from honest errors, but a deliberate attempt to defraud [62]. They base their conclusion on the contention that a great deal of thought and effort must have gone into drafting the paper to achieve the results he wanted, since the discrepancies all led in one direction and the misreporting was gross [62].

7. Consequences of the "MMR causes autism" fraud

Nearly 12 years after its original publication, the journal *Lancet* fully retracted the article by Wakefield *et al.*, based on several elements of the paper being proven to be false [67]. The GMC completed its longest-ever "fitness to practice" hearing, and based upon it withdrew Dr. Wakefield's license to practice medicine [68]. Andrew Wakefield was branded as being "dishonest," "unethical," and "callous" [69]. His associate Professor John Walker-Smith, the senior clinician in the project, was found to have presided over "high risk" research without clinical indication or ethical approval, and also struck off the medical register [70]. Further details about this controversy and autism research have been published in a couple of recent books [71, 72].

The Wakefield case and its aftermath have resulted in a reevaluation of how biomedical research is regulated. There have been calls for a national Health Research Agency to be established in the UK to oversee the regulation and governance of health research [73]. Others have advocated for public access to raw data, arguing that the apparent discrepancies between the patient records and the data in the article by Wakefield *et al.* [8] might have come to light sooner, perhaps even before publication, had the raw data been available for public scrutiny [74]. Opel et al. propose that as part of an effort to improve research integrity traditional hierarchies and authority gradients need to be bypassed in order to empower everyone in the research enterprise-especially those on the front lines, such as research assistants, data analysts, and project managers-to raise questions and be able to report suspected misconduct without fear of reprisal [75]. They suggest that the ability to investigate research incidents needs to be strengthened using the best tools and techniques available to protect the safety of research subjects [75]. They also assert that the customs and culture around biomedical research need rethinking and reform. They point to the disastrous impact that Wakefield's flawed study has had on vaccine coverage, recrudescence of vaccine-preventable diseases and erosion of the public's trust in science, and exhort rapid action to remedy the current system of ensuring research integrity [75].

Based on the above referenced body of knowledge, few people could deny that Wakefield *et al.*'s paper was fatally flawed both scientifically and ethically, if not outright fraudulent. Unfortunately, an allegation is remembered long after it has been disproved. In a postal survey of parent's decisions, attitudes and use of information about MMR immunization in the UK, Casiday *et al.* report that both MMR-accepting and refusing parents were supportive of immunization, but had a high level of concern about the safety of vaccines [76]. A web-based survey of parents conducted in 2009 in the US, showed that while most parents agreed that vaccines protect their child(ren) from diseases, more than half of the respondents also expressed concerns regarding serious adverse effects of vaccines [77]. Overall, 11.5% of the parents had refused at least 1 vaccine that their doctor had recommended for their child(ren), with 17.7% refusing the MMR vaccine [77]. A quarter of the survey responders either strongly agreed or agreed with the statement "Some vaccines cause autism in healthy children" [77]. Wakefield's legacy promises to live on.

8. Origins of the thimerosal and autism controversy

Another hot button issue that has been debated in relationship to the onset of ASDs is exposure to thimerosal, a preservative that has been present in vaccines since the 1930s [78]. Multidose vaccine vials have the antibacterial agent thimerosal added to preserve the sterility of the contents. Thimerosal contains 49.6% mercury by weight and metabolizes into *ethylmercury* and thiosalicylate. Towards the end of the 20th century, the US government became aware of and concerned about mercury exposure in the general population [79] and the US Environmental Protection Agency (EPA) published standards of safe limits of oral *methylmercury* exposure particularly from fish and shellfish [80, 81]. Statements from the EPA clearly indicate that people in the U.S. are mainly exposed to organic *methylmercury*, when they eat fish and shellfish that contain it. The EPA identifies factors that determine how severe the health effects are from mercury exposure including:

- the chemical form of mercury
- the dose
- the age of the person exposed (the fetus is the most susceptible)
- the duration of exposure
- the route of exposure -- inhalation, ingestion, dermal contact, etc.
- the health of the person exposed.

Various agencies have developed guidelines for "safe" exposure to *methylmercury*, including the EPA [82, 83], U.S. Agency for Toxic Substances and Disease Registry (ATSDR) [84], the U.S. Food and Drug Administration (FDA) [85], and the World Health Organization (WHO) [86]. These exposure levels ranged from 0.1 μg/kg body weight/day (EPA) to 0.47 μg/kg body weight/day (WHO), and while clearly different, were within the same order of magnitude. The various mercury guidelines were based on epidemiological and laboratory studies of *methylmercury*, whereas thimerosal as noted above is a derivative of *ethylmercury*. Because they are different chemical entities i.e. *ethyl* versus *methylmercury*, different toxicological profiles are expected for the two compounds. It should be recognized that there was uncertainty in applying the *methylmercury*-based guidelines to thimerosal. The FDA has noted that these guidelines may be used as screening tools in risk assessment to evaluate the "safety" of mercury exposures, but are not meant to be bright lines above which toxicity *will* occur [87].

In 1997, Frank Pallone, a U.S. congressman from New Jersey, added an amendment to a (FDA) reauthorization bill which gave the FDA 2 years to "compile a list of drugs and foods that contain intentionally introduced mercury compounds and provide a quantitative and qualitative analysis of the mercury compounds in the list" [88]. The bill was signed into law as the FDA Modernization Act of 1997, and garnered little public or press attention at the time. To abide by this law, the FDA conducted a comprehensive review of the use of thimerosal in childhood vaccines in 1999, and notably, found *no evidence of harm* from the use of thimerosal as a vaccine preservative, other than local hypersensitivity reactions [89]. The maximum cumulative exposure to mercury from vaccines in the recommended childhood immunization schedule at the time, was found to be within acceptable limits for the *methylmercury* exposure guidelines set by FDA, ATSDR, and WHO. However, depending on the vaccine formulations used and the weight of the infant, some infants could have been exposed to cumulative levels of mercury during the first six months of life that exceeded EPA recommended guidelines for safe intake of *methylmercury*.

As more thimerosal-containing vaccines were added to the recommended infant and child immunization schedule, theoretical concerns based on the cumulative amounts of thimerosal that a child was receiving in the first two years of life were raised. As a precautionary measure, the US Public Health Service (USPHS) (which includes the FDA, National Institutes of Health (NIH), Centers for Disease Control and Prevention (CDC) and the Health Resources and Services Administration (HRSA)) and the American Academy of Pediatrics issued two Joint Statements, urging vaccine manufacturers to reduce or eliminate thimerosal in vaccines as soon as possible [90, 91]. This was done through an "abundance of caution," even though there was no evidence that thimerosal-containing vaccines contributed to toxic mercury levels in children, and in fact *ethylmercury* does not have the neurotoxic effects of *methylmercury*. The action taken by the USPHS and AAP had significant, unintended ripple effects on the general public's concerns about vaccine safety in young children. Beginning in 2000, many parents founded advocacy groups based on the belief that thimerosal had caused their children's autism [92]. The birth dose of hepatitis B vaccine which in 1999 contained thimerosal was subsequently withheld from many children and the hepatitis B vaccination campaign in the US experienced a serious setback [93]. Although thimerosal-free hepatitis B vaccines became available shortly thereafter, the effort to vaccinate infants at birth remains a challenge in some areas.

9. Insufficient scientific evidence linking thimerosal with autism

The signs and symptoms of autism are clearly distinct from those of mercury poisoning. Children with mercury poisoning show characteristic motor, speech, sensory, psychiatric, visual, and head circumference changes that are fundamentally different from those of or absent in children with autism. Concerns about mercury as a cause of autism therefore seemed biologically implausible [94]. Nevertheless, it began to be suggested that there may be adverse neurological effects including autism due to *ethylmercury* exposure from the use of thimerosal in vaccines [95-103]. Notably, most of the studies that reported an association of thimerosal with neurodevelopmental disorders including autism were performed by the same group of researchers [98-103], using the Vaccine Adverse Events Reporting System (VAERS) as their data source. The VAERS is a passive reporting system to which anyone can report adverse events that are purported to be associated with vaccines. Goodman and Nordin have shown that most reports to the VAERS system in recent years regarding thimerosal were influenced by litigation, and are therefore unsuitable for scientific study [104]. In other words, most of the adverse reports regarding thimerosal and autism were related to pending lawsuits for vaccine injury. This severely biased the dataset, and it should not have been used to assign causality.

Meanwhile, several other studies performed by various groups of researchers did not support the postulated association between thimerosal and ASDs [33, 105-112]. Three ecological studies were performed in 3 different countries comparing the incidence of autism with thimerosal exposure from vaccines [33, 109, 110]. In each case, thimerosal had been removed from childhood vaccines, allowing robust comparisons of vaccination with thimerosal-containing and thimerosal-free products. A large study from Denmark showed no difference in the incidence of autism among children who had received vaccines containing different amounts of thimerosal [109]. Despite the removal of thimerosal from vaccines in 1992 in Sweden and Denmark, the incidence of autism increased steadily from 1990 to 2000 [110]. Thimerosal exposure and pervasive developmental disorder diagnosis were found to be independent variables in a study from Canada [33]. In this study, the highest rates of pervasive developmental disorder were found in children who had received thimerosal-free vaccines [33].

Additional epidemiologic studies also failed to show any association between thimerosol exposure and ASDs. In a large study from Denmark, researchers found that the risk of autism did not differ between children vaccinated with thimerosal-containing vaccines and those vaccinated with thimerosal-free vaccines or between children who received larger or smaller quantities of thimerosal [107]. They also reported that the rates of autism increased after the removal of thimerosal from all vaccines. In the United States, researchers at the CDC used the Vaccine Safety Data Link to examine the health records of 140,887 children born during 1991–1999, including over 200 children diagnosed with autism [111]. They found no relationship between receipt of thimerosal-containing vaccines and autism. In a large study conducted in the UK, researchers evaluated the vaccination records of 100,572 children born during 1988–1997, 104 of whom were affected with autism [105]. No relationship between thimerosal exposure and developmental disorders was observed. In a separate study from the UK, researchers prospectively followed 12,810 children born during 1991–1992, for whom they had complete vaccination records, and again found no relationship between early thimerosal exposure and subsequent adverse neurological or psychological outcomes [106].

A long-term follow-up study by the CDC showed that early thimerosal exposure from vaccines did not cause adverse neuropsychological outcomes after 7-10 years [113]. In another long-term follow-up study performed in Italy, 2 groups of children with exposure to different doses of thimerosol were examined [114]. Among the 24 neuropsychological outcomes that were evaluated, only 2 were significantly associated with thimerosal exposure. The authors noted that due to the large number of statistical comparisons performed, the few associations found between thimerosal exposure and neuropsychological development might be attributable to chance [114]. The associations found, although statistically significant, were based on small differences in mean test scores, and their clinical relevance could not be determined. A case-control study was conducted in 3 managed care organizations (MCOs) in the US that included 256 children with ASD and 752 matched controls [115]. The authors report that in their study, prenatal and early-life exposure to *ethylmercury* from thimerosal-containing vaccines and immunoglobulin preparations was not related to increased risk of ASDs [115]. Several scientific and public policy review committees have carefully evaluated the existing data and concluded that there was not sufficient evidence of a link between autism and thimerosal in vaccines [59, 87, 116]. In fact, the Institute of Medicine's 2004 evaluation included a strong statement that rejected the idea that thimerosal-containing vaccines cause autism, concluding that "...epidemiological evidence favors rejection of a causal relationship between thimerosal-containing vaccines and autism" [116].

Interestingly, comparisons of *methylmercury* and *ethylmercury* tissue distribution following exposure in young mice [117] and monkeys [118] both reported significantly less mercury deposited in the brain following *ethylmercury* or thimerosal exposure, as compared to *methylmercury* exposure. The authors of these studies concluded that the clearance and tissue distribution of the two compounds differ significantly in animal models [117, 118]. The route of exposure (injection versus ingestion) to *methylmercury* also resulted in differences in the amount of mercury deposited in the brain in mice, with exposure via intramuscular injection resulting in *less* mercury deposition than via ingestion [117]. In a study by Pichichero ME, *et al.*, mercury levels in blood and other samples from infants who had received routine immunizations with thimerosal-containing vaccines were measured [119]. Blood levels of mercury did not exceed safety guidelines for *methylmercury* for all infants in this study. Further, mercury was cleared from the blood in infants exposed to thimerosal faster than would be predicted for *methylmercury*. Infants excreted significant amounts of mercury in stool after thimerosal exposure, thus removing mercury from their bodies. These results suggest that there are differences in the way that thimerosal and *methylmercury* are distributed, metabolized, and excreted. Thimerosal appears to be removed from the blood and body more rapidly than *methylmercury* in young children. Due to the differences in the biological behavior of these two compounds, the flaws in extrapolating the risk assessment of thimerosal by direct comparison with *methylmercury* are well described in a review by Aschner and Ceccatelli [120]. Another review article summarizes the studies investigating thimerosal exposure and the development of ASDs (Table 2) [121].

	Type of Study	Outcome Measure	Association with Thimerosal Exposure
Andrews et al., 2004	Cohort	Autism	No
Croen et al., 2008	Case-Control	Autism	No
Geier and Geier, 2007	Case-Control	Autism	Yes
Heron et al., 2004	Cohort	Developmental Disorders	No
Hviid et al., 2003	Cohort	Autism	No
Madsen et al., 2003	Ecologic	Autism	No
Miles and Takahashi, 2007	Cross-Sectional	Autism	No
Thompson et al., 2007	Cohort	Neuropschological Functioning	No
Verstraeten et al., 2003	Cohort	Autism	No
Young, Geier, and Geier, 2008	Ecologic	Autism	Yes

Table 2. Studies Investigating Thimerosal Exposure with Autism and Other Developmental Outcomes.

Thimerosal has been removed from all childhood vaccines in the US, but this has also increased production costs which are ultimately passed on to the consumer. Only some preparations of influenza vaccine still contain thimerosal (See Table 3). However, largely unfounded concerns about the adverse effects of thimerosal have made many parents reluctant to have their children receive this vaccine. What goes unrecognized by the lay public and even many health care providers is that influenza virus causes hundreds of thousands of hospitalizations and an average of 100 deaths among children every year. Mistakenly attempting to protect their children from a theoretical risk, these parents inadvertently place them at the real risk of being hospitalized or killed by influenza. An alarming recent trend has been that physicians, scientists, government policy advisors, and child advocates who publicly state that vaccines do not cause neurologic problems or autism have been harassed, threatened, and vilified, receiving hate mail and occasionally even death threats [92].

Vaccine Brand Name Manufacturer Thimerosal Concentration1				Mercury mcg/0.5 ml
Anthrax	BioThrax	BioPort Corp	**0**	**0**
DTaP	Tripedia	sanofi pasteur	*	*
	Infanrix	GlaxoSmithKline	**0**	**0**
	DAPTACEL	sanofi pasteur	**0**	**0**
DTaP-HepB-IPV	Pediarix	GlaxoSmithKline	**0**	**0**
DTaP-IPV-Hib	Pentacel	sanofi pasteur	**0**	0
DTaP-Hib	TriHIBit	sanofi pasteur	*	*
DTwP	All Products		.01%	25
DT	Diphtheria & Tetanus Toxoids Adsorbed USP multi-dose single dose	sanofi pasteur	.01%	25
			*	*
Td	DECAVAC	sanofi pasteur	*	*
	Tetanus and Diphtheria Toxoids Adsorbed	sanofi pasteur	*	*
Tdap	ADACEL	sanofi pasteur	**0**	**0**
	Boostrix	GlaxoSmithKline	**0**	**0**
Tetanus Toxoid	Tetanus Toxoid Adsorbed USP	sanofi pasteur	.01%	25
	Tetanus Toxoid Adsorbed Adult Use		.01%	25
	Booster		.01%	25
Hib	ActHIB	sanofi pasteur	**0**	**0**
	Hiberix	GlaxoSmithKline	**0**	**0**
	HibTITER	Wyeth-Ayerst	**0**	**0**
	PedvaxHIB liquid (2)	Merck	**0**	**0**
Hib/HepB	Comvax (3)	Merck	**0**	**0**
Hepatitis A	Havrix	GlaxoSmithKline	**0**	**0**
	Vaqta adult/pediatric	Merck	**0**	**0**
Hepatitis B	Engerix-B preservative free	GlaxoSmithKline	**0**	**0**
	Recombivax HB preservative free	Merck	**0**	**0**
Hep A-B	Twinrix	GlaxoSmithKline	**0**	**0**
HPV	Cervarix	GlaxoSmithKline	**0**	**0**
	Gardasil	Merck	**0**	**0**
Influenza 2009/10 Formula	Afluria multi-dose single dose	CSL Limited	.01%	24.5
			0	**0**
	Agriflu	Novartis	**0**	**0**
	Fluarix	GlaxoSmithKline		≤1
	FluLaval	GlaxoSmithKline	.01%	25
	FluMist	MedImmune	**0**	**0**
	Fluvirin	Novartis	.01%	24.5
	Fluzone 5 mL vial No Preservative	sanofi pasteur	.01%	25
			0	**0**

Vaccine	Brand Name	Manufacturer	Thimerosal Concentration1	Mercury mcg/0.5 ml
Influenza A H1N1 2009	Influenza A (H1N1) 2009 Monovalent Vaccine multi-dose single dose	CSL Limited	.01%	24.5
			0	0
	Influenza A (H1N1) 2009 Monovalent Vaccine multi-dose single dose	Novartis	.01%	24.5
				≤1
	Influenza A (H1N1) 2009 Monovalent Vaccine multi-dose single dose	sanofi pasteur	.01%	25
			0	0
	Influenza A (H1N1) 2009 Monovalent Vaccine	GlaxoSmith Kline	.01%	25
	Influenza A (H1N1) 2009 Monovalent Vaccine Live, Intranasal	MedImmune	0	0
IPV Japanese Encephalitis	IPOL Ixiaro commercial | military	sanofi pasteur Intercell Bio	0	0
	JE-Vax	sanofi pasteur	0.007%	
Meningococcal	Menactra	sanofi pasteur	0	0
	MENOMUNE-A/C/Y/W-135 multi-dose single dose	sanofi pasteur	.01%	25
			0	0
	Menveo	Novartis	0	0
MMR	M-M-R II	Merck	0	0
MMR-Varicella	ProQuad	Merck	0	0
Polio	IPOL	sanofi pasteur	0	0
Pneumococcal	Prevnar	Wyeth-Ayerst	0	0
	Pneumovax 23	Merck	0	0
Rabies	RabAvert	Chiron	0	0
	IMOVAX	sanofi pasteur	0	0
Rotavirus	RotaTeq	Merck	0	0
Typhoid Fever	Typhim Vi	sanofi pasteur	0	0
	Vivotif	Berna Biotch	0	0
Varicella Zoster	Varivax	Merck	0	0
	Zostavax	Merck	0	0
Yellow Fever	YF-VAX	sanofi pasteur	0	0

1. A concentration of 1:10,000 is equivalent to a 0.01% concentration. Thimerosal is approximately 50% Hg by weight. A 1:10,000 concentration contains 25 mcg of Hg per 0.5 mL. (2). A previously marketed lyophilized preparation contained 0.005% thimerosal. (3). COMVAX is not approved for use under 6 weeks of age because of decreased response to the Hib component * This product should be considered equivalent to thimerosal-free products. This vaccine may contain trace amounts (<0.3 mcg) of mercury left after post-production thimerosal removal; these amounts have no biological effect. JAMA 1999;282(18) and JAMA 2000;283(16).

Institute for Vaccine Safety, Johns Hopkins Bloomberg School of Public Health. Available at www.vaccinesafety.edu. Accessed 4/17/11.

Table 3. Thimerosal Concentration in Licensed Vaccines.

10. If vaccines are not to blame, why are ASDs increasing?

The increase in prevalence of the ASDs may be explained by three reasons as described by Scahill *et al.* [122]. First, in 1994, with the release of the *Diagnostic and Statistical Manual of Mental Disorders*, 4th edition (*DSM–IV*), there was a broadening of the diagnostic criteria for autism. The *DSM–IV* also added criteria for Asperger's syndrome and clarified the criteria for Pervasive Developmental Disorders-Not Otherwise Specified (PDD-NOS). Finally, better assessment methods have provided clarification across PDD diagnoses and improved the demarcation between PDD and non-PDD cases. Thus, more recent studies have overcome the systematic undercount of PDD cases in previous studies. The perceived increase in ASDs is therefore likely driven by broadened diagnostic criteria and increased awareness.

11. Role of multiple vaccines and emergence of alternative schedules

With the revelations of the likely fraudulent claims about the MMR vaccine causing ASDs and well-designed studies of thimerosal-containing vaccines failing to show an association with autism, alternative theories about the role of vaccines in causing ASDs have been proposed. The most prominent among these is that the simultaneous administration of multiple vaccines "overwhelms" or "weakens" the immature immune system in young children and through some interaction with the nervous system "triggers" autism in a susceptible host. Sensationalized cases in the media have given credence to this theory. The case that has garnered the most attention is that of a 9-year-old girl with a mitochondrial enzyme deficiency whose encephalopathy, which included features of ASD, was judged to have worsened following the receipt of multiple vaccines at age 19 months [123]. Her family was able to successfully obtain compensation through the US Vaccine Injury Compensation Program (VICP) which was developed in the 1980s to fairly compensate individuals who feel they have been harmed by a vaccine. In the wake of this case, despite reassurances by the CDC that the VICP's action should not be interpreted as scientific evidence that vaccines cause autism, the theory that multiple vaccines given simultaneously can trigger autism has gained credence among the lay press and public.

The idea that multiple vaccines given to young children might either overwhelm an immature immune system or generate a pathologic, autism-inducing autoimmune response is flawed for several reasons. Although the infant immune system is relatively naive, it is capable of generating a vast array of protective responses, starting at birth. In fact, vaccines represent a small fraction of the challenges to a young child's immune system. For example, the average child is infected with 4-6 viruses per year [124], exposing its immune system to numbers of antigens that far exceed those present in simultaneously administered childhood vaccines. Proponents of the theory point to the increasing number of vaccines that are administered to young children. However, most people do not recognize that although the number of recommended childhood vaccines has increased during the past 30 years, with advanced technologies that are used to manufacture modern vaccines, the immunologic load has actually decreased. The childhood vaccines given today contain <200 bacterial and viral antigens, compared with >3000 of these immunological components in the vaccines administered to children in 1980 [125]. In fact, combinations of vaccines are actually known to induce immune responses that are comparable to those given individually [126].

Susceptibility to non-vaccine-preventable infections does not differ in vaccinated and unvaccinated children [127–129]. Put in another way, vaccination does not appear to suppress the immune system in young children in a clinically relevant manner. On the contrary, infections with some vaccine-preventable diseases are known to predispose children to severe, invasive infections with other pathogens [130,131]. Therefore, the available data suggest that vaccines do not "weaken" the immune system. Furthermore, it should be recognized that autism is not an immune-mediated disease such as multiple sclerosis. There is no evidence of immune activation or inflammatory lesions in the brains of autistic people [116]. Instead, new research suggests that genetic variation in neuronal circuitry that affects synaptic development in the brain might in part account for the symptoms of autism [132]. Therefore, the theory that an exaggerated or inappropriate immune response to vaccination results in autism is at variance with current scientific data that address the pathogenesis of autism.

In 2007, Dr. Robert Sears, a pediatrician from Southern California published *The Vaccine Book: Making the Right Decision for Your Child* [133]. In it he offers 2 alternative schedules (that are not approved or endorsed by any public health authority or professional physician group) to parents who are concerned about vaccines so that they may delay, withhold, separate, or space out vaccines for their children. Dr. Sears has publicly stated that he isn't against vaccinations [134]. Instead, his book suggests an untraditional "alternative" schedule that delays vaccines or spaces them further apart. If parents are unwilling to vaccinate at all, he offers a separate "selective" schedule to encourage them to give their child(ren) at least the "bare minimum" of vaccinations. Healthcare providers are facing many parents who are questioning the need for immunization and insisting that their children receive vaccines according to Dr. Sears' schedule, rather than that recommended by the CDC, the American Academy of Pediatrics, and the American Academy of Family Physicians. Most parents are unaware that no research studies have compared the incidence of autism in vaccinated, unvaccinated, or alternatively vaccinated children (i.e., schedules that spread out vaccines, avoid combination vaccines, or include only select vaccines) [135]. The problem with Dr. Sears' schedules is the fact that it can take up to 5-6 years for children to complete their immunizations, during which some children will be at risk for contracting vaccine-preventable diseases due to lack of adequate immunity. Dr. Sears' book has been described as dangerous by some, because it validates the pervasive myths that are currently scaring parents into making ill-informed decisions for their children [136].

12. Legal repercussions

Perhaps inevitably in a litiginous society, the question of whether childhood vaccines cause autism has moved from the scientific into the legal realm [137-139]. Parents of children with autism have submitted thousands of claims to the federal VICP, seeking damages because they believe that their child's autism was caused by vaccines. In 2002, to resolve such claims more expeditiously, the VICP announced that some "test cases" would examine the general causation question, putting aside the question of harm to any particular child. On February 12, 2009, the U.S. Court of Federal Claims published decisions about these claims, which were considered as a group under the Omnibus Autism Proceeding. The Court, after reviewing 5,000 pages of transcripts, 939 medical articles, 50 expert reports, and hearing testimony from 28 experts, found that the MMR and thimerosal-containing vaccines, independently or together, were not causal factors in the development of autism or ASD [48, 50, 140]. Finally, in *Bruesewitz v. Wyeth* (No. 09-152), the Supreme Court of the United States has held that the National Childhood Vaccine Injury Act "preempts all design-defect claims against vaccine manufacturers brought by plaintiffs who seek compensation for injury or death caused by vaccine side effects" [141]. In so doing, it likely closed the door on thousands of claims by parents alleging a link between vaccines and childhood autism.

13. Role of the media

As described above, many parents are hesitant about vaccinating their children. Vaccine hesitancy can be explained in part by a lack of trust in those who make vaccine recommendations; a suspicion of profit motive driven by pharmaceutical companies; misinformation on the Internet; failure to appreciate the seriousness of vaccine-preventable diseases, given their low rates; and constant stories in the media claiming that vaccines cause a variety of illnesses, ranging from allergies to autism [135]. In spite of overwhelming scientific evidence to the contrary, the debate over vaccines and ASDs rages on, with media reports fueling the general public's fear. The disconnect between the scientific community and the popular media is clear in a study published by researchers at the Stanford University School of Medicine [142]. They found that while 41 percent of research funding and published scientific papers on autism dealt with brain and behavior research, only 11 percent of newspaper stories in the United States, United Kingdom and Canada dealt with those issues. Instead, 48 percent of the media coverage dealt with environmental causes of autism, particularly the childhood MMR vaccine [142]. However, in a study by Smith *et al.*, there was a significant increase in selective MMR non-receipt in the US that was temporally associated with the publication of the original scientific literature suggesting a link between MMR and autism. This decline in MMR vaccination preceded media coverage of the MMR-autism controversy and suggests a limited influence of mainstream media on MMR immunization in the United States [20].

Poland and Jacobson note that there has been opposition to vaccination published in newspapers since the introduction of the first vaccine for smallpox over 200 years ago [143]. According to them, little has changed since that time, although now the antivaccinationists' media of choice are typically television and the Internet, including its social media outlets, which are used to sway public opinion and distract attention from scientific evidence [143]. The authors propose various remedies to the misinformation about vaccines that may be presented in the media. Chief among these is enhanced public education and public persuasion, with increasing scientific literacy at all levels of education. They also recommend public-private partnerships of scientists and physicians be developed to make accurate vaccine information accessible to the public in multiple languages, on a range of reading levels, and through various media outlets.

14. Conclusions

Due to the vaccine discoveries of the 20th century and implementation of successful immunization programs around the world, many infectious diseases such as smallpox, polio and measles have either been eradicated or become rare. Parents and many health care providers of the 21st century have limited or no experience with the devastating effects of these diseases. In parallel, over the last few decades there has been an alarming increase in the number of children diagnosed with autism spectrum disorders (ASDs). Why this increase has occurred is not entirely known, although some explanations have been offered by the medical and scientific community. ASDs are often diagnosed in children at about the same chronologic age as the peak time for vaccine delivery. This congruence in time of two separate but important health issues, has led to the peculiar situation where fear of disease has shifted to concerns about vaccine safety, particularly ASDs among some members of the public. Although scientific evidence has refuted many of the misconceptions regarding vaccines and ASDs, this information has not been disseminated sufficiently among the lay public. The unfortunate result has been an erosion of public confidence in vaccines. Consequently, some vaccine-preventable diseases such as measles and polio have reappeared in parts of the world where they had been nearly eliminated. In order to restore the public's trust, all stakeholders including parents, healthcare providers and public health authorities need to ensure that rigorously researched scientific information on the issue of vaccines and autism be accurately collected and disseminated.

15. References

[1] Demirjian A, Levy O. Novel vaccines: bridging research, development and production. Expert Rev Vaccines 2008;7:1321-4.

[2] Oxford JS. The end of the beginning: vaccines for the next 25 years. Vaccine 2008;26:6179-82.

[3] Plotkin SL, Plotkin SA. A short history of vaccination. In: Vaccines. Ed. Plotkin SA, Orenstein W, Offit P. Saunders-Elsevier, 5th Edition, 2008, pp1-16.

[4] National Immunization Program, CDC. Achievements in Public Health, 1900-1999 Impact of Vaccines Universally Recommended for Children -- United States, 1990-1998. MMWR 1999; 48(12);243-248.

[5] Jana L. Don't take the benefits of vaccines for granted. Omaha World Herald, April 12, 2011.

[6] Gupta VB. Communicating with parents of children with autism about vaccines and complementary and alternative approaches. J Dev Behav Pediatr 2010;31:343–345.

[7] Coulter HL, Fisher BL. A shot in the dark. Why the P in the DPT vaccination may be hazardous to your child's health. Garden City Park, N.Y. : Avery Pub. Group, 1991.

[8] Wakefield AJ, Murch SH, Linnell J *et al.* Ileal-lymphoid-nodular hyperplasia, non-specific colitis, and pervasive developmental disorder in children [retracted]. Lancet 1998;351:637–641.

[9] Kawashima H, Mori T, Kashiwagi Y, Takekuma K, Hashika A, Wakefield A. Detection and sequencing of measles virus from peripheral mononuclear cells from patients with inflammatory bowel disease and autism. Dig Dis Sci 2000;45:723–729.

[10] BBC News. MMR immunisation rate falls again. Available at: http://news.bbc.co.uk/2/hi/health/3683930.stm. Accessed 4/17/11.

[11] McIntyre P, Leask J. Improving uptake of MMR vaccine. BMJ 2008;336:729–30.

[12] Asaria P, MacMahon E. Measles in the United Kingdom: can we eradicate it by 2010? BMJ 2006;333:890–5.

[13] Gupta RK, Best J, MacMahon E. Mumps and the UK epidemic 2005. BMJ 2005;330:1132–5.

[14] Centers for Disease Control and Prevention (CDC). Mumps epidemic--United Kingdom, 2004-2005. MMWR 2006;55:173-5.

[15] Editorial team. Measles once again endemic in the United Kingdom. Euro Surveill 2008;13(27):pii=18919. Available at:

http://www.eurosurveillance.org/ViewArticle.aspx?ArticleId=18919. Accessed 4/10/11.

[16] Pepys MB. Science and serendipity. Clin Med 2007;7:562–78.

[17] McBrien J, Murphy J, Gill D, Cronin M, O'Donovan C, Cafferkey MT. Measles outbreak in Dublin, 2000. Pediatr Infect Dis J 2003 Jul;22(7):580-4.

[18] Murphy JF. Fallout of the enterocolitis, autism, MMR vaccine paper. Ir Med J 2011;104(2):36.

[19] BBC News. MMR vaccine uptake rise 'stalls'. Available at: http://news.bbc.co.uk/2/hi/health/7633115.stm. Accessed 4/10/11.

[20] Smith MJ, Ellenberg SS, Bell LM, Rubin DM. Media coverage of the measles-mumps-rubella vaccine and autism controversy and its relationship to MMR immunization rates in the United States. Pediatrics 2008;121:e836–43.

[21] Centers for Disease Control and Prevention. Update: Measles --- United States, January--July 2008. MMWR 2008;57:893-896.

[22] Chen RT, DeStefano F. Vaccine adverse events: causal or coincidental? Lancet 1998;351:611-2.

[23] Payne C, Mason B. Autism, inflammatory bowel disease, and MMR vaccine. Lancet 1998;351:907.

[24] Black C, Kaye JA, Jick H. Relation of childhood gastrointestinal disorders to autism: nested case-control study using data from the UK General Practice Research Database. BMJ 2002;325:419-21.

[25] Taylor B, Miller E, Lingam R, Andrews N, Simmons A, Stowe J. Measles, mumps, and rubella vaccination and bowel problems or developmental regression in children with autism: population study. BMJ 2002;324:393-6.

[26] Madsen KM, Hviid A, Vestergaard M, Schendel D, Wohlfahrt J, Thorsen P, *et al.* A population-based study of measles, mumps, and rubella vaccination and autism. N Engl J Med 2002;347:1477-82.

[27] Honda H, Shimizu Y, Rutter M. No effect of MMR withdrawal on the incidence of autism: a total population study. J Child Psychol Psychiatry 2005;46:572-9.

[28] Chen W, Landau S, Sham P, Fombonne E. No evidence for links between autism, MMR and measles virus. Psychol Med 2004;34:543–553.

[29] Dales L, Hammer SJ, Smith NJ. Time trends in autism and in MMR immunization coverage in California. JAMA 2001;285:1183–1185.

[30] DeStefano F, Bhasin TK, Thompson WW, Yeargin-Allsopp M, Boyle C. Age at first measles-mumps-rubella vaccination in children with autism and school-matched control subjects: a population-based study in metropolitan Atlanta. Pediatrics 2004;113:259–266.

[31] Farrington CP, Miller E, Taylor B. MMR and autism: further evidence against a causal association. Vaccine 2001;19:3632–3635.

[32] Fombonne E, Chakrabarti S. No evidence for a new variant of measles-mumps-rubella-induced autism. Pediatrics 2001;108:E58.

[33] Fombonne E, Zakarian R, Bennett A, Meng L, McLean-Heywood D. Pervasive developmental disorders in Montreal, Quebec, Canada: prevalence and links with immunizations. Pediatrics 2006;118:e139–150.

[34] Kaye JA, del Mar Melero-Montes M, Jick H. Mumps, measles, and rubella vaccine and the incidence of autism recorded by general practitioners: a time trend analysis. BMJ 2001;322: 460–463.

[35] Lingam R, Simmons A, Andrews N, Miller E, Stowe J, et al. Prevalence of autism and parentally reported triggers in a north east London population. Arch Dis Child 2003;88: 666–670.

[36] Makela A, Nuorti JP, Peltola H. Neurologic disorders after measles-mumps-rubella vaccination. Pediatrics 2002;110:957–963.

[37] Patja A, Davidkin I, Kurki T, Kallio MJ, Valle M, et al. Serious adverse events after measles-mumps-rubella vaccination during a fourteen-year prospective follow-up. Pediatr Infect Dis J 2000;19:1127–1134.

[38] Peltola H, Patja A, Leinikki P, Valle M, Davidkin I, et al. No evidence for measles, mumps, and rubella vaccine-associated inflammatory bowel disease or autism in a 14-year prospective study. Lancet 1998;351:1327–1328.

[39] Richler J, Luyster R, Risi S, Hsu WL, Dawson G, et al. Is there a 'regressive phenotype' of Autism Spectrum Disorder associated with the measles-mumps-rubella vaccine? A CPEA Study. J Autism Dev Disord 2006;36:299–316.

[40] Smeeth L, Cook C, Fombonne E, Heavey L, Rodrigues LC, et al. MMR vaccination and pervasive developmental disorders: a case-control study. Lancet 2004;364:963–969.

[41] Smeeth L, Hall AJ, Fombonne E, Rodrigues LC, Huang X, et al. A case-control study of autism and mumps-measles-rubella vaccination using the general practice research database: design and methodology. BMC Public Health 2001;1:2.

[42] Takahashi H, Arai S, Tanaka-Taya K, Okabe N. Autism and infection/immunization episodes in Japan. Jpn J Infect Dis 2001;54:78–79.

[43] Taylor B, Miller E, Farrington CP, Petropoulos MC, Favot-Mayaud I, *et al.* Autism and measles, mumps, and rubella vaccine: no epidemiological evidence for a causal association. Lancet 1999;353:2026–2029.

[44] Mroz˙ek-Budzyn D, Kiełtyka A, Majewska R. Lack of association between measles-mumps-rubella vaccination and autism in children. A case-control study. Pediatr Infect Dis J 2010;29:397–400.

[45] Klein K, Diehl E. Relationship between MMR vaccine and autism. Ann Pharmacother 2004;38:1297–1300.

[46] Uchigama T, Kurosawa M, Inaba Y. MMR-vaccine and regression in autism spectrum disorders: negative results presented from Japan. J Autism Dev Disord 2007;37:210 –217.

[47] DeStefano F. Vaccines and autism: evidence does not support a causal association. Clin Pharmacol Ther 2007;82:756–759.

[48] Miller L, Reynolds J. Autism and vaccination-The current evidence. J Spec Pediatr Nurs 2009;14:166-172.

[49] Gerber JS, Offit A. Vaccines and autism: A tale of shifting hypotheses. Clin Infect Dis 2009;48(4):456–461.

[50] Hensley E, Briars L. Closer look at autism and the measles–mumps–rubella vaccine. J Am Pharm Assoc 2010;50:736–741.

[51] Wakefield AJ, Montgomery SM. Measles, mumps, rubella vaccine: Through a glass, darkly. Adverse Drug React Toxicol Rev 2000;19:265-83.

[52] Weibel RE, Caserta V, Benor DE, Evans G. Acute encephalopathy followed by permanent injury or death associated with further attenuated measles vaccines: A review of claims submitted to the National Vaccine Injury Compensation Program. Pediatrics 1998;101:383-7.

[53] Geier MR, Geier DA. Pediatric MMR vaccination safety. International Pediatrics 2003;18:203-208.

[54] Murch SH, Anthony A, Casson DH, Malik M, Berelowitz M, Dhillon AP, Thomson MA, Valentine A, Davies SE, Walker-Smith JA. Retraction of an interpretation. Lancet 2004; 363:750.

[55] D'Souza Y, Fombonne E, Ward BJ. No evidence of persisting measles virus in peripheral blood mononuclear cells from children with autism spectrum disorder. Pediatrics 2006;118:1664–1675.

[56] Afzal MA, Ozoemena LC, O'Hare A, Kidger KA, Bentley ML, et al. Absence of detectable measles virus genome sequence in blood of autistic children who have had their MMR vaccination during the routine childhood immunization schedule of UK. J Med Virol 2006;78: 623–630.

[57] Baird G, Pickles A, Simonoff E, Charman T, Sullivan P, Chandler S, Loucas T, Meldrum D, Afzal M, Thomas B, Jin L, Brown D. Measles vaccination and antibody response in autism spectrum disorders. Arch Dis Child 2008;93(10):832-7.

[58] Hornig M, Briese T, Buie T, Bauman ML, Lauwers G, Siemetzki U, Hummel K, Rota PA, Bellini WJ, O'Leary JJ, Sheils O, Alden E, Pickering L, Lipkin WI. Lack of association between measles virus vaccine and autism with enteropathy: a case-control study. PLoS One 2008 4;3(9):e3140.

[59] Institute of Medicine. Immunization safety review: vaccines and autism. Available at www.nap.edu/catalog/10997.html. Accessed 4/15/2011.

[60] Fisher BL. Vaccines, autism & chronic inflammation: The new epidemic. PB Industries, Inc. 2008.

[61] Office of Research Integrity. Definition of research misconduct. Available at http://ori.hhs.gov/misconduct/definition_misconduct.shtml. Accessed 4/17/11.

[62] Godlee F, Smith J, Markovitch H. Wakefield's article linking MMR vaccine and autism was fraudulent. Clear evidence of falsification of data should now close the door on this damaging vaccine scare. BMJ 2011;342:c7452.

[63] Deer B. Secrets of the MMR scare. How the case against the MMR vaccine was fixed. BMJ 2011;342:c5347.

[64] Deer B. Secrets of the MMR scare. How the vaccine crisis was meant to make money. BMJ 2011;342:c5258.

[65] Deer B. Secrets of the MMR scare. The Lancet's two days to bury bad news. BMJ 2011;342:c7001.

[66] Deer B. Revealed: MMR research scandal. *Sunday Times* 2004 February 22. Available at www.timesonline.co.uk/tol/life_and_style/health/article1027636.ece. Accessed 4/17/11.

[67] The editors of the Lancet. Retraction—Ileal-lymphoid-nodular hyperplasia, non-specific colitis, and pervasive developmental disorder in children. Lancet 2010;375:445.

[68] Transcripts of hearings of fitness to practice panel (misconduct) in the case of Wakefield, Walker-Smith, and Murch, 16 July 2007 to 24 May 2010. GMC; 2010.

[69] General Medical Council. Dr Andrew Jeremy Wakefield. Determination on serious professional misconduct (SPM) and sanction. 24 May 2010. Available at: www.gmc-uk.org/Wakefield_SPM_and_SANCTION.pdf_32595267.pdf. Accessed: 4/17/11.

[70] General Medical Council. Professor John Angus Walker-Smith. Determination on serious professional misconduct (SPM) and sanction. 24 May 2010. http://www.gmc-uk.org/Professor_Walker_Smith_SPM.pdf_32595970.pdf. Accessed: 4/17/11.

[71] Offit PA. Autism's false prophets. Bad science, risky medicine, and the search for a cure. Columbia University Press, NY, USA 2008.

[72] Offit PA. Deadly choices. How the anti-vaccine movement threatens us all. Basic Books, NY, USA 2011.

[73] Dillner L. Regulating research. BMJ 2010;341:c5128.

[74] Wager L. Wakefield: should we demand public access to raw data? BMJ 2011;342:135.

[75] Opel DJ, Diekema DS, Marcuse EK. Assuring research integrity in the wake of Wakefield. BMJ 2011;342:179-180.

[76] Casiday R, Cresswell T, Wilson D, Panter-Brick C. A survey of UK parental attitudes to the MMR vaccine and trust in medical authority. Vaccine 2006;24(2):177-84.

[77] Freed GL, Clark SJ, Butchart AT, Singer DC, Davis MM. Parental vaccine safety concerns in 2009. Pediatrics 2010;125(4):654-9.

[78] Baker JP. Mercury, vaccines, and autism: one controversy, three histories. Am J Public Health 2008;98:244–53.

[79] United States Enviromental Protection Agency. Implementation of the Mercury-Containing and Rechargeable Battery Management Act. Available at http://www.epa.gov/osw/hazard/recycling/battery.pdf. Accessed 4/17/11.

[80] United States Enviromental Protection Agency. Methyl mercury. Available at http://www.epa.gov/iris/subst/0073.htm#reforal. Accessed 4/17/11.

[81] United States Enviromental Protection Agency. Health Effects. Mercury. Available at http://www.epa.gov/hg/effects.htm#comp. Accessed 4/17/11.

[82] Mahaffey KR, Rice G, et al. An assessment of exposure to mercury in the United States: Mercury study report to Congress. Washington, DC: U.S. Environmental Protections Agency; 1997. Document EPA-452/R097-006.

[83] Mahaffey KR. Methylmercury: A new look at the risks. Public Health Rep 1999;114:397-413.

[84] Agency for Toxic Substances and Disease Registry. Toxicological profile for mercury. Atlanta, GA: Agency for Toxic Substances and Disease Registry;1999.

[85] Federal Register, January 19, 1979;44;3990.

[86] World Health Organization. Trace elements and human nutrition and health. Geneva: World Health Organization;1996:209.

[87] United States Food and Drug Administration. Thimerosal in vaccines. Available at http://www.fda.gov/BiologicsBloodVaccines/SafetyAvailability/VaccineSafety/UCM096228#guid. Accessed 4/17/11.

[88] 21 USC 397 Section 413, 1997.

[89] Ball LK, Ball R, Pratt RD. An assessment of thimerosal use in childhood vaccines. Pediatrics 2001;1147-1154.

[90] Centers for Disease Control and Prevention. Notice to Readers: Thimerosal in Vaccines: A Joint Statement of the American Academy of Pediatrics and the Public Health Service. MMWR 1999;48:563-565.

[91] Centers for Disease Control and Prevention. Notice to Readers: Summary of the Joint Statement on Thimerosal in Vaccines. MMWR 2000;49(27):622-631.

[92] Offit PA. Thimerosal and vaccines--a cautionary tale. N Engl J Med 2007; 357(13):1278-9.

[93] Biroscak BJ, Fiore AE, Fasano N, Fineis P, Collins MP, Stoltman G. Impact of the thimerosal controversy on hepatitis B vaccine coverage of infants born to women of unknown hepatitis B surface antigen status in Michigan. Pediatrics 2003;11:e645-9.

[94] Nelson KB, Bauman ML. Thimerosal and autism? Pediatrics 2003;111:674–9.

[95] Morbidity and Mortality Weekly Report. 2001. Blood and hair mercury levels in young children and women of childbearing age--United States, 1999. MMWR 50(8):140-143.

[96] Institute of Medicine. Thimerosal-containing vaccines and neurodevelopmental disorders. National Academy Press. Washington, D.C. 2001.

[97] Mutter J, Naumann J, Schneider R, Walach H, Haley B. Mercury and autism: accelerating evidence? Neuro Endocrinol Lett 2005;26(5):439-446.

[98] Geier MR, Geier DA. Neurodevelopmental disorders after thimerosal-containing vaccines: a brief communication. Exp Biol Med 2003;228:660–664.

[99] Geier DA, Geier MR. A comparative evaluation of the effects of MMR immunization and mercury doses from thimerosal-containing childhood vaccines on the population prevalence of autism. Med Sci Monit 2004;10(3):I33-I39.

[100] Geier DA, Geier MR. An assessment of the impact of thimerosal on childhood neurodevelopmental disorders. Pediatr Rehabil 2003;6(2):97-102.

[101] Geier DA, Geier MR. Neurodevelopmental disorders following thimerosal-containing childhood immunizations: a follow-up analysis. Int J Toxicol 2004;23(6):369-376.

[102] Geier DA, Geier MR. An evaluation of the effects of thimerosal on neurodevelopmental disorders reported following DTP and Hib vaccines in comparison to DTPH vaccine in the United States. J Toxicol Environ Health 2006;69:1481–1495.

[103] Geier DA, Geier MR. A prospective study of thimerosal-containing Rho(D)-immune globulin administration as a risk factor for autistic disorders. J Matern Fetal Neonatal Med 2007;20:358–390.

[104] Goodman MJ, Nordin J. Vaccine Adverse Event Reporting System source: a possible source of bias in longitudinal studies. Pediatrics 2006;117: 387–390.

[105] Andrews N, et al. Thimerosal exposure in infants and developmental disorders: a retrospective cohort study in the United Kingdom does not support a causal association. Pediatrics 2004;114(3):584-591.

[106] Heron J, Golding J. Thimerosal exposure in infants and developmental disorders: a prospective cohort study in the United Kingdom does not support a causal association. Pediatrics 2004;114(3):577-583.

[107] Hviid A, Stellfeld M, Wohlfahrt J, Melbye M. Association between thimerosal-containing vaccine and autism. JAMA 2003;290(13):1763-1766.

[108] Jick H, Kaye JA. Autism and DPT vaccination in the United Kingdom. N Engl J Med 2004;350(26):2722-2723.

[109] Madsen KM, Lauritsen MB, Pedersen CB, Thorsen P, Plesner AM, Andersen PH, Mortensen PB. Thimerosal and the occurrence of autism: negative ecological evidence from Danish population-based data. Pediatrics 2003;112(3 Pt 1):604-606.

[110] Stehr-Green P, Tull P, Stellfeld M, Mortenson PB, Simpson D. Autism and thimerosal-containing vaccines: lack of consistent evidence for an association. Am J Prev Med 2003;25(2):101-106.

[111] Verstraeten T, Davis RL, DeStefano F, Lieu TA, Rhodes PH, Black SB, Shinefield H, Chen RT; Vaccine Safety Datalink Team. Safety of thimerosal-containing vaccines: a two-phased study of computerized health maintenance organization databases. Pediatrics 2003;112(5):1039-1048.

[112] Clements CJ. The evidence for the safety of thiomersal in newborn and infant vaccines. Vaccine 2004;22:1854-1861.

[113] Thompson WW, Price C, Goodson B, et al. Early thimerosal exposure and neuropsychological outcomes at 7 to 10 years. N Engl J Med 2007;357:1281–92.

[114] Tozzi AE, Bisiacchi P, Tarantino V, De Mei B, D'Elia L, Chiarotti F, Salmaso S. Neuropsychological performance 10 years after immunization in infancy with thimerosal-containing vaccines. Pediatrics 2009;123(2):475-82.

[115] Price CS, Thompson WW, Goodson B, Weintraub ES, Croen LA, Hinrichsen VL, Marcy M, Robertson A, Eriksen E, Lewis E, Bernal P, Shay D, Davis RL, DeStefano F. Prenatal and infant exposure to thimerosal from vaccines and immunoglobulins and risk of autism. Pediatrics 2010;126(4):656-64.

[116] Immunization Safety Review Committee, Board on Health Promotion and Disease Prevention. Immunization Safety Review: Vaccines and Autism. Institute of Medicine of the National Academies. The National Academies Press. Washington, D.C. 2004.

[117] Harry GJ, Harris MW, Burka LT. Mercury concentrations in brain and kidney following ethylmercury, methylmercury and thimerosal administration to neonatal mice. Toxicol Lett 2004;154(3):183-189.

[118] Burbacher TM, et al. Comparison of blood and brain mercury levels in infant monkeys exposed to methylmercury or vaccines containing thimerosal. Environ Health Perspect 2005;113(8):1015-1021.

[119] Pichichero ME, et al. Lancet 2002;360:1737-1741.

[120] Aschner M, Ceccatelli S. Are neuropathological conditions relevant to ethylmercury exposure? Neurotox Res 2010;18:59–68.

[121] Shultz ST. Does thimerosal or other mercury exposure increase the risk for autism? A review of current literature. Acta Neurobiol Exp 2010, 70: 187–195.

[122] Scahill L, Bearss K. The rise in autism and the mercury myth. J Child Adolesc Psychiatr Nurs 2009;22:51-53.

[123] Offit PA. Vaccines and autism revisited-the Hannah Poling case. N Engl J Med 2008;358:2089-91.

[124] Dingle JH, Badger GF, Jordan WS. Illness in the home: a study of 25,000 illnesses in a group of Cleveland families. Press of Western Reserve University; Cleveland: 1964.

[125] Offit PA, Quarles J, Gerber MA, *et al.* Addressing parents' concerns: do multiple vaccines overwhelm or weaken the infant's immune system? Pediatrics 2002;109:124–9.

[126] King GE, Hadler SC. Simultaneous administration of childhood vaccines: an important public health policy that is safe and efficacious. Pediatr Infect Dis J 1994;13:394–407.

[127] Black SB, Cherry JD, Shinefield HR, Fireman B, Christenson P, Lampert D. Apparent decreased risk of invasive bacterial disease after heterologous childhood immunization. Am J Dis Child 1991;145:746–9.

[128] Davidson M, Letson GW, Ward JI, *et al.* DTP immunization and susceptibility to infectious diseases: is there a relationship? Am J Dis Child 1991;145:750–4.

[129] Storsaeter J, Olin P, Renemar B, *et al.* Mortality and morbidity from invasive bacterial infections during a clinical trial of acellular pertussis vaccines in Sweden. Pediatr Infect Dis J 1988;7:637–45.

[130] O'Brien KL, Walters MI, Sellman J, *et al.* Severe pneumococcal pneumonia in previously healthy children: the role of preceding influenza infection. Clin Infect Dis 2000;30:784–9.

[131] Laupland KB, Davies HD, Low DE, Schwartz B, Green K, McGeer A. Invasive group A Streptococcal disease in children and association with varicella-zoster virus infection. Ontario Group A Streptococcal Study Group. Pediatrics 2000;105:e60.

[132] Morrow EM, Yoo SY, Flavell SW, et al. Identifying autism loci and genes by tracing recent shared ancestry. Science 2008;321:218–23.

[133] Sears R. The Vaccine Book: making the right decision for your child. Little, Brown, 2007.

[134] Kam K. What is the Alternative Vaccine Schedule? Web MD. Children's Vaccines Health Center. Available at: http://children.webmd.com/vaccines/features/robert-sears-alternative-vaccine-schedule. Accessed 4/19/11.

[135] Offit PA, Moser CA. The problem with Dr. Bob's alternative vaccine schedule. Pediatrics 2009;123:e164–e169.

[136] Snyder J. Cashing in on fear: The danger of Dr. Sears. Science-Based Medicine. Available at: http://www.sciencebasedmedicine.org/?p=512. Accessed 4/19/11.

[137] Sugarman SD. Cases in Vaccine Court-Legal battles over vaccines and autism. N Engl J Med 2007;357:1275-1277.

[138] Stewart AM. When vaccine injury claims go to court. N Engl J Med 2009;360(24):2498-500.

[139] Silverman RD. Litigation, regulation, and education--protecting the public's health through childhood immunization. N Engl J Med 2009;360(24):2500-1.

[140] Riedmann EM. Court finds no Vaccines-autism link. Human Vaccines 2010;6:368–372.

[141] Supreme Court of the United States. Russell Bruesewitz et al v. Wyeth et al. No. 09-152. Argued October 12, 2010 - Decided February 22, 2011.

[142] Singh J, Hallmayer J, Illes J. Interacting and paradoxical forces in neuroscience and society. Nat Rev Neurosci 2007 Feb;8(2):153-60.

[143] Poland GA, Jacobson RM. The age-old struggle against the antivaccinationists. N Engl J Med 2011; 364:97-99.

18

Immune Dysfunction in Autism Spectrum Disorder

Jonna B. Westover[3], Thayne L. Sweeten[1], Michael Benson[3], Patricia Bray-Ward[2] and Anthony R. Torres[3]
[1]Utah State University, Department of Biology,
[2]University of Hawai'i Cancer Center,
[3]Utah State University, Center for Persons with Disabilities,
U.S.A.

1. Introduction

Autism was first described in 1943 (Kanner, 1943) after examining a group of children with abnormal social characteristics and obsessive behavior. Initially considered a rare disorder, it did not receive much attention in the media or research circles until the latter end of the 1900's when an increasing number of cases were being diagnosed. The term Autism Spectrum Disorder (ASD) combines a group of syndromes with fundamental deficits in reciprocal social communications and repetitive stereotyped verbal and non-verbal behaviors (Volkmar et al., 2009). The severity spectrum varies widely and more severe forms can be accompanied by language regression, seizures and lower IQ. The diversity of specific behavioral deficits in different individuals suggests that autism is not a single disorder, but rather a collection of variants each with its own characteristics and perhaps etiologies (Altevogt et al., 2008). Despite the recognition that there has been an increase in public awareness, an improvement in patient ascertainment and a broadening of diagnosis, it is generally accepted that the incidence of autism and related ASDs is increasing (CDC, 2009) with current estimates being 1 in 110 children in the United States have ASD. There is much debate around the world, both in the research and parent communities, on the etiology of autism. Everything from childhood vaccines, exposure to mercury, maternal viral infections and autoimmune disorders has been blamed. Unfortunately, despite extensive research over the last decade, there is very little that is universally accepted about the etiology of autism except, of course, that autism results from abnormal brain function arising in either the prenatal period or infancy stage of life.

2. Infection

In the 1960's and early 70's a number of case studies associating autism with congenital infection appeared in the literature. Many of these cases were reports of autism associated with congenital rubella infection. In fact, at least 43 cases associating autistic symptoms with congenital rubella have been described (Hwang & Chen, 2010; Rimland, 1964). Congenital rubella is caused by first trimester infection with German measles. Resultant symptoms in the infant may include deafness, developmental delay, mental retardation and seizures. In 1964 a rubella epidemic swept through the United States leaving behind a large group of rubella children that also suffered a higher than expected incidence of autism (Chess, 1971). It is estimated that the incidence of autism in children with congenital rubella syndrome is between 4-7%. Although these were clearly cases of congenital rubella with resultant autistic symptoms that represented a unique subset of patients, the association of a congenital infection resulting in autistic behaviors sparked interest in the theory that the immune system may play a vital role in protecting against or perhaps contributing to the onset of some cases of autism. After decades of research, evidence of a broader infectious role in this disorder remains elusive, yet findings of altered immune system parameters continually appear in the literature.

3. Familial studies

Familial clustering and twin studies indicate a strong genetic component to autism predisposition. Twin studies show the concordance rates of monozygotic twins at 36-96%, whereas dizygotic twins are 0-24% concordant, resulting in an estimated heritability of autism at >90% (Bailey et al., 1995; Folstein & Rutter, 1977; Steffenburg et al., 1989). Additionally, family studies have shown autism to have familial aggregation with 3-8% of subsequently born siblings either being autistic or showing some form of pervasive developmental disorder (PDD) (Bolton et al., 1994; Ritvo et al., 1989). This is a 30-80% increase in risk for siblings over the general population. At that time it was thought the data from family genetics studies and the phenotypic variability indicated that the likely cause of autism was due to specific genes in certain combinations (Freitag, 2007; Rutter, 2000). It became apparent that genetics was another area to study along with the immune system.

4. Genetics

When the ability to genotype one or more alleles became routine in the 1990s, candidate gene association studies became common. Many of these candidate gene association studies were unreliable and often contradictory, casting doubt on this approach (State, 2010). It is now realized that the probability of any candidate gene or allele being associated with a particular phenotype, let alone a spectrum of phenotypes, is extremely low. Modern microarray based technology allows for the genotyping of thousands of genetic polymorphisms on a single chip, however, other issues such as case/control selection and statistical evaluation are especially important when examining large amounts of subjects and polymorphisms.

Several large studies examining hundreds of thousands of single nucleotide polymorphisms (SNPs) have been exhaustively tested in thousands of cases and thousands of controls in an effort to find disease associations (Szatmari et al., 2007; Veenstra-VanderWeele et al., 2004). Three genome-wide association studies (GWAS) each with over 1,000 subjects did not replicate each other's findings for candidate genes (Anney et al., 2010; Wang et al., 2009; Weiss et al., 2009). At this stage it appears that none of the studies had a sufficient number of subjects to replicate findings for alleles having a small to moderate effect. Weiss et al. (2009) examined over 500,000 SNPs in 1,553 autistic subjects using the Affymetrix 5.0 microarray platform. No genome-wide associations were identified until single SNP genotyping was done on the most suggestive regions. Even after genotyping, only one SNP on chromosome

5p15 had a strong autism association. Although a number of loci causative of early childhood psychiatric disorders have been identified using GWAS, few of them have been validated in additional studies. The lack of success with this approach may be due to the fact that while GWAS are well suited for the discovery of common, low impact alleles causing a clinically well-defined disease, it is less effective for gene discovery when diseases with clinical and biochemical heterogeneity are caused by rare high impact genetic changes in many different genes.

Another method for looking at genetic differences involves copy number variation (CNV) by comparative genomic hybridization (Sebat et al., 2007) or by microarray analysis of single samples using SNP/CN chips, with observed CNVs being compared to the CNVs found in genomic databases. Normal CNV includes deletion and amplification of segments of the genome that occur very frequently without apparent phenotypic effect other than producing normal phenotypic variation. Deletions and amplifications of normally diploid genes can cause higher or lower levels of gene expression, or truncated or abnormal fusion gene products, producing pathological phenotypes. CNV is an area of considerable interest in autism as the effects of gene dosage may be important in conferring outcome (Toro et al., 2010). Difficulties with detection of abnormal CNVs by array methodology (Craddock et al., 2010; Sebat et al., 2007) include the confounding presence of extensive CNVs in all genomes, limitations in the size of genomic regions detected (typically 1 Mb or more) (Magi et al., 2011), and the variation in clinical phenotypes that can result from CNVs (Ching et al., 2010). Many copy number mutation differences have been reported in autism including genes involved in neuronal cell adhesion and ubiquitin degradation (Glessner et al., 2009). Mutations and CNVs in the neurexin-1 gene have been discovered that are associated with a variety of developmental disorders including autism, ASD and schizophrenia (Ching et al., 2010). Point mutations and deletions have been associated with ASD by several research groups (reviewed by Ching et al., 2010). Neurexin-1 (2p16.3) is one of the largest genes in the human genome with 24 exons in 1.1Mb. The neurexin-1 gene has two independent promotors that encode alpha and beta neurexin-1 proteins and alternate splicing of the 24 exons can result in over 1,000 possible neurexin isoforms with differential expression of isoforms in different cells or tissues. Similarly, CNVs and mutations have been identified in contactin associated protein-like 2 (CNTNAP2), first described as having a role in developmental delay in Old-Order Amish subjects with intractable epilepsy and autism (Strauss et al., 2006). Three other groups have now replicated the involvement of the CNTNAP2 gene in autism (Alarcón et al., 2008; Arking et al., 2008; Bakkaloglu et al., 2008). Both CNTNAP2 and neurexin-1 are illustrative of an additional hallmark of genes causing psychiatric illness; clinical phenotypes of identical rare variants suggest that ASD, intellectual disability, seizure disorder, schizophrenia, ADHD, Tourette syndrome, OCD, or a combination of these may result from functionally identical genetic changes (State, 2010).

For many years, there was a consensus that the "common variant common disease" hypothesis would best explain the genetics behind autism. This model suggests that each common risk variant (greater than 5% in the population) only confers a small degree of risk and that disease is the result of the combination of many common variants (Risch & Merikangas, 1996). This genetic disease hypothesis has been entertained for over 100 years. After a decade of genetic research using modern microarray and sequencing technology it is now clear that "common risk variants" cannot explain the vast majority of genetic heritability for any human disease (Manolio et al., 2009). Both SNP and GWAS studies have found that rare alleles can be major contributors to complex diseases like ASD. However, the "rare allele variant" can only explain a small proportion (about 10%) of the genetic risk for ASD (Goldstein, 2009; Manolio et al., 2009; State, 2010). It is now apparent that the rare allele is a major contributor not a minor contributor in diseases like breast and ovarian cancers (McClennen & King, 2010). The 90% "missing inheritance" may be explained, at least in part, by rare alleles that have not yet been detected (Goldstein, 2009). Current research suggests that rare mutations are important in neurodevelopmental disorders like autism and schizophrenia (Bucan et al., 2009; Guilmatre et al., 2009). It also appears that diseases with significant phenotype differences are characterized by marked genetic heterogeneity. Whole genome sequencing is becoming a viable experimental approach to attempt to clarify the issue of "missing inheritance" and determine whether it results from mutations in many different genes associated with CNS development and function, or whether it results from differences in the highly complex genes that regulate immune responses, linking genetics with inflammatory responses to infections or environmental insults.

Stubbs & Magenis (1980) first suggested over 30 years ago that the human leukocyte antigen (HLA) region might be involved in autism. Unraveling the relationship between the immune system and autism is difficult, not in the least because of the extraordinary complexity of the immune system. Analysis of the HLA class I and II genomic regions is challenging due to the abundance of genetic polymorphisms, which often occur within a few base pairs of each other. Genetic HLA allelic typing by standard polymerase chain reaction methods followed by DNA sequencing for questionable regions is very precise at a reasonable cost for classical HLA alleles. However, there may be genes that are being missed, as there are close to 100 genes in this region. Research in our laboratory has noted differences in certain HLA class I and class II alleles as well as C4 complement genes in the class III region (Torres et al., 2002; Torres et al., 2006; Warren et al., 1991; Warren et al., 1996).

5. Immune system

The word immune comes from the Latin "immunis" meaning free or exempt. It was noticed in about 1500 that survivors of a certain disease did not get sick after a second exposure. They were considered "exempt from" or "immune" from that disease. The field of immunology was solidified in 1796 with the observations of an English physician, Edward Jenner, when he noted that dairymaids and farmers lacked the characteristic pock-marked complexion of those who had been infected with smallpox. It was determined that individuals who acquired cowpox from cattle were immune from smallpox. In the late 1800s Louis Pasteur coined the term vaccination which means "derived from cows" in honor of Edward Jenner.

Modern immunology is the study of a system of cells, tissue and molecules that recognize and attack pathogens and tumor cells that endanger the individual. Physical barriers prevent most bacteria and viruses from entering the body. Once foreign organisms are inside the body the immune system has two arms of antigen response: innate and adaptive immunity. The innate immune system provides a non-specific response that is built into the organism and therefore always available, but does not elicit an immunological memory. The adaptive immune response is a slower system that is tailored for the individual to fight a specific pathogen (Figure 1). Both systems depend on the organism's ability to recognize self from non-self molecules. Self molecules are components of the individual organism and non-self molecules are defined as substances that are recognized by the organism's immune receptors and elicit an immune response. There is an ever changing immunological memory in the adaptive immune response. For example, when Edward Jenner vaccinated individuals with cowpox an immunological memory was developed that protected them from infection with smallpox at a later date.

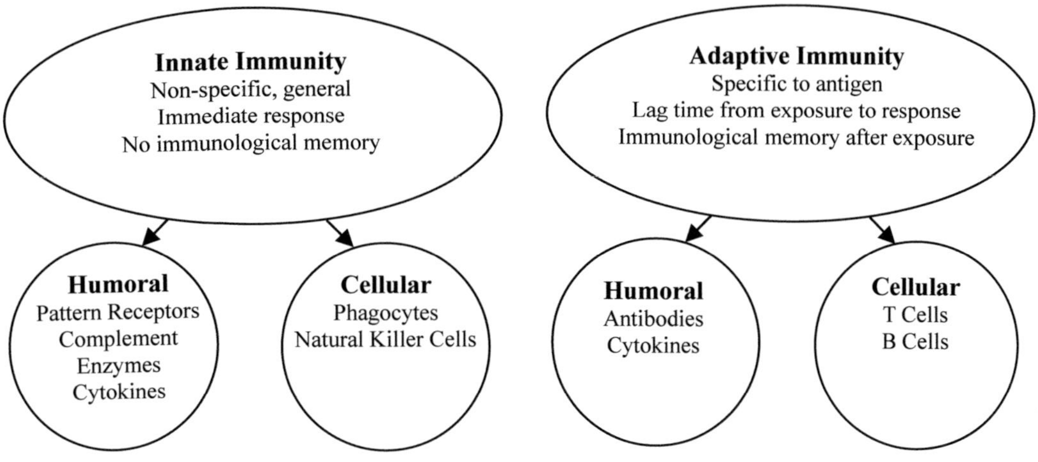

Fig. 1. Innate and Adaptive Immune Systems. The human immune system is divided into two main branches: innate and adaptive. These two branches are further divided into humoral and cellular immunity. Innate immunity provides a non-specific response that is built into the organism and therefore always available, but does not elicit an immunological memory. The adaptive immune response is a slower system that is designed to protect against foreign antigens and develops memory.

5.1 Innate immune system

Microorganisms that get past surface barriers such as skin, tears and mucus will encounter the cells and mechanisms of the innate immune system. Microorganisms are targeted by pattern recognition receptors (PRR) on the surface of innate immune cells such as granulocytes, macrophages and dendritic cells which recognize and bind to pathogen-associated molecular patterns (PAMPs). These PAMPs include common molecules of bacterial carbohydrates and peptides, lipopolysaccharide, bacterial and viral RNA and DNA, and other microbial molecules that are non-self molecules.

Certain PRRs are secreted and these proteins bind to a variety of bacteria, fungi, viruses and protozoa. The complement system contains over 25 proteins and protein fragments that are mainly synthesized in the liver, but also exist in the brain. These proteins in a complicated manner help or "complement" antibodies and phagocyte cells to clear pathogens from the organism. The C4 complement proteins (C4A and C4B) are encoded by two separate genes in the class III HLA region. Although similar, C4A and C4B proteins have different binding characteristics and each have distinct associations with autoimmune diseases.

There are several classes of PRRs according to their ligand binding specificity. Signaling PRRs include the membrane bound Toll-like receptor (TLR) family. There are 11 TLRs with different binding characteristics and, like many molecules, were discovered in *Drosophila.* A series of events including the production of cytokines as well as the activation of cellular pathways occurs upon the engagement of PAMPs by membrane bound TLRs. After binding to a PAMP, cytokine molecules are released to signal other innate and adaptive cells. After binding, endocytosis and phagocytosis of the infectious agent is triggered by the complement system. The second arm of the innate immune system involves phagocytes (macrophages, dendritic cells, neutrophils, eosinophils and basophils), mast cells and natural killer (NK) cells. These cells recognize and kill pathogens by engulfing them or by rupturing the invader's cell membrane and have the ability to send signals to the adaptive immune system. NK cells are a subset of lymphocytes with the ability to produce cytokines and kill target cells without prior sensitization and are essential for self-tolerance. NK cells thus participate in early responses against virally infected and transformed cells by recognizing the lack of HLA class I proteins on cell surfaces, a phenomenon called "missing self".

5.2 Adaptive immune system

A slower immune reaction is elicited by the adaptive than the innate immune system, but more importantly, immunological memory is invoked. Memory involves the recognition of the signature invading organism for extended periods of time, often decades. This is done by the generation of memory cells that are tailor-made to each antigen.

The first arm of the adaptive immune system involves soluble antibodies and cytokines. Antibodies are glycoproteins from the immunoglobulin superfamily produced by memory B cells called plasma cells. Memory cells survive in the body for years and become activated to produce the respective antibody upon exposure to the particular antigen. There are an unlimited number of antibodies that can be produced in response to antigens. This is accomplished by recombination of genes in B cells followed by a complicated cellular selection process to create antibodies to particular antigens.

The second arm of the adaptive immune system involves T cells and B cells. T cells are important in cell-mediated immunity and are distinguished by the T cell receptors which recognize short peptides bound to HLA class I and class II proteins. Several subsets of T cells with distinct functions exist. Helper T cells assist B cells to mature into antibody producing memory B cells. Cytotoxic T cells seek and destroy virally infected cells and tumor cells. Memory T cells exist long term after an infection is resolved. Upon exposure to their particular antigen they quickly expand into effector T cells. Suppressor or regulatory T cells suppress auto-reactive T cells to limit tissue damage. In addition to all this complexity, while attacking pathogens, the immune system must maintain a balance and not attack self molecules.

Helper T cells have an important role in activating and directing other immune cells by the release of cytokines. These cytokines include interleukins, interferons, tumor necrosis factors, chemokines and growth factors. Helper T cells fall into two different subsets (Th1 and Th2) based on the cytokines they release and how they regulate the immune system. There is signaling or significant cross-talk that occurs between the innate and adaptive immune system which is accomplished through cytokines, chemokines and other small molecules. For example, IL-10 will act to inhibit macrophage activation and TNF-β can kill activated B cells. Also, HLA class I proteins bind to receptors on various adaptive immune T cells as well as innate immune NK cells.

6. Immunogenetics

The immune system must distinguish a wide variety of microorganisms including viruses, bacteria and parasitic worms from the organism's own cells and tissues and, therefore, has a complex genetic basis. The HLA region on chromosome 6 (Figure 1) is the most complex region in the human genome and central to many immunological reactions, both innate and adaptive. There are over 100 genes and pseudogenes in this region. The HLA genes are grouped into three classes based on their function: class I, class II, and class III. Class I molecules are expressed on all nucleated cells and are responsible for presenting viral antigens to cytotoxic T cells which are specially designed to kill targeted cells. Class II molecules are typically expressed on B lymphocytes where they present antigens to helper T cells to stimulate the B cells to produce antibodies and they are also expressed on macrophages which are then activated to destroy the pathogen they have engulfed. Class III genes encode for numerous proteins, including C4 complement, involved in immune interactions. Complement C4 proteins are involved in numerous immune functions, including lysing pathogens and marking pathogens for clearance by immune cells.

There are over 3,000 genetic alleles contained in the class I and class II HLA regions that encode proteins important in cellular discrimination of 'self' and 'non-self'. The class I region is important in cellular immunology and the class II region is important in antibody production. For example, there are about 30 common HLA-B alleles and several hundred uncommon ones. This diversity endows the immune system with great plasticity in adaptive immune responses and allows researchers to precisely follow the inheritance of specific alleles.

Haplotype from the Greek means one-fold, single, or simple. In genetics a haplotype refers to a set of closely linked DNA markers present on one chromosome. The HLA region was recognized in 1983 as consisting of several distinct linkage disequilibrium blocks (Dawkins et al., 1983). This information was helpful for the construction of large haplotypes of over 4 million base pairs containing over 150 protein-coding genes. Smaller ancestral haplotypes (AHs) that encode about 100 proteins and the classic HLA alleles (1.2 million base pairs) fit within these blocks. Haplotypes in the HLA region are probably the best characterized in the human genome due to the fact that there are so many alleles. By typing one or both parents as well as the subject of interest one can assign the inheritance as it is not common for both parents to have the same HLA class I and II alleles. Complete DNA sequences have been published for 8 of the most common AHs in an effort to expedite disease association research (Horton et al., 2008). It is difficult by current diploid genotyping methods to determine haplotypes.

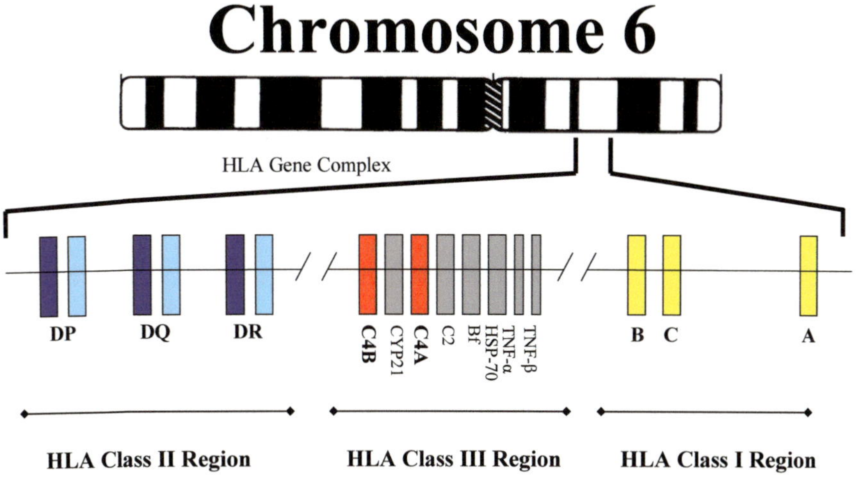

Fig. 2. A simplified map of the HLA region on human chromosome 6.

Some of the genes in these AHs encode for HLA class I and class II proteins which are important in antigen presentation to cytotoxic T cell ($CD8^+$) and helper T cell ($CD4^+$) lymphocytes, respectively. The HLA region of the human genome also encodes for important innate immune proteins, such as the complement C4A and C4B proteins, as well as many proteins vital in immune regulation like HSP-70 and TNF-α. Complement C4 proteins are involved in numerous immune functions, including lysing pathogens and marking pathogens for clearance by immune cells.

7. Cytokines

In Greek the term cytokine means cell movement. All cytokines are peptides/proteins/glycoproteins that serve as immune signals from one cell to regulate other cells. They are very potent molecules with systemic concentrations in the picomolar (10^{12}) range. Virtually all nucleated cells produce cytokines and the target cells have highly specific receptors to the respective cytokine. Cytokines have pronounced effects on many cellular functions in all branches of the immune system, most importantly inflammation and brain chemistry. Since cytokines bind to specific receptors on the membrane of target cells, triggering signal transduction pathways that ultimately alter gene expression in the target cells, they can significantly alter neural development and maintenance, especially in the developing fetus.

Th1 cytokines such as interferon-γ, TNF-α, TNF-β, IL-3 and GM-CSF are considered to be pro-inflammatory. They are involved in the cellular immune system and act to maximize the killing efficacy of macrophages and the proliferation of cytotoxic T cells. Th2 cytokines are non-inflammatory and promote the humoral immune system with cytokines IL-4, IL-5, IL-6, IL-10, IL-13 and TGF-β. IL-6 for example, can determine B cell antibody class switching by driving the proliferation and differentiation of B cells into antibody-secreting plasma cells and is also an endogenous pyrogen so it can cause a fever to help eliminate infections by acting in the hypothalamus. The adverse effects of unusually high levels of certain cytokines have been documented in several disease states like cancer, depression and Alzheimer's disease. The therapeutic application of cytokines such as IL-2 and interferon-γ can result in severe systemic reactions referred to as "cytokine storm".

8. Immune system abnormalities in autism

8.1 Humoral innate system

The complement C4A and C4B genes exist in a region of the genome that demonstrates a particularly complex morphology including CNV. Previous research in our laboratory utilizing a protein immunoelectrophoresis assay demonstrated a significant link to a deficiency of C4B protein and autism (Warren et al., 1991). It was reported that a C4B null allele (C4B gene deletion) was significantly more common in subjects with autism than in subjects without autism. This initial finding was supported by additional research in our laboratory that showed a relative risk of 4 for subjects with autism who have a C4B null allele (Odell et al., 2005). Interestingly, Mostafa & Shehab (2010) supported this observation with an odds ratio of 6 for autism subjects with a C4B null allele and an odds ratio 6.26 for children with autism and a family history of autoimmune diseases in an Egyptian population. Relative risk and odds ratio are different mathematical models for describing the chances of developing a certain disease.

Other humoral innate system genes/proteins have not been investigated as closely as C4 CNV. However, Torres et al. (2001) examined the neighboring TNF-α gene and found that TNF-α microsatellite markers in autistics favored a microsatellite known for lower TNF-α production. Accordingly, plasma TNF-α levels were lower in the subjects with autism compared to controls.

8.2 Cellular innate system

8.2.1 Natural killer cells

Warren et al. (1987) noted reduced NK cell killing (or cytotoxicity) in subjects with autism when using a standard K562 target cell assay. The K562 tumor cell line is used to stimulate NK cells because it lacks HLA class I surface proteins (missing-self) and is targeted by NK cells for destruction. This deficiency in NK cell killing in subjects with autism has been observed by at least two more research groups (Enstrom et al., 2009; Vojdani et al., 2008). Enstrom et al. (2009) also noted decreased NK cell killing upon stimulation with K562 cells in ASD subjects. They examined resting NK cell killing and RNA expression and observed an increase in killing in resting NK cells from ASD subjects as determined by higher levels of interferon-γ and the enzymes perforin and granzyme B. The microarray experiment demonstrated the increased expression of inhibitory killer-cell immunoglobulin-like (KIR) receptors in ASD subjects, which is in agreement with the decreased killing of K562-stimulated NK cells. In the resting state this increased expression of interferon-γ in ASD subjects could have profound long term effects, especially in the brain.

8.2.2 Monocytes

Progenitor cells called monocytes circulate in the blood and differentiate into macrophages upon leaving the blood and entering other body tissues. Macrophages are a type of phagocyte that engulf and digest microorganisms then present the antigens to lymphocytes. Some macrophages are resident cells within specific organs taking on different names depending on the organ they inhabit, for example, in the liver they are referred to as Kupffer cells and in the brain microglial cells. When stimulated by interferon-γ, monocytes and macrophages produce increased amounts of the molecule neopterin. High levels of neopterin serve as an indicator of monocyte/macrophage activation and are indicative of pro-inflammatory immune status. Increased monocyte counts and neopterin levels were observed in autistic children compared to gender and age-matched healthy controls (Sweeten et al., 2003a) suggesting that the immune system is over-activated in the ASD group. Enstrom et al. (2010) isolated monocytes from ASD and control subjects then stimulated the monocytes with several different TLR ligands. The researchers noted marked increases or decreases in pro-inflammatory cytokines following TLR stimulation depending on the particular ligand used.

Since interferon-γ normally activates monocytes/macrophages, an increase in the resting state production of interferon-γ by NK cells in autistic subjects could be over stimulating monocytes/macrophages and altering their function. If microglial cells in the brain are activated this way, it could have profound effects on neurochemistry, brain development and ultimately behavior. For instance, one mechanism whereby microglial cells could alter brain chemistry and development is by increased production of nitric oxide (NO). Interferon-γ activated immune cells such as microglia produce high levels of NO as a means to kill pathogens by oxidative stress. In the brain NO is typically produce by neurons where it acts as an intercellular messenger modulating synaptogenesis, dendrite and axonal growth, and neuronal release of various neurotransmitters (Hess et al., 1993; Lizasoain et al., 1996; Lonart et al., 1992). Increased NO production by interferon-γ responsive immune cells in the brain would likely alter these processes.

NO production has been shown to be elevated in the blood of autistic subjects and was positively correlated with plasma interferon-γ concentration (Sweeten et al., 2004). Vargas et al. (2005) showed that microglia as well as monocyte and macrophages are activated and produce high levels of interferon-γ in the brain tissue of autistic patients. Interferon-γ levels in cerebrospinal fluid were increased over 200-fold in the autistic subjects versus controls.

8.3 Humoral adaptive system

Since autism is a neurodevelopmental disorder, one hypothesis proposes that autism could be caused by an antibody attack against one or more brain proteins. By examining total serum proteins and serum concentrations of antibodies, Croonenberghs et al. (2002) found that children with autism had significantly increased concentrations of albumin and gamma globulin as well as significantly increased levels of IgG subclasses IgG2 and IgG4. Autoantibodies have been detected against several brain proteins including myelin basic protein (Jyonouchi et al., 2001), neuron-axon filament protein and glial fibrillary acidic protein (Singh et al., 1997), nerve growth factor (Bashina et al., 1997), cell nuclei (Stubbs, 1988), brain endothelium (Connolly et al., 1999), Purkinje cells of the cerebellum (Zimmerman et al., 1993), cerebellar proteins (Goines et al., 2011), serotonin 5-HT receptors (Todd & Ciaranello, 1985), and most recently to tissue transglutaminase-2 which is present in the brain and involved in cell adhesion and synaptic stabilization (Rosenspire et al., 2011). Although autoantibodies have been detected against several proteins important in neuronal function, there is no evidence of pathological brain lesions in autism.

8.4 Cellular adaptive system

The cellular arm of the adaptive immune system involves T cells and B cells, both of which have been reported to be altered in subjects with autism. For example, decreased numbers of T lymphocytes and an altered ratio of helper T cells to suppressor T cells were observed as early as 1986 (Denney et al., 1996; Warren et al., 1986). More recently, alterations in lymphocyte subsets was observed when data showed a significant decrease in $CD4^+$ naïve and an increase in $CD4^+$ memory T cells in autistic subjects who also bore the HLA-A*02 and HLA-DRβ1*11 alleles (Ferrante et al., 2003). In addition, the production of numerous autoantibodies in autism that was discussed above provides evidence that indirectly points to B cells playing a role in autism.

Alterations in cytokine levels or production can indicate abnormal cellular immunity. A study that compared the production of several cytokines in peripheral blood mononuclear cells found that levels of both Th2 and Th1 cytokines were significantly higher in the autism group (Molloy et al., 2006). However, when 12 children with autism were compared to children with other neurological disorders, only TNF receptor II was elevated (Zimmerman et al., 2005). When children with ASD were examined for cytokine production, Ashwood et al. (2011) found an association between increased levels of some Th1 pro-inflammatory cytokines and severity of clinical behavioral outcomes, specifically, aberrant behaviors and impaired communication. As previously mentioned, Vargas et al., (2005) found elevated cerebrospinal fluid interferon-γ levels in autistic subjects. This finding was accompanied by an elevated pro-inflammatory profile of cytokines in the brain tissue and cerebrospinal fluid, including macrophage chemoattractant protein-1 and TGF-β1. A prominent theory for the pathophysiology of autism is that increased cytokines produced as a result of inflammation or other aberrant immune processes in the brain may alter neurochemistry and neurodevelopment resulting in the behavioral profile characteristic of autism.

9. HLA associations in autism

There are numerous associations between autism and genes located in the HLA region (Table 1). Researchers in our laboratory demonstrated a significant association between autism and the allele HLA-A2 (Torres et al., 2006). This may be important as HLA class I molecules have been shown to be involved in brain development (Boulanger & Shatz, 2004). HLA class II molecules have been shown to be associated with numerous autoimmune disorders. In a very elegant paper Warren et al. (1996) reported that the third hypervariable region of DRβ1, which is part of the shared epitope binding pocket (DRβ1*0401, *0404, and *0101), has a strong association with autism. This is an interesting observation as the shared epitope has been associated with certain autoimmune diseases (De Alemeida et al., 2010). Torres et al. (2002) confirmed the association of the HLA-DR4 allele and also found that the DR13 and DR14 alleles occurred less often in subjects with autism, suggesting a possible protective mechanism. Interestingly, the DR13 allele was inherited less frequently than expected from the mothers. Associations with autism and the DR4 allele have since been confirmed in two additional laboratories. Lee et al. (2006) demonstrated that boys with autism and their mothers had a significantly higher frequency of DR4 than normal control subjects (odds ratios 4.20 and 5.54, respectively), suggesting that a maternal-fetal immune interaction could be involved in autism. Johnson et al. (2009) reported significant transmission disequilibrium for HLA-DR4 (odds ratio 4.67) from maternal grandparents to mothers of children with autism which also suggests a maternal-fetal interaction for HLA-DR4. The HLA class III region encodes for many components involved in the immune system such as the C4 complement proteins, cytokines (TNF-α), and heat shock proteins. Although C4A and C4B proteins are encoded in the HLA class III region, they are considered part of the humoral innate system as they are not classical HLA proteins.

In addition to single genes and alleles, several publications have shown haplotypes to be associated with autism (Table 1). Warren et al. (1991) first reported that the AH 44.1-SC30-DR4 was associated with autism with a relative risk of 7.9. That result was confirmed in a new case/control population (Daniels et al., 1995). Interestingly, the individual components of AH 44.1-SC30-DR4 include C4B null and HLA-DR4, both of which have been shown independently to be significantly associated with autism, although not in the population reported on by Daniels et al. (1995). In autistic children bearing both the HLA-A2 and DR11 alleles, Ferrante et al. (2003) observed a significant decrease in CD4+ naïve and an increase in CD4+ memory T cells compared to the HLA-A2, DR11 negative autistic children. Two HLA haplotypes, TNF-MIB-B*38-Cw12 and D6S265-A*38-MOG, were shown to be transmitted more often to subjects with ASD than their unaffected siblings after performing linkage analysis (Guerini et al., 2011). Guerini et al. (2011) also examined the individual components of these two haplotypes using a family based TDT analysis and showed that the D6S2239*105 allele and the MOGc*131 allele were more often transmitted to children with ASD from the fathers.

HLA Region	Gene/Allele	Reference
HLA Class I	A2	Ferrante et al., 2003 Torres et al., 2006
HLA Class II	Third hypervariable region of DRβ1	Warren et al., 1996
	DR4	Torres et al., 2002
	DR13,14	Torres et al., 2002
	DR4	Lee et al., 2006
	DR4	Johnson et al., 2009
HLA Class III	C4B null allele	Warren et al., 1991
	TNF-α microsatellites	Torres et al., 2001
	C4B null allele	Odell et al., 2005
	C4B null allele	Mostofa & Shehab, 2010
HLA Haplotypes	AH 44.1-SC30-DR4	Warren et al., 1992
	AH 44.1-SC30-DR4	Daniels et al., 1995
	TNF-238, TNF-308-MIB*332-HLA-B*38-HLA-Cw12	Guerini et al., 2011
	D6S265*218-HLA-A*23-MOGc*131-rs2857766	Guerini et al., 2011

Table 1. HLA alleles, genes and haplotypes associated with autism.

10. Autoimmunity

Autoimmunity occurs when a person's immune system treats its own DNA, proteins, cells or tissues as non-self. About 5% of the population has an autoimmune disease such as celiac disease, type 1 diabetes, systemic lupus erythematosus (SLE), Graves' disease, multiple sclerosis (MS) and rheumatoid arthritis (RA). Autoimmune diseases involve both innate and adaptive systems and often cause severe tissue damage. For example, in celiac disease there is destruction of villi in the small intestines, psoriasis and SLE both have skin lesions, RA has joint destruction and diabetes has destruction of the beta cells of the pancreas. Human C4 complement genes have strong associations with several autoimmune diseases such as SLE and MS (Pickering & Walport, 2000; Tegla et al., 2009) as well as central nervous system disorders like Alzheimer's (Kolev et al., 2009). HLA class I and class II alleles have been shown to be associated with various autoimmune diseases. For example, HLA-DR4 (DR4) has an association with RA with a relative risk of 4 and HLA-DR3 (DR3) is associated with diabetes with a relative risk of 5. If an individual has both DR3 and DR4 the relative risk for diabetes increases to 15. A large scale study found that the frequency of autoimmune disorders in the families with autistic children was found to be higher than in control subjects, especially mothers of autistic children (Comi et al., 1999). Another group showed that autoimmunity was increased significantly in families with ASD compared with those of healthy control subjects (Sweeten et al., 2003b), suggesting a link between the two disorders.

Some very interesting research has surfaced where physicians/ researchers treated Crohn's disease and ulcerative colitis using nematode parasites. This is consistent with the increase in the rate of autoimmune disease in more developed countries and in urban environments (Grant, 2011). Therefore, it has been hypothesized that the lack of parasite infection may lead to autoimmune diseases (Summers et al., 2003).

11. Childhood infections

A number of retrospective studies have investigated the histories of infections in autistic children. One study found a significant difference in rates of ear infections in children with autism compared to normal controls. No difference was found in the frequency of colds or fevers unrelated to ear infections in the two groups (Kostantareas & Homatidis, 1987). However, another survey did not find increased rates of ear infection, glue ear or hearing grommets in autistic children (Fombonne et al., 1997). Researchers in Japan compared the histories of 145 autistic children and 224 normal children for various diseases including pneumonia, seizures, high fever, measles, meningitis, significant diarrhea, hydrocephalus, rubella, varicella, mumps and otitis media in the first 18 months of life. No significant difference in rates of illness between groups was found (Tanoue et al., 1988). No difference was found for overall infections in the first 2 years of life when records of 403 children with autism and 2100 matched controls were analyzed in a California study (Rosen et al., 2007). A study of past illness utilizing a questionnaire found no significant difference between autistics and healthy controls in the occurrence of chickenpox, reactions to immunizations, meningitis, encephalitis, influenza or repeated infections (Comi et al., 1999).

The role of infection in the child as a possible autism trigger remains unclear. As stated earlier, the literature does support that congenital rubella and perhaps some herpes family viruses may contribute to the onset of autistic behavior in a small percentage of cases. Yet consistent evidence of a broader infectious role in this disorder remains elusive. It is yet to be seen if future research in this area will reveal a larger role for infection in autism and other neuropsychiatric disorders (Yolken & Torrey, 2008).

12. Maternal immune factors

Since there are many children who show signs of autistic behavior at a very early age and are thus thought to be born with autism, it is reasonable to assume that something acting on the developing fetus contributes to the disorder in these children. While a mother and fetus do not normally share or even mix blood during pregnancy, hormones, cytokines and antibodies do cross the placenta. An atypical maternal immune response during pregnancy could have a profound impact on the developing fetal brain, thus, contributing to autism. Maternal infection during pregnancy has been shown to increase the risk of several neurodevelopmental disorders in the fetus. Since associations with numerous infectious agents have been observed, it has been proposed that the maternal induction of pro-inflammatory cytokines may be the link between maternal infection and adverse effects on the fetal brain (Meyer et al., 2009).

In rats, maternal exposure to bacterial infection led to altered cytokine levels in the fetal environment by significantly increasing placental IL-1β, IL-6 and TNF-α which may have profound effects on the developing fetal brain (Urakubo et al., 2001). Patterson (2002) suggested that maternal immune response to human influenza virus respiratory infections during pregnancy can have both short-term and long-lasting deleterious effects on developing brain structure in the progeny through increased levels of circulating cytokines. Hsiao & Patterson (2011) used a mouse model to investigate how the maternal immune system plays a role in the development of autism by mimicking a virus infection and activating the maternal immune response. They observed increases in IL-6 mRNA (pro-inflammatory cytokine) as well as maternally-derived IL-6 protein in the placenta.

To test the hypothesis that ASD is caused by exposure of the fetal brain to maternal autoantibodies during pregnancy, Martin et al. (2008) studied rhesus monkeys gestationally exposed to IgG class antibodies from mothers of children with ASD. These exposed monkeys consistently demonstrated increased whole-body stereotypies across multiple testing paradigms and were also hyperactive compared to controls. Monkeys exposed to IgG antibodies purified from mothers of typically developing children did not display stereotypical or hyperactive behaviors. These findings support the potential for a maternal immune system etiology in patients with autism.

Another theory is that antibodies present in the mother's sera react directly against fetal proteins. Warren et al. (1990) found that mothers with an autistic child are more likely to have plasma antibodies that react against their child's lymphocytes. Zimmerman et al. (2007) tested the serum reactivity of mothers and their autistic children as well as control mothers and unaffected children to adult rat brain proteins. They found antibody reactivity against myelin basic protein and glial fibrillary acidic protein in the sera from mothers with autistic children. Both of these proteins are expressed in the brain of fetuses suggesting a mechanism by which antibodies that cross the placenta could affect fetal brain development.

Maternal autoimmune disease, immune dysfunction and other immune-related disorders have been reported to be associated with autism. Croen et al. (2005) showed that maternal psoriasis diagnosed around the time of pregnancy is significantly associated with a subsequent diagnosis of autism in the child. Additionally, they showed a 2-fold increase in risk for a child having ASD if the mother was diagnosed with asthma or allergies during pregnancy. An association between a family history of type 1 diabetes and infantile autism as well as a significant association between maternal histories of either RA or celiac disease and ASDs was observed by Atladóttir et al. (2009) who stated that these associations between familial autoimmunity and ASDs/infantile autism are probably attributable to a combination of a common genetic background and a possible prenatal antibody exposure or alteration in fetal environment during pregnancy.

Comi et al. (1999) also found that during pregnancy 43% of the mothers of autistic children, compared to 26% of the control mothers had an influenza-like illness, upper respiratory tract infection, urinary tract infection or vaginal infection. An investigation of children born in Denmark from 1980 through 2005 revealed no association between maternal infection over the entire pregnancy and autism in the child; however, admission into the hospital due to maternal viral infection in the first trimester and maternal bacterial infection in the second trimester were found to be associated with diagnosis of ASD in the offspring (Atladóttir et al., 2010). Another study found a trend toward increased infections during pregnancy in mothers bearing autistic children (Gillberg & Gillberg, 1983), and two studies found a significant increase in viral infections including rubella (Harper & Williams, 1974; Wilkerson et al., 2002), yet at least seven other studies have found no such correlations (Juul-Dam et al., 2001; Lobascher et al., 1970).

13. Vaccines

Over the past decade an energetic debate has brood around the possible role of the measles-mumps-rubella (MMR) vaccine in the development of autism. Although this debate has surfaced in headlines and news reports, the body of scientific data overwhelmingly fails to show any association between autism and the MMR vaccine. However, many parents of children with autism report that their child was normal until receiving vaccinations and they therefore felt there was something to this story: Is there a government or vaccine company cover-up? The initial paper proposing a correlation between the MMR vaccine and autism onset (Wakefield et al., 1998) has been retracted amid reports of unethical research (Full retraction 2010). At least 15 epidemiological studies from around the world have found no connection between autism and vaccines (Honda et al., 2005; Madsen et al., 2002; Mäkelä et al., 2002). These epidemiological findings have been substantiated by biological studies finding no evidence of persistent measles virus infection in autistic individuals (Afzal et al., 2006; Hornig et al., 2008). Interestingly, Pangborn (2002) reported that autism onset-at-18 months of age has increased more than 10 times the 1980 levels while the onset-at-birth cases has increased 3-4 times.

One common response to infection in the animal kingdom is the induction of fever. A vaccination is basically an artificial infection and it is common to give fever suppressors such as acetaminophen after a vaccination to suppress the fever and discomfort. Torres (2003) proposed that perhaps the acetaminophen given to suppress fever was disrupting a normal immune mechanism that in certain subjects could lead to autism. It is known that acetaminophen usage has increased greatly in the last three decades. In 1980 the CDC reported that aspirin was associated with Reye's syndrome, a rare but often fatal disease in children. This resulted in a sharp decrease in the purchase of aspirin between 1980 and 1985 and a reciprocal increase in the purchase of acetaminophen-containing drugs (Arrowsmith et al., 1987). Schultz et al. (2008) found that children given acetaminophen for reactions to the MMR vaccines were more likely to become autistic than children given ibuprofen. The use of acetaminophen by pregnant mothers has also increased over the last few decades.

14. Future directions

Despite several decades of research using cutting-edge technologies, the etiology of autism is unknown. This is not to say that progress has not been made. However, we will use the term used by most researchers that "more research is needed". We anticipate additional advances in autism genetics and perhaps more importantly research that ties together genetics and immune dysfunction in autism. The phenotypic heterogeneity among children with autism who participate as subjects in research studies has added to the inconsistent results. There is increasing interest in using phenotypic subgroups of psychiatric disorders in the detection of susceptibility genes (Tsuang, 2001). Szatmari et al. (2007) discusses ways to increase sample size and identify genetically informative phenotypes which will segregate with susceptibility loci and ultimately lead to a causative gene.

Recent advances in genetic analysis methods are positively impacting autism research. Researchers in the genetics field are very excited about the findings of "rare alleles" in the neurexin superfamily of genes so one can expect to see a great amount of research in this area. As stated above, the neurexin genes are incredibly complex and it will take years to understand how these genes are involved in the etiology of autism. However, one must remember that the "common variant common disease" model has not held up after 15 years of scrutiny leaving the genetic area ripe for discovery.

Whole genome sequencing will result in the identification of mutations or deletions/ insertions in many additional genes involved in neurologic development and function as causes of autism. The impact of single DNA molecule sequencing by Third-Generation instruments should have a significant impact in genetics as single molecule sequences result in haplotypes. Current methods cannot determine haplotypes with any certainty. This is reminiscent of gold companies extracting considerable amounts of gold from the tailing of old mining operations. Perhaps current genetic data that identifies regions of interest will prove golden for single molecule haplotype sequencing.

It is entirely possible that unknown genes in the HLA region that are in linkage disequilibrium with classical genes including C4 are involved in the etiology of autism. For example, there are several AHs that have been associated with autism by Warren et al. (1992). The AH 44.1 (B44, C4B null allele, DR4) has the strongest association. The AH 8.1 (A1, C7, B8, C4A null allele, DR3, DQ2) also referred to as COX has been entirely sequenced providing information on all of the genes in this haplotype. The AH 44.1 has the DR4 allele that has been associated with RA and the DR3 DQ2 portion of the AH 8.1 is associated with autoantibodies to tissue transglutiminase-2 as is seen in celiac disease (Rosenspire et al., 2011). The C4 null alleles in these two ancestral haplotypes associate with several autoimmune diseases like SLE. Although the current methods work very well for typing classical HLA alleles, newer microarray beads such those from Illumina will find autism associations when interrogating the 100 non-classical HLA genes.

Another area of developing interest in autism is epigenetics (reviewed by Grafodatskaya et al., 2010). Epigenetics is the field of research focused on the interaction between the environment and genetics. Epigenetic effects cause alterations in gene expression, resulting in phenotypic changes, without changes in the primary DNA sequence. Epigenetic mechanisms include DNA methylation, histone modifications, nucleosome repositioning, higher-order chromatin remodeling, non-coding RNAs, and RNA and DNA editing. One known example of an epigenetic cause of ASD is the alteration in DNA methylation caused by a mutation in the methyl CpG binding protein 2 (Rett syndrome). Most areas of epigenetic research in autism are largely unexplored. In addition, the expression of genes in the many cells in the brain is an incredibly complex process and it will take years to understand how these genes are involved in the etiology of autism.

Evidence continues to mount implicating an involvement of immune molecules in the etiology of autism. The evidence for pathogen involvement in autism does not appear to be strong as all children get infections and children with autism did not appear out of the norm. On the other hand, immune differences continue to mount and it is our belief that discoveries will be made that tie together seemingly unrelated immune findings. Autoimmune associations, differences in NK cells, cytokine levels and maternal influences appear to be the most fruitful areas for research. One legitimate question raised about autoimmunity in autism is, "Why is there no tissue damage as seen in typical autoimmune diseases?". Autism does not appear to be an autoimmune disease in a traditional sense, but it has similar immune associations. Some differences are that most autoimmune diseases appear late in life whereas autism appears very early and autoimmune diseases generally affect more females than males where autism disproportionately affects more males.

Another interesting research area involves NK cells in autism. It is fairly well established that NK cells from subjects with autism have decreased killing. Perhaps more importantly, resting NK cells from autistic subjects have higher cytolytic activity and produce higher levels of interferon-γ. Higher baseline levels of interferon-γ have been measured in CSF and the brain in subjects with autism. Vargas et al. (2005) noted an increase of over 200-fold in interferon-γ in CSF in subjects with autism. At this time, these levels have unknown consequences, however, cytokines have powerful effects on cells including those in the brain. There are two major subsets of NK cells (CD56^{dim} and CD^{bright}) as determined by flow cytometry and they have somewhat different functions. NK cells CD56^{dim} have more cytolytic activity but produce less interferon-γ than CD56^{bright}. How and which type of NK cell enters the brain could be significant. Also, since the NK cell is the major immune cell at maternal-fetal interfaces during pregnancy, the characterization of NK cells from mothers who have borne children with autism may be important. The NK cell is sentinel in pregnancy and environmental factors can change their phenotype by epigenetic pathways (Karimi & Arck, 2010). Therefore, the maternal/fetal NK cell population could be of major importance in autism.

15. References

Afzal, M.A., Ozoemena, L.C., O'Hare, A., Kidger, K.A., Bentley, M.L., & Minor, P.D. (2006). Absence of detectable measles virus genome sequence in blood of autistic children who have had their MMR vaccination during the routine childhood immunization schedule of UK. *Journal of Medical Virology*, Vol.78, No.5, (May 2006), pp. 623-630, ISSN 0146-6615

Alarcón, M., Abrahams, B.S., Stone, J.L., Duvall, J.A., Perederly, J.V., Bomar, J.M., et al. (2008). Linkage association, and gene-expression analyses identify CNTNAP2 as an autism susceptibility gene. *American Journal of Human Genetics*, Vol.82, No.1, (January 2008), pp. 150-159, ISSN 0002-9297

Atladóttir, H.O., Pedersen, M.G., Thorsen, P., Mortensen, P.B., Deleuran, B., Eaton, W.W., et al. (2009). Association of family history of autoimmune diseases and autism spectrum disorders. *Pediatrics*, Vol.124, No.2, (August 2009), pp. 687-694, ISSN 0031-4005

Atladóttir, H.O., Thorsen, P., Østergaard, L., Schendel, D.E., Lemcke, S., Abdallah, M., et al. (2010). Maternal infection requiring hospitalization during pregnancy and autism spectrum disorders. *Journal of Autism and Developmental Disorders*, Vol.40, No.12, (December 2010), pp. 1423-1430, ISSN 0162-3257

Altevogt, B.M., Hanson, S.L., & Leshner, A.I. (2008). Autism and the environment: challenges and opportunities for research. *Pediatrics*, Vol.121, No.6, (June 2008), pp. 1225-1229, ISSN 0031-4005

Anney, R., Klei, L., Pinto, D., Regan, R., Conroy, J., Magalhaes, T.R. et al. (2010). A genome-wide scan for common alleles affecting risk for autism. *Human Molecular Genetics*, Vol.19, No.20, (October 2010), pp. 4072-4082, ISSN 0964-6906

Arking, D.E., Cutler, D.J., Brune, C.W., Teslovich, T.M., West, K., Ikeda, M., et al. (2008). A common genetic variant in the neurexin superfamily member CNTNAP 2 increases familial risk of autism. *American Journal of Human Genetics*, Vol.82, No.1, (January 2008), pp. 160-164, ISSN 0002-9297

Arrowsmith, J.B., Kennedy, D.L., Kuritsky, J.N., & Faich, G.A. (1987). National patterns of aspirin use and Reye syndrome reporting, United States, 1980 to 1985. *Pediatrics*, Vol.79, No.6, (June 1987), pp. 858-863, ISSN 0031-4005

Ashwood, P., Krakowiak, P., Hertz-Picciotto, I., Hansen, R., Pessah, I., & Van de Water, J. (2011). Elevated plasma cytokines in autism spectrum disorders provide evidence of immune dysfunction and are associated with impaired behavioral outcome. *Brain, Behavior, and Immunity*, Vol.25, No.1, (January 2011), pp. 40-45, ISSN 0889-1591

Bailey, A., Le Couteur, A., Gottesman, I., Bolton, P., Simonoff, E., Yuzda, E., et al. (1995). Autism as a strongly genetic disorder: evidence from a British twin study. *Psychological Medicine,* Vol.25, No.1, (January 1995), pp. 63-77, ISSN 0033-2917

Bailey, A.R., Giunta, B.N., Obregon, D., Nikolic, W.V., Tian, J., Sanberg, C.D., et al. (2008). Peripheral biomarkers in autism: secreted amyloid precursor protein-alpha as a probable key player in early diagnosis. *International journal of clinical and experimental medicine,* Vol.1, No.4, (October 2008), pp. 338-344, ISSN 1940-5901

Bakkaloglu, B., O'Roak, B.J., Louvi, A., Gupta, A.R., Abelson, J.F., Morgan, T.M., et al. (2008). Molecular cytogenetic analysis and resequencing of contactin associated protein-like 2 in autism spectrum disorders. *American Journal of Human Genetics,* Vol.82, No.1, (January 2008), pp. 165-173, ISSN 0002-9297

Bashina, V.M., Kozlova, I.A., Kliushnik, T.P., Simashkova, N.V., Danilovskaia, E.V., Gorbachevskaia, N.L., et al. (1997). An elevation in the level of autoantibodies to nerve-growth factor in the blood serum of schizophrenic children. *Zhurnal nevrologii i psikhiatrii imeni S.S. Korsakova,* Vol.97, No.1, (1997), pp. 47-51, ISSN 1997-7298

Bolton, P., Macdonald, H., Pickles, A., Rios, P., Goode, S., Crowson, M., et al. (1994). A case-control family history study of autism. *Journal of Child Psychology and Psychiatry,* Vol.35, No.5, (July 1994), pp. 877-900, ISSN 0021-9630

Boulanger, L.M. & Shatz, C.J. (2004). Immune signaling in neural development, synaptic plasticity and disease. *Nature reviews. Neuroscience,* Vol.5, No.7, (July 2004), pp. 521-531, ISSN 1471-003X

Bucan, M., Abrahams, B.S., Wang, K., Glessner, J.T., Herman, E.I., Sonnenblick, L.I., et al. (2009). Genome-wide analyses of exonic copy number variants in a family-based study point to novel autism susceptibility genes. *PloS Genetics,* Vol.5, No.6, (June 2009), pp. e1000536, ISSN 1553-7390

Centers for Disease Control and Prevention (CDC). (2009). Prevalence of autism spectrum disorders-Autism and Developmental Disabilities Monitoring Network, United States, 2006. *Morbidity and mortality weekly report. Surveillance summaries / CDC,* Vol.58, No.10, (December 2009), pp. 1-20, ISSN 1546-0738

Chess, S. (1971). Autism in children with congenital rubella. *Journal of Autism and Childhood Schizophrenia,* Vol.1, No.1, (January-March 1971), pp. 33-47, ISSN 0021-9185

Ching, M.S., Shen, Y., Tan, W.H., Jeste, S.S., Morrow, E.M., Chen, X., et al. (2010). Deletions of NRXN1 (neurexin-1) predispose to a wide spectrum of developmental disorders. *American Journal of Medical Genetics. Part B, Neuropsychiatric Genetics,* Vol.153B, No.4, (June 2010), pp. 937-947, ISSN 1552-4841

Comi, A.M., Zimmerman, A.W., Frye, V.H., Law, P.A., & Peeden, J.N. (1999). Familial clustering of autoimmune disorders and evaluation of medical risk factors in autism. *Journal of Child Neurology,* Vol.14, No.6, (June 1999), pp. 388-394, ISSN 0883-0738

Connolly, A.M., Chez, M.G., Pestronk, A., Arnold, S.T., Mehta, S., & Deuel, R.K. (1999). Serum autoantibodies to brain in Landau-Kleffner variant, autism, and other neurologic disorders. *The Journal of Pediatrics,* Vol.134, No.5, (May 1999), pp. 607-13, ISSN 0022-3476

Craddock, N., Hurles, M.E., Cardin, N., Pearson, R.D., Plagnol, V., Robson, S., et al. (2010). Genome-wide association study of CNV's in 16,000 cases of eight common diseases and 3,000 shared controls. *Nature*, Vol.464, No.7289, (April 2010), pp. 713-720, ISSN 0028-0836

Croen, L.A., Grether, J.K., Yoshida, C.K., Odouli, R., & Van de Water, J. (2005). Maternal autoimmune diseases, asthma and allergies, and childhood autism spectrum disorders: a case-control study. *Archives of Pediatric & Adolescent Medicine*, Vol.159, No.2, (February 2005), pp. 151-157, ISSN 1072-4710

Croonenberghs, J., Wauters, A., Devreese, K., Verkerk, R., Scharpe, S., Bosmans, E., et al. (2002). Increased serum albumin, gamma globulin, immunoglobulin IgG, and IgG2 and IgG4 in autism. *Psychological Medicine*, Vol.32, No.8, (November 2002), pp. 1457-1463, ISSN 0033-2917

Daniels, W.W., Warren, R.P., Odell, J.D., Maciulis, A., Burger, R.A., Warren, W.L., et al. (1995). Increased frequency of the extended or ancestral haplotype B44-SC30-DR4 in autism. *Neuropsychobiology*, Vol.32, No.3, (1995), pp. 120-123, ISSN 0302-282X

Dawkins, R.L., Christiansen, F.T., Kay, P.H., Garlepp, M., McCluskey, J., Hollingsworth, P.N., et al. (1983). Disease associations with complotypes, supratypes and haplotypes. *Immunological Reviews*, Vol.70, No.1, (February 1983), pp. 5-22, ISSN 0105-2896

De Almeida, D.E., Ling, S., Pi, X., Hartman-Scruggs, A.M., Pumpens, P., & Holoshitz, J. (2010). Immune dysregulation by the rheumatoid arthritis shared epitope. *Journal of Immunology*, Vol.185, No.3, (August 2010), pp. 1927-1934, ISSN 0022-1767

Denney, D.R., Frei, B.W., & Gaffney, G.R. (1996). Lymphocyte subsets and interleukin-2 receptors in autistic children. *Journal of Autism and Developmental Disorders*, Vol.26, No.1, (February 1996), pp. 87-97, ISSN 0162-3257

Enstrom, A.M., Lit, L., Onore, C.E., Gregg, J.P., Hansen, R.L., Pessah, I.N., et al. (2009). Altered gene expression and function of peripheral blood natural killer cells in children with autism. *Brain, Behavior, and Immunity*, Vol.23, No.1, (January 2009), pp. 24-33, ISSN 0889-1591

Enstrom, A.M., Onore, C.E., Van de Water, J.A., & Ashwood, P. (2010). Differential monocyte responses to TLR ligands in children with autism spectrum disorders. *Brain, Behavior, and Immunity*, Vol.24, No.1, (January 2010), pp. 64-71, ISSN 0889-1591

Ferrante, P., Saresella, M., Guerini, F.R., Marzorati, M., Musetti, M.C., & Cazzullo, A.G. (2003). Significant association of HLA A2-DR11 with CD4 naïve decrease in autistic children. *Biomedicine and Pharmacotherapy*, Vol.57, No.8, (October 2003), pp. 372-374, ISSN 0753-3322

Folstein, S. & Rutter, M. (1977). Infantile autism: a genetic study of 21 twin pairs. *Journal of child psychology and psychiatry*, Vol.18, No.4, (September 1877), pp. 297-321, ISSN 0021-9630

Fombonne, E., Du Masaubrun, C., Cans, C., & Grandjean, H. (1997). Autism and associated medical disorders in a French epidemiological survey. *Journal of the American Academy of Child and Adolescent Psychiatry*, Vol.36, No.11, (November 1997), pp. 1561-1569, ISSN 0890-8567

Freitag, C.M. (2007). The genetics of autistic disorders and its clinical relevance: a review of the literature. *Molecular Psychiatry*, Vol.12, No.1, (January 2007), pp. 2-22, ISSN 1359-4184

Gillberg, C. & Gillberg, I.C. (1983). Infantile autism: A total population study of reduced optimality in the pre-, peri-, and neonatal period. *Journal of Autism and Developmental Disorders*, Vol.13, No.2, (June 1983), pp. 153-166, ISSN 0162-3257

Glessner, J.T., Wang, K., Cai, G., Korvatska, O., Kim, C.E., Wood, S., et al. (2009). Autism genome-wide copy number variation reveals ubiquitin and neuronal genes. *Nature*, Vol.459, No.7246, (May 2009), pp. 459-573, ISSN 0028-0836

Goines, P., Haapanen, L., Boyce, R., Duncanson, P., Braunschweig, D., Delwiche, L., et al., (2011). Autoantibodies to cerebellum in children with autism associated with behavior. *Brain, Behavior, and Immunity*, Vol. 25, No.3, (March 2001), pp. 514-523, ISSN 0889-1591

Goldstein, D.B. (2009). Common genetic variation and human traits. *The New England Journal of Medicine*, Vol.360, No.17, (April 2009), pp. 1696-1698, ISSN 0028-4793

Grafodatskaya, D., Chung, B., Szatmari, P., & Weksberg, R. (2010). Autism spectrum disorders and epigenetics. *Journal of the American Academy of Child and Adolescent Psychiatry*, Vol.49, No.8, (August 2010), pp. 794-809, ISSN 0890-8567

Grant, B. (2011). Opening a can of worms. *The Scientist*, Vol.25, No.2, (February 2011), pp. 42-47

Guerini, F.R., Bolognesi, E., Chiappedi, M., De Silvestri, A., Ghezzo, A., Zanette, M., et al. (2011). HLA polymorphisms in Italian children with autism spectrum disorders: Results of a family based linkage study. *Journal of Neuroimmunology*, Vol.230, No.1-2, (January 2011), pp. 135–142, ISSN 0165-5728

Guilmatre, A., Dubourg, C., Mosca, A.L., Legallic, S., Goldenberg, A., Drouin-Garrard, V., et al. (2009). Recurrent rearrangements in synaptic and neurodevelopment genes and shared biologic pathways in schizophrenia, autism and mental retardation. *Archives of General Psychiatry*, Vol.66, No.9, (September 2009), pp. 947-956, ISSN 0003-990X

Harper, J. & Williams, S. (1974). Early environmental stress and infantile autism. *The Medical Journal of Australia*, Vol.1, No.10, (March 1974), pp. 341-346, ISSN 0025-729X

Hess, D.T., Patterson, S.I., Smith, D.S., & Skene, J.H. (1993). Neuronal growth cone collapse and inhibition of protein fatty acylation by nitric oxide. *Nature*, Vol.366, No.6455, (December 1993), pp. 562-565, ISSN 0028-0836

Honda, H., Shimizu, Y., & Rutter, M. (2005). No effect of MMR withdrawal on the incidence of autism: a total population study. *Journal of Child Psychology and Psychiatry*, Vol.46, No.6, (June 2005), pp. 572-579, ISSN 0021-9630

Hornig, M., Briese, T., Buie, T., Bauman, M.L., Lauwers, G., Siemetzki, U., et al. (2008). Lack of association between measles virus vaccine and autism with enteropathy: a case-control study. *PLoS One*, Vol.3, No.9, (September 2008), pp. e3140, ISSN 1932-6203

Horton, R., Gibson, R., Coggill, P., Miretti, M., Allcock, R.J., Almeida, J., et al. (2008). Variation analysis and gene annotation of eight MHC haplotypes: The MHC Haplotype Project. *Immunogenetics*, Vol.60, No.1, (January 2008), pp. 1-18, ISSN 0093-7711

Hsiao, E.Y. & Patterson, P.H. (2011). Activation of the maternal immune system induces endocrine changes in the placenta via IL-6. *Brain, Behavior, and Immunity*, (in press 2011), doi:10.1016/j.bbi.2010.12.017, ISSN 0889-1591

Hwang, S.J. & Chen, Y.S. (2010). Congenital rubella syndrome with autistic disorder. *Journal of the Chinese Medical Association*, Vol.73, No.2, (February 2010), pp. 104-107, ISSN 1726-4901

Johnson, W.G., Buyske, S., Mars, A.E., Sreenath, N., Stenroos, E.S., Williams, T.A., et al. (2009). HLA-DR4 as a risk allele for autism acting in mothers of probands possibly during pregnancy. *Archives of Pediatrics & Adolescent Medicine*, Vol.163, No.6, (June 2009), pp. 542-546 ISSN 1072-4710

Juul-Dam, N., Townsend, J., & Courchesne, E. (2001). Prenatal, perinatal, and neonatal factors in autism, pervasive developmental dosorder-not otherwise specified, and the general population. *Pediatrics*, Vol.107, No.4, (April 2001), pp. e63, ISSN 0031-4005

Jyonouchi, H., Sun, S., & Le, H. (2001). Proinflammatory and regulatory cytokine production associated with innate and adaptive immune responses in children with autism spectrum disorders and developmental regression. *Journal of Neuroimmunology*, Vol.120, No.1-2, (November 2001), pp. 170-179, ISSN 0165-5728.

Kanner L. (1943) Autistic disturbances of affective contact. *The Nervous Child*, Vol.2, (1943), pp. 217-250.

Karimi, K., & Arck, P.C. (2010). Natural Killer cells: keepers of pregnancy in the turnstile of the environment. *Brain, Behavior, and Immunity*, Vol.24, No.3, (March 2010), pp. 339-47, ISSN 0889-1591

Kolev, M.V., Ruseva, M.M., Harris, C.L., Morgan, B.P., & Donev, R.M. (2009). Implication of complement system and its regulators in Alzheimer's disease. *Current Neuropharmacology*, Vol.7, No.1, (March 2009), pp. 1-8, ISSN 1570-159X

Konstantareas, M.M. & Homatidis, S. (1987). Ear infections in autistic and normal children. *Journal of Autism and Developmental Disorders*, Vol.17, No.4, (December 1987), pp. 585-594, ISSN

Lee, L.C., Zachary, A.A., Leffell, M.S., Newschaffer, C.J., Matteson, K.J., Tyler, J.D., et al. (2006). HLA-DR4 in families with autism. *Pediatric Neurology*, Vol.35, No.5, (November 2006), pp. 303-307, ISSN 0887-8994

Lizasoain, I., Weiner, C.P., Knowles, R.G., & Moncada, S. (1996). The ontogeny of cerebral and cerebellar nitric oxide synthase in the guinea pig and rat. *Pediatric Research*, Vol.39, No.5, (May 1996), pp. 779-783, ISSN 0031-3998

Lobascher, M.E., Kingerlee, P.E., & Gubbay, S.S. (1970). Childhood autism: An investigation of aetiological factors in twenty-five cases. *The British Journal of Psychiatry*, Vol.117, No.540, (November 1970), pp. 525-529, ISSN 0007-1250

Lonart, G., Wang, J., & Johnson, K.M. (1992). Nitric oxide induces neurotransmitter release from hippocampal slices. *European Journal of Pharmacology*, Vol.220, No.2-3, (September 1992), pp. 271-272, ISSN 0014-2999

Madsen, K.M., Hviid, A., Vestergaard, M., Schendel, D., Wohlfahrt, J., Thorsen, P., et al. (2002). A population-based study of measles, mumps, and rubella vaccination and autism. *The New England Journal of Medicine*, Vol.347, No.19, (November 2002), pp. 1477-1482, ISSN 0028-4793

Magi, A., Benelli, M., Yoon, S., Roviello, F., & Torricelli, F. (2011). Detecting common copy number variants in high-throughput sequencing data by using JointSLM algorithm. *Nucleic Acids Research*, Vol.39, No.10, (May 2011), pp. e65, ISSN 0305-1048

Mäkelä, A., Nuorti, J.P., & Peltola, H. (2002). Neurologic disorders after measles-mumps-rubella vaccination. *Pediatrics* Vol.110, No.5, (November 2002), pp. 957-963, ISSN 0031-4005

Manolio, T.A., Collins, F.S., Cox, N.J., Goldstein, D.B., Hindorff, L.A., Hunter, D.J., et al. (2009). Finding the missing heritability of complex diseases. *Nature*, Vol.461, No.7265, (October 2009), pp. 747-753, ISSN 0028-0836

Martin, L.A., Ashwood, P., Braunschweig, D., Cabanlit, M., Van de Water, J., & Amaral, D.G. (2008). Stereotypies and hyperactivity in rhesus monkeys exposed to IgG from mothers of children with autism. *Brain, Behavior, and Immunity*, Vol.22, No.6, (August 2008), pp. 806-816, ISSN 0889-1591

McClellan, J. & King, M.C. (2010). Genetic heterogeneity in human disease. *Cell*, Vol.141, No.2, (April 2010), pp. 210-217, ISSN 0092-8674

Meyer, U., Feldon, J., & Yee, B.K. (2009). A review of the fetal brain cytokine imbalance hypothesis of schizophrenia. *Schizophrenia Bulletin*, Vol.35, No.5, (September 2009), pp. 959-972, ISSN 0586-7614

Molloy, C.A., Morrow, A.L., Meinzen-Derr, J., Schleifer, K., Dienger, K., Manning-Courtney, P., et al. (2006). Elevated cytokine levels in children with autism spectrum disorder. *Journal of Neuroimmunology*, Vol.172, No.1-2, (March 2006), pp. 198-205, ISSN 0165-5728

Mostafa, G.A. & Shehab, A.A. (2010). The link of C4B null allele to autism and to a family history of autoimmunity in Egyptian autistic children. *Journal of Neuroimmunology*, Vol.223, No.1-2, (June 2010), pp. 115-119, ISSN 0165-5728

Odell, D., Maciulis, A., Cutler, A., Warren, L., McMahon, W.M., Coon, H., et al. (2005). Confirmation of the association of the C4B null allele in autism. *Human Immunology*, Vol.66, No.2, (February 2005), pp. 140-145, ISSN 0198-8859

Pangborn, J.B. (2002). Introduction to the diseases of autism and laboratory testing options, In: *Biomedical Assessment Options for Children with Autism and Related Problems*, Pangborn, J.B. & Baker, S.M. (Eds.), pp. 1-130, Autism Research Institute, San Diego, CA.

Patterson, P.H. (2002). Maternal infection: window on neuroimmune interactions in fetal brain development and mental illness. *Current Opinion in Neurobiology*, Vol.12, No.1, (February 2002), pp. 115-118, ISSN 0959-4388

Pickering, M.C. & Walport, M.J. (2000). Links between complement abnormalities and systemic lupus erythematosus. *Rheumatology (Oxford)*, Vol.39, No.2, (February 2000), pp. 133-141, ISSN 1462-0324

Rimland, B. (1964). *Infantile Autism*, Prentice-Hall, Inc., New Jersey

Risch, N., & Merikangas, K. (1996). The future of genetic studies of complex human diseases. *Science*, Vol.273, No.5281, (September 1996), pp. 1516-1517, ISSN 0193-4511

Ritvo, E.R., Jorde, L.B., Mason-Brothers, A., Freeman, B.J., Pingree, C., Jones, M.B., et al. (1989). The UCLA-University of Utah epidemiologic survey of autism: recurrence risk estimates and genetic counseling. *The American Journal of Psychiatry*, Vol.146, No.8, (August 1989), pp. 1032-1036, ISSN 0002-953X

Rosen, N.J., Yoshida, C.K., & Croen, L.A. (2007). Infection in the first 2 years of life and autism spectrum disorders. *Pediatrics*, Vol.119, No.1, (January 2007), pp. e61-69, ISSN 0031-4005

Rosenspire, A., Yoo, W., Menard, S., & Torres, A.R. (2011). Autism spectrum disorders are associated with an elevated autoantibody response to tissue transglutaminase-2. *Autism Research*, (April 2011), 4: n/a. doi: 10.1002/aur.194, ISSN 1939-3792

Rutter, M. (2000). Genetic studies of autism: from the 1970s into the millennium. *Journal of Abnormal Child Psychology*, Vol.28, No.1, (February 2000), pp. 3-14, ISSN 0091-0627

Schultz, S.T., Klonoff-Cohen, H.S., Wingard, D.L., Akshoomoff, N.A., Macera, C.A., & Ji, M. (2008). Acetaminophen (paracetamol) use, measles-mumpsrubella vaccination, and autistic disorder: the results of a parent survey. *Autism*, Vol.12, No.3, (May 2008), pp. 293-307, ISSN 1362-3613

Sebat, J., Lakshmi, B., Malhotra, D., Troge, J., Lese-Martin, C., Walsh, T., et al. (2007). Strong association of de novo copy number mutations with autism. *Science*, Vol.316, No.5823, (April 2007), pp. 445-449, ISSN 0193-4511

Singh, V.K., Warren, R., Averett, R., & Ghaziuddin, M. (1997). Circulating autoantibodies to neuronal and glial filament proteins in autism. *Pediatric Neurology*, Vol.17, No.1, (July 1997), pp. 88-90, ISSN 0887-8994

State, M.W. (2010). The genetics of child psychiatric disorder: focus on autism and tourette syndrome. *Neuron*, Vol.68, No.2, (October 2010), pp. 254-269, ISSN 0896-6273

Steffenburg, S., Gillberg, C., Hellgren, L., Anderson, L., Gillberg, I.C., Jakobsson, G., et al. (1989). A twin study of autism in Denmark, Finland, Iceland, Norway and Sweden. *Journal of Child Psychology and Psychiatry*, Vol.30, No.3, (May 1989), pp. 405-416, ISSN 0021-9630

Strauss, K.A., Puffenberger, E.G., Huentelman, M.J., Gottlieb, S., Dibrin, S.E., Parod, J.M., et al. (2006). Recessive symptomatic focal epilepsy and mutant contactin-associated-like 2. *The New England Journal of Medicine*, Vol.354, No.13, (March 2006), pp. 1370-1377, ISSN 0028-4793

Stubbs, E.G. & Magenis, R.E. (1980). HLA and autism. *Journal of Autism and Developmental Disorders*, Vol.10, No.1, (March 1980), pp. 15-19, ISSN 0162-3257

Stubbs, E.G. (1988). Does intrauterine cytomegalovirus plus autoantibodies contribute to autism?, In: *Aspects of Autism: Biological Research*, Wing, L. (Ed.), pp. 91-101, Alden, Oxford

Summers, R.W., Qadir, K., Urban, J.F., Jr., Thompson, R., & Weinstock, J.V. (2003). Trichuris suis seems to be save and possibly effective in the treatment of inflammatory bowel disease. *The American Journal of Gastroenterology*, Vol.98, No.9, (September 2003), pp. 2034-2041, ISSN 0002-9270

Sweeten, T.L., Posey, D.J., & McDougle, C.J. (2003a). High blood monocyte counts and neopterin levels in children with autistic disorder. *The American Journal of Psychiatry*, Vol.160, No.9, (September 2003), pp. 1691-1693, ISSN 0002-953X

Sweeten, T.L., Bowyer, S.L., Posey, D.J., Halberstadt, G.M., & McDougle, C.J. (2003b). Increased prevalence of familial autoimmunity in probands with pervasive developmental disorders. *Pediatrics*, Vol.112, No.5, (November 2003), pp. e420-e424, ISSN 0031-4005

Sweeten, T.L., Posey, D.J., Shankar, S., & McDougle, C.J. (2004). High nitric oxide production in autistic disorder: a possible role for interferon-gamma. *Biological Psychiatry*, Vol.55, No.4, (February 2004), pp. 434-437, ISSN 0006-3223

Szatmari, P., Maziade, M., Zwaigenbaum, L., Mérette, C., Roy, M.A., Joober, R., et al. (2007). Informative phenotypes for genetic studies of psychiatric disorders. *American Journal of Medical Genetics. Part B, Neuropsychiatric Genetics*, Vol.144B, No.5, (July 2007), pp. 581-588, ISSN 1552-4841

Tanoue, Y., Oda, S., Asano, F., & Kawashima, K. (1988). Epidemiology of infantile autism in Southern Ibaraki, Japan: differences in prevalence in birth cohorts. *Journal of Autism and Developmental Disorders*, Vol.18, No.2, (June 1988), pp. 155-166, ISSN 0162-3257

Tegla, C.A., Cudrici, C., Rus, V., Ito, T., Vlaicu, S., Singh, A., et al. (2009). Neuroprotective effects of the complement terminal pathway during demyelination: implications for oligodendrocyte survival. *Journal of Neuroimmunology*, Vol.213, No.1-2, (August 2009), pp. 3-11, ISSN 0165-5728

Todd, R.D. & Ciaranello, R.D. (1985). Demonstration of inter- and intraspecies differences in serotonin binding sites by antibodies from an autistic child. *Proceedings of the National Academy of Sciences of the United States of America*, Vol.82, No.2, (January 1985), pp. 612-616, ISSN 0027-8424

Toro, R., Konyukh, M., Delorma, R., Leblond, C., Chase, P., Fauchereau, F., et al. (2010). Key role for gene dosage and synaptic homeostasis in autism spectrum disorders. *Trends in Genetics*, Vol.26, No.8, (August 2010), pp. 363-372, ISSN 0168-9525

Torres, A.R., Maciulis, A., & Odell, D. (2001). The association of MHC genes with autism. *Frontiers in bioscience: a journal and virtual library*, Vol.6, (August 2001), pp. D936-943, ISSN 1093-9946

Torres, A.R., Maciulis, A., Stubbs, E.G., Cutler, A., & Odell, D. (2002). The transmission disequilibrium test suggests that HLA-DR4 and DR13 are linked to autism spectrum disorder. *Human Immunology*, Vol.63, No.4, (April 2002), pp. 311-316, ISSN 0198-8859

Torres, A.R. (2003). Is fever suppression involved in the etiology of autism and neurodevelopmental disorders? *BMC Pediatrics*, Vol.3, No.9, (September 2003), pp. 1-6, ISSN 1471-2431

Torres, A.R., Sweeten, T.L., Cutler, A., Bedke, B.J., Fillmore, M., Stubbs, E.G., et al. (2006). The association and linkage of the HLA-A2 class I allele with autism. *Human Immunology*, Vol.67, No.4-5, (April-May 2006), pp. 346-351, ISSN 0198-8859

Tsuang, M.T. (2001). Defining alternative phenotypes for genetic studies: what can we learn from studies of schizophrenia? *American Journal of Medical Genetics*, Vol.105, No.1, (January 2001), pp. 8-10, ISSN 0148-7299

Urakubo, A., Jarskog, L.F., Lieberman, J.A., & Gilmore, J.H. (2001). Prenatal exposure to maternal infection alters cytokine expression in the placenta, amniotic fluid and fetal brain. *Schizophrenia Research*, Vol.47, No.1, (January 2001), pp. 27-36, ISSN 0920-9964

Vargas, D.L., Nascimbene, C., Krishnan, C., Zimmerman, A.W., & Pardo, C.A. (2005). Neuroglial activation and neuroinflammation in the brain of patients with autism. *Annals of Neurology*, Vol.57, No.1, (January 2005), pp. 67-81, ISSN 0364-5134

Veenstra-VanderWeele, J., Christian, S.L., & Cook, E.H., Jr. (2004). Autism as a paradigmatic complex genetic disorder. *Annual Review of Genomics and Human Genetics*, Vol.5, (September 2004), pp. 379-405, ISSN 1527-8204

Vojdani, A., Mumper, E., Granpeesheh, D., Mielke, L., Traver, D., Bock, K., et al. (2008). Low natural killer cell cytotoxic activity in autism: the role of glutathione, IL-2 and IL-15. *Journal of Neuroimmunology*, Vol.205, No.1-2, (December 2008), pp. 148-154, ISSN 0165-5728

Volkmar, F.R., State, M., & Klin, A. (2009). Autism and autism spectrum disorders: diagnostic issues for the coming decade. *Journal of Child Psychology and Psychiatry*, Vol.50, No.1-2, (January 2009), pp. 108-115, ISSN 0021-9630

Wakefield, A.J., Murch, S.H., Anthony, A., Linnell, J., Casson, D.M., Malik, M., et al. (1998). Illeal-lymphoid-nodular hyperplasia, non-specific colitis, and pervasive developmental disorder in children. *Lancet*, Vol.351, No.9103, (February 1998), pp. 637-641, ISSN 0140-6736. Partial retraction: *Lancet*, Vol.363, No.9411, (March 2004), pp. 750. Full retraction: *Lancet*, Vol.375, No.9713, (February 2010), pp. 445.

Wang, K., Zhang, H., Ma, D., Bucan, M., Glessner, J.T., Abrahams, B.S., et al. (2009). Common genetic variants on 5p14.1 associate with autism spectrum disorder. *Nature*, Vol.459, No.7246, (May 2009), pp. 528-533, ISSN 0028-0836

Warren, R.P., Margaretten, N.C., Pace, N.C., & Foster, A. (1986). Immune abnormalities in patients with autism. *Journal of Autism and Developmental Disorders*, Vol.16, No.2, (June 1986), pp. 189-97, ISSN 0162-3257

Warren, R.P., Foster, A., & Margaretten, N.C. (1987). Reduced natural killer cell activity in autism. *Journal of the American Academy of Child and Adolescent Psychiatry*, Vol.26, No.3, (May 1987), pp. 333-335, ISSN 0890-8567

Warren, R.P., Cole, P., Odell, J.D., Pingree, C.B., Warren, W.L., White, E., et al. (1990). Detection of maternal antibodies in infantile autism. *Journal of the American Academy of Child and Adolescent Psychiatry*, Vol.29, No.6, (November 1990), pp. 873-877, ISSN 0890-8567

Warren, R.P., Singh, V.K., Cole, P., Odell, J.D., Pingree, C.B., Warren, W.L., et al. (1991). Increased frequency of the null allele at the complement C4B locus in autism. *Clinical and Experimental Immunology*, Vol.83, No.3, (March 1991), pp. 438-440, ISSN 0009-9104

Warren, R.P., Singh, V.K., Cole, P., Odell, J.D., Pingree, C.B., Warren, W.L., et al. (1992). Possible association of the extended MHC haplotype B44-SC30-DR4 with autism. *Immunogenetics*, Vol.36, No.4, (July 1992), pp. 203-207, ISSN 0093-7711

Warren, R.P., Odell, J.D., Warren, W.L., Burger, R.A., Maciulis, A., Daniels, W.W., et al. (1996). Strong association of the third hypervariable region of the HLA-DRβ1 with autism. *Journal of Neuroimmunology*, Vol.67, No.2, (July 1996), pp. 97-102, ISSN 0165-5728

Weiss, L.A., Arking, D.E. Gene Discovery Project of Johns Hopkins & the Autism Consortium, Daly, M.J., & Chakravarti, A. (2009). A genome-wide linkage and association scan reveals novel loci for autism. *Nature*, Vol.461, No.7265, (October 2009), pp. 802-808, ISSN 0028-0836

Wilkerson, D.S., Volpe, A.G., Raymond, S.D., & Titus, J.B. (2002). Perinatal complications as predictors of infantile autism. *The International Journal of Neuroscience*, Vol.112, No.9, (September 2002), pp. 1085-1098, ISSN 0020-7454

Yolken, R.H. & Torrey, E.F. (2008). Are some cases of psychosis caused by microbial agents? A review of the evidence. *Molecular Psychiatry*, Vol.13, No.5, (May 2008), pp. 470-479, ISSN 1359-4184

Zimmerman, A.W., Brashear, H.R., Frye, V.H., & Potter, N.T. (1993). Anticerebellar antibodies in autism. *Annals of Neurology*, Vol.34, No.3, (September 1993), pp. 498, ISSN 0364-5134

Zimmerman, A.W., Jyonouchi, H., Comi, A.M., Connors, S.L., Milstien, S., Varsou A, et al. (2005). Cerebrospinal fluid and serum markers of inflammation in autism. *Pediatric Neurology*, Vol.33, No.3, (September 2005), pp. 195-201, ISSN 0887-8994

Zimmerman, A.W., Connors, S.L., Matteson, K.J., Lee, L.C., Singer, H.S., Castaneda, J.A., et al. (2007). Maternal antibrain antibodies in autism. *Brain, Behavior, and Immunity*, Vol.21, No.3, (March 2007), pp. 351-357, ISSN 0889-1591

19

Impaired Oral Tolerance in ASD Children with Food Protein Induced Enterocolitis Syndrome (FPIES) – Altered Innate and Adaptive Immune Responses in ASD Children Recovered from FPIES in Comparison with non-ASD/FPIES and ASD/non-FPIES Children

Harumi Jyonouchi, Lee Geng, Pamella Kattouf, Deanna L. Streck and Gokce Toruner
University of Dentistry and Medicine of New Jersey, New Jersey Medical School, Newark, NJ, USA

1. Introduction

The gut mucosal immune system has to maintain an intricate immune homeostasis by maintaining tolerance to macronutrients, commensal flora, and other harmless molecules in the gut lumen while exerting effective immune defense against pathogenic microbes. It takes the first few years of life to establish this intricate immune homeostasis and until then the gut mucosal immune system is rather error-prone. This is reflected in the fact that young infants and children often suffer from temporary intolerance to common food proteins (FP), e.g. food allergy (FA). During this period, most FA manifests as delayed type FA, mediated by cellular immune responses - often called FP induced enterocolitis syndrome (FPIES) (Jyonouchi, 2008; Nowak-Wegrzyn and Muraro, 2009; Sicherer et al., 1998). Although FPIES seldom cause fatal anaphylaxis, as opposed to IgE-mediated FA, this condition is often under-diagnosed and under-treated. This is, in part, due to the delayed onset of symptoms that renders clinical connection more difficult, as well as, the lack of commercially available diagnostic measures. This is in contrast to IgE mediated FA in which a causal relationship is apparent in many cases and reactivity to food allergens is easily detected with prick skin test (PST) and/or presence of FA specific IgE antibody. The delay in the onset of symptoms of FPIES makes it especially difficult to diagnose FPIES clinically in infants and young children, as well as those with limited expressive language including children with autism spectrum disorders (ASD). In general, FPIES has an excellent prognosis provided there is timely implementation of avoidance measures against offending food (Jyonouchi, 2008; Nowak-Wegrzyn and Muraro, 2009). However, delayed diagnosis/treatment of FPIES can lead to failure to thrive (FTT), protein losing enteropathy, and possibly other irreversible complications.

GI symptoms are frequently observed in ASD (Buie et al., 2010). A role of FPIES has been suspected to play a role in GI symptoms observed in ASD children, since many parents of ASD children with GI symptoms report favorable responses to the casein-free, gluten-free (cf/gf) diet. We have reported previously that FPIES accounts for the GI symptoms experienced in many young ASD children (Jyonouchi et al., 2005). We have also observed that resolution of GI symptoms in ASD/FPIES children following implementation of avoidance of offending food was associated with improvement of certain behavioral symptoms, (Jyonouchi H, 2007).

However, in our observation, even after recovering from FPIES, some ASD children continue to suffer from recurrent or persistent GI symptoms. We also observed that flare ups of GI symptoms in these ASD/FPIES children are usually triggered by GI insults such as microbial gastroenteritis and prolonged oral antibiosis. Such 'treatment-resistant' GI symptoms after implementation of appropriate avoidance measures are less frequently observed in non-ASD/FPIES children. Since diagnosis of FPIES in ASD children are typically delayed partly due to their limited expressive language, it may be argued that this is simply reflecting a longer period of GI inflammation in ASD/FPIES children than in non-ASD/FPIES children. However, ASD/FPIES children with persistent GI symptoms also often exhibit other co-morbid conditions such as recurrent respiratory infection, adverse reaction to multiple medications, and seizure disorders, in our observation. Thus we hypothesized that 1) oral tolerance to FP and commensal flora is fragile and easily broken in ASD/FPIES children with persistent GI symptoms, and 2) this is associated with altered innate immune and adaptive immune responses in these 'treatment-resistant' ASD/FPIES children. We also hypothesized that evidence of impaired oral tolerance can be detected by studying peripheral blood mononuclear cells (PBMCs). These hypotheses are based on the facts that 1) aberrant immune responses to commensal flora is implicated with onset of chronic inflammatory conditions in the gut (Schirbel and Fiocchi, 2010), 2) plasticity of T-helper (Th) cell differentiation is not as definite as initially thought and aberrant innate immune responses and altered gut microbiota can hinder development of gut immune homeostasis (Lee and Mazmanian, 2010; Zhou et al., 2009), and 3) FP specific Th cells circulate in the peripheral blood (PB) (Karlsson et al., 2004).

This study focused on ASD/FPIES children and non-ASD/FPIES children who have already been treated for FPIES. In these subjects, we assessed innate immune responses, Th cell polarization, and T cell functions in comparison with normal control children. Our results indicate that a subset of ASD/FPIES children with persistent GI symptoms revealed a different pattern of innate and adaptive immune responses as compared to both ASD/FPIES or non-ASD/FPIES children.

2. Materials and methods

2.1 The study subjects

The study subjects were recruited following the study protocols approved by the Institutional Review Board, University of Medicine and Dentistry of New Jersey-New Jersey Medical School (UMDNJ-NJMS). Blood samples were collected after obtainment of signed parental consent forms. Signed assent forms were also obtained, if applicable, in children older than 7 years of age.

A total of 45 ASD/FPIES children who have been treated for FPIES were included in this study in addition to control ASD/non-FPIES (N=24), non-ASD/FPIES (N=26), and typically growing, healthy control children without FPIES (N=43). Among 45 ASD/FPIES children, 16 children revealed persistent or recurrent GI symptoms with suboptimal responses to dietary intervention measures (avoidance of offending food). These children are categorized as ASD-immune subtype (ASD-IS), since their behavioral and GI symptoms flared up repeatedly following immune insults such as viral infection. The demographics of the study subjects are summarized in Table 1. Non-ASD/FPIES children recruited to the study were also already treated for FPIES with implementation of appropriate avoidance measures. ASD and FPIES children were recruited in the Pediatric Allergy/Immunology Clinic at our institution. Normal healthy control and ASD/non-FPIES children were recruited in the Subspecialty Clinic at UMDNJ-NJMS where subspecialties include allergy/immunology, cardiology, developmental pediatrics, endocrinology, genetics, gastroenterology, nephrology, pulmonology, and general pediatrics. In most cases, blood samples were obtained when they were medically indicated to have venipuncture for routine blood work or general health screening. At the time of sample obtainment, all the subjects were examined to ensure absence of active infection.

	Age (yr) median (range)	Sex (male: female)	Ethnicity
ASD-IS (N=16)	9.7 (5-15.9)	16 : 0	13 W, 1 AA, 1 mixed, 1 Asian
ASD/FPIES (N=29)	6.2 (3-17.3)	25 : 4	24 W, 1 AA, 1 mixed, 3 Asians
ASD/non-FPIES (N-24)	6.5 (3-15.9)	20 : 4	16 W, 1 AA, 2 mixed, 5 Asians
Non-ASD/FPIES (N=26)	4.2 (1.7-15.9)	17 : 9	22 W, 2 mixed, 2 Asians
Normal control (N=43)	8.5 (1-17.8)	29 : 14	33 W, 6 AA, 4 mixed

[1]Abbreviations used: AA (African Americans), W (Caucasians)

Table 1. Demographics of the pediatric study subjects.

2.1.1 ASD diagnosis

ASD diagnosis was made or ascertained by DSM-IV (Diagnostic and Statistical Manual of Mental Disorders IV) criteria, ADI-R (Autism Diagnostic Interview-Revised), and/or ADOS (Autism Diagnostic Observational Schedules). All the ASD children recruited to this study were those with established autism diagnosis from established autism diagnostic centers including ours at UMDNJ.

2.1.2 Diagnosis of atopic disorders

Allergic rhinitis (AR), allergic conjunctivitis (AC) were diagnosed with positive prick skin test reactivity and/or presence of allergen-specific IgE accompanied by clinical features consistent with AR and AC (Butrus and Portela, 2005; Nassef et al., 2006). Asthma diagnosis was based on NIH guideline criteria (National Heart, Lung, Blood Institute, 2007). Asthma without prick skin test reactivity and/or allergen-specific IgE antibody was categorized as non-atopic asthma (Nassef et al., 2006)

2.1.3 Diagnosis of FPIES

FPIES to common food proteins (FPs; cow's milk protein, wheat, and soy) was diagnosed using the following diagnostic criteria: 1) presence of objective GI symptoms (diarrhea, loose stool, and constipation) which resolved with avoidance of causative FPs (Sicherer and Sampson, 2006), 2) delayed (more than 6 h) onset of GI symptoms following exposure to offending FPs after resolution of GI symptoms, and 3) cellular immune reactivity to offending FPs defined as the production of more than 1 standard deviation (SD) + control mean value of TNF-α and/or IL-12 by PBMCs with stimuli of causative DPs (Jyonouchi et al., 2005).

Diagnoses of other GI conditions were ascertained by reviewing medical charts and previous laboratory findings. Persistent GI symptoms are defined as lack of complete resolution of GI symptoms following introduction of appropriate restricted diet (avoidance of offending food) with persistent unformed stools, loose stool, and constipation alternating with diarrhea or loose stool. It is of note that these ASD/FPIES children with persistent GI symptoms (e.g. ASD-IS children) often revealed beneficial effects from oral anti-fungal medications (fluconazole or nystatin) and antibiotics (usually metronidazole) with improvement of GI symptoms. However, the beneficial effects of anti-fungals and antibiotics are generally short-lived and repeated courses of these therapies are often required, along with the persistent use of probiotics to control GI symptoms.

2.2 Cultures of PBMCs

PBMCs were isolated by Ficoll-Hypaque density gradient centrifugation. Innate immune responses were assessed by incubating PBMCs (10^6 cells/ml) overnight with TLR4 agonist (LPS; 0.1 μg/ml, GIBCO-BRL, Gaithersburg, MD), TLR2/6 agonist (zymosan; 50 μg/ml, Sigma-Aldrich, St. Luis, Mo), TLR3 agonist (Poly I:C, 0.1 μg/ml, Sigma-Aldrich), and TLR7/8 agonist (CL097, water-soluble derivative of imidazoquinoline, 20 μM, InvivoGen, San Diego, CA) in RPMI 1640 with additives as previously described (Jyonouchi et al., 2001). Overnight incubation was adequate to induce the optimal responses in this setting. Levels of proinflammatory [tumor necrosis factor-α (TNF-α), interleukin (IL)-1β, IL-6, IL-12p40, and IL-23] and counter-regulatory [IL-10, transforming growth factor-ß (TGF-ß) and soluble TNF receptor II (sTNFRII)] cytokines in culture supernatant were then measured by an enzyme-linked immunosorbent assay (ELISA).

Cellular reactivity to T cell stimulants was assessed by incubating PBMCs (10^6 cells/ml) with T cell mitogens [Con A (2 μg/ml) and PHA (5 μg/ml)], recall antigens (Ags)[soy protein (100 μg/ml), ß-lactoglobulin (ßLG; 10 μg/ml) Sigma-Aldrich, candida Ag (5 μg/ml), dust mite (5 μl/ml) Greer, Lenoir, NC, tetanus toxoid (1:5000)], and IFN-γ inducing cytokines [IL-12p70 (0.2 ng/ml, BD Biosciences, San Diego, CA), IL-18 (1 ng/ml, BD Biosciences) for 4 days and measuring levels of IFN-γ, TNF-α, IL-5, IL-10, IL-12p40, and IL-17 in the culture supernatant (Jyonouchi et al., 2005). Initial titration studies showed that a four day incubation period resulted in the optimal production of these cytokines, in this setting.

Cytokine levels were measured by ELISA, using OptEIA™ Reagent Sets (BD Biosciences) for IFN-γ, IL-1ß, IL-5, IL-6, IL-10, IL-12p40, and TNF-α, and ELISA reagent set (R & D, Minneapolis, MN) for sTNFRII, IL-17 (IL-17A), and TGF-ß. IL-23 ELISA kit was purchased from eBiosciences, San Diego, CA. Intra- and inter-variations of cytokine levels were less than 5%.

2.3 Flow cytometry

For intracellular cytokine staining in $CD4^+$ T cells, the following fluorochrome-conjugated monoclonal antibodies were used: CD4-PerCp, IFN-γ-PE-Cy7, IL-17-PE, IL-4-FITC, IL-10-Pacific Blue (all from eBiosicences), and TGF-ß-APC (R & D, Minneapolis, MN). PBMCs were incubated at 37°C overnight (16 h) with medium alone, Staphylococcal enterotoxin B (5 μg/mL, Sigma-Aldrich), or candida Ag (5 μg/ml, Greer) in the presence of Brefeldin A (BFA; 5 μg/ml, Sigma-Aldrich), anti-CD28 (1 μg/ml, eBiosciences), and anti-CD49 (1 μg/ml, eBiosciences). The same culture medium used for the cytokine production assay was utilized. Then PBMCs were permeabilized (permeabilization buffer, BD Biosciences) and stained with the above described antibodies. All flow cytometry was conducted by using FACSVantage SE TM (BD Biosciences) and the data were analyzed with the CellQuest software (BD Biosciences) and FlowJo (TreeStar, Ashland, OR).

2.4 Transcription profiling

PB monocytes were purified using an immuno-affinity column following the company's instructions (MACS monocytes isolation kit, Miltenyi Biotec, Auburn, CA). Total RNA were extracted by the RNA easy kit (Quiagen , Valencia, CA). RNA labelling and hybridizations on Agilent Human 4x44K arrays (Agilent, Lexington, MA) were done using the Agilent One-Color Microarray-Based Gene Expression Analysis Ver 5.5 protocol (Agilent). All slides were scanned by an Agilent Scanner and normalized numerical data were obtained by Agilent Feature extraction software 9.5.

2.5 Statistical analysis

For comparison of test values with control values, a Wilcoxon signed rank test was used. For comparison of values of multiple groups, a Kruskall-Wallis test was used. A Chi square (χ^2) test was used to examine the difference in frequency. These tests were performed using R.2.10.1 (R-Development Core Team 2009). A p value of <0.05 was considered to be statistically significant. For the analysis of microarrays experiments, Gene Spring GX v11 software (Agilent) was used. After filtering for "present" calls in at least 20% of samples, fold change analysis were performed for group for comparisons on 26992 probes. Genes with at least two fold changes, as compared to controls, are determined to be either up-regulated or down-regulated. Using a specific module of GeneSpring software (Agilent), pathways enrichment analysis on those genes was performed to see if there is a statistically significant enrichment ($p<0.05$) for specific BioPax pathways.

3. Results

3.1 Clinical features

Prevalence of common childhood disorders in the study groups is shown in Table 2. The prevalence of allergic rhinoconjunctivitis and asthma in ASD-IS, ASD/FPIES, ASD/non-FPIES, as well as in non-ASD/FPIES children was found to be similar to what is reported in general population (Table 2) (Akinbami et al., 2011; Singh et al., 2010). However, ASD-IS children revealed a higher prevalence of recurrent infection [recurrent otitis media (ROM) and chronic rhinosinusitis (CRS)] than other study groups (Table 2). It is of note, that 3 subjects out of these 6 ASD-IS children with recurrent infection were diagnosed with specific polysaccharide antibody deficiency (SPAD). In contrast, non-ASD/FPIES children seldom revealed chronic infection. Around 10% of ASD/FPIES and ASD/non-FPIES children had history of ROM but they did not suffer from CRS and they are responsive to the first-line antibiotics such as amoxicillin. It remains to be seen whether apparent higher prevalence of ROM in the ASD children in our study is associated with under-diagnosis or under-treatment, secondary to their limited expressive language.

	AR+AC	Asthma	ROM/CRS	Seizure disorders
ASD-IS (N=16)	4 (25.0%)	3 (18.8%)[4]	6[2] (37.5%)	2 (12.5%)
ASD/FPIES (N=29)	4 (13.8%)	3 (10.3%)	4[3] (13.8%)	1 (3.4%)
ASD/non-FPIES (N=24)	5 (20.8%)	2/24 (8.3%)	2 (8.3%)	0
Non-ASD/FPIES (N=26)	6 (23.1%)	3/26 (11.5%)	1 (0.4%)	0
Normal control (N=43)	8 (18.6%)	5 (11.6%)	0	0

[1]Abbreviations used: AR+AC, allergic rhinoconjunctivitis, ROM, recurrent otitis media, CRS, chronic rhinosinusitis
[2]Among 6 subjects with recurrent ROM and CRS, 3 subjects were diagnosed with specific polysaccharide antibody deficiency (SPAD).
[3]One patient was diagnosed with SPAD.
[4]All 3 asthma patients are diagnosed with SPAD and asthma is considered as non-atopic asthma triggered by infection.

Table 2. Prevalence of co-morbid condition in the study subjects.

3.2 Innate immune responses

In response to a TLR4 agonist (LPS), ASD/FPIES PBMCs revealed lower TNF-α and IL-12 production than normal controls (Fig. 1). Non-ASD/FPIES PBMCs also revealed similar tendencies, however, the ASD/non-FPIES and ASD-IS cells did not differ from normal controls in the production of these cytokines (Fig. 1). IL-23 production with a TLR4 agonist

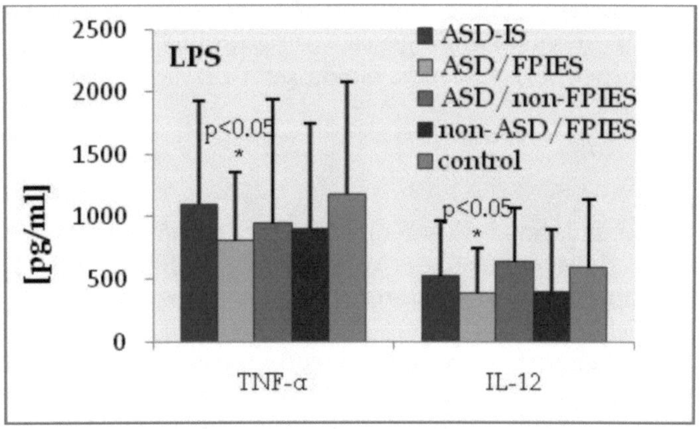

Fig. 1. Production of TNF-α and IL-12 by PBMCs (10^6 cells/ml) from the study groups following overnight incubation with a TLR4 agonist (LPS). *; lower than normal controls by Wilcoxon signed rank test.

was lower in both the ASD-IS and ASD/FPIES groups than normal controls (Fig. 2). We also observed lower production of IL-12 (with a TLR9 agonist) and IL-1ß (in the absence of stimuli and with a dectin 1 agonist) in the ASD-IS group. This was not observed in any other study groups (Figs. 2-3). IL-6 production without stimuli and in response to a TLR9 agonist were also the lowest in the ASD-IS group (Fig.3). ASD/FPIES but not non-ASD/FPIES PBMCs revealed a similar tendency. However, this was less evident than in the ASD-IS cells (Fig. 3). Non-ASD/FPIES PBMCs revealed higher TGF-ß production with a TLR2/6 agonist than normal controls (Fig. 3). Taken together, our results indicate that ASD-IS, ASD/FPIES, and non-ASD/FPIES children reveals different patterns of cytokine production in response to TLR agonists.

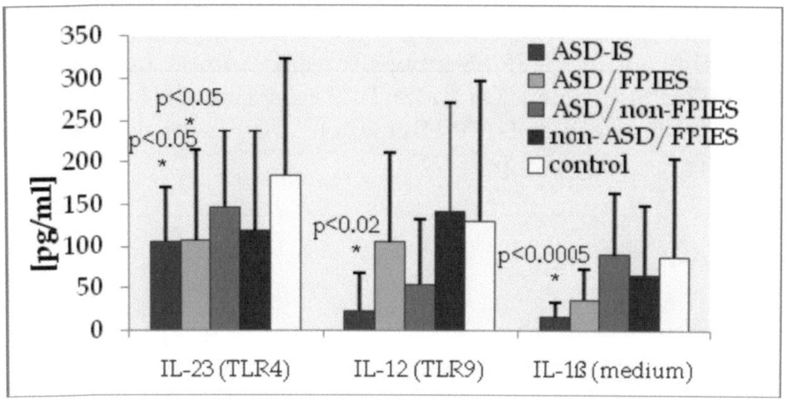

Fig. 2. Production of IL-23, IL-12, and IL-1β by PBMCs (10^6 cells/ml) in responses to TLR agonists as shown in the figure. *; significantly lower than normal controls by Wilcoxon signed rank test. IL-1β production was obtained when cells were cultured in the medium without a stimulus.

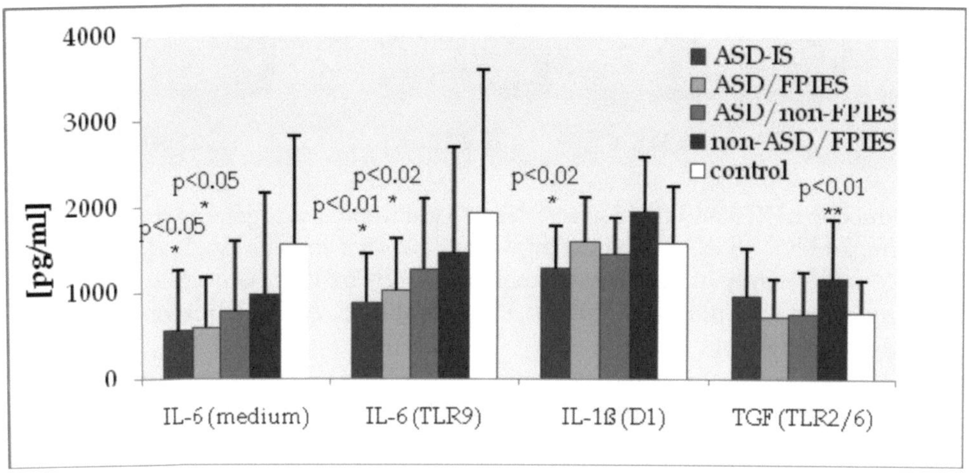

Fig. 3. Production of IL-6, IL-1β, and TGF-β by PBMCs (10^6 cells/ml) when incubated overnight with medium only (IL-6), TLR9 agonist (IL-6), Dectin-1 agonist (D1 - IL-1β), and TLR2/6 agonist (TGF-β). *; significantly lower as compared to normal controls by Wilcoxon signed rank test . **; significantly higher than normal controls by Wilcoxon signed rank test.

3.3 Adaptive immune responses

No significant differences were observed among the study groups in responses to stimuli of T cell mitogens or recall antigens via vaccination (tetanus toxoid) or respiratory tract (dust mite). However, when PBMC responses to gut luminal Ags were tested, we observed a higher IL-5 production (with candida Ag) in non-ASD/FPIES children, while IL-17 production with ß-LG and candida Ag were higher in the ASD-IS group as compared to the non-ASD/FPIES group (Fig. 4). Such increase in IL-17 production was not observed in ASD/FPIES children (Fig. 4). Moreover, IL-10 production was lower in the ASD-IS children without stimuli as well as in response to candida Ag (Fig. 5). We also assessed frequency of Th cell subsets following stimulation of PBMCs with a polyclonal T cell stimulant (SEB) overnight by measuring intracellular expression of Th-lineage specific cytokines in $CD4^+$ T cells. Both the ASD-IS and ASD/FPIES groups revealed a lower frequency of IFN-γ^+ Th1 cells than controls (Fig. 6). Frequency of IL-17^+ Th17 cells were also lower in the ASD/FPIES group but not in ASD-IS or non-ASD/FPIES children. Our results also indicated differences in T cell responses in the study groups.

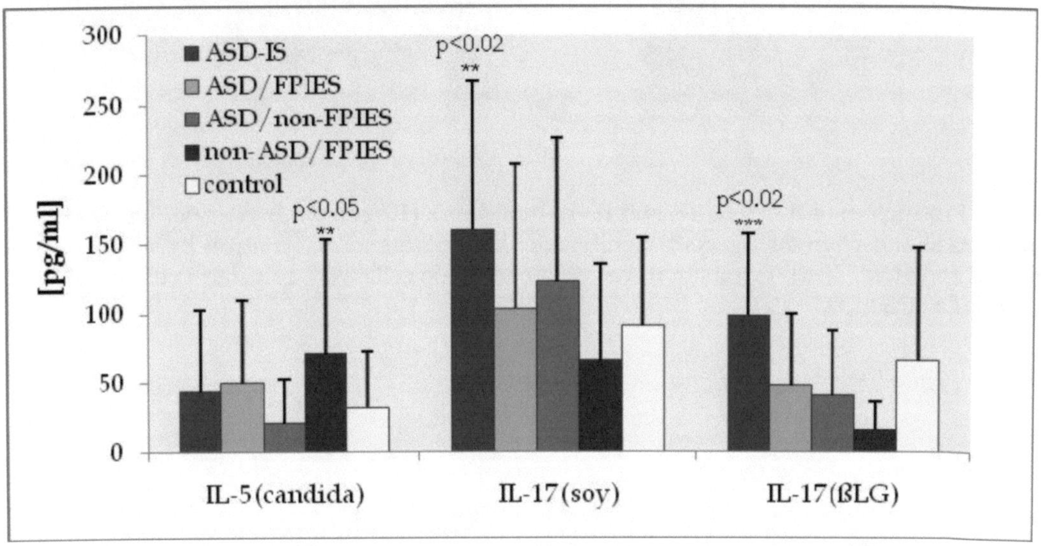

Fig. 4. Production of IL-5 and IL-17 in response to gut luminal antigens as shown in Figure. βLG; β-lactoglobulin. PBMCs (10^6 cells/ml) were incubated for 4 days with these luminal antigens. **; significantly higher than normal controls by by Wilcoxon signed rank test. ***; significantly higher than non-ASD/FPIES, ASD/non-FPIES, ASD/FPIES controls by Wilcoxon signed rank test.

3.4 Transcription profiling results

Transcript profiles of PB monocytes were tested in 16 ASD-IS, 14 ASD/FPIES, 16 ASD/non-FPIES, and 26 normal control children. The changes of transcript expression are summarized in Table 3. As compared to ASD/FPIES, ASD/non-FPIES, and normal control groups, ASD-IS PB monocytes revealed that a large numbers of genes are up- or down-regulated over 2-fold and the difference was most significant when compared to normal controls (Table 3). ASD/FPIES children also revealed changes in gene expression as compared to ASD/non-

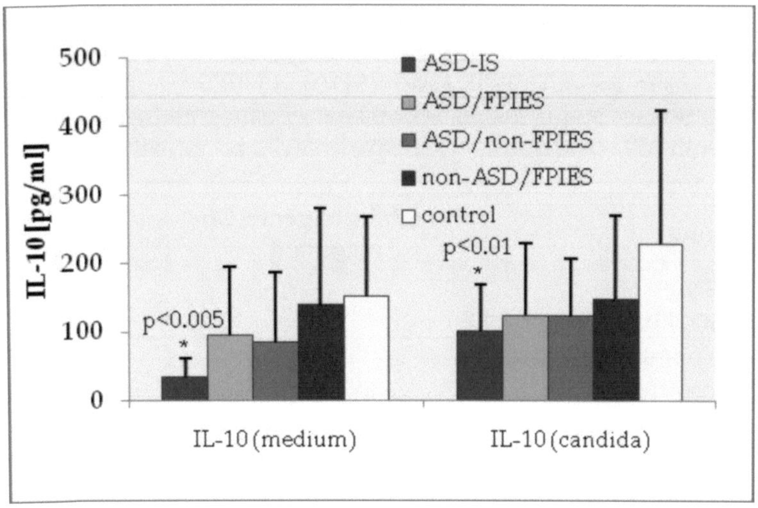

Fig. 5. IL-10 production when PBMCs (10^6 cells/ml) were cultured for 4 days with candida Ag or medium only in the study groups. *; significantly lower than normal controls by Wilcoxon signed rank test.

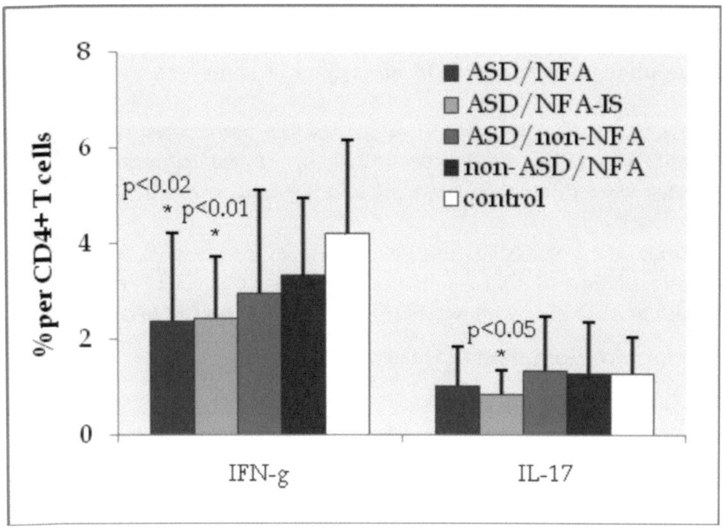

Fig. 6. Percent of IFN-γ^+ and IL-17$^+$ CD4$^+$ T-helper cells per total CD4$^+$ cells in PBMCs obtained from the study groups. PBMCs were incubated overnight with SEB and intracellular cytokine expression was assessed by flow cytometry as detailed in the Materials and Methods section. *; lower than normal controls by Wilcoxon signed rank test.

FPIES and normal controls. However, the numbers of genes up- or down-regulated were not as many as seen in ASD-IS children (Table 3). ASD/non-FPIES children also revealed changes in gene expression in a large number of genes as compared to controls (Table 3). Notable findings are the down-regulated expression of cytokines and chemokines in ASD-IS, as well as ASD/FPIES children, as compared to ASD/non-FPIES children. This finding is consistent with those from the bioassays. CCL7 expression was down regulated in ASD/non-FPIES children as compared to all the study groups. The pathway analysis did not reveal any significant enrichment of genes in known signaling pathways. GO analysis revealed significant differences in the ASD/FPIES and ASD/non-FPIES groups; changes were evident in inflammatory processes and cytokine/chemokine signaling pathways. In comparison with normal controls, both ASD-IS and ASD/FPIES patients did not reveal significant changes.

Group	Numbers of genes Up-regulated	Numbers of genes down-regulated
ASD-IS		
Vs. ASD/FPIES	231[1]	61
Vs. ASD/non-FPIES	784	148[2]
Vs. control	1341	617
ASD/FPIES		
Vs. ASD/non-FPIES	251	92[2]
Vs. control	223	191
ASD/non-FPIES		
Vs. control	723[3]	1152[3]

[1]CCR7 and CCL7 are up-regulated in ASD-IS monocytes as compared to ASD/FPIES monocytes (>2 fold)

[2]Transcripts of CCL3, IL-1ß, IL-23A, and IL-6 in ASD-IS monocytes were all down-regulated as compared to the ASD/non-FPIES monocytes (>2 fold). Gene expression of IL-23A and IL-6 in ASD/FPIES monocytes were down regulated 6.4 and 4.9 fold, respectively, than ASD/non-FPIESPB monocytes.

[3]As compared to normal controls, ASD/non-FPIES monocytes revealed up-regulation of IL-23A (5.4 fold) as well down-regulation of CCL7 (chemockine- 5.4 fold). CCL7 expression was also down-regulated as compared to ASD-IS as well as ASD/FPIES monocytes (>2 fold).

Table 3. Transcription profiling data in the study subjects.

4. Discussion

FPIES was first reported in 1940's and has been recognized as a relatively benign condition that shows good responses to avoidance measures. However, delayed onset of symptoms and lack of readily (commercially) available diagnostic measures hinders early diagnosis and early intervention. This is especially true for children with impaired expressive language, such as ASD children. For example, it has been our observation that behavioral changes associated with GI discomfort and pain are often attributed to be just 'being autistic', despite presence of objective GI symptoms. In previous years, we have attempted to sort out whether FPIES is prevalent in ASD children and if so, how FPIES affects their behavioral symptoms. Our previous results indicated a high prevalence of FPIES in ASD children, probably partly due to delayed diagnosis and treatment (Jyonouchi, 2010; Jyonouchi H, 2007). We also observed changes of behaviors after resolution of GI symptoms when behavioral changes were assessed using the Aberrant Behavior Check list (ABC) (Jyonouchi, 2010; Jyonouchi H, 2007).

Group	GO ID	GO Term	Corrected p value
ASD-IS vs. ASD/FPIES	3290	Oxygen transporter activity	0.02578
	3593	Hemoglobin complex	0.02578
	8215	Gas transport	0.01368
	8217	Oxygen transport	0.02578
ASD-IS vs. ASD/non-FPIES	4528	Response to stress	0.06276
	6431	Response to wounding	0.06276
	19583	Positive regulation of cell adhesion	0.02184
ASD-IS vs. controls			NS
ASD/FPIES vs. ASD/non-FPIES	801	G-protin-coupled receptor binding	0.06068
	936	Positive regulation of cytokine production	0.06644
	1402	Immune system process	0.03348
	1465	Chronic inflammatory response to antigenic stimulus	0.05304
	3128	Receptor binding	0.05304
	3145	Cytokine activity	3.75E-04
	3430	Extracellular space	0.03299
	4518	Chemotaxis	0.00644
	4528	Response to stress	2.43E-05
	4529	Defense response	2.45E-08
	4531	Inflammatory response	2.14E-08
	4532	Immune response	3.75E-04
	5082	Behavior	0.05386
	5098	Locomotry behavior	0.03956
	5114	Chemokine activity	0.00822
	6425	Response to external stimulus	1.25E-05
	6431	Response to wounding	2.15E-08
	7346	Regulation of vascular endotheial growth factor production	0.06512
	10575	Cytokine and chemokine mediated signaling pathway	0.03138
		Nitric oxide transport	
	12106	Positive regulation of synaptic plasticity	0.05304
	13736	Positive regulation of heterotype cell-cell adhesion	0.05304
	15920	Regulation of cytokine biosynthetic process	0.05304
	16941	Positive regulation of cytokine biosynthetic process	0.05304
	17002	Taxis	0.05613
	17201	Chemokine receptor binding	0.00644
	17250	Regulation of nitric oxide biosynthetic process	0.00918
	19242	Positive regulation of nitric oxide biosynthetic process	0.06068
	19243	Positive regulation of mitosis	0.03956
		Regulation of smooth muscle cell proliferation	
	19628	Positive regulation of smooth muscle cell proliferation	0.05304
	22266	Response to stimulus	0.03956
	22267	Positive regulation of nitrogen compound metabolic process	0.00822
	23421	Regulation of muticellular organismal process	0.07731
	23697	Positive regulation of muticellular organismal process	0.05304
	23761		0.05304
	23762		0.03348

Table 4. GO analysis results in the study subjects

However, among ASD children with FPIES, there exists a subset of children who are also vulnerable to recurrent infection and have history of adverse reactions to multiple medications and other substances. These children are sensitive to a variety of food proteins and often require the intake of extremely hydrolyzed hypoallergic formulas to get sufficient nutrition, in our experience. Parents of these children also reports fluctuating behavioral symptoms and cognitive skills following immune insults, such as viral syndrome. Our previous studies indicated that these children, with fluctuating behaviors, often exhibit distinct innate immune abnormalities (Jyonouchi et al., 2008). Further analysis of this population led to the finding that ASD children with these clinical phenotypes, as well as 'treatment-resistant' FPIES, do reveal more significant changes in innate immunity which can be detected using bioassays as well as transcript profiles in PB monocytes (manuscript in press). We have categorized this group of ASD children as ASD-immune subtype (ASD-IS), since they appear to reveal different immune abnormalities than ASD/FPIES as well as ASD/non-FPIES children. These children also seem to be vulnerable to dysbiosis, since their GI symptoms, as well as behavioral symptoms, tend to improve following anti-fungal treatment and/or antibiosis targeting pathogenic microbes in the gut. Unfortunately, as reported by others (Sandler et al., 2000), these effects are generally transient.

It remains unclear as to how such clinical phenotypes are associated with innate and adaptive immune responses and how their immune responses are different from those observed in ASD/FPIES and non-ASD/FPIES children. To address this question, we conducted detailed studies of innate and adaptive immune responses in ASD-IS children in comparison with ASD/FPIES, non-ASD/FPIES, ASD/non-FPIES, and normal controls. Since the presence of active GI inflammation is likely to affect the bioassay results, we conducted these assays after FPIES subjects were appropriately treated with avoidance measures and nutritional supplements if required. Co-morbid conditions summarized in Table 2 in our study groups revealed that ASD-IS children do seem to have a higher prevalence of COM/CRS than other study groups, although prevalence of atopic disorders does not appear to be altered significantly among the study groups, indicating that atopy or Th2 deviated responses are unlikely to be associated with their clinical characteristics.

When we assessed innate immune responses in the study groups, our results revealed lower TNF-α and IL-12 production with a TLR4 agonist (LPS) than normal controls in the ASD/FPIES group. In the colon where microbes reside at the highest concentration, responses to endotoxin such as LPS are suppressed – so-called LPS desensitization (Abreu et al., 2005; Michalek et al., 1982; Smith and Nagler-Anderson, 2005; Wannemuehler et al., 1982). Tendency for lower responses to LPS in ASD/FPIES children may be beneficial for maintaining oral tolerance after recovering from FPIES. Although it was not significant, the same tendency was observed in non-ASD/FPIES children.

In ASD-IS children, while their LPS responses are equivalent to those of normal controls, spontaneous production of IL-1ß and IL-12 production with a TLR9 agonist was markedly lower. IL-1ß production with a dectin 1 agonist (heat killed candida) was also lower in the ASD-IS children. Lower production of IL-6 (without stimuli and with TLR9 agonist) and IL-23 (with a TLR4 agonist) were observed in both ASD/FPIES and ASD-IS children but this was more evident in ASD-IS children. On the other hand, non-ASD/FPIES children were noted to have higher TGF-ß with a TLR2/6 agonist (zymosan). IL-1ß, IL-6, and IL-23 are all important for differentiation and maintenance of Th17 cells as a part of their pleiotropic biological actions (Kimura and Kishimoto, 2010; Zhou et al., 2009). While IL-12 has a key role in Th1 cell differentiation. Given these findings, it may be questioned whether ASD-IS, as well as ASD/FPIES children, have impaired development of Th17 cells. When we tested frequency of IL-17^{+} Th cells after stimulation of PBMCs with SEB overnight, frequency of

Th17 cells was lower in ASD/FPIES children than normal controls, but this was not evident in the ASD-IS children. Both ASD-IS and ASD/FPIES children revealed a little lower frequency of IFN-γ^+ Th1 cells. However, to our surprise, IL-17 production in response to food proteins was higher in the ASD-IS children. Comparative frequency of Th17 cells, but higher IL-17 production in the ASD-IS children indicate that committed Th17 cells may keep producing a large amount of IL-17. These findings also indicate that excessive IL-17 responses in ASD-IS children may be associated with dysregulated development and maintenance of Th17 cells. In addition, persistent IL-17 responses will not aid in and may hinder establishment of gut immune homeostasis consistent with clinical phenotype of 'treatment-resistant' FPIES in the ASD-IS children.

In contrast to the ASD-IS group, non-ASD/FPIES children recovering from FPIES revealed higher TGF-ß with a TLR2/6 agonist (zymosan). These children did not reveal an increase in frequency of Th17 cells or IL-17 production. Since TGF-ß regulates (down-regulates) various immune responses and also serves as a key differentiation factor for regulatory T (Treg) cells, especially in the absence of inflammatory cytokines (Burgler et al., 2009; Zhou et al., 2009), it may be wondered whether ASD-IS children and in some-degree, ASD/FPIES children have impaired development or function of regulatory T cells. In the GI mucosa, IL-10 and TGF-ß producing Treg cells are believed to have a major role in the gut immune homeostasis (Lee and Mazmanian, 2010). In that regard, non-ASD/FPIES children and perhaps ASD/FPIES children who did not reveal excessive IL-17 responses may be in the process of establishing gut mucosal immune homeostasis, while recovering from FPIES.

In the gut mucosa, it became known that Th17 cell develop prior to antigen exposure, bearing an important role in mucosal immune defense. However, following antigen-exposure, regulatory T cells develops, developing immune homeostasis in the gut. During this process, the gut microbiota is believed to have an pivotal role in developing gut immune homeostasis affecting plasticity of Th cell and regulatory T cell development (Atarashi et al., 2011; Lee and Mazmanian, 2010; Lochner et al., 2011; Zhou et al., 2009). Consistent with this assumption, we also found that spontaneous IL-10 production, as well as that in response to luminal antigens (food proteins and candida Ag), were equivalent in non-ASD/FPIES children in comparison with normal controls. While ASD-IS children revealed lower IL-10 production, in the absence of stimuli and with candida Ag. These results again indicate that tolerance induction or establishment of immune homeostasis may be impaired in the ASD-IS children. However, it needs to be cautioned that further studies are necessary to explore this possibility, including testing gut mucosal expression of Th lineage cells in these patients. As previously stated, ASD-IS children are clinically characterized with markedly fluctuating behavioral and cognitive activity. Our findings also indicate the possibility that in addition to impairment of gut immune homeostasis, systemic immune homeostasis may also be dysregulated in the ASD-IS children. This may be the results of chronic gut inflammation or uncontrolled inflammatory responses.

To further assess changes in innate immunity, we also studies transcript profiles of PB monocytes in the study groups. However, partly due to the restriction of approved protocol, we were not able to conduct such a study in the non-ASD/FPIES children. Nevertheless, our results indicated a significantly altered gene expression in PB macrophages in the ASD-IS children. The numbers of genes with >2 fold up- or down-regulated expression was the highest when compared to normal controls and least when compared to ASD/FPIES children. This is not surprising, given the fact that ASD-IS and ASD/FPIES children both suffered from FPIES and gut inflammation. However, what was surprising is that the significant differences in transcript profiles of PB monocytes between ASD/non-FPIES and normal controls. These results also indicate that ASD/non-FPIES children may have altered immune responses or other changes that can be reflected in PB monocytes. This seems to be consistent with the results of the GO analysis, which revealed significant changes between ASD/FPIES and ASD/non-FPIES children. Involved pathways revealed differences with GO analysis are largely associated with immune/inflammatory responses as well as cytokines and chemokines, possibly reflecting fundamental differences in the 2 study groups. It should be noted that recent genetic studies have been accumulating evidence that ASD is a behavioral syndrome encompassing markedly heterogeneous populations (Bale et al., 2010; Rudan, 2010; Toro et al., 2010). Our findings also support results of previous genetic studies. However, we have to be cautious about the results given the low number of study subjects that underwent this analysis.

With transcription profiling, we also found that certain cytokines, notably IL-6, IL-1ß, and IL-23, were down regulated in ASD-IS children, as compared to ASD/non-FPIES children. ASD/FPIES children also revealed down-regulation of IL-6 and IL-23. Interestingly, down-regulation of IL-6 was more prominent in ASD/FPIES children, despite the fact that in vitro IL-6 protein production was lower in the ASD-IS children. In addition, GO analysis revealed significant differences between ASD/FPIES and ASD/non-FPIES groups, although given clinical features, one can expect that differences may be more prominent between ASD/IS and ASD/non-FPIES children. Our findings may indicate an importance of post-transcriptional regulation including ones exerted by microRNA (Baltimore et al., 2008).

The notable weakness of this study is lack of data of transcription profiling in the non-ASD/FPIES children. This was secondary to the restriction imposed by the IRB and lack of funding. In the future, it will be interesting to compare transcript profiles between ASD-IS/ASD-FPIES and non-ASD/FPIES children which may yield further important information.

5. Conclusion

In summary, our results indicate that altered adaptive and innate immune responses are observed in both ASD-IS and ASD/FPIES children but difference exist between the ASD-IS and ASD/FPIES children. In addition, these changes also differed from those observed in non-ASD/FPIES children. Our findings thus indicate that changes in innate and adaptive immunity observed in ASD-IS children and in some degree even in ASD/FPIES children, are not likely to be attributed to FPIES but may be associated with impaired establishment of immune homeostasis, perhaps, affected by aberrant innate immune responses.

6. Acknowledgment

This study was partly funded by Jonty Foundation, St. Paul, MN, and Autism Research Institute, San Diego, CA. We are also thankful for critical review by Dr. Lisa Huguienin.

7. References

Abreu, M.T.; Fukata, M., & Arditi, M. (2005). TLR signaling in the gut in health and disease. Journal of Immunolology, Vol.174, No. 8, (April 2005), pp. 4453-4460, ISSN 0022-1767

Akinbami, L.J.; Moorman, J.E., & Liu, X. (2011). Asthma prevalence, health care use, and mortality: United States, 2005-2009. *National Health State Report,* No. 32, (January 2011), pp. 1-14, http://www.ncbi.nlm.nih.gov/pubmed/21355352?dopt=Citation

Atarashi, K., Tanoue, T., Shima, T., Imaoka, A., Kuwahara, T., Momose, Y., Cheng, G., Yamasaki, S., Saito, T., Ohba, Y.,

Bale, T.L., Baram, T.Z., Brown, A.S., Goldstein, J.M., Insel, T.R., McCarthy, M.M., Nemeroff, C.B., Reyes, T.M., Simerly, R.B.,

Baltimore, D., Boldin, M.P., O'Connell, R.M., Rao, D.S., & Taganov, K.D. (2008). MicroRNAs: new regulators of immune cell development and function. *Nature Immunolology*, Vol. 9, No. 8, (August 2008), pp. 839-845, ISSN 1529-2916

Buie, T., Campbell, D.B., Fuchs, G.J., 3rd, Furuta, G.T., Levy, J., Vandewater, J., Whitaker, A.H., Atkins, D., Bauman, M.L.,

Beaudet, A.L., Carr, E. G., Gershon, M. D., Hyman, S. L., Jirapinyo, P., Jyonouchi, H., Kooros, K., Kushak, R., Levitt, P., Levy, S. E., Lewis, J. D., Murray, K. F., Natowicz, M. R., Sabra, A., Wershil, B. K., Weston, S. C., Zeltzer, L., Winter, H. (2010). Evaluation, diagnosis, and treatment of gastrointestinal disorders in individuals with ASDs: a consensus report. Pediatrics Vol. 125, Suppl 1, (January 2010), pp. S1-18 ISSN 1098-4275

Burgler, S., Ouaked, N., Bassin, C., Basinski, T.M., Mantel, P.Y., Siegmund, K., Meyer, N., Akdis, C.A. & Schmidt-Weber, C.B. (2009). Differentiation and functional analysis of human T(H)17 cells. Journal of Allergy and Clinical Immunology, Vol. 123, No. 3, (March 2009), pp. 588-595, ISSN 1097-6825

Butrus, S, & Portela, R. (2005). Ocular allergy: diagnosis and treatment. Ophthalmology Clininics of North America, Vol. 18,No. 4, (December 2005), pp. 485-492, ISSN 896-1549

Jyonouchi, H. (2008). Non-IgE mediated food allergy. *Inflammation and Allergy Drug Targets,* Vol. 7, No. 3, (September 2008),pp. 173-180, ISSN 1871-5281

Jyonouchi, H. (2010). Autism spectrum disorders and allergy: observation from a pediatric allergy/immunology clinic. Expert Review of Clininical Immunology, Vol. 6, No. 3, (May 2010), pp. 397-411, ISSN 1774-8409

Jyonouchi, H.; Geng, L., Cushing-Ruby, A., and Quraishi, H. (2008). Impact of innate immunity in a subset of children with autism spectrum disorders: a case control study. *J Neuroinflammation* Vol. 5, No. 5, (November, 2008), pp. 52, ISSN 1742-2094

Jyonouchi, H.; Geng, L., Ruby, A., Reddy, C., & Zimmerman-Bier, B. (2005). Evaluation of an association between gastrointestinal symptoms and cytokine production against common dietary proteins in children with autism spectrum disorders. Journal of Pediatrics, Vol. 146, No. 5, (May 2005), pp. 605-610, ISSN 0022-3476

Jyonouchi, H.; Geng, L., Ruby, A. & Reddy, C. (2007). Suboptimal responses to dietary intervention in children with autism spectrum disorders and Non-IgE mediated food allergy. In: *Autism Research Advance,* L B Zhao (Ed), pp. 169-184.Nova Science Publishers, ISBN 1-60021-523-8.New York, NY

Jyonouchi, H.; Sun, S., & Le, H. (2001). Proinflammatory and regulatory cytokine production associated with innate and adaptive immune responses in children with autism spectrum disorders and developmental regression. Journal of Neuroimmunology, Vol. 120, No. 1-2, (November, 2001), pp. 170-179, ISSN 0165-5728

Karlsson, M.R.; Rugtveit, J. & Brandtzaeg, P. (2004). Allergen-responsive CD4+CD25+ regulatory T cells in children who have outgrown cow's milk allergy. Journal of Experimental Medicine, Vol. 199, No. 12, (June 2004), pp. 1679-1688, ISSN 0022-1007

Kimura, A. & Kishimoto, T. (2010). Th17 cells in inflammation. International Immunopharmacology, Vol. 11, No. 3, (March, 2011), pp. 319-322, ISSN 1878-1705

Lee, Y.K. & Mazmanian, S.K. (2010). Has the microbiota played a critical role in the evolution of the adaptive immune system? *Science,* Vol. 330, No. 6012, (December 2010), pp. 1768-1773, ISSN 1095-9203

Lochner, M.; Berard, M., Sawa, S., Hauer, S., Gaboriau-Routhiau, V., Fernandez, T.D., Snel, J., Bousso, P., Cerf-Bensussan, N. & Eberl, G. (2011). Restricted microbiota and absence of cognate TCR antigen leads to an unbalanced generation of Th1 cells. *Journal of Immunolology*, Vol. 186, No. 3, (February 2011), pp. 1531-1537, ISSN 1550-6606

Michalek, S.M.; Kiyono, H., Wannemuehler, M.J., Mosteller, L.M. & McGhee, J.R. (1982). Lipopolysaccharide (LPS) regulation of the immune response: LPS influence on oral tolerance induction. *Journal of Immunology*, Vol. 128, No. 5, (May 1982), pp. 1992-1998, ISSN 0022-1767

Nassef, M.; Shapiro, G. & Casale, T.B. (2006). Identifying and managing rhinitis and its subtypes: allergic and nonallergic components--a consensus report and materials from the Respiratory and Allergic Disease Foundation. *Current Medical Research and Opinion*, Vol. 22, No. 12, (December 2006), pp. 2541-2548, ISSN 1473-4877

National Heart, Lung, and Blood Institute; National Asthma Education and Prevention Program (2007). Expert Panel Report 3 (ER3) Guidelines for the diagnosis and management of asthma, L. National Heart, and Blood Institute, ed. No. 08-4051, (Bethesda, NIH Publication).
http://www.nhlbi.nih.gov/guidelines/asthma/asthgdln.htm

Nowak-Wegrzyn, A. & Muraro, A. (2009). Food protein-induced enterocolitis syndrome. *Current Opinions in Allergy and Clinical Immunology*, Vol. 9, No. 4, (August, 2009), pp. 371-377, ISSN 1473-6322

Rudan, I. (2010). New technologies provide insights into genetic basis of psychiatric disorders and explain their co-morbidity. *Psychiatria Danubina*, Vol. 22, No. 2, (June 2010), pp. 190-192, ISSN 0353-5053

Sandler, R.H.; Finegold, S.M., Bolte, E.R., Buchanan, C.P., Maxwell, A.P., Vaisanen, M.L., Nelson, M.N. & Wexler, H.M (2000). Short-term benefit from oral vancomycin treatment of regressive-onset autism. Journal of Child Neurology, Vol. 15, No. 7, (July 2000), pp. 429-435, ISN 0883-0738

Schirbel, A. and Fiocchi, C. (2010). Inflammatory bowel disease: Established and evolving considerations on its etiopathogenesis and therapy. *Journal of Digestive Diseases*, Vol. 11, No. 5, (October, 2010), pp. 266-276, ISSN 1751-2980

Sicherer, S.H.; Eigenmann, P.A. & Sampson, H.A. (1998). Clinical features of food protein-induced enterocolitis syndrome. *Journal of Pediatrics*, Vol. 133, No. 2, (August, 1998), pp. 214-219, ISSN 0022-3496

Sicherer, S.H. & Sampson, H.A. (2006). 9. Food allergy. Journal of Allergy and Clinical Immunology, Vol. 117, 2 suppl Mini-Primer, pp. S470-475, ISSN 0091-6749

Singh, K.; Axelrod, S. & Bielory, L. (2010). The epidemiology of ocular and nasal allergy in the United States, 1988-1994. Journal of Allergy and Clinical Immunology, Vol. 126, No. 4, (October, 2010), pp. 778-783 e776, ISSN 1097-6825

Smith, D.W. & Nagler-Anderson, C. (2005). Preventing intolerance: the induction of nonresponsiveness to dietary and microbial antigens in the intestinal mucosa. Journal of Immunology, Vol. 174, No. 7, (April, 2005), pp. 3851-3857, ISSN 0022-1767

Toro, R.; Konyukh, M., Delorme, R., Leblond, C., Chaste, P., Fauchereau, F., Coleman, M., Leboyer, M., Gillberg, C. & Bourgeron, T. (2010). Key role for gene dosage and synaptic homeostasis in autism spectrum disorders. *Trends in Genetics*, Vol. 26, No. 8, (August, 2010), pp. 363-372, ISSN 0168-9525

Wannemuehler, M.J.; Kiyono, H., Babb, J.L., Michalek, S.M. & McGhee, J.R. (1982). Lipopolysaccharide (LPS) regulation ofthe immune response: LPS converts germfree mice to sensitivity to oral tolerance induction. Journal of Immunolology, Vol. 129, No. 3, (September, 1982), pp. 959-965, ISSN 0022-1767

Zhou, L.; Chong, M.M. & Littman, D.R. (2009). Plasticity of CD4+ T cell lineage differentiation. Immunity Vol. 30, No. 5,(May, 2009), pp. 646-655, ISSN 1097-4180

20

Clinical Evaluations on the Diagnosis of Autism

Michaela Dobre[1] and Ileana Stoica[2]
[1]*"Lower Danube" University of Galati, Faculty of Medicine and Pharmacy*
[2]*University of Bucharest, Faculty of Biology*
Romania

1. Introduction

Autism spectrum disorders are currently formally diagnosed after pursuing two levels of investigation (Blackwell, 2001; Filipek et al., 1999, 2000). The first consists of monitoring the child's development and specific autism signal detection, which implies parent and physician attention, while the second involves actual diagnosis and evaluation, performed by specialized medical staff. Both levels of diagnosis rely on the careful observation of specific changes in child development, concerning movement, discovery and interaction with objects and people, games and language, which exhibit atypical variations in autism spectrum disorders. Intelligence tests, which aim to be a conventional estimate for the severity of mental retardation that may accompany autism, can serve as a relatively suitable approximation. There are many specific tests to diagnose autism, none ubiquously applied, and the final diagnosis depends on an overall estimation of brain function and, to a great extent, on the competence of the diagnostician. Therefore, the evaluation of a clinical diagnosis of autism through paraclinical methods - whether imaging, functional or laboratory work (metabolic, immunological, cytogenetic or molecular tests) - that would justify a medical conclusion, represents a necessity that is still unattainable. For autism spectrum disorders no specific paraclinical methods have as yet been identified, despite serious efforts invested. Traditional medicine was overcome in discriminate between different patients, thus opening the door for interdisciplinary studies which developed into new sciences, like psychoneuroendocrinology, psychoneuroimmunology and their corresponding insertions in genetics.
Over the years, some of the elements suspected to be involved in causing autism spectrum disorders were studied, providing mixed results. The present study will list converging results, and will attempt to portray autism from the perspective of modern investigation methods.

2. Neuroimaging

Imaging methods, from the classic X-ray to the CT-scan (computerized axial tomography), the PET (positron emission tomography), the SPECT (simple proton emission computerized tomography) and the MRI (magnetic resonance imaging), along with their functional variations - fMRI (functional Magnetic Resonance Imaging), sMRI (structural MRI), MEG (magnetoencephalography), DTI (diffusion-tensor tracking) - aim to highlight anatomical and physiological modifications in the central nervous system and associate them with autism spectrum disorders via physical or physiological parameter evaluations.

Imaging is useful in diagnosing autism first of all by measuring brain component dimensions because in most autistic children, MRI studies showed a particular dynamics of brain and cranium development. Development peaks occur between 2 and 5 years of age, when cranial circumferences are above average and cerebral volume increases, and are followed, in adolescent and adult ages, by reduced cerebral dimension and function (Aylward et al., 2002; Courchesne et al., 2010; Hazlett et al., 2011). The evaluation and monitoring of child development dynamics may be performed using *WHO Anthro* software (World Health Organization-WHO, 2011), made in accordance with WHO standards (WHO, 2007) for children aged between 0 and 5. For children above 5, statistical data is available from the *Centers for Disease Control and Prevention – USA (*Kuczmarski et al., 2002).

An in-depth and detailed analysis of the brain revealed some anomalies in component dimensions and especially in the ratio of white to grey matter: the corpus callosum is diminished (Frazier & Hardan, 2009; Just et al., 2007); the frontal lobes are enlarged due to an excess of white matter (Cheng et al., 2010); the temporal lobes have a reduced grey matter content; the cerebellum has an over 30% excess of white matter (Carper et al., 2002; Courchesne et al., 2010). Microscopical *post-mortem* imagery revealed Purkinje cell loss in the cerebellum. These particular cells are selectively vulnerable to hypoxia, ischemia, G-protein dysfunctions, viral infections, heavy metals, as well as to a wide range of metabolites and chemicals (Kern & Jones, 2006). Histological studies also showed numerical reductions in amygdala axons (Schumann & Amaral, 2006) and reduced cell sizes in the hippocampus, similar to what is seen in the case of a precocious maturation (Bauman & Kemper, 2005). The two cerebral formations are involved in emotional and memory process development, two aspects typically targeted by autism spectrum disorders (Kleinhans et al., 2009; Saitoh et al., 2001). The use of MRI-DTI for the microstructural analysis of pathways interconnecting hippocampus and amygdala with the mid-fusiform gyrus in the temporal lobe (which participates in number, word, color and physiognomy recognition) revealed dysfunctionalities of these connections in autistic subjects which took a face recognition test (Conturo et al., 2008). The cause for this diminished response to stimuli was considered to be the reduction of axon diameters as a result of hypermielinisation, which yields a lower neuron conduction speed. Hippocampus and amygdala dimensions, which vary with the age and functionality of the investigated subjects, suggest the incongruence of cerebral components in the developmental dynamics.

A disharmonious development, a reduced corpus callosum volume and a small quantity of grey matter might justify the reduction in inter-hemispherical connectivity specific to autism (Hardan et al., 2009). Investigations using fMRI placed the causes for poor connectivity in the frontal insula, the superior temporal gyrus, the primary sensorimotor cortex, the lateral inferior premotor cortex and the superior parietal lobe (Anderson J.S. et al., 2010). The aforementioned areas are associated with social interaction management, social intelligence, new stimulus identification, sound processing and language linking, fine and gross motor activities, thereby encompassing all areas damaged by autism.

Specific modifications in verbal communication in autism spectrum disorders incriminate the reduced volume of grey matter in the Broca area (where the primary motor complex is responsible for talking motions in the left hemisphere), considering the fact that insufficient neural connectivity between the Wernicke and Broca areas lead to difficulties in lexical and semantic processing (Yamasaki et al., 2010). The comprehension and elaboration of language also depends, however, on auditory sensorial integration, which serves to perceive and analyze sound patterns. In the Knaus et al. study, linguistic scores were correlated with grey matter deficits found in frontal and temporal lobes, autistic subjects manifesting a typical lateralization of language (Knaus et al., 2010).

Temporal auditory associative areas, where mono- and multimodal auditory sensor integration crosses, are involved in the transfer of information to the internal structures of the limbic and frontoparietal integration systems. By measuring cerebral blood flow using high resolution PET as well as high resolution SPECT, the superior temporal region of the autistic child's brain was shown to have bilaterally localized hypoperfusions in the superior temporal gyrus and superior temporal sulcus, centered on the multimodal associative temporal cortex and auditory associative areas (Boddaert & Zilbovicius, 2002; Ohnishi et al., 2000; Zilbovicius et al., 2000). This hypoperfusion was presumed to impair the transfer of auditory information and social perception (eye contact and facial expression), relational difficulties characteristic of autistic children. Ulterior studies, which undertook a neurofunctional perspective, reported diminished responses to auditory stimuli (Gomot et al., 2006) and visual stimuli (Bölte et al., 2008), along with hemodynamic modifications (Zilbovicius et al., 2006). The autistic brain reacts faintly to stimuli variations, a fact which agrees with some of the basic diagnosis criteria, such as a reduced interest for all surroundings, resistance to change and stereotypical activities.

Admitting that a diminished neural connectivity, due to an excess of white matter, and some cerebral area hypoperfusions are responsible for the functional afflictions specific to autism, it would be expected that a normalization of the blood flow would improve subject behavior. Positive developments as concerning hyperactivity, irritability, stereotyping as well as improvements in verbal communication were observed during prolonged periods of pathological fever, presumably responsible for an increased flow of blood to the cortex (Curran et al., 2007). Hyperbaric oxygen therapy (HBOT), as proposed by Rossignol (Rossignol & Rossignol, 2006) also registered temporary improvements for autistic behavior (Chungpaibulpatana et al., 2008), oxygen supplementation techniques being well known also for their anti-inflammatory effects and their ability to regulate tissue perfusion and reduce oxidative stress.

Grey matter is not only present in reduced volume in the autistic brain, but also badly distributed, as it is found in a thin layer in cerebral areas associated with socialization and all that which it entails: audio-visual and olfactory stimuli processing, face recognition, imitation and language (Hadjikhani et al., 2006). This situation corresponds to the mirror neuron system, activated in the normal brain when observing and imitating an action, and also considered responsible for empathy and mentalization. The mirror neuron system is dysfunctional in autistic subjects, as enforced by fMRI, MEG and EEG findings (Dapretto et al., 2006; Oberman et al., 2005; Perkins et al., 2010).

Modern neuroimaging supports neuropsychological hypotheses concerning the causes of autism, i.e. the limbic system hypothesis, the weak central coherence hypothesis, the executive function hypothesis and the mentalization theory, but without being able to validate them (Bade-White et al., 2008; Joseph, 1999).

The limbic system, presumed responsible for memory deficits that negatively impact communication and social interaction, was associated through imaging with structural and functional changes especially in the hypothalamus and amygdala, known to be correlated with affective and emotional behavior, sense of smell and long term memory.

Weak central coherence, which implies a difficulty to integrate separately perceived elements into a whole, is associated with restricted interest and behaviour. Poor central coherence may be supported by a disproportion between white and grey matter and diminished neural connectivity but, considering the fact that the central coherence mechanism is postulated rather than demonstrated, imagistic information such as activity lateralization in the autistic brain and predominant involvement of the left hemisphere may be admitted as atypical, compensatory connections (Pierce et al., 2001).

The executive function hypothesis, which concerns intelligence and flexibility when establishing priorities, is associated with the prefrontal cortex and, especially, the dorso-lateral areas. It has been proven that lesions in the aforementioned areas substantially impair executive functions, with imagistic evidence available (Levin & Hanten, 2005). In the case of autism, executive dysfunction manifests in the absence of lesions and exhibits heterogeneous cortically distribution, associable with cyto-architectural defects and/or atypical connectivity (Gilbert et al., 2008).

To oppose the executive dysfunction and weak central coherence hypotheses, which aim to explain obsessive, repetitive autistic behavior as a blockage in information processing resulting from neuroanatomical and neurofunctional causes, Baron-Cohen et al. (Baron-Cohen et al, 2009) suggest repetitiveness as a consequence of hyper-systematization. This aspect, remarkable in gifted high-functioning autism, is associated with sensory hypersensitivity which favors an exceptional perception of visual, auditory and tactile details and may be due to a special type of informational processing, correlated with various neurotransmitters through modifications in receptor density and sensitivity or inhibition/stimulation mechanisms.

The mentalization theory concerns one's capacity to understand another's thoughts, engaging intention, visual contact and emotional communication. The existence of deficiencies in empathy and mentalization, associated with autism using psychological tests (Baron-Cohen, 1985), was endorsed by functional imaging, which revealed reduced activity in mirror neurons when recognizing a motor act preformed by another (Iacoboni & Dapretto, 2006). Admitting that mentalization is more than imitation, the same imagistic tests performed when intentionality was separated from the motor act (*why-use vs. why-place*), showed that autistic patients recognize the intention behind an action but can not anticipate the action using the motions that make it up (Boria et al., 2009). Therefore, mentalization theory can not be limited to the incrimination of mirror neurons, as these seem to be included into an extended cortical motor system (Rizzolatti & Fabbri-Destro, 2010). Probable causes for this system's malfunctions are presumed to also be neural connectivity defects and white substance distribution.

Imaging, along with specific marker neurochemicals, helped detect areas activated by some cerebrally distributed hormones, impossible to dose *in vivo*. Thusly, in the presence of a marked substitute of tryptophane, PET revealed reduced serotonin activity in the thalamus and frontal cortex (Chugani et al., 1999). By analyzing active serotonin and dopamine transport in the high-functioning autistic brain through a combination of MRI and PET, a significant decrease in serotonin distribution volumes was noticed throughout the brain, along with a volume increase for dopamine, in the medial frontal area covering the orbital-frontal complex (Nakamura et al., 2010). Serotonin deficiencies in the thalamus were in this case associated with repetitive/obsessive manifestations of autism.

MEG, another modern imagistic technique, proved to be much more sensitive, compared to EEG, in detecting forms of epilepsy, and contributed to the reevaluation of the frequency of these being associated with autism (Lewine et al., 1999).

3. Biomarkers

Clinical laboratory tests aim to reveal measurable parameter variations in blood, urine, CSF or tissue samples, suggesting possibly altered metabolic processes in patients suffering from autism spectrum disorders.

Anemia, revealed by blood tests and iron and transferrin level assessments, support the hypothesis of cerebral tissue hypoxia and are frequently associated with autism spectrum disorders. It is believed that anemia has an irrevocably negative effect on intellectual function, especially when it occurs in pre-, peri- and postnatal life, even if hemoglobin levels subsequently normalize (Hurtado et al., 1999). Anemia tends to be chronic in autism spectrum disorders and is usually correlated with iron (Latif et al., 2002), folate (Moretti et al., 2005), and possibly ferritin (Dosman et al., 2007), transferrin and associated receptor deficiencies (Chauhan et al., 2004).

Common iron deficits may be due to insufficient food intake or absorption problems caused by the various gastrointestinal diseases that may accompany autism. A significant competitor to iron absorption from food is lead, classified as a neurotoxic environmental pollutant. Studies suggest that intoxication may occur even in pre- or postnatal periods, lead being delivered to the child through the mother's blood or milk (Gulson et al., 1998). Research has yet to accurately identify the origin of the lead or its role in autism spectrum disorders.

Iron deficits were also associated with non-alimentary consumption (pica), sometimes apparent in autism spectrum disorders (Barton et al., 2010).

Due to its ability to easily modify its oxidation state, iron participates in a large number of metabolic reactions, is included in hemoprotein structures and acts as an enzyme co-factor. As a component of hem, iron bound in hemoproteins takes part in vital metabolic reactions, such as oxygen transportation and storage (hemoglobin, mioglobin, neuroglobin), electron transfer in the respiratory chain and energy generation in the mitochondria (cytochromes b and c) and is a component of catalases, peroxidases and hemothyolate proteins (NO synthetase, P_{450} proteins/cytochromes), which act to bind carbon monoxide, nitrogen monoxide and produce free radical oxygen. Qualitative or quantitative anomalies in these proteins, together with specific cognitive process impairments led to the hypothesis that metabolic causes may be behind autism, among these tissular hypoxia, mitochondrial function damage and oxidative stress.

Iron deficits associated with low hemoglobin levels explain the reduced levels of transported oxygen and plead for the cerebral hypo-oxygenation theory. However, there were cases when iron deficiencies accompany cognitive impairments without the presence of anemia, the cause of such occurrences being unknown (Hulthén, 2003). Recently discovered hemoprotein, neuroglobin, (Burmester et al., 2000), was attributed a neuroprotective role in cases of hypoxia and also a regulatory function for G-protein signaling systems (Wakasugi et al., 2005), but no related studies have yet been undertaken in autism spectrum disorders.

On the other hand, iron deficiencies do not only reflect low hemoglobin, iron being also well distributed in tissues. A recent study (Fukunaga et al., 2010), used immunofluorescence to prove that iron bound in ferritine and basic myelinic proteins is non-uniformly distributed between cortical layers, a fact that may be interpreted as the quality of contrast on MRI imagery. Iron deposits were frequently identified in collocation with myelin. Considering the predominantly lipoidic structure of the white matter, such positioning might suggest iron placement is opportune for insertion, via cytochrome P_{450}, into the biosynthesis of cholesterol and other myelinic lipids that envelop the nervous fiber. Ferritin, in its two forms, (H and L), its adequately placed receptors and its different expressions in microglia and oligodendrocytes (Sammarco et al., 2008) might play an as yet undeciphered part in myelin regularization, impaired in autism. Considering the excessive white matter synthesis in the early years of an autistic child's life and the frequency of anemia in the same period, Hay et al. (Hay et al., 2007) performed a study on normal children of the same age group in order to assess the quantity of transferring receptors. Receptor concentrations proved to vary, dropping in the first two years of life, but being higher in boys ever in intrauterine life, possibly as an influence of testosterone.

Excessive testosterone in the womb has also been incriminated, based on typical psychological differences between the sexes as a cause of autism, in what was called the "extreme male brain theory" forwarded by Baron-Cohen in 2002. This theory places autistic behavior in the male extremity of a human behavior curve built upon two criteria: systematization (as a mark of masculinity) and empathy (as a mark of femininity). In this theory, testosterone is the only hormone made responsible for sexual dimorphism. Excessive testosterone, infiltrated prenatally, has supposedly exacerbated systematic capabilities, as well as dramatically reduced empathy, thereby creating an extremely masculine individual. At the same time, increased levels of fetal testosterone may lead to more dramatic behavior alterations within the male population, thus justifying the predominance of autism in males (sex ratio 4:1). The validity of this theory is still to be demonstrated, with large scale future experiments aiming to clarify the role of fetal testosterone in the delimitation between normal and pathological psychoneurological development. The few studies hitherto carried out used radioimmunochemical methods to evaluate fetal testosterone in the amniotic fluid (Auyeung et al., 2009; Chura et al., 2010).

The "extreme male brain" theory is not the only one to place forward a hormonal cause to autism. More and more studies link autism-specific behavior alterations with levels of hormones derived from aminoacids (melatonin, serotonin, adrenalin/epinephrine, dopamine, noradrenalin/norepinephrine, GABA) or neuropeptides with hormonal activity (arginine vasopressin/ADH, oxytocin), making up the foundation of the hypothesis that autism is rooted in neuropsychoendocrinology.

Serum half-life for some hormones can be very low, so that measuring them in biological fluids is difficult and subject to operator bias.

Hormones derived from amino acids are biosynthesized by oxidizing enzymes with iron cofactors. Identifying hormonal deficiencies linked to altered psycho-social behavior in autism may also correlate with the low iron levels.

Tryptophane 5-monooxygenase (TPH) is part of the tryptophane metabolism enzyme chain and generates precursors for serotonin and melatonin. Serotonin, which occurs cerebrally as well as periferically (more than 95%), is excreted as 5-hydroxy-indolil-acetic acid (5-HIAA). Serotonin concentrations may be measured directly or assessed by measuring its metabolite in spinal fluid or blood and urine, taking into account the fact that significant differences exist between periferical and cerebral media. Central serotonin deficiencies have been metabolically associated with autism using neuro-pharmacological methods. Prenatal stress induced by serotonin enhancing drugs, such as valproic acid, was associated with serotonergic neuron damage and is considered to be a potential cause of serotonin deficiencies in autism (Miyazaki et al., 2005). Low serotonin may explain the panic, anxiety, (auto)aggression as well as the repetitive/obsessive activities manifested in some forms of autistic behavior.

Tyrozine-3-monooxigenase (TH) is involved in the synthesis of L-dihydroxyphenylalanin (L-DOPA), a precursor of dopamine and, consequently, of noradrenaline and adrenaline. Phenylalanine-4-monooxygenase (PAH) participates in the L-phenylalanine metabolism as a precursor of tyrosine and, consequently, chatecholamines.

Dopamine is a neuromediator stored in the hypothalamus, tightly linked to the central reward system, which associates certain actions with a sense of pleasure, and controls the actions necessary to fulfill the action. Through its responding to the novelty stimulus, dopamine is also involved in the learning by experience process. Levels of dopamine may be assessed by measuring the concentration of its metabolite, homovanilic acid (HVA). In some cases, autism was pharmacologically associated with elevated levels of dopamine, as its agonists actually accentuated stereotypical behavior and increased aggression, while dopamine antagonists produced slight alleviations of such behavior (Tsai, 1999).

From the perspective of hormonal activity, repetitive/stereotypical behavior may be regarded as an obstinate search for an unattained pleasure (lack of serotonin). However, the subject anticipates satisfaction and understands the motric details that should lead up to it, and therefore performs them repeatedly (high dopamine levels). This interpretation is consistent with the hyper-systemizing hypothesis, but may also explain short-term memory impairment which prevents learning and obtaining ultimate satisfaction from goal achievement.

The appliance of immunohistochemistry on animal brains lead to the observation that dopamine, as well as one of its receptors (D_1), accentuates the development of GABA neurons, which play an important role in neuro-epithelial cell growth modulation, neuron migration and circuit forming (Crandall et al., 2007). The release of GABA in the synaptic space is equivalent to a cease of information transmission and depends on specific receptors, included in the neural membrane. One of these receptors ($GABA_A$) is inhibited by zinc ions, located cerebrally using fluoro- and radiomarkers (Smart et al., 2004). When released into the synaptic space, zinc ions can block some post-synaptic glutamate receptors (NMDA type), thereby becoming a neuromodulator. Physiological zinc concentrations take on neuroprotective and anticonvulsive roles. In zinc deficiencies, excessive glutamate may lead to prolonged neural hyperexcitability, displayed through epileptic seizures (Takeda et al., 2003), often associated with autism spectrum disorders. Additionally, proton magnetic resonance imaging found elevated glutamate levels in the hippocampus and amygdala (Page et al., 2006). Zinc pools mainly in the hippocampus and amygdala, areas tightly linked to autism by imaging (Szewczyk et al., 2010).

GABA neurons are close to acetylcholine receptors, part of the cholinergic system, which coordinate autonomous functions using sympathetic and parasympathetic pathways. Upon stimulation of the acetylcholine receptor, the dopaminergic system was found to react by releasing GABA (Lester et al., 2010). Therefore, cognitive process alteration may reside in the dopaminergic system (neuroanatomical modifications, zinc metabolism, iron deficiencies) as well as in the cholinergic system (Youdim, 2008). This logic justifies autonomous function impairments that accompany autism, such as afflictions of the digestive tract (difficult nutritional uptake and swallowing, defective transit and vomiting), arrhythmias, fever episodes, altered pain perception, problems with urination, anxiety, emotional lability and impaired socialization.

Various numbers of studies were carried out on L-DOPA, adrenalin and noradrenalin and did not reveal significant correlations with autism.

Low levels of melatonin link to frequent sleep abnormalities registered in autism (Melke et al., 2008). Melatonin uptake improves nightly sleep for autistic children as well as for other categories of people struggling with circadian rhythms. Circadian preferences may be associated with autistic hyper-sensitivity to auditory or other disruptive environmental stimuli, unnoticeable to normal people.

Neuropeptides involved in neural signaling, influence cerebral activity and are linked to complex processes such as learning and memory, essential to create personality and adaptability. Studies suggest arginine vasopressin and oxytocin may influence neural connections related to individuals relation forming therefore, recently come important for autism correlations.

Arginine vasopressin is synthesized in the hypothalamus, stored in the posterior hypophysis and released in the supraoptical nuclei according to the circadian rhythm. Oxytocin is secreted by the hypophysis and also acts remotely, like arginine vasopressin, its action modulated by receptors. Animal studies showed that hormonal action completion depends both on the concentration of neuropeptides and respective receptor density. In 1987, Le Moal and Dantzer carried out studies on mice to prove associative behavioral differences correlate with antidiuretic hormone levels. Excess arginine vasopressin was linked with associative relation stability, i.e. monogamy. Winslow et al. studies associated arginine vasopressin with male voice modulations in a social context, in some species of monkeys, and also with increases in motor activity upon central hormone administration (Winslow & Insel 1991; Winslow et al., 1993). Increased levels of arginine vasopressin in the hypothalamus was recorded post-partum, in both males and females of certain rodent families, hinting to the hormone's role in affiliation, attachment and sexual and parental behavior manifestations (Wang et al., 2000). Increased levels of ADH found in human CSF and murine brains were correlated with aggression as well as with social memory. Autistic subjects showed increased levels of arginine vasopressin (Boso et al., 2007) and reduced levels of oxytocin (Modahl et al., 1998). Resulting anxiety was relieved by oxytocin nasal spray administration, which appeared to also contribute to an improvement in verbal and affective communication. Ferguson et al. hypothesized arginine vasopressin involvement in the consolidation (rather than acquisition) of social memory (Ferguson et al., 2002). All these studies revealed a more predominant influence of vasopressin in males, the equivalent of which being oxytocin in females (Lim & Young, 2006). Protective behavior towards offspring and the pursuit of social integration (*"tend-and-befriend"*) are linked to oxytocin and the female sex, while the male is attributed adaptive-aggressive (*„fight-or-flight"*) behavior. Specific receptor distribution was found to be influenced by steroid sexual hormones and differentiated by sex and displayed level of affinitive behavior (Hammock et al., 2006). Some dopamine receptors are laminar distributed with oxytocin receptors, making their interaction more likely in social attachment behavior, while arginine vasopressin receptors are disfavored (Smeltzer et al., 2006).

The hypothetic hormonal cause to autism intersects with the oxidative stress hypothesis when it comes to glutamate exocytotoxicity. Glutamate release into the extracelluar space leads to increased intracytoplasmatic calcium ion levels but, when concentrations above 10^{-5} M are reached, calcium is pumped through membrane transport proteins into the interior of mitochondria (Krauss, 2003), where it triggers metabolism through matrix dehydrogenases. Besides contributing to ATP production, the shift of iron state of oxidation generates free radicals that are detoxified, in physiological conditions, by the mitochondrial antioxidation systems: superoxid dismutase, catalase, glutathione peroxidase, glutathione reductase –

enzymes lacking in autism (Yorbik et al., 2002). Given time and non-physiological conditions, reactive oxygen species generated cause mitochondrial damage, decreased ATP production and, eventually, apoptosis. Considering the fact that apoptosis is integral to development, morphogenesis and homeostasis, as it removes excessive or irrecoverable cells from the cellular cycle arrest, it is admissible that endogenous and/or exogenous stress factors in prenatal, postnatal periods and early years influence cerebral architecture and development, vulnerable to oxidative stress.

Some substances, such as lactate, piruvate, aspartate, found in excess levels in blood and urine also plead for the hypothesis that mitochondrial functions may be impaired (Giulivi et al., 2010; Weissmann et al., 2008). Lactate, which appears in blood as a secondary product of anaerobic glycolysis, points to tissular hypoxia caused by mitochondrial oxidative phosphorilation impairment and, in association with anion gaps, may lead to metabolic acidosis, encountered in some cases of autism (Lombardi, 1998). In some patients, MRI and PET scans showed diminished rates of glucose metabolization in the striatum and thalamus (Haznedar et al., 2006). Lactatemia also registers as a side effect of anticonvulsive treatment, administered when autism is aggravated by epilepsy (Willmore et al., 2006).

Oxidative stress occurs when antioxidant capacity is overwhelmed, either because insufficient free radical scavenging enzymes are produced, because glutathione levels drop, or there are excessive reactive oxygen species (ROS) induced by exogenous factors (e.g. toxic heavy metals such as lead or mercury).

In antioxidant enzyme deficits, mitochondrial permeability transition increases in astrocytes for an endogenous toxic, difficult to annihilate: ammonia (Seyan et al., 2010). Excessive ammonia, found in some autistic subjects, represents the consequence of urea cycle defects (ornitin transcarbamilase/OTC, argininosuccinic acid/ASA syntethase deficiencies) or is a side-effect of anticonvulsive medication. Chauhan et al. (2004) associates oxidative stress in autism with low transferrin and plasmatic ceruloplasmin coupled with accentuated lipid peroxidation. Normal zinc levels would not allow excessive lipid peroxidation but, in autists, zinc levels were usually found to be low. Zinc deficiencies may also generate shortages of superoxide dismutase (a Cu-Zn enzyme).

Glutathione insufficiencies were revealed by HPLC in a study carried out on autistic subjects by James et al. and were associated (via homocystein) with decreased methyonin cycle turnover (James et al., 2004). This, in turn, implies a diminished synthesis of S-adenosyl methyonin (SAM), which is required to methylate proteins and DNA (Waterlow, 2006). The redox/methylation hypothesis (Deth et al., 2008), put forward as an expression of the response to environmental stress factors, is, as yet, insufficiently supported by ample genetic studies to investigate methylation profiles in autism and their roles in neural network synchronization.

The possibility exists that this scenario correlates with various types of anemia encountered in autism either through folate deficiencies (cysteine is obtained from homocysteine by means of a folate-dependent enzyme) or through the iron lost by high urinary porphyrin elimination, as a result of oxidative stress induced by heavy metal poisoning (Geier et al., 2009). Urinary porphyrin concentrations are comparable to those found in adults exposed to mercury poisoning, in which the presence of an atypical porphyrin in urine (precoproporphyrin) is associated with an inhibitory effect of mercury on kidney uroporphyrinogen decarboxilase (Woods et al., 2010). Since most investigated subjects did not have significant (organic/inorganic) mercury levels, the mechanism of this disturbance in the hemoglobin metabolism in autism is still unclear.

Glutathione deficiencies also correlate with gastrointestinal dysfunctions that often accompany autism. Oxidative stress is emphasized upon intake of a large gamut of foods, while the intestinal wall opposes the passage of toxic agents into the body. The maintenance of high levels of antioxidants (glutathione, vitamin C, tocopherols) and also of antioxidant enzymes represents two cellular integrity defense mechanisms. Should these fail, the intestinal cell is directed towards apoptosis or necrosis. (Aw, 1999).

A preliminary response to oxidative stress is white cell proliferation and subsequent signal cascade, followed by neutrophyle infiltration into intestinal cells, through diapedesis. Cellular migration is associated with an increased synthesis of proinflammatory cytokines (e.g.: IL-1β, IL-6, IL-8 and TNF-α), which act as a "first response" in acute inflammatory reactions. Anti-inflammatory response is quick to appear, as evident through IL-4, IL-13 and TGF-β cytokine synthesis. Ingesting food high in Maillard reaction final products activates macrophages and spreads intestinal inflammation to the entire organism (Georgescu, 2005; Muscat, 2007). Cytokines, as well as free radicals, activate the intranuclear translocation of the NF-kB transcription factor which, when bound to DNA, modulates numerous genes involved in inflammation and immunity. Intestinal mucose inflammation, histologically investigated through biopsies, was associated in autism with increased TNF-α, IL-2 and IL-4 in the duodenum and elevated TNF-α in the colon, with a more pronounced proinflammatory activity in subjects whose diets included casein and gluten (Ashwood et al., 2004). Proinflammatory (IL-1β, IL-8 and especially TNF- α) and anti-inflammatory (IL-4, IL-10, IL-12) cytokines, found in plasma, CSF, culture cells and in *post-mortem* cerebral tissue were found to be in high concentrations in autism (Ashwood et al., 2006; Chez et al., 2007). It is noteworthy that cytokines can cross the blood-brain barrier, affect neuron function and accentuate metabolism for hormones such as noradrenalin, dopamine and serotonin (Pickering et al., 2005; Zhao B. & Schwartz, 1998). Some cytokines were associated with autism specific aspects such as sleep, cognitive function abnormalities or depression (IL-2, TNF-α) (Larson, 2002). Additionally, anti-cerebral peptide autoantibodies (antimyelin, anti-Purkinje cells and anti-caudate nucleus) were identified by Western blots in serum, which speaks of a vulnerability of the blood brain barrier in autism (Singer et al., 2006). On the other hand, plasma harvested from autistic children's mothers was found to contain antibodies to human fetal brain of approximately 37 kD and 73 kD, which may represent a possible contribution of the maternal immune system to autism etiology (Braunschweig et al, 2008). Since oxidative stress and/or immunological vulnerability is manifested in autistic children from a very early age, it is legitimate to suspect that these may be consequences of environmental conditions in the mother's womb (Dietert & Dietert, 2008; Kinney et al, 2008).

4. Genetic and molecular testing

Any attempt to evaluate a diagnosis of autism can not escape genetic and molecular investigation. The heterogeneity of the psychological, morphological and functional aspects which contributed to the classification of the forms of autism undoubtedly has a genetic background that is currently being investigated, each organism being in fact an expression of its genetic pattern adapting to its environment. It is presumed that autism spectrum disorders have a major genetic component, with a complex and as yet undiscovered means of transmission. Twin studies revealed a rate of correlation of between 60 and 90% for monozygote twins and of 0 to 10% for dizygote twins (Bailey et al., 1995).

It is unlikely, given the great diversity of symptoms and particularities of autism, that a single gene may mutate with pleiotropic effects. On the other hand, the same reasons make it very likely that the diverse manifestations may be due to environmental factors, which occur with various intensities and cause different responses, according to genetically dictated, individual tolerances. All things considered, autism is outlined as a complex manifestation, posing great difficulties when it comes to pinpointing its origins.

Identification of candidate genes was performed using methodology both direct and indirect (in the case of secondary autism, associated with other syndromes characterized by neuropsychiatric afflictions). The discovery of genes linked to Rett syndrome (MECP2), Joubert syndrome (AHI1), neurofibromatosis type 1 (NF1), tuberous sclerosis (TSC1) and fragile-X syndrome (FMR1) allowed for differential diagnoses for these illnesses and opened the door to further genetic association for autism.

The direct approach to identifying the genetic components of autism used three categories of methods: chromosomal analysis, linkage and association studies and, respectively, candidate gene direct analysis. The numerous and various outcomes called for the development of specific databases. The Centre for Applied Genomics, established in 2004, in Toronto, maintains the *Autism Chromosome Rearrangement Database*, which comprises both cytogenetic as well as array comparative genomic hybridization/aCGH data. The *Autism Genetic Database (AGD)*, established in 2009 at the University of Kansas, offers a generous list of autism-susceptible copy-number variations/CNVs (743) and genes (243), as well as a series of fragile sites (120) and non-coding RNA genes which act as regulators and are intensely expressed in the nervous system (more than 650.000) (Matuszek & Talabizadeh, 2009). Also, the Stritch School of Medicine database from the Loyola University of Chicago, offers an eloquent mapping of candidate genes (238), CNVs, SNP polymorphisms and microinsertion/deletion sites linked to autism (*Autism Candidate Gene Map (ACGMAP) Database*).

Structural variations, recorded through cariotyping and SNP polymorphism analysis, are found in many chromosomal sites throughout the haploid genome: 1p, 1q, 2q, 3p, 3q, 4p, 4q, 5p, 5q, 6q, 7q, 8q, 11p, 12p, 13q, 15q, 16p, 17q, 19p, 19q, Xp şi Xq (Xu et al., 2004). The most frequent rearrangements appear in 2p, 7p (translocations in the 7q22-q33 regions), 15q (duplications in the 15q11-q13 regions), 16p and X. Studies suggested that the modifications encountered are most often hereditary, but may also appear spontaneously - as previously documented or indeed, as novel forms - therefore demanding the taking into account of additional risk factors, whether genetic, epigenetic or environmental.

The loci that presented high CNVs comprise genes involved in neural development and synaptic activity, genes associated with oxidative processes, immunity or transcriptional processes, genes encoding various metabolic pathway enzymes, hormones and receptors, zones associated with mental retardation (15q24 and 16p11.2) as well as other genes with modified expressions. Locus 16p11.2 stood out with the highest CNVs frequency (1%), correspondent to structural modifications that were mostly microdeletions and microduplications (Marshall et al., 2008; Weiss et al., 2008).

The most sought-after candidate genes were the ones most likely to be involved in central synaptic formation and modeling, cell adhesion and position and neural migration, such as: NLGN1 (3q26.31), NLGN3 (Xq13.1), X-linked NLGN4 (Xp22.32/33) and Y-linked NLGN4 (Yq11.221)/encoding for neuroligins, SHANK3 (22q13.3)/ankirin, PCDH9 (13q21.32)/protocadherin, NRXN1 (2p16.3), CNTNAP2 (7q35)/neurexins, RELN (7q22)/reelin (Glessner et al., 2009).

After a study which looked at a great number of families with autism, Yonan et al. attribute significance to gene locus SLC6A4 (17q11.2) around the sequence exhibiting the maximum LOD score for 17q chromosome linkage (Yonan et al., 2003). The gene codifies a membrane protein for serotonin transport which carries the neurohormone from synaptic spaces into presynaptic neurons and which correlates with aggressive behavior and attention deficits. Deciphering the ever more important hormone contribution to psycho-behavioral modelling triggered a reorientation of studies towards genes encoding hormones, neuropeptides and membrane receptor cells like: GABRA3 (Xq28), GABRB3 (15q11.2-q12), GABBR2 (9q22.1-q22.3)/encoding GABA receptor-proteins, which participate in the inhibition of excitation transmission, AVPR1A (12q14.2)/encoding the 1a arginine vasopressin receptor, OXT (20p13)/coding for oxytocin and neurophyzine 1, OXTR (3p25.3)/coding for oxytocin receptor, DRD1 and 3 (5q35 and 3q13.3)/coding for dopamine receptors (Anderson B.M. et al., 2008; Harony & Wagner, 2010; Muhler et al., 2004; Yirmiya et al., 2006).

Some genes from autism linked-CNVs have metabolically correlated expressions, which amplify their candidate gene status. For example, 10 genes encode proteins from the phosphatidylinositol signal pathway and the glutamatergic synapse which are correlated through phosphatidylinositol-3-OH kinase (Cuscó et al., 2009). Genes like UBE3A (15q11.2), PARK2 (6q25.2-q27), RFWD2 (1q25.1-q25.2) and FBXO40 (3q13.33), which encode similar function proteins (ubiquitin protein ligases), are associated with ubiquitin degradation as well as neuron cell adhesion (Glessner et al., 2009). Also, the SLC6A4 gene, involved in serotonin metabolic and signal pathways, was associated, by epistasis, with the ITGB3 (17q21.32) gene, encoding an integrin (Coutinho et al., 2007). An interesting correlation was found in astrocytes, between a V1a arginine vasopressin receptor antagonist and the hormone's anti-inflammatory and immunomodulating activities, by suppressing gene expression for IL-1α, IL-1β, IL-2 and TNF-α cytokines using a CREB transcription factor-dependant mechanism (Zhao L. & Brinton, 2004). The CREB1 (2q34), along with SHANK3, DLG3 (Xq13.1) and DLG4 (17p13.1) genes, were also studied in association with memory and learning processes, afflicted by autism.

As deciphering genetic mechanisms is still a distant goal, suspicion falls, for the moment on genes like FOXP2 (7q31), involved, in embryogenesis, in the developing cerebral regions associated with language, CNTNAP2 (7q35-q36), encoding a neurexin-type protein or DYX1C1 (15q21.3), which gives susceptibility to dyslexia.

The contribution of these genes to autism remains disputed, as association frequencies are low. Without any firm links between identified genetic modifications and autism, more recent studies aimed to characterize polymorphisms in regulatory zones - an undertake made easier by the completion of human genome sequencing. Also, animal model studies additionally augment data acquisition. An eloquent example is the analysis of AVPR1A gene promoter polymorphisms: by observing affiliative behavior in mice, it was concluded that arginine vasopressin levels are not the only factors that influence affiliation, and that the distribution of membrane hormone receptors, encoded by the AVPR1A gene, also plays a part. The three polymorphisms identified in the gene promoter were presumed to influence gene expression and, in consequence, were associated with types of affiliative behavior. Considering the genetic similarities between mice and humans, these polymorphisms were studied in connection with autism, which exhibits a powerful alteration of affiliative impulses. Studies revealed a transmission imbalance of RS1 and RS3 polymorphisms linked to autism (Kim et al., 2002), associated with willing socialization (Yirmiya et al., 2006) and prepulse inhibition in auditory stimulus response as stress-adaptive reaction (Levin et al., 2009). Another pathway for research are the epigenetic studies, from among which global methylation profile investigation revealed significant modifications in two genes: RORA (15q22.2) and BCL-2 (18q21.3), whose expressions are subdued in the autistic brain (Nguyen et al., 2010). A recent study by Sarachana et al. stresses the importance of RORA as a candidate gene, because of the correlation between its expression and testosterone levels. RORA-encoded protein synthesis is stimulated by female hormones and inhibited by male ones. In the autistic brain, besides reduced RORA-protein levels, low aromatase levels were also found (Sarachana et al., 2011). Aromatase is an enzyme from the P_{450} cytochrome family which acts to maintain an elevated testosterone level. The molecular mechanism proposed by this study might explain the elevated frequency of autism among the male population and also add an argument to support Baron-Cohen's "extreme male brain" theory.

5. Conclusion

There is no concrete data on autism before Kanner (1943). For years, autism was considered a rare illness consisting of emotional contact impairments in children. As psychologists and psychiatrists looked more closely into the matter, autism was scrutinized, classified and linked with a triad of modified behavioral aspects involving socialization, language and obsessive/repetitive actions. During the last 30-odd years, the frequency of autism began to rise dramatically, up to alarming levels in some countries. Whatever the cause of this increase (better information dissemination among the medical community and the general population and/or a real increase in the number of afflicted children) autism was boosted to the peak of modern research interest and major resources were devoted to discovering the underlying causes of this developmental disorder.

The study of the brain, as the material basis for psychic processes, emerged as a prioritized research direction and the discovery of new means of imagistic investigation helped make a significant leap for normal and pathological cerebral anatomy and function understanding. Studies of the autistic brain converge to a general unitary conclusion of cerebral spatial and temporal development impairment. The defining elements for autistic anatomo-functional defects are excessive white substance and low neural connectivity, which may be linked to anomalous neuron migration, axon and synaptic development, neural impulse transmission. Although modifications were described in many areas of the brain, the most frequent associations lead to the corpus callosum, thalamus, amygdala and cerebellum. Considering the behavioral triad afflicted by autism, deficient socialization was linked to the orbitofrontal, anterior cingulate and posterior parietal cortexes, the fusiform and inferior frontal gyruses, the amygdala and mirror neurons, impaired communication was traced to the Broca area, the superior temporal sulcus, the thalamus and the basal ganglia, while repetitive behavior was found to be reflected in the orbitofrontal and anterior cingulated cortexes, the basal ganglia ant the thalamus. For temporal anomalies, it remains to be said if excessive brain dimensions recorded in the early years of life is caused by prolonged neurogenesis or glial cell overproduction, perhaps in conjunction with delayed neuron death (Verhoeven et al., 2010). Performing neuroimagistic investigation on children with idiopathic autism is difficult. Functional study subjects require a former period of introduction and accommodation and need to exhibit a certain degree of comprehension and participation, therefore unavoidably implying older test subjects, which are often cited as "high-functioning autists" or Asperger syndrome sufferers. Moreover, correlating functional imaging with psycho-behavioral studies concerning affection entails, even for normal individuals, a certain degree of imprecision derived from the multiple variables that concur in the momentary psychic status-quo. In order to obtain relevant results, all applied techniques require synchronization between the time that the brain responds to a stimulus and the time at which the response is recorded, a reason for using professional actors as a control group (Perreau-Linck et al., 2007). Therefore, neuroimagistic studies have not yet been able to tell *"where"* autism-specific structural and functional modifications occur in the brain, but are breaking new ground for general human brain exploration, cerebral stimulus processing and reactive response.

Another perspective for evaluating autistic diagnoses involves investigating biochemical and immunological profiles. This path is however subject to limitations imposed by the inherent differences between the presence and/or concentration of various substances in blood, CSF and brain tissue as a result of blood-brain barrier activity - while under suspicion that even this mechanism might be modified by autism. As access to brain tissue for biochemical or immunological studies is restricted and spinal punctures are invasive procedures to which parents seldom consent, the great majority of tests have to focus on blood or urine, in an attempt to identify products relevant for a diagnosis.

For patients diagnosed with autism spectrum disorders, laboratory testing indicated a series of very diverse blood or urine parameter alterations, which, even if associable with metabolic process dysfunctions, did not yield a stable marker for autism, but instead generated new plausible causalities materialized into theories based on the oxidative stress, mitochondrial, hormonal and neuro-transmission dysfunctions, signal pathway alterations, immune dysregulation, maternal-fetal effects or subsequent combinations thereof. Unfortunately, no complex study has yet to use a single large study lot with good representation for all types of autism in order to simultaneously check for biochemical and immunological anomalies otherwise fractionally identified (a metabolomic approach) such as: oxidative enzyme and glutathione deficiencies, metal poisoning, excess lactate and pyruvate, disturbances in the urea cycle, iron and zinc deficits, low levels of serotonin and melatonin, excessive dopamine, the arginine vasopressin/oxytocin balance, atypical porphyrins in urine or excessive pro- and anti-inflammatory cytokines. For the moment, test results show little uniformity, as the polymorphic and complex spectrum of autism demands a quasi-individual approach for each patient. Understanding the need for a systematic approach to autism, the National Institute of Health (USA) launched an ambitious attempt to create a national, interactive network (*The Interactive Autism Network)* to facilitate data flow. *The National Database for Autism Research (NDAR),* established in November 2010, also aids future studies by making available a wide database of reference imagery as well as autistic genotypes and phenotypes. It functions in close collaboration with the Autism Genetic Resource Exchange (AGRE), which offers DNA, cell lineages and serum from families with autistic children. The aforementioned institutions are also currently considering the future establishment of a brain tissue database.

While biochemical and immunological analysis can not definitively answer *„what"* causes autistic behavior or *„how"*, neither can genetic testing indicate *"who"* is responsible for the modifications observed. Hundreds of studies sought associations between abnormal protein levels (structural proteins, enzymes, hormones, receptors, immunoproteins), their encoding genes and their corresponding regulatory regions. Results encountered the same inconsistencies regarding autism causality, but the amount of new data brought into behavioral genetics is by no means unimportant. New genomic analysis methods (aCGH) point to chromosomal areas most likely to be associated with autism, which reduce gene selection areas, admitting that their numbers may mean to autism either the contribution of more than one locus, genetic heterogeneity, epigenetic effects, or the result of environmental influence on genetic givens.

The evaluation of a diagnosis of autism has to arrive at answering *"when?"* and *"how?"* development impairment occurs on such a debilitating scale. Mental illnesses - therefore also autism - are regarded as multifactorial disorders. It is therefore plausible that an environmental factor, even in small quantities, is capable of setting in motion a series of molecular and cellular alterations which induce major behavioral modifications compatible with autism. It can be also inferred that this trigger for anatomical and functional brain damage penetrates the womb, where it may act upon the central nervous system throughout its long development, thus accounting for the permanence of the affliction. In order to reach the cerebral level, the presumed environmental factor, whether biological or chemical in origin, should be capable of crossing the open blood-brain barrier (while in uterus) or, if contamination occurs after the barrier has closed, to navigate it with or without inflicting structural changes. Some studies suggest that oxidative stress may affect blood brain barrier function (Lochhead et al., 2010) and others point to certain inefficiencies in the barrier itself in autism.

The fact that oxidative stress is capable of influencing gene expression by modifying DNA methylation patterns enforces the suspicion that it plays a critical role in autism. Oxidative stress on the organism is maintained by heavy metals, pesticides or Maillard reaction final products. These are brought into the body through aliments and may exert neurotoxic effects even in the womb, provoke redox imbalances, alter metabolic pathways, change membrane permeability or cause inflammation and immunological vulnerability. Perhaps it is not by chance that countries that use large quantities of organo-phosphate pesticides have the highest frequencies of autism, considering the fact that the enzyme that breaks down these products, paraoxonase 1, is deficient in autism (Deth et al., 2008; Paşca et al., 2006). Also, career-making women, who come to consume food that is nutritionally altered by packaging, coloring and flavoring, develop greater risks of delivering autistic children. Maillard reaction products intervene, among other things, in iron and zinc complexation (Zhang et al., 2009) - which may explain why these ions are scarce in autistic children, even before birth. On the other hand, from the hormonal perspective, whether modern women suffer from excessive testosterone and, furthermore, whether that can be linked to autism via the "extreme male brain" justification remains to be proven by further, more consistent, studies.

The effort to identify the cause behind autism spectrum disorders is coming more and more to resemble a worldwide, multifrontal assault on an impregnable labyrinthine territory. Current research implies an interdisciplinary approach to the complexity of the autistic phenomenon in order to arrive at relevant genotype-phenotype-environment correlations that offer a chance of early diagnosis and/or therapeutic approach.

6. References

Anderson, B. M., Schnetz-Boutaud, N., Bartlett, J., Wright, H. H., Abramson, R. K., Cuccaro, M. L., Gilbert, J. R., Pericak-Vance, M. A., & Haines, J. L. (2008). Examination of association to autism of common genetic variationin genes related to dopamine. *Autism Res*, Vol. 1, No. 6, (December 2008), pp. 364-369, ISSN: 1939-3792

Anderson, J. S., Druzgal, T. J., Froehlich, A., DuBray, M. B., Lange, N., Alexander, A. L., Abildskov, T., Nielsen, J. A., Cariello, A. N., Cooperrider, J. R., Bigler, E. D., & Lainhart, J. E. (2010). Decreased Interhemispheric Functional Connectivity in Autism. *Cerebral Cortex*. Advance online publication. Accessed on February 7, 2011, Available from: http://cercor.oxfordjournals.org/content/early/2010/10/12/ cercor.bhq190.long

Ashwood, P., Anthony, A., Torrente, F., & Wakefield, A. J. (2004). Spontaneous Mucosal Lymphocyte Cytokine Profiles in Children with Autism and Gastrointestinal Symptoms: Mucosal Immune Activation and Reduced Counter Regulatory Interleukin-10. *Journal of Clinical Immunology*, Vol. 24, No. 6, (November 2004), pp. 664-673, ISSN: 0271-9142

Ashwood, P., Wills, S., & Van de Water, J. (2006). The immune response in autism: a new frontier for autism research. *Journal of Leukocyte Biology*, Vol. 80, No. 1, (July 2006), pp. 1-15, ISSN: 0741-5400

Auyeung, B., Baron-Cohen, S., Ashwin, E., Knickmeyer, R., Taylor, K., & Hackett, G. (2009). Fetal testosterone and autistic traits. *British Journal of Psychology*, Vol. 100, No 1, (February 2009), pp. 1-22, ISSN: 2044-8295

Aw, T. Y. (1999). Molecular and cellular responses to oxidative stress and changes in oxidation-reduction imbalance in the intestine. *Am J Clinl Nutr*, Vol. 70, No. 4, (October 1999), pp. 557-565, ISSN: 0002-9165

Aylward, E. H., Minshew, N. J., Field, K., Sparks, B. F., & Singh, N. (2002). Effects of age on brain volume and head circumference in autism. *Neurology*, Vol. 59, No.2, (July 2002), pp. 175-183, ISSN: 0028-3878

Bade-White, P. A., Obrzut, J. E., & Randall, P. P. (2008). Neuropsychological Aspects of Pervasive Development and Autism Spectrum Disorders, In: *Handbook of Clinical Child Neuropsychology*, 3-rd Ed., Reynolds, C.R. & Fletcher-Janzen, E., pp. 765-783, Springer-Verlag, ISBN: 0387707085, New York, LLC

Bailey, A., Le Couteur, A., Gottesman, I., Bolton, P., Simonoff, E., Yuzda, E., & Rutter, M. (1995). Autism as a strongly genetic disorder: evidence from a British twin study. *Psychological Medicine*, Vol. 25, No. 01, (January 1995), pp. 63-77, ISSN: 0033-2917

Baron-Cohen, S., Leslie, A. M., & Frith, U. (1985). Does the autistic child have a "theory of mind"?. *Cognition*, Vol. 21, No. 1, (October 1985), pp. 37-46, ISSN: 0010-0277

Baron-Cohen, S. (2002). The extreme male brain theory of autism. *Trends Cogn Sci*. Vol. 6, No. 6, (June 2002), pp. 248-254. ISSN: 1364-6613

Baron-Cohen, S., Ashwin, E., Ashwin, C., Tavassoli, T., & Chakrabarti, B. (2009). Talent in autism: hyper-systemizing, hyper-attention to detail and sensory hypersensitivity. *Philos Trans R Soc Lond B Biol Sci*, Vol. 364, No. 1522, (May 2009), pp. 1377-1383, ISSN: 1471-2970

Barton, J., Barton, J. C., & Bertoli, L. (2010). Pica associated with iron deficiency or depletion: clinical and laboratory correlates in 262 non-pregnant adult outpatients. *BMC Blood Disorders*, Vol.10, No. 1, (December 2010), pp. 9, ISSN: 1471-2326

Bauman, M. L., & Kemper, T. L. (2005). Neuroanatomic observations of the brain in autism: A review and future directions. *Int J Dev Neurosc*, Vol. 23, No. 2-3, (April-May 2005), pp. 183–187, ISSN: 0736-5748

Blackwell, J. (2001). Clinical practice guideline: screening and diagnosing autism, *J Am Acad Nurs Pract*, Vol. 13, No. 12, (December 2001), pp. 534-6, ISSN: 1041-2972

Boddaert, N., & Zilbovicius, M. (2002). Functional neuroimaging and childhood autism, *Pediatr Radiol,* Vol.32, No.1, (January 2002), pp. 1-7, ISSN: 0301-0449

Bölte, S., Hubl, D., Dierks, T., Holtmann, M., & Poustka, F. (2008). An fMRI-study of locally oriented perception in autism: altered early visual processing of the block design test. *J Neural Transm,* Vol 115, No.3, (February 2008), pp. 545-552. ISSN: 0300-9564.

Boria, S., Fabbri-Destro, M., Cattaneo, L., Sparaci, L., Sinigaglia, C., Santelli, E., Cossu, G., & Rizzolatti, G. (2009). Intention understanding in autism. *PLoS One*. Vol. 4, No. 5, (May 2009), e5596. eISSN: 1932-6203. doi:10.1371/journal.pone.0005596

Boso, M., Emanuele, E., Politi, P., Pace, A., Arra, M., Ucelli di Nemi, S., & Barale, F. (2007). Reduced Plasma Apelin Levels in Patients with Autistic Spectrum Disorder. *Arch Medl Res,* Vol. 38, No. 1, (January 2007), pp. 70-74, ISSN: 0188-4409

Braunschweig, D., Ashwood, P., Krakowiak, P., Hertz-Picciotto, I., Hansen, R., Croen, L. A., Pessah, I. N., & Van de Water, J. (2008). Autism: Maternally derived antibodies specific for fetal brain proteins. *Neurotoxicology,* Vo. 29, No. 2, (March 2008), pp. 226-231, ISSN: 0161-813X

Burmester, T., Weich, B., Reinhardt, S., & Hankeln, T. (2000). A vertebrate globin expressed in the brain. *Nature,* Vol. 407, No. 6803, (September 2000), pp. 520-523, ISSN: 0028-0836

Carper, R. A., Moses, P., Tigue, Z. D., & Courchesne, E. (2002). Cerebral lobes in autism: early hyperplasia and abnormal age effects. *Neuroimage,* Vol. 16, No.4, (August 2002), pp. 1038–1051. ISSN: 1053-8119

Chauhan, A., Chauhan, V., Brown, W. T., & Cohen, I. (2004). Oxidative stress in autism: increased lipid peroxidation and reduced serum levels of ceruloplasmin and transferrin--the antioxidant proteins. *Life Sci,* Vol. 75, No. 21, (October 2004), pp. 2539-2549, ISSN: 0024-3205

Cheng, Y., Chou, K. H., Chen, I. Y., Fan, Y. T., Decety, J., & Lin, C. P. (2010). Atypical development of white matter microstructure in adolescents with autism spectrum disorders. *Neuroimage,* Vol 50, No.3, (April 2010), pp. 873-882, ISSN: 1053-8119

Chez, M. G., Dowling, T., Patel, P. B., Khanna, P., & Kominsky, M. (2007). Elevation of Tumor Necrosis Factor-Alpha in Cerebrospinal Fluid of Autistic Children. *Pediatric neurology,* Vol. 36, No. 6, (June 2007), pp. 361-365, ISSN: 0887-8994

Chugani, D. C., Muzik, O., Behen, M., Rothermel, R., Janisse, J. J., Lee, J., & Chugani, H. T. (1999). Developmental changes in brain serotonin synthesis capacity in autistic and nonautistic children. *Ann Neurol,* Vol. 45, No. 3, (March 1999), pp. 287–95, ISSN: 0364-5134

Chungpaibulpatana, J., Sumpatanarax, T., Thadakul, N., Chantharatreerat, C., Konkaew, M., & Aroonlimsawas, M. (2008). Hyperbaric oxygen therapy in Thai autistic children. *J Med Assoc Thai.,* Vol. 91, No. 8, (August 2008), pp. 1232-8, ISSN: 0125-2208

Chura, L. R., Lombardo, M. V., Ashwin, E., Auyeung, B., Chakrabarti, B., Bullmore, E. T., & Baron-Cohen, S. (2010). Organizational effects of fetal testosterone on human corpus callosum size and asymmetry. *Psychoneuroendocrinology,* Vol. 35, No. 1, (January 2010), pp. 122-132, ISSN: 0306-4530

Conturo, T. E., Williams, D. L., Smith, C. D., Gultepe, E., Akbudak, E., & Minshew, N. J. (2008). Neuronal fiber pathway abnormalities in autism: an initial MRI diffusion tensor tracking study of hippocampo-fusiform and amygdalo-fusiform pathways. *J Int Neuropsychol Soc.,* Vol. 14, No. 6, (November 2008), pp. 933-46, ISSN: 1355-6177

Courchesne, E., Campbell, K., & Solso, S. (2010). Brain growth across the life span in autism: Age-specific changes in anatomical pathology. *Brain Res.*, Vol. In Press, Corrected Proof, (Oct, 2010), ISSN 0006-8993, doi:10.1016/j.brainres.2010.09.101 Article in Press.

Coutinho, A., Sousa, I., Martins, M., Correia, C., Morgadinho, T., Bento, C., Marques, C., Ataíde, A., Miguel, T., Moore, J., Oliveira, G., & Vicente, A. (2007). Evidence for epistasis between SLC6A4 and ITGB3 in autism etiology and in the determination of platelet serotonin levels. *Hum Genet,* Vol. 121, No. 2, (April 2007), pp. 243-256, ISSN: 0340-6717

Crandall, J. E., McCarthy, D. M., Araki, K. Y., Sims, J. R., Ren, J.-Q., & Bhide, P. G. (2007). Dopamine Receptor Activation Modulates GABA Neuron Migration from the Basal Forebrain to the Cerebral Cortex. *J. Neurosci.*, Vol. 27, No. 14, (April 2007), pp. 3813-3822, ISSN: 0270-6474

Curran, L. K., Newschaffer, C. J., Lee, L. C., Crawford, S. O., Johnston, M. V., & Zimmerman, A. W. (2007). Behaviors associated with fever in children with autism spectrum disorders. *Pediatrics,* Vol. 120, No. 6, (December 2007), pp. 1386-1392, ISSN: 1098-4275

Cuscó, I., Medrano, A., Gener, B., Vilardell, M., Gallastegui, F., Villa, O., González, E., Rodriguez-Santiago, B., Vilella, E., Del Campo, M., & Pérez-Jurado, L. A. (2009). Autism-specific copy number variants further implicate the phosphatidylinositol signaling pathway and the glutamatergic synapse in the etiology of the disorder. *Hum Mol Genet,* Vol. 18, *No.* 10, (February 2009), pp. 1795-1804, ISSN: 0964-6906

Dapretto, M., Davies, M. S., Pfeifer, J. H., Scott, A. A., Sigman, M., Bookheimer, S. Y., & Iacoboni, M. (2006). Understanding emotions in others: mirror neuron dysfunction in children with autism spectrum disorders. *Nat Neurosci.*, Vol. 9, No. 1, (January 2006), pp. 28-31, ISSN: 1097-6256

Deth, R., Muratore, C., Benzecry, J., Power-Charnitsky, V. A., & Waly, M. (2008). How environmental and genetic factors combine to cause autism: A redox/methylation hypothesis. *Neurotoxicology,* Vol. 29, No. 1, (January 2008), pp. 190-201, ISSN: 0161-813X

Dietert, R. R., & Dietert, J. M. (2008). Potential for early-life immune insult including developmental immunotoxicity in autism and autism spectrum disorders: focus on critical windows of immune vulnerability. *J Toxicol Environ Health B Crit Rev,* Vol. 11, No. 8, (October 2008), pp. 660-680, ISSN: 1521-6950

Dosman, C. F., Brian, J. A., Drmic, I. E., Senthilselvan, A., Harford, M. M., Smith, R. W., Sharieff, W., Zlotkin, S. H., Moldofsky, H., & Roberts, S. W. (2007). Children with autism: effect of iron supplementation on sleep and ferritin. *Pediatr Neurol,* Vol 36, No. 3, (March 2007), pp. 152-158, ISSN: 0887-8994

Ferguson, J. N., Young, L. J., & Insel, T. R. (2002). The Neuroendocrine Basis of Social Recognition. *Front Neuroendocrinol,* Vol. 23, No. 2, (April 2002), pp. 200-224, ISSN: 0091-3022

Filipek, P. A., Accardo, P. J., Ashwal, S., Baranek, G. T., Cook, E. H. Jr., Dawson, G., Gordon, B., Gravel, J. S., Johnson, C. P., Kallen, R. J., Levy, S. E., Minshew, N. J., Ozonoff, S., Prizant, B. M., Rapin, I., Rogers, S. J., Stone, W. L., Teplin, S. W., Tuchman, R. F., & Volkmar, F. R. (2000). Practice parameter: screening and diagnosis of autism: report of the Quality Standards Subcommittee of the American Academy of Neurology and the Child Neurology Society. *Neurology,* Vol. 55, No. 4, (August 2000), pp. 468-79, ISSN: 0028-3878

Frazier, T. W., & Hardan, A. Y. (2009). A meta-analysis of the corpus callosum in autism. *Biol Psychiatry,* Vol. 66, No. 10, (November 2009), pp. 935-941. ISSN: 1873-2402.

Fukunaga, M., Li, T-Q., van Gelderen, P., de Zwart, J. A., Shmueli, K., Yao, B., Lee, J., Maric, D., Aronova, M. A., Zhang, G., Leapman, R. D., Schenck, J. F., Hellmut Merkle, H., & Duyn, J. H. (2010). Layer-specific variation of iron content in cerebral cortex as a source of MRI contrast. *PNAS,* Vol. 107, No. 8, (February 23), pp. 3834-3839, ISSN: 0027-8424

Geier, D. A., Kern, J. K., Garver, C. R., Adams, J. B., Audhya, T., Nataf, R., & Geier, M. R. (2009). Biomarkers of environmental toxicity and susceptibility in autism. *J Neurol Sci,* Vol. 280, No. 1-2, (May 2009), pp. 101-108, ISSN: 1878-5883

Georgescu, L. P. (2005). Reacţia de degradare oxidativă şi radicalică, In: *Inhibarea degradărilor radicalice din sisteme model cu utilizarea substanţelor naturale,* Georgescu P. Lucian, pp. 9-57, Editura Fundaţiei Universitare "Dunărea de Jos" din Galaţi, ISBN: 973-627-242-7, Galaţi.

Gilbert, S. J., Bird, G., Brindley, R., Frith, C. D., & Burgess, P. W. (2008). Atypical recruitment of medial prefrontal cortex in autism spectrum disorders: an fMRI study of two executive function tasks. *Neuropsychologia.* Vol. 46, No. 9, (April 2008), pp. 2281-91, ISSN: 0028-3932

Giulivi, C., Zhang, Y.-F., Omanska-Klusek, A., Ross-Inta, C., Wong, S., Hertz-Picciotto, I., Tassone, F., & Pessah, I. N. (2010). Mitochondrial Dysfunction in Autism. *JAMA,* Vol. 304, No. 21, (December 2010), pp. 2389-2396, ISSN: 0098 7484

Glessner, J. T., Wang, K., Cai, G., Korvatska, O., Kim, C. E., Wood, S., Zhang, H., Estes, A., Brune, C. W., Bradfield, J. P., Imielinski, M., Frackelton, E. C., Reichert, J., Crawford, E. L., Munson, J., Sleiman, P. M. A., Chiavacci, R., Annaiah, K., Thomas, K., Hou, C., Glaberson, W., Flory, J., Otieno, F., Garris, M., Soorya, L., Klei, L., Piven, J., Meyer, K. J., Anagnostou, E., Sakurai, T., Game, R. M., Rudd, D. S., Zurawiecki, D., McDougle, C. J., Davis, L. K., Miller, J., Posey, D. J., Michaels, S., Kolevzon, A., Silverman, J. M., Bernier, R., Levy, S. E., Schultz, R. T., Dawson, G., Owley, T., McMahon, W. M., Wassink, T. H., Sweeney, J. A., Nurnberger, J. I., Coon, H., Sutcliffe, J. S., Minshew, N. J., Grant, S. F. A., Bucan, M., Cook, E. H., Buxbaum, J. D., Devlin, B., Schellenberg, G. D., & Hakonarson, H. (2009). Autism genome-wide copy number variation reveals ubiquitin and neuronal genes. *Nature,* Vol. 459, No. 7246, (February 2009), pp. 569-573, ISSN: 0028-0836

Gomot, M., Bernard, F. A., Davis, M. H., Belmonte, M. K., Ashwin, C., Bullmore, E. T., & Baron-Cohen S. (2006). Change detection in children with autism: An auditory event-related fMRI study, *NeuroImage,* Vol.29, No.2, (January 2006), pp. 475–484, ISSN: 1095-9572.

Gulson, B. L., Jameson, C. W., Mahaffey, K. R., Mizon, K. J., Patison, N., Law, A. J., Korsch, M. J., & Salter, M. A. (1998). Relationships of lead in breast milk to lead in blood, urine, and diet of the infant and mother. *Environ Health Perspect.* Vol. 106, No. 10, (October 1998), pp. 667-74, ISSN: 0091-6765

Hadjikhani, N., Joseph, R. M., Snyder, J., & Tager-Flusberg, H. (2006). Anatomical Differences in the Mirror Neuron System and Social Cognition Network in Autism. *Cerebral Cortex,* Vol. 16, No. 9, (September 2006), pp. 1276-1282, ISSN: 1047-3211

Hammock, E. A. D., & Young, L. J. (2006). Oxytocin, vasopressin and pair bonding: implications for autism. *Philos Trans R Soc Lond B Biol Sci,* Vol. 361, No. 1476, (December 2006), pp. 2187-2198, ISSN: 0962-8436

Hardan, A. Y., Pabalan, M., Gupta, N., Bansal, R., Melhem, N. M., Fedorov, S., Keshavan, M. S., & Minshew, N. J. (2009). Corpus Callosum Volume in Children with Autism. *Psychiatry Res.*, Vol. 174, No. 1, (October 2009), pp. 57–61. ISSN: 0925-4927

Harony, H., & Wagner, S. (2010). The contribution of oxytocin and vasopressin to Mammalian social behavior: potential role in autism spectrum disorder. *Neurosignals,* Vol. 18, No. 2, (February 2011), pp. 82-97, ISSN: 1424-862X

Hay, G., Refsum, H., Whitelaw, A., Melbye, E. L., Haug, E., & Borch-Iohnsen, B. (2007). Predictors of serum ferritin and serum soluble transferrin receptor in newborns and their associations with iron status during the first 2 y of life. *Am J Clin Nutr.* 2007 Jul; Vol. 86, No. 1, (July 2007), pp. 64-73, ISSN: 0002-9165

Hazlett, H. C., Poe, M. D., Gerig, G., Styner, M., Chappell, C., Smith, R. G., Vachet, C., & Piven, J. (2011). Early brain overgrowth in autism associated with an increase in cortical surface area before age 2 years. *Arch Gen Psychiatry,* Vol. 68, No. 5, (May 2011), p. 467-476, ISSN:1538-3636

Haznedar, M. M., Buchsbaum, M. S., Hazlett, E. A., LiCalzi, E. M., Cartwright, C., & Hollander, E. (2006). Volumetric analysis and three-dimensional glucose metabolic mapping of the striatum and thalamus in patients with autism spectrum disorders. *Am J Psychiatry,* Vol. 163, No. 7, (July 2006), pp. 1252-1263, ISSN: 0002-953X

Hulthén, L. (2003). Iron deficiency and cognition. *Scan J Nutr,* Vol. 47, No. 3, (September 2003), pp. 152-156, ISSN: 1102-6480

Hurtado E. K., Claussen A. H., & Scott K. G. (1999). Early childhood anemia and mild or moderate mental retardation. *Am J Clin Nutr,* Vol 69, No. 1, (January 1999), pp. 115–9, ISSN: 0002-9165

Iacoboni, M., & Dapretto, M. (2006). The mirror neuron system and the consequences of its dysfunction. *Nat Rev Neurosci* Vol. 7, No. 12, (December 2006), pp. 942–951, ISSN: 1471-003X

James, S. J., Cutler, P., Melnyk, S., Jernigan, S., Janak, L., Gaylor, D. W., & Neubrander, J. A. (2004). Metabolic biomarkers of increased oxidative stress and impaired methylation capacity in children with autism. *Am J of Clin Nutr,* Vol. 80, No. 6, (December 2004), pp. 1611-1617, ISSN: 0002-9165

Joseph, R. M. (1999). Neuropsychological Frameworks for Understanding Autism. *Int Rev Psychiatry,* Vol. 11, No. 4, (November 1999), pp. 309–324, ISSN: 0954-0261

Just, M. A., Cherkassky, V. L., Keller, T. A., Kana R. K., & Minshew N. J. (2007). Functional and Anatomical Cortical Underconnectivity in Autism: Evidence from an fMRI Study of an Executive Function Task and Corpus Callosum Morphometry. *Cereb. Cortex* 17 (4): 951-961. ISSN: 1047-3211.

Kern, J. K., & Jones, A.M. (2006). Evidence of toxicity, oxidative stress, and neuronal insult in autism. *J Toxicol Environ Health B Crit Rev,* Vol. 9, No. 6, (June 2006), pp. 485–499, ISSN: 1093–7404

Kim, S. J., Young, L. J., Gonen, D., Veenstra-VanderWeele, J., Courchesne, R., Courchesne, E., Lord, C., Leventhal, B. L., Cook, E. H., Jr., & Insel, T. R. (2002). Transmission disequilibrium testing of arginine vasopressin receptor 1A (AVPR1A)

polymorphisms in autism. *Mol Psychiatry,* Vol. 7, No. 5, (June 2002), pp. 503-507, ISSN: 1359-4184

Kinney, D. K., Munir, K. M., Crowley, D. J., & Miller, A. M. (2008). Prenatal stress and risk for autism. *Neurosci Biobehav Rev,* Vol. 32, No. 8, (October 2008), pp. 1519-1532, ISSN: 0149-7634

Kleinhans, N. M., Johnson, L. C., Richards, T., Mahurin, R., Greenson, J., Dawson, G., & Aylward, E. (2009). Reduced neural habituation in the amygdala and social impairments in autism spectrum disorders. *Am J Psychiatry,* 166(4), 467-475. ISSN: 1535-7228.

Knaus, T. A., Silver, A. M., Kennedy, M., Lindgren, K. A., Dominick, K. C., Siegel, J., & Tager-Flusberg, H. (2010). Language laterality in autism spectrum disorder and typical controls: a functional, volumetric, and diffusion tensor MRI study. *Brain Lang,* Vol. 112, No. 2, (February 2010), pp. 113-120, ISSN: 0093-934X

Krauss, G. (2003). Intracellular Messenger Substances: "Second Messengers", In *Biochemistry of Signal Transduction and Regulation.* 3rd Edition. Gerhard Krauss, pp. 245-6, WILEY-VCH Verlag GmbH & Co. KGaA, ISBN: 3-527-30591-2, Weinheim

Kuczmarski, R. J., Ogden, C. L., Guo, S. S., Grummer-Strawn, L. M., Flegal, K. M., Mei, Z., Wei, R., Curtin, L. R., Roche, A. F., & Johnson, C. L. (2002). 2000 CDC growth charts for the United States: Methods and development. *Vital Health Stat,* Series 11, No. 246, (May 2002), pp. 39-40, ISBN 0-8406-0575-7

Larson, S. J. (2002). Behavioral and motivational effects of immune-system activation. *J. Gen. Psychol.* Vol. 129, No. 4, (October 2002), pp. 401-14, ISSN: 0022-1309

Latif, A., Heinz, P., & Cook, R. (2002). Iron deficiency in autism and Asperger syndrome. *Autism,* Vol. 6, No. 1, (March 2002), pp. 103-114, ISSN: 1362-3613

Lester, D. B., Rogers, T. D., & Blaha, C. D. (2010). Acetylcholine–Dopamine Interactions in the Pathophysiology and Treatment of CNS Disorders. *CNS Neuroscience & Therapeutics,* Vol. 16, No. 3, (June 2010), pp. 137–162, ISSN: 1755-5930

Levin, H. S., & Hanten, G. (2005). Executive Functions After Traumatic Brain Injury in Children. *Pediatric Neurology,* Vol. 33, No. 2, (August 2005), pp. 79-93, ISSN: 0887-8994

Levin, R., Heresco-Levy, U., Bachner-Melman, R., Israel, S., Shalev, I., & Ebstein, R.P. (2009). Association between arginine vasopressin 1a receptor (AVPR1a) promoter region polymorphisms and prepulse inhibition. *Psychoneuroendocrinology,* Vol. 34, No. 6, (July 2009), pp. 901-908, ISSN: 0306-4530

Lewine, J. D., Andrews, R., Chez, M., Patil, A. A., Devinsky, O., Smith, M., Kanner, A., Davis, J. T., Funke, M., Jones, G., Chong, B., Provencal, S., Weisend, M., Lee, R. R., & Orrison, W. W., Jr. (1999). Magnetoencephalographic patterns of epileptiform activity in children with regressive autism spectrum disorders. *Pediatrics,* Vol. 104, No. 3 Pt 1, (September 1999), pp. 405-418, ISSN: 0031-4005

Lim, M. M., & Young, L. J. (2006). Neuropeptidergic regulation of affiliative behavior and social bonding in animals. *Hormones and Behavior,* Vol. 50, No. 4, (November 2006), pp. 506-517, ISSN: 0018-506X

Lochhead, J. J., McCaffrey, G., Quigley, C. E., Finch, J., DeMarco, K. M., Nametz, N., & Davis, T.P. (2010). Oxidative stress increases blood-brain barrier permeability and induces alterations in occludin during hypoxia-reoxygenation. *J Cereb Blood Flow Metab,* Vol. 30, No. 9, (September 2010), pp. 1625-1636, ISSN: 0271-678X

Lombard, J. (1998). Autism: a mitochondrial disorder? *Med Hypotheses,* Vol. 50, No. 6, (June 1998), pp. 497-500, ISSN: 0306-9877

Marshall, C. R., Noor, A., Vincent, J. B., Lionel, A. C., Feuk, L., Skaug, J., Shago, M., Moessner, R., Pinto, D., Ren, Y., Thiruvahindrapduram, B., Fiebig, A., Schreiber, S., Friedman, J., Ketelaars, C. E. J., Vos, Y. J., Ficicioglu, C., Kirkpatrick, S., Nicolson, R., Sloman, L., Summers, A., Gibbons, C. A., Teebi, A., Chitayat, D., Weksberg, R., Thompson, A., Vardy, C., Crosbie, V., Luscombe, S., Baatjes, R., Zwaigenbaum, L., Roberts, W., Fernandez, B., Szatmari, P., & Scherer, S. W. (2008). Structural Variation of Chromosomes in Autism Spectrum Disorder. *American journal of human genetics,* Vol. 82, No. 2, (February 2008), pp. 477-488, ISSN: 0002-9297

Matuszek, G., & Talabizadeh Z. (2009). Autism Genetic Database: A comprehensive database for autism susceptibility gene-CNVs integrated with known noncoding RNAs and fragile sites. *BMC Medical Genetics.,* Vol. 10, No. 1, (September 2009), pp. 102, ISSN: 1471-2350

Melke, J., Botros, H. G., & Chaste, P. (2008). Abnormal melatonin synthesis in autism spectrum disorders. *Mol Psychiatry,* Vol. 13, No. 1, (January 2008), pp. 90–8, ISSN: 1359-4184.

Miyazaki, K., Narita, N., & Narita, M. (2005). Maternal administration of thalidomide or valproic acid causes abnormal serotonergic neurons in the offspring: implication for pathogenesis of autism. *Int J Dev Neurosc,i* Vol. 23, No. 2-3, (April-May 2005), pp. 287–97, ISSN: 0736-5748

Modahl, C., Green, L. A., Fein, D., Morris, M., Waterhouse, L., Feinstein, C., & Levin, H. (1998). Plasma oxytocin levels in autistic children. *Biological psychiatry,* Vol. 43, No. 4, (February 1998), pp. 270-277, ISSN: 0006-3223

Moretti, P., Sahoo, T., Hyland, K., Bottiglieri, T., Peters, S., del Gaudio, D., Roa, B., Curry, S., Zhu, H., Finnell, R. H., Neul, J. L., Ramaekers, V. T., Blau, N., Bacino, C. A., Miller, G., & Scaglia, F. (2005). Cerebral folate deficiency with developmental delay, autism, and response to folinic acid. *Neurology,* Vol. 64, No. 6, (March 2005), pp. 1088-1090, ISSN: 0003-9942

Muhle, R., Trentacoste, S. V., & Rapin, I. (2004). The Genetics of Autism. *Pediatrics,* Vol. 113, No. 5, (May 2004), pp. e472-486, ISSN: 0031 4005

Muscat, S. (2007). Influence of Food-Derived Maillard Reaction Products on the Cellular Response of Macrophages. Dissertation. *Friedrich-Alexander-Universität Erlangen-Nürnberg,* (March 2011), Available from http://www.opus.ub.uni-erlangen.de/opus/volltexte/2007/604/

Nakamura, K., Sekine, Y., Ouchi, Y., Tsujii, M., Yoshikawa, E., Futatsubashi, M., Tsuchiya, K. J., Sugihara, G., Iwata, Y., Suzuki, K., Matsuzaki, H., Suda, S., Sugiyama, T., Takei, N., & Mori, N. (2010). Brain Serotonin and Dopamine Transporter Bindings in Adults With High-Functioning Autism. *Arch Gen Psychiatry,* Vol. 67, No. 1, (January 2010), pp. 59-68, ISSN: 0003-990x

Nguyen, A., Rauch, T. A., Pfeifer, G. P., & Hu, V. W. Global methylation profiling of lymphoblastoid cell lines reveals epigenetic contributions to autism spectrum disorders and a novel autism candidate gene, RORA, whose protein product is reduced in autistic brain. *The FASEB Journal,* Vol 24, No. 8, (August 2010), pp. 3036-51, ISSN: 0892-6638

Oberman, L. M., Hubbard, E. M., McCleery, J. P., Altschuler, E. L., Ramachandran, V. S., & Pineda, J. A. (2005). EEG evidence for mirror neuron dysfunction in autism spectrum disorders. *Cog. Brain Res*, Vol. 24, No. 2, (July 2005), pp. 190–198, ISSN: 0926-6410

Ohnishi, T., Matsuda, H., Hashimoto, T., Kunihiro, T., Nishikawa, M., Uema, T., & Sasaki, M. 2000. Abnormal regional cerebral blood flow in childhood autism. *Brain*, Vol.123, No.9, (September 2000), pp.1838-1844, ISSN: 1838-1844

Page, L. A., Daly, E., Schmitz, N., Simmons, A., Toal, F., Deeley, Q., Ambery, F., McAlonan, G. M., Murphy, K. C., & Murphy, D. G. M. (2006). In Vivo 1H-Magnetic Resonance Spectroscopy Study of Amygdala-Hippocampal and Parietal Regions in Autism. *Am J Psychiatry*, Vol. 163, No. 12, (December 2006), pp. 2189-2192, ISSN: 0002-953X

Paşca, S. P., Nemeş, B., Vlase, L., Gagyi, C. E., Dronca, E., Miu, A. C., & Dronca, M. (2006). High levels of homoczsteine and low serum paraoxonase 1 arylesterase activity in children with autism. *Life Sciences*, Vol. 78, No. 19, (April 2006), pp. 2244-48, ISSN: 0024-3205

Perkins, T., Stokes, M., McGillivray, J., & Bittar, R. (2010). Mirror neuron dysfunction in autism spectrum disorders. *J Clin Neurosci*, Vol. 17, No. 10, (October 2010), pp. 1239-1243, ISSN: 1532-2653

Perreau-Linck, E., Beauregard, M., Gravel, P., Paquette, V., Soucy, J. P., Diksic, M., & Benkelfat, C. (2007). In vivo measurements of brain trapping of C-labelled alpha-methyl-L-tryptophan during acute changes in mood states. *J Psychiatry Neurosci.*, Vol. 32, No. 6, (November 2007), pp. 430-4, ISSN: 1180-4882

Pickering, M., Cumiskey, D., & O'Connor, J. J. (2005). Actions of TNF-alpha on glutamatergic synaptic transmission in the central nervous system. *Experimental Physiology*, Vol. 90, No. 5, (September 2005), pp. 663–670, ISSN: 0958-0670

Pierce, K., Müller, R. A., Ambrose, J., Allen, G., & Courchesne, E. (2001). Face processing occurs outside the fusiform "face area" in autism: evidence from functional MRI. *Brain*, Vol. 124, No. 10, (October 2001), pp. 2059-2073, ISSN: 0006-8950

Rizzolatti, G., & Fabbri-Destro, M. (2010). Mirror neurons: from discovery to autism. *Exp Brain Res*, Vol. 200, No. 3-4, (January 2010), pp. 223-237, ISSN: 0014-4819

Rossignol, D. A., & Rossignol L. W. (2006). Hyperbaric oxygen therapy may improve symptoms in autistic children. *Med Hypotheses*, Vol 67, No.2, (March 2006), pp. 216-28, ISSN: 0306-9877

Saitoh, O., Karns, C. M., & Courchesne, E. (2001). Development of the hippocampal formation from 2 to 42 years: MRI evidence of smaller area dentata in autism. *Brain*, Vol. 124, No. Pt 7, (July 2001), pp. 1317-1324, ISSN: 0006-8950

Sammarco, M. C., Ditch, S., Banerjee, A., & Grabczyk, E. (2008). Ferritin L and H Subunits Are Differentially Regulated on a Post-transcriptional Level. *Journal of Biological Chemistry*, Vol. 283, No. 8, (February 2008), pp. 4578-4587, ISSN: 0021-9258

Sarachana, T, Xu, M, Wu, R. C., & Hu, V. W. (2011). Sex hormones in autism: androgens and astrogens differentially and reciprocally regulate RORA, a novel candidate gene for autism. *PLoS One*, Vol l6, No. 2, (February 2011), pp. e17116, ISSN: 1932-6203

Schumann, C. M., & Amaral, D. G. (2006). Stereological analysis of amygdala neuron number in autism. *J Neurosci.*, Vol 26, No. 29, (July 2006), pp. 7674–7679, ISSN: 0270-6474

Seyan, A. S., Hughes, R. D., & Shawcross, D. L. (2010). Changing face of hepatic encephalopathy: role of inflammation and oxidative stress. *World J Gastroenterol.* Vol. 16, No. 27, (July 2010), pp. 3347-57, ISSN: 1007-9327

Singer, H. S., Morris, C. M., Williams, P. N., Yoon, D. Y., Hong, J. J., & Zimmerman, A. W. (2006). Antibrain antibodies in children with autism and their unaffected siblings. *J Neuroimmunol,* Vol. 178, No. 1, (December 2006), pp 149 -55, ISSN: 0165-5728

Smart, T. G., Hosie, A. M., & Miller, P.S. (2004). Zn2+ Ions: Modulators of Excitatory and Inhibitory Synaptic Activity. *The Neuroscientist,* Vol. 10, No. 5, (October 2004), pp. 432-442, ISSN: 1073-8584

Smeltzer, M. D., Curtis, J. T., Aragona, B. J., & Wang, Z. (2006). Dopamine, oxytocin, and vasopressin receptor binding in the medial prefrontal cortex of monogamous and promiscuous voles. *Neuroscience Letters,* Vol. 394, No. 2, (February 2006), pp. 146-151, ISSN: 0304-3940

Stritch School of Medicine. (2009). *Autism Candidate Gene Map (ACGMAP) Database,* Accessed on 13.03.2011, Available from:
http://www.meddean.luc.edu/node/375

Stromswold, K. (2008). The Genetics of Speech and Language Impairments. *NEJM,* Vol. 359, No. 22, (November 2008), pp. 2381-83, ISSN: 0028-4793

Szewczyk, B., Poleszak, E., Pilc, A., & Nowak, G. (2010). Ionic Glutamate Modulators in Depression (Zinc, Magnesium), In *Glutamate-based Therapies for Psychiatric Disorders, Milestones in Drug Therapy,* P. Skolnick (ed.), p. 21-38, Springer, Retrieved from: DOI 10.1007/978-3-0346-0241-9_2

Takeda, A., Hirate, M., Tamano, H., Nisibaba, D., & Oku, N. (2003). Susceptibility to kainate-induced seizures under dietary zinc deficiency. *Journal of Neurochemistry,* Vol. 85, No. 6, (June 2003), pp. 1575-1580, ISSN: 1471-4159

Tsai, L. Y. (1999). Psychopharmacology in Autism. *Psychosomatic Medicine* Vol. 61, No. 5, (September 1999), pp.651-665, ISSN: *0033-3174*

Verhoeven, J., De Cock, P., Lagae, L., & Sunaert, S. Neuroimaging of autism. *Neuroradiology,* Vol. 52, No. 1, (January 2010), pp. 3-14, ISSN: 0028-3940

Wakasugi, K., Kitatsuji, C., & Morishima, I. (2005). Possible neuroprotective mechanism of human neuroglobin. *Ann N Y Acad Sci,* Vol. 1053, (August 2005), pp. 220-230, ISSN: 0077-8923

Wang, Z. X., Liu, Y., Young, L. J., & Insel, T. R. (2000). Hypothalamic vasopressin gene expression increases in both males and females postpartum in a biparental rodent. *J Neuroendocrinol.,* Vol. 12, No. 2, (February 2000), pp. 111–120, ISSN: 0953-8194

Waterlow, J. C. (2006). *Protein turnover,* CABI Publishing, ISBN: 0851996132, UK-USA.

Weiss, L. A., Shen, Y., Korn, J. M., Arking, D. E., Miller, D. T., Fossdal, R., Saemundsen, E., Stefansson, H., Ferreira, M. A. R., Green, T., Platt, O. S., Ruderfer, D. M., Walsh, C. A., Altshuler, D., Chakravarti, A., Tanzi, R. E., Stefansson, K., Santangelo, S. L., Gusella, J. F., Sklar, P., Wu, B.-L., & Daly, M. J. (2008). Association between Microdeletion and Microduplication at 16p11.2 and Autism. *NEJM,* Vol. 358, No. 7, (February 2008), pp. 667-675, ISSN: 0028-4793

Weissman, J. R., Kelley, R. I., Bauman, M. L., Cohen, B. H., Murray, K. F., Mitchell, R. L., Kern, R. L., & Natowicz, M. R. (2008). Mitochondrial disease in autism spectrum disorder patients: a cohort analysis. *PLoS One,* Vol. 3, No. 11, (December 2008), pp. e3815, ISSN: 1932-6203

WHO Multicentre Growth Reference Study Group. (2007). WHO Child Growth Standards: Head circumference-for-age, arm circumference-for-age, triceps skinfold-for-age and subscapular skinfold-for-age: Methods and development. World Health Organization, ISBN-13: 9789241547185, Geneva, Retrived from http://www.who.int/childgrowth/standards/second_set/technical_report_2/en/index.html

WHO. (2011). WHO Anthro (version 3.2.2, January 2011) and macros. Accesses on 2011-02-05, Available from: http://www.who.int/childgrowth/software/en/

Willmore, L. J., Pickens IV, A., & Pellok, J. M. (2006). Monitoring for Adverse Effects of Antiepileptic Drugs. In: *The treatment of epilepsy: principles & practice,* Elaine Wyllie, Ajay Gupta, Deepak K. Lachhwani, pp. 735-747, Lippincott Williams & Wilkins, ISBN: 9780781749954, Hardcover

Winslow, J., & Insel, T. (1991). Social status in pairs of male squirrel monkeys determines the behavioral response to central oxytocin administration. *J Neurosci.,* Vol. 11, No. 7. (July 1991), pp. 2032-2038, ISSN: 0270-6474

Winslow, J. T., Hastings, N., Carter, C. S., Harbaugh, C. R., & Insel T. R. (1993). A role for central vasopressin in pair bonding in monogamous prairie voles. *Nature,* Vol. 365, No. 6446, (October 1993), pp. 545 – 548, ISSN: 0028-0836

Woods, J. S., Armel, S. E., Fulton, D. I., Allen, J., Wessels, K., Simmonds, P. L., Granpeesheh, D., Mumper, E., Bradstreet, J. J., Echeverria, D., Heyer, N. J., & Rooney, J. P. K. (2010). Urinary Porphyrin Excretion in Neurotypical and Autistic Children. *Environ Health Perspect.* Vol. 118, No. 10, (October 2010), ISSN: 0091-6765

Xu, J., Zwaigenbaum, L., Szatmari, P., & Scherer, S. W. (2004). Molecular Cytogenetics of Autism. Current Genomics Vol. 5, No. 4, (May 2004), pp. 347-364, ISSN: 1389-2029

Yamasaki, S., Yamasue, H., Abe, O., Suga, M., Yamada, H., Inoue, H., Kuwabara, H., Kawakubo, Y., Yahata, N., Aoki, S., Kano, Y., Kato, N., & Kasai, K. (2010). Reduced Gray Matter Volume of Pars Opercularis Is Associated with Impaired Social Communication in High-Functioning Autism Spectrum Disorders. *Biol Psychiatry,* Vol 68, No. 12, (December 2010), pp. 1141-7, ISSN: 0006-3223

Yirmiya, N., Rosenberg, C., Levi, S., Salomon, S., Shulman, C., Nemanov, L., Dina, C., & Ebstein, R. P. (2006). Association between the arginine vasopressin 1a receptor (AVPR1a) gene and autism in a family-based study: mediation by socialization skills. *Mol Psychiatry,* Vol. 11, No. 5, (May 2006), pp. 488-494, ISSN: 1359-4184

Yonan, A. L., Alarcon, M., Cheng, R., Magnusson, P. K., Spence, S. J., Palmer, A. A., Grunn, A., Juo, S. H., Terwilliger, J. D., Liu, J., Cantor, R. M., Geschwind, D. H., & Gilliam, T. C. (2003). A genomewide screen of 345 families for autism-susceptibility loci. *Am J Hum Genet,* Vol. 73, No. 4, (October 2003), pp. 886-897, ISSN: 0002-9297

Yorbik, O., Sayal, A., Akay, C., Akbiyik, D. I., & Sohmen, T. (2002). Investigation of antioxidant enzymes in children with autistic disorder. *Prostaglandins, Leukotrienes and Essential Fatty Acids,* Vol. 67, No. 5, (November 2002), pp. 341-343, ISSN: 0952-3278

Youdim, M.B. (2008). Brain iron deficiency and excess; cognitive impairment and neurodegeneration with involvement of striatum and hippocampus. *Neurotox Res,* Vol. 14, No. 1, (August 2008), pp. 45-56, ISSN: 1029-8428

Zhang, Q., Ames, J. M., Smith, R. D., Baynes, J. W., & Metz, T. O. (2009). A perspective on the Maillard reaction and the analysis of protein glycation by mass spectrometry: probing the pathogenesis of chronic disease. *J Proteome Res,* Vol. 8, No. 2, (February 2009), pp. 754-769, ISSN: 1535-3893

Zhao, B., & Schwartz, J. P. (1998). Involvement of cytokines in normal CNS development and neurological diseases: recent progress and perspectives. *J. Neurosci. Res.* Vol. 52, No. 1, (April 1998), pp. 7-16, ISSN: 0360-4012

Zhao, L., & Brinton, R. D. (2004). Suppression of Proinflammatory Cytokines Interleukin-1β and Tumor Necrosis Factor-α in Astrocytes by a V1 Vasopressin Receptor Agonist: A cAMP Response Element-Binding Protein-Dependent Mechanism. *J Neurosci.,* Vol. 24, No. 9, (March 2004), pp. 2226-2235, ISSN: 0270-6474

Zilbovicius, M., Boddaert, N., Belin, P., Poline, J. B., Remy, P., Mangin, J.F., Thivard, L., Barthelemy, C., & Samson, Y. (2000). Temporal lobe dysfunction in childhood autism: a PET study. *Am J Psychiatry,* Vol. 157, No. 12, (December 2000), pp.1988-93, ISSN: 0002-953X

Zilbovicius, M., Meresse, I., & Boddaert, N. (2006). Autism: neuroimaging. *Rev Bras Psiquiatr,* Vol.28, Suppl 1, (May 2006), pp. S21-28, ISSN: 1516-4446

21

Loneliness and Silence in Autism - Implications for Psychotherapy

Anna Bieniarz
Medical University of Lublin
Poland

1. Introduction

The aim of the following chapter is to present the two phenomena (loneliness and silence) that can be seen in the autistics and non-autistic subjects. At first there were presented characteristics of each of the mentioned experiences. The main hypothesis in the psychotherapy with subject suffering from autism is connected with the existence of the healthy and disordered part of the personality. The therapist should refer to the healthy part and make efforts to find the meaning of the observed pathological symptoms.

This content shows a phenomenological description of loneliness and silence and theirs connections with autism. It also considers how these two phenomena might influence the treatment.

The notion "autism" was used for the first time by E. Bleur in the beginning of the 20th century. The name originates from the Greek word "autos' (alone) and refers to the psychopathological tendency of turning away from the outside world and focusing on the inner world. Initially, autism was treated as the result of or a part of schizophrenic detachment. The term 'autism' entered the official classification in 1943, when Leo Kanner used it to describe a disorder whose primarily feature was the significant abnormalities of social, cognitive and behavioral function in a few children (APA, 1994; WHO, 2000).

The short description of autism indicates the importance of the loneliness seen as a behavioral factor noticed by the researchers.

2. Phenomenon of loneliness

Loneliness is a phenomenon perfectly known to every human being. It is a universal, subjective and difficult to articulate sensation typically associated with negative emotions. Yet loneliness can take the shape of solitude and the positive aspects should not be underestimated. It enables humans to reflect upon themselves, their lives and the surrounding world. Undoubtedly, a significant intensity of loneliness/ solitude leads to physical (Hawkley&Cacioppo, 2010) and mental (Heinrich& Gullone, 2006) dysfunctions. In social understanding loneliness is perceived as a pejorative phenomenon and only unwillingly is it acknowledged by the lonely person. He or she usually fears being stigmatized and becoming a social outcast with the label of a scapegoat or a failure.

Consequently, loneliness is often accompanied by shame and remorse (Dill&Anderson, 1999). In his analysis of the literature, Rotenberg enumerates two elements of loneliness: 1).

cognitive, referring to the disparity between the desired and real social relations in respect to quantity and quality, and 2) affective, highlighting the negative experience of being disoriented, lonely and lost (Rotenberg, 1999).

2.1 Definition

Researchers have not arrived at a coherent definition concerning loneliness. Generally, they distinguish two approaches to the phenomenon: 1) social needs theory and 2) cognitive process theory. The first approach represented by Bowlby, Sullivan and Weiss defines loneliness as a *response to a relational deficit that gives rise to a learning for the insufficient relationship* (Terrel-Deutsch, 1999, p.11). It underlines the human need of social relationships as a means for proper emotional and physical development. Here, loneliness is perceived affectively. The second approach, the cognitive process theory, developed by Peplau and Perlman, focuses on analyzing loneliness from the inner perspective. This theory revolves around cognition and the social relations of an individual. It defines loneliness as a lack of satisfaction resulting from an insufficient network of social relations. There are two factors which may be responsible for such a situation. The first one is the changes within the interpersonal relations of an individual and the second is the changes in the desired, or expected social relations (Terrell-Duetsch, 1999).

A slightly different definition of loneliness, rooted in the philosophical approach, was given by McGraw. He claimed: *loneliness pertains primarily to desired but unattained personal or personified relationships for which one yearns; secondarily, to the absence of current relationships which one misses; and, tertiarily, to the loss of past relationships which one mourns or grieves. Furthermore, the lack or loss of such relationships refers principally to a defectiveness in their quality and, in a subsidiary manner, to a deficit in their quality. (…) Loneliness is a deficiency of the needs and metaneeds of intimacy/meaning, specifically of that kind of intimacy which is meaningful and of that kind of meaning which is intimate* (McGraw,1995, p.44).

The above definition indicates that in order to avoid the negative experience associated with loneliness, the person's metaneeds should be fulfilled.

2.2 Types of loneliness

One of the most known typologies of loneliness was created by Weiss who described it as the most distressful experience in a human life. He distinguished two subcategories of loneliness: emotional and social. According to him emotional loneliness appears when a person undergoes lack of a close emotional relation with another human being. Consequently, anxiety and isolation appear which can be overcome only by the establishment of a new relationship involving attachment. Social loneliness, as Weiss referred to it, is caused by the lack of the network of social relations. It mainly influences people who enter a new and unknown social environment, and have not created the typical network of relations yet. A possible solution to the problem of social loneliness is the involvement in friendships enabling social interactions (Weiss, 1985).

Basing on the time criterion, Young subdivided loneliness into: short-term, incidental, situational and chronic loneliness. These subcategories may all be conditioned by subjective personal characteristics, measured with the Eysenck Personality Questionnaire (Young, 1978 cited in: Hojat, 1983). Short-term loneliness can be experienced by everybody as a reaction to the occurrence of an undesirable event in the life of every human being. Chronic loneliness is linked with such personal characteristics as: elevated neuroticism and anxiety, introversion, low self-esteem, low level of openness, increased depressive mood, and external locus of control (Hojat, 1982).

Some researchers examining loneliness indicate the necessity of a clear-cut division between pathological type of loneliness and the common type known to every human being (Moustakas, 1961; Booth, 1997). Pathological loneliness may endanger human health, and especially mental health. It originates in existential loneliness exhibited by all people and closely connected with human sense of fundamental emptiness and disconnectedness as well as the consciousness of the unknown, the feeling of transience, and the inevitability of death (Booth, 1997).

Applebaum's psychodynamic approach to the phenomenon of loneliness is based on the feeling of loss caused by the process of separation-individuation (Applebaum, 1978). She indicated 4 types of loneliness: 1) existential, 2) reactive, 3) pervasive nonspecific and 4) psychotic. The first category that is existential loneliness, results in a process of individuation and differentiation leading to a greater degree of autonomy and independent functioning. The second type, the reactive loneliness, appears in response to specific loss. As Applebaum notices:

It can include lonely reactions to any recent loss, as those endured during enforced physical isolation, separation from loved ones, loss of a body part, loss of self-esteem, loss of a vocation or avocation, loss of fantasy, or loss of attachment figures for any reason or cause (Applebaum, 1978, p.17). The third category of loneliness, namely, pervasive nonspecific, refers to people capable of creating social relations but whose dysfunctional habits influence the formation of these relations. According to Applebaum, this category consists of neurotics and people suffering from personality disorders, and particularly those presenting borderline personality disorder (BPD). The key characteristics of psychotic loneliness are: prolonged regression to the unity of infancy and primary narcissism. It can be defined as specific vulnerability typical for schizophrenic patients. The main feature distinguishing psychotic loneliness from all the other types is its degree, duration and defenses.

A very interesting idea of loneliness was put forward by John McGraw who, reaching for its philosophical sources, subdivided it into 10 categories: 1) metaphysical, 2) epistemological, 3) communicative, 4) ontological (or intrapersonal), 5) ethical (or moral), 6) existential, 7) emotional (linked with love and sexual instinct), 8) social (linked with friendship), 9) cultural and 10) cosmic (McGraw, 1995).

All the aboved mentioned forms of loneliness affect healthy people with no mental disorders. The classification is cited to indicate the significance of the phenomenon of loneliness, also observed in patients with the Autistic Spectrum Disorders.

McGraw claims that all types of loneliness emerge from the metaphysical loneliness (1). He states: *it denotes the lack of intimate/ meaningful solidarity with other beings and bespeaks an entitative-emotional longing for their plenitude and connectedness* (McGraw, 1995, p.46). This type of loneliness arises from the consciousness of individual distinctiveness from others resulting in insecurity and instability. Seemingly, it can be compared with the psychodynamic type of loneliness. The core of the epistemological loneliness (2) is defined by McGraw as following: *one is too close to one's self to intimately grasp the self and too far from others to know or be known by them, with the result that such knowledge is ephemeral and superficial* (McGraw,1995, p.50). This assumption postulates the incapability of knowing profoundly both oneself and other human beings.

Communicative loneliness (3) consists of 3 subcategories. The first occurs in reaction to the inability to communicate effectively with others because of inadequate social skills, communicative deficits or lack of social intelligence. The second subcategory of communicative loneliness refers to the failure to develop relationships with desired intimates. The last subcategory designates the incompetence to express negative or self-devaluating emotions connected with isolation resulting in all types of loneliness.

Ontological loneliness (intrapersonal) (4) *is felt threat to obtaining or maintaining self identity and self-integrity, a menace that occurs due to a deficiency of intimacy/meaning which the self seeks within itself* (McGraw,1995, p.53). This type of loneliness is caused by the absence of *the desired other.*

Ethical (moral) loneliness (5) is defined as: *loneliness inherent in freedom, choice and responsibility as well as in value formation, enactment and commitment. It entails the formidable moral task of facing one's loneliness in its diverse forms and of converting it into ethically constructive attitudes and actions* (McGraw, 1995, p.55). Such description of loneliness can also be found in Sartre.

Existential loneliness (6) *is found in the inevitable ruptures of intimacy/meaning within the separate self as it journeys by way of its individuation and socialization through its checkered history of involution / evolution and disintegration / integration* (McGraw, 1995, p.57). This form of loneliness is the most frequent in the so-called *Jasperian* "borderline situation," a situation of radical "*loneliness*".

Emotional loneliness (7), according to McGraw's typology, refers to the physical lack of sexual partner or a person we can feel emotionally attached to. Whereas social loneliness (8) is prescribed to experienced lack of company, friendship or sense of community. The two types of loneliness are borrowed from classification created by Weiss.

Cultural loneliness (9) appears in the case of the lack of the sense of belonging to the mainstream society. It affects people disconnected from the surrounding society. This type of loneliness is typically characterized by lack of cohesion or identity and individuals suffering from cultural loneliness are often perceived as outsiders. Consequently, they may develop an alternative subgroup or 'tribe' of a negative identity.

People sensing cosmic loneliness (10) are under the impression that there is no human being or force, they can refer to or identify with. It is a feeling of a complete and universal loneliness. A sufferer of cosmic loneliness sees the world as unfriendly and hostile and unable to fulfill human needs, and especially emotional needs.

2.3 Developmental stages and loneliness

According to Moustakas' theory, loneliness is the main force motivating human actions (Moustakas, 1961). In order to avoid being lonely and protect themselves from helplessness, humans are capable of making great efforts to connect with other human beings.

Humans experience loneliness for the first time during infancy (at the age of 3-8 months) while the process of differentiation form the outside world represented by the mother's breast begins. Because of experiencing frustration resulting from unfulfilled nutritional needs, the baby may painfully experience his/ her own loneliness (Mahler, 19962 cited in: Dorr & Friedenberg, 1993). The beginning of the process of separation-individuation reaches its peak in adolescence, when the subjective enforcement of loneliness seems to be the greatest. The particularly strong deprivation of the need for contact during infancy may result in the creation of analytical depression, endangering both health and life of the baby (Reis & Grenyer, 2002).

Loneliness in pre-school children is difficult to diagnose. In his 3 to 5 years old, a child is not able to identify and define sensed emotional states. Recently, the research has proved that typically developed pre-school children often confuse loneliness with boredom (Kirova 2004). The appearance of loneliness in healthy children does not cause any reasons for worry and consideration as the child is only beginning to learn how to interact socially through a process of socialization. Playing with mates constitutes an attractive alternative to everyday life and boosts the cognitive development.

During the school period, children enter a group of peers, make friends and become more sensitive to rejection which may lead to loneliness. A prolonged instance of being alone has proved to show a positive correlation with a depressive mood and an elevated level of anxiety in adolescence (Fontaine et al.,2009; Qualter et al. ,2010). Adolescence is a period during which the company and acceptance of peers begin to have a significant meaning. In the background of maturing, there is a clearly visible dichotomy between the need for community, contacts with peers, and the need for seclusion and isolation from the surrounding environment. Lack of acceptance on the part of peers may cause loneliness and inhibit the fulfillment of major social needs (Pretty et al., 1994). A teenager tends to reflect on his/ her place in the world, to experiment on his/ her own body, that is to change the image as well as try the effects of using psychoactive substances and alcohol. Such activities are especially frequent when accompanied by the acceptance of the group. Sexual drive is also awakened and the possible difficulties in finding a partner can result in the feeling of loneliness. Moreover, adolescence is an important phase in a life of a youth both because of crossing the line ending childhood and because of the need to fulfill specific social requirements characteristics for adulthood.

Adult failures in the sphere of social relationships may result in the appearance of different addictions masking loneliness. The focus on productivity and effectiveness in the professional sphere overlaps the social expectance to set up a family, exerting pressure on the young adult who often cannot deal with.

Senility is characterized by the greatest intensity of social loneliness, when the senior person is particularly exposed to the experience of loss of the peers, friends, and neighbours because of death (Weiss, 1995). Loneliness may become a real threat for the life of the elderly, who not only is not able to deal with the daily activities as their health starts to decline, but also has difficulties in communicating their needs. The creation of a social network of support seems to be necessary in order to reduce the distance resulting in senility.

2.4 Autism and loneliness

Loneliness among autistic patients, suffering from neurodevelopmental deficiencies, is tightly connected with their inclination for isolating from the environment as the consequence of a lesser degree of understanding of the social life. As research has proved, when compared with healthy individuals , the intensity of loneliness is elevated in autistics (Bauminger et al., 2004, Bauminger et al., 2003, Bauminger & Kasari ,2000). The results of an examination of a group of 22 autistics aged from 8 to 14 indicated that an elevated degree of loneliness (measured with the Loneliness Rating Scale) co-occurred with a poorer quality of created friendships when the following ratios were considered: companionship, security and help (Bauminger & Kasari, 2000). Importantly, the ASD children did not indicate the connection of sadness, fear, emptiness or depression with an intensified sensation of loneliness. In these subjects the emotional dimension did not aid the explanation of experiencing loneliness. It was interpreted as a cognitive-social category (Weiss, 1985) caused by the cognitive evaluation of oneself, and then compared with other children in respect of the quality of created friendships. Nevertheless, having friends did not reduce the sensation of loneliness among autistic children due to the poorer quality of these friendships (Bauminger & Kasari, 2000).

ASD adolescents when contrasted with typically developing teenagers, report an elevated degree of temporal or constant loneliness (Lasgaard et al., 2010) which positions them in a group of chronically lonely subjects (Young, 1978 cited in: Hojat, 1983). It is important to highlight that chronic loneliness in addition to an inefficient social support and a general social dysfunction among teenagers with ASD, may lead to the risk of depression and suicidal moods. Another factor impairing the functioning and deepening loneliness in ASD teenagers is the coexisting anxiety disorder (White et al., 2009). This disorder contributes to the social withdrawal of individuals with autism manifested by unsatisfactory functioning in a group of peers. Researchers draw attention to the fact that the psychopathological picture of both depression and the anxiety disorder in autistics, may considerably differ from the typical picture. When fear is experienced it may include such symptoms as for example repetitive behaviors (White et al., 2009).

High-functioning autistic adults report a pervasive sensation of loneliness in reference to their own dissimilarity and lack of belonging to the surrounding world (Davidson, 2007; Jones et al., 2001). This feeling may be compared to cultural loneliness (McGraw, 1995). Seemingly, the incapability to communicate with others with the use of a language is also a type of loneliness.

As it was reported in the case of depression and anxiety disorders, the picture of loneliness in individuals with autism may differ considerably from the picture in non-autistics. Indisputably, loneliness among autistic patients manifests itself by an intensification of autistic symptoms such as repetitive behaviors, or self- destructive behaviors (e.g. self harm). Episodes of tantrums understood as attempts to communicate with family, can be particularly difficult for the parents of autistics. In adolescence and adulthood loneliness is disguised by different addictions (addiction to computer games, excessive internet use, alcohol abuse). A dedication to one's hobby or work may function as a substitute to inner loneliness in autistics. An additional phenomenon closely linked with loneliness is silence.

3. Phenomenon of silence

Qui nescit tacere, nescit et loqui is an ancient proverb meaning that if someone cannot be silent he cannot speak either and highlights the existence of the phenomenon of silence in the consciousness of the great minds of the ancient times, who indicated its salience as a means of social effects. Not surprisingly, great attention is paid to silence in communication. Linguists distinguish different types of silence in verbal communication (common silence, lack of response, and pause) highlighting its links with language, speech, culture and indicating its presence in philosophy, literature and theory of literature, rhetoric and stylistics (Śniatkowski, 2002). All these phenomena (silence, lack of response and pause) are instances of language use. The main feature aiding the distinction of silence from the other phenomena is the lack of any sounds. Lack of response and the pause are instances of silencer in reference to particular matters, and speaking is acknowledging something else (Rokoszowa, 1994). Silence in its broad context is defined as lack of speech, lack of verbalization, and triggers the non-verbal communication (Śniatkowski, 2002).

From the philosophical perspective represented by Wittgeinstein there is an equation between silence and the inexpressible, stressing the lack of verbal competence to describe a given experience. The following division of silence into three subcategories is close to Wittgeinstein's approach. 1. Referring to the unuttered/ unvoiced – in the sense of sound production and its lack. 2. Referring to the unexpressed - in the sense of reference to reality. 3. Referring to the unspoken /undeclared - in the sense of conscious strategic decisions (Rokoszowa, 1994, p.36). The phenomenon of silence is also exploited during therapy, and its acknowledgement indicates a shift in the therapeutic process.

3.1 Definition

The formulation of one coherent definition of the phenomenon of silence seems to be a complex task since different researchers tend to underline different features depending on the field of their primary interest.

Thus, the selection of definitions referring to the notion relating to autism appears the most adequate.

Paul Watzlawick, a famous researcher of human communication in the beginning of the 70s, noticed that in a social interaction it is impossible not to communicate, consequently, even silence provides certain information. This observation was particularly important in the context of miscommunication understood in terms of lack of significance for the typical verbal means. Silence provided evidence for dysfunctional interpersonal relationships (Watzlawick, 1972 cited in: Rokoszowa, 1994).

Linguists argue that *silence is understood as such a communicative behaviour that makes the alternative or the frames of verbal utterances. The meaning silence (unlike the transcendent silence) can also be called 'silent language behaviour'*. Whereas a pause is *a formally and functionally differentiated real or potential break/gap within verbal utterances* (Śniatkowski, 2002, p. 97). Silence treated as a framework for language has entered linguistics (Rokoszowa, 1994) and is defined as a necessary background phenomenon of language in its all instances. Since silence constitutes the frame, it may not remain neutral, conversely, it is constant and salient (Rokoszowa, 1994, p.28). The importance of silence as a necessary feature of language functioning has been evaluated, and Rokoszowa (1994, p.29) claims that language in its all functions is shaped and penetrated by silence.

Such picture of silence constitutes the basis for distinguishing the so-called transcendental silence compared to the inexpressible as well as the meaningful silence as a part of verbal communication. Schmitz (cited in: Roszkowa, 1994, p.36) subcategorizes silence into the following types: 1) conspicuous silence – remaining silent when sound/ speech is expected, 2) fitted silence – remaining silent when expected, 3) side-silence – remaining silent when it is not perceived as silence, but remains unnoticed (i.e. pausing in speech).

3.2 Functions of silence

Silence may play various roles depending on the context of appearance. In situations when expected, e.g. in a library, it remains unnoticed constituting the frame for the overlapping activity of acquiring knowledge. In this context, silence, is a factor inevitable for being attentive. In the context of a communication act, silence may disguise both positive and negative emotions.

In the context of some company or public gathering, silence may evoke either a dignified atmosphere, or be a closing element, or the sign of embarrassment. A noticeable silence differentiates itself from other communicative behaviors. It carries meanings which can be interpreted differently depending on the subjective characteristics of the silent person.

Sabbadini describes silence in the following way: *Silence can be a barrier. It can be a shield. It can be a bridge. It can be a way of avoiding saying something and it can be a way of saying what no words could ever tell. It can express anger, excitement, despair, gratitude, emptiness, joy, shame, helplessness or indeed any other emotion* (Sabbadini, 1991, p. 232).

According to Crafoord, silence can take different shapes and he distinguishes: 1).the searching silence (that no words can express), 2).the grey silence (emphasizing the lack of words in a given individual), 3). the passionate silence (carrying strong impulses, inclining towards the danger zone, e.g. sexual desires), 4). the pondering silence (referring to the wordless, mutual conviction about the meaning of the particular kind of silence), 5).the creative silence (which should not be disturbed since it is part of a creative process), 6). the threatening silence (disguising perseverance and obstinacy, protest and separation and being the reaction to anger, outrage, or desire for revenge), 7). the black silence (synonymous with complete rejection and self-destruction, or presence of death) (Ronningstan, 2006, p. 1279).

3.3 Psychotherapy and silence

Psychodynamic psychotherapy may be called "taking cure" since it emphases the impact of words in reference to psychopathological symptoms of personality in patients. From the perspective of psychodynamic psychotherapy silence is interpreted as emptiness and vacuum and rooted in fear of death and annihilation (Lane et al., 2002).

Importantly, silence as experienced in therapeutic environment does not necessarily indicate emptiness, but may be the manifestation to a variety of emotions. It is typically analyzed in terms of preverbal communication and delineates regression in the course of therapy. Silence can be explained as a mechanism aiding the withdrawal to the earlier stages of development. As such it may be interpreted in terms of the desire for control over a object or the desire for fusion with the object in control (Levitt, 2001).

From the traditional perspective, silence equals defensive mechanism of the ego when facing change (Freud 1912 cited in: Lane et al., 2002). It can also be the patient's reaction to wrong therapeutic intervention. According to Sabbadini, silence is a kind of compromise formation, that is the result of a conflict between different mental organizations, or the result of the tensions between different impulses. She claims that most cases of silence in therapy appear in response to experiencing an unconscious fear (Sabbadini, 1991). The basic therapeutic task, in the moments of silence on the part of the patient, is to decide whether or not to break it. And professional experience, sensitivity, tactfulness and the right timing all play a great role in the therapeutic process (Sabbadini, 1991).

Silence from the perspective of the psychodynamic psychotherapy might be analyzed in respect to inner conflicts of a patient, his defense mechanisms and the style of social functioning (Lane et al., 2002). Silence, on the part of the patient, may evoke the feelings of emptiness, helplessness and anger in the therapist caused by the lack of visible progress in the treatment. It is believed that the role of the therapist is to help the patient overcome silence and verbalize his/ her thoughts and fantasies concerning both the current therapeutic situation and the earlier experiences with objects in the life of the patient. The appearance of silence may convey such feelings as anger, fear, sadness, boredom, withdrawal from the contact, or lack of emotions in the patient. It also serves the function of censorship on what may and may not be verbalized (e.g. aggressive or sexual content). When examined from the point of view of the unconscious, it may be associated with the need to return to a safe place, such as the womb of the mother or the cradle or sleep (Lane et al., 2002). In the conscious surface, the patient may explain his silence in terms of unwillingness to discuss some difficult topics or in reference to difficulty in naming matters adequately (Storr, 1990 cited in: Lane et al., 2002). Nevertheless, it is important to remember that an appearance of silence during the course of the therapeutic process may indicate working through of analyzed experiences and feelings in the patient, providing space for the afterthought concerning his/her behavior. In the final phase of the therapy silence on the part of the patient may indicate a fear of separation and the requirement to begin an autonomic functioning. Additionally, it denotes sadness and pain caused by the necessity of partition from a meaningful object (i.e. the therapist). Silence occurring at the beginning of the treatment may be caused by feelings of transference. Nacht (1964 cited in: Lane et al., 2002) suggests that silence becomes a factor unifying the patient and the therapist (as an ideal object) and the inner integration. Sabbadini indicates that *silence is not, or not just, an absence (of words) but an active presence* (Sabbadini, 1991, p.232). She also claims that silence creates space for words which are impossible to express (Sabbadini, 1991).

In silence the therapist may both focus on observing the mental condition of the patient and his behavior and look for an explanation for a particular reaction of the patient to the given words. It is a time for recess contributing to the precise recognition, observation and analysis of countertransference. Importantly, it should be underlined that a prolonged silence on the part of the patient is undesirable because it indicates the retention of the experienced sensations inhibiting the therapist from any therapeutic intervention. It is crucial to recognize the type of silence appearing during the therapeutic session. In the view employing the communicative function of silence supported by Levitt, it is referred to as a pause. He distinguishes a few kinds of pauses: disengaged pauses and interactional pauses. The first one deals with the active avoidance of difficult emotions on the part of the patients, whereas the second type tracks the communicational process in respect to disorientation, the positive self-presentation, the threat to the therapeutic relationships (Levitt 2001). In some cases, silence is a vivid manifestation of an aggressive attitude towards the therapist and if this remains un-interpreted, may hinder therapeutic work in these two personality types (personality of a patient and the therapist) (Sabbadini, 1991). The protective function of silence should not be disregarded. It shields the patient from what he/ she perceives as the destructive influence of the outside environment.

Silence may also become a therapeutic method introduced by therapists in specific situation, e.g. in the phenomenon of holding and containing. The therapist contains the information provided by the patient, transforms it and gives it back in a processed form which becomes bearable for the patient. Furthermore, such remodeled information enables the patients to confront his or her fears and other rejected emotions. The therapist may use silence to frustrate the patient when the need for reflection on inner states is not fulfilled in the words of the analyst. This method may help the patient increase the frustration tolerance level (Lane et al., 2002). Wordlessness on the part of the therapist may both trigger the acceptance of the rejected thoughts, impulses and fantasies, and result in the deprivation of needs, or even in rejection or negation of the therapeutic alliance. Generally speaking, this type of silence creates detachment, endangers the patient's feeling of security and undermines trust, although all three are inevitable during the treatment (Lane et al., 2002). In patients with serious disorders (e.g. psychotics), who do not present a good verbal communication with the therapist, a silent mirroring of the patients' movements is recommended (Blumenson, 1993 cited in: Lane et al. 2002). This method aids the development of the non-verbal communication and has been practiced with autistics for a long time (Alvarez & Reid, 1991). Silence understood as the countertransference reaction has been interpreted in terms of anger, desire for punishing the patient, or rejecting him, often in return to an improper and nonfunctional response (Lane et al., 2002).

The introduction of silence in the course of the therapeutic treatment must be carefully analyzed and interpreted. Thoughtless therapeutic interventions concerning silence first of all may indicate strong countertransference reactions on the part of the therapist. Secondly, they may threaten further therapeutic cooperation by weakening the therapeutic relation between the patient and the therapist. Should such situation occur, listening to the silence, recognizing its type and identifying of the emotions it disguises along with the presentation of its bearable interpretation to the patient, seem to the best solution. The main issue is to encourage the patient to present his/ her own interpretation of the instances of silences in the course of the therapy. Additionally, the therapist should together with his patient learn how to perceive the silence maturely in social interactions. Silence interpreted in this way would provide the patients with reflecting tool and equip the patient in skills for distancing from others.

3.4 Autism and silence

In the case of autistic patients silence is linked with the lack of or with inadequate verbal competences resulting in disturbances in the exchange of information with the environment. Presumably, language does not serve the autistics as a means of communication, thus in their analysis of the role of silence in reference to verbal communication is out of question. Yet, this assumption does not recognize the strenuous attempts of the high-functioning autistics to communicate with others.

Silence in autism may appear in different degrees. The highest degree of silence is manifested in mutism and is interpreted as complete silence, creating a barrier separating autistics from the competent speech users. The silence surrounding autistics possesses the greatest therapeutic value; it reduces the number of stimuli from the outside environment. In these cases silence serves as a defense mechanism, protecting the unformed intrapsychic space from the chaos, and disarray challenging with the disintegration of the personality.

A case of silence of a lower intensity appears in autistics whose verbal communication is practiced only for self-stimulation and not for interchange with the environment. In these cases pauses are observed and their analysis aids the observation of the behavior of autistics in terms of their needs. Basing on the examination of pauses researchers made an attempt to engage the autistic child in the outside world.

The least intensity of silence is present in high-functioning autistics as well as in Asperger Syndrome patients. In their case silence equals with lack of sound that has its cause, goal and meaning and is suitable for profound psychotherapeutic analysis. The insight into the mechanisms joined with silence in autistics (i.e. social rejection, peer stigmatization, limited interpersonal contacts and loneliness) may aid the formulation of functional ways to deal with other people.

In the course of treatment of autistics psychotherapists often have to bend the fixed analytical rules in order to initiate contact with the autistic child, teenager or adult.

Silence is an inevitable part of the treatment of autistics individuals. Only occasionally does it function as an instance of inner censorship; more often it is connected with the search for temporal harmony, integration and feeling of security especially in HFA and AS patients. Silence not only aids the non-verbal communication with the autistic but also the recognition of the needs, talents and personal interests. While listening to the silence in autistics, we may discover their personality, pay attention to their emotions, tensions and repetitive behaviors. Silence itself may be analyzed during the therapy, although very few therapists focus on it. Personality features of a therapist play a vital role in examining silence since it can either predispose its retention or interruption. The analyst usually interrupts the period of silence when he senses tensions, fears, lack of tolerance for speechlessness, anger and helplessness. In the case of high-functioning autistics the retention of silence in a therapeutic environment may result in the feeling of rejection, being punished or blamed for prolonging the silence. In the course of psychodynamic group treatment, an autistic individual typically acts as a silent observer, and does not participate in the exchanges between members of the group. Adoption of the role of the observer by autistics commonly results in the feeling of self-inadequacy and discrepancy from the group members. The group may react destructively to the silence of one of its members, fearing negative evaluation and rejection. The increase of frustration within the group as well as the need for cohesion and completeness, may either cause an attack on the silent participant or result in appointing him a scapegoat (Brown, 2008). The task of the therapist is to broaden the scope of tolerance towards instances of discrepancies within the group.

Children, teenagers and adults interpret silence differently. It is most burdensome in the case of the therapy of autistic children, who understand silence in terms of withdrawal and lack of affective relationship with other people. The social communication in autistic children represents a specific problem when the speech disturbances are primal and autistics symptoms are subsequent to them (Rutter, 1988 cited in: Jaklewicz, 1993). A different scenario takes place in the case of patients whose initial decent verbal competencies deteriorate with time weakening their verbal abilities for communicating with the environment. In children who after a period of normal development acquired clinical symptoms of autism still constitute the more promising cases. Nevertheless, in these cases a grueling, long-term, multispecialty assistance of professionals is inevitable in order to rebuild the verbal competences for interaction with the outside world. The clinical picture of autism tends to present a milder case. The interpretation of silence changes in autistic adolescents. The speechlessness of the peers is a more frequently noticeable instance of silence than the recognition of personal withdrawal from speech in company. Autistic adolescents do not sense the discomfort that results from the silence in the context of the peer group. They do not notice that silence may increase tensions, provoke the need for closing the interaction and finally cause the exclusion from the peer group. It is very important during a social training to emphasize the bond between silence and physical and emotional distancing in other people. In HFA and AS patients, in adolescence the need for silence on the one hand and the precise formulation of information on the other, may be interpreted as a particularly economical form of communication with people. This form aims at accumulation of energy necessary for interacting with inanimate objects, ideas and numbers. Silence provides a perfect condition for self-development and learning. In adults silence is perceived mainly in terms of security, though different experiences may change this perception. Consequently, it conveys diverse emotions depending on the context of occurrence.

The link between autism, loneliness and silence is very close. Loneliness can be experienced in silence, and silence may deepen the feeling of loneliness. Both can be accompanied by different emotions (sadness, fear, tension, anger, remorse as well as joy and optimism). Loneliness is sensed not only by the autistic child, teenager or adult but also by their family (Park 1982) struggling to socialize the disabled relative.

4. Conclusion

The aim of the article was to analyze the two factors -loneliness and silence- common both for autistics and people typically developed. Loneliness as an inseparable feature of every human life seems to predominate in the experiences of autistics (Bauminger et al., 2003). The form and the strategies of loneliness as the means for protection against unpleasant emotions are still in question. The tools exploited in the evaluation of the level of loneliness help determine only its two subcategories, namely, emotional and social. The other subcategories of loneliness appearing in autism, that is cultural and loneliness in the interpersonal communication (communicative loneliness) cannot be measured with these tools. The presented characteristics of loneliness and silence aims at a greater understanding of these phenomena as well as a more precise positioning of these features in the clinical treatment of autistics. Significantly, the sensation of loneliness may be inadequately recognized in patients with autism for two reasons. Firstly, the patients with autism have difficulties in identifying and communicating their own emotions and the emotions of others. Secondly, the acknowledgement of loneliness may seem socially stigmatizing in healthy subjects. They may subconsciously generate negative impulses concerning the articulation of loneliness in the disabled. There seems to exist a need for common social dispute regarding the familiarizing of the issue of loneliness on the one hand, and a need for a scientific research of the hypotheses concerning the shape of loneliness in psychotherapy, on the other. Only then the verification of the assumptions may be possible. The number of tools necessary for access to the phenomenon of loneliness is still insufficient. There are no tools enabling the identification of the different shades of loneliness. What is more, autistics and healthy people manifest loneliness in significantly different ways (similarly to the manifestations of anxiety and depression). Loneliness in patients with autism may be disguised by intensified typical pathological symptoms (tantrums, self-injury, self-stimulation, aggression, elevated echolalia). The excessive control of autistics on the part of the caretakers or parents may also result in loneliness. The dichotomy of the needs of the patient and the caretaker may lead to loneliness in both. Moreover, peer rejection separates the autistic from the social environment and threatens with the appearance of additional psychopathological symptoms (i.e. fear, depression - in their clinical picture).The phenomenon of loneliness is often accompanied by silence. Nevertheless, on the contrary to its origin in healthy individuals, in autistics, silence usually designates positive emotions. The reduction of external stimuli may result in both quietening the patient and deepening of the social isolation.

In autistics, the search for the source and meaning of silence seems to be a major issue both during the course of the psychotherapeutic treatment and in aiding the design and selection of adequate therapeutic methods.

5. References

American Psychiatric Association (1994). *Diagnostic and statistical manual of mental disorders* (fourth edition). ISBN 0-89-042-062-9, Washington, DC: Author.

Applebaum F. (1978). Loneliness: a taxonomy and psychodynamic view. *Clinical Social Work Journal*. Vol. 6, No. 1, pp. 13-20, ISSN 1573-3343.

Alvarez A., Reid S. (1999). *Autism and personality. Findings from the Tavistock Autism Workshop.* Routledge Taylor & Francis Group, ISBN 978-0-415-14602-9, London & New York.

Bauminger N., Kasari C. (2000). Loneliness and friendship in high-functioning children with autism. *Child development*. Vol. 71, No. 2, pp. 447-456, ISSN 1467-8624.

Bauminger N., Shulman C., Agam G. (2003). Peer interaction and loneliness in high-functioning children with autism. *Journal of Autism and Developmental Disorders*. Vol. 33, No. 5, pp. 489-507, ISSN 1573-3432.

Bauminger N., Shulman C., Agam G. (2004). The link between perceptions of self and of social relations in high-functioning children. *Journal of Developmental and Physical Disabilities*. Vol.16, No. 2, pp.193-214, ISSN 1573-3580.

Booth R.(1997) Existential loneliness: the other side of the void. *The International Journal of Interpersonal Studies*. Vol. 16, No.1, pp. 23-32, available on 20 march 2011 on : https://securehost85.hrwebservices.net/~affectp/docs/existentialloneliness.doc

Brown N.W. (2008). Troubling silences in therapy groups. *Journal of Contemporary Psychychotherapy*. Vol. 38, No. 2, pp. 81-85, ISSN 1573-3564.

Davidson J. (2007). "In a World of her Own…":Re-presenting alienation and emotion in the lives and writings of women with autism. *Gender, Place and Culture*. Vol. 14, No. 6, pp. 659-677, ISSN 1360-0524.

Dill J. C., Anderson C.A. (1999). Loneliness, shyness and depression: the etiology and interrelationships of everyday problems in living. In: *The interactional nature of depression: Advances in interpersonal approaches*, Joiner T., Coyne J.C.(Eds), pp. 93-125., APA, ISBN 978-1-55798-534-7, Washinghton D.C.

Dorr D., Friedenberg L.(1983). Mothering and the young child. *Issues in Mental Health Nursing*. Vol. 5, No. 1-4, pp. 45-60, ISSN 1096-4673.

Fontaine R.G., Yang Ch., Burks V.S., Dodge K.A., Price J.M., Pettit G.S., Bates J.E. (2009). Loneliness as a partial mediator of the relation between low social preference in childhood and anxious/depressed symptoms in adolescence. *Development and Psychopathology*. Vol. 21, No 2, pp. 479-491, ISSN 1469-2198.

Hawkley L.C., Cacioppo J.T. (2010). Loneliness matters: a theoretical and empirical review of consequences and mechanisms. *Annals of Behavioral Medicine*. Vol. 40, No. 2, pp. 218-227, ISNN 1532-4796.

Heinrich L.M., Gullone E. (2006).The clinical significance of loneliness: a literature review. *Clinical Psychology Review*. Vol. 26, No.6, pp. 695-718, ISSN 0272-7358.

Hojat M. (1983). Comparison of transitory and chronic loners on selected personality variables. *British Journal of Psychology*. Vol. 74.No.2., pp. 199-202, ISSN 2044-8295.

Hojat M. (1982). Loneliness as function of selected personality variables. *Journal of Clinical Psychology*. Vol. 38, No. 1, pp. 137-141, ISSN 0021-9762.

Jaklewicz H. (1993). *Autyzm wczesnodziecięcy. Diagnoza, przebieg, leczenie* [Early childhood autism. Diagnosis, course, therapy].Gdańskie Wydawnictwo Psychologiczne, ISBN 83-85416-09-9, Gdańsk, Poland.

Jones R.S.P., Zahl A., Huws J.C. (2001). First-hand accounts of emotional experiences in autism: a qualitative analysis. *Disability and Society*. Vol. 16, No. 3, pp. 393-401, ISSN 1360-0508.

Kirova A. (2004). Lonely or bored: children's lived experiences reveal the difference. *Interchange. A Quarterly Review of Education*. Vol. 35, No. 2, pp. 243-268, ISSN 1573-1790.

Lane R.C., Koetting M.G., Bishop J. (2002). Silence as communication in psychodynamic psychotherapy. *Clinical Psychology Review*. Vol. 22, No 7, pp. 1091-1104, ISSN 0272-7358.

Lasgaard M., Nielsen A., Eriksen M.E., Goossens L. (2010). Loneliness as social support in adolescent boys with autism spectrum disorders. *Journal of Autism and Developmental Disorders*. Vol. 40, No. 2, pp. 218-226, ISSN 1573-3432.

Levitt H.M. (2001). Client's experiences of obstructive silence: Integrating conscious reports and analytic theories. Journal of Contemporary Psychotherapy. Vol. 31, No. 4, pp. 221-244, ISSN 1573-3564.

McGraw J. G. (1995). Loneliness, its nature and forms: an existential perspective. *Man and World*. Vol. 28, No.1, pp. 43-64, ISSN 0025-1534.

Moustakas C.E. (1961). *Loneliness*. Englewood Cliffs, ISBN 076570028X, Prentice-Hall, New York.

Park C.C. (1982). *Oblężenie. Rodzinna wyprawa w świat dziecka autystycznego* [The siege. A family's journey into the world of an autistic child]. Wydawnictwo Literackie. ISBN 83-08-03412-8, Kraków, Poland.

Pretty G.M.H., Andrewes L., Collett Ch. (1994). Exploring adolescents' sense of community and its relationship to loneliness. *Journal of Community Psychology*. Vol. 22, No. 4, pp. 346-358, ISSN 0090-4392.

Qualter P., Brown S.L., Munn P., Rotenberg K.J. (2010). Childhood loneliness as a predictor of adolescent depressive symptoms: an 8-year longitudinal study. *European Child and Adolescent Psychiatry*. Vol. 19, No. 6, pp. 493-501, ISSN 1435-165X

Reis S., Grenyer B.F.S. Pathways to anaclitic and introjective depression. (2002). *Psychology and Psychotherapy: Theory, Research and Practice*. Vol. 75, No. 4, pp. 445-459, ISSN 2044-8341.

Rokoszowa J. (1994). Milczenie jako fakt językowy [Silence as a linguistic fact]. *Bulletin de la Societe Polonaise de Linguisique*. Vol. L, pp.27-47, ISSN 0032-3802.

Ronningstam E. (2006). Silence. Cultural function and psychological transformation in psychoanalysis and psychoanalytic psychotherapy. *International Journal of Psychoanalysis*. Vol. 87, pp. 1277-1295, ISSN 1745-8315.

Rotenberg K. J.(1999). Childhood and adolescent loneliness: an introduction. In: *Loneliness in childhood and adolescence*. Rotenberg K.J., Hymel S. (Eds), pp. 3-8, Cambridge University Press, ISBN 0-521-56135-3, Cambridge, United Kingdom.

Sabbadini A. (1991). Listening to silence. *British Journal of Psychotherapy*. Vol. 7. No. 4, pp. 229-240, ISSN 1752-0118.

Śniatkowski S. (2002). *Milczenie i pauza w gramatyce nadawcy i odbiorcy. Ujęcie lingwoedukacyjne* [Silence and pausing in the sender's and receiver's grammar. Educational linguistics approach]. Wydawnictwo Naukowe Akademii Pedagogicznej. ISBN 83-7271-141-0, Kraków, Poland.

Terrel-Deutsch B. (1999). The conceptualization and measurement of childhood loneliness. In: *Loneliness in childhood and adolescence*. Rotenberg K.J., Hymel S. (Eds), pp.11-33, Cambridge University Press, ISBN 0-521-56135-3, Cambridge, United Kingdom.

White S.W., Oswald D., Ollendick Th., Scahill L. (2009). Anxiety in children and adolescents with autism spectrum disorders. *Clinical Psychology Review*. Vol. 29, No. 3, pp. 216-229, ISSN 0272-7358.

Weiss R.S. (1985). *Loneliness. The experience of emotional and social isolation*, The MIT Press Classics, ISBN 0-262-73041-3, Cambridge, Massachusetts and London, England.

World Health Organization (1992). *Manual of the International Statistical Classification of the Diseases and Related Health Problems: Diagnostic criteria for research* (10th edition). ISBN 92-4-154649-2, Geneva, Switzerland: Author.

22

Mnesic Imbalance and the Neuroanatomy of Autism Spectrum Disorders

Miguel Ángel Romero-Munguía
Hospital Psiquiátrico "Dr. Samuel Ramírez Moreno", Secretaría de Salud
México

1. Introduction

Currently, the diagnostic and statistical manual of the American Psychiatric Association (APA) defines autistic disorder together with Rett syndrome, childhood disintegrative disorder, Asperger syndrome and Pervasive Developmental Disorder Not Otherwise Specified (PDD-NOS) as Pervasive Developmental Disorders (APA, 2000). This term implies disorders that vary in several ways but the term Autism Spectrum Disorders (ASD), which excludes Rett syndrome and childhood disintegrative disorder, implies that there are shared clinical and etiological factors among diagnoses (Gabis et al., 2008). It has even been thought that the diagnosis of PDD-NOS, which does not require all three diagnostic symptoms of autism, does not deserve to be included as an ASD (Mercadante et al., 2006). On the other hand, the assumption of a normal early language development in Asperger syndrome has been challenged (Howlin, 2003).

Three major cognitive theories (theory of mind deficit, weak central coherence and executive dysfunction) have unsuccessfully attempted to explain the core triadic symptoms of ASD (impairments in social interaction, communicative capacity and behavioural flexibility) (Pisula, 2010). On the other hand, the mnesic imbalance theory seems to be a very promising solution to this problem. This account proposes that all three diagnostic symptoms of autism may be explained by an imbalance between a faulty procedural memory and a relatively preserved declarative memory (Romero-Munguía, 2008). The theory of mind is a system that enables one to infer mental states, central coherence is a tendency to create higher meanings from samples of data and the executive function is a set of mental processes that help us control our actions. Besides, it has recently been proposed that the alterations described by the three dominant theories can be explained by the mnesic imbalance theory. However, this theory can only be convincing if it is in accordance with data available from the neurobiological literature. In keeping with this view, this paper begins by reviewing the neurobiological basis of declarative and procedural memories. Next, it presents neuropathological, structural and functional imaging data of patients with ASD in order to support the mnesic imbalance theory.

2. Memory systems

Learning can be defined as a process for acquiring a behavioral change based on experience; this change is termed memory (Okano et al., 2000). Memory can be divided into declarative (explicit) and non-declarative (implicit) memory systems. Declarative memory can be subdivided into working memory, semantic memory and episodic memory, whereas non-declarative memory includes procedural memory, conditioned reflexes, emotional conditioning and priming. Declarative memory involves information that is subject to conscious recollection and verbal reflection, whereas procedural memory involves behavioural algorithms that operate at an unconscious level (Budson & Price, 2001). Besides, lesions limited to the hippocampus, located within the medial temporal lobe, impair the acquisition of declarative knowledge sparing more remote declarative memory (Squire et al., 2004), whereas lesions limited to the cerebellum impair the acquisition of procedural knowledge, sparing more remote implicit knowledge (Bracha et al., 1997; Molinari et al., 1997; Quintero-Gallego et al., 2006). There are patients with severe deficits of declarative memory whose procedural memory is spared (Budson & Price, 2001; Okano et al., 2000), while there are patients whose procedural learning is significantly below their declarative learning (Molinari et al., 1997; Quintero-Gallego et al., 2006). For instance, in a study utilizing the Serial Response Time Task (SRTT), a procedural learning task, the data suggest that in comparison with control group (C), children and adolescents with acquired pathology confined to the cerebellum (CE), and children and adolescents with additional damage due to the chemotherapy and radiotherapy used (CE+) had faulty procedural learning, while the California Verbal Learning Test (CVLT) was used to measure declarative learning, which was preserved in the two groups with cerebellar damage (Fig. 1).

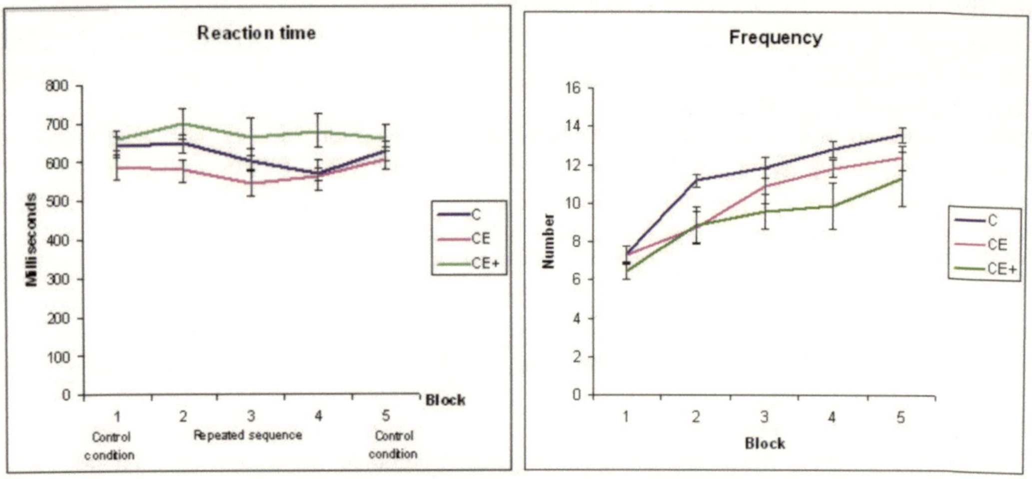

Fig. 1. Progressive reduction of reaction time during repeated sequence was an index of procedural learning, while the frequency of words remembered was an index of declarative learning. The standard error of measurement is also represented. (From Quintero-Gallego et al., 2006).

It is also important to note that there are neurobiological changes that occur with the development of memories in healthy human individuals. For instance, it is now accepted that neurogenesis occurs in the adult human hippocampus and it has been surmised that adding new neurons helps in acquiring new knowledge without disrupting knowledge that has already been stored (Appleby et al., 2011), whereas neurogenesis occurs in the human cerebellum mainly before the end of the first postnatal year, but its cytoarchitectonical development is completed subsequently (Ábrahám et al., 2001; Rakic & Sidman, 1970). On the other hand, a postmortem study of persons who had developed a complex declarative memory showed significant reductions of minicolumnar width in six Brodmann areas (BA) of interests: primary motor cortex (BA 4), prefrontal association cortex (BA 9), primary visual cortex (BA 17), visual inferotemporal area (BA 21), higher order auditory cortex (BA 22), and parietal-temporal-occipital association cortex (BA 40) (Casanova et al., 2007). Furthermore, a three-dimensional magnetic resonance imaging study (Fig. 2) in people who had developed a complex procedural memory showed significant morphological enlargement in the vermal lobules VI–VII (declive, folium, and tuber), which might reflect an increased glial volume per purkinje cell, a larger volume of the molecular layer, a greater number of synapses, and/or a dendritic hypertrophy of stellate cells in the cerebellum (Park et al., 2009).

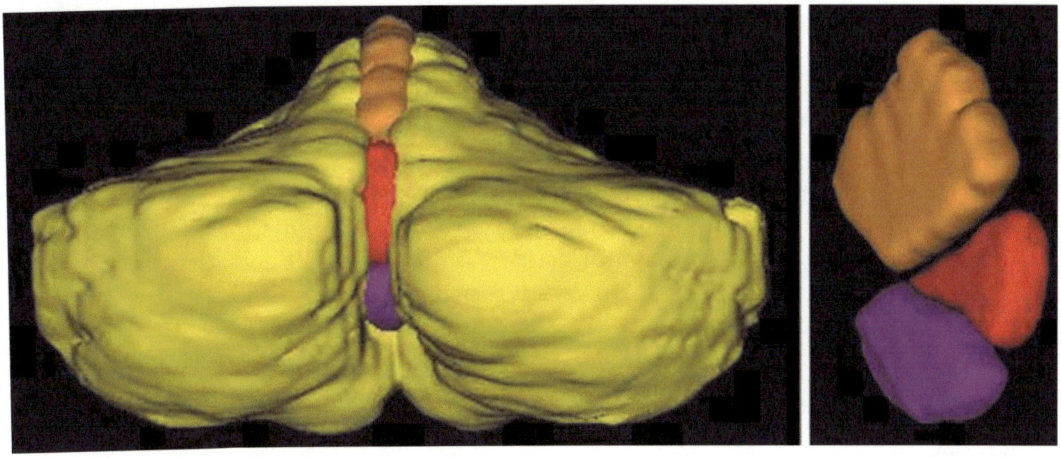

Fig. 2. 3-D model of the cerebellum. *Yellow* cerebellar hemisphere, *orange* vermian lobules I–V, *red* vermian lobules VI–VII, *purple* vermian lobules VIII–X (From Park et al., 2009), with permission.

Several studies have shown that declarative learning results in gray matter changes. One of them demonstrates that students in intensive declarative learning showed a significant gray matter volume increase in the dorsomedial frontal cortex (DMPFC), the orbitofrontal cortex, and the precuneus (Ceccarelli et al., 2009). Furthermore, after a 3 month period of daily study sessions, students in extensive declarative learning showed a significant increase in gray matter volume in the posterior parietal cortex, inferior parietal cortex, and the hippocampus (Draganski et al., 2006). Another study has reported gray matter volume increases in temporal lobe, occipital lobe, inferior parietal lobe, middle frontal gyrus, and middle temporal gyrus after motor procedural learning, while individuals with procedural learning of motor sequences with objects had gray matter volume increases in the hippocampus; individuals with procedural learning of complex motor stereotypies had gray matter volume increases of the insula, and the inferior parietal lobe (Filippi et al., 2010).

Procedural learning is associated with changes in brain activation patterns, such as initial increases of activity in the prefrontal, bilateral somatosensory and parietal cortices, caudate nucleus and the ipsilateral cerebellar hemisphere. This activity progressively decreases during the subsequent acquisition of procedural knowledge, while there is an increase in activity in the cerebellar dentate, thalamus and putamen (Floyer-Lea & Matthews, 2004). In another study, this process was divided into three distinct phases. A first phase, called the cognitive phase, has been associated with activation of the right frontal cortex (BA 10/11/45/46), right and left posterior cerebellum, left precuneus (BA 7/19), right angular region (BA 7/19/39), anterior cingulate cortex (ACC), and left frontal regions (BA 9/10). A second phase, called the associative phase, has shown bilateral activation of the lingual and calcarine regions (BA 17/18), right middle and right orbitofrontal region (BA 10/11/46), right thalamus and right caudate nucleus, left occipital region (BA 19), and right posterior cerebellum (Crus 1 and 2). A last phase, called the autonomous phase, has shown increased activity of the lingual and calcarine regions (BA 17/18), right anterior cerebellum (vermis VI-VII/hemisphere IV-VI), bilateral superior and middle frontal areas (BA 10/11) as well as the left thalamus (Hubert et al., 2007). In addition, data from an experimental study suggest that during procedural learning, cerebellar stimulation resulted in selective facilitation of contralateral excitability in the primary motor cortex, but it does not occur if procedural knowledge has already been acquired (Torriero et al., 2011). Similarly, if a stimulus to be stored into declarative memory occurs, then a period of increased activation will be produced in the dorsolateral prefrontal cortex (DLPFC), but there is a decrease in activation once the declarative knowledge has already been acquired (León-Carrión et al., 2010). The practice in working memory tasks is associated with progressive decrease of activation mainly in the right superior frontal gyrus/DLPFC (BA 8/9/46), the middle frontal gyrus bilaterally (BA 10/11), the left precentral gyrus (BA 4/6), and the dorsal part of the right ACC (BA 32), whereas there is a progressive increase of activation in the posterior cingulated cortex (PCC) adjacent to the corpus callosum (Koch et al., 2006). These temporal changes in neural activation might be linked to the efficiency of the declarative memory function; indeed, the activity of the posterior midline region, which includes PCC and precuneus, and the activity of the ventral part of lateral posterior parietal cortex, which includes the supramarginal gyrus and the angular gyrus, have shown reduced activity for hits relative to misses during encoding, whereas they showed increased activity for hits relative to misses during retrieval (Daselaar et al., 2009). Moreover, the lateral prefrontal cortex (LPFC) activity tends to increase with age, which is positively correlated with gains in declarative memory for details of experiences and reduction of gray matter volume in children and young adults, whereas medial temporal lobe activations appeared fully development by eight years of age (Ofen et al., 2007).

In functional magnetic resonance imaging, a number of studies have shown brain regions that are typically more active during resting states than during demanding cognitive tasks. This ensemble of cortical regions is called the default mode network (DMN) and it has been observed that DMN regions are not only more strongly deactivated with a higher cognitive load relative to a lower cognitive load during a short-term memory (working memory) task, but also the activation of the DMN is more pronounced at rest after a cognitively challenging task relative to an easier task, as illustrated in Fig. 3 (Pyka et al., 2009). In addition, DMN functional connectivity is progressively increased in the medial prefrontal cortex (MPFC) regions and progressively decreased in PCC regions when moving from lower to higher short-term declarative memory loads (Esposito et al., 2009).

In the first clinical description of children with autism, Kanner wrote, referring to their excellent declarative memory, “It is difficult to know for certain whether the stuffing as such

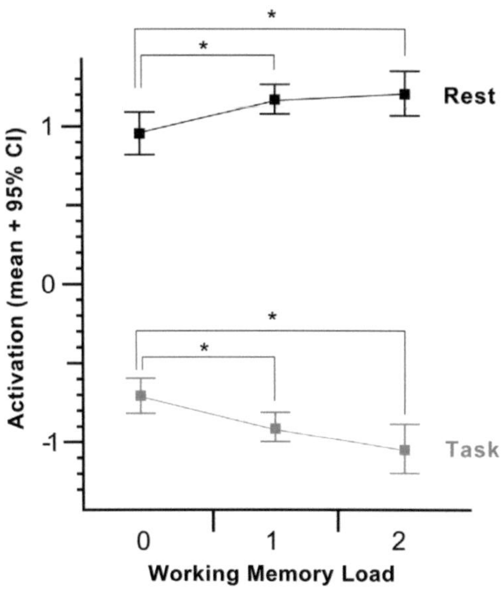

Fig. 3. Means for activation in the DMN were calculated for each n-back (lower values: 0-back, 1-back, 2-back) and subsequent resting block (upper values: R0-back, R1-back, R2-back) x-Axis; working memory load, y-Axis (From Pyka et al., 2009).

has contributed essentially to the course of the psychopathologic condition" (Kanner, 1943). Thereafter, the notion of faulty procedural memory that is compensated for by relatively preserved declarative memory has been proposed as an account of the savant abilities, as well as of the symptoms in autism (Goldberg, 1987; Klinger et al., 2007; Mostofsky et al., 2000; Romero-Munguía, 1998, 2002, 2008). For instance, in a study utilizing the Behavioral Summarized Evaluation (BSE) to evaluate autistic symptomatology in children with autistic disorder with non-functional verbal language (NFV) or no spoken language (NSL), as well as children with mixed receptive-expressive language disorder (RLD). All participants solved tests for reception of gestural language (Ges), vocabulary (Voc), and verbal commands (Com). A significant procedural knowledge deficit and a significant positive correlation between autism symptoms and declarative knowledge were observed only among the overall group of children with autistic disorder, which indicates an imbalance between declarative and procedural memory in autism rather than mere faulty procedural memory, as suggested in Fig. 4 (From Romero-Munguía, 2002).

However, in recent studies the view of a procedural learning deficit has been challenged (Barnes et al., 2008; Brown et al., 2010, Nemeth et al., 2010), but this discrepancy seems to be more apparent than real, because each author may be referring to different types of memories although using the same terms (Romero-Munguía, 2009). For instance, papers that supports the mnesic imbalance theory have been described as being based on a general deficit in implicit learning (Brown et al., 2010), but this is inaccurate because procedural memory is only one type of implicit memory (Budson & Price, 2001). Indeed, papers that argue against the presence of faulty implicit learning have shown significant decreases in reaction time, but this was independent of learning about sequences (Barnes et al., 2008; Brown et al., 2010, Nemeth et al., 2010), which suggests any other type of preserved implicit learning rather than intact procedural learning (Budson & Price, 2001); however,

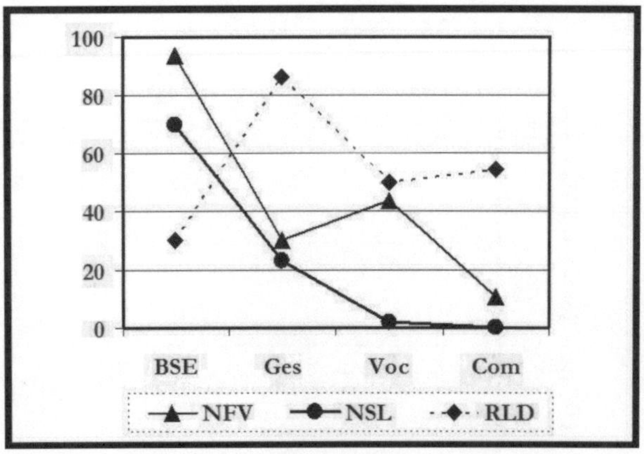

Fig. 4. Percent of the highest score achieved by any participant on every test, x-Axis; the reception of gestural language was an index of procedural knowledge, while the reception of vocabulary was an index of declarative knowledge, y-Axis (From Romero-Munguía, 2002).

interestingly, it has been surmised that this faster reaction time may imply an enhancement of declarative (explicit) memory in autism (Gidley Larson & Mostofsky, 2008; Molinari et al., 1997; Walenski et al., 2008). Nevertheless, it is important to note that the present paper does not suggest that procedural learning is entirely absent in autism, but rather its development is significantly lower in comparison to declarative learning (Gordon & Stark, 2007; Romero-Munguía, 2008).

According to the mnesic imbalance theory, a cerebellar maldevelopment may cause faulty procedural memory, as with relatively intact declarative memory as well as a reciprocal causal interaction, in which some neuroanatomical changes might result from this mnesic imbalance (Romero-Munguía, 2007). New information has appeared that has increased knowledge of how this interaction may occur; the present paper is devoted to this matter.

3. Neurobiological findings in autism

At present, it is difficult to avoid the view that a single cognitive theory can explain neither the neurobiological heterogeneity nor the core triadic symptoms of ASD (Pisula, 2010; Rajendran & Mitchell, 2007; Waterhouse, 2008). However, this section offers arguments based on neurobiological findings in favor of the mnesic imbalance theory.

3.1 Neuropathological findings

Purkinje cells are the only output neurons of the cerebellar cortex and loss of them within the cerebellum is the most replicated finding (72%) according to a review on the neuropathology of autism (Palmen, 2004). Indeed, some of the negative cases included in that review might be false negatives. For instance, a study included a case with encephalopathy of phenylketonuria (Williams et al., 1980), and other case could be diagnosed as childhood disintegrative disorder (APA, 2000; Guerin et al., 1996); both disorders may mimic ASD. In a study of six cases, five cases had low Purkinje cell counts, but one did not. However, only in that case, eosinophilic inclusions in the perikaryon and proximal dendrites of 30-40% of Purkinje cells were seen in the cerebellar vermis; they were less frequent in the cerebellar hemispheres, and they were not observed elsewhere (Bailey et al., 1998). These inclusions might have been associated with cerebellar dysfunction. Moreover, basket and stellate cells (GABAergic interneurons) in the molecular layer of the cerebellar cortex may influence the output of Purkinje cells (Bao et al., 2010; Rancillac & Crépel, 2004), so the increase in GAD67 mRNA levels in cerebellar interneurons of persons with ASD might be related to Purkinje cell dysfunction (Yip et al., 2008).

Abnormally enlarged neurons have been observed in the inferior olive of the brainstem in young individuals with autism, whereas these neurons are small and pale in adults with autism (Bauman & Kemper, 2005), which might be explained by the need to keep the scant procedural knowledge that has been acquired. This assumption is in agreement with experimental observations involving stimulation of the nucleo-olivary pathway just before an unconditioned stimulus, which leads to extinction of the conditioned response (Bengtsson et al., 2007), hence, an interruption in this pathway might prevent procedural memory loss, but this disruption of neuronal signal to the inferior olive may lead to its hypertrophy, as well as to olivary atrophy after several years (Akar et al., 2008).

A subgroup of individuals with ASD has an apparently normal early development, followed by a loss of verbal and/or non-verbal skills prior to 2 years of age. Such a regression has been found in 30% of children with autistic disorder, whereas it has been found in 14% of those with PDD-NOS, and none with Asperger syndrome (Meilleur & Fombonne, 2009). A study found that children with autistic regression show more frequent use of words and babble compared with typical infants at 12 months of age (Werner & Dawson, 2005). This might be explained by assuming that basket and stellate cells established precocious synaptic contacts which could improve the procedural learning, although thereafter, an increased number of interneurons relative to Purkinje cells could impair this procedural learning. This hypothesis is in accordance with the report of autistic cases with a lower density of Purkinje cells in comparison to cerebellar interneurons (Whitney et al., 2009). In addition, it has been suggested that during the normal development of the human cerebellum, granule cells influence the survival of interneurons, they regulate their migration to the molecular layer, and induce their differentiation into basket and stellate cells after birth, whereas before birth Purkinje cells would induce differentiation of granular layer interneurons, as proposed in Fig. 5 (Leto et al., 2008). The aforementioned might be clinically relevant because the source of cerebellar granule cells is exclusively the external granular layer, which disappears between the end of the first and the second postnatal year (Ábrahám et al.,, 2001; Rakic & Sidman, 1970), hence a delay in the disappearance of external granular layer might cause the autistic regression since a reduced density of Purkinje cells in presence of a normal or high density of cerebellar interneurons has been reported in at least two of six autistic cases (Whitney et al., 2009). The latter might be associated with dysfunctional synaptic transmission between cerebellar interneurons and Purkinje cells.

The aforementioned review reported increased cell packing density, reduced cell size, and simplified dendritic pattern in the hippocampus in 64% of cases with autism (Palmen, 2004). According to the mnesic imbalance theory, declarative memory replaces faulty procedural memory in some of its functions in individuals with ASD, which implies an overload for the declarative memory (Romero-Munguía, 2008). Therefore, it is necessary to use a best adaptation strategy, and additive neurogenesis seems better than either conventional synaptic plasticity or neuronal turnover (Appleby et al., 2011),, so this best strategy might lead to the particularly large number of likely immature cells (newly born neurons) depicted

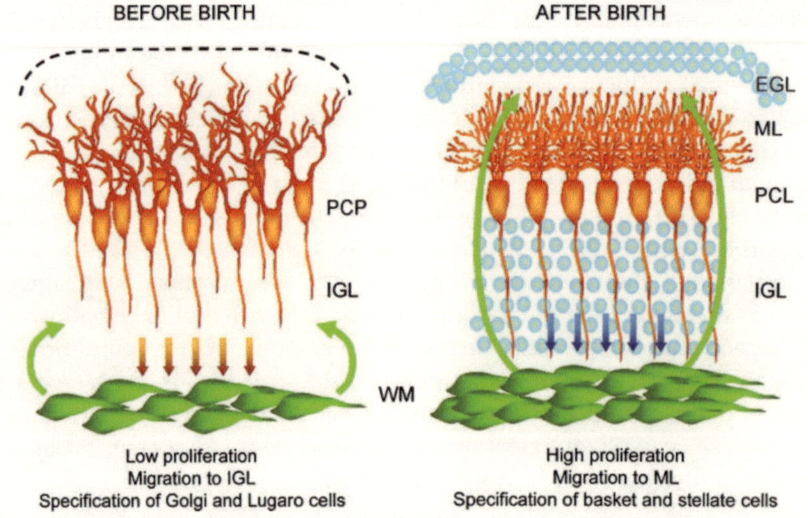

Fig. 5. Schematic illustration of the hypothesis that neurons adjacent to the prospective white matter influence the fate choices of interneuron precursors during prenatal and postnatal cerebellar development. *PCP* Purkinje cell plate, *WM* white matter, *IGL* internal granular layer, *PCL* Purkinje cell layer, *ML* molecular layer, *EGL* external granular layer (From Leto et al., 2008), with permission.

above. Indeed, an increased cell packing density (minicolumns per cortical area) with reduced cell size also has been reported in several areas of the cerebral cortex of non-autistic individuals with complex declarative memory, as well as of individuals with ASD (Casanova, 2007; Casanova et al., 2007). The possibility that the human neocortex produces neurons after birth has been suggested (Gould, 2007). Likewise, increased microglial cell density has been reported in gray matter of the DLPFC of patients with autism compared to control cases (Morgan et al., 2010); activity in DLPFC has been associated with declarative memory (Galea et al., 2010; Koch et al., 2006; León-Carrión et al., 2010). In addition, physiological stimuli such as environmental enrichment and physical activity lead to production of new glial cells (Ehninger & Kempermann, 2003). Therefore, it is possible that an overload for the declarative memory is related to an increase in this type of cells. Furthermore, contrary to other disorders involving mental retardation, which are associated with loss of dendritic spines, higher spine densities were found in temporal (BA 21), frontal (BA 9), and parietal (BA 7) lobes of autistic patients who had suffered moderate to severe mental retardation, as suggested in Fig. 6 (Hutsler & Zhang, 2010), that is, in some cerebral cortical regions that have been associated with complex declarative memory (Casanova et al., 2007).

3.2 Structural neuroimaging

A meta-analysis of structural magnetic resonance imaging studies from over 800 individuals with autism reported generalized enlargement of cerebral hemispheres, cerebellum and caudate nucleus, whereas reductions in the size were observed in corpus callosum, vermal lobules VIII-X, midbrain, and vermal lobules VI-VII; a normalization in the size of vermal lobules VI-VII with increasing age was also reported (Stanfield et al., 2008). The latter is consistent with a more recent analysis of abnormality in vermis lobules VI-VII, as shown in Fig. 7 (Courchesne et al., 2011). This increase in vermis size might be a consequence of the

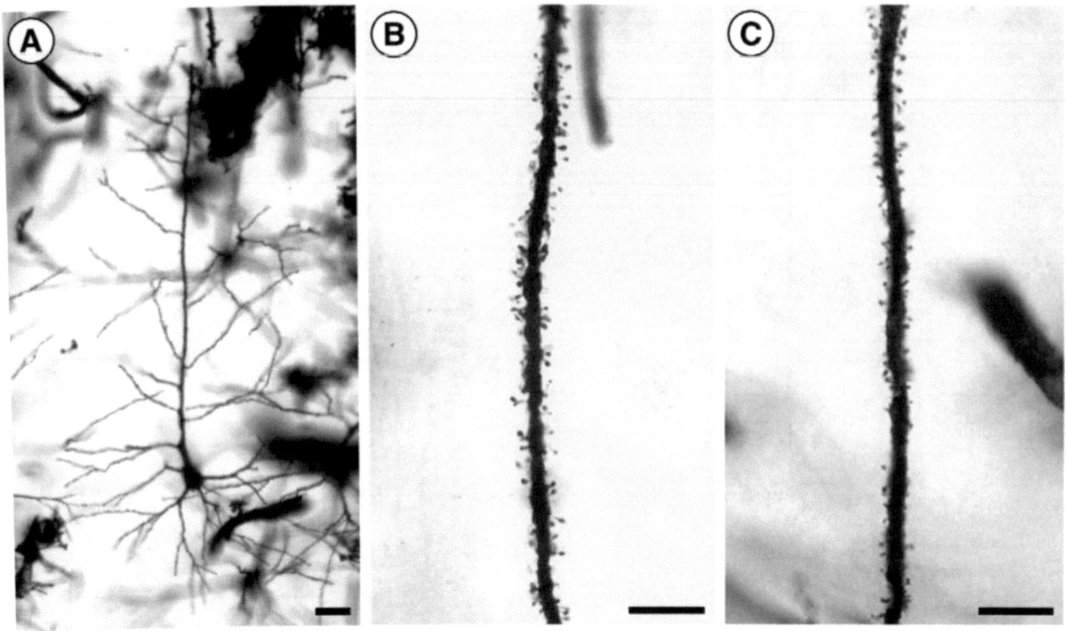

Fig. 6. (A) An example of a Golgi-impregnated pyramidal cell in layer III of the superior frontal gyrus (BA 9) in an ASD case. (B and C) Apical dendrite segments showing spine densities of layer V pyramidal cells in BA 9 of ASD (B) and control subjects (C). Both examples are approximately 300 μm from the cell soma. Scale bar in A=25 μm; Scale bars in B and C=20 μm (From Hutsler & Zhang, 2010), with permission.

slow development of procedural memory, a hypothesis that is consistent with previous research and the mnesic imbalance theory (Gordon & Stark, 2007; Park et al., 2009; Romero-Munguía, 2008).

A volumetric analysis of the cerebellum in children and adolescents with ASD showed that total vermis volume was decreased in ASD, but when the ASD groups were examined separately, only the children and adolescents with higher functioning autism showed a statistically significant smaller volume of the vermis, compared with that of the typically developing controls. However, the data showed a trend towards a larger midsagittal area of vermal lobules VI-VII of individuals with higher functioning autism, compared with individuals with lower functioning autism, as well as with Asperger syndrome. This suggests a smaller vermal width in people with higher functioning autism (Scott et al, 2009). A previous study found a similar trend towards a larger midsagittal area of vermal lobules VI-VII of preschool children with higher functioning autism, compared with that of preschool children with lower functioning autism (Akshoomoff et al., 2004). Therefore, a larger midsagittal area of vermal lubules might be the vestige of an enlargement in vermal lobules VI–VII that could improve the procedural learning (Park et al., 2009), although thereafter, the lobules might show a decrease in volume without damaging remote procedural knowledge (Bracha et al., 1997; Molinari et al., 1997; Quintero-Gallego et al., 2006). In addition, an inverse correlation between the area of cerebellar vermis lobules VI-VII and frontal gray matter was found in preschool children with autistic disorder (Carper & Courchesne, 2000). In concordance with this, a study of monozygotic twins

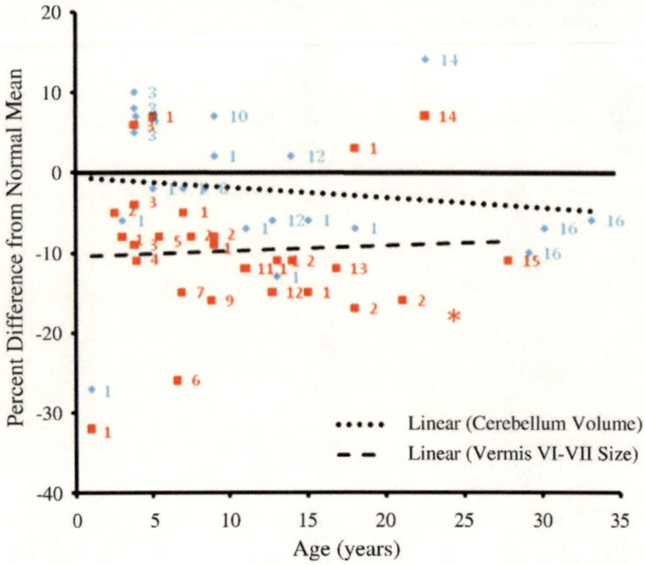

Fig. 7. Cerebellum and vermis size from infancy to adulthood in autism. Percent differences from normal average in each study of autistic subjects are plotted for cerebellar volume (blue diamonds) and vermis lobule VI–VII cross-sectional area (red squares) against age. Regression lines are shown for cerebellar volume and vermis lobule VI–VII area. The solid line represents normal average. Each number represents a separate study that has been cited in the paper from where this figure has been obtained (From Courchesne et al, 2011), with permission.

discordant for autism reported that DLPFC, amygdala, and vermal lobules VI-VII volumes were significantly associated with the severity of autistic symptomatology based on scores from the Autism Diagnostic Observation Schedule-Generic (Mitchell et al., 2009). All these findings are in accordance with the proposal of overload for declarative memory (Galea et al., 2010; Koch et al., 2006; León-Carrión et al., 2010), as well as faulty procedural memory (Hubert et al., 2007; Park et al., 2009). On the other hand, brain overgrowth is most pronounced in the frontal and temporal lobes; this occurs in people with ASD mainly during the first four years of life, but it is followed by arrest of growth, as illustrated in Fig. 8 (Courchesne et al., 2011). According to the mnesic imbalance theory, an overload for declarative memory might lead to a particularly precocious brain growth that is followed by a precocious arrest of growth (Romero-Munguía, 2007, 2008).

A recent study has estimated that cortical thickness provided the best classification between adults with ASD and controls: a pattern of reduced gray matter in frontal regions and increased gray matter in temporal regions (Ecker et al., 2010). This pattern resembles an exaggeration of the typical development of declarative memory in the human brain (Ceccarelli et al., 2009; Ofen et al., 2007). Another study has reported significant correlations between the volumes of the caudate, frontal lobe, temporal lobe, cerebellum and a measure of repetitive behaviors, while social and communication deficits in autism were associated with caudate, cerebellar, precuneus, frontal and temporal lobe regional volumes (Rojas et al., 2006). The overload for the declarative memory, as well as procedural learning of sequences and stereotypies might lead to these regional gray matter volumetric changes in autism (Ceccarelli et al., 2009; Draganski et al., 2006; Filippi et al., 2010). Moreover, a study involving

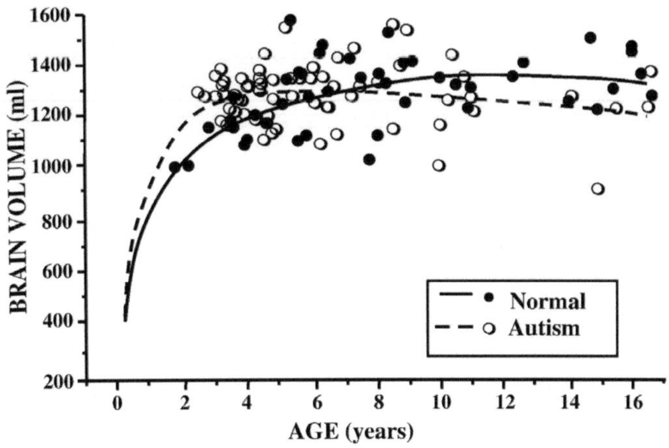

Fig. 8. Brain growth in autism through 16 years. Data plot shows larger MRI-based volumes in autistic 2- to 4-year-old males as compared to normal 2- to 4-year-old males and smaller overall brain volumes in autistic 8–16 year olds as compared to normal (From Courchesne et al, 2011), with permission.

boys with high-functioning autism compared to boys with developmental disorder reported that radiate but not deep white matter volume is increased in autism (Herbert, 2005). This might imply not only local over-connectivity, but also long-distance disconnection in the frontal lobe, as well as in other lobes (Courchesne & Pierce, 2005). All of this is in accordance with the assumption that a faulty procedural memory may complicate the simultaneous application of procedural knowledge (Romero-Munguía, 2008). So for instance, a study of children with autism showed that greater gray matter volume was associated with better communication skills in several regions of the frontal lobe, whereas increased total gray matter volume was correlated with more severe symptoms of autism (Parks et al., 2009); another study has reported that amygdala enlargement is correlated with better joint attention ability at age 4 years in children with autism (Mosconi et al., 2009), whereas this amygdala enlargement has been associated with more severe social and communication impairments in toddlers with autism (Schumann et al., 2009). However, the relationship between neuroanatomy and symptoms may be different across ASD due to cognitive differences among patients with low-functioning autism, high-functioning autism, and Asperger syndrome. So, the presence of a history of early speech delay in autistic disorder may reflect a greater deficit of procedural memory, compared with that in Asperger syndrome, which may lead to greater overload for declarative memory that might lead to an increase in gray matter that is smaller in Asperger syndrome (Ceccarelli et al., 2009; Draganski et al., 2006; Toal et al., 2010), while in high-functioning autism, the cerebral gray matter volume might have a negative correlation, not only with performance IQ but also with gains in declarative memory for details of experiences (Lotspeich et al., 2004; Ofen et al., 2007).

3.3 Functional neuroanatomy

The mnesic imbalance theory suggests a significantly slow development of procedural learning in ASD (Gordon & Stark, 2007; Romero-Munguía, 2008), which must be in accordance with data available on functional neuroanatomy of ASD. So for instance, in a study utilizing functional magnetic resonance imaging, the data showed that in comparison with the control group, adults with high-functioning autism had increased activity in the primary somatosensory cortex and the primary motor cortex during the late stages of a procedural learning task (Müller et al., 2004), which suggest a delay in procedural learning (Floyer-Lea & Matthews, 2004; Torriero et al., 2011). In another study, regional cerebral blood flow was measured in men with high-functioning autism during an emotion-recognition task, the results showed higher regional cerebral blood flow in the right anterior temporal lobe, ACC and right thalamus in individuals with autism (Hall et al., 2003); that is, the autistic group has shown activity of the first phases of procedural learning whereas the control group has shown activity of its last phase (Hubert et al., 2007). A study found that in comparison with the control group, males with high-functioning autism had diffusely increased connectivity of the caudate nuclei during simple visuomotor coordination tasks (Turner et al., 2006); an experimental study showed that during procedural learning, the striatum (caudate nucleus and putamen) has a diffuse functional connectivity, but this is selective if procedural knowledge has already been acquired (Barnes et al., 2005). Furthermore, the mnesic imbalance theory suggests that procedural memory is required to infer what a person is feeling from the facial expression, or some other perceptual category (Romero-Munguía, 2008); hence its deficit might lead to the utilization of alternate neural pathways dedicated to that task. For instance, a study examined whether successful facial affect recognition training is associated with increased activation of the fusiform gyrus in autism. Trained participants showed behavioral improvements with higher activity in the superior parietal lobe and maintained activation in the occipital cortex, but no significant activation changes in the fusiform gyrus were shown (Bölte et al., 2006). In contrast, the results of a study involving expert and novices radiologists during the interpreting of radiographic images, suggest that the fusiform gyrus becomes more engaged and the occipital cortex less engaged in interpreting radiographic images over the course of training (Harley et al., 2009).

Studies of the DMN in autism show several abnormalities. One of them demonstrated that ASD subjects fail to show the deactivation effect (Kennedy et al., 2006). The overload for the declarative memory might explain the fail to show the deactivation effect in ASD because the activation of the DMN is more pronounced at rest, after a cognitively challenging task (Pyka et al., 2009). In addition, the mnesic imbalance theory has argued that persons with ASD use explicit mental representations to resolve systemizing problems, while normally developing individuals do not (Romero-Munguía, 2008; Sahyoun et al., 2010). These explicit mental representations might not only prevent the deactivation effect but also increasing the activation of the DMN since the DMN is significantly activated during mental imagery (Daselaar et al., 2010). Other studies have observed decreased functional connectivity between posterior regions of DMN and other regions (Assaf et al., 2010; Monk et al., 2009). This might also be explained by overload for the declarative memory because DMN functional connectivity is progressively decreased in PCC regions during high short-term declarative memory loads (Esposito et al., 2009). In addition, association between delays in the decrease of amygdala activity and the reduction in the number of errors during a second set of facial recognition memory task was observed in adults with ASD, which was not present in healthy comparison subjects (Kleinhans et al, 2009). Albeit, a previous study reported that increased amygdala activation is associated with better face memory performance in healthy individuals (Kleinhans et al., 2007). Moreover,a significant increase of choline/creatine and a significant decrease of N-acetyl-aspastate/creatine in the hippocampus and amygdala was observed in children with ASD (Gabis et al., 2008), while in another study healthy elderly subjects had these same results at the hippocampus after they completed declarative memory training (Valenzuela et al, 2003). Finally, it is interesting to note that the disruption of the DLPFC improves the consolidation of procedural knowledge (Galea et al., 2010), which might shed light on its preliminary results in ASD (Baruth et al., 2010; Sokhadze et al., 2010).

4. Conclusion

The mnesic imbalance theory surmises that a cerebellar dysfunction causes faulty procedural memory with relatively intact declarative memory, as well as a reciprocal causal interaction, in which some neuroanatomical changes may result from this mnesic imbalance. The present work has offered arguments based on neurobiological findings in favour of this cognitive theory.

5. Acknowledgment

The author would like to thank Edith Monroy for reviewing the language of the manuscript.

6. References

Ábrahám, H.; Tornóczky, T.; Kosztolányi, G. & Seress, L. (2001). Cell formation in the cortical layers of the developing human cerebellum. *International journal of developmental neuroscience,* Vol. 19, No. 1, pp. 53-62, ISSN: 0736-5748.

Akar, S.; Drappatz, J.; Hsu, L.; Blinder, R.A.; Black, P.M. & Kesari, S. (2008). Hypertrophic olivary degeneration after resection of a cerebellar tumor. *Journal of neuro-oncology,* Vol. 87, No. 3, pp. 341-345. ISSN: 1573-7373.

Akshoomoff, N.; Lord, C.; Lincoln, A.J.; Courchesne, R.Y.; Carper, R.A.; Townsend, J. & Courchesne, E. (2004). Outcome classification of preschool children with autism spectrum disorders using MRI brain measures. *Journal of the American Academy of Child and Adolescent Psychiatry,* Vol. 43, No. 3, pp. 349-357, ISSN: 1527-5418.

American Psychiatric Association. (2000). *Diagnostic and statistical manual of mental disorders* (4th edition, text revision), American Psychiatric Publishing, ISBN: 978-0890420256, Washington, DC.

Appleby, P.A.; Kempermann, G. & Wiskott, L. (2011). The role of additive neurogenesis and synaptic plasticity in a hippocampal memory model with grid-cell like input. *PLoS computational biology,* Vol. 7, No. 1, pp. e1001063, ISSN: 1553-7358.

Assaf, M.; Jagannathan, K.; Calhoun, V.D.; Miller, L.; Stevens, M.C.; Sahl, R.; O'Boyle, J.G.; Schultz, R.T. & Pearlson, G.D. (2010). Abnormal functional connectivity of default mode sub-networks in autism spectrum disorder patients. *NeuroImage,* Vol. 53, No. 1, 247-256, ISSN: 1095-9572.

Bailey, A.; Luthert, P.; Dean, A.; Harding, B.; Janota, I.; Montgomery, M.; Rutter, M. & Lantos, P. (1998). A clinicopathological study of autism. *Brain : a journal of neurology,* Vol. 121, No. 5, pp. 889-905, ISSN: 0006-8950.

Bao, J.; Reim, K. & Sakaba. T. (2010). Target-dependent feedforward inhibition mediated by short-term synaptic plasticity in the cerebellum. *The Journal of neuroscience,* Vol. 30, No. 24, pp. 8171-8179, ISSN: 1563-5279.

Barnes, T.D.; Kubota, Y.; Hu, D.; Jin, D.Z. & Graybiel, A.M. (2005). *Nature,*Vol. 437, No. 7062, pp. 1158-1161, ISSN: 1476-4687.

Barnes, K.A.; Howard, J.H. Jr.; Howard, D.V.; Gilotty, L.; Kenworthy, L.; Gaillard, W.D. & Vaidya, C.J. (2008). Intact implicit learning of spatial context and temporal sequences in childhood autism spectrum disorder. *Neuropsychology,* Vol. 22, No. 5, pp. 563-570, ISSN: 1931-1559.

Baruth, J.M.; Casanova, M.F.; El-Baz, A.; Horrell, T.; Mathai, G.; Sears, L. & Sokhadze, E. (2010). Low-Frequency Repetitive Transcranial Magnetic Stimulation (rTMS) Modulates Evoked-Gamma Frequency Oscillations in Autism Spectrum Disorder (ASD). *Journal of neurotherapy,* Vol. 14, No. 3, pp. 179-194, ISSN: 1530-017X.

Bauman, M.L. & Kemper, T.L. (2005). Neuroanatomic observations of the brain in autism: a review and future directions. *International journal of developmental neuroscience,* Vol. 23, No. 2-3, pp. 183-187, ISSN: 0736-5748.

Bengtsson, F.; Jirenhed, D.A.; Svensson, P. & Hesslow, G. (2007). Extinction of conditioned blink responses by cerebello-olivary pathway stimulation. *Neuroreport,* Vol. 18, No.14, pp. 1479-1482, ISSN: 1473-558X.

Bölte, S.; Hubl, D.; Feineis-Matthews, S.; Prvulovic, D.; Dierks, T. & Poustka, F. (2006). Facial affect recognition training in autism: can we animate the fusiform gyrus? *Behavioral neuroscience,* Vol. 120, No. 1, pp. 211-216, ISSN: 1939-0084.

Bracha, V.; Zhao, L.; Wunderlich, D.A.; Morrissy, S.J. & Bloedel, J.R. (1997).Patients with cerebellar lesions cannot acquire but are able to retain conditioned eyeblink reflexes. *Brain : a journal of neurology,* Vol. 120, No. 8, pp. 1401-1413, ISSN: 0006-8950.

Brown, J.; Aczel, B.; Jiménez, L.; Kaufman, S.B. & Grant, K.P. (2010). Intact implicit learning in autism spectrum conditions. *The quarterly journal of experimental psychology,* Vol. 63, No. 9, pp. 1789-1812, ISSN: 1747-0226.

Budson, A.E. & Price, B.H. (2001). Memory: clinical disorders. In: *Nature encyclopedia of life sciences,* Vol. 11, Nature Publishing Group. (Ed.), pp. 529-536, Macmillan Publishers Ltd,London, UK. Available from *http://people.bu.edu/abudson/budson-memory.pdf*

Carper, R.A. & Courchesne, E. (2000). Inverse correlation between frontal lobe and cerebellum sizes in children with autism. *Brain : a journal of neurology,* Vol. 123, No. 4, pp. 836-844, ISSN: 0006-8950.

Casanova, M.F. (2007). The neuropathology of autism. *Brain pathology,* Vol. 17, No.4, pp.422-433, ISSN: 1750-3639.

Casanova, M.F.; Switala, A.E.; Trippe, J. & Fitzgerald, M. (2007). Comparative minicolumnar morphometry of three distinguished scientists. *Autism : the international journal of research and practice,* Vol. 11, No. 6, pp. 557-569, ISSN: 1461-7005.

Ceccarelli, A.; Rocca, M.A.; Pagani, E.; Falini, A.; Comi, G. & Filippi, M. (2009). Cognitive learning is associated with gray matter changes in healthy human individuals: a tensor-based morphometry study. *NeuroImage,* Vol. 48, No. 3, pp. 585-589, ISSN:1095-9572.

Courchesne, E. & Pierce, K. (2005). Why the frontal cortex in autism might be talking only to itself: local over-connectivity but long-distance disconnection. *Current opinion in neurobiology*, Vol. 15, No. 2, pp. 225-230, ISSN: 1873-6882.

Courchesne, E.; Campbell, K. & Solso, S. (2011). Brain growth across the life span in autism: Age-specific changes in anatomical pathology. *Brain research*, Vol. 1380, pp. 138-45, ISSN: 1872-6240.

Daselaar, S.M.; Prince, S.E.; Dennis, N.A.; Hayes, S.M.; Kim, H. & Cabeza, R. (2009). Posterior midline and ventral parietal activity is associated with retrieval success and encoding failure. *Frontiers in human neuroscience*, Vol. 3, pp. e13, ISSN: 1662-5161.

Daselaar, S.M.; Porat, Y.; Huijbers, W. & Pennartz, C.M. (2010). Modality-specific and modality-independent components of the human imagery system. *NeuroImage*, Vol. 52, No. 2, pp. 677-685, ISSN: 1095-9572.

Draganski, B.; Gaser, C.; Kempermann, G.; Kuhn, H.G.; Winkler, J.; Büchel, C. & May, A. (2006). Temporal and spatial dynamics of brain structure changes during extensive learning. *The Journal of neuroscience: the official journal of the Society for Neuroscience*,Vol. 26, No. 23, pp. 6314-6317, ISSN: 1529-2401.

Ecker, C.; Marquand, A.; Mourão-Miranda, J.; Johnston, P.; Daly, E.M.; Brammer, M.J.; Maltezos, S.; Murphy, C.M.; Robertson, D.; Williams, S.C. & Murphy, D.G. (2010). Describing the brain in autism in five dimensions--magnetic resonance imaging-assisted diagnosis of autism spectrum disorder using a multiparameter classification approach. *The Journal of neuroscience: the official journal of the Society for Neuroscience*, Vol. 30, No. 32, pp. 10612-10623, ISSN: 1529-2401.

Ehninger, D. & Kempermann, G. (2003). Regional effects of wheel running and environmental enrichment on cell genesis and microglia proliferation in the adult murine neocortex. *Cerebral cortex*, Vol. 13, No. 8, pp. 845-851, ISSN: 1460-2199.

Esposito, F.; Aragri, A.; Latorre, V.; Popolizio, T.; Scarabino, T.; Cirillo, S.; Marciano, E.; Tedeschi, G. & Di Salle, F. (2009). Does the default-mode functional connectivity of the brain correlate with working-memory performances? *Archives italiennes de biologie*, Vol. 147, No. 1-2, pp. 11-20, ISSN: 0003-9829.

Filippi, M.; Ceccarelli, A.; Pagani, E.; Gatti, R.; Rossi, A.; Stefanelli, L.; Falini, A.; Comi, G. & Rocca, M.A. (2010). Motor learning in healthy humans is associated to gray matter changes: a tensor-based morphometry study. *PloS one*, Vol. 5, No. 4, pp. e10198, ISSN: 1932-6203.

Floyer-Lea, A. & Matthews, P.M. (2004). Changing brain networks for visuomotor control with increased movement automaticity. *Journal of neurophysiology*, Vol.92, No. 4, pp. 2405-2412, ISSN: 1522-1598.

Gabis, L.; Huang, W.; Azizian, A.; Devincent, C.; Tudorica, A.; Kesner-Baruch, Y.; Roche, P. & Pomeroy, J. (2008). 1H-magnetic resonance spectroscopy markers of cognitive and language ability in clinical subtypes of autism spectrum disorders. *Journal of child neurology*, Vol. 23, No. 7, pp. 766-774, ISSN: 1708-8283.

Galea, J.M.; Albert, N.B.; Ditye, T. & Miall, R.C. (2010). Disruption of the dorsolateral prefrontal cortex facilitates the consolidation of procedural skills. *Journal of cognitive neuroscience*, Vol. 22, No. 6, pp. 1158-1164, ISSN: 1530-8898.

Gidley Larson, J.C. & Mostofsky, S.H. (2008). Evidence that the pattern of visuomotor sequence learning is altered in children with autism. *Autism research : official journal of the International Society for Autism Research,* Vol. 1, No. 6, pp. 341-353, ISSN: 1939-3806.

Goldberg, T.E. (1987). On hermetic reading abilities. *Journal of autism and developmental disorders,* Vol. 17, No. 1, pp. 29-44, ISSN: 1573-3432.

Gordon, B. & Stark S. (2007). Procedural Learning of a Visual Sequence in Individuals with Autism. *Focus on autism and other developmental disabilities,* Vol.22, No. 1, pp. 14-22, ISSN: 1088-3576.

Gould, E. (2007). How widespread is adult neurogenesis in mammals? *Nature reviews. Neuroscience,* Vol. 8, No. 6, pp. 481-488, ISSN: 1471-0048.

Guerin, P.; Lyon, G.; Barthelemy, C.; Sostak, E.; Chevrollier, V.; Garreau, B. & Lelord, G. (1996). Neuropathological study of a case of autistic syndrome with severe mental retardation. *Developmental medicine and child neurology,* Vol. 38, No. 3, pp. 203-211, ISSN: 0012-1622.

Hall, G.B.; Szechtman, H. & Nahmias, C. (2003). Enhanced salience and emotion recognition in Autism: a PET study. *The American journal of psychiatry,* Vol. 160, No. 8, pp. 1439-1441, ISSN: 1535-7228.

Harley, E.M.; Pope, W.B.; Villablanca, J.P.; Mumford, J.; Suh, R.; Mazziotta, J.C.; Enzmann, D. & Engel S.A. (2009). Engagement of fusiform cortex and disengagement of lateral occipital cortex in the acquisition of radiological expertise. *Cerebral cortex,* Vol. 19, No. 11, pp. 2746-2754, ISSN: 1460-2199.

Herbert, M.R. (2005). Large brains in autism: the challenge of pervasive abnormality. *The Neuroscientist : a review journal bringing neurobiology, neurology and psychiatry,* Vol. 11, No. 5, pp. 417-440, ISSN: 1089-4098.

Howlin, P. (2003). Outcome in high-functioning adults with autism with and without early language delays: implications for the differentiation between autism and Asperger syndrome. *Journal of autism and developmental disorders,* Vol. 33, No. 1, pp. 3-13, ISSN: 1573-3432.

Hubert, V.; Beaunieux, H.; Chételat, G.; Platel, H.; Landeau, B. ; Danion, J.M. ; Viader, F. ; Desgranges, B. & Eustache, F. (2007). The dynamic network subserving the three phases of cognitive procedural learning. *Human brain mapping,* Vol.28, No.12, pp.1415-1429, ISSN: 1097-0193.

Hutsler, J.J. & Zhang, H. (2010). Increased dendritic spine densities on cortical projection neurons in autism spectrum disorders. *Brain research,* Vol. 1309, pp. 83-94, ISSN: 1872-6240.

Kanner, L. (1943). Autistic disturbances of affective contact. *The nervous child,* Vol. 2, pp. 217-250, ISSN: 0099-4286.

Kennedy, D.P.; Redcay, E. & Courchesne, E. (2006). Failing to deactivate: resting functional abnormalities in autism. *Proceedings of the National Academy of Sciences of the United States of America,* Vol. 103, No. 21, pp. 8275-8280, ISSN: 1091-6490.

Kleinhans, N.M.; Johnson, L.C.; Mahurin, R.; Richards, T.; Stegbauer, K.C.; Greenson, J.; Dawson, G. & Aylward, E. (2007). Increased amygdala activation to neutral faces is associated with better face memory performance. *Neuroreport,* Vol. 18, No. 10, 987-991, ISSN: 1473-558X.

Kleinhans, N.M.; Johnson, L.C.; Richards, T.; Mahurin, R.; Greenson, J.; Dawson, G. & Aylward, E. (2009). Reduced neural habituation in the amygdala and social impairments in autism spectrum disorders. *The American journal of psychiatry,* Vol. 166,No. 4, pp. 467-475, ISSN: 1535-7228.

Klinger, L.G.; Klinger, M.R. & Pohlig, R. (2007). Implicit learning impairments in autism spectrum disorders: Implications for treatment. In: *New developments in autism: The future is today,* Pérez, J.M.; González, P.M.; Comí, M.L. & Nieto, C. (Eds.), pp. 76–103, Jessica Kingsley Publishers, ISBN: 978-1-84310-449-0, London, UK.

Koch, K.; Wagner, G.; von Consbruch, K.; Nenadic, I.; Schultz, C.; Ehle, C.; Reichenbach, J.; Sauer, H. & Schlösser, R. (2006). Temporal changes in neural activation during practice of information retrieval from short-term memory: an fMRI study. *Brain research,* Vol. 1107, No. 1, pp. 140-150, ISSN: 1872-6240.

León-Carrión, J.; Izzetoglu, M.; Izzetoglu, K.; Martín-Rodríguez, J.F.; Damas-López, J.; Barroso y Martin, J.M. & Domínguez-Morales, M.R. (2010). Efficient learning produces spontaneous neural repetition suppression in prefrontal cortex. *Behavioural brain research,* Vol. 208, No. 2, pp. 502-508, ISSN: 1872-7549.

Leto, K.; Bartolini, A. & Rossi, F. (2008). Development of cerebellar GABAergic interneurons: origin and shaping of the "minibrain" local connections. *The cerebellum,* Vol. 7, No. 4, pp. 523-529, ISSN: 1473-4230.

Lotspeich, L.J.; Kwon, H.; Schumann, C.M.; Fryer, S.L.; Goodlin-Jones, B.L.; Buonocore, M.H.; Lammers, C,R.; Amaral, D.G. & Reiss, A.L. (2004). Investigation of neuroanatomical differences between autism and Asperger syndrome. *Archives of general psychiatry,* Vol. 61, No. 3, pp. 291-298, ISSN: 1538-3636.

Meilleur, A.A. & Fombonne, E. (2009). Regression of language and non-language skills in pervasive developmental disorders. *Journal of intellectual disability research,* Vol. 53, No. 2, pp. 115-124, ISSN: 1365-2788.

Mercadante, M.T.; Van der Gaag, R.J. & Schwartzman, J.S. (2006). Non-Autistic Pervasive Developmental Disorders: Rett's syndrome, childhood disintegrative disorder and pervasive developmental disorder not otherwise specified. *Revista brasileira de psiquiatria,* Vol. 28, Suppl. 1, pp. 12-20, ISSN: 1809-452X.

Mitchell SR, Reiss AL, Tatusko DH, Ikuta I, Kazmerski DB, Botti JA, Burnette CP, Kates WR. (2009). Neuroanatomic alterations and social and communication deficits in monozygotic twins discordant for autism disorder. *The American journal of psychiatry,* Vol. 166, No. 8, pp. 917-925, ISSN: 1535-7228.

Molinari, M.; Leggio, M.G.; Solida, A.; Ciorra, R.; Misciagna, S.; Silveri, MC. & Petrosini, L. (1997). Cerebellum and procedural learning: evidence from focal cerebellar lesions. *Brain : a journal of neurology,* Vol. 120, No. 10, pp. 1753-1762, ISSN: 1460-2156.

Monk, C.S.; Peltier, S.J.; Wiggins, J.L.; Weng, S.J.; Carrasco, M.; Risi, S. & Lord, C. (2009). Abnormalities of intrinsic functional connectivity in autism spectrum disorders. *NeuroImage,* Vol. 47, No. 2, pp. 764-772, ISSN: 1095-9572.

Morgan, J.T.; Chana, G.; Pardo, C.A.; Achim, C.; Semendeferi, K.; Buckwalter, J.; Courchesne, E. & Everall, I.P. (2010). Microglial activation and increased microglial density observed in the dorsolateral prefrontal cortex in autism. *Biological psychiatry,* Vol. 68, No. 4, pp. 368-376, ISSN: 1873-2402.

Mosconi, M.W.; Cody-Hazlett, H.; Poe, M.D.; Gerig, G.; Gimpel-Smith, R. & Piven, J. (2009). Longitudinal study of amygdala volume and joint attention in 2- to 4-year-old children with autism. *Archives of general psychiatry,* Vol. 66, No. 5, pp. 509-516, ISSN: 1538-3636.

Mostofsky, S.H.; Goldberg, M.C.; Landa R.J. & Denckla M.B. (2000). Evidence for a deficit in procedural learning in children and adolescents with autism: implications for cerebellar contribution. *Journal of the International Neuropsychological Society,* Vol. 6, No. 7, pp. 752-759, ISSN: 1469-7661.

Müller, R.A.; Cauich, C.; Rubio, M.A.; Mizuno, A. & Courchesne, E. (2004). Abnormal activity patterns in premotor cortex during sequence learning in autistic patients, *Biological psychiatry,* Vol. 56, No. 5, pp. 323-332, ISSN: 1873-2402.

Nemeth, D.; Janacsek, K.; Balogh, V.; Londe, Z.; Mingesz, R.; Fazekas, M.; Jambori, S.; Danyi, I. & Vetro, A (2010). Learning in autism: implicitly superb. *PloS one,* Vol. 5, No. 7, pp. 11731, ISSN: 1932-6203.

Ofen, N.; Kao, Y.C.; Sokol-Hessner, P.; Kim, H.; Whitfield-Gabrieli, S. & Gabrieli, J.D. (2007). Development of the declarative memory system in the human brain. *Nature neuroscience,* Vol. 10, No. 9, pp. 1198-1205, ISSN: 1546-1726.

Okano, H.; Hirano, T. & Balaban, E. Learning and memory. (2000). *Proceedings of the National Academy of Sciences of the United States of America,* Vol. 97, No. 23, pp. 12403-12404, ISSN: 1091-6490.

Park, I.S.; Lee, K.J.; Han, J.W.; Lee, N.J.; Lee, W.T.; Park, K.A. & Rhyu, I.J. (2009). Experience-dependent plasticity of cerebellar vermis in basketball players. *The cerebellum,* Vol. 8, No. 3, pp. 334-339, ISSN: 1473-4230.

Parks, L.K.; Hill, D.E.; Thoma, R.J.; Euler, M.J.; Lewine, J.D. & Yeo, R.A. (2009). Neural correlates of communication skill and symptom severity in autism: A voxel-based morphometry study. *Research in Autism Spectrum Disorders,* Vol. 3, No. 2, pp. 444-454, ISSN: 1750-9467.

Pisula, E. (2010). The autistic mind in the light of neuropsychological studies. *Acta neurobiologiae experimentalis,* Vol. 70, No. 2, pp. 119-130, ISSN: 1689-0035.

Pyka, M.; Beckmann, C.F.; Schöning, S.; Hauke, S.; Heider, D.; Kugel, H.; Arolt, V. & Konrad, C. (2009). Impact of working memory load on FMRI resting state pattern in subsequent resting phases. *PloS one,* Vol. 4, No. 9, pp. e7198, ISSN: 1932-6203.

Quintero-Gallego, E.A.; Gómez, C.M.; Vaquero Casares, E.; Márquez, J. & Pérez-Santamaría, F.J. (2006). Declarative and procedural learning in children and adolescents with posterior fossa tumours. *Behavioral and brain functions,* Vol. 2, No. 1, pp. e9, ISSN: 1744-9081.

Rojas, D.C.; Peterson, E.; Winterrowd, E.; Reite, M.L.; Rogers, S.J. & Tregellas J.R. (2006). Regional gray matter volumetric changes in autism associated with social and repetitive behavior symptoms. *BMC psychiatry,* Vol. 6, pp. e56, ISSN: 1471-244X.

Rajendran, G. & Mitchell, P. (2007). Cognitive Theories of Autism. *Developmental Review,* Vol. 27, No. 2, pp. 224-260, ISSN: 0273-2297.

Rancillac, A. & Crépel, F. (2004). Synapses between parallel fibres and stellate cells express long-term changes in synaptic efficacy in rat cerebellum. *The Journal of physiology,* Vol. 554, No. 3, pp. 707-720. ISSN: 1469-7793.

Rakic, P. & Sidman, R.L. (1970). Histogenesis of cortical layers in human cerebellum, particularly the lamina dissecans. *Journal of comparative neurology and psychology,* Vol. 139, No. 4, pp. 473-500, ISSN: 0021-9967.

Romero-Munguía, M.A. (1998). ¿Es la memoria procesal deficiente la causa del comportamiento estereotipado en el autismo? *Psiquiatría, Órgano Oficial de Difusión de la Asociación Psiquiátrica Mexicana, A.C.,* Vol. 14, No. 2 pp. 62-65, ISSN: 0187-4543.

Romero-Munguía, M.A. (2002). Memoria procesal deficiente y alteraciones de la comunicación en la patogenia del autismo infantil. *Neurología, Neurocirugía y Psiquiatría,* Vol. 35, No. 4, pp.203-208, ISSN: 0028-3851.

Romero-Munguía, M.A. (2007). Sistemas de memoria y neuroanatomía en los trastornos del espectro autista, In: *Psiquiatría y Neurociencias,* Castro-Román, R. & Velázquez-Moctezuma, J. (Eds.), pp. 247-276, Hospital Psiquiátrico "Dr. Samuel Ramírez Moreno" - Universidad Autónoma Metropolitana, ISBN: 978-970-31-0983-8, México, DF.

Romero-Munguía, M.A. (2008). Mnesic imbalance: a cognitive theory about autism spectrum disorders. *Annals of general psychiatry,* Vol. 7, No. 1, pp. e20, ISSN: 1744-859X.

Romero-Munguía, M.A. (2009). Trastorno de aprendizaje no verbal frente a trastorno del espectro autista: el papel del aprendizaje procesal. *Revista de Neurología,* Vol. 49, No. 8, pp. 448, ISSN: 1576-6578.

Sahyoun, C.P.; Belliveau, J.W.; Soulières, I.; Schwartz, S. & Mody, M. (2010). Neuroimaging of the functional and structural networks underlying visuospatial vs. linguistic reasoning in high-functioning autism. *Neuropsychologia,* Vol . 48, No. 1, pp. 86-95, ISSN: 1873-3514.

Schumann, C.M.; Barnes, C.C.; Lord, C. & Courchesne, E. (2009). Amygdala enlargement in toddlers with autism related to severity of social and communication impairments. *Biological psychiatry,* Vol. 66, No. 10, pp. 942-949, ISSN: 1873-2402.

Scott, J.A.; Schumann, C.M.; Goodlin-Jones, B.L. & Amaral, D.G. (2009). A comprehensive volumetric analysis of the cerebellum in children and adolescents with autism spectrum disorder. *Autism research : official journal of the International Society for Autism Research,* Vol. 2, No. 5, pp. 246-257, ISSN: 1939-3806.

Sokhadze, E.; Baruth, J.; Tasman, A.; Mansoor, M.; Ramaswamy, R.; Sears, L.; Mathai, G.; El-Baz, A. & Casanova, M.F. (2010). Low-frequency repetitive transcranial magnetic stimulation (rTMS) affects event-related potential measures of novelty processing in autism. *Applied psychophysiology and biofeedback,* No. 35, No. 2, pp. 147-161, ISSN: 1573-3270.

Squire, L.R.; Stark, C.E. & Clark, R.E. (2004). The medial temporal lobe. *Annual review of neuroscience,* Vol. 27, pp. 279-306, ISSN: 1545-4126.

Stanfield, A.C.; McIntosh, A.M.; Spencer, M.D.; Philip, R.; Gaur, S. & Lawrie, S.M. (2008). Towards a neuroanatomy of autism: a systematic review and meta-analysis of structural magnetic resonance imaging studies. *European psychiatry : the journal of the Association of European Psychiatrists,* Vol. 23, No. 4, pp. 289-299, ISSN: 1778-3585.

Toal, F.; Daly, E.M.; Page, L.; Deeley, Q.; Hallahan, B.; Bloemen, O.; Cutter, W.J.; Brammer, M.J.; Curran, S.; Robertson, D.; Murphy, C.; Murphy, K.C. & Murphy, D.G. (2010). Clinical and anatomical heterogeneity in autistic spectrum disorder: a structural MRI study. *Psychological medicine,* Vol. 40, No. 7, pp. 1171-1181, ISSN: 1469-8978.

Torriero, S.; Oliveri, M.; Koch, G.; Lo Gerfo, E.; Salerno, S.; Ferlazzo, F.; Caltagirone, C. & Petrosini, L. (2011). Changes in cerebello-motor connectivity during procedural learning by actual execution and observation. *Journal of cognitive neuroscience,* Vol. 23, No. 2, pp. 338-348, ISSN: 1530-8898.

Turner, K.C.; Frost, L.; Linsenbardt, D.; McIlroy, J.R. & Müller, R.A. (2006). Atypically diffuse functional connectivity between caudate nuclei and cerebral cortex in autism. *Behavioral and brain functions,* Vol. 2, No. 1, pp. e34, ISSN: 1744-9081.

Valenzuela MJ, Jones M, Wen W, Rae C, Graham S, Shnier R, Sachdev P. (2003). Memory training alters hippocampal neurochemistry in healthy elderly. *Neuroreport,* Vol. 14, No. 10, pp. 1333-1337, ISSN: 1473-558X.

Walenski, M.; Mostofsky, S.H.; Gidley-Larson, J.C. & Ullman M.T (2008). Brief report: enhanced picture naming in autism. *Journal of autism and developmental disorders,* Vol. 38, No. 7, pp.1395-1399, ISSN: 1573-3432.

Waterhouse, L. (2008). Autism overflows: increasing prevalence and proliferating theories. *Neuropsychology review,* Vol. 18, No. 4, pp. 273-286, ISSN: 1573-6660.

Werner, E. & Dawson, G. (2005). Validation of the phenomenon of autistic regression using home videotapes. *Archives of general psychiatry,* Vol. 62, No. 8, pp. 889-895, ISSN: 1538-3636.

Whitney, E.R.; Kemper, T.L.; Rosene, D.L.; Bauman, M.L. & Blatt, G.J. (2009). Density of cerebellar basket and stellate cells in autism: evidence for a late developmental loss of Purkinje cells. *Journal of neuroscience research,* Vol. 87, No. 10, pp. 2245-2254, ISSN: 1097-4547.

Williams, R.S.; Hauser, S.L.; Purpura, D.P.; DeLong, G.R. & Swisher, C.N. (1980). Autism and mental retardation: neuropathologic studies performed in four retarded persons with autistic behavior. *Archives of neurology,* Vol. 37, No. 12, pp. 749-753, ISSN: 0003-9942.

Yip, J.; Soghomonian, J.J. & Blatt, G.J. (2008). Increased GAD67 mRNA expression in cerebellar interneurons in autism: implications for Purkinje cell dysfunction. *Journal of Neuroscience Research,* Vol. 86, No. 3, pp. 525-530, ISSN: 1097-4547.

23

Behavioral and Electrophysiological Characterization of Induced Neural Plasticity in the Autistic Brain

Jaime A. Pineda[1,2] et al.*
[1]*Departments of Cognitive Science, University of California,*
[2]*Neuroscience Group, University of California,*
San Diego,
USA

1. Introduction

Despite extensive research, the causes of Autistic Spectrum Disorders (ASD) are still unknown and no single explanation has been proposed that can account for the heterogeneous profile (Muller, 2007). The DSM-IV diagnostic criteria for ASD include deficits in social and communicative skills such as imitation, empathy, and shared attention, as well as restricted interests and repetitive patterns of behaviors. Elucidating the neuroetiology of these symptoms has been a challenge because the behavioral manifestations vary both in severity (high, medium, or low) as well as expression (Autistic Disorder, Asperger Syndrome, or Pervasive Developmental Disorder-Not Otherwise Specified) (Matson, 2006; Volkmar et al., 1994). Although a growing body of work has raised questions about the role of mirror neurons in human social behavior (Hickok, 2009; Lingnau et al., 2009), many recent studies suggest that a dysfunction in the frontal human mirror neuron system (hMNS) could potentially account for the social deficits in autism, including problems with imitation, understanding actions, emotion recognition, and empathy (Williams et al., 2001; Oberman et al., 2005; Dapretto et al., 2006; Williams et al., 2006).
Mirror neurons were initially reported by Rizzolatti and colleagues (di Pellegrino et al., 1992) in the premotor cortex of macaque monkeys (area F5), an area thought to be analogous to Broca's area (Brodmann's area 44) in humans (Buccino et al., 2001; Buccino et al., 2004; Petrides et al., 2005). Some cells in this region increase firing during the execution of an action as well as during observation of a similar action performed by others. This execution/observation matching feature is hypothesized to provide a mechanism for

*Heather Pelton[1], Oriana Aragon[3], Jia-Min Bai[4], Matt Erhart[1], Dane Chambers[1], Burcu Darst[1], Ernesto Enrique[5], Steven Gilmore[1], Stephen Johnson[1], Albert Anaya[1], Alicia Trigeiro[1], Dan T. Lotz[1], Nicholas Pojman[1], Tom Gamage[6] and David Linderman[1]
3 *Department of Psychology, Yale University, USA*
4 *Department of Psychology, University of Illinois at Urbana-Champaign, USA*
5 *Departments of Biology, San Diego State University, USA*
6 *Department of Psychology, Virginia Commonwealth University, USA*

mapping seeing into doing and vice versa - a mechanism capable of internally mirroring the action it perceives and performing a "simulation" of the action without accompanying motor execution (Rizzolatti & Craighero, 2004). Furthermore, the existence of auditory mirror neurons in the same region make it likely that a system exists that responds to implied action or attention to movement in the absence of discrete visual perception of the action (Umilta et al., 2001; Kohler et al., 2002; Iacoboni et al., 2005).

While individual mirror neurons have not been studied directly in humans, the existence of an analogous system has been supported by indirect population-level measures including transcranial magnetic stimulation (TMS: (Fadiga et al., 1999), functional magnetic resonance imaging (fMRI: (Iacoboni et al., 1999), and electroencephalogram/magnetoencephalogram (EEG/MEG: (Pineda, 2005; Muthukumaraswamy et al., 2004; Muthukumaraswamy & Singh, 2008; Hari et al., 1997). These and many other studies strongly support the existence of a mirror neuron system consisting of interconnected regions, in the ventral premotor area (PMv) of the IFG, parietal frontal (PF) in the rostral cortical convexity of the inferior parietal lobule (IPL), and the superior temporal sulcus (Rizzolatti & Craighero, 2004; Iacoboni et al., 2005; Buccino et al., 2004). MEG and EEG studies have further suggested that mirroring activity is reflected in the mu frequency oscillations (8-13 Hz, 13-15 Hz, and 15-25 Hz) measured over the sensorimotor cortex (Hari et al., 1997; Cochin et al., 1999; Pineda et al., 2000; Muthukumaraswamy et al., 2004; Muthukumaraswamy & Johnson, 2004). Although some of these frequency bands overlap with the traditional alpha frequency band, the evidence argues for distinct neural sources (Niedermeyer, 1997). More specifically, traditional alpha oscillations reflect visual information processing in the occipital lobes and are readily affected by the opening and closing of the eyes. In contrast, mu rhythms are generated in more anterior sources and are not susceptible to eye closure. It is also assumed that their sources are in sensorimotor cortices, where neurons fire synchronously while at rest to produce high amplitude oscillations measured at the scalp. It is argued that input from premotor areas, presumably correlated with mirroring or simulation activity, produces asynchronous firing in the sensorimotor circuits during self, observed, and imagined movement resulting in suppressed or desynchronized EEG activity (Gastaut & Bert, 1954; Pfurtscheller & Aranibar, 1979; Pineda et al., 2000; Pineda, 2005). Such suppression to observed movement in the absence of self movement is hypothesized to reflect downstream modulation of sensorimotor circuits by premotor hMNS (Altschuler et al., 1998; Oberman et al., 2005; Pineda, 2005).

In short, hMNS is hypothesized to be engaged during the observation of actions and thought to be one of the neural mechanisms by which we comprehend such actions (Gallese et al., 1996; Rizzolatti & Fabbri-Destro, 2009), understand the goal or intentions of those actions (Blakemore et al., 2001), learn through imitation (Williams et al., 2001; Wohlschlager & Bekkering, 2002), interpret facial expressions (Carr et al., 2003), and exhibit empathy (Leslie et al., 2004). Given these relationships, Williams et al. (Williams et al., 2001) posited that a developmental deficiency in hMNS could lead to problems in imitation learning and account for the type of theory of mind deficits thought to occur in ASD individuals (Baron-Cohen et al., 1985; Baron-Cohen et al., 1997; Baron-Cohen, 2009). That is, an inability to "form and coordinate social responses of self and others via amodal or cross-modal representation processes" could impede early affective, social, and communication development (Ozonoff et al., 1991; Rogers et al., 2003).

Impairments in individuals with autism include deficits in behaviors that parallel the presumed functions of hMNS, (Rogers et al., 2003; Leslie et al., 2004; Williams et al., 2004; Buxbaum et al., 2005). Anatomical evidence provides support for such a link. Villalobos et al (Villalobos et al., 2005) found reduced functional connections between the inferior frontal cortex and primary visual area (V1) while Just et al (Just et al., 2004) found reduced functional connections between the inferior frontal cortex and other areas during language tasks. Recent neurophysiological studies provide further evidence for an hMNS dysfunction in individuals with ASD (Just et al., 2004). Oberman et al (Oberman et al., 2005) found that children with ASD, when compared to matched typically-developing controls, exhibit mu suppression for self but not observed action. In a study by Dapretto et al (Dapretto et al., 2006) comparing children with ASD to typically developing children it was found that ASD children did not show activation of the IFG during imitation of facial expressions.

Recent studies argue that many aspects of autism arise from atypical anatomical connections and therefore produce altered activity between different areas in the brains (Courchesne & Pierce, 2005). Such a 'disconnection syndrome' as a function of hypoconnectivity could lead to desynchronization and ineffective intra- and interhemispheric communication impacting many neural circuits. If hMNS is dysfunctional because of altered connectivity, it seems appropriate to consider whether it is possible to change cortical dynamics in this circuit to induce neural plastic changes that would normalize those connections and *inter alia* alleviate symptoms of the disorder. One approach in this regard is plasticity-inducing rehabilitation training (PIRT), which utilizes neurofeedback training of EEG signals as a form of operant conditioning (Lubar 1997). This type of training has been used clinically for many years and provides real-time feedback to the user allowing for alteration and enhancement of brain function (Hirshberg et al., 2005). One explanation of how such training produces its behavioral and electrophysiological effects is by gaining access to and control over regulatory mechanisms that increase or decrease synchronous or desynchronous activity in thalamocortical networks (Lilienfeld, 2005). This leads to the induction of neural plastic changes.

To verify and extend the positive changes in behavior reported in previous autism work with neurofeedback (Pineda et al., 2008; Coben et al., 2009; Jarusiewicz, 2003), the following study was designed to compare the effects of this type of training methodology on high-functioning children with ASD compared to typically-developing matched controls. Participants received a total of 30 hrs or training and were assessed with a large variety of cognitive assessment tools.

2. Methods

2.1 Participants

Eighteen individuals diagnosed with autism (17 male, 1 female; right-handed), as well as twelve typically developing (TD) individuals (10 males; 2 females) were initially recruited for the study. One participant was diagnosed as low-functioning and was not included in the analyses. An additional autistic child and one TD child completed pretraining assessments but did not complete the 20 weeks of training. Hence, post-training analyses included 16 participants with autism (age 6-17; M = 9.8 $\pm$ 3.3 years) and 11 TD (age 6-17; M = 11.2 $\pm$ 3.5 years). Autism participants were recruited through Valerie's List, a listserv of families and professionals in the San Diego autism community. Parents were asked to provide an outside diagnosis, which was verified by a licensed clinical psychologist not associated with the research. In most cases this involved administration of the Autism Diagnostic Observation Schedule - Generic (ADOS-G), the Autism Diagnostic Interview-Revised (ADI-R), and the Wechsler Abbreviated Scale of Intelligence (WASI) (see Table 1). Based on the results of these assessments and clinical judgment all 16 children met criteria for *Autism Spectrum Disorder*. All were considered high-functioning, defined as having age appropriate verbal comprehension abilities and an IQ greater than 80. No children were diagnosed with attention deficit hyperactivity disorder (ADHD) and only one child had experienced epileptic-like seizures prior to the study. TD participants scored within the normal range on a standardized test of intelligence, had no neurological or psychological disorder, and were matched on chronological age and gender with a participant in the ASD group. All participants and their guardians (for children under 18 years of age) signed the informed consent form before participating. The protocol was approved by the University of California, San Diego's Institutional Review Board and the study has been performed in accordance with the ethical standards described in the 1964 Declaration of Helsinki.

	IQ Verbal	IQ Non Verbal	IQ Full	ADI COM	ADI SOC	ADI REP	ADOS COM	ADOS SOC	ADOS C+S	ADOS REP
ASD	105.4	108.1	107.5	12.5	20.2	4.8	3.7	9.2	12.8	2.2
TD	114.7	109.7	116	-	-	-	-	-	-	-

Table 1. Mean scores on the WASI, ADI, and ADOS

2.2 Cognitive assessments

All participants received a series of electrophysiological, cognitive, behavioral, and parental-assessments before and after 30 hours of PIRT over a 20 week period. These included a mini-Quantitative EEG (mini-QEEG), Suppression Indices for all frequencies, including mu suppression index (MSI), evaluation of imitation and sustained attention, and parental assessments.

2.2.1 Mini-QEEG

Participants were asked to remain still and to relax for one minute intervals while EEG was recorded from two sites at a time over six intervals for a total of twelve EEG sites. Participants were given a short break after each minute of recording. Mini-QEEG is used to assess the asymmetry and coherence of the scalp-recorded EEG at rest and provides an electrophysiological profile indicating the degree of functional connectivity between different pairs of electrodes that reflect processing in distinct cortical areas.

2.2.2 Suppression indices

The mu suppression index (MSI) is an electrophysiological tool developed by Oberman et al (Oberman et al., 2005) to measure the changes in mu power during the observation of biological or non-biological movement that is either goal- or non-goal oriented. Similar indices for other EEG frequencies (delta, theta, SMR, beta, and gamma) were computed in addition to the MSI. Recordings were taken while participants viewed silent action videos (120 seconds each) on a computer monitor while performing a continuous performance attention task (counting the number of pauses in the action). A baseline, non-biological "Ball" condition consisted of two light gray balls (32.9 cd/m^2) on a black background (1.0 cd/m^2) moving vertically towards each other, touching in the middle of the screen, and then moving apart to their initial starting position. The ball stimulus subtended 2° of visual angle when touching in the middle of the screen and 5° at its maximal point of separation. Experimental, biological movement conditions incorporated simple and complex non-goal and goal-directed movements. Simple non-goal directed movement included a hand opening and closing (Hand). This motion was visually equivalent to the trajectory taken by the bouncing ball in the "Ball" video and subtended 5° of visual angle when open and 2° when closed. Simple goal-directed movement included a hand extracting a crayon from a crayon box using a precision grip (Crayon), and complex goal-directed movement included three individuals playing catch with a small ball (Social). Videos were presented at a distance of 48 cm. The hand, crayon, and crayon box were medium gray (8.6 cd/m^2) on a black background (3.5 cd/m^2). The social video was in color.

2.2.3 Imitation tasks

Imitation abilities were assessed using the Apraxia Imitation Scale (AIS), developed by De Renzi and normed in the general public (De Renzi E. et al., 1980). This test measures imitation ability of arm/hand, finger, and general movements of varying complexity. The experimenter demonstrates each movement and participants are instructed to attempt to copy the movement exactly. Each movement is repeated three times.

2.2.4 Sustained attention

Participants were administered the visual form of the Test of Variables of Attention (TOVA) to measure sustained attention. The TOVA is a computerized visual continuous performance test for the diagnosis and treatment of children and adults with attentional disorders (Greenberg & Waldman, 1993). The visual form of the TOVA has been normed in the general population.

2.2.5 Parental assessment

One of the parents for each of the ASD participants completed the Autism Treatment Evaluation Checklist before and after training. ATEC calculates four subscale scores in which ASD individuals have known deficits: speech/language communication, sociability, sensory/cognitive awareness, and health/physical/behavior, as well as a total score. These are weighted according to the response and the corresponding subscale. The higher the subscale and total scores, the more impaired the individual. Participants in the TD group were not administered the ATEC.

2.3 EEG recording

The resting EEG from twelve scalp electrode sites was recorded for the mini-QEEG using a Brainmaster 2.5W system and Mini-Q software. The sites were recorded in pairs with a sampling rate of 256 Hz referenced to mastoids and grounded at Fpz. The MSI was extracted from EEG recorded from the C3 site over the left hemisphere and C4 over the right hemisphere at a sampling rate of 256 Hz, referenced at the mastoids and grounded at Fpz. PIRT required the use of BioExplorer software, a Brainmaster 2.5W system, and a five electrode configuration. EEG from the C4 over the right hemisphere was recorded using a sampling rate of 256 Hz, referenced to the right ear lobe, and grounded at the left ear lobe. EMG was recorded from the right trapezius muscle of the shoulder and referenced to the left trapezius muscle. The EMG was bandpass filtered for 30-60 Hz, a frequency range previously found to be sensitive to movement artifact.

2.4 Plasticity-Inducing Rehabilitation Training (PIRT)

Both ASD and TD groups received a total of 30 hours of PIRT in either 30 minute sessions three times a week or 45 minute sessions twice a week for approximately 20 weeks. Children played a variety of video games, such as racecar, asteroids, and jigsaw puzzle completions during this training. Both groups received feedback focused on the high mu band (10-13 Hz) recorded over the right sensorimotor area (C4) and referenced to the right earlobe, as well as on EMG activity (30-60 Hz) recorded from the right trapezius muscle referenced to the left trapezius muscle. Feedback was given based on satisfying two conditions: 1) power in the mu band exceeded a specified threshold, and 2) power from the muscle electrodes fell below a specified threshold. When both criteria were met, the video game progressed (e.g., car moved along the track) and a pleasant tone sounded. When the criteria were not met, visual and auditory feedbacks paused. Thresholds for both EEG and EMG channels were set as a function of the levels in the initial mini QEEG analysis and subsequently raised as a function of the preceding session. All participants viewed a computer screen displaying two threshold bars on the left and right side of a video game window. The left threshold bar corresponded to power in the 10-13 Hz band and the right threshold bar corresponded to power in the EMG band. Participants were instructed to make the mu band display larger in order to exceed a threshold bar, while making the EMG band display fall below a threshold bar. In order to help children stay focused, the experimenter encouraged them to pay attention to the game to meet these goals.

EMG feedback was included in the design for two reasons. First, it ensured that children could not advance in the game by producing movement-induced power increases in the entire EEG spectrum. Second, it allowed us to distinguish improvement effects as a function of EEG modulation, modulation of autonomic nervous system activity, or placebo effects. Because the comparison was between ASD and TD groups, nonspecific effects, such as child-trainer interaction would not explain any significant differences.

3. Results

Mixed ANOVAs were used to analyze all the data, including training performance, EEG power, suppression indices, imitation, TOVA, ATEC, and mini-QEEG results. Step down ANOVAs and post hoc comparisons with Bonferroni corrections were performed on data with significant effects. The Greenhouse-Geisser correction for degrees of freedom was applied. A summary of the results can be seen in Table 2.

3.1 Training performance

A behavioral performance measure was computed during training based on the number of hits achieved during each session multiplied by the mu rhythm threshold level established

Assessment tool	Effects of training
TOVA	No change
ATEC	• Total scores ↓ • Scores of sociability subscale ↓ • Scores of sensory/cognitive awareness subscale ↓
Imitation	• Under imitation errors ↓ • No imitation errors ↓ • Over imitation errors ↑
Absolute power	• Overall absolute power ↓ • Beta and gamma frequencies ↓ • Theta and mu frequencies ↑
Mu Suppression	• Social > Hand > Crayon • All frequencies ↓
EEG coherence	• Delta, theta and mu coherence ↑ • SMR, beta, high beta and gamma coherence ↓

Table 2. Summary of Results

for that session (Hits/Mins*MuThr). Sessions were then aggregated into phases with Phase1 including the first 5 training sessions as a baseline, Phase 2 included sessions 6-20 and Phase3 included all sessions after session 20. These data were analyzed using phases (3) as a within subjects factor and group (ASD, TD) as a between subjects factor.

A highly significant main effect of phase occurred, with Phase1 exhibiting the smallest hit rate (96.4), Phase2 with a higher hit rate (132.5), and Phase3 showing the highest rate (150.6), $F(2,52) = 20.8$, $p < 0.001$. There was also a significant interaction between phases x group, $F(2,26) = 5.04$, $p < 0.05$. Step-down one-way ANOVAs confirmed that the only significant difference between groups occurred in Phase3 ($p < 0.01$). As can be observed from Fig. 1, both groups demonstrated increases in their performance with PIRT, although the TD group profiles a steeper learning curve, while the ASD group appears to plateau in the later sessions.

3.2 Behavioral assessments

3.2.1 Sustained attention

The various subscales of the TOVA (attention deficit hyperactivity disorder or ADHD scores, errors of omission, errors of commission, time response, variability reaction time, and signal detection) were analyzed using a repeated measures ANOVAs including training (pre, post) and subscales (6) as within subjects factor and group (ASD, TD) as a between subjects factor.

A highly significant main effect of subscales indicated that standard scores for errors of commission (95.3) and reaction times (95.1) were larger than errors of omission (87.6), reaction time variability (84.2), and signal detection (88.8), $F(4,104) = 7.15$, $p < 0.001$. Stepdown ANOVAs for each subscale indicated that the only one reaching statistical significance between groups were errors of omission where the mean ASD score was lower (82) compared to that for the TD group (93), $F(1,26) = 4.57$, $p < 0.05$. Analysis of z-scores produced a highly significant main effect of subscales, $F(5,130) = 12.24$, $p < 0.001$ indicating that the ADHD (-1.87) and reaction time variability (-1.05) had the largest z scores compared

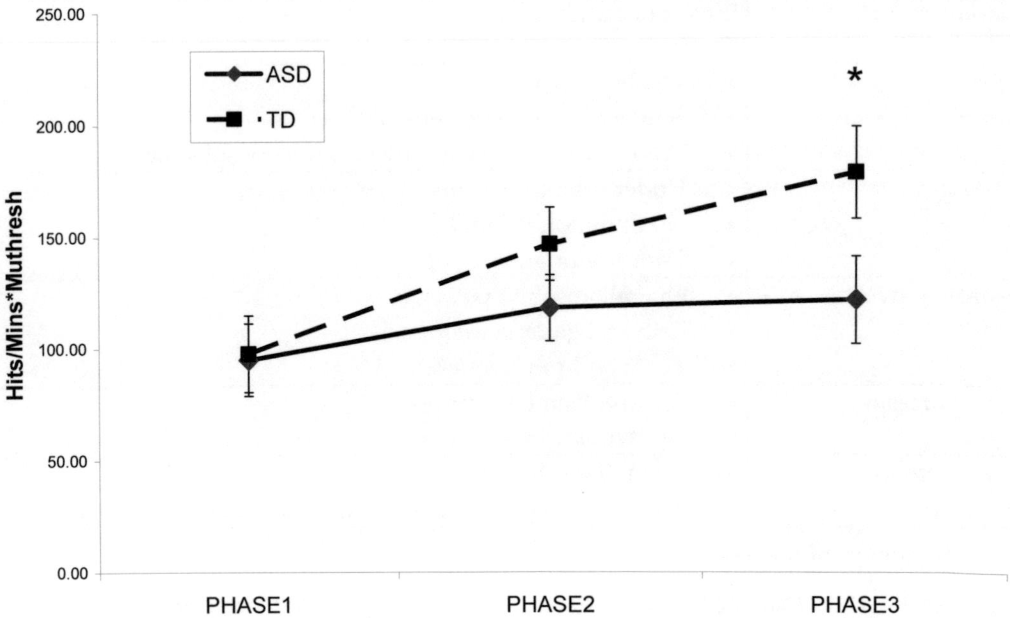

Fig. 1. Behavioral learning curves for ASD and TD groups during plasticity-inducing rehabilitation training. Performance measure was computed based on the number of hits achieved during each session multiplied by the threshold level of the mu rhythm for that session (Hits/Mins*MuThr). Illustrates scores aggregated into sessions 1-5 (Phase1), sessions 6-20 (Phase 2) and sessions after session 20 (Phase 3).

to intermediate scores for errors of omission (-0.82) and signal detection (-0.75), while reaction time (-0.32) and errors of commission (-0.19) showed the lowest scores. Errors of omission was the only subscale that produced a statistically significant z-score difference, with the ASD group exhibiting a z-score further from the mean (-1.2) than the TD group (-.5), $F(1,26) = 4.56$, $p < 0.05$.

3.2.2 Imitation

Scores on the AIS were used to compute four levels of accuracy: no imitation, under imitation, correct imitation and over imitation. Sessions were videotaped and later analyzed by three independent raters. Lack of imitation of a movement was scored as a 0 (no imitation), partial imitation was given a 0.5 score (a score $\geq$ 0.5 but < 1 was considered under imitation), accurate imitation was scored as a 1, while excessive imitation was given a 1.5 score (anything > 1.0 was considered over imitation). These scores were then subjected to a non parametric analysis using the Kruskal-Wallis analysis of ranks with factors of imitation type (no imitation, under imitation, correct imitation, over imitation) and training (pre, post) as within subjects factors and group (ASD, TD) as a grouping factor.

Analysis of the AIS indicated that pre-training ranking of no imitation scores was significantly different between groups, with ASD showing a larger number of no responses (17.97) compared to the TD group (9.88), χ^2 (1, N = 28) = 7.12, $p < .01$. These differences disappeared post-training, χ^2 (1, N = 28) = 0.164, $p = .69$. A similar difference occurred pre-training for correct imitation scores. ASD children had less correct responses (11.83) compared to TD children (18.33), χ^2 (1, N = 28) = 4.78, p < .05. In contrast, following training the differences were not statistically significant in that the ASD (13.47) group had about the same level of responding as the TD group (15.88), χ^2 (1, N = 28) = 0.64, p = .42. No differences occurred between groups for under- and over-imitation responses.

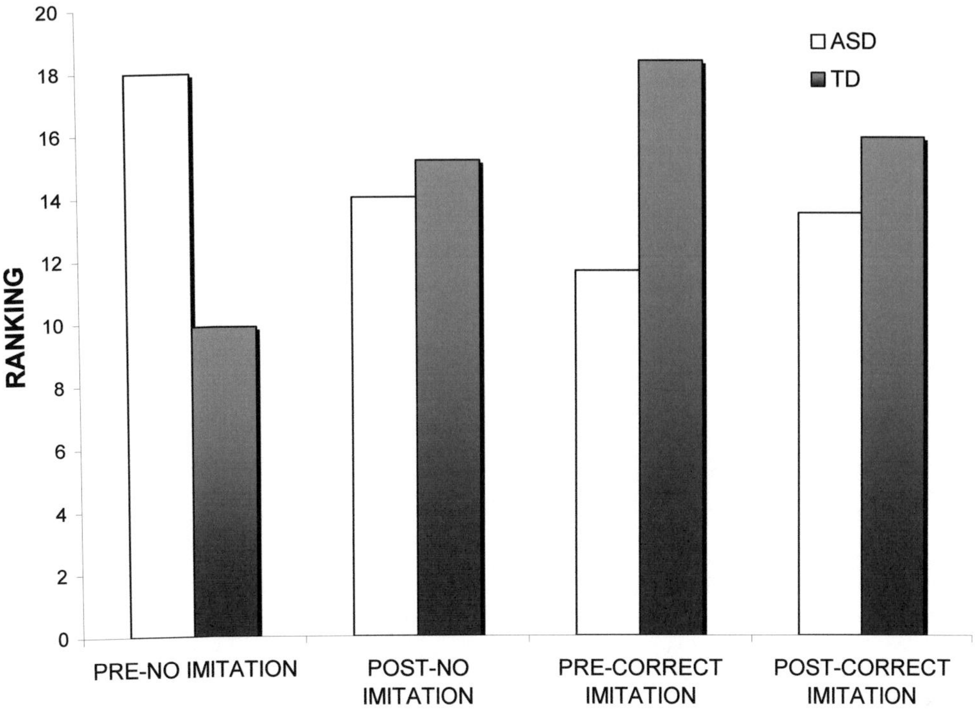

Fig. 2. Performance during the Imitation tasks for ASD and TD groups. Note that ASD had more non responses (no imitation errors) than the TD group before but not following training. Similarly, ASD had statistically fewer correct imitation responses before but not following training.

3.2.3 ATEC parental assessments

Each of the four dimensions of symptoms measured by the ATEC, which includes speech/language communication, sociability, sensory/cognitive awareness, health/physical/behavior contain descriptions of behaviors that are rated on a scale of 1-5. Each score on these dimensions, as well as the total score, were converted to a percentage of the highest possible score for that dimension. These percentile scores were then analyzed using a repeated measures ANOVA with training (pre,post) and subscales (5) as within subjects factors.

There was a highly significant main effect of subscales in which scores on the sociability subscale were the highest and therefore indicated the most impairment, while speech/language communication were the lowest and indicated the least impairment, $F(4,60) = 33.0$, $p < 0.001$. There was also a significant main effect of training on all scores such that post-training produced a lowering of scores (0.20) relative to pre-training (0.25), $F(1,15) = 4.73$, $p < 0.05$. Finally, there was a significant training x subscales interaction, $F(4,60)$ =

3.78, $p < 0.05$. As illustrated in Fig. 3, training produced a reduction in all scores, with the greatest changes occurring in the sociability and sensory/cognitive awareness dimensions.

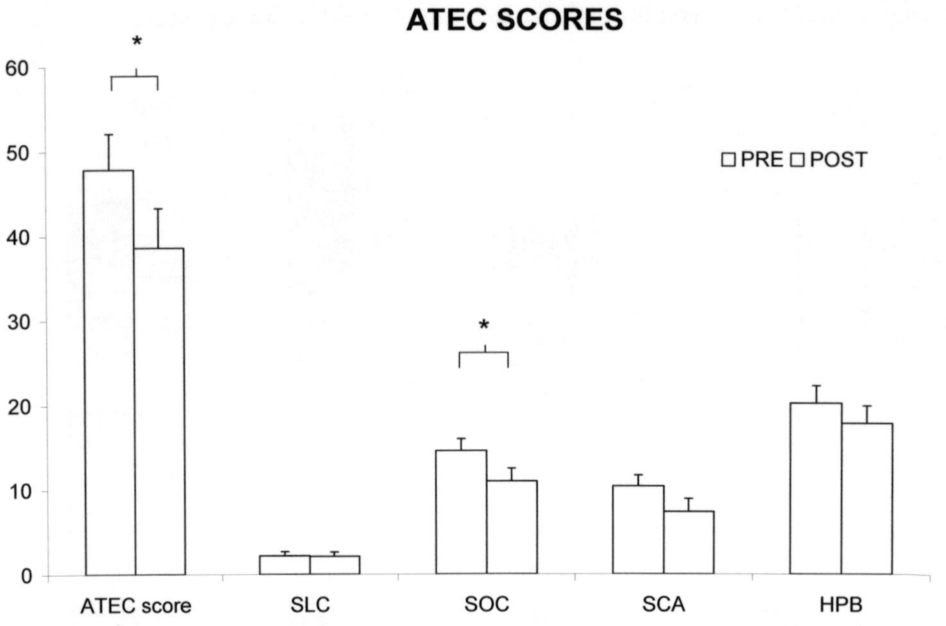

Fig. 3. Scores on the Autism Treatment Evaluation Checklist (ATEC). Questionnaire was filled out by the same parent before and following the 20 weeks of training. Graph depicts the total score (ATEC score), as well as scores on the speech/language communication (SLC), sociability (SOC), sensory/cognitive awareness (SCA), and health/physical/behavior (HPB) subscales.

3.3 Electrophysiological results

3.3.1 Absolute power

The first and last ten seconds of EEG recorded during the observation of movement in the various video conditions were eliminated in order to remove attentional transients due to initiation and termination of the stimulus. The remaining one-minute segments were combined with data from the same conditions resulting in two, one-minute segments of data per condition. Eye blinks and movement artifacts were removed automatically as well as manually prior to analyses. A Fast Fourier Transform was performed on the edited data set (1024 points) using cosine windowing to control for artifacts resulting from data splicing. Absolute power at scalp locations corresponding to premotor cortex (C_3, C_Z, and C_4) was compared to baseline condition. Absolute power was analyzed using training (pre, post-), video type (crayon, hand, social), and frequencies (delta, theta, mu, SMR, beta, gamma) as within subjects factors and group (ASD, TD) as a between subjects factor.

As illustrated in Fig. 4, there was a main effect of video type such that experimental conditions (Hand, Crayon, Social) were significantly reduced in power from baseline (Ball), $F(3,78) = 8.24$, $p < 0.05$. Pairwise comparisons confirmed that responses to the Social video varied significantly from baseline ($p < 0.01$), while Hand ($p = 0.06$) and Crayon ($p = 0.92$) were not significant. Additionally, a video type x training interaction indicated that while the effect of training was to reduce power, the largest reduction occurred in the baseline condition, $F(3,78) = 3.65$, $p < 0.05$. No differences occurred between groups. Step down analysis for each video type revealed that the responsiveness to the Ball condition reflected a marginally significant effect of training, $F(1,26) = 4.08$, $p = 0.054$. That is, participants displayed reduced mu power to the baseline Ball condition post-training. A highly significant training x frequencies interaction, $F(5,130) = 14.7$, $p < 0.001$ indicated that the decrease was primarily in the beta and gamma frequencies. Highly significant training x frequencies interactions also occurred for responses to the Hand ($F(5,130) = 9.68$, $p < 0.001$), Crayon ($F(5,130) = 10.3$, $p < 0.001$), and Social videos ($F(5,130) = 8.55$, $p < 0.001$). In all these cases, power increased post-training, primarily for theta and mu frequencies, while decreases occurred primarily for beta and gamma bands.

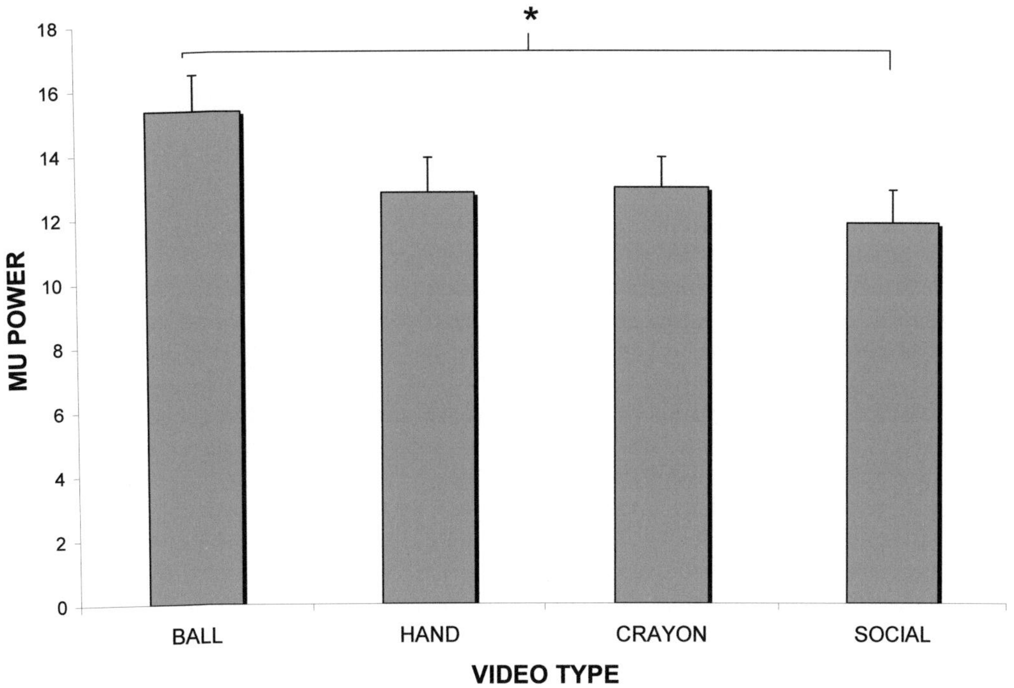

Fig. 4. Mu power in response to the observation of movement in the various videos (Ball, Hand, Crayon, and Social). The Ball video was used as the baseline condition. Note the decrease in mu power in the experimental conditions relative to baseline.

3.3.2 Suppression indices

In order to control for individual differences in scalp thickness and electrode impedance, a ratio of absolute power between the experimental and baseline conditions was calculated. A log transform of these ratios was used since ratio data are inherently non-normal as a result of lower bounding. A log ratio less than zero indicates suppression, a value of zero indicates no suppression and a value greater than zero indicates enhancement. Suppression indices were analyzed using training (pre, post-), video type (crayon, hand, social), and frequencies (delta, theta, mu, SMR, beta, gamma) as within subjects factors and group (ASD, TD) as a between subjects factor.

There was an overall main effect of training on suppression indices with more suppression observed pre-training (-.17) compared to post-training (-.028), $F(1,26) = 6.09$, $p < 0.05$. There was also a main effect of video type in which the largest suppression occurred to the social video (-.129), followed by the hand (-.091) and then the crayon (-.079), $F(2,52) = 4.06$, $p < 0.05$. A main effect of frequencies occurred ($F(5,130) = 3.12$, $p < 0.05$), although pairwise comparisons did not show any that differed significantly. Nonetheless, a highly significant training x frequencies interaction, $F(5,130) = 7.43$, $p < 0.001$ indicated that all frequencies were reduced following training.

3.3.3 Mini QEEG

Covariance of power at two sites (amplitude coherence), asymmetry, and attention indices (theta/alpha, theta/SMR, theta/beta, mu/beta ratios) were measured as part of the mini-QEEG analysis and analyzed using repeated measures ANOVAs with frequency bands (delta, theta, mu, SMR, beta, high beta, gamma), electrode pairs (C3/C4, F3/F4, Fz/Cz, O1/O2, P3/P4, T3/T4) and training (pre, post) as within subjects factors and group (ASD, TD) as a between subjects factor.

In terms of asymmetry, there was an overall (across group and training) marginally significant effect of frequency bands, $F(6,156) = 2.95$, $p = 0.058$, with none of the pairwise comparisons reaching significance. Individual ANOVAs for each frequency band showed that only the delta frequency exhibited a main effect of electrode pair, $F(5,130) = 4.13$, $p < 0.05$ such that the main difference occurred between F3/F4 and both T3/T4 ($p < 0.01$) and O1/O2 ($p < 0.05$). In terms of amplitude coherence, there was a highly significant main effect of electrode pairs, $F(5,130) = 18.1$, $p < 0.001$ such that coherence at temporal (T3/T4) and parietal (P3/P4) sites were significantly different from all other pairs of electrodes. These differences varied across groups since there was a group x electrode pairs interaction, $F(5,26) = 2.79$, $p < 0.05$. There was also a highly significant main effect of frequencies, $F(6,156) = 478.5$, $p < 0.001$ as well as a frequencies x training interaction, $F(6,156) = 6.4$, $p < 0.01$. Step down analyses of the individual frequencies revealed that delta, theta, and mu coherence increased or tended to increase with training but SMR, beta, high beta, and gamma decreased or tended to decrease. Only the SMR data for both groups, as illustrated in Fig. 5, displayed a marginally significant training x group interaction, $F(1,26)=3.96$, $p = 0.057$. That is, following training the ASD group exhibited a statistically significant decrease in coherence compared to the TD group.

3.3.4 Frequency band ratios

Analysis of the ratios of the asymmetry data for the various frequencies revealed a main effect, with the theta/alpha ratio being closest to 1, while theta/SMR (1.05) and theta/beta (1.08) were both significantly larger, $F(3,78) = 4.41$, $p < 0.05$. These asymmetry ratios varied as a function of location since there was an electrode pair x ratio interaction, $F(15,390) = 3.57$, $p < 0.05$.

4. Discussion

Results from this study provide evidence for significant positive effects of plasticity-inducing rehabilitation training (PIRT) on the behavioral, cognitive, and electrophysiological indices of children with autistic spectrum disorders (ASD). The effectiveness of this type of operant training is supported by the fact that both ASD and typically-developing (TD)

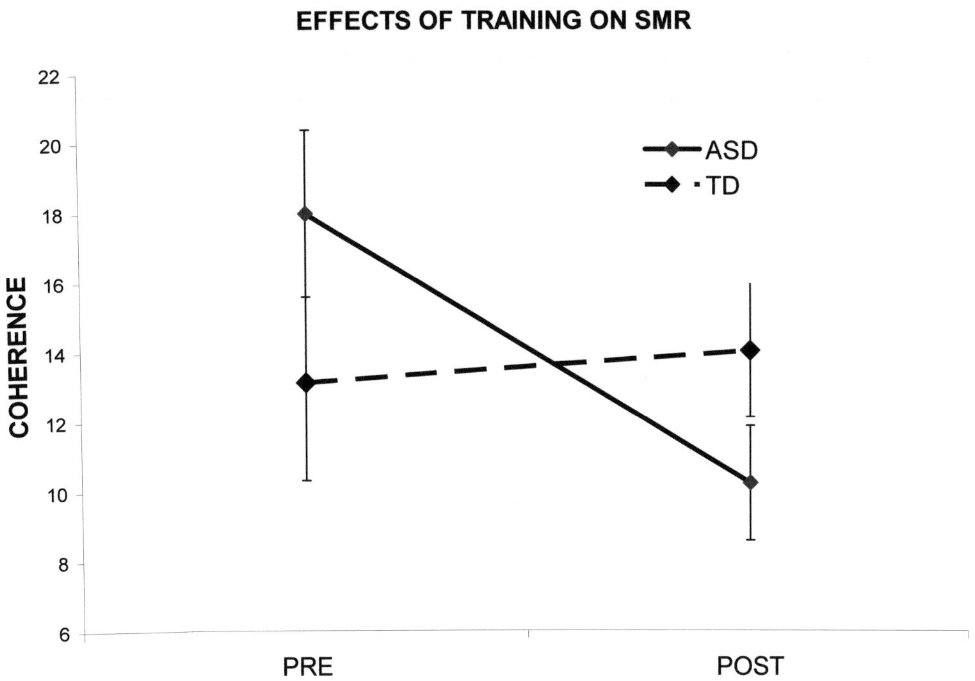

Fig. 5. Line graph showing the effects of training on the SMR.

children learned to modulate the amplitude of EEG mu rhythm oscillations (10-13 Hz) across the 20 weeks of training. Learning rates differed between groups, with ASD children exhibiting a more gradual learning curve and ultimately a leveling of their performance during the later stages of training. Such learning impacted a number of behavioral, cognitive and electrophysiological measures. Some of these measures, however, like the assessment of sustained attention using the Test of Variable Attention (TOVA) scale showed group differences in terms of errors of omission and response times but were not significantly affected by the training. While ASD children were significantly more impaired in sustained attention compared to the TD group, with larger differences from the mean than the control group, sustained attention was insensitive to PIRT. In contrast, in terms of the assessment of imitation skills, results indicated that ASD children exhibited significantly more no imitation responses than the TD group, as well as reduced levels of correct responses prior to training. Training had a significant impact on reducing the lack of imitation behaviors while increasing the tendency to correctly imitate. Positive behavioral changes were also noted by parents using the Autism Treatment Evaluation Checklist (ATEC) scale of participant's behaviors following training. Results suggest that the greatest impairment measured in the ASD group before training was in the sociability subscale. Training significantly reduced those scores, as well as scores on the sensory/cognitive awareness subscale. Reduction of such scores indicates behavioral improvement.

The present results are consistent with a variety of studies that have provided evidence that EEG neurofeedback training can be an effective form of intervention for ASD symptoms (Jarusiewicz, 2003; Pineda et al., 2008). All studies in the recent literature using this methodology have reported significant reductions in autistic symptoms following such training. In a single case study, Sichel et al. (Sichel, 1995) reported positive changes in all

DSM-IV-R diagnostic criteria for autism. Jarusiewicz (Jarusiewicz, 2003) reported an average of 26% improvement in the ATEC in 12 children diagnosed with autism, compared to 3% improvement in a control group. Pineda et al. (Pineda et al., 2008) reported decreased mu power and coherence, increased sustained attention ability, improved imitation ability, and improved scores on the Sensory/Cognitive subscale of the ATEC in children with ASD compared to a placebo group following training. Similarly, Coben et al. (Coben et al., 2009) validated EEG neurofeedback as a therapeutic modality for autistic children and argued that changes in connectivity anomalies may be related to the mechanism of action.

There was no distinction in the present study between ASD and TD groups in terms of electrophysiological responses to the variety of observed movement in the videos. Both groups displayed a gradient in such suppression with respect to the type of movements observed. The social video elicited the largest suppression, followed by the hand and then the crayon video. This is consistent with previous studies indicating a normal gradient of mu rhythm responsiveness as a function of sociability (Oberman et al., 2006). However, one difference between the current and previous results was the effect of training on mu rhythm suppression. Whereas in the earlier study, mu rhythm suppression increased with training, it decreased in the current study. One possible explanation for these differences may be that in the present study all frequencies, including the mu rhythm, showed an overall amplitude reduction in ASD children following training. Because the decreases were seen for all frequencies, it suggests a more general explanation of the results. That is, since suppression is typically computed as the ratio of experimental to the baseline condition, a training difference could result from changes in either one. The results indicated that the effect of training was to reduce power and the largest reduction occurred in response to the baseline Ball video. Additionally, while absolute power in response to Ball was reduced, responses to Hand, Crayon, and Social videos exhibited increases in power with training, especially for theta and mu frequencies, while decreases occurred primarily in the beta and gamma frequencies.

EEG asymmetry measures differed significantly between groups, but only delta frequencies showed a significant asymmetry between cortical areas, specifically those in the frontal and temporo-occipital regions. In contrast, measures of coherence indicated more widespread effects across all frequencies and especially between regions in temporo-parietal cortex and other electrode sites. PIRT also had a significant impact on amplitude coherence. Slower EEG oscillations (delta, theta, mu) tended to increase in coherence, while faster oscillations (SMR, beta, gamma) tended to decrease. Overall, the most significant difference occurred for the SMR frequency band.

EEG rhythmic oscillations are assumed to be generated by thalamocortical circuits and modulated by a variety of motivational, attentional, motor, and cognitive factors (Sterman, 1996). The PIRT intervention is grounded in the theory of operant conditioning and reinforcement in which participants are taught to modulate and volitionally control specific EEG frequencies. One explanation of how this methodology produces its behavioral and electrophysiological effects is by gaining access to and control over regulatory mechanisms that increase or decrease synchronous or desynchronous activity in thalamocortical networks (Lubar, 1997), leading to the induction of neural plastic changes.

Egner et al. (Egner et al., 2004) have argued that an assumption in the neurofeedback field that motivates clinical practice, especially the use of pre-training quantitative EEG assessments to formulate training protocols, is that "spectral EEG variables related to the operant learning of EEG frequency modulation mediate observed behavioural effects."

Although this assumption has had little support from the clinical research literature, several studies have provided evidence of beneficial effects of neurofeedback training on healthy human participants (Egner et al., 2002; Egner et al., 2004; Egner & Gruzelier, 2004). The present results are consistent with the general beneficial effects of such training on children on the autism spectrum but do not support the assumption that operant training contingencies result in corresponding enhancement/suppression of only the trained frequency components. Indeed, our results suggest a widespread nonspecific effect of training on most other EEG frequencies.

4.1 Conclusion

Sixteen high-functioning autistic spectrum disorder (ASD) participants and twelve typically-developing (TD) participants received twenty weeks of plasticity-inducing rehabilitation training (PIRT) centered on the high mu frequency band (10-13 Hz). Training had a significant impact on reducing the lack of imitation behaviors in ASD children while increasing correct imitation responses. While pre-training assessments indicated that the greatest impairment in the ASD group occurred in the sociability subscale of the Autism Treatment Evaluation Checklist, training reduced those scores as well as scores on the sensory/cognitive awareness subscale, indicating behavioral improvement. Furthermore, PIRT had significant and widespread effects on EEG coherence measures with slower EEG rhythms (delta, theta, and mu) increasing coherence with training, while faster rhythms (SMR, beta, and gamma) decreased coherence. These results suggest that operant conditioning of EEG mu rhythms can induce global neural changes with positive consequences on both the electrophysiology and behavior in high-functioning children on the autism spectrum.

5. References

Altschuler, E. L., Vankov, A., Wang, V., Ramachandran, V. S., & Pineda, J. A. (1998). Person see, person do: Human cortical electrophysiological correlates of monkey see monkey do cells. *J.Cogn Neurosci.*.

Baron-Cohen, S. (2009). Autism: the empathizing-systemizing (E-S) theory. *Ann.N.Y.Acad.Sci., 1156,* 68-80.

Baron-Cohen, S., Jolliffe, T., Mortimore, C., & Robertson, M. (1997). Another advanced test of theory of mind: evidence from very high functioning adults with autism or asperger syndrome. *J.Child Psychol.Psychiatry, 38,* 813-822.

Baron-Cohen, S., Leslie, A. M., & Frith, U. (1985). Does the autistic child have a "theory of mind"? *Cognition, 21,* 37-46.

Blakemore, S. J., Frith, C. D., & Wolpert, D. M. (2001). The cerebellum is involved in predicting the sensory consequences of action. *Neuroreport, 12,* 1879-1884.

Buccino, G., Binkofski, F., Fink, G. R., Fadiga, L., Fogassi, L., Gallese, V. et al. (2001). Action observation activates premotor and parietal areas in a somatotopic manner: an fMRI study. *Eur.J.Neurosci., 13,* 400-404.

Buccino, G., Binkofski, F., & Riggio, L. (2004). The mirror neuron system and action recognition. *Brain Lang, 89,* 370-376.

Buxbaum, L. J., Kyle, K. M., & Menon, R. (2005). On beyond mirror neurons: internal representations subserving imitation and recognition of skilled object-related actions in humans. *Brain Res.Cogn Brain Res., 25,* 226-239.

Carr, L., Iacoboni, M., Dubeau, M. C., Mazziotta, J. C., & Lenzi, G. L. (2003). Neural mechanisms of empathy in humans: a relay from neural systems for imitation to limbic areas. *Proc.Natl.Acad.Sci.U.S.A, 100,* 5497-5502.

Coben, R., Sherlin, S., Hudspeth, W. J., & McKeon, K. (2009). Connectivity-guided EEG biofeedback for autism spectrum disorder: evidence of neurophysiological changes. *Journal of Autism and Developmental Disorders.*

Cochin, S., Barthelemy, C., Roux, S., & Martineau, J. (1999). Observation and execution of movement: similarities demonstrated by quantified electroencephalography. *Eur.J.Neurosci., 11,* 1839-1842.

Courchesne, E. & Pierce, K. (2005). Why the frontal cortex in autism might be talking only to itself: local over-connectivity but long-distance disconnection. *Curr.Opin.Neurobiol., 15,* 225-230.

Dapretto, M., Davies, M. S., Pfeifer, J. H., Scott, A. A., Sigman, M., Bookheimer, S. Y. et al. (2006). Understanding emotions in others: mirror neuron dysfunction in children with autism spectrum disorders. *Nat.Neurosci., 9,* 28-30.

De Renzi E., Motti, F., & Nichelli, P. (1980). Imitating gestures. A quantitative approach to ideomotor apraxia. *Arch.Neurol., 37,* 6-10.

di Pellegrino, G., Fadiga, L., Fogassi, L., Gallese, V., & Rizzolatti, G. (1992). Understanding motor events: a neurophysiological study. *Exp.Brain Res., 91,* 176-180.

Egner, T. & Gruzelier, J. H. (2004). EEG biofeedback of low beta band components: frequency-specific effects on variables of attention and event-related brain potentials. *Clin.Neurophysiol., 115,* 131-139.

Egner, T., Strawson, E., & Gruzelier, J. H. (2002). EEG signature and phenomenology of alpha/theta neurofeedback training versus mock feedback. *Appl.Psychophysiol.Biofeedback, 27,* 261-270.

Egner, T., Zech, T. F., & Gruzelier, J. H. (2004). The effects of neurofeedback training on the spectral topography of the electroencephalogram. *Clin.Neurophysiol., 115,* 2452-2460.

Fadiga, L., Buccino, G., Craighero, L., Fogassi, L., Gallese, V., & Pavesi, G. (1999). Corticospinal excitability is specifically modulated by motor imagery: a magnetic stimulation study. *Neuropsychologia, 37,* 147-158.

Gallese, V., Fadiga, L., Fogassi, L., & Rizzolatti, G. (1996). Action recognition in the premotor cortex. *Brain, 119 (Pt 2),* 593-609.

Gastaut, H. J. & Bert, J. (1954). EEG changes during cinematographic presentation. *Electroencephalogr.Clin.Neurophysiol., 6,* 433-444.

Hari, R., Salmelin, R., Makela, J. P., Salenius, S., & Helle, M. (1997). Magnetoencephalographic cortical rhythms. *Int.J.Psychophysiol., 26,* 51-62.

Hickok, G. (2009). Eight problems for the mirror neuron theory of action understanding in monkeys and humans. *J.Cogn Neurosci., 21,* 1229-1243.

Hirshberg, L. M., Chiu, S., & Frazier, J. A. (2005). Emerging brain-based interventions for children and adolescents: overview and clinical perspective. *Child Adolesc.Psychiatr.Clin.N.Am., 14,* 1-19, v.

Iacoboni, M., Molnar-Szakacs, I., Gallese, V., Buccino, G., Mazziotta, J. C., & Rizzolatti, G. (2005). Grasping the intentions of others with one's own mirror neuron system. *PLoS.Biol., 3,* e79.

Iacoboni, M., Woods, R. P., Brass, M., Bekkering, H., Mazziotta, J. C., & Rizzolatti, G. (1999). Cortical mechanisms of human imitation. *Science, 286,* 2526-2528.

Jarusiewicz, B. (2003). Efficacy of neurofeedback for children in the autistic spectrum: A pilot study. *Applied Psychophysiology and Biofeedback, 28,* 311.

Just, M. A., Cherkassky, V. L., Keller, T. A., & Minshew, N. J. (2004). Cortical activation and synchronization during sentence comprehension in high-functioning autism: evidence of underconnectivity. *Brain, 127,* 1811-1821.

Kohler, E., Keysers, C., Umilta, M. A., Fogassi, L., Gallese, V., & Rizzolatti, G. (2002). Hearing sounds, understanding actions: action representation in mirror neurons. *Science, 297,* 846-848.

Leslie, K. R., Johnson-Frey, S. H., & Grafton, S. T. (2004). Functional imaging of face and hand imitation: towards a motor theory of empathy. *Neuroimage., 21,* 601-607.

Lilienfeld, S. O. (2005). Scientifically unsupported and supported interventions for childhood psychopathology: a summary. *Pediatrics, 115,* 761-764.

Lingnau, A., Gesierich, B., & Caramazza, A. (2009). Asymmetric fMRI adaptation reveals no evidence for mirror neurons in humans. *Proc.Natl.Acad.Sci.U.S.A, 106,* 9925-9930.

Lubar, J. F. (1997). Neocortical dynamics: implications for understanding the role of neurofeedback and related techniques for the enhancement of attention. *Appl.Psychophysiol.Biofeedback, 22,* 111-126.

Matson, J. L. (2006). Current status of differential diagnosis for children with autism spectrum disorders. *Res.Dev.Disabil.*.

Muller, R. A. (2007). The study of autism as a distributed disorder. *Ment.Retard.Dev.Disabil.Res.Rev., 13,* 85-95.

Muthukumaraswamy, S. D. & Johnson, B. W. (2004). Changes in rolandic mu rhythm during observation of a precision grip. *Psychophysiology, 41,* 152-156.

Muthukumaraswamy, S. D., Johnson, B. W., & McNair, N. A. (2004). Mu rhythm modulation during observation of an object-directed grasp. *Brain Res.Cogn Brain Res., 19,* 195-201.

Muthukumaraswamy, S. D. & Singh, K. D. (2008). Modulation of the human mirror neuron system during cognitive activity. *Psychophysiology, 45,* 896-905.

Niedermeyer, E. (1997). Alpha rhythms as physiological and abnormal phenomena. *Int.J.Psychophysiol., 26,* 31-49.

Oberman, L. M., Hubbard, E. M., McCleery, J. P., Altschuler, E. L., Ramachandran, V. S., & Pineda, J. A. (2005). EEG Evidence for Mirror Neuron Dysfunction in Autism Spectrum Disorders. *Cognitive Brain Research.*

Oberman, L. M., Pineda, J. A., & Ramachandran, V. S. (2006). The Human Mirror Neuron System: A Link Between Action Observation and Social Skills. *Soc.Cog.Affect.Neurosci.*.

Ozonoff, S., Pennington, B. F., & Rogers, S. J. (1991). Executive function deficits in high-functioning autistic individuals: relationship to theory of mind. *J.Child Psychol.Psychiatry, 32,* 1081-1105.

Petrides, M., Cadoret, G., & Mackey, S. (2005). Orofacial somatomotor responses in the macaque monkey homologue of Broca's area. *Nature, 435,* 1235-1238.

Pfurtscheller, G. & Aranibar, A. (1979). Evaluation of event-related desynchronization (ERD) preceding and following voluntary self-paced movement. *Electroencephalogr.Clin.Neurophysiol., 46,* 138-146.

Pineda, J. A. (2005). The functional significance of mu rhythms: translating "seeing" and "hearing" into "doing". *Brain Res.Brain Res.Rev., 50,* 57-68.

Pineda, J. A., Allison, B. Z., & Vankov, A. (2000). The effects of self-movement, observation, and imagination on mu rhythms and readiness potentials (RP's): toward a brain-computer interface (BCI). *IEEE Trans.Rehabil.Eng, 8,* 219-222.

Pineda, J. A., Brang, D., Hecht, E., Edwards, L., Carey, S., Bacon, M. et al. (2008). Positive behavioral and electrophysiological changes following neurofeedback training in children with autism. *Research in Austism Spectrum Disorders, 2,* 557-581.

Rizzolatti, G. & Craighero, L. (2004). The mirror-neuron system. *Annu.Rev.Neurosci., 27,* 169-192.

Rizzolatti, G. & Fabbri-Destro, M. (2009). Mirror neurons: from discovery to autism. *Exp.Brain Res.*.

Rogers, S. J., Hepburn, S. L., Stackhouse, T., & Wehner, E. (2003). Imitation performance in toddlers with autism and those with other developmental disorders. *J.Child Psychol.Psychiatry, 44,* 763-781.

Sichel, A. G. (1995). Positive outcome with neurofeedback treatment in a case of mild autism. *Journal of Neurotherapy, Vol 1 (1),* p. 60-p. 64.

Sterman, M. B. (1996). Physiological origins and functional correlates of EEG rhythmic activities: implications for self-regulation. *Biofeedback Self Regul., 21,* 3-33.

Umilta, M. A., Kohler, E., Gallese, V., Fogassi, L., Fadiga, L., Keysers, C. et al. (2001). I know what you are doing. a neurophysiological study. *Neuron, 31,* 155-165.

Villalobos, M. E., Mizuno, A., Dahl, B. C., Kemmotsu, N., & Muller, R. A. (2005). Reduced functional connectivity between V1 and inferior frontal cortex associated with visuomotor performance in autism. *Neuroimage, 25,* 916-925.

Volkmar, F. R., Klin, A., Siegel, B., Szatmari, P., Lord, C., Campbell, M. et al. (1994). Field trial for autistic disorder in DSM-IV. *Am.J.Psychiatry, 151,* 1361-1367.

Williams, J. H., Waiter, G. D., Gilchrist, A., Perrett, D. I., Murray, A. D., & Whiten, A. (2006). Neural mechanisms of imitation and 'mirror neuron' functioning in autistic spectrum disorder. *Neuropsychologia, 44,* 610-621.

Williams, J. H., Whiten, A., & Singh, T. (2004). A systematic review of action imitation in autistic spectrum disorder. *J.Autism Dev.Disord., 34,* 285-299.

Williams, J. H., Whiten, A., Suddendorf, T., & Perrett, D. I. (2001). Imitation, mirror neurons and autism. *Neurosci.Biobehav.Rev., 25,* 287-295.

Wohlschlager, A. & Bekkering, H. (2002). Is human imitation based on a mirror-neuron system? Some behavioural evidence. *Exp.Brain Res., 143,* 335-341.

24

Connecting Electroencephalography Profiles with the Gamma-Amino-Butyric Acid (GABA) Neuropathology of Autism as a Prelude to Treatment

Jane Yip[1,2], Sara Davis[2] and Oliver Wendt[3]
[1]*Boston University*
[2]*ABA Inc*
[3]*Purdue University*
USA

1. Introduction

The nervous system has the ability to adjust in the face of disease by reorganizing its molecular, cellular, and systems function for survival through the mechanism of neuroplasticity. Recently, the National Institutes of Health Blueprint for Neuroscience Research, USA, gathered experts in various fields of neural disorders to look into the translation of neuroplasticity and circuit retraining research: the message is that the future ahead is very bright for effective clinical therapies (Cramer et al., 2011). When research studies are driven by basic science conceived alongside therapeutic disciplines designed in a congruent manner, there is the exciting potential that the innate mechanism of neuroplasticity could be shaped more precisely and more thoughtfully than is currently available. The vision for the future is patient-centered therapy aimed at the rewiring of brain circuits for successful function with long term changes in the molecular and genetic level.

The transcriptional machinery is a site where neuropathology can occur. Alterations in gamma-amino-butyric acid (GABA) through transcriptional machinery may be one of the factors that underlie the neuropathology of autism. The level of transcription via the synthesizing enzymes for GABA plus the neuroplasticity response of interneurons in the brain will be discussed in this chapter. The implication on the afferent-efferent circuitry in the cerebellum with the promising application of a neuropathological understanding of autism to treatment and behavior will be presented.

Autism is not a single disease entity. It is often referred to as autism spectrum disorder (ASD). The condition is behaviorally defined, described, and diagnosed by a triad of deficits which include the following characteristics: impaired social interaction, impaired communication, restricted interests with stereotyped and/or repetitive behaviors. Another behavioral feature is sensory input sensitivity. Autistic individuals are hypersensitive to stimuli regarding sound, light, touch, smell, temperature, and pressure. These sensations can limit the individual's ability adapt to the regular noise and activity levels typically encountered in shopping malls, school classrooms, and other public gathering sites.

Additionally, the sensory overload severely affects one's ability to screen out noise and process relevant signals. This reactive hypersensitivity significantly impacts the family's lifestyle as each member has to adjust to the disease manifestations on a day-to-day basis. Paradoxically, sensitivity to pain (Tordjman et al., 1999) and the ability to be sedated (Marrosu et al., 1987) may be diminished in certain autistic individuals as there exist in certain subtypes, an alteration in pain and sedation neuroreceptors.

The behavioral features of autism remain a mystery. Kanner's (1943) original description of autism stated that it is a disorder of "nervousness". "Nervousness" is regarded as the staunch refusal of the autistic individual to change his or her routine in order to "go with the flow. This is a common behavior that caregivers have to grapple with routinely. Both parties become frustrated as a result of the emotional intensity that is generated. A favored treatment recommendation, recognized by the health system and the public, consists of behavior modification on the part of the autistic individual (i.e., the child).

To date, preferred treatments of the neurodevelopmental disorder defined as autism are the behaviorally-oriented refinements of ABA. However, there is a paucity of data linking behavior with its neurobiological correlates that could have caused the behavior. Behavior is the final product of a cascade of events beginning with genetic and transcriptional coding of genomic information to the culmination of neurochemical transmitters which intersect on neuronal circuits in the brain to finally give rise to behavior (Figure 1). It is necessary to link the multiple systems into a hierarchy of events so that a big picture could be constructed lest we become like the legend of "the blind men and the elephant", where a group of blind men were each feeling a different part the elephant and each man came up with a different description depending on the part of the elephant he has touched. In the case of autism, each researcher- through the lense of his or her discipline- feels one part of the puzzle and assumes that one or another represents the whole.

2. The neurotransmitter γ-Amino butyric acid (GABA)

This journey begins with a single neuro-transmitter. GABA is selected because considerable evidence points towards a hypothesis of GABA deficiency occurring in autism. For example, a reduced density of $GABA_A$ and benzodiazepine receptors in the hippocampus (Blatt et al., 2001), a reduction in the level for GABA synthesizing enzymes (GAD65 and GAD67) in the parietal and cerebellar cortex (Fatemi et al., 2002; Yip et al., 2007, 2008, 2009), and the effectiveness of the $GABA_A$ receptor agonists in treating seizure and anxiety disorders in autistic patients (Askalan et al., 2003; Acosta, 2004) reported in autistic patients all point towards a lowered GABAergic function. Genetic studies have found interstitial duplication in the chromosome 15q11.2-q13, a region containing three GABAA receptor subunits, in autism phenotypes (Schroer et al., 1998; Shao et al., 2003; Menold et al., 2001; Ashley-Koch et al., 2006).

The perturbation of GABA in autism can lead to neuropathological changes at a cellular level that resides in a particular circuit within the brain. The main components governing GABAergic neurotransmission lies in its two isoforms of synthesizing enzyme- glutamic acid decarboxylase (GAD), which are GAD65 and GAD67. GAD is the precursor through which GABA transmission is carried out. Results from our study implies that an *in vivo* cerebellar circuit of cell-type specific afferent-efferent network leads to a GABA neuropathological mechanism in autism that are likewise compensated through the same circuitry via GAD through neuroplasticity (Yip et al., 2007, 2008, 2009; Blatt et al., 2010).

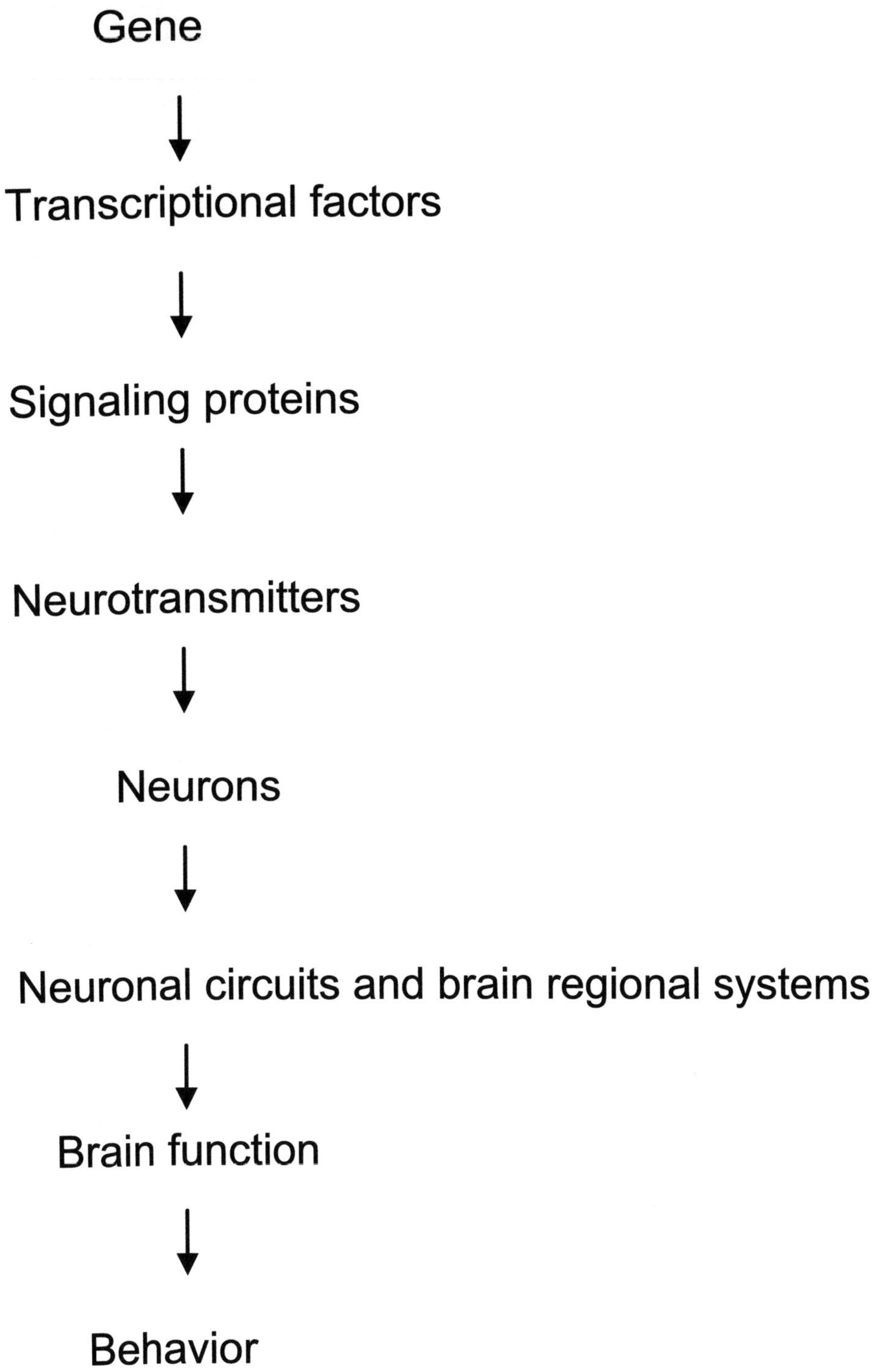

Fig. 1. The hierarchy of biological systems giving rise to behavior.

The next step is to extend the study GABA dysregulation to behavior. An autistic child often finds informational input, particularly sensory data, overwhelming. Autistic children have stated that they feel as if the physical world is too difficult or confusing to handle sometimes. In short, their comfort zone is disrupted. Since one of the major brain centers of sensory information processing is located in the cerebellar circuitry, its study is beneficial in order to identify the source of their confusion. Furthermore, it is necessary to correlate the child's brain function within one's natural environment (i.e., home, schools, clinician's office) as they relate various activity states such as resting, performing a mental task, and being subjected through a bombardment of sensory input that surrounds an autistic child on a daily basis. The authors propose that through a non-invasive portable method of testing one's brain function in the child's natural environment (i.e., *in situ*) such as the electroencephalography (EEG) technologies, the field of autism research and treatment could be further enhanced.

3. Physiological regulation requires glutamic acid decarboxylase synthesis

GABA is a bountiful inhibitory amino acid neurotransmitter catalyzed by the enzymatic action of GAD from glutamate. GAD activity is altered in many disease states such as stiff-person syndrome (Meinck et al., 2002), epilepsy (Vianello et al., 2002), and Parkinson's disease (Guridi et al., 1996).

GAD exists as two isoforms encoded by GAD65 and GAD67 that exist in mammals and are highly conserved. GAD67 is acted upon by coenzyme pyridoxal-5'-phosphate (Martin et al., 1991; Martin and Rimvall, 1993) for the synthesis of GABA via ribonucleic acid (RNA) regulation whereas GAD65 is constituitively expressed by a promoter regulated by coenzyme-saturation (Rimvall et al., 1993). GAD65 is considered to be mainly involved in regulated GABA synthesis. These two GAD isoforms presumably originated in vertebrates following gene duplication approximately 450 million years ago before the emergence of sharks and rays (Lariviere et al., 2002) and differ in subcellular distribution and have different roles in the regulation of GABA. The relative amounts of the isoforms expressed may reflect the functional adaptations of that particular cell. There is a third novel form of GAD, GAD3, that has been shown in certain fish species: *Coryphaenoides artmatus*, the deep sear armed grenadier, and *Carassius auratus*, the gold fish (Lariviere et al., 2002).

Martin and Rimvall (1993) first suggested that GAD65 is involved in GABA synthesis for vesicular release whereas GAD67 primarily functions in cytoplasmic GABA for metabolic purposes. Soghomonian and Martin (1998) extended this work and proposed that GAD65 synthesized most of the neuro-transmitter GABA under normal conditions. Based on the characteristic cellular localization, differing N-terminal domains, and interaction with co-factors, each isoform regulates the level of GABA. A differing role in the synthesis of GABA in the two isoforms have been observed in knock out mice studies. For example, the deletion of GAD65 gene does not significantly alter brain GABA levels in young mice (Asada et al., 1996). Contrastingly, GAD67 gene deletion resulted in mice born with low GABA contents that failed to survive due to cleft palate deficiency (Asada et al., 1997).

Studies from immunohistochemistry demonstrated considerable differences in the level of individual GAD isoform among cell types and regional brain distribution (Esclapez et al., 1993, 1994). Esclapez et al (1994) reported that the levels of GAD67 mRNA are greater than GAD65 mRNA in most brain regions. There is now reproducible evidence for a decreased expression of GAD67 mRNA in multiple brain regions such as the cortico-limbic regions of patients with schizophrenia, bipolar and major depressive disorders (Torrey et al., 2005).

4. The roles of GAD65 and GAD67 in neural development and its implication in autism

GABA is involved in the development and plasticity of the postnatal nervous system (McLean et al., 1996; Pisu et al., 2004). In the human cerebellum, both GAD65 and GAD67 mRNA are strongly expressed during development yet the two isoforms differ in their timing of expression. GAD65 and GAD67 mRNA are both present by gestational week 12 (Chan et al., 1997). For GAD67, the messenger level remains high and following a slight decrease maintains its abundance for the rest of the gestational period. In contrast, GAD65 mRNA levels decreases rapidly from GW12 and becomes undetectable by GW19 (Chan et al., 1997). Despite the apparent low activity of GAD65 in subsequent fetal development, GAD65 is important in maintaining GABA levels. While both GAD65 and GAD67 maintain GABA levels *in vivo*, GAD65 could act to powerfully offset perturbation to GABA synthesis in the event of a GAD67 deficiency. This was substantiated in an organotypic culture where the presence of GAD65 and GABA synthesis as well as the proliferation of GABAergic neurons was supported without GAD67. Interestingly, mutant mice lacking GAD67 had maintained their GABA contents. Additionally, there was a markedly increased number of GABA-containing neurites. In the absence of GAD67, the development of cerebellar Purkinje cells (PC) was sustained (Ji & Obata, 1999). In young mice, GAD65 may not be required (Asada et al., 1996) though it is indispensable in adult mice (Stork et al., 2000). GAD65 deficiency could lead to serious behavioral impairment as experimental adult rats who have reduced GABA content, to nearly 50% of normal values, in the hypothalamus and amygdala exhibited severe anxiety (Ji & Obata, 1999).

In humans, development of the visual cortex coincides with an increase in GAD65 expression. The additional GAD65 expression provides a larger pool of synaptic GAD for GABA release for the purpose of short term changes in neuronal activity (Feldblum et al., 1993; 1995) and further provides an agency for neuroplasticity. Electrophysiologically, GAD65 provides tonic inhibition and the short term effect balances more rapid fluctuation in excitation during development (Walls et al., 2010) and confers the balance between excitation and inhibition in development.

The failure of an adaptive control of GAD65 and GAD67, especially in circuits modulating emotions, may be related to defective emotional coping in times of stress. Dysregulated GAD67 have been linked to psychosis (Kalkman & Loetscher, 2003). One interesting note is that there appears to be a parallel between psychosis and autism. This is especially true in relation to aggressive behaviors such as temper tantrums, property damage, self injurious behavior, and so on. Some individuals with autism have an increase risk of hyperactive responses to emotional distress with a predisposition to acting out aggressively towards oneself and others. However, in these autistic individuals- who exhibit aggression- the relationship between emotional distress and the actual mechanism of autism is less clear. Certainly, an inherent abnormality of the GABA system could dampen the nervous systems' ability to shut off stimuli once it has been turned on. When GABA malfunctions start at the GAD65 loci where its developmental trajectory may have been perturbed during fetal development, it could likely lead to a long standing state of reduced GABAergic inhibition and increased neuronal hyperexcitability of stimuli.

Current autism neuropathology literature shows that there is a decreased level of GAD65 and GAD67 proteins in the cerebellum (Fatemi et al., 2002). Our study revealed that there is a also a decrease (51%) in transcript level for GAD65 (GAD65 mRNA) in a select population of neurons neurons located in the dentate nuclei (i.e. the larger dentate neurons) -which projected into the thalamo-cortical circuit- compared to no change in the smaller dentate neurons- which are likely interneurons- in adult autism cases (Yip et al., 2009; figure 2 and 3). A deficiency of GAD65 in the larger dentate neuron population in the absence of a compensatory increase in neighboring GABAergic neuronal populations suggests serious deficiency in the ability of GAD65 to maintain GABA levels. To compound the GAD65 mRNA deficiency, there was a 40% reduction of GAD67 mRNA levels in PC that formed reciprocal connections to the dentate nuclei, inferior olivary complex, and the dentato-thalamic-cortical pathways; the PC is ultimately responsible for timing and gating of incoming hyperpolarizing impulse (Yip et al., 2007; figure 4). A decrease of GAD67 mRNA level in the PC of the cerebellar cortex in our subjects examined is consistent with a deficiency in the protein level of GAD67 in whole cerebellar homogenates (Fatemi et al., 2002) suggesting a direct relationship between protein and transcript level. Overall, this suggests that there is a lowered level of GAD67 in autism which can compromise the ability of neurons to maintain baseline GABA levels.

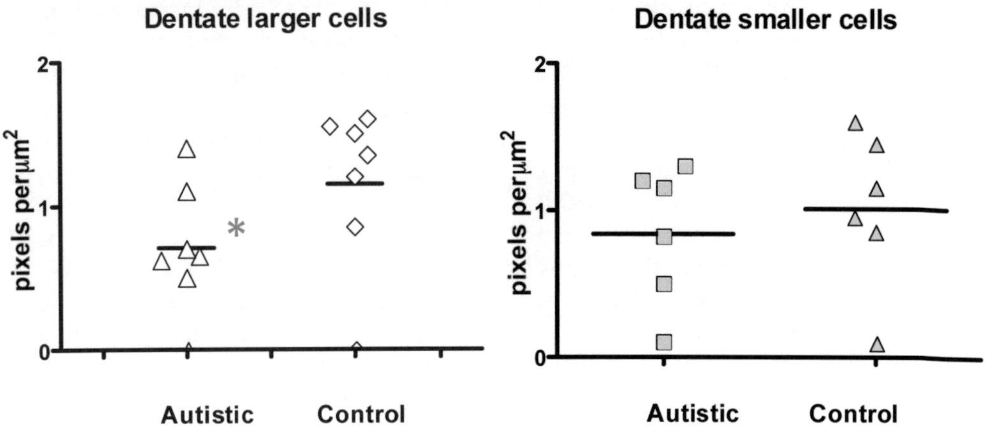

p=0.03, students t test

Fig. 2. Comparison of the level of GAD65 mRNA level in the dentate nuclei in the larger versus the smaller cells. There was a statistically significant difference reduction in the level of GAD65 mRNA (*p*=0.03; student's t test) between the larger dentate cells compared to the smaller cells in the autistic population.

5. Mechanism of neuroplasticity in the cerebellum in subjects with autism

Neuroplasticity balances deficiency of GABA through the modulation of the two isoforms. For example: GAD67 in contrast to GAD65, is strongly experience driven. The activity of GAD coincides with regulation of protein and mRNA levels during intense neuronal activity: stress stimulation (Uchida et al., 2011), seizure events (Walls et al., 2010), and chronic psychotropic drug treatment (Fatemi et al., 2009). GAD67 balances the cast of inhibitory plasticity in a dynamic manner as the developing organism acquired neuronal inputs from the environment. GAD67 is important for long term regulation of phasic neuronal activity, while GAD65 is recruited during tonic activity and as needed (Walls et al., 2010).

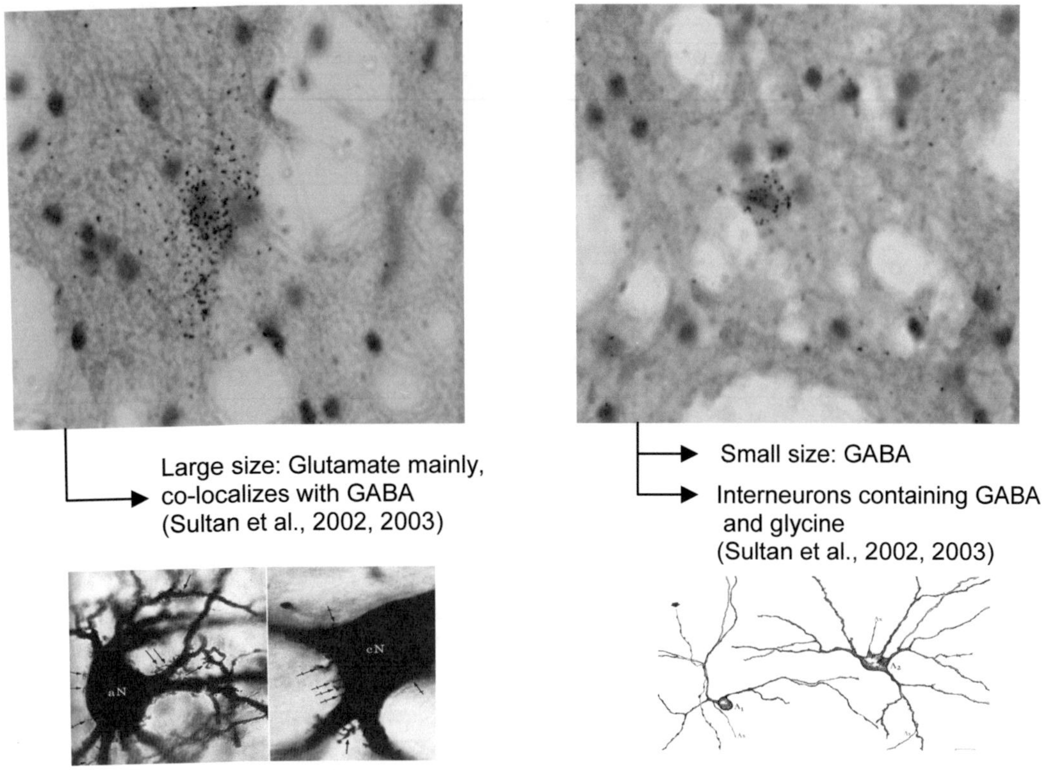

Fig. 3. The differential sizes of the dentate nuclei co-localize with different neurotransmitters. The larger-size dentate nuclei contains mainly GABA whereas the smaller-size dentate nuclei are GABAergic mainly as well as colocalizes GABA with Glycine in the interneuron population.

The inter-neurons in the molecular layer are agents responsive in plasticity mechanisms to balance GAD deficiency. Indeed, in the event of a GAD67mRNA decrease in PC, the authors observed a compensatory increase in basket cells in the cerebellar cortex of autistic individuals (Yip et al., 2008; figure 5). Inhibitory interneurons in the cerebellar cortex consist of basket and stellate cells (figure 6). The inhibitory nature of basket cells and stellate cells was clarified in the 1960s (Eccles et al., 1966) when antibodies became available for GAD. Each class of inhibitory interneuron contained GAD, and the synaptogenesis of these GABAergic neurons marks the differentation of basket and stellate cells (Simat et al., 2007).

6. The cerebellar circuitry in autism and conditioned learning: implications for behavioral treatment

Based on the study of GABA regulation described in previous sections, the authors propose a summary of a putative mechanism of "behavior in autism"- linking what is well known about the cerebellar circuitry to current understanding of behavioral output as being governed by a system of memory called long term potentiation (LTP) and long term depression (LTD) of memory. Afterall, it is the memory system that determines the behavior of an individual because each behavior is generated as a response to a pool of learned responses whether conditioned or not.

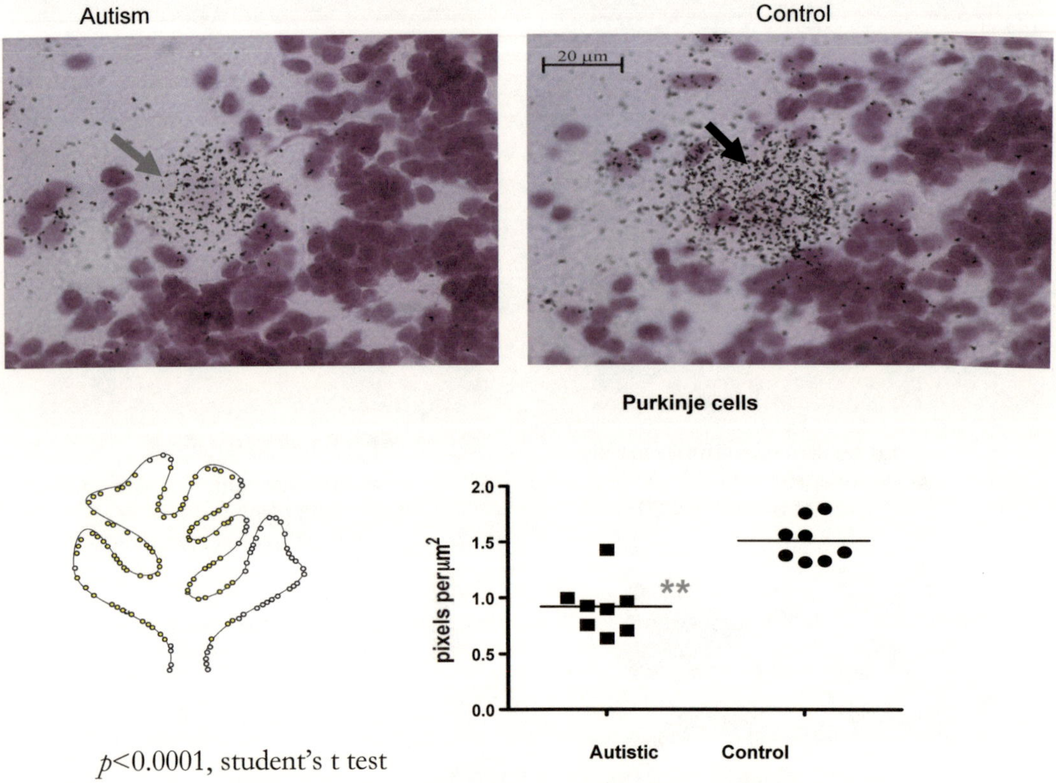

Fig. 4. Reduction in the level of GAD67 mRNA in Purkinje cells (PC) in the autistic compared to control (t(degrees of freedom)= …, $p<0.0001$). Note the sparseness of speckled labeling of GAD67 mRNA-positive PC soma in the autism compared to the control case.

The cerebellum has a role pertaining to complex behaviors such as cognition and emotional processing that include planning and impulse control (Schmahmann, 2010; Stoodley, 2011) and is more than a motor and sensory processing device as previously believed to be. Its simple, modular structure holds the greatest promise for uncovering the "holy grail" as it relates to the neurological system. With less uncertainties, cerebellar circuitry offers a more viable model to derive an understanding of how (almost directly from molecules and cells) synapses and circuits could influence one's behavior (i.e., learning and memory) as compared to the immense complexity of the neuronal configuration within the cerebral cortex. One of the pioneers who elaborated on the idea of cerebellar neuronal structure and its circuit connections as it related to motor learning recently expressed strong support for a plausible, if not important, role of cerebellar circuitry in the dysfunction within information processing (Ito, 2008). Specifically, perturbation of GAD in the cerebellum of our subjects with autism is presented in relation to the circuit connection according to Ito in figure 7.

The cerebellum consists of a well defined structure containing a unique, compartmental, modular structure with complex signal transduction processes in distinct cerebellar neurons. The study of neuronal circuits has attracted a great deal of interest in understanding how these specific entities operate to generate one's mental activities and will continue to be the forefront of modern neuroscience. Factors that allow for this included an unusually well

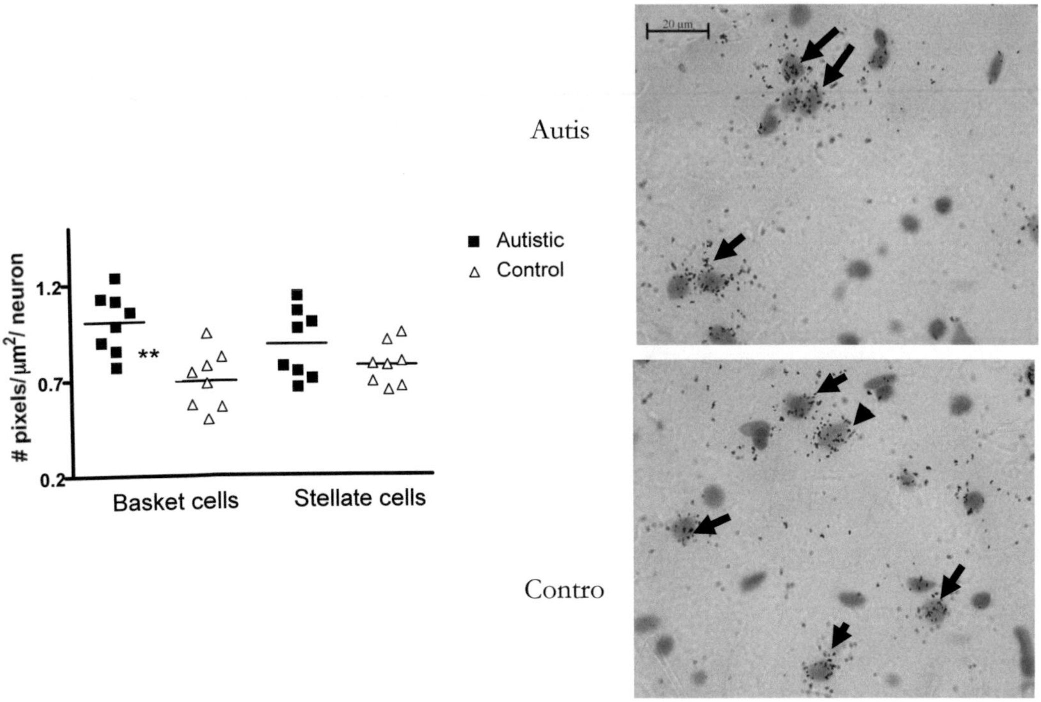

(Simat et al., 2007).

Fig. 5. Increased level of GAD67 mRNA in the basket cell interneurons of the cerebellar cortex in the autistic versus the control group as evidence of neuroplasticity mechanism (-$p<0.0001$). Speckled labeling represents GAD67-mRNA-positive cells.

defined circuit diagram. The cerebellar circuitry is essentially composed of a relay station in the deep cerebellar nucleus (DCN) and a cortical 'side loop'. Cerebellar output to pre-motor centers originate in the DCN which in turn are driven by direct excitatory input from the mossy fibers (MFs). Additionally, it is modulated by the inhibitory input from PC axons, which conveys computations and interactions in the PC. These computations will be performed upon a matrix of subtle and informationally rich excitatory Parallel fiber(PF) input (~200,000 axons), massive and synchronous excitation produced by the one climbing fiber (CF) axon innervating each mature PC, an input from inhibitory inter-neurons. This unusual anatomical configuration inspired a model of motor learning proposed that the PF-PC synapses could provide contextual information. Additionally, the CF-PC synapses could signal an error in motor performance that required alterations of subsequent behavior, and that the conjunction of these two signals could strengthen the PF-PC synapse to create a memory trace for motor learning (Ito, 2002). Compared to other brain areas, the cerebellum has the most organized cellular arrangement consisting of PCs, basket cells, stellate cells, granule cells, and interneurons along with mossy, parallel and climbing fibers. The precision of the cerebellar circuitry has inspired numerous proposals to emulate the cerebellum as a universal learning machine (Ito, 2006, Welsh et al., 2001, Raymond et.al., 1996).

The emerging view of cerebellar circuitry pushes strongly in the direction of regarding this structure as a real-time processing device, whose output is governed strictly by the pattern of inputs received from the other nervous system areas for sensory, motor, or cognitive task

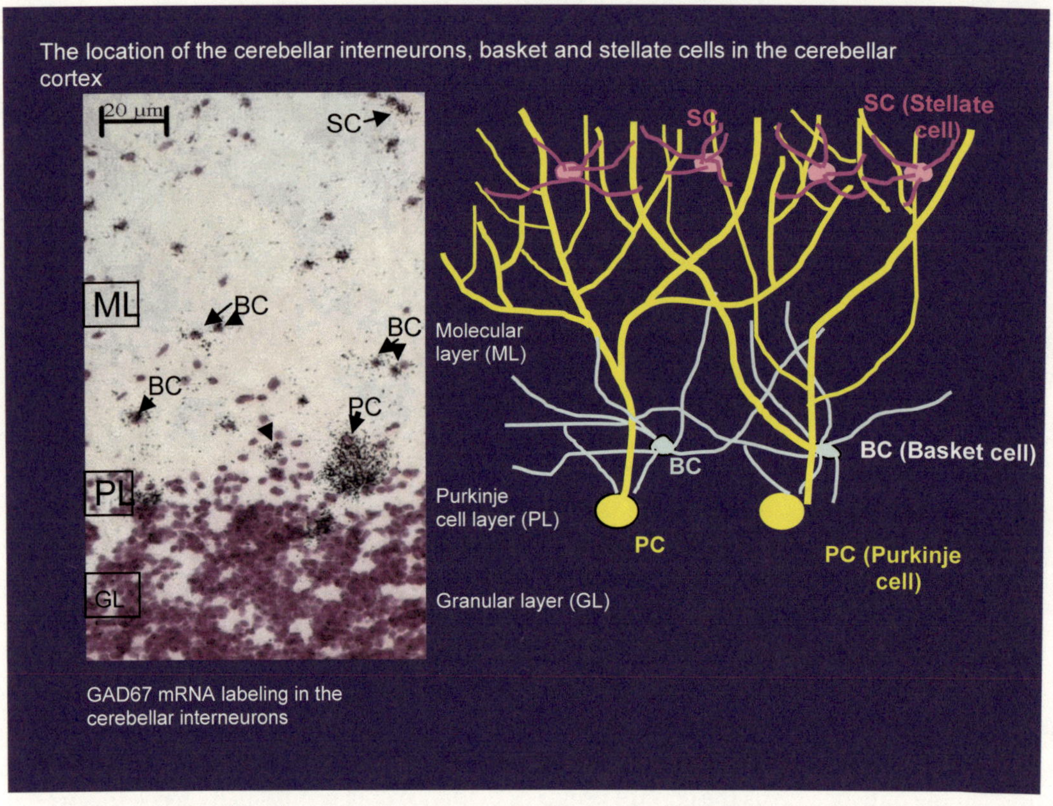

Fig. 6. **The relative location of interneurons in the cerebellar cortex.** Basket cells are located at the PC layer (PCL) whereas stellate cells (SC) are locate at the PC dendritic tree. Speckled labelling represent GAD67 mRNA-positive cells.

completion (Bower, 2002). The following diagram attempts to summarize current literature views of the cerebellar circuitry in LTP and LTD. The general consensus of those in the field of LTP learning in the cerebellum is that it operates precisely upon moment-by-moment task demand that can be conditioned. Recently, using an eye-blink conditioning mode, stimulation of the mossy fibers in the cerebellum elicited a potentiation of memory, although short lasting, at the granule cell-to-PC synapse indicating and that plasticity can be induced through training in animals (D'Angelo et al., 2005; Ohyama et al., 2010). The literature supporting plasticity changes in the brain, in particular the cerebellum, in response to conditioning is extensive (Aiba et al., 1994; Coesmans et al., 2004; Weber et al., 2003; Wadiche & Jahr, 2005) This type of plasticity conditioning has been observed mainly in animals, and may have correlates in human learning since both the substrate and the structure of the brain are the same. It may be for this reason that ABA-based treatments proved to be successful in improving the behavior of individuals with autism (Glen et al., 2005 in the Wisconsin Early Project). In this project, 24 children with autism were trained according to an early intensive behavioral treatment developed by at the University of California-Los Angeles (UCLA) and the outcomes after 4 years of treatment were that 48% of all children showed rapid learning including cognitive, language, adaptive, social, and by age 7, were succeeding in regular education classrooms. It is tempting to speculate whether training could have the effect of modifying brain circuitry and adjusting the deficient transcriptional machinery. Answering a question like this would provide substantiation to clinical professionals working with individuals so that redirection of behavior is congruent with the underlying brain (dys)function and the molecular mechanisms of the disorder, as shown in figure 1. Also, it will be an initial step into a much needed interdisciplinary collaboration between biomedical and applied behavioral research.

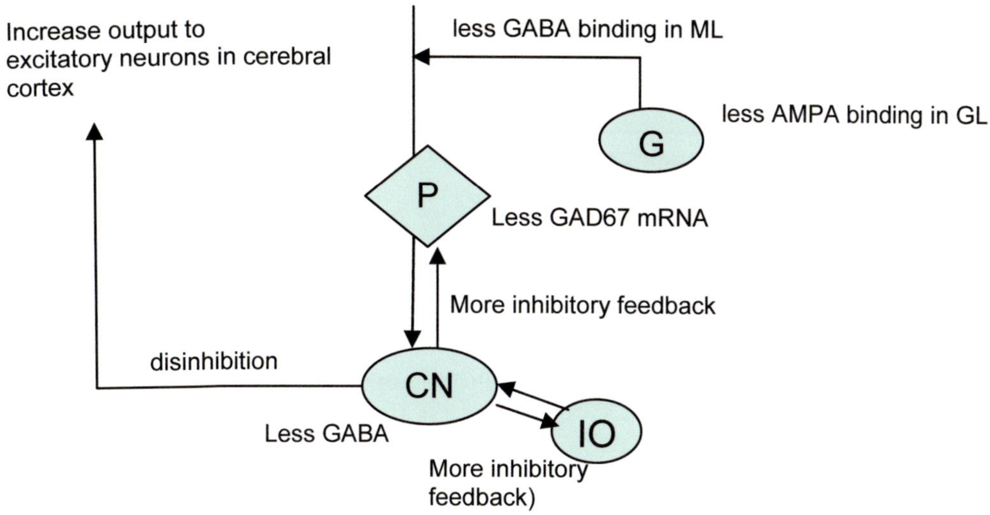

Fig. 7. A diagrammatic summary of the findings on the GABAergic system in the autistic cerebellar cortex from the laboratory of Dr Gene Blatt, Boston University, in which the neurobiological work discussed was performed (some of the results not presented in this chapter). Overall, our results showed that there is a down-regulation in inhibitory GABAergic input to Purkinje cells (PCs) and a down-regulation to the dentate nuclei (CN). The resulting cascading event lies in the circuitry connection such that an overall disinhibition occurs in the cerebellar circuitry from the feedback circuit of the PC-CN and the CN to the inferior olive (IO). Ultimately, lowered inhibition will result in increased output to excitatory neurons in the cerebral cortex, which the dentate nuclei project to, since the balance between inhibition and excitation is perturbed. A background inhibitory synaptic drive in addition to the excitation is necessary to maintain normal PC output. Therefore, an altered balance of inhibition in the PCs is likely to adversely affect glutamate receptor functions essential for the normal maintenance of LTP/LTD.

7. Future studies - how to connect brain function to the child's performance in the natural environment

The sum total of neuronal output is the balance between excitatory and inhibitory neurotransmitter signals. A reduction of inhibitory GABA signals predicts an overall perturbation of neuronal signals that favors a more "excited output which is reflected in numerous parameters of brainwave signals in autism that can be observed from infancy (Ahmadlou et al., 2010; Bosl et al., 2011). An illustration of a simple EEG tracing obtained in a preliminary study is shown in figure 9 (collaboration with Dr. Sara Davis). Briefly, the brain waves of a typical child versus an autistic child were measured using EEG during the following conditions: mathematical processing task, reading task, block building task, and at rest. Preliminary results reveal that there is a significant difference between a typical child and an autistic child during these four monitored conditions. It is interesting to note that the autistic child exhibited a tendency towards "normalization" when building the blocks as compared to the other three conditions.

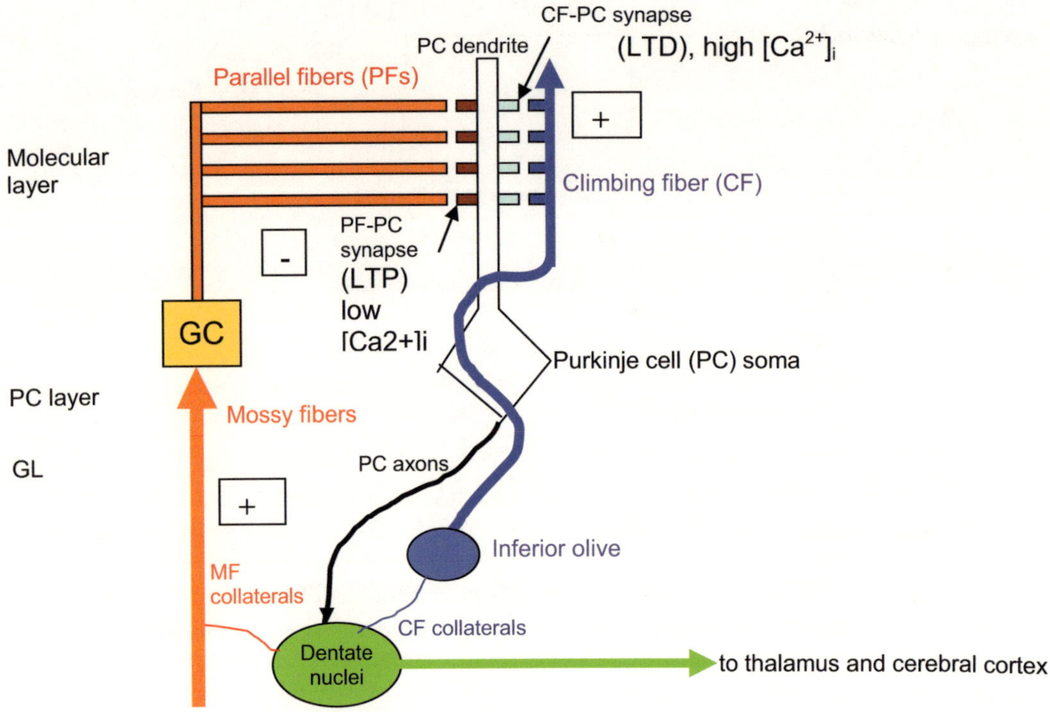

Fig. 8. The converging synaptic site of CF-PC and PF-PC on PC dendrites in the molecular layer is the location for the mediation of LTD or LTP. The components of the cerebellar circuitry belong to a larger neuronal output that is responsible for LTD by amplifying somatic responses through increasing the activity of the voltage-gated [Ca^{2+}] channels or decreasing responses through lowering the influx of [Ca^{2+}] into PCs. Therefore, an imbalance will occur if the ratio between LTD/LTP changes in the molecular layer. Since the granular layer (GL) is also involved in LTP in an activity-dependent manner, the granular layer LTP that is mediated through NMDA receptor activation will increase as a neuroplasticity mechanism. Abbreviations: CF, Climbing fibers; PC, Purkinje Cell; PF, Parallel fiber; LTP, long term potentiation of memory; LTD, long term depression of memory, GL, Granular layer; GC, Granule Cell

To date, many advances have been made in developing an electrophysiological recording device that is portable, reasonably accurate for documenting brainwaves, and has the ability to estimate neural activity from various brain regions. The quantitative electroencephalograph (qEEG) holds such as a promise. QEEG assessment allows to identify anomalies in brain function. QEEG maps (such as the ones shown in Figure 10 are typically constructed using

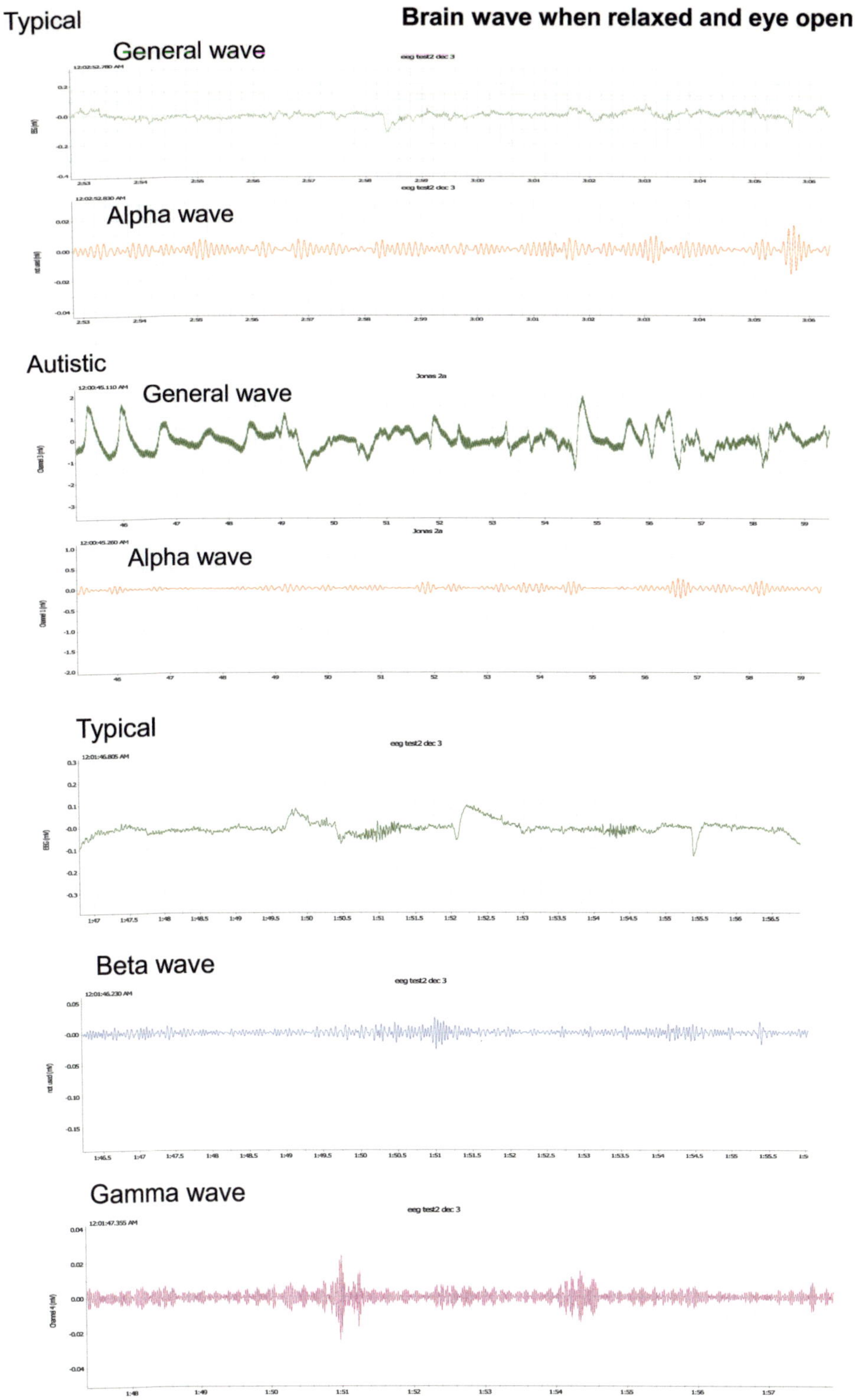
Typical
Brain wave when relaxed and eye open
General wave
eeg test2 dec 3
Alpha wave
eeg test2 dec 3
Autistic
General wave
Jonas 2a
Alpha wave
Jonas 2a
Typical
eeg test2 dec 3
Beta wave
eeg test2 dec 3
Gamma wave
eeg test2 dec 3

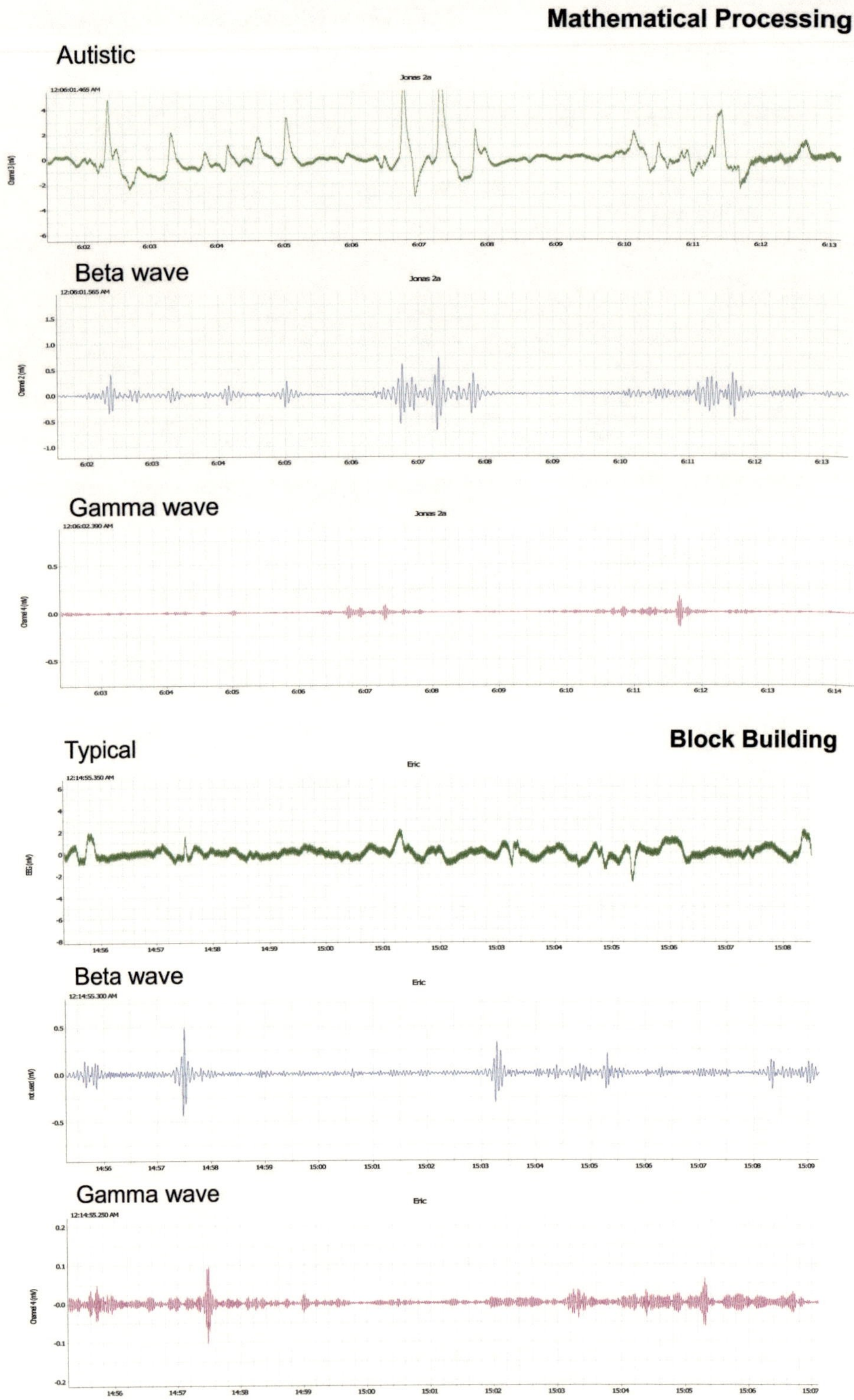
Mathematical Processing
Autistic
Jonas 2a
Beta wave
Jonas 2a
Gamma wave
Jonas 2a
Typical
Block Building
Eric
Beta wave
Eric
Gamma wave
Eric

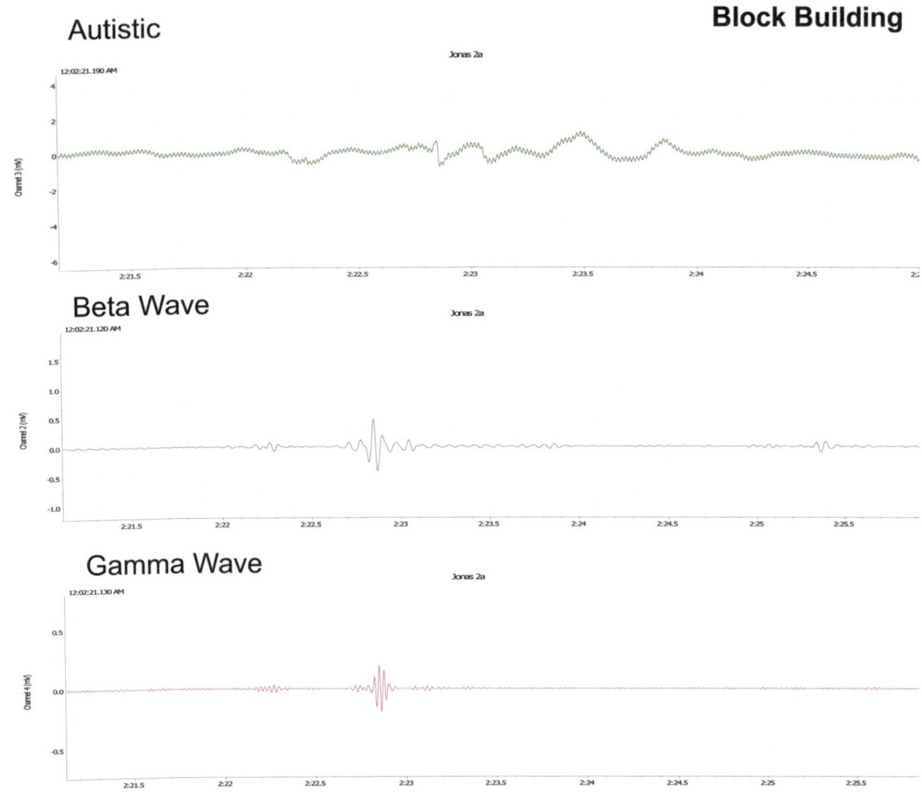

Fig. 9. The brain waves of a typical child versus an autistic child were measured using EEG at rest and while performing three tasks: mathematical processing, reading, and block building. Preliminary results reveal that there is a significant difference between a typical child and an autistic child during these tasks. It is interesting to note that the autistic child studied exhibited a tendency towards "normalization" in general of brain waves when building blocks as compared to the other tasks.

19 electrodes based on the International 10-20 system (Jasper, 1958). These maps are quantitative summaries of EEG characteristics such as frequency, amplitude and coherence emmitted during different conditions or tasks. Software programs along with the qEEG devices that are United States Food and Drug Administration (FDA) approved can then be used to calculate various indices of brain function such as connectivity, amplitude, phase lag, and brain performance index (Thatcher et al., 2005). One advantage of this method is that it can be widely applied, is less costly as Magnetic Resonance Imaging (MRI), and less invasive than Positron Emission Tomography (PET). Thus far, the qEEG analyses in autistic individuals showed aberrant activity in the frontal lobe (Pop-Jordanova et al., 2010).

The authors have collected qEEG brain maps for both an autistic and a control subject during resting states using the same system as Pop-Jordanova et al (2010)(i.e., BrainMaster amplifier and NeuroGuide software) . Although slight individual differences are typical in qEEG measurement these maps illustrate the more fundamental disparities of brain activity in autistic individuals (Figure 10). Our aim for future studies is to connect neurological deficits to treatment regimens Further investigations will look at the effects of language and

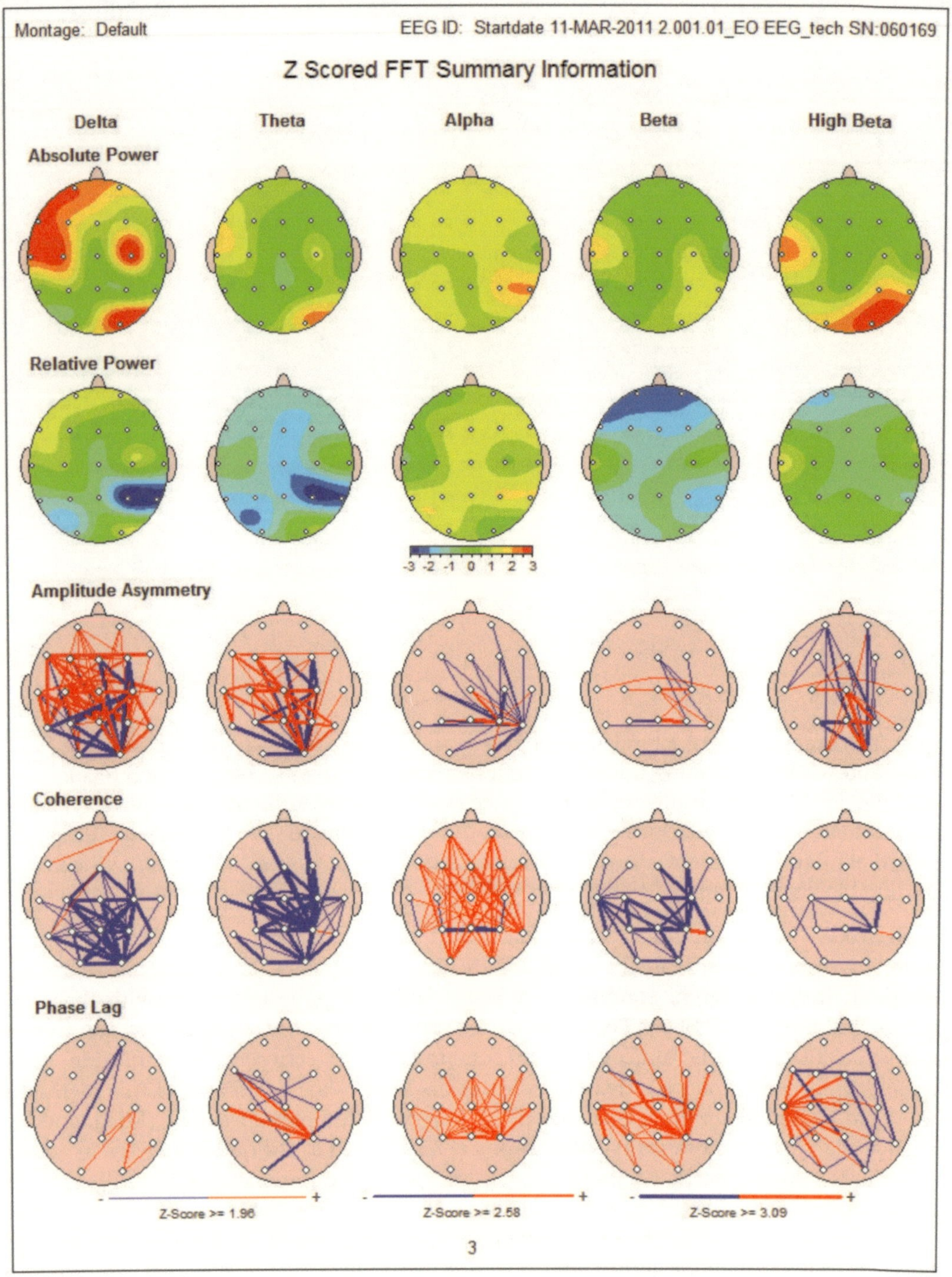

Fig. 10A. A brain map using qEEG obtained from Brainmaster and NeuroGuide of a volunteer with autism (A) and a control (B) during eye open relaxed state.

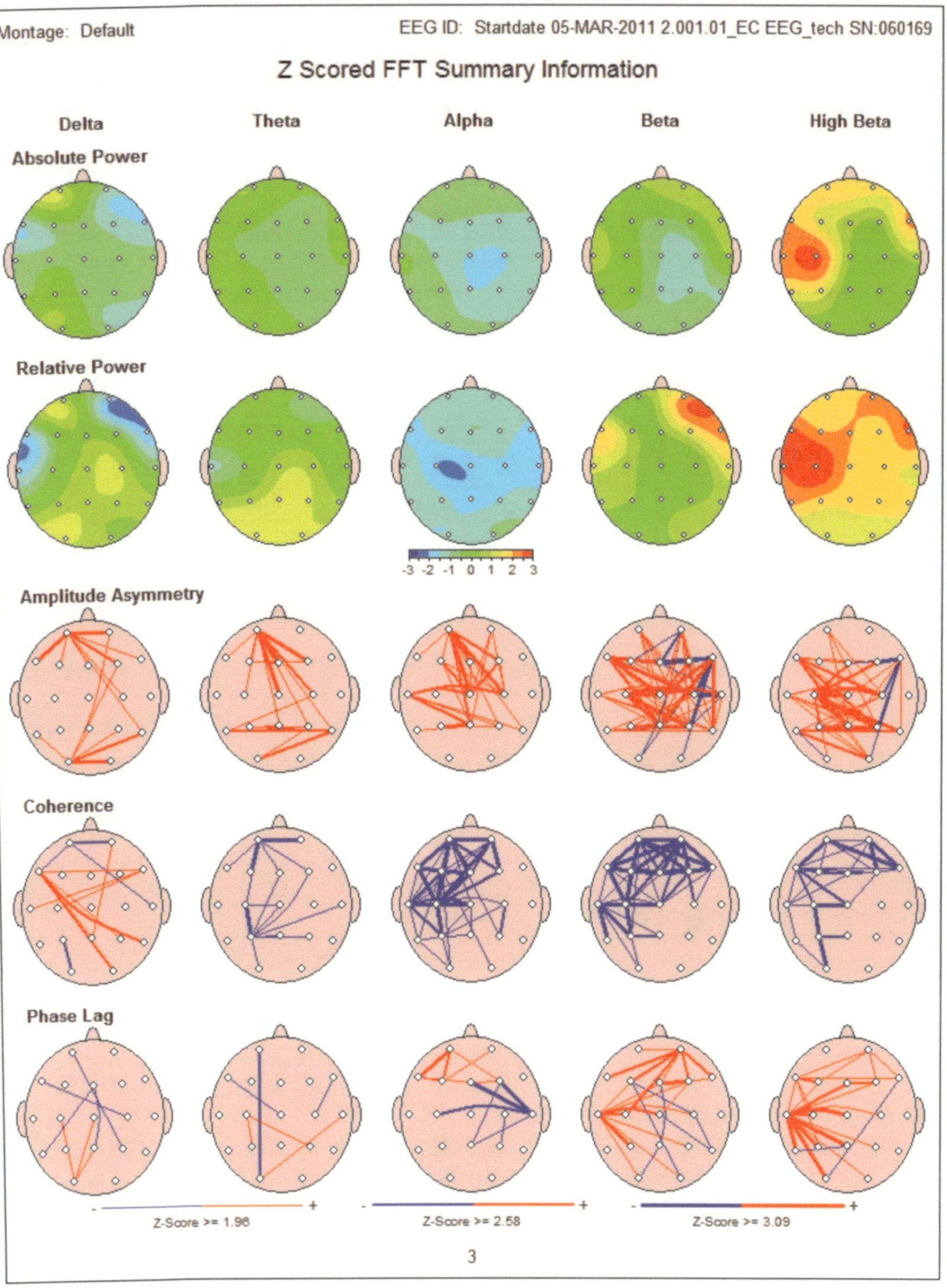

Fig. 10B. A brain map using qEEG obtained from Brainmaster and NeuroGuide of a volunteer with autism (A) and a control (B) during eye open relaxed state.

social communication treatments on brain connectivity, combining behavioral assessment with electrophysiological measurement. Such an approach has been implemented successfully in evaluations of neurofeedback training (Pineda et al., 2008) and acupuncture treatments (Chan, 2009) with subjects mostly diagnosed as high-functioning autism. Our goal is to extend these initial efforts to a wider range of autism spectrum disorders and treatments, particularly focusing on the domains of verbal and non-verbal language and social-communication in those diagnosed with moderate to severe autism.

A widely implemented ABA-based treatment to target communication skills innon-verbal, severely autistic children is the Picture Exchange Communication System (PECS). PECS follows a manualized treatment protocol for prelinguistic communicators that involves six phases: In *Phase I: Physical Exchange*, children are instructed to exchange a graphic symbol for a desired item (often a snack item or toy). In *Phase II: Expanding Spontaneity*, children learn to exchange a symbol with a communication partner who is not in the immediate surrounding. In *Phase III: Picture Discrimination* the child is taught to discriminate among symbols for requesting. Subsequently, in *Phase IV: Sentence Structure*, the child learns to put an "I want" symbol onto a blank sentence strip, along with the symbol for a desired item, and to exchange the sentence strip with a partner. In *Phase V: Responding to "What do you want,"* the child is trained to respond to a direct question. Finally, *Phase VI: Responsive and Spontaneous Commenting* uses previously acquired skills to elicit a response to additional questions (i.e., "What do you see?") and encourage spontaneous commenting (Bondy & Frost, 1994)

An autistic child's ability to recognize pictures is relatively intact due to preserved visuo-spatial skills (Mirenda & Brown, 2009). It is expected that successful intervention will be reflected in enhanced brain connectivity and overall gain in cognitive capacity. Additionally, it is expected that individuals who are slower to learn- and/ are unable to proceed to later PECS phases -have brain connectivity patterns that are different from their more successful counterparts, and hence could reveal both a brain and performance marker in the more challenged subtypes. QEEG data will be critical in measuring treatment effects at the neurophysiological level, individualizing treatments for participants and making refinements to the treatment protocol as necessary.

8. References

Acosta MT (2004) Pharmacotherpy in autism: where to start. Drug Discov. Today 9(11): 474

Ahmadlou M, Adeli H, Adeli A (2010) Fractality and a wavelet-chaos-neural network methodology for EEG-based diagnosis of autistic spectrum disorder. J Clin Neurophysiol. 2010 Oct;27(5):328-33

Aiba A, Kano M, Chen C, Stanton ME, Fox, GD, Herrup K, zwingman TA, Tonegawa S (1994) Deficient cerebellar long term depression and imparied motor learning in mGluR1 mutant mice. Cell 79: 377-388

Asada H, Kawamura Y, Maruyama K, Kume H, Ding R, Ji FY, Kanbara N, Kuzume H, Sanbo M, Yagi T, Obata K (1996) Mice lacking the 65 kDa isoform of glutamic acid decarboxylase (GAD65) maintain normal levels of GAD67 and GABA in their brains but are susceptible to seizures. Biochem Biophys Res Commun. Dec 24;229(3):891-5

Asada H, Kawamura Y, Maruyama K, Kume H, Ding RG, Kanbara N, Kuzume H, SanboM, Yagi T, Obata K (1997) Cleft palate and decrased brain gamma-aminobutyric acid in mice lacking the 67-kDa isoform of glutamic acid decarboxylase. Proc Natl Acad Sci USA 94(12): 6496-6499

Ashley-Koch AE, Mei H, Jaworski J, Ma DQ, Ritchie MD, Menold MM, Delong GR, Abramson RK, Wright HH, Hussman JP, Cuccaro ML, Gilbert JR, Martin ER, Pericak-Vance MA. An analysis paradigm for investigating multi-locus effects in complex disease: examination of three GABA receptor subunit genes on 15q11-q13 as risk factors for autistic disorder. Ann Hum Genet. 2006 May;70(Pt 3):281-92

Askalan R, Mackay M, Brian J, Otsubo H, McDermott C, Bryson S, Boyd J, Snead C 3rd, Roberts W, Weiss S (2003). Prospective preliminary analysis of the development of autism and epilepsy in children with infantile spasms. J Child Neurol. 2003 Mar;18(3):165-70.

Bao J, Reim K, Sakaba T (2010) Target-dependent feedforward inhibition mediated by short-term synaptic plasticity in the cerebellum. J Neurosci. Jun 16;30(24):8171-9

Blatt GJ, Fitzgerald CM, Guptill JT, Booker AB, Kemper TL, Bauman ML (2001). Density and distribution of hippocampal neurotransmitter receptors in autism: an autoradiographic study. J Autism Dev Disord. 2001 Dec;31(6):537-43

Blatt GJ, Soghomonian JJ, Yip J (2010). Glutamic Acid Decarboxylase (GAD) as a Biomarker of GABAergic Activity in Autism: Impact on Cerebellar Circuitry and Function. In Springer, G.J. Blatt (ed.), *The Neurochemical Basis of Autism*

Bondy, A., & Frost, L. (1994). The Picture Exchange Communication System. *Focus on Autistic Behavior, 9*, 1-19.

Bosl W, Tierney A, Tager-Flusberg H, Nelson C (2011) EEG complexity as a biomarker for autism spectrum disorder risk. BMC Med. 2011 Feb 22;9:18

Bower JM. (2002) The organization of cerebellar cortical circuitry revisited: implications for function. Ann N Y Acad Sci. 978:135-55

Chan, AS, Cheung, MC, Sze, SL, &Leung, W.W. (2009). Seven-star needle stimulation improves language and social interaction of children with autistic spectrum disorders. American Journal of Chinese Medicine, 37, 495-504.

Chan SO, Lyman WD, Chiu FC (1997) Temporal and spatial expression of glutamic acid decarboxylases in human fetal brain. Brain Res Mol Brain Res. 1997 Jun;46(1-2):318-20.

Coesmans M, Weber JT, De Zeeuw CI, Hansel C (2004) Bidirectional parallel fiber plasticity in the cerebellum under climbing fiber control. Neuron 44: 691-700

Cramer SC, Sur M, Dobkin BH, O'Brien C, Sanger TD, John Q. Trojanowski, Rumsey JM, Hicks R, Cameron J, Chen D, Wen G. Chen,Cohen LG, deCharms C, Duffy CJ, Eden GF, Fetz EE, Filart R, Freund M, Grant SJ, Haber S, Kalivas PW, Bryan Kolb B, Kramer AF, Lynch M, Helen S. Mayberg, McQuillen PS, Nitkin R, Pascual-Leone A, Reuter-Lorenz P, Schiff N, Sharma A, Shekim L, Stryker M, Sullivan EV and Vinogradov S. (2011) Harnessing neuroplasticity for clinical applications. Brain 134; 1591–1609

D'Angelo E, Rossi P, Gall D, Prestori F, Nieus T, Maffei A, Sola E (2005) Long-term potentiation of synaptic transmission at the mossy fiber-granule cell relay of the cerebellum. Prog Brain Res 148: 69-80

Eccles JC, Llinás R, Sasaki K. (1966) The inhibitory interneurones within the cerebellar cortex. Exp Brain Res 1(1):1-16

Esclapez M, Tillakaratne NJ, Kaufman DL, Tobin AJ, Houser CR (1994) Comparative localization of two forms of glutamic acid decarboxylase and their mRNAs in rat brain supports the concept of functional differences between the forms. J Neurosci 14(3 Pt 2): 1834-1855

Esclapez M, Tillakaratne NJK, Tobin AJ, Houser CR (1993) Comparative localization of mRNAs encoding two forms of glutamic acid decarboxylase with non radioactive in situ hybridization methods. J Comp Neurol 331:339-362

Fatemi SH, Halt AR, Stary JM, Kanodia R, Schulz SC, Realmuto GR (2002) Glutamic acid decarboxylase 65 and 67 kDa proteins are reduced in autistic parietal and cerebellar cortices. Biol Psychiatry 52(8):805-10

Fatemi SH, Reutiman TJ, Folsom TD (2009) Chronic psychotropic drug treatment causes differential expression of Reelin signaling system in frontal cortex of rats. Schizophr Res. Jun;111(1-3):138-52

Feldblum S, Dumoulin A, Anoal M, Sandillon F, Privat A. (1995) Comparative distribution of GAD65 and GAD67 mRNAs and proteins in the rat spinal cord supports a differential regulation of these two glutamate decarboxylases in vivo. J Neurosci Res. 1995 Dec 15;42(6):742-57

Feldblum S, Erlander MG, Tobin AJ (1993) Different distributions of GAD65 and GAD67 mRNAs suggest that the two glutamate decarboxylases play distinctive functional roles. J Neurosci Res 34(6): 689-706

Blatt GJ, Soghomonian JJ, and Yip J (2010) Glutamic Acid Decarboxylase (GAD) as a Biomarker of GABAergic Activity in Autism: Impact on Cerebellar Circuitry and Function. In Blatt GJ(ed.), *The Neurochemical Basis of Autism*, 95. Springer 2010

Glen O. Sallows and Tamlynn D. Graupner (2005) Intensive Behavioral Treatment for Children With Autism: Four-Year Outcome and Predictors. Wisconsin Early Autism Project (Madison). VOLUME 110, NUMBER 6: 417–438 z NOVEMBER 2005 AMERICAN JOURNAL ON MENTAL RETARDATION

Gray, C. A., & Garand, J. D. (1993). Social stories: Improving responses of students with autism with accurate social information. Focus *on autistic behavior, 8*(1), 1-10.

Guridi J, Herrero MT, Luquin MR, Guillén J, Ruberg M, Laguna J, Vila M, Javoy-Agid F, Agid Y, Hirsch E, Obeso JA (1996) Subthalamotomy in parkinsonian monkeys. Behavioural and biochemical analysis. Brain. 119 (Pt 5):1717-27

Ito M (2002) The molecular organization of cerebellar long-term depression. Nat Rev Neurosci 3: 896-902

Ito M (2008) Control of mental activities by internal models in the cerebellum. Nat Rev Neurosci. Apr;9(4):304-13

Ito, M (2006) Cerebellar circuitry as a neuronal machine. Prog Neurobiol 78: 272-303

Ji F, Obata K (1999) Development of the GABA system in organotypic culture of hippocampal and cerebellar slices from a 67-kDa isoform of glutamic acid decarboxylase (GAD67)-deficient mice. Neurosci Res. 1999 Mar;33(3):233-7

Kalkman HO, Loetscher E. (2003) GAD(67): the link between the GABA-deficit hypothesis and the dopaminergic- and glutamatergic theories of psychosis. J Neural Transm. 2003 Jul;110(7):803-12

Kanner, L . (1943) Autistic disturbances of affective contact. Nervous Child. Vol 2: 217-250

Koegel, R. L., & Koegel. L. K. (2006). Pivotal Response Treatment for Autism: Communication, Social, & Academic Development. Baltimore: Paul Brookes.

Lariviere K, MacEachern L, Greco V, Majchrzak G, Chiu S, Drouin G, Trudeau VL. (2002) GAD(65) and GAD(67) isoforms of the glutamic acid decarboxylase gene originated before the divergence of cartilaginous fishes. Mol Biol Evol. 19(12):2325-9

Lovaas, O. I. (1977). The autistic child: Language development through behavior modification. New York: Irvington.

Lovaas, O. I. (2003). Teaching individuals with developmental delays: Basic intervention techniques. Austin, TX: Pro-Ed.

Marrosu F, Marrosu G, Rachel MG, Biggio G (1987) Paradoxical reactions elicited by diazepam in children with classic autism. Funct Neurol. 2(3):355-61

Martin DL, Rimvall K (1993) Regulation of gamma-aminobutyric synthesis in the brain. J Neurochem 60(2): 395-407

Martin, D.L., Martin, S.B., Wu, S.J., & Espina, N. (1991). Cofactor interactions on the regulation of glutamate decarboxylase activity. Neurochem Res 16, 243-249.

McLean HA, Caillard O, Ben-Ari Y, Gaiarsa JL (1996) Bidirectional plasticity expressed by GABAergic synapses in the neonatal rat hippocampus. J Physiol. 1996 Oct 15;496 (Pt 2):471-7

Meinck HM. (2002) Stiff man syndrome. CNS Drugs. 15(7):515-26. Review.

Menold MM, Shao Y, Wolpert CM, Donnelly SL, Raiford KL, Martin ER, Ravan SA, Abramson RK, Wright HH, Delong GR, Cuccaro ML, Pericak-Vance MA, Gilbert JR (2001) Association analysis of chromosome 15 gabaa receptor subunit genes in autistic disorder. J Neurogenet. 2001;15(3-4):245-59

Mirenda, P., & Brown, K. E. (2009). A picture is worth a thousand words: Using visual supports for augmented input with individuals with autism spectrum disorders. In P. Mirenda & T. Iacono (Eds.), *Autism Spectrum Disorders and AAC* (pp. 303-332). Baltimore: Paul H. Brookes.

Ohyama T, Voicu H, Kalmbach B, Mauk MD (2010) A decrementing form of plasticity apparent in cerebellar learning. J Neurosci. 2010 Dec 15;30(50):16993-7003.

Pineda, JA, Brang, D., Hecht, E., Edwards, L., Carey, S., Bacon, M., Futagaki, C., Suk, D., Tom, J., Birnbaum, C., & Rork, A. (2008). Positive behavioral and neurophysiological changes following neurofeedback training in children with autism. Research in Autism Spectrum Disorders, 2, 557-581.

Pisu MB, Roda E, Avella D, Bernocchi G (2004) Developmental plasticity of rat cerebellar cortex after cisplatin injury: inhibitory synapses and differentiating Purkinje neurons. Neuroscience. 2004;129(3):655-64

Pop-Jordanova N, Zorcec T, Demerdzieva A, Gucev Z (2010) QEEG characteristics and spectrum weighted frequency for children diagnosed as autistic spectrum disorder. Nonlinear Biomed Phys. Sep 30;4(1):4

Raymond JL, Lisberger SG, Mauk MD (1996) The cerebellum: a neuronal learning machine? Science 272(5265): 1126-1131

Rimvall K, Sheikh SN, Martin DL (1993) Effects of increased γ-aminobutyric acid levels on GAD67 protein and mRNA levels in rat cerebral cortex. J Neurochem 60: 714-720

Schmahmann JD (2010) The role of the cerebellum in cognition and emotion: personal reflections since 1982 on the dysmetria of thought hypothesis, and its historical evolution from theory to therapy. Neuropsychol Rev 20(3): 236-60

Schmahmann JD, Kiliany RJ, Moore TL, DeMong C, MacMore JP, Moss MB (2004) Cerebellar dentate nucleus lesions impair flexibility but not motor function in monkeys. Soc Neurosci Abstr 34: 254-312

Schroer RJ, Phelan MC, Michaelis RC, Crawford EC, Skinner SA, Cuccaro M, Simensen RJ, Bishop J, Skinner C, Fender D, Stevensone RE (1998) Autism and maternally derived aberrations of chromosome 15q. Amer J Med Genet 76: 327-336

Shao Y, Wolpert CM, Raiford KL, Menold MM, Donnelly SL, Ravan SA, Bass MP, McClain C, Von Wendt L, Vance JM, Abramson RH, Wright HH, Ashley-Koch A, Gilbert JR, DeLong RG, Cuccaro ML, Pericak-Vance MA (2002) Genomic screen and follow-up analysis for autistic disorder. Am J Med Genetics 114: 99-105

Simat M, Ambrosetti L, Lardi-Studler B, Fritschy JM (2007) GABAergic synaptogenesis marks the onset of differentiation of basket and stellate cells in mouse cerebellum. Eur J Neurosci 26(8):2239-56.

Soghomonian JJ, Martin DL (1998) Two isoforms of glutamate decarboxylase: why? Trends Pharmacol Sci 19(12): 500-505

Stoodley CJ. (2011) The Cerebellum and Cognition: Evidence from Functional Imaging Studies. Cerebellum. Mar 4.

Stork O, Ji FY, Kaneko K, Stork S, Yoshinobu Y, Moriya T, Shibata S, Obata K. (2000) Postnatal development of a GABA deficit and disturbance of neural functions in mice lacking GAD65. Brain Res. May 19;865(1):45-58

Thatcher RW, North D, Biver C (2005) EEG and intelligence: Relations between EEG coherence, EEG phase delay and power. Clinical Neurophysiology 116 (2005) 2129–2141

Tordjman S, Antoine C, Cohen DJ, Gauvain-Piguard A, Carlier M, Roubertoux P, Ferrari P (1999) Study of the relationships between self-injurious behavior and pain reactivity in infantile autism. Encephale. 1999 Mar-Apr;25(2):122-34

Torrey EF, Barci BM, Webster MJ, Bartko JJ, Meador-Woodruff JH, Knable MB (2005) Neurochemical markers for schizophrenia, bipolar disorder, and major depression in postmortem brains. Biol Psychiatry. Feb 1;57(3):252-60

Vianello M, Tavolato B, Giometto B (2002) Glutamic acid decarboxylase autoantibodies and neurological disorders. Neurol Sci. 23(4):145-51.

Wadiche JI, Jahr CE (2005) Patterned expression of Purkinje cell glutamate transporters controls synaptic plasticity. Nat Neurosci 8(10): 1329-1334

Walls AB, Nilsen LH, Eyjolfsson EM, Vestergaard HT, Hansen SL, Schousboe A, Sonnewald U, Waagepetersen HS. (2010) GAD65 is essential for synthesis of GABA destined for tonic inhibition regulating epileptiform activity. J Neurochem. Dec;115(6):1398-408

Weber JT, De Zeeuw CI, Linden DJ, Hansel C (2003) Long-term depression of climbing fiber-evoked calcium transients in Purkinje cell dendrites. Proc Natl Acad Sci 100(5): 2878-2883

Welsh JP, Schwarz C, Garbourg Y (2001) The cerebellum as a neuronal prosthesis machine. Prog Brain Res 130: 297-315

Yip J, Soghomonian JJ, Blatt GJ (2007) Decreased GAD67 mRNA levels in cerebellar Purkinje cells in autism: pathophysiological implications. Acta Neuropathol. 113(5):559-68

Yip J, Soghomonian JJ, Blatt GJ (2008). Increased GAD67 mRNA expression in cerebellar interneurons in autism: implications for Purkinje cell dysfunction. J Neurosci Res. 86(3):525-30

Yip J, Soghomonian JJ, Blatt GJ (2009) Decreased GAD65 mRNA levels in select subpopulations of neurons in the cerebellar dentate nuclei in autism: an in situ hybridization study. Autism Res. 2(1):50-9